Springer Proceedings in Mathematics & Statistics

Volume 311

Springer Proceedings in Mathematics & Statistics

This book series features volumes composed of selected contributions from workshops and conferences in all areas of current research in mathematics and statistics, including operation research and optimization. In addition to an overall evaluation of the interest, scientific quality, and timeliness of each proposal at the hands of the publisher, individual contributions are all refereed to the high quality standards of leading journals in the field. Thus, this series provides the research community with well-edited, authoritative reports on developments in the most exciting areas of mathematical and statistical research today.

More information about this series at http://www.springer.com/series/10533

Mercedes Siles Molina · Laiachi El Kaoutit ·
Mohamed Louzari · L'Moufadal Ben Yakoub ·
Mohamed Benslimane
Editors

Associative and Non-Associative Algebras and Applications

3rd MAMAA, Chefchaouen, Morocco, April 12–14, 2018

Editors
Mercedes Siles Molina
Departamento Álgebra
Geometría y Topología
Universidad de Málaga
Málaga, Spain

Mohamed Louzari
Department of Mathematics
Abdelmalek Essaâdi University
Tetouan, Morocco

Mohamed Benslimane
Department of Mathematics
Abdelmalek Essaâdi University
Tetouan, Morocco

Laiachi El Kaoutit
Department of Algebra
University of Granada
Ceuta, Spain

L'Moufadal Ben Yakoub
Department of Mathematics
Abdelmalek Essaâdi University
Tetouan, Morocco

ISSN 2194-1009 ISSN 2194-1017 (electronic)
Springer Proceedings in Mathematics & Statistics
ISBN 978-3-030-35258-5 ISBN 978-3-030-35256-1 (eBook)
https://doi.org/10.1007/978-3-030-35256-1

Mathematics Subject Classification (2010): 01A30, 11Y40, 13P10, 16D10, 16D90, 16E30, 16Y60, 16U90, 17A01, 17A99, 17B63, 17D92, 18D10, 18D20, 20K40, 35J93, 37K30, 46B04, 46B06, 46L70, 46K70, 47B15

This Springer imprint is published by the registered company Springer Nature Switzerland AG
The registered company address is: Gewerbestrasse 11, 6330 Cham, Switzerland

Dedicated to professors Amin Kaidi and Ángel Rodríguez Palacios, without their long scientific relationship, this proceeding had never seen the light.

Foreword

The scientific cooperation relationships between Moroccan and Andalusian universities in the field of Mathematics began more than 30 years ago. Initially between the universities of Granada and Rabat, thanks to the efforts of Profs. Ángel Rodríguez Palacios and Amin Kaidi. These relationships were extended to the Faculty of Sciences in Tetouán with the valuable support of the dean Layachi Imlahi, who encouraged the Algebra and its Applications Research Team of the Department of Mathematics, with many other Andalusian universities, to start the first scientific collaboration projects in the context of institutional cooperation. These projects were mainly supported by the "Junta de Andalucía" and/or by "La Agencía Española de Cooperación Internacional". Under the sponsorship of these institutional cooperation, the first Moroccan Andalusian meeting on algebras and their Applications took place at Tetouán in September 2001. This first meeting gathered many researchers from Andalusian and Moroccan universities, and other European universities (Oviedo, Barcelona, Zaragoza, Murcia, Paris, Reims and Montpellier). Subsequently, in collaboration with the universities of Almería, Cádiz, Granada, and Málaga, The Algebra and its Applications Research Team of Tetuán, organized the second Moroccan Andalusian meeting on Algebras and their Applications, celebrated at Tetouán in July 2003.

This proceeding conference reports on *The 3rd Moroccan Andalusian Meeting on Algebras and their Applications* conference held in the city of Chefchaouen, from April 12–14, 2018, actually with contributors from different Universities of Morocco, Europe, and Occidental Africa.

Ceuta, Spain; Tetouan, Morocco; Málaga, Spain
July 2019

The Editors

Preface

This is the proceedings book that reports on the contributors research activities, who attend *The 3rd Moroccan Andalusian Meeting on Algebras and their Applications* conference, which took place in Chefchaouen (Morocco), April 12–14, 2018.

The subject of the conference, of course, is too vague as well as the participants' research fields. So, the purpose of the book is to offer the participants an opportunity to report on their recent research results, in a way that the proceedings can encompass the wide range of their interdisciplinary research domains. In fact, these multidisciplinary research areas could make the task almost impossible, as the research fields, in most cases, seem to have absolutely no interrelation between one contribution and the other. Nevertheless, for future reasons, especially for young Occidental African researchers, it is, in our opinion, worthwhile to make the effort in assembling these contributions with such scattered contents.

It is for that reason that we have decided to divide this book into three parts, although, we hasten to remark that these divisions are hardly exact!

The first part entitled *Algebraic and Analytic Methods in Associative and Non-associative Structures. Applications*, aggregates, to a large extent, contributions whose methods are mostly based on elementary techniques by using either algebraic or analytic tools employing associative or non-associative algebraic structures. For example, Poisson and evolution algebras, operators algebras, Lipschitz spaces, rings, modules, etc.

The second part is called *Homological and Categorical Methods in Algebra*, and contains contributions that employ homological and/or categorical techniques in different branches of the general framework of Algebra, like tensor categories, triples, Hom-Lie algebras, derived functors, localization in categories of modules, etc.

The last part entitled *History of Mathematics*, contains only one contribution and perhaps could be seen as the "melancholic" component of the Moroccan Andalusian mathematical collaborations carried out over the last 30 years.

This book can be addressed to postgraduate students and young researchers, who are either interested in the field of algebra and/or of mathematical analysis.

Ceuta, Spain
Tetouan, Morocco
Málaga, Spain
July 2019

Laiachi El Kaoutit
Mohamed Louzari
Mercedes Siles Molina

Acknowledgements

We would like to thank Abdelmalek Essaâdi University and the Faculty of Science of Tetuán for their logistical support and for providing such an excellent scientific ambience in organizing the meeting. The conference was in fact held at the Cultural Complex Mohamed VI and Youth House in the city of Chefchaouen (Morocco) from April 12 to 14, 2018. The choice of Chefchaouen to host this event was not coincidental. It is a city with a great historical load and a diverse and rich cultural heritage. We extend our thanks and gratitude to all the people of this region. We also acknowledge the financial support of the following corporations:

- Ministry of Youth and Sports,
- Abdelmalek Essaâdi University,
- Regional Council of Tangier-Tetuán-Al Hoceima,
- Faculty of Sciences of Tetuán,
- National School of Applied Sciences of Tetuán,
- Hassan II Academy of Science and Technology,
- Province of Chefchaouen,
- Provincial Council of Chefchaouen,
- Urban Municipality of Chefchaouen,
- Urban Municipality of Tetuán.

We would like also to thank the anonymous referees for their excellent job, as well as Springer for the facilities that were offered to us, during the editorial process of this book.

Contents

Contributors

Seddik Abdelalim Department of Mathematical and Computer Sciences, Faculty of Sciences Ain Chock, Laboratory of Topology, Algebra, Geometry and Discrete Mathematics, Hassan II University of Casablanca, Maarif, Casablanca, Morocco

Mostafa Alaoui Abdallaoui Department of Mathematics, Chouaib Doukkali University, El jadida, Morocco

Faouzi Ammar Sfax University, Sfax, Tunisia

Imen Ayadi IRIMAS - Département de Mathématiques, Mulhouse, France

Abdelmalek Azizi ACSA Laboratory, Faculty of Sciences, Universite Mohammed Premier, Oujda, Morocco

Antonio Behn Departamento de Matemática, Facultad de Ciencias, Universidad de Chile, Santiago, Chile

Daniel Bulacu Faculty of Mathematics and Computer Science, University of Bucharest, Bucharest 1, Romania

Yolanda Cabrera Casado Departamento de Matemática Aplicada, Universidad de Málaga, Málaga, Spain

Miguel Cabrera García Departamento de Análisis Matemático, Facultad de Ciencias, Universidad de Granada, Granada, Spain

M. G. Cabrera-Padilla Departamento de Matemáticas, Universidad de Almería, Almería, Spain

Antonio J. Calderón Martín Department of Mathematics, University of Cádiz, Cádiz, Spain

José Carmona Departamento de Matemáticas, Universidad de Almería, Almería, Spain

Abdelhak Chaichaa Department of Mathematical and Computer Sciences, Faculty of Sciences Ain Chock, Laboratory of Topology, Algebra, Geometry and Discrete Mathematics, Hassan II University of Casablanca, Maarif, Casablanca, Morocco

Bassirou Dembele Laboratory of Algebra, Codes and Cryptography Applications (LACCA) of UFR of Applied Sciences and Technologies of Gaston Berger University, Saint-Louis (UGB), Senegal

Boubacar Dieme Department of Mathematics, University Cheikh Anta Diop of Dakar, Dakar, Senegal

Yatma Diop Cheikh Anta Diop University of Dakar, Dakar, Senegal

Mohammed El Badry Department of Mathematics, Chouaib Doukkali University, El jadida, Morocco

Mostafa El garn Department of Mathematical and Computer Sciences, Faculty of Sciences Ain Chock, Laboratory of Topology, Algebra, Geometry and Discrete Mathematics, Hassan II University of Casablanca, Maarif, Casablanca, Morocco

Laiachi El Kaoutit Departamento de Álgebra and IEMath-Granada, Universidad de Granada, Granada, Spain;
Facultad de Educación, Econonía y Tecnología de Ceuta, Ceuta, Spain

Mohamed Ben Faraj ben Maaouia Laboratory of Algebra, Codes and Cryptography Applications (LACCA) of UFR of Applied Sciences and Technologies of Gaston Berger University, Saint-Louis (UGB), Senegal

Babacar Gaye Department of Mathematics, University Cheikh Anta Diop of Dakar, Dakar, Senegal

Abdelfattah Haily Department of Mathematics, Chouaib Doukkali University, El jadida, Morocco

Mbarek Haynou Faculty of Sciences and Technology, Moulay Ismail University, Errachidia, Morocco

A. Jiménez-Vargas Departamento de Matemáticas, Universidad de Almería, Almería, Spain

Salvador López-Martínez Departamento de Análisis Matemático, Universidad de Granada, Facultad de Ciencias, Granada, Spain

Mohamed Louzari Department of Mathematics, Abdelmalek Essaadi University, Tetouan, Morocco

Sami Mabrouk Faculty of Sciences, University of Gafsa, Gafsa, Tunisia

Abdenacer Makhlouf IRIMAS - Département de Mathématiques, Mulhouse, France

Pedro J. Martínez-Aparicio Departamento de Matemáticas, Universidad de Almería, Almería, Spain

Laila Mesmoudi Cheikh Anta Diop University of Dakar, Dakar, Senegal

Francisco J. Navarro Izquierdo Department of Mathematics, University of Cádiz, Cádiz, Spain

Moussa Ouattara Département de Mathématiques et Informatique, Université Joseph KI-ZERBO, Ouagadougou 03, Burkina Faso

Mourad Oudghiri Département de Mathématiques, Labo LAGA, Faculté des Sciences, Université Mohammed Premier, Oujda, Morocco

Armando Reyes Departamento de Matemáticas, Universidad Nacional de Colombia, Bogotá, Colombia

Ángel Rodríguez Palacios Departamento de Análisis Matemático, Facultad de Ciencias, Universidad de Granada, Granada, Spain

Mamadou Sanghare Cheikh Anta Diop University of Dakar (UCAD), Doctoral School of Mathematics-Computer Sciences, UCAD, Senegal

Souleymane Savadogo Université Norbert Zongo, Koudougou, Burkina Faso

Mercedes Siles Molina Departamento de Álgebra, Geometría y Topología, Universidad de Málaga, Málaga, Spain

Khalid Souilah Département de Mathématiques, Labo LAGA, Faculté des Sciences, Université Mohammed Premier, Oujda, Morocco

Djiby Sow Cheikh Anta Diop University of Dakar, Dakar, Senegal

Mohammed Taous Faculty of Sciences and Technology, Moulay Ismail University, Errachidia, Morocco

Blas Torrecillas Department of Mathematics, Universidad de Almería, Almería, Spain

Moisés Villegas-Vallecillos Departamento de Matemáticas, Universidad de Cádiz, Puerto Real, Spain

Algebraic and Analytic Methods in Associative and Non-associative Structures. Applications

With the exception of one contribution, that reports on the quasilinear elliptic operator, this part contains contributions from different aspects in associative and non-associative algebras. It includes new results on the classification of evolution algebras and two exhaustive surveys. The first one reports on isometries in Lipschitz spaces and the second deals with multiplication algebras and their structure. Apart from these, we found others' contributions to ring theory and their modules, Poisson algebras, operator algebras, algebraic number theory, as well as to Gröbner bases.

Isomorphisms of Four Dimensional Perfect Non-simple Evolution Algebras

Antonio Behn, Yolanda Cabrera Casado and Mercedes Siles Molina

Abstract In this paper we complete the classification of four dimensional perfect non-simple evolution algebras (under mild conditions on the based field), started in Casado et al. (Linear Algebra and its Applications, [1]). We consider the different parametric families of evolution algebras appearing in the classification and study which algebras in the same family are isomorphic.

Keywords Evolution algebra · Perfect evolution algebra

1 Introduction

Evolution algebras appear for the first time in the 2006 paper [7] by Tian and Vojtechovsky. This work is part of the doctoral dissertation of the first author, published in [8]. After this monography there has been a flurry of activity. A systematic study of evolution algebras was started in [3], considering not only finite dimensional evolution algebras as in [8]. In the paper [6] the two dimensional evolution algebras

Y. Cabrera Casado and M. Siles Molina—have been supported by the Junta de Andalucía and Fondos FEDER, jointly, through project FQM-336 and also by the Spanish Ministerio de Economía y Competitividad and Fondos FEDER, jointly, through project MTM2016-76327-C3-1-P.

A. Behn
Departamento de Matemática, Facultad de Ciencias, Universidad de Chile,
Casilla 653, Santiago, Chile
e-mail: abehn@uchile.cl

Y. Cabrera Casado (✉)
Departamento de Matemática Aplicada, Universidad de Málaga,
Campus de Teatinos s/n., 29071 Málaga, Spain
e-mail: yolandacc@uma.es

M. Siles Molina
Departamento de Álgebra, Geometría y Topología, Universidad de Málaga,
Campus de Teatinos s/n., 29071 Málaga, Spain
e-mail: msilesm@uma.es

M. Siles Molina et al. (eds.), *Associative and Non-Associative Algebras and Applications*, Springer Proceedings in Mathematics & Statistics 311,
https://doi.org/10.1007/978-3-030-35256-1_1

are classified (on the complex field); over a general field (with mild restrictions), there are 10 mutually non-isomorphic families of evolution algebras (see [5]). The three dimensional evolution algebras over a field having characteristic different from 2 and in which there are roots of orders 2, 3 and 7 were classified in [2, 5], being listed 116 non-isomorphic families. As for the classification of the four dimensional evolution algebras, it was initiated in [1]. The reducible case (i.e. when the evolution algebra is the sum of two non-trivial evolution ideals) can be trivially completed taking into account the classification of the two and three dimensional evolution algebras, achieved in [2, 5]. The case of four dimensional irreducible evolution algebras, has been treated in [1], where a classification into families of the four dimensional perfect non-simple evolution algebras was given (an algebra A is said to be perfect if $A^2 = A$). The simple four dimensional evolution algebras will appear in the forthcoming paper [4].

In the mentioned paper [1] the authors classify in families of mutually non-isomorphic evolution algebras. Each family is determined by some parameters (this is the reason because of which we speak about parametric families). The aim of this article is to study which algebras lying in the same parametric family are isomorphic. By a family of evolution algebras we mean all the algebras whose structure matrices have the same number of zeros and these zeros are in the same place.

We have divided the paper into four sections, being the first one this introduction. The other sections (as well as the subsections) coincide in name with Sects. 3, 4 and 5 in [1] (and the subsections with the corresponding ones).

Every four dimensional perfect evolution algebra A contains a maximal evolution ideal which has a natural basis that can be extended to a natural basis of the whole algebra. Such an ideal is called basic ideal (see [1, Lemma 2.5]). Each of the sections that appears in the classification given in [1] corresponds to each of the possible dimensions that maximal basic ideals in A can have (1, 2 or 3). Maximal basic ideals are unique when their dimension is 1 or 2.

The study we carry out relays on three facts. The first one is the relation between structure matrices of the same evolution algebra relative to different natural basis given in [2, Theorem 2.2]. The second one is the description of the structure matrix into blocks, which appear in [1, Proposition 2.13]. The third and last one is that the only change of basis matrices which are possible are the elements in $S_4 \rtimes (\mathbb{K}^\times)^4$ (see [1, Proposition 2.13 (iv)]).

For our purposes we have designed a program with the free software SageMath whose output is a triple consisting of: the first and the third component of the triple are the structure matrices in the same family which are isomorphic and the second component is the corresponding change of basis matrix.

Recall (from [2]) that given two basis B and B' of an evolution algebra A, the structure matrices M_B and $M_{B'}$ relative to the bases B and B', respectively, are related by the formula:

$$M_{B'} = P^{-1} M_B P^{(2)},$$

where P is the change of basis matrix from B to B'.

Now we explain how the program works. In the particular case of a four-dimensional perfect evolution algebra, given a matrix M in a family (which corresponds to an structure matrix), the program computes all the matrices $P^{-1}MP^{(2)}$, where $P \in S_4 \rtimes (\mathbb{K}^\times)^4$ and shows only those which are in the same family (i.e. they have the same form).

The list of routines that we have used, together with a brief description of them, follows:

- **new_M**: Computes the new structure matrix associated to P.
- **ring_definition**: Defines the Polynomial Ring of the variables used for structure matrix and change of basis matrix.
- **reduce_fraction**: Simplify an algebraic fraction.
- **isomorphism_M**: Computes the structure matrices that belong to the same family.

The *SageMath* code of the routines is the following:

```
def new_M(M, P) :
    m2 = matrix([M * v.pairwise_product(v) for v in P.columns])
    return P.inverse() * (m2.transpose())
def reduce_fraction(I1) :
    def Mreducefraction(x) :
        return I1.reduce(x.numerator())/I1.reduce(x.denominator())
    return Mreducefraction
def ring_definition(r) :
    WW = ['w' + Ã,Â´str(k) + str(j) for k in range (1, r + 1) for j in range (1, r + 1)]
    XX = ['x' + str(k) for k in range(1, r + 1)]
    Rax = PolynomialRing(QQ, r^2 + r + 1, names = XX + ['z'] + AA, order =' lex')
    Rax.inject_variables()
    return Rax
def isomorphism_M(M) :
    r = M.nrows()
    Sn = SymmetricGroup(r)
    Rax = M.base_ring()
    WW = Rax.gens()[r + 1 :]
    XX = Rax.gens()[: r]
    Z = Rax.gens()[r]
    ones = [[i, j] for i in range(r) for j in range(r) if M[i, j] == 1]
    mx = diagonal_matrix(XX)
    T = []
    for g in Sn :
        m1 = permutation_action(g, identity_matrix(r))
        M1 = new_M(M, mx * m1)
        L1 = [mx.det() * Z - 1]
        for k in ones :
            L1.append(M1[k].numerator() - M1[k].denominator())
            I1 = Ideal(L1).radical()
            G1 = I1.groebner_basis()
            p = I1.reduce(mx.det())
            if M.nonzero_positions() == M1.nonzero_positions() and p! = 0 :
                m2 = (mx * m1).apply_map(I1.reduce)
                M2 = M1.apply_map(reduce_fraction(I1))
                M0 = M.apply_map(I1.reduce)
                T.append([M0, m2, M2])
    return pretty_print(table(T))
```

In the forthcoming sections we will consider that the evolution algebra A has a maximal basic ideal I. Note that, since A is non-simple, the dimension of I is at least one.

Assume that $B = \{e_1, e_2, e_3, e_4\}$ is a natural basis of A, decompose $\{1, 2, 3, 4\} = \Lambda_1 \sqcup \Lambda_2$ in such a way that $\{e_j\}_{j\in\Lambda_1}$ is a natural basis for I. There is no loss in generality if we suppose that $\Lambda_1 = \{1, \ldots, s\}$. Then,

$$M_B = \begin{pmatrix} W & U \\ 0 & Y \end{pmatrix}. \tag{1}$$

Note that W is the structure matrix of I relative to the natural basis $\{e_1, \ldots, e_s\}$. By [1, Proposition 2.10], A/I is a basic simple evolution algebra and Y can be seen as the structure matrix of A/I relative to the natural basis $\{\overline{e_{s+1}}, \ldots, \overline{e_4}\}$.

The classification is divided into three blocks, depending on s, which can be 1, 2 or 3. Each block is divided into different pieces depending on the number of zeros in the matrix U, which is invariant by [1, Theorem 2.15].

2 The Maximal Basic Ideal Is One-Dimensional

In this section we assume that the maximal basic ideal I is one-dimensional. If we write

$$M_B = \left(\begin{array}{c|ccc} 1 & \omega_{12} & \omega_{13} & \omega_{14} \\ \hline 0 & & & \\ 0 & & Y & \\ 0 & & & \end{array}\right) \tag{2}$$

then the matrix U is $(\omega_{12}\ \omega_{13}\ \omega_{14})$. Observe that ω_{12}, ω_{13} and ω_{14} cannot be zero at the same time because otherwise the evolution algebra will be reducible.

2.1 *The Matrix U Has Two Zero Entries*

M_B	$P_{B'B}$	M'_B	M_B	$P_{B'B}$	M'_B	M_B	$P_{B'B}$	M'_B
$\begin{pmatrix}1&1&0&0\\0&0&0&w_{24}\\0&w_{32}&0&0\\0&0&w_{43}&0\end{pmatrix}$	$\begin{pmatrix}1&0&0&0\\0&x_3x_4z&0&0\\0&0&x_3&0\\0&0&0&x_4\end{pmatrix}$	$\begin{pmatrix}1&1&0&0\\0&0&0&\frac{x_4w_{24}}{x_3z}\\0&\frac{w_{32}}{x_3}&0&0\\0&0&\frac{x_3^2w_{43}}{x_4}&0\end{pmatrix}$	$\begin{pmatrix}1&1&0&0\\0&0&w_{23}&0\\0&w_{32}&0&w_{34}\\0&0&w_{43}&w_{44}\end{pmatrix}$	$\begin{pmatrix}1&0&0&0\\0&x_3x_4z&0&0\\0&0&x_3&0\\0&0&0&x_4\end{pmatrix}$	$\begin{pmatrix}1&1&0&0\\0&0&\frac{x_3w_{23}}{x_4z}&0\\0&\frac{w_{32}}{x_3}&0&\frac{x_4^2w_{34}}{x_3}\\0&0&\frac{x_3^2w_{43}}{x_4}&x_4w_{44}\end{pmatrix}$	$\begin{pmatrix}1&1&0&0\\0&0&w_{23}&0\\0&0&w_{33}&w_{34}\\0&w_{42}&w_{43}&0\end{pmatrix}$	$\begin{pmatrix}1&0&0&0\\0&x_3x_4z&0&0\\0&0&x_3&0\\0&0&0&x_4\end{pmatrix}$	$\begin{pmatrix}1&1&0&0\\0&0&\frac{x_3w_{23}}{x_4z}&0\\0&0&x_3w_{33}&\frac{x_4^2w_{34}}{x_3}\\0&\frac{w_{42}}{x_4}&\frac{x_3^2w_{43}}{x_4}&0\end{pmatrix}$
$\begin{pmatrix}1&1&0&0\\0&w_{22}&0&w_{24}\\0&w_{32}&0&0\\0&0&w_{43}&0\end{pmatrix}$	$\begin{pmatrix}1&0&0&0\\0&x_3x_4z&0&0\\0&0&x_3&0\\0&0&0&x_4\end{pmatrix}$	$\begin{pmatrix}1&1&0&0\\0&x_3x_4zw_{22}&0&\frac{x_4w_{24}}{x_3z}\\0&\frac{w_{32}}{x_3}&0&0\\0&0&\frac{x_3^2w_{43}}{x_4}&0\end{pmatrix}$	$\begin{pmatrix}1&1&0&0\\0&0&w_{23}&w_{24}\\0&w_{32}&0&0\\0&w_{42}&0&w_{44}\end{pmatrix}$	$\begin{pmatrix}1&0&0&0\\0&x_3x_4z&0&0\\0&0&x_3&0\\0&0&0&x_4\end{pmatrix}$	$\begin{pmatrix}1&1&0&0\\0&0&\frac{x_3w_{23}}{x_4z}&\frac{x_4w_{24}}{x_3z}\\0&\frac{w_{32}}{x_3}&0&0\\0&\frac{w_{42}}{x_4}&0&x_4w_{44}\end{pmatrix}$	$\begin{pmatrix}1&1&0&0\\0&0&w_{23}&w_{24}\\0&0&0&w_{34}\\0&w_{42}&0&w_{44}\end{pmatrix}$	$\begin{pmatrix}1&0&0&0\\0&x_3x_4z&0&0\\0&0&x_3&0\\0&0&0&x_4\end{pmatrix}$	$\begin{pmatrix}1&1&0&0\\0&0&\frac{x_3w_{23}}{x_4z}&\frac{x_4w_{24}}{x_3z}\\0&0&0&\frac{x_4^2w_{34}}{x_3}\\0&\frac{w_{42}}{x_4}&0&x_4w_{44}\end{pmatrix}$
$\begin{pmatrix}1&1&0&0\\0&0&w_{23}&0\\0&0&w_{33}&w_{34}\\0&w_{42}&0&0\end{pmatrix}$	$\begin{pmatrix}1&0&0&0\\0&x_3x_4z&0&0\\0&0&x_3&0\\0&0&0&x_4\end{pmatrix}$	$\begin{pmatrix}1&1&0&0\\0&0&\frac{x_3w_{23}}{x_4z}&0\\0&0&x_3w_{33}&\frac{x_4^2w_{34}}{x_3}\\0&\frac{w_{42}}{x_4}&0&0\end{pmatrix}$	$\begin{pmatrix}1&1&0&0\\0&w_{22}&w_{23}&0\\0&w_{32}&0&w_{34}\\0&0&w_{43}&0\end{pmatrix}$	$\begin{pmatrix}1&0&0&0\\0&x_3x_4z&0&0\\0&0&x_3&0\\0&0&0&x_4\end{pmatrix}$	$\begin{pmatrix}1&1&0&0\\0&x_3x_4zw_{22}&\frac{x_3w_{23}}{x_4z}&0\\0&\frac{w_{32}}{x_3}&0&\frac{x_4^2w_{34}}{x_3}\\0&0&\frac{x_3^2w_{43}}{x_4}&0\end{pmatrix}$	$\begin{pmatrix}1&1&0&0\\0&0&w_{23}&w_{24}\\0&w_{32}&0&w_{34}\\0&0&w_{43}&0\end{pmatrix}$	$\begin{pmatrix}1&0&0&0\\0&x_3x_4z&0&0\\0&0&x_3&0\\0&0&0&x_4\end{pmatrix}$	$\begin{pmatrix}1&1&0&0\\0&0&\frac{x_3w_{23}}{x_4z}&\frac{x_4w_{24}}{x_3z}\\0&\frac{w_{32}}{x_3}&0&\frac{x_4^2w_{34}}{x_3}\\0&0&\frac{x_3^2w_{43}}{x_4}&0\end{pmatrix}$
$\begin{pmatrix}1&1&0&0\\0&0&w_{23}&0\\0&0&0&w_{34}\\0&w_{42}&0&w_{44}\end{pmatrix}$	$\begin{pmatrix}1&0&0&0\\0&x_3x_4z&0&0\\0&0&x_3&0\\0&0&0&x_4\end{pmatrix}$	$\begin{pmatrix}1&1&0&0\\0&0&\frac{x_3w_{23}}{x_4z}&0\\0&0&0&\frac{x_4^2w_{34}}{x_3}\\0&\frac{w_{42}}{x_4}&0&x_4w_{44}\end{pmatrix}$	$\begin{pmatrix}1&1&0&0\\0&w_{22}&w_{23}&w_{24}\\0&w_{32}&0&0\\0&0&w_{43}&0\end{pmatrix}$	$\begin{pmatrix}1&0&0&0\\0&x_3x_4z&0&0\\0&0&x_3&0\\0&0&0&x_4\end{pmatrix}$	$\begin{pmatrix}1&1&0&0\\0&x_3x_4zw_{22}&\frac{x_3w_{23}}{x_4z}&\frac{x_4w_{24}}{x_3z}\\0&\frac{w_{32}}{x_3}&0&0\\0&0&\frac{x_3^2w_{43}}{x_4}&0\end{pmatrix}$	$\begin{pmatrix}1&1&0&0\\0&0&w_{23}&w_{24}\\0&w_{32}&0&w_{34}\\0&w_{42}&0&0\end{pmatrix}$	$\begin{pmatrix}1&0&0&0\\0&x_3x_4z&0&0\\0&0&x_3&0\\0&0&0&x_4\end{pmatrix}$	$\begin{pmatrix}1&1&0&0\\0&0&\frac{x_3w_{23}}{x_4z}&\frac{x_4w_{24}}{x_3z}\\0&\frac{w_{32}}{x_3}&0&\frac{x_4^2w_{34}}{x_3}\\0&\frac{w_{42}}{x_4}&0&0\end{pmatrix}$
$\begin{pmatrix}1&1&0&0\\0&0&w_{23}&w_{24}\\0&w_{32}&0&0\\0&0&w_{43}&0\end{pmatrix}$	$\begin{pmatrix}1&0&0&0\\0&x_3x_4z&0&0\\0&0&x_3&0\\0&0&0&x_4\end{pmatrix}$	$\begin{pmatrix}1&1&0&0\\0&0&\frac{x_3w_{23}}{x_4z}&\frac{x_4w_{24}}{x_3z}\\0&\frac{w_{32}}{x_3}&0&0\\0&0&\frac{x_3^2w_{43}}{x_4}&0\end{pmatrix}$	$\begin{pmatrix}1&1&0&0\\0&0&w_{23}&0\\0&w_{32}&w_{33}&w_{34}\\0&w_{42}&0&0\end{pmatrix}$	$\begin{pmatrix}1&0&0&0\\0&x_3x_4z&0&0\\0&0&x_3&0\\0&0&0&x_4\end{pmatrix}$	$\begin{pmatrix}1&1&0&0\\0&0&\frac{x_3w_{23}}{x_4z}&0\\0&\frac{w_{32}}{x_3}&x_3w_{33}&\frac{x_4^2w_{34}}{x_3}\\0&\frac{w_{42}}{x_4}&0&0\end{pmatrix}$	$\begin{pmatrix}1&1&0&0\\0&0&w_{23}&0\\0&w_{32}&0&w_{34}\\0&w_{42}&w_{43}&0\end{pmatrix}$	$\begin{pmatrix}1&0&0&0\\0&x_3x_4z&0&0\\0&0&x_3&0\\0&0&0&x_4\end{pmatrix}$	$\begin{pmatrix}1&1&0&0\\0&0&\frac{x_3w_{23}}{x_4z}&0\\0&\frac{w_{32}}{x_3}&0&\frac{x_4^2w_{34}}{x_3}\\0&\frac{w_{42}}{x_4}&\frac{x_3^2w_{43}}{x_4}&0\end{pmatrix}$
$\begin{pmatrix}1&1&0&0\\0&0&w_{23}&0\\0&w_{32}&0&w_{34}\\0&w_{42}&0&0\end{pmatrix}$	$\begin{pmatrix}1&0&0&0\\0&x_3x_4z&0&0\\0&0&x_3&0\\0&0&0&x_4\end{pmatrix}$	$\begin{pmatrix}1&1&0&0\\0&0&\frac{x_3w_{23}}{x_4z}&0\\0&\frac{w_{32}}{x_3}&0&\frac{x_4^2w_{34}}{x_3}\\0&\frac{w_{42}}{x_4}&0&0\end{pmatrix}$	$\begin{pmatrix}1&1&0&0\\0&0&w_{23}&0\\0&0&0&w_{34}\\0&w_{42}&w_{43}&w_{44}\end{pmatrix}$	$\begin{pmatrix}1&0&0&0\\0&x_3x_4z&0&0\\0&0&x_3&0\\0&0&0&x_4\end{pmatrix}$	$\begin{pmatrix}1&1&0&0\\0&0&\frac{x_3w_{23}}{x_4z}&0\\0&0&0&\frac{x_4^2w_{34}}{x_3}\\0&\frac{w_{42}}{x_4}&\frac{x_3^2w_{43}}{x_4}&x_4w_{44}\end{pmatrix}$	$\begin{pmatrix}1&1&0&0\\0&w_{22}&w_{23}&0\\0&0&w_{33}&w_{34}\\0&w_{42}&0&w_{44}\end{pmatrix}$	$\begin{pmatrix}1&0&0&0\\0&x_3x_4z&0&0\\0&0&x_3&0\\0&0&0&x_4\end{pmatrix}$	$\begin{pmatrix}1&1&0&0\\0&x_3x_4zw_{22}&\frac{x_3w_{23}}{x_4z}&0\\0&0&x_3w_{33}&\frac{x_4^2w_{34}}{x_3}\\0&\frac{w_{42}}{x_4}&0&x_4w_{44}\end{pmatrix}$
$\begin{pmatrix}1&1&0&0\\0&0&w_{23}&0\\0&0&0&w_{34}\\0&w_{42}&w_{43}&0\end{pmatrix}$	$\begin{pmatrix}1&0&0&0\\0&x_3x_4z&0&0\\0&0&x_3&0\\0&0&0&x_4\end{pmatrix}$	$\begin{pmatrix}1&1&0&0\\0&0&\frac{x_3w_{23}}{x_4z}&0\\0&0&0&\frac{x_4^2w_{34}}{x_3}\\0&\frac{w_{42}}{x_4}&\frac{x_3^2w_{43}}{x_4}&0\end{pmatrix}$	$\begin{pmatrix}1&1&0&0\\0&w_{22}&0&w_{24}\\0&w_{32}&w_{33}&0\\0&0&w_{43}&0\end{pmatrix}$	$\begin{pmatrix}1&0&0&0\\0&x_3x_4z&0&0\\0&0&x_3&0\\0&0&0&x_4\end{pmatrix}$	$\begin{pmatrix}1&1&0&0\\0&x_3x_4zw_{22}&0&\frac{x_4w_{24}}{x_3z}\\0&\frac{w_{32}}{x_3}&x_3w_{33}&0\\0&0&\frac{x_3^2w_{43}}{x_4}&0\end{pmatrix}$	$\begin{pmatrix}1&1&0&0\\0&w_{22}&w_{23}&w_{24}\\0&w_{32}&0&0\\0&w_{42}&0&w_{44}\end{pmatrix}$	$\begin{pmatrix}1&0&0&0\\0&x_3x_4z&0&0\\0&0&x_3&0\\0&0&0&x_4\end{pmatrix}$	$\begin{pmatrix}1&1&0&0\\0&x_3x_4zw_{22}&\frac{x_3w_{23}}{x_4z}&\frac{x_4w_{24}}{x_3z}\\0&\frac{w_{32}}{x_3}&0&0\\0&\frac{w_{42}}{x_4}&0&x_4w_{44}\end{pmatrix}$
$\begin{pmatrix}1&1&0&0\\0&0&w_{23}&0\\0&w_{32}&0&w_{34}\\0&w_{42}&0&w_{44}\end{pmatrix}$	$\begin{pmatrix}1&0&0&0\\0&x_3x_4z&0&0\\0&0&x_3&0\\0&0&0&x_4\end{pmatrix}$	$\begin{pmatrix}1&1&0&0\\0&0&\frac{x_3w_{23}}{x_4z}&0\\0&\frac{w_{32}}{x_3}&0&\frac{x_4^2w_{34}}{x_3}\\0&\frac{w_{42}}{x_4}&0&x_4w_{44}\end{pmatrix}$	$\begin{pmatrix}1&1&0&0\\0&w_{22}&w_{23}&0\\0&0&w_{33}&w_{34}\\0&w_{42}&0&0\end{pmatrix}$	$\begin{pmatrix}1&0&0&0\\0&x_3x_4z&0&0\\0&0&x_3&0\\0&0&0&x_4\end{pmatrix}$	$\begin{pmatrix}1&1&0&0\\0&x_3x_4zw_{22}&\frac{x_3w_{23}}{x_4z}&0\\0&0&x_3w_{33}&\frac{x_4^2w_{34}}{x_3}\\0&\frac{w_{42}}{x_4}&0&0\end{pmatrix}$	$\begin{pmatrix}1&1&0&0\\0&0&w_{23}&0\\0&w_{32}&w_{33}&w_{34}\\0&0&w_{43}&w_{44}\end{pmatrix}$	$\begin{pmatrix}1&0&0&0\\0&x_3x_4z&0&0\\0&0&x_3&0\\0&0&0&x_4\end{pmatrix}$	$\begin{pmatrix}1&1&0&0\\0&0&\frac{x_3w_{23}}{x_4z}&0\\0&\frac{w_{32}}{x_3}&x_3w_{33}&\frac{x_4^2w_{34}}{x_3}\\0&0&\frac{x_3^2w_{43}}{x_4}&x_4w_{44}\end{pmatrix}$
$\begin{pmatrix}1&1&0&0\\0&0&w_{23}&w_{24}\\0&w_{32}&0&0\\0&0&w_{43}&w_{44}\end{pmatrix}$	$\begin{pmatrix}1&0&0&0\\0&x_3x_4z&0&0\\0&0&x_3&0\\0&0&0&x_4\end{pmatrix}$	$\begin{pmatrix}1&1&0&0\\0&0&\frac{x_3w_{23}}{x_4z}&\frac{x_4w_{24}}{x_3z}\\0&\frac{w_{32}}{x_3}&0&0\\0&0&\frac{x_3^2w_{43}}{x_4}&x_4w_{44}\end{pmatrix}$	$\begin{pmatrix}1&1&0&0\\0&0&w_{23}&0\\0&0&w_{33}&w_{34}\\0&w_{42}&0&w_{44}\end{pmatrix}$	$\begin{pmatrix}1&0&0&0\\0&x_3x_4z&0&0\\0&0&x_3&0\\0&0&0&x_4\end{pmatrix}$	$\begin{pmatrix}1&1&0&0\\0&0&\frac{x_3w_{23}}{x_4z}&0\\0&0&x_3w_{33}&\frac{x_4^2w_{34}}{x_3}\\0&\frac{w_{42}}{x_4}&0&x_4w_{44}\end{pmatrix}$	$\begin{pmatrix}1&1&0&0\\0&w_{22}&0&w_{24}\\0&0&0&w_{34}\\0&w_{42}&w_{43}&w_{44}\end{pmatrix}$	$\begin{pmatrix}1&0&0&0\\0&x_3x_4z&0&0\\0&0&x_3&0\\0&0&0&x_4\end{pmatrix}$	$\begin{pmatrix}1&1&0&0\\0&x_3x_4zw_{22}&0&\frac{x_4w_{24}}{x_3z}\\0&0&0&\frac{x_4^2w_{34}}{x_3}\\0&\frac{w_{42}}{x_4}&\frac{x_3^2w_{43}}{x_4}&x_4w_{44}\end{pmatrix}$
$\begin{pmatrix}1&1&0&0\\0&w_{22}&0&w_{24}\\0&w_{32}&0&w_{34}\\0&0&w_{43}&0\end{pmatrix}$	$\begin{pmatrix}1&0&0&0\\0&x_3x_4z&0&0\\0&0&x_3&0\\0&0&0&x_4\end{pmatrix}$	$\begin{pmatrix}1&1&0&0\\0&x_3x_4zw_{22}&0&\frac{x_4w_{24}}{x_3z}\\0&\frac{w_{32}}{x_3}&0&\frac{x_4^2w_{34}}{x_3}\\0&0&\frac{x_3^2w_{43}}{x_4}&0\end{pmatrix}$	$\begin{pmatrix}1&1&0&0\\0&w_{22}&0&w_{24}\\0&w_{32}&0&0\\0&w_{42}&w_{43}&0\end{pmatrix}$	$\begin{pmatrix}1&0&0&0\\0&x_3x_4z&0&0\\0&0&x_3&0\\0&0&0&x_4\end{pmatrix}$	$\begin{pmatrix}1&1&0&0\\0&x_3x_4zw_{22}&0&\frac{x_4w_{24}}{x_3z}\\0&\frac{w_{32}}{x_3}&0&0\\0&\frac{w_{42}}{x_4}&\frac{x_3^2w_{43}}{x_4}&0\end{pmatrix}$	$\begin{pmatrix}1&1&0&0\\0&w_{22}&w_{23}&w_{24}\\0&w_{32}&0&0\\0&0&w_{43}&w_{44}\end{pmatrix}$	$\begin{pmatrix}1&0&0&0\\0&x_3x_4z&0&0\\0&0&x_3&0\\0&0&0&x_4\end{pmatrix}$	$\begin{pmatrix}1&1&0&0\\0&x_3x_4zw_{22}&\frac{x_3w_{23}}{x_4z}&\frac{x_4w_{24}}{x_3z}\\0&\frac{w_{32}}{x_3}&0&0\\0&0&\frac{x_3^2w_{43}}{x_4}&x_4w_{44}\end{pmatrix}$

M_B	$P_{B'B}$	M'_B
$\begin{pmatrix}1&1&0&0\\0&0&w_{23}&0\\0&w_{32}&w_{33}&w_{34}\\0&w_{42}&0&w_{44}\end{pmatrix}$	$\begin{pmatrix}1&0&0&0\\0&x_3x_4z&0&0\\0&0&x_3&0\\0&0&0&x_4\end{pmatrix}$	$\begin{pmatrix}1&1&0&0\\0&0&\frac{x_3w_{23}}{x_4z}&0\\0&\frac{w_{32}}{x_3}&x_3w_{33}&\frac{x_4^2w_{34}}{x_3}\\0&\frac{w_{42}}{x_4}&0&x_4w_{44}\end{pmatrix}$
$\begin{pmatrix}1&1&0&0\\0&w_{23}&w_{32}&0\\0&0&0&w_{33}\\0&w_{34}&w_{42}&w_{44}\end{pmatrix}$	$\begin{pmatrix}1&0&0&0\\0&x_3x_4z&0&0\\0&0&x_3&0\\0&0&0&x_4\end{pmatrix}$	$\begin{pmatrix}1&1&0&0\\0&x_3x_4zw_{23}&\frac{x_3w_{32}}{x_4z}&0\\0&0&0&\frac{x_4^2w_{33}}{x_3}\\0&\frac{w_{34}}{x_4}&\frac{x_3^2w_{42}}{x_4}&x_4w_{44}\end{pmatrix}$
$\begin{pmatrix}1&1&0&0\\0&0&w_{23}&w_{24}\\0&w_{32}&w_{33}&0\\0&w_{42}&0&w_{44}\end{pmatrix}$	$\begin{pmatrix}1&0&0&0\\0&x_3x_4z&0&0\\0&0&x_3&0\\0&0&0&x_4\end{pmatrix}$	$\begin{pmatrix}1&1&0&0\\0&0&\frac{x_3w_{23}}{x_4z}&\frac{x_4w_{24}}{x_3z}\\0&\frac{w_{32}}{x_3}&x_3w_{33}&0\\0&\frac{w_{42}}{x_4}&0&x_4w_{44}\end{pmatrix}$
$\begin{pmatrix}1&1&0&0\\0&0&w_{23}&w_{24}\\0&w_{32}&w_{33}&0\\0&w_{42}&0&w_{44}\end{pmatrix}$	$\begin{pmatrix}1&0&0&0\\0&x_3x_4z&0&0\\0&0&0&x_3\\0&0&x_4&0\end{pmatrix}$	$\begin{pmatrix}1&1&0&0\\0&0&\frac{x_4w_{24}}{x_3z}&\frac{x_3w_{23}}{x_4z}\\0&\frac{w_{42}}{x_4}&x_4w_{44}&0\\0&\frac{w_{32}}{x_3}&0&x_3w_{33}\end{pmatrix}$
$\begin{pmatrix}1&1&0&0\\0&w_{22}&w_{23}&0\\0&w_{32}&0&w_{34}\\0&0&w_{43}&w_{44}\end{pmatrix}$	$\begin{pmatrix}1&0&0&0\\0&x_3x_4z&0&0\\0&0&x_3&0\\0&0&0&x_4\end{pmatrix}$	$\begin{pmatrix}1&1&0&0\\0&x_3x_4zw_{22}&\frac{x_3w_{23}}{x_4z}&0\\0&\frac{w_{32}}{x_3}&0&\frac{x_4^2w_{34}}{x_3}\\0&0&\frac{x_3^2w_{43}}{x_4}&x_4w_{44}\end{pmatrix}$
$\begin{pmatrix}1&1&0&0\\0&w_{22}&w_{23}&0\\0&w_{32}&0&w_{34}\\0&0&w_{43}&w_{44}\end{pmatrix}$	$\begin{pmatrix}1&0&0&0\\0&x_3x_4z&0&0\\0&0&x_3&0\\0&0&0&x_4\end{pmatrix}$	$\begin{pmatrix}1&1&0&0\\0&x_3x_4zw_{22}&\frac{x_3w_{23}}{x_4z}&0\\0&\frac{w_{32}}{x_3}&0&\frac{x_4^2w_{34}}{x_3}\\0&0&\frac{x_3^2w_{43}}{x_4}&x_4w_{44}\end{pmatrix}$
$\begin{pmatrix}1&1&0&0\\0&w_{22}&w_{23}&0\\0&w_{32}&0&w_{34}\\0&w_{42}&0&w_{44}\end{pmatrix}$	$\begin{pmatrix}1&0&0&0\\0&x_3x_4z&0&0\\0&0&x_3&0\\0&0&0&x_4\end{pmatrix}$	$\begin{pmatrix}1&1&0&0\\0&x_3x_4zw_{22}&\frac{x_3w_{23}}{x_4z}&0\\0&\frac{w_{32}}{x_3}&0&\frac{x_4^2w_{34}}{x_3}\\0&\frac{w_{42}}{x_4}&0&x_4w_{44}\end{pmatrix}$
$\begin{pmatrix}1&1&0&0\\0&w_{22}&w_{23}&0\\0&0&w_{33}&w_{34}\\0&w_{42}&w_{43}&0\end{pmatrix}$	$\begin{pmatrix}1&0&0&0\\0&x_3x_4z&0&0\\0&0&x_3&0\\0&0&0&x_4\end{pmatrix}$	$\begin{pmatrix}1&1&0&0\\0&x_3x_4zw_{22}&\frac{x_3w_{23}}{x_4z}&0\\0&0&x_3w_{33}&\frac{x_4^2w_{34}}{x_3}\\0&\frac{w_{42}}{x_4}&\frac{x_3^2w_{43}}{x_4}&0\end{pmatrix}$
$\begin{pmatrix}1&1&0&0\\0&0&w_{23}&w_{24}\\0&w_{32}&0&w_{34}\\0&w_{42}&0&w_{44}\end{pmatrix}$	$\begin{pmatrix}1&0&0&0\\0&x_3x_4z&0&0\\0&0&x_3&0\\0&0&0&x_4\end{pmatrix}$	$\begin{pmatrix}1&1&0&0\\0&0&\frac{x_3w_{23}}{x_4z}&\frac{x_4w_{24}}{x_3z}\\0&\frac{w_{32}}{x_3}&0&\frac{x_4^2w_{34}}{x_3}\\0&\frac{w_{42}}{x_4}&0&x_4w_{44}\end{pmatrix}$
$\begin{pmatrix}1&1&0&0\\0&0&w_{23}&w_{24}\\0&w_{32}&0&w_{34}\\0&0&w_{43}&w_{44}\end{pmatrix}$	$\begin{pmatrix}1&0&0&0\\0&x_3x_4z&0&0\\0&0&x_3&0\\0&0&0&x_4\end{pmatrix}$	$\begin{pmatrix}1&1&0&0\\0&0&\frac{x_3w_{23}}{x_4z}&\frac{x_4w_{24}}{x_3z}\\0&\frac{w_{32}}{x_3}&0&\frac{x_4^2w_{34}}{x_3}\\0&0&\frac{x_3^2w_{43}}{x_4}&x_4w_{44}\end{pmatrix}$

M_B	$P_{B'B}$	M'_B
$\begin{pmatrix}1&1&0&0\\0&w_{22}&0&w_{24}\\0&w_{32}&0&w_{34}\\0&w_{42}&w_{43}&0\end{pmatrix}$	$\begin{pmatrix}1&0&0&0\\0&x_3x_4z&0&0\\0&0&x_3&0\\0&0&0&x_4\end{pmatrix}$	$\begin{pmatrix}1&1&0&0\\0&x_3x_4zw_{22}&0&\frac{x_4w_{24}}{x_3z}\\0&\frac{w_{32}}{x_3}&0&\frac{x_4^2w_{34}}{x_3}\\0&\frac{w_{42}}{x_4}&\frac{x_3^2w_{43}}{x_4}&0\end{pmatrix}$
$\begin{pmatrix}1&1&0&0\\0&0&w_{23}&w_{24}\\0&w_{32}&0&0\\0&w_{42}&w_{43}&w_{44}\end{pmatrix}$	$\begin{pmatrix}1&0&0&0\\0&x_3x_4z&0&0\\0&0&x_3&0\\0&0&0&x_4\end{pmatrix}$	$\begin{pmatrix}1&1&0&0\\0&0&\frac{x_3w_{23}}{x_4z}&\frac{x_4w_{24}}{x_3z}\\0&\frac{w_{32}}{x_3}&0&0\\0&\frac{w_{42}}{x_4}&\frac{x_3^2w_{43}}{x_4}&x_4w_{44}\end{pmatrix}$
$\begin{pmatrix}1&1&0&0\\0&0&w_{23}&0\\0&w_{32}&0&w_{34}\\0&w_{42}&w_{43}&w_{44}\end{pmatrix}$	$\begin{pmatrix}1&0&0&0\\0&x_3x_4z&0&0\\0&0&x_3&0\\0&0&0&x_4\end{pmatrix}$	$\begin{pmatrix}1&1&0&0\\0&0&\frac{x_3w_{23}}{x_4z}&0\\0&\frac{w_{32}}{x_3}&0&\frac{x_4^2w_{34}}{x_3}\\0&\frac{w_{42}}{x_4}&\frac{x_3^2w_{43}}{x_4}&x_4w_{44}\end{pmatrix}$
$\begin{pmatrix}1&1&0&0\\0&w_{22}&w_{23}&w_{24}\\0&0&0&w_{34}\\0&w_{42}&w_{43}&0\end{pmatrix}$	$\begin{pmatrix}1&0&0&0\\0&x_3x_4z&0&0\\0&0&x_3&0\\0&0&0&x_4\end{pmatrix}$	$\begin{pmatrix}1&1&0&0\\0&x_3x_4zw_{22}&\frac{x_3w_{23}}{x_4z}&\frac{x_4w_{24}}{x_3z}\\0&0&0&\frac{x_4^2w_{34}}{x_3}\\0&\frac{w_{42}}{x_4}&\frac{x_3^2w_{43}}{x_4}&0\end{pmatrix}$
$\begin{pmatrix}1&1&0&0\\0&w_{22}&w_{23}&w_{24}\\0&w_{32}&w_{33}&0\\0&0&w_{43}&0\end{pmatrix}$	$\begin{pmatrix}1&0&0&0\\0&x_3x_4z&0&0\\0&0&x_3&0\\0&0&0&x_4\end{pmatrix}$	$\begin{pmatrix}1&1&0&0\\0&x_3x_4zw_{22}&\frac{x_3w_{23}}{x_4z}&\frac{x_4w_{24}}{x_3z}\\0&\frac{w_{32}}{x_3}&x_3w_{33}&0\\0&0&\frac{x_3^2w_{43}}{x_4}&0\end{pmatrix}$
$\begin{pmatrix}1&1&0&0\\0&w_{22}&w_{23}&0\\0&w_{32}&w_{33}&w_{34}\\0&w_{42}&0&0\end{pmatrix}$	$\begin{pmatrix}1&0&0&0\\0&x_3x_4z&0&0\\0&0&x_3&0\\0&0&0&x_4\end{pmatrix}$	$\begin{pmatrix}1&1&0&0\\0&x_3x_4zw_{22}&\frac{x_3w_{23}}{x_4z}&0\\0&\frac{w_{32}}{x_3}&x_3w_{33}&\frac{x_4^2w_{34}}{x_3}\\0&\frac{w_{42}}{x_4}&0&0\end{pmatrix}$
$\begin{pmatrix}1&1&0&0\\0&0&w_{23}&0\\0&0&w_{33}&w_{34}\\0&w_{42}&w_{43}&w_{44}\end{pmatrix}$	$\begin{pmatrix}1&0&0&0\\0&x_3x_4z&0&0\\0&0&x_3&0\\0&0&0&x_4\end{pmatrix}$	$\begin{pmatrix}1&1&0&0\\0&0&\frac{x_3w_{23}}{x_4z}&0\\0&0&x_3w_{33}&\frac{x_4^2w_{34}}{x_3}\\0&\frac{w_{42}}{x_4}&\frac{x_3^2w_{43}}{x_4}&x_4w_{44}\end{pmatrix}$
$\begin{pmatrix}1&1&0&0\\0&w_{22}&w_{23}&w_{24}\\0&w_{32}&0&0\\0&w_{42}&w_{43}&0\end{pmatrix}$	$\begin{pmatrix}1&0&0&0\\0&x_3x_4z&0&0\\0&0&x_3&0\\0&0&0&x_4\end{pmatrix}$	$\begin{pmatrix}1&1&0&0\\0&x_3x_4zw_{22}&\frac{x_3w_{23}}{x_4z}&\frac{x_4w_{24}}{x_3z}\\0&\frac{w_{32}}{x_3}&0&0\\0&\frac{w_{42}}{x_4}&\frac{x_3^2w_{43}}{x_4}&0\end{pmatrix}$
$\begin{pmatrix}1&1&0&0\\0&0&w_{23}&0\\0&w_{32}&w_{33}&w_{34}\\0&w_{42}&w_{43}&0\end{pmatrix}$	$\begin{pmatrix}1&0&0&0\\0&x_3x_4z&0&0\\0&0&x_3&0\\0&0&0&x_4\end{pmatrix}$	$\begin{pmatrix}1&1&0&0\\0&0&\frac{x_3w_{23}}{x_4z}&0\\0&\frac{w_{32}}{x_3}&x_3w_{33}&\frac{x_4^2w_{34}}{x_3}\\0&\frac{w_{42}}{x_4}&\frac{x_3^2w_{43}}{x_4}&0\end{pmatrix}$
$\begin{pmatrix}1&1&0&0\\0&0&w_{23}&w_{24}\\0&0&0&w_{34}\\0&w_{42}&w_{43}&w_{44}\end{pmatrix}$	$\begin{pmatrix}1&0&0&0\\0&x_3x_4z&0&0\\0&0&x_3&0\\0&0&0&x_4\end{pmatrix}$	$\begin{pmatrix}1&1&0&0\\0&0&\frac{x_3w_{23}}{x_4z}&\frac{x_4w_{24}}{x_3z}\\0&0&0&\frac{x_4^2w_{34}}{x_3}\\0&\frac{w_{42}}{x_4}&\frac{x_3^2w_{43}}{x_4}&x_4w_{44}\end{pmatrix}$

M_B	$P_{B'B}$	M'_B
$\begin{pmatrix}1&1&0&0\\0&0&w_{23}&w_{24}\\0&w_{32}&0&w_{34}\\0&w_{42}&w_{43}&0\end{pmatrix}$	$\begin{pmatrix}1&0&0&0\\0&x_3x_4z&0&0\\0&0&x_3&0\\0&0&0&x_4\end{pmatrix}$	$\begin{pmatrix}1&1&0&0\\0&0&\frac{x_3w_{23}}{x_4z}&\frac{x_4w_{24}}{x_3z}\\0&\frac{w_{32}}{x_3}&0&\frac{x_4^2w_{34}}{x_3}\\0&\frac{w_{42}}{x_4}&\frac{x_3^2w_{43}}{x_4}&0\end{pmatrix}$
$\begin{pmatrix}1&1&0&0\\0&0&w_{23}&w_{24}\\0&w_{32}&0&w_{34}\\0&w_{42}&w_{43}&0\end{pmatrix}$	$\begin{pmatrix}1&0&0&0\\0&x_3x_4z&0&0\\0&0&0&x_3\\0&0&x_4&0\end{pmatrix}$	$\begin{pmatrix}1&1&0&0\\0&0&\frac{x_4w_{24}}{x_3z}&\frac{x_3w_{23}}{x_4z}\\0&\frac{w_{42}}{x_4}&0&\frac{x_3^2w_{43}}{x_4}\\0&\frac{w_{32}}{x_3}&\frac{x_4^2w_{34}}{x_3}&0\end{pmatrix}$
$\begin{pmatrix}1&1&0&0\\0&w_{22}&w_{23}&w_{24}\\0&w_{32}&w_{33}&0\\0&w_{42}&0&w_{44}\end{pmatrix}$	$\begin{pmatrix}1&0&0&0\\0&x_3x_4z&0&0\\0&0&x_3&0\\0&0&0&x_4\end{pmatrix}$	$\begin{pmatrix}1&1&0&0\\0&x_3x_4zw_{22}&\frac{x_3w_{23}}{x_4z}&\frac{x_4w_{24}}{x_3z}\\0&\frac{w_{32}}{x_3}&x_3w_{33}&0\\0&\frac{w_{42}}{x_4}&0&x_4w_{44}\end{pmatrix}$
$\begin{pmatrix}1&1&0&0\\0&w_{22}&w_{23}&w_{24}\\0&w_{32}&w_{33}&0\\0&w_{42}&0&w_{44}\end{pmatrix}$	$\begin{pmatrix}1&0&0&0\\0&x_3x_4z&0&0\\0&0&0&x_3\\0&0&x_4&0\end{pmatrix}$	$\begin{pmatrix}1&1&0&0\\0&x_3x_4zw_{22}&\frac{x_4w_{24}}{x_3z}&\frac{x_3w_{23}}{x_4z}\\0&\frac{w_{42}}{x_4}&x_4w_{44}&0\\0&\frac{w_{32}}{x_3}&0&x_3w_{33}\end{pmatrix}$
$\begin{pmatrix}1&1&0&0\\0&w_{22}&w_{23}&0\\0&w_{32}&w_{33}&w_{34}\\0&0&w_{43}&w_{44}\end{pmatrix}$	$\begin{pmatrix}1&0&0&0\\0&x_3x_4z&0&0\\0&0&x_3&0\\0&0&0&x_4\end{pmatrix}$	$\begin{pmatrix}1&1&0&0\\0&x_3x_4zw_{22}&\frac{x_3w_{23}}{x_4z}&0\\0&\frac{w_{32}}{x_3}&x_3w_{33}&\frac{x_4^2w_{34}}{x_3}\\0&0&\frac{x_3^2w_{43}}{x_4}&x_4w_{44}\end{pmatrix}$
$\begin{pmatrix}1&1&0&0\\0&w_{22}&0&w_{24}\\0&0&w_{33}&w_{34}\\0&w_{42}&w_{43}&w_{44}\end{pmatrix}$	$\begin{pmatrix}1&0&0&0\\0&x_3x_4z&0&0\\0&0&x_3&0\\0&0&0&x_4\end{pmatrix}$	$\begin{pmatrix}1&1&0&0\\0&x_3x_4zw_{22}&0&\frac{x_4w_{24}}{x_3z}\\0&0&x_3w_{33}&\frac{x_4^2w_{34}}{x_3}\\0&\frac{w_{42}}{x_4}&\frac{x_3^2w_{43}}{x_4}&x_4w_{44}\end{pmatrix}$
$\begin{pmatrix}1&1&0&0\\0&w_{22}&w_{23}&w_{24}\\0&w_{32}&w_{33}&0\\0&0&w_{43}&w_{44}\end{pmatrix}$	$\begin{pmatrix}1&0&0&0\\0&x_3x_4z&0&0\\0&0&x_3&0\\0&0&0&x_4\end{pmatrix}$	$\begin{pmatrix}1&1&0&0\\0&x_3x_4zw_{22}&\frac{x_3w_{23}}{x_4z}&\frac{x_4w_{24}}{x_3z}\\0&\frac{w_{32}}{x_3}&x_3w_{33}&0\\0&0&\frac{x_3^2w_{43}}{x_4}&x_4w_{44}\end{pmatrix}$
$\begin{pmatrix}1&1&0&0\\0&w_{22}&w_{23}&0\\0&w_{32}&w_{33}&w_{34}\\0&w_{42}&0&w_{44}\end{pmatrix}$	$\begin{pmatrix}1&0&0&0\\0&x_3x_4z&0&0\\0&0&x_3&0\\0&0&0&x_4\end{pmatrix}$	$\begin{pmatrix}1&1&0&0\\0&x_3x_4zw_{22}&\frac{x_3w_{23}}{x_4z}&0\\0&\frac{w_{32}}{x_3}&x_3w_{33}&\frac{x_4^2w_{34}}{x_3}\\0&\frac{w_{42}}{x_4}&0&x_4w_{44}\end{pmatrix}$
$\begin{pmatrix}1&1&0&0\\0&w_{22}&w_{23}&0\\0&0&w_{33}&w_{34}\\0&w_{42}&w_{43}&w_{44}\end{pmatrix}$	$\begin{pmatrix}1&0&0&0\\0&x_3x_4z&0&0\\0&0&x_3&0\\0&0&0&x_4\end{pmatrix}$	$\begin{pmatrix}1&1&0&0\\0&x_3x_4zw_{22}&\frac{x_3w_{23}}{x_4z}&0\\0&0&x_3w_{33}&\frac{x_4^2w_{34}}{x_3}\\0&\frac{w_{42}}{x_4}&\frac{x_3^2w_{43}}{x_4}&x_4w_{44}\end{pmatrix}$
$\begin{pmatrix}1&1&0&0\\0&w_{22}&w_{23}&w_{24}\\0&w_{32}&0&0\\0&w_{42}&w_{43}&w_{44}\end{pmatrix}$	$\begin{pmatrix}1&0&0&0\\0&x_3x_4z&0&0\\0&0&x_3&0\\0&0&0&x_4\end{pmatrix}$	$\begin{pmatrix}1&1&0&0\\0&x_3x_4zw_{22}&\frac{x_3w_{23}}{x_4z}&\frac{x_4w_{24}}{x_3z}\\0&\frac{w_{32}}{x_3}&0&0\\0&\frac{w_{42}}{x_4}&\frac{x_3^2w_{43}}{x_4}&x_4w_{44}\end{pmatrix}$

M_B	$P_{B'B}$	M'_B
$\begin{pmatrix}1&1&0&0\\0&0&w_{23}&0\\0&w_{32}&w_{33}&w_{34}\\0&w_{42}&w_{43}&w_{44}\end{pmatrix}$	$\begin{pmatrix}1&0&0&0\\0&x_3x_4z&0&0\\0&0&x_3&0\\0&0&0&x_4\end{pmatrix}$	$\begin{pmatrix}1&1&0&0\\0&0&\frac{x_3w_{23}}{x_4z}&0\\0&\frac{w_{32}}{x_3}&x_3w_{33}&\frac{x_4^2w_{34}}{x_3}\\0&\frac{w_{42}}{x_4}&\frac{x_3^2w_{43}}{x_4}&x_4w_{44}\end{pmatrix}$
$\begin{pmatrix}1&1&0&0\\0&w_{22}&w_{23}&w_{24}\\0&0&0&w_{34}\\0&w_{42}&w_{43}&w_{44}\end{pmatrix}$	$\begin{pmatrix}1&0&0&0\\0&x_3x_4z&0&0\\0&0&x_3&0\\0&0&0&x_4\end{pmatrix}$	$\begin{pmatrix}1&1&0&0\\0&x_3x_4zw_{22}&\frac{x_3w_{23}}{x_4z}&\frac{x_4w_{24}}{x_3z}\\0&0&0&\frac{x_4^2w_{34}}{x_3}\\0&\frac{w_{42}}{x_4}&\frac{x_3^2w_{43}}{x_4}&x_4w_{44}\end{pmatrix}$
$\begin{pmatrix}1&1&0&0\\0&w_{22}&w_{23}&w_{24}\\0&w_{32}&0&w_{34}\\0&w_{42}&0&w_{44}\end{pmatrix}$	$\begin{pmatrix}1&0&0&0\\0&x_3x_4z&0&0\\0&0&x_3&0\\0&0&0&x_4\end{pmatrix}$	$\begin{pmatrix}1&1&0&0\\0&x_3x_4zw_{22}&\frac{x_3w_{23}}{x_4z}&\frac{x_4w_{24}}{x_3z}\\0&\frac{w_{32}}{x_3}&0&\frac{x_4^2w_{34}}{x_3}\\0&\frac{w_{42}}{x_4}&0&x_4w_{44}\end{pmatrix}$
$\begin{pmatrix}1&1&0&0\\0&0&w_{23}&w_{24}\\0&w_{32}&w_{33}&w_{34}\\0&0&w_{43}&w_{44}\end{pmatrix}$	$\begin{pmatrix}1&0&0&0\\0&x_3x_4z&0&0\\0&0&x_3&0\\0&0&0&x_4\end{pmatrix}$	$\begin{pmatrix}1&1&0&0\\0&0&\frac{x_3w_{23}}{x_4z}&\frac{x_4w_{24}}{x_3z}\\0&\frac{w_{32}}{x_3}&x_3w_{33}&\frac{x_4^2w_{34}}{x_3}\\0&0&\frac{x_3^2w_{43}}{x_4}&x_4w_{44}\end{pmatrix}$
$\begin{pmatrix}1&1&0&0\\0&w_{22}&0&w_{24}\\0&w_{32}&0&w_{34}\\0&w_{42}&w_{43}&w_{44}\end{pmatrix}$	$\begin{pmatrix}1&0&0&0\\0&x_3x_4z&0&0\\0&0&x_3&0\\0&0&0&x_4\end{pmatrix}$	$\begin{pmatrix}1&1&0&0\\0&x_3x_4zw_{22}&0&\frac{x_4w_{24}}{x_3z}\\0&\frac{w_{32}}{x_3}&0&\frac{x_4^2w_{34}}{x_3}\\0&\frac{w_{42}}{x_4}&\frac{x_3^2w_{43}}{x_4}&x_4w_{44}\end{pmatrix}$
$\begin{pmatrix}1&1&0&0\\0&0&w_{23}&w_{24}\\0&w_{32}&w_{33}&w_{34}\\0&w_{42}&0&w_{44}\end{pmatrix}$	$\begin{pmatrix}1&0&0&0\\0&x_3x_4z&0&0\\0&0&x_3&0\\0&0&0&x_4\end{pmatrix}$	$\begin{pmatrix}1&1&0&0\\0&0&\frac{x_3w_{23}}{x_4z}&\frac{x_4w_{24}}{x_3z}\\0&\frac{w_{32}}{x_3}&x_3w_{33}&\frac{x_4^2w_{34}}{x_3}\\0&\frac{w_{42}}{x_4}&0&x_4w_{44}\end{pmatrix}$
$\begin{pmatrix}1&1&0&0\\0&w_{22}&w_{23}&w_{24}\\0&w_{32}&0&w_{34}\\0&0&w_{43}&w_{44}\end{pmatrix}$	$\begin{pmatrix}1&0&0&0\\0&x_3x_4z&0&0\\0&0&x_3&0\\0&0&0&x_4\end{pmatrix}$	$\begin{pmatrix}1&1&0&0\\0&x_3x_4zw_{22}&\frac{x_3w_{23}}{x_4z}&\frac{x_4w_{24}}{x_3z}\\0&\frac{w_{32}}{x_3}&0&\frac{x_4^2w_{34}}{x_3}\\0&0&\frac{x_3^2w_{43}}{x_4}&x_4w_{44}\end{pmatrix}$
$\begin{pmatrix}1&1&0&0\\0&w_{22}&0&w_{24}\\0&w_{32}&w_{33}&w_{34}\\0&w_{42}&w_{43}&0\end{pmatrix}$	$\begin{pmatrix}1&0&0&0\\0&x_3x_4z&0&0\\0&0&x_3&0\\0&0&0&x_4\end{pmatrix}$	$\begin{pmatrix}1&1&0&0\\0&x_3x_4zw_{22}&0&\frac{x_4w_{24}}{x_3z}\\0&\frac{w_{32}}{x_3}&x_3w_{33}&\frac{x_4^2w_{34}}{x_3}\\0&\frac{w_{42}}{x_4}&\frac{x_3^2w_{43}}{x_4}&0\end{pmatrix}$
$\begin{pmatrix}1&1&0&0\\0&0&w_{23}&w_{24}\\0&w_{32}&0&w_{34}\\0&w_{42}&w_{43}&w_{44}\end{pmatrix}$	$\begin{pmatrix}1&0&0&0\\0&x_3x_4z&0&0\\0&0&x_3&0\\0&0&0&x_4\end{pmatrix}$	$\begin{pmatrix}1&1&0&0\\0&0&\frac{x_3w_{23}}{x_4z}&\frac{x_4w_{24}}{x_3z}\\0&\frac{w_{32}}{x_3}&0&\frac{x_4^2w_{34}}{x_3}\\0&\frac{w_{42}}{x_4}&\frac{x_3^2w_{43}}{x_4}&x_4w_{44}\end{pmatrix}$
$\begin{pmatrix}1&1&0&0\\0&w_{22}&w_{23}&w_{24}\\0&w_{32}&0&w_{34}\\0&w_{42}&w_{43}&0\end{pmatrix}$	$\begin{pmatrix}1&0&0&0\\0&x_3x_4z&0&0\\0&0&x_3&0\\0&0&0&x_4\end{pmatrix}$	$\begin{pmatrix}1&1&0&0\\0&x_3x_4zw_{22}&\frac{x_3w_{23}}{x_4z}&\frac{x_4w_{24}}{x_3z}\\0&\frac{w_{32}}{x_3}&0&\frac{x_4^2w_{34}}{x_3}\\0&\frac{w_{42}}{x_4}&\frac{x_3^2w_{43}}{x_4}&0\end{pmatrix}$

M_B	$P_{B'B}$	M'_B
$\begin{pmatrix}1&1&0&0\\0&w_{22}&w_{23}&w_{24}\\0&w_{32}&0&w_{34}\\0&w_{42}&w_{43}&0\end{pmatrix}$	$\begin{pmatrix}1&0&0&0\\0&x_3x_4z&0&0\\0&0&0&x_3\\0&0&x_4&0\end{pmatrix}$	$\begin{pmatrix}1&1&0&0\\0&x_3x_4zw_{22}&\frac{x_4w_{24}}{x_3z}&\frac{x_3w_{23}}{x_4z}\\0&\frac{w_{42}}{x_4}&0&\frac{x_3^2w_{43}}{x_4}\\0&\frac{w_{32}}{x_3}&\frac{x_4^2w_{34}}{x_3}&0\end{pmatrix}$
$\begin{pmatrix}1&1&0&0\\0&w_{22}&w_{23}&w_{24}\\0&w_{32}&w_{33}&w_{34}\\0&w_{42}&0&w_{44}\end{pmatrix}$	$\begin{pmatrix}1&0&0&0\\0&x_3x_4z&0&0\\0&0&x_3&0\\0&0&0&x_4\end{pmatrix}$	$\begin{pmatrix}1&1&0&0\\0&x_3x_4zw_{22}&\frac{x_3w_{23}}{x_4z}&\frac{x_4w_{24}}{x_3z}\\0&\frac{w_{32}}{x_3}&x_3w_{33}&\frac{x_4^2w_{34}}{x_3}\\0&\frac{w_{42}}{x_4}&0&x_4w_{44}\end{pmatrix}$
$\begin{pmatrix}1&1&0&0\\0&w_{22}&w_{23}&w_{24}\\0&w_{32}&w_{33}&w_{34}\\0&0&w_{43}&w_{44}\end{pmatrix}$	$\begin{pmatrix}1&0&0&0\\0&x_3x_4z&0&0\\0&0&x_3&0\\0&0&0&x_4\end{pmatrix}$	$\begin{pmatrix}1&1&0&0\\0&x_3x_4zw_{22}&\frac{x_3w_{23}}{x_4z}&\frac{x_4w_{24}}{x_3z}\\0&\frac{w_{32}}{x_3}&x_3w_{33}&\frac{x_4^2w_{34}}{x_3}\\0&0&\frac{x_3^2w_{43}}{x_4}&x_4w_{44}\end{pmatrix}$
$\begin{pmatrix}1&1&0&0\\0&w_{22}&0&w_{24}\\0&w_{32}&w_{33}&w_{34}\\0&w_{42}&w_{43}&w_{44}\end{pmatrix}$	$\begin{pmatrix}1&0&0&0\\0&x_3x_4z&0&0\\0&0&x_3&0\\0&0&0&x_4\end{pmatrix}$	$\begin{pmatrix}1&1&0&0\\0&x_3x_4zw_{22}&0&\frac{x_4w_{24}}{x_3z}\\0&\frac{w_{32}}{x_3}&x_3w_{33}&\frac{x_4^2w_{34}}{x_3}\\0&\frac{w_{42}}{x_4}&\frac{x_3^2w_{43}}{x_4}&x_4w_{44}\end{pmatrix}$
$\begin{pmatrix}1&1&0&0\\0&w_{22}&w_{23}&w_{24}\\0&w_{32}&w_{33}&w_{34}\\0&w_{42}&w_{43}&0\end{pmatrix}$	$\begin{pmatrix}1&0&0&0\\0&x_3x_4z&0&0\\0&0&x_3&0\\0&0&0&x_4\end{pmatrix}$	$\begin{pmatrix}1&1&0&0\\0&x_3x_4zw_{22}&\frac{x_3w_{23}}{x_4z}&\frac{x_4w_{24}}{x_3z}\\0&\frac{w_{32}}{x_3}&x_3w_{33}&\frac{x_4^2w_{34}}{x_3}\\0&\frac{w_{42}}{x_4}&\frac{x_3^2w_{43}}{x_4}&0\end{pmatrix}$
$\begin{pmatrix}1&1&0&0\\0&0&w_{23}&w_{24}\\0&w_{32}&w_{33}&w_{34}\\0&w_{42}&w_{43}&w_{44}\end{pmatrix}$	$\begin{pmatrix}1&0&0&0\\0&x_3x_4z&0&0\\0&0&x_3&0\\0&0&0&x_4\end{pmatrix}$	$\begin{pmatrix}1&1&0&0\\0&0&\frac{x_3w_{23}}{x_4z}&\frac{x_4w_{24}}{x_3z}\\0&\frac{w_{32}}{x_3}&x_3w_{33}&\frac{x_4^2w_{34}}{x_3}\\0&\frac{w_{42}}{x_4}&\frac{x_3^2w_{43}}{x_4}&x_4w_{44}\end{pmatrix}$
$\begin{pmatrix}1&1&0&0\\0&0&w_{23}&w_{24}\\0&w_{32}&w_{33}&w_{34}\\0&w_{42}&w_{43}&w_{44}\end{pmatrix}$	$\begin{pmatrix}1&0&0&0\\0&x_3x_4z&0&0\\0&0&0&x_3\\0&0&x_4&0\end{pmatrix}$	$\begin{pmatrix}1&1&0&0\\0&0&\frac{x_4w_{24}}{x_3z}&\frac{x_3w_{23}}{x_4z}\\0&\frac{w_{42}}{x_4}&x_4w_{44}&\frac{x_3^2w_{43}}{x_4}\\0&\frac{w_{32}}{x_3}&\frac{x_4^2w_{34}}{x_3}&x_3w_{33}\end{pmatrix}$
$\begin{pmatrix}1&1&0&0\\0&w_{22}&w_{23}&w_{24}\\0&w_{32}&w_{33}&w_{34}\\0&w_{42}&w_{43}&w_{44}\end{pmatrix}$	$\begin{pmatrix}1&0&0&0\\0&x_3x_4z&0&0\\0&0&x_3&0\\0&0&0&x_4\end{pmatrix}$	$\begin{pmatrix}1&1&0&0\\0&x_3x_4zw_{22}&\frac{x_3w_{23}}{x_4z}&\frac{x_4w_{24}}{x_3z}\\0&\frac{w_{32}}{x_3}&x_3w_{33}&\frac{x_4^2w_{34}}{x_3}\\0&\frac{w_{42}}{x_4}&\frac{x_3^2w_{43}}{x_4}&x_4w_{44}\end{pmatrix}$
$\begin{pmatrix}1&1&0&0\\0&w_{22}&w_{23}&w_{24}\\0&w_{32}&w_{33}&w_{34}\\0&w_{42}&w_{43}&w_{44}\end{pmatrix}$	$\begin{pmatrix}1&0&0&0\\0&x_3x_4z&0&0\\0&0&0&x_3\\0&0&x_4&0\end{pmatrix}$	$\begin{pmatrix}1&1&0&0\\0&x_3x_4zw_{22}&\frac{x_4w_{24}}{x_3z}&\frac{x_3w_{23}}{x_4z}\\0&\frac{w_{42}}{x_4}&x_4w_{44}&\frac{x_3^2w_{43}}{x_4}\\0&\frac{w_{32}}{x_3}&\frac{x_4^2w_{34}}{x_3}&x_3w_{33}\end{pmatrix}$

2.2 The Matrix U Has One Zero Entry

M_B	$P_{B'B}$	M'_B
$\begin{pmatrix}1&0&1&1\\0&0&0&w_{24}\\0&w_{32}&0&0\\0&0&w_{43}&0\end{pmatrix}$	$\begin{pmatrix}1&0&0&0\\0&x_2&0&0\\0&0&x_3&0\\0&0&0&x_4\end{pmatrix}$	$\begin{pmatrix}1&0&1&1\\0&0&0&\frac{w_{24}}{x_2}\\0&\frac{x_2^2w_{32}}{x_3}&0&0\\0&0&\frac{w_{43}}{x_4}&0\end{pmatrix}$
$\begin{pmatrix}1&0&1&1\\0&0&w_{23}&0\\0&0&w_{33}&w_{34}\\0&w_{42}&0&0\end{pmatrix}$	$\begin{pmatrix}1&0&0&0\\0&x_2&0&0\\0&0&x_3&0\\0&0&0&x_4\end{pmatrix}$	$\begin{pmatrix}1&0&1&1\\0&0&\frac{w_{23}}{x_2}&0\\0&0&x_3w_{33}&\frac{w_{34}}{x_3}\\0&\frac{x_2^2w_{42}}{x_4}&0&0\end{pmatrix}$
$\begin{pmatrix}1&0&1&1\\0&0&w_{23}&0\\0&0&0&w_{34}\\0&w_{42}&0&w_{44}\end{pmatrix}$	$\begin{pmatrix}1&0&0&0\\0&x_2&0&0\\0&0&x_3&0\\0&0&0&x_4\end{pmatrix}$	$\begin{pmatrix}1&0&1&1\\0&0&\frac{w_{23}}{x_2}&0\\0&0&0&\frac{w_{34}}{x_3}\\0&\frac{x_2^2w_{42}}{x_4}&0&x_4w_{44}\end{pmatrix}$
$\begin{pmatrix}1&0&1&1\\0&0&w_{23}&w_{24}\\0&w_{32}&0&0\\0&0&w_{43}&0\end{pmatrix}$	$\begin{pmatrix}1&0&0&0\\0&x_2&0&0\\0&0&x_3&0\\0&0&0&x_4\end{pmatrix}$	$\begin{pmatrix}1&0&1&1\\0&0&\frac{w_{23}}{x_2}&\frac{w_{24}}{x_2}\\0&\frac{x_2^2w_{32}}{x_3}&0&0\\0&0&\frac{w_{43}}{x_4}&0\end{pmatrix}$
$\begin{pmatrix}1&0&1&1\\0&0&w_{23}&0\\0&w_{32}&0&w_{34}\\0&w_{42}&0&0\end{pmatrix}$	$\begin{pmatrix}1&0&0&0\\0&x_2&0&0\\0&0&x_3&0\\0&0&0&x_4\end{pmatrix}$	$\begin{pmatrix}1&0&1&1\\0&0&\frac{w_{23}}{x_2}&0\\0&\frac{x_2^2w_{32}}{x_3}&0&\frac{w_{34}}{x_3}\\0&\frac{x_2^2w_{42}}{x_4}&0&0\end{pmatrix}$
$\begin{pmatrix}1&0&1&1\\0&0&w_{23}&0\\0&0&0&w_{34}\\0&w_{42}&w_{43}&0\end{pmatrix}$	$\begin{pmatrix}1&0&0&0\\0&x_2&0&0\\0&0&x_3&0\\0&0&0&x_4\end{pmatrix}$	$\begin{pmatrix}1&0&1&1\\0&0&\frac{w_{23}}{x_2}&0\\0&0&0&\frac{w_{34}}{x_3}\\0&\frac{x_2^2w_{42}}{x_4}&\frac{w_{43}}{x_4}&0\end{pmatrix}$
$\begin{pmatrix}1&0&1&1\\0&0&w_{23}&0\\0&w_{32}&0&w_{34}\\0&w_{42}&0&w_{44}\end{pmatrix}$	$\begin{pmatrix}1&0&0&0\\0&x_2&0&0\\0&0&x_3&0\\0&0&0&x_4\end{pmatrix}$	$\begin{pmatrix}1&0&1&1\\0&0&\frac{w_{23}}{x_2}&0\\0&\frac{x_2^2w_{32}}{x_3}&0&\frac{w_{34}}{x_3}\\0&\frac{x_2^2w_{42}}{x_4}&0&x_4w_{44}\end{pmatrix}$
$\begin{pmatrix}1&0&1&1\\0&0&w_{23}&w_{24}\\0&w_{32}&0&0\\0&0&w_{43}&w_{44}\end{pmatrix}$	$\begin{pmatrix}1&0&0&0\\0&x_2&0&0\\0&0&x_3&0\\0&0&0&x_4\end{pmatrix}$	$\begin{pmatrix}1&0&1&1\\0&0&\frac{w_{23}}{x_2}&\frac{w_{24}}{x_2}\\0&\frac{x_2^2w_{32}}{x_3}&0&0\\0&0&\frac{w_{43}}{x_4}&x_4w_{44}\end{pmatrix}$
$\begin{pmatrix}1&0&1&1\\0&w_{22}&0&w_{24}\\0&w_{32}&0&w_{34}\\0&0&w_{43}&0\end{pmatrix}$	$\begin{pmatrix}1&0&0&0\\0&x_2&0&0\\0&0&x_3&0\\0&0&0&x_4\end{pmatrix}$	$\begin{pmatrix}1&0&1&1\\0&x_2w_{22}&0&\frac{w_{24}}{x_2}\\0&\frac{x_2^2w_{32}}{x_3}&0&\frac{w_{34}}{x_3}\\0&0&\frac{w_{43}}{x_4}&0\end{pmatrix}$

M_B	$P_{B'B}$	M'_B
$\begin{pmatrix}1&0&1&1\\0&0&w_{23}&0\\0&w_{32}&0&w_{34}\\0&0&w_{43}&w_{44}\end{pmatrix}$	$\begin{pmatrix}1&0&0&0\\0&x_2&0&0\\0&0&x_3&0\\0&0&0&x_4\end{pmatrix}$	$\begin{pmatrix}1&0&1&1\\0&0&\frac{w_{23}}{x_2}&0\\0&\frac{x_2^2w_{32}}{x_3}&0&\frac{w_{34}}{x_3}\\0&0&\frac{w_{43}}{x_4}&x_4w_{44}\end{pmatrix}$
$\begin{pmatrix}1&0&1&1\\0&0&w_{23}&w_{24}\\0&w_{32}&0&0\\0&w_{42}&0&w_{44}\end{pmatrix}$	$\begin{pmatrix}1&0&0&0\\0&x_2&0&0\\0&0&x_3&0\\0&0&0&x_4\end{pmatrix}$	$\begin{pmatrix}1&0&1&1\\0&0&\frac{w_{23}}{x_2}&\frac{w_{24}}{x_2}\\0&\frac{x_2^2w_{32}}{x_3}&0&0\\0&\frac{x_2^2w_{42}}{x_4}&0&x_4w_{44}\end{pmatrix}$
$\begin{pmatrix}1&0&1&1\\0&w_{22}&w_{23}&0\\0&w_{32}&0&w_{34}\\0&0&w_{43}&0\end{pmatrix}$	$\begin{pmatrix}1&0&0&0\\0&x_2&0&0\\0&0&x_3&0\\0&0&0&x_4\end{pmatrix}$	$\begin{pmatrix}1&0&1&1\\0&x_2w_{22}&\frac{w_{23}}{x_2}&0\\0&\frac{x_2^2w_{32}}{x_3}&0&\frac{w_{34}}{x_3}\\0&0&\frac{w_{43}}{x_4}&0\end{pmatrix}$
$\begin{pmatrix}1&0&1&1\\0&w_{22}&w_{23}&w_{24}\\0&w_{32}&0&0\\0&0&w_{43}&0\end{pmatrix}$	$\begin{pmatrix}1&0&0&0\\0&x_2&0&0\\0&0&x_3&0\\0&0&0&x_4\end{pmatrix}$	$\begin{pmatrix}1&0&1&1\\0&x_2w_{22}&\frac{w_{23}}{x_2}&\frac{w_{24}}{x_2}\\0&\frac{x_2^2w_{32}}{x_3}&0&0\\0&0&\frac{w_{43}}{x_4}&0\end{pmatrix}$
$\begin{pmatrix}1&0&1&1\\0&0&w_{23}&0\\0&w_{32}&w_{33}&w_{34}\\0&w_{42}&0&0\end{pmatrix}$	$\begin{pmatrix}1&0&0&0\\0&x_2&0&0\\0&0&x_3&0\\0&0&0&x_4\end{pmatrix}$	$\begin{pmatrix}1&0&1&1\\0&0&\frac{w_{23}}{x_2}&0\\0&\frac{x_2^2w_{32}}{x_3}&x_3w_{33}&\frac{w_{34}}{x_3}\\0&\frac{x_2^2w_{42}}{x_4}&0&0\end{pmatrix}$
$\begin{pmatrix}1&0&1&1\\0&0&w_{23}&0\\0&0&0&w_{34}\\0&w_{42}&w_{43}&w_{44}\end{pmatrix}$	$\begin{pmatrix}1&0&0&0\\0&x_2&0&0\\0&0&x_3&0\\0&0&0&x_4\end{pmatrix}$	$\begin{pmatrix}1&0&1&1\\0&0&\frac{w_{23}}{x_2}&0\\0&0&0&\frac{w_{34}}{x_3}\\0&\frac{x_2^2w_{42}}{x_4}&\frac{w_{43}}{x_4}&x_4w_{44}\end{pmatrix}$
$\begin{pmatrix}1&0&1&1\\0&w_{22}&0&w_{24}\\0&w_{32}&w_{33}&0\\0&0&w_{43}&0\end{pmatrix}$	$\begin{pmatrix}1&0&0&0\\0&x_2&0&0\\0&0&x_3&0\\0&0&0&x_4\end{pmatrix}$	$\begin{pmatrix}1&0&1&1\\0&x_2w_{22}&0&\frac{w_{24}}{x_2}\\0&\frac{x_2^2w_{32}}{x_3}&x_3w_{33}&0\\0&0&\frac{w_{43}}{x_4}&0\end{pmatrix}$
$\begin{pmatrix}1&0&1&1\\0&w_{22}&w_{23}&0\\0&0&w_{33}&w_{34}\\0&w_{42}&0&0\end{pmatrix}$	$\begin{pmatrix}1&0&0&0\\0&x_2&0&0\\0&0&x_3&0\\0&0&0&x_4\end{pmatrix}$	$\begin{pmatrix}1&0&1&1\\0&x_2w_{22}&\frac{w_{23}}{x_2}&0\\0&0&x_3w_{33}&\frac{w_{34}}{x_3}\\0&\frac{x_2^2w_{42}}{x_4}&0&0\end{pmatrix}$
$\begin{pmatrix}1&0&1&1\\0&0&w_{23}&0\\0&0&w_{33}&w_{34}\\0&w_{42}&0&w_{44}\end{pmatrix}$	$\begin{pmatrix}1&0&0&0\\0&x_2&0&0\\0&0&x_3&0\\0&0&0&x_4\end{pmatrix}$	$\begin{pmatrix}1&0&1&1\\0&0&\frac{w_{23}}{x_2}&0\\0&0&x_3w_{33}&\frac{w_{34}}{x_3}\\0&\frac{x_2^2w_{42}}{x_4}&0&x_4w_{44}\end{pmatrix}$
$\begin{pmatrix}1&0&1&1\\0&w_{22}&0&w_{24}\\0&w_{32}&0&0\\0&w_{42}&w_{43}&0\end{pmatrix}$	$\begin{pmatrix}1&0&0&0\\0&x_2&0&0\\0&0&x_3&0\\0&0&0&x_4\end{pmatrix}$	$\begin{pmatrix}1&0&1&1\\0&x_2w_{22}&0&\frac{w_{24}}{x_2}\\0&\frac{x_2^2w_{32}}{x_3}&0&0\\0&\frac{x_2^2w_{42}}{x_4}&\frac{w_{43}}{x_4}&0\end{pmatrix}$

M_B	$P_{B'B}$	M'_B
$\begin{pmatrix}1&0&1&1\\0&0&w_{23}&0\\0&0&w_{33}&w_{34}\\0&w_{42}&w_{43}&0\end{pmatrix}$	$\begin{pmatrix}1&0&0&0\\0&x_2&0&0\\0&0&x_3&0\\0&0&0&x_4\end{pmatrix}$	$\begin{pmatrix}1&0&1&1\\0&0&\frac{w_{23}}{x_2}&0\\0&0&x_3w_{33}&\frac{w_{34}}{x_3}\\0&\frac{x_2^2w_{42}}{x_4}&\frac{w_{43}}{x_4}&0\end{pmatrix}$
$\begin{pmatrix}1&0&1&1\\0&0&w_{23}&w_{24}\\0&0&0&w_{34}\\0&w_{42}&0&w_{44}\end{pmatrix}$	$\begin{pmatrix}1&0&0&0\\0&x_2&0&0\\0&0&x_3&0\\0&0&0&x_4\end{pmatrix}$	$\begin{pmatrix}1&0&1&1\\0&0&\frac{w_{23}}{x_2}&\frac{w_{24}}{x_2}\\0&0&0&\frac{w_{34}}{x_3}\\0&\frac{x_2^2w_{42}}{x_4}&0&x_4w_{44}\end{pmatrix}$
$\begin{pmatrix}1&0&1&1\\0&0&w_{23}&w_{24}\\0&w_{32}&0&w_{34}\\0&0&w_{43}&0\end{pmatrix}$	$\begin{pmatrix}1&0&0&0\\0&x_2&0&0\\0&0&x_3&0\\0&0&0&x_4\end{pmatrix}$	$\begin{pmatrix}1&0&1&1\\0&0&\frac{w_{23}}{x_2}&\frac{w_{24}}{x_2}\\0&\frac{x_2^2w_{32}}{x_3}&0&\frac{w_{34}}{x_3}\\0&0&\frac{w_{43}}{x_4}&0\end{pmatrix}$
$\begin{pmatrix}1&0&1&1\\0&0&w_{23}&w_{24}\\0&w_{32}&0&w_{34}\\0&w_{42}&0&0\end{pmatrix}$	$\begin{pmatrix}1&0&0&0\\0&x_2&0&0\\0&0&x_3&0\\0&0&0&x_4\end{pmatrix}$	$\begin{pmatrix}1&0&1&1\\0&0&\frac{w_{23}}{x_2}&\frac{w_{24}}{x_2}\\0&\frac{x_2^2w_{32}}{x_3}&0&\frac{w_{34}}{x_3}\\0&\frac{x_2^2w_{42}}{x_4}&0&0\end{pmatrix}$
$\begin{pmatrix}1&0&1&1\\0&0&w_{23}&0\\0&w_{32}&0&w_{34}\\0&w_{42}&w_{43}&0\end{pmatrix}$	$\begin{pmatrix}1&0&0&0\\0&x_2&0&0\\0&0&x_3&0\\0&0&0&x_4\end{pmatrix}$	$\begin{pmatrix}1&0&1&1\\0&0&\frac{w_{23}}{x_2}&0\\0&\frac{x_2^2w_{32}}{x_3}&0&\frac{w_{34}}{x_3}\\0&\frac{x_2^2w_{42}}{x_4}&\frac{w_{43}}{x_4}&0\end{pmatrix}$
$\begin{pmatrix}1&0&1&1\\0&w_{22}&w_{23}&0\\0&0&w_{33}&w_{34}\\0&w_{42}&0&w_{44}\end{pmatrix}$	$\begin{pmatrix}1&0&0&0\\0&x_2&0&0\\0&0&x_3&0\\0&0&0&x_4\end{pmatrix}$	$\begin{pmatrix}1&0&1&1\\0&x_2w_{22}&\frac{w_{23}}{x_2}&0\\0&0&x_3w_{33}&\frac{w_{34}}{x_3}\\0&\frac{x_2^2w_{42}}{x_4}&0&x_4w_{44}\end{pmatrix}$
$\begin{pmatrix}1&0&1&1\\0&w_{22}&w_{23}&w_{24}\\0&w_{32}&0&0\\0&w_{42}&0&w_{44}\end{pmatrix}$	$\begin{pmatrix}1&0&0&0\\0&x_2&0&0\\0&0&x_3&0\\0&0&0&x_4\end{pmatrix}$	$\begin{pmatrix}1&0&1&1\\0&x_2w_{22}&\frac{w_{23}}{x_2}&\frac{w_{24}}{x_2}\\0&\frac{x_2^2w_{32}}{x_3}&0&0\\0&\frac{x_2^2w_{42}}{x_4}&0&x_4w_{44}\end{pmatrix}$
$\begin{pmatrix}1&0&1&1\\0&0&w_{23}&0\\0&w_{32}&w_{33}&w_{34}\\0&0&w_{43}&w_{44}\end{pmatrix}$	$\begin{pmatrix}1&0&0&0\\0&x_2&0&0\\0&0&x_3&0\\0&0&0&x_4\end{pmatrix}$	$\begin{pmatrix}1&0&1&1\\0&0&\frac{w_{23}}{x_2}&0\\0&\frac{x_2^2w_{32}}{x_3}&x_3w_{33}&\frac{w_{34}}{x_3}\\0&0&\frac{w_{43}}{x_4}&x_4w_{44}\end{pmatrix}$
$\begin{pmatrix}1&0&1&1\\0&w_{22}&0&w_{24}\\0&0&0&w_{34}\\0&w_{42}&w_{43}&w_{44}\end{pmatrix}$	$\begin{pmatrix}1&0&0&0\\0&x_2&0&0\\0&0&x_3&0\\0&0&0&x_4\end{pmatrix}$	$\begin{pmatrix}1&0&1&1\\0&x_2w_{22}&0&\frac{w_{24}}{x_2}\\0&0&0&\frac{w_{34}}{x_3}\\0&\frac{x_2^2w_{42}}{x_4}&\frac{w_{43}}{x_4}&x_4w_{44}\end{pmatrix}$
$\begin{pmatrix}1&0&1&1\\0&w_{22}&w_{23}&w_{24}\\0&w_{32}&0&0\\0&0&w_{43}&w_{44}\end{pmatrix}$	$\begin{pmatrix}1&0&0&0\\0&x_2&0&0\\0&0&x_3&0\\0&0&0&x_4\end{pmatrix}$	$\begin{pmatrix}1&0&1&1\\0&x_2w_{22}&\frac{w_{23}}{x_2}&\frac{w_{24}}{x_2}\\0&\frac{x_2^2w_{32}}{x_3}&0&0\\0&0&\frac{w_{43}}{x_4}&x_4w_{44}\end{pmatrix}$

M_B	$P_{B'B}$	M'_B
$\begin{pmatrix}1&0&1&1\\0&0&w_{23}&0\\0&w_{32}&w_{33}&w_{34}\\0&w_{42}&0&w_{44}\end{pmatrix}$	$\begin{pmatrix}1&0&0&0\\0&x_2&0&0\\0&0&x_3&0\\0&0&0&x_4\end{pmatrix}$	$\begin{pmatrix}1&0&1&1\\0&0&\frac{w_{23}}{x_2}&0\\0&\frac{x_2^2w_{32}}{x_3}&x_3w_{33}&\frac{w_{34}}{x_3}\\0&\frac{x_2^2w_{42}}{x_4}&0&x_4w_{44}\end{pmatrix}$
$\begin{pmatrix}1&0&1&1\\0&w_{22}&w_{23}&0\\0&0&0&w_{34}\\0&w_{42}&w_{43}&w_{44}\end{pmatrix}$	$\begin{pmatrix}1&0&0&0\\0&x_2&0&0\\0&0&x_3&0\\0&0&0&x_4\end{pmatrix}$	$\begin{pmatrix}1&0&1&1\\0&x_2w_{22}&\frac{w_{23}}{x_2}&0\\0&0&0&\frac{w_{34}}{x_3}\\0&\frac{x_2^2w_{42}}{x_4}&\frac{w_{43}}{x_4}&x_4w_{44}\end{pmatrix}$
$\begin{pmatrix}1&0&1&1\\0&0&w_{23}&w_{24}\\0&w_{32}&w_{33}&0\\0&w_{42}&0&w_{44}\end{pmatrix}$	$\begin{pmatrix}1&0&0&0\\0&x_2&0&0\\0&0&x_3&0\\0&0&0&x_4\end{pmatrix}$	$\begin{pmatrix}1&0&1&1\\0&0&\frac{w_{23}}{x_2}&\frac{w_{24}}{x_2}\\0&\frac{x_2^2w_{32}}{x_3}&x_3w_{33}&0\\0&\frac{x_2^2w_{42}}{x_4}&0&x_4w_{44}\end{pmatrix}$
$\begin{pmatrix}1&0&1&1\\0&0&w_{23}&w_{24}\\0&w_{32}&w_{33}&0\\0&w_{42}&0&w_{44}\end{pmatrix}$	$\begin{pmatrix}1&0&0&0\\0&x_2&0&0\\0&0&0&x_3\\0&0&x_4&0\end{pmatrix}$	$\begin{pmatrix}1&0&1&1\\0&0&\frac{w_{24}}{x_2}&\frac{w_{23}}{x_2}\\0&\frac{x_2^2w_{42}}{x_4}&x_4w_{44}&0\\0&\frac{x_2^2w_{32}}{x_3}&0&x_3w_{33}\end{pmatrix}$
$\begin{pmatrix}1&0&1&1\\0&w_{22}&w_{23}&0\\0&w_{32}&0&w_{34}\\0&0&w_{43}&w_{44}\end{pmatrix}$	$\begin{pmatrix}1&0&0&0\\0&x_2&0&0\\0&0&x_3&0\\0&0&0&x_4\end{pmatrix}$	$\begin{pmatrix}1&0&1&1\\0&x_2w_{22}&\frac{w_{23}}{x_2}&0\\0&\frac{x_2^2w_{32}}{x_3}&0&\frac{w_{34}}{x_3}\\0&0&\frac{w_{43}}{x_4}&x_4w_{44}\end{pmatrix}$
$\begin{pmatrix}1&0&1&1\\0&0&w_{23}&w_{24}\\0&w_{32}&w_{33}&0\\0&0&w_{43}&w_{44}\end{pmatrix}$	$\begin{pmatrix}1&0&0&0\\0&x_2&0&0\\0&0&x_3&0\\0&0&0&x_4\end{pmatrix}$	$\begin{pmatrix}1&0&1&1\\0&0&\frac{w_{23}}{x_2}&\frac{w_{24}}{x_2}\\0&\frac{x_2^2w_{32}}{x_3}&x_3w_{33}&0\\0&0&\frac{w_{43}}{x_4}&x_4w_{44}\end{pmatrix}$
$\begin{pmatrix}1&0&1&1\\0&w_{22}&w_{23}&0\\0&w_{32}&0&w_{34}\\0&w_{42}&0&w_{44}\end{pmatrix}$	$\begin{pmatrix}1&0&0&0\\0&x_2&0&0\\0&0&x_3&0\\0&0&0&x_4\end{pmatrix}$	$\begin{pmatrix}1&0&1&1\\0&x_2w_{22}&\frac{w_{23}}{x_2}&0\\0&\frac{x_2^2w_{32}}{x_3}&0&\frac{w_{34}}{x_3}\\0&\frac{x_2^2w_{42}}{x_4}&0&x_4w_{44}\end{pmatrix}$
$\begin{pmatrix}1&0&1&1\\0&w_{22}&w_{23}&0\\0&0&w_{33}&w_{34}\\0&w_{42}&w_{43}&0\end{pmatrix}$	$\begin{pmatrix}1&0&0&0\\0&x_2&0&0\\0&0&x_3&0\\0&0&0&x_4\end{pmatrix}$	$\begin{pmatrix}1&0&1&1\\0&x_2w_{22}&\frac{w_{23}}{x_2}&0\\0&0&x_3w_{33}&\frac{w_{34}}{x_3}\\0&\frac{x_2^2w_{42}}{x_4}&\frac{w_{43}}{x_4}&0\end{pmatrix}$
$\begin{pmatrix}1&0&1&1\\0&0&w_{23}&w_{24}\\0&w_{32}&0&w_{34}\\0&w_{42}&0&w_{44}\end{pmatrix}$	$\begin{pmatrix}1&0&0&0\\0&x_2&0&0\\0&0&x_3&0\\0&0&0&x_4\end{pmatrix}$	$\begin{pmatrix}1&0&1&1\\0&0&\frac{w_{23}}{x_2}&\frac{w_{24}}{x_2}\\0&\frac{x_2^2w_{32}}{x_3}&0&\frac{w_{34}}{x_3}\\0&\frac{x_2^2w_{42}}{x_4}&0&x_4w_{44}\end{pmatrix}$
$\begin{pmatrix}1&0&1&1\\0&0&w_{23}&w_{24}\\0&w_{32}&0&w_{34}\\0&0&w_{43}&w_{44}\end{pmatrix}$	$\begin{pmatrix}1&0&0&0\\0&x_2&0&0\\0&0&x_3&0\\0&0&0&x_4\end{pmatrix}$	$\begin{pmatrix}1&0&1&1\\0&0&\frac{w_{23}}{x_2}&\frac{w_{24}}{x_2}\\0&\frac{x_2^2w_{32}}{x_3}&0&\frac{w_{34}}{x_3}\\0&0&\frac{w_{43}}{x_4}&x_4w_{44}\end{pmatrix}$

M_B	$P_{B'B}$	M'_B
$\begin{pmatrix}1&0&1&1\\0&w_{22}&0&w_{24}\\0&w_{32}&0&w_{34}\\0&w_{42}&w_{43}&0\end{pmatrix}$	$\begin{pmatrix}1&0&0&0\\0&x_2&0&0\\0&0&x_3&0\\0&0&0&x_4\end{pmatrix}$	$\begin{pmatrix}1&0&1&1\\0&x_2w_{22}&0&\frac{w_{24}}{x_2}\\0&\frac{x_2^2w_{32}}{x_3}&0&\frac{w_{34}}{x_3}\\0&\frac{x_2^2w_{42}}{x_4}&\frac{w_{43}}{x_4}&0\end{pmatrix}$
$\begin{pmatrix}1&0&1&1\\0&0&w_{23}&w_{24}\\0&w_{32}&0&0\\0&w_{42}&w_{43}&w_{44}\end{pmatrix}$	$\begin{pmatrix}1&0&0&0\\0&x_2&0&0\\0&0&x_3&0\\0&0&0&x_4\end{pmatrix}$	$\begin{pmatrix}1&0&1&1\\0&0&\frac{w_{23}}{x_2}&\frac{w_{24}}{x_2}\\0&\frac{x_2^2w_{32}}{x_3}&0&0\\0&\frac{x_2^2w_{42}}{x_4}&\frac{w_{43}}{x_4}&x_4w_{44}\end{pmatrix}$
$\begin{pmatrix}1&0&1&1\\0&0&w_{23}&0\\0&w_{32}&0&w_{34}\\0&w_{42}&w_{43}&w_{44}\end{pmatrix}$	$\begin{pmatrix}1&0&0&0\\0&x_2&0&0\\0&0&x_3&0\\0&0&0&x_4\end{pmatrix}$	$\begin{pmatrix}1&0&1&1\\0&0&\frac{w_{23}}{x_2}&0\\0&\frac{x_2^2w_{32}}{x_3}&0&\frac{w_{34}}{x_3}\\0&\frac{x_2^2w_{42}}{x_4}&\frac{w_{43}}{x_4}&x_4w_{44}\end{pmatrix}$
$\begin{pmatrix}1&0&1&1\\0&w_{22}&w_{23}&w_{24}\\0&0&0&w_{34}\\0&w_{42}&w_{43}&0\end{pmatrix}$	$\begin{pmatrix}1&0&0&0\\0&x_2&0&0\\0&0&x_3&0\\0&0&0&x_4\end{pmatrix}$	$\begin{pmatrix}1&0&1&1\\0&x_2w_{22}&\frac{w_{23}}{x_2}&\frac{w_{24}}{x_2}\\0&0&0&\frac{w_{34}}{x_3}\\0&\frac{x_2^2w_{42}}{x_4}&\frac{w_{43}}{x_4}&0\end{pmatrix}$
$\begin{pmatrix}1&0&1&1\\0&w_{22}&w_{23}&w_{24}\\0&w_{32}&w_{33}&0\\0&0&w_{43}&0\end{pmatrix}$	$\begin{pmatrix}1&0&0&0\\0&x_2&0&0\\0&0&x_3&0\\0&0&0&x_4\end{pmatrix}$	$\begin{pmatrix}1&0&1&1\\0&x_2w_{22}&\frac{w_{23}}{x_2}&\frac{w_{24}}{x_2}\\0&\frac{x_2^2w_{32}}{x_3}&x_3w_{33}&0\\0&0&\frac{w_{43}}{x_4}&0\end{pmatrix}$
$\begin{pmatrix}1&0&1&1\\0&w_{22}&w_{23}&0\\0&w_{32}&w_{33}&w_{34}\\0&w_{42}&0&0\end{pmatrix}$	$\begin{pmatrix}1&0&0&0\\0&x_2&0&0\\0&0&x_3&0\\0&0&0&x_4\end{pmatrix}$	$\begin{pmatrix}1&0&1&1\\0&x_2w_{22}&\frac{w_{23}}{x_2}&0\\0&\frac{x_2^2w_{32}}{x_3}&x_3w_{33}&\frac{w_{34}}{x_3}\\0&\frac{x_2^2w_{42}}{x_4}&0&0\end{pmatrix}$
$\begin{pmatrix}1&0&1&1\\0&0&w_{23}&0\\0&0&w_{33}&w_{34}\\0&w_{42}&w_{43}&w_{44}\end{pmatrix}$	$\begin{pmatrix}1&0&0&0\\0&x_2&0&0\\0&0&x_3&0\\0&0&0&x_4\end{pmatrix}$	$\begin{pmatrix}1&0&1&1\\0&0&\frac{w_{23}}{x_2}&0\\0&0&x_3w_{33}&\frac{w_{34}}{x_3}\\0&\frac{x_2^2w_{42}}{x_4}&\frac{w_{43}}{x_4}&x_4w_{44}\end{pmatrix}$
$\begin{pmatrix}1&0&1&1\\0&w_{22}&w_{23}&w_{24}\\0&w_{32}&0&0\\0&w_{42}&w_{43}&0\end{pmatrix}$	$\begin{pmatrix}1&0&0&0\\0&x_2&0&0\\0&0&x_3&0\\0&0&0&x_4\end{pmatrix}$	$\begin{pmatrix}1&0&1&1\\0&x_2w_{22}&\frac{w_{23}}{x_2}&\frac{w_{24}}{x_2}\\0&\frac{x_2^2w_{32}}{x_3}&0&0\\0&\frac{x_2^2w_{42}}{x_4}&\frac{w_{43}}{x_4}&0\end{pmatrix}$
$\begin{pmatrix}1&0&1&1\\0&0&w_{23}&0\\0&w_{32}&w_{33}&w_{34}\\0&w_{42}&w_{43}&0\end{pmatrix}$	$\begin{pmatrix}1&0&0&0\\0&x_2&0&0\\0&0&x_3&0\\0&0&0&x_4\end{pmatrix}$	$\begin{pmatrix}1&0&1&1\\0&0&\frac{w_{23}}{x_2}&0\\0&\frac{x_2^2w_{32}}{x_3}&x_3w_{33}&\frac{w_{34}}{x_3}\\0&\frac{x_2^2w_{42}}{x_4}&\frac{w_{43}}{x_4}&0\end{pmatrix}$
$\begin{pmatrix}1&0&1&1\\0&0&w_{23}&w_{24}\\0&0&0&w_{34}\\0&w_{42}&w_{43}&w_{44}\end{pmatrix}$	$\begin{pmatrix}1&0&0&0\\0&x_2&0&0\\0&0&x_3&0\\0&0&0&x_4\end{pmatrix}$	$\begin{pmatrix}1&0&1&1\\0&0&\frac{w_{23}}{x_2}&\frac{w_{24}}{x_2}\\0&0&0&\frac{w_{34}}{x_3}\\0&\frac{x_2^2w_{42}}{x_4}&\frac{w_{43}}{x_4}&x_4w_{44}\end{pmatrix}$

M_B	$P_{B'B}$	M'_B
$\begin{pmatrix}1&0&1&1\\0&0&w_{23}&w_{24}\\0&w_{32}&0&w_{34}\\0&w_{42}&w_{43}&0\end{pmatrix}$	$\begin{pmatrix}1&0&0&0\\0&x_2&0&0\\0&0&x_3&0\\0&0&0&x_4\end{pmatrix}$	$\begin{pmatrix}1&0&1&1\\0&0&\frac{w_{23}}{x_2}&\frac{w_{24}}{x_2}\\0&\frac{x_2^2w_{32}}{x_3}&0&\frac{w_{34}}{x_3}\\0&\frac{x_2^2w_{42}}{x_4}&\frac{w_{43}}{x_4}&0\end{pmatrix}$
$\begin{pmatrix}1&0&1&1\\0&0&w_{23}&w_{24}\\0&w_{32}&0&w_{34}\\0&w_{42}&w_{43}&0\end{pmatrix}$	$\begin{pmatrix}1&0&0&0\\0&x_2&0&0\\0&0&0&x_3\\0&0&x_4&0\end{pmatrix}$	$\begin{pmatrix}1&0&1&1\\0&0&\frac{w_{24}}{x_2}&\frac{w_{23}}{x_2}\\0&\frac{x_2^2w_{42}}{x_4}&0&\frac{w_{43}}{x_4}\\0&\frac{x_2^2w_{32}}{x_3}&\frac{w_{34}}{x_3}&0\end{pmatrix}$
$\begin{pmatrix}1&0&1&1\\0&w_{22}&w_{23}&w_{24}\\0&w_{32}&w_{33}&0\\0&w_{42}&0&w_{44}\end{pmatrix}$	$\begin{pmatrix}1&0&0&0\\0&x_2&0&0\\0&0&x_3&0\\0&0&0&x_4\end{pmatrix}$	$\begin{pmatrix}1&0&1&1\\0&x_2w_{22}&\frac{w_{23}}{x_2}&\frac{w_{24}}{x_2}\\0&\frac{x_2^2w_{32}}{x_3}&x_3w_{33}&0\\0&\frac{x_2^2w_{42}}{x_4}&0&x_4w_{44}\end{pmatrix}$
$\begin{pmatrix}1&0&1&1\\0&w_{22}&w_{23}&w_{24}\\0&w_{32}&w_{33}&0\\0&w_{42}&0&w_{44}\end{pmatrix}$	$\begin{pmatrix}1&0&0&0\\0&x_2&0&0\\0&0&0&x_3\\0&0&x_4&0\end{pmatrix}$	$\begin{pmatrix}1&0&1&1\\0&x_2w_{22}&\frac{w_{24}}{x_2}&\frac{w_{23}}{x_2}\\0&\frac{x_2^2w_{42}}{x_4}&x_4w_{44}&0\\0&\frac{x_2^2w_{32}}{x_3}&0&x_3w_{33}\end{pmatrix}$
$\begin{pmatrix}1&0&1&1\\0&w_{22}&w_{23}&0\\0&w_{32}&w_{33}&w_{34}\\0&0&w_{43}&w_{44}\end{pmatrix}$	$\begin{pmatrix}1&0&0&0\\0&x_2&0&0\\0&0&x_3&0\\0&0&0&x_4\end{pmatrix}$	$\begin{pmatrix}1&0&1&1\\0&x_2w_{22}&\frac{w_{23}}{x_2}&0\\0&\frac{x_2^2w_{32}}{x_3}&x_3w_{33}&\frac{w_{34}}{x_3}\\0&0&\frac{w_{43}}{x_4}&x_4w_{44}\end{pmatrix}$
$\begin{pmatrix}1&0&1&1\\0&w_{22}&0&w_{24}\\0&0&w_{33}&w_{34}\\0&w_{42}&w_{43}&w_{44}\end{pmatrix}$	$\begin{pmatrix}1&0&0&0\\0&x_2&0&0\\0&0&x_3&0\\0&0&0&x_4\end{pmatrix}$	$\begin{pmatrix}1&0&1&1\\0&x_2w_{22}&0&\frac{w_{24}}{x_2}\\0&0&x_3w_{33}&\frac{w_{34}}{x_3}\\0&\frac{x_2^2w_{42}}{x_4}&\frac{w_{43}}{x_4}&x_4w_{44}\end{pmatrix}$
$\begin{pmatrix}1&0&1&1\\0&w_{22}&w_{23}&w_{24}\\0&w_{32}&w_{33}&0\\0&0&w_{43}&w_{44}\end{pmatrix}$	$\begin{pmatrix}1&0&0&0\\0&x_2&0&0\\0&0&x_3&0\\0&0&0&x_4\end{pmatrix}$	$\begin{pmatrix}1&0&1&1\\0&x_2w_{22}&\frac{w_{23}}{x_2}&\frac{w_{24}}{x_2}\\0&\frac{x_2^2w_{32}}{x_3}&x_3w_{33}&0\\0&0&\frac{w_{43}}{x_4}&x_4w_{44}\end{pmatrix}$
$\begin{pmatrix}1&0&1&1\\0&w_{22}&w_{23}&0\\0&w_{32}&w_{33}&w_{34}\\0&w_{42}&0&w_{44}\end{pmatrix}$	$\begin{pmatrix}1&0&0&0\\0&x_2&0&0\\0&0&x_3&0\\0&0&0&x_4\end{pmatrix}$	$\begin{pmatrix}1&0&1&1\\0&x_2w_{22}&\frac{w_{23}}{x_2}&0\\0&\frac{x_2^2w_{32}}{x_3}&x_3w_{33}&\frac{w_{34}}{x_3}\\0&\frac{x_2^2w_{42}}{x_4}&0&x_4w_{44}\end{pmatrix}$
$\begin{pmatrix}1&0&1&1\\0&w_{22}&w_{23}&0\\0&0&w_{33}&w_{34}\\0&w_{42}&w_{43}&w_{44}\end{pmatrix}$	$\begin{pmatrix}1&0&0&0\\0&x_2&0&0\\0&0&x_3&0\\0&0&0&x_4\end{pmatrix}$	$\begin{pmatrix}1&0&1&1\\0&x_2w_{22}&\frac{w_{23}}{x_2}&0\\0&0&x_3w_{33}&\frac{w_{34}}{x_3}\\0&\frac{x_2^2w_{42}}{x_4}&\frac{w_{43}}{x_4}&x_4w_{44}\end{pmatrix}$
$\begin{pmatrix}1&0&1&1\\0&w_{22}&w_{23}&w_{24}\\0&w_{32}&0&0\\0&w_{42}&w_{43}&w_{44}\end{pmatrix}$	$\begin{pmatrix}1&0&0&0\\0&x_2&0&0\\0&0&x_3&0\\0&0&0&x_4\end{pmatrix}$	$\begin{pmatrix}1&0&1&1\\0&x_2w_{22}&\frac{w_{23}}{x_2}&\frac{w_{24}}{x_2}\\0&\frac{x_2^2w_{32}}{x_3}&0&0\\0&\frac{x_2^2w_{42}}{x_4}&\frac{w_{43}}{x_4}&x_4w_{44}\end{pmatrix}$

M_B	$P_{B'B}$	M'_B	M_B	$P_{B'B}$	M'_B
$\begin{pmatrix}1&0&1&1\\0&0&w_{23}&0\\0&w_{32}&w_{33}&w_{34}\\0&w_{42}&w_{43}&w_{44}\end{pmatrix}$	$\begin{pmatrix}1&0&0&0\\0&x_2&0&0\\0&0&x_3&0\\0&0&0&x_4\end{pmatrix}$	$\begin{pmatrix}1&0&1&1\\0&0&\frac{w_{23}}{x_2}&0\\0&\frac{x_2^2w_{32}}{x_3}&x_3w_{33}&\frac{w_{34}}{x_3}\\0&\frac{x_2^2w_{42}}{x_4}&\frac{w_{43}}{x_4}&x_4w_{44}\end{pmatrix}$	$\begin{pmatrix}1&0&1&1\\0&w_{22}&w_{23}&w_{24}\\0&w_{32}&0&w_{34}\\0&w_{42}&w_{43}&0\end{pmatrix}$	$\begin{pmatrix}1&0&0&0\\0&x_2&0&0\\0&0&0&x_3\\0&0&x_4&0\end{pmatrix}$	$\begin{pmatrix}1&0&1&1\\0&x_2w_{22}&\frac{w_{24}}{x_2}&\frac{w_{23}}{x_2}\\0&\frac{x_2^2w_{42}}{x_4}&0&\frac{w_{43}}{x_4}\\0&\frac{x_2^2w_{32}}{x_3}&\frac{w_{34}}{x_3}&0\end{pmatrix}$
$\begin{pmatrix}1&0&1&1\\0&w_{22}&w_{23}&w_{24}\\0&0&0&w_{34}\\0&w_{42}&w_{43}&w_{44}\end{pmatrix}$	$\begin{pmatrix}1&0&0&0\\0&x_2&0&0\\0&0&x_3&0\\0&0&0&x_4\end{pmatrix}$	$\begin{pmatrix}1&0&1&1\\0&x_2w_{22}&\frac{w_{23}}{x_2}&\frac{w_{24}}{x_2}\\0&0&0&\frac{w_{34}}{x_3}\\0&\frac{x_2^2w_{42}}{x_4}&\frac{w_{43}}{x_4}&x_4w_{44}\end{pmatrix}$	$\begin{pmatrix}1&0&1&1\\0&w_{22}&w_{23}&w_{24}\\0&w_{32}&w_{33}&w_{34}\\0&w_{42}&0&w_{44}\end{pmatrix}$	$\begin{pmatrix}1&0&0&0\\0&x_2&0&0\\0&0&x_3&0\\0&0&0&x_4\end{pmatrix}$	$\begin{pmatrix}1&0&1&1\\0&x_2w_{22}&\frac{w_{23}}{x_2}&\frac{w_{24}}{x_2}\\0&\frac{x_2^2w_{32}}{x_3}&x_3w_{33}&\frac{w_{34}}{x_3}\\0&\frac{x_2^2w_{42}}{x_4}&0&x_4w_{44}\end{pmatrix}$
$\begin{pmatrix}1&0&1&1\\0&w_{22}&w_{23}&w_{24}\\0&w_{32}&0&w_{34}\\0&w_{42}&0&w_{44}\end{pmatrix}$	$\begin{pmatrix}1&0&0&0\\0&x_2&0&0\\0&0&x_3&0\\0&0&0&x_4\end{pmatrix}$	$\begin{pmatrix}1&0&1&1\\0&x_2w_{22}&\frac{w_{23}}{x_2}&\frac{w_{24}}{x_2}\\0&\frac{x_2^2w_{32}}{x_3}&0&\frac{w_{34}}{x_3}\\0&\frac{x_2^2w_{42}}{x_4}&0&x_4w_{44}\end{pmatrix}$	$\begin{pmatrix}1&0&1&1\\0&w_{22}&w_{23}&w_{24}\\0&w_{32}&w_{33}&w_{34}\\0&0&w_{43}&w_{44}\end{pmatrix}$	$\begin{pmatrix}1&0&0&0\\0&x_2&0&0\\0&0&x_3&0\\0&0&0&x_4\end{pmatrix}$	$\begin{pmatrix}1&0&1&1\\0&x_2w_{22}&\frac{w_{23}}{x_2}&\frac{w_{24}}{x_2}\\0&\frac{x_2^2w_{32}}{x_3}&x_3w_{33}&\frac{w_{34}}{x_3}\\0&0&\frac{w_{43}}{x_4}&x_4w_{44}\end{pmatrix}$
$\begin{pmatrix}1&0&1&1\\0&0&w_{23}&w_{24}\\0&w_{32}&w_{33}&w_{34}\\0&0&w_{43}&w_{44}\end{pmatrix}$	$\begin{pmatrix}1&0&0&0\\0&x_2&0&0\\0&0&x_3&0\\0&0&0&x_4\end{pmatrix}$	$\begin{pmatrix}1&0&1&1\\0&0&\frac{w_{23}}{x_2}&\frac{w_{24}}{x_2}\\0&\frac{x_2^2w_{32}}{x_3}&x_3w_{33}&\frac{w_{34}}{x_3}\\0&0&\frac{w_{43}}{x_4}&x_4w_{44}\end{pmatrix}$	$\begin{pmatrix}1&0&1&1\\0&w_{22}&0&w_{24}\\0&w_{32}&w_{33}&w_{34}\\0&w_{42}&w_{43}&w_{44}\end{pmatrix}$	$\begin{pmatrix}1&0&0&0\\0&x_2&0&0\\0&0&x_3&0\\0&0&0&x_4\end{pmatrix}$	$\begin{pmatrix}1&0&1&1\\0&x_2w_{22}&0&\frac{w_{24}}{x_2}\\0&\frac{x_2^2w_{32}}{x_3}&x_3w_{33}&\frac{w_{34}}{x_3}\\0&\frac{x_2^2w_{42}}{x_4}&\frac{w_{43}}{x_4}&x_4w_{44}\end{pmatrix}$
$\begin{pmatrix}1&0&1&1\\0&0&w_{23}&w_{24}\\0&w_{32}&w_{33}&w_{34}\\0&w_{42}&0&w_{44}\end{pmatrix}$	$\begin{pmatrix}1&0&0&0\\0&x_2&0&0\\0&0&x_3&0\\0&0&0&x_4\end{pmatrix}$	$\begin{pmatrix}1&0&1&1\\0&0&\frac{w_{23}}{x_2}&\frac{w_{24}}{x_2}\\0&\frac{x_2^2w_{32}}{x_3}&x_3w_{33}&\frac{w_{34}}{x_3}\\0&\frac{x_2^2w_{42}}{x_4}&0&x_4w_{44}\end{pmatrix}$	$\begin{pmatrix}1&0&1&1\\0&w_{22}&w_{23}&w_{24}\\0&w_{32}&w_{33}&w_{34}\\0&w_{42}&w_{43}&0\end{pmatrix}$	$\begin{pmatrix}1&0&0&0\\0&x_2&0&0\\0&0&x_3&0\\0&0&0&x_4\end{pmatrix}$	$\begin{pmatrix}1&0&1&1\\0&x_2w_{22}&\frac{w_{23}}{x_2}&\frac{w_{24}}{x_2}\\0&\frac{x_2^2w_{32}}{x_3}&x_3w_{33}&\frac{w_{34}}{x_3}\\0&\frac{x_2^2w_{42}}{x_4}&\frac{w_{43}}{x_4}&0\end{pmatrix}$
$\begin{pmatrix}1&0&1&1\\0&w_{22}&w_{23}&w_{24}\\0&w_{32}&0&w_{34}\\0&0&w_{43}&w_{44}\end{pmatrix}$	$\begin{pmatrix}1&0&0&0\\0&x_2&0&0\\0&0&x_3&0\\0&0&0&x_4\end{pmatrix}$	$\begin{pmatrix}1&0&1&1\\0&x_2w_{22}&\frac{w_{23}}{x_2}&\frac{w_{24}}{x_2}\\0&\frac{x_2^2w_{32}}{x_3}&0&\frac{w_{34}}{x_3}\\0&0&\frac{w_{43}}{x_4}&x_4w_{44}\end{pmatrix}$	$\begin{pmatrix}1&0&1&1\\0&0&w_{23}&w_{24}\\0&w_{32}&w_{33}&w_{34}\\0&w_{42}&w_{43}&w_{44}\end{pmatrix}$	$\begin{pmatrix}1&0&0&0\\0&x_2&0&0\\0&0&x_3&0\\0&0&0&x_4\end{pmatrix}$	$\begin{pmatrix}1&0&1&1\\0&0&\frac{w_{23}}{x_2}&\frac{w_{24}}{x_2}\\0&\frac{x_2^2w_{32}}{x_3}&x_3w_{33}&\frac{w_{34}}{x_3}\\0&\frac{x_2^2w_{42}}{x_4}&\frac{w_{43}}{x_4}&x_4w_{44}\end{pmatrix}$
$\begin{pmatrix}1&0&1&1\\0&w_{22}&0&w_{24}\\0&w_{32}&w_{33}&w_{34}\\0&w_{42}&w_{43}&0\end{pmatrix}$	$\begin{pmatrix}1&0&0&0\\0&x_2&0&0\\0&0&x_3&0\\0&0&0&x_4\end{pmatrix}$	$\begin{pmatrix}1&0&1&1\\0&x_2w_{22}&0&\frac{w_{24}}{x_2}\\0&\frac{x_2^2w_{32}}{x_3}&x_3w_{33}&\frac{w_{34}}{x_3}\\0&\frac{x_2^2w_{42}}{x_4}&\frac{w_{43}}{x_4}&0\end{pmatrix}$	$\begin{pmatrix}1&0&1&1\\0&0&w_{23}&w_{24}\\0&w_{32}&w_{33}&w_{34}\\0&w_{42}&w_{43}&w_{44}\end{pmatrix}$	$\begin{pmatrix}1&0&0&0\\0&x_2&0&0\\0&0&0&x_3\\0&0&x_4&0\end{pmatrix}$	$\begin{pmatrix}1&0&1&1\\0&0&\frac{w_{24}}{x_2}&\frac{w_{23}}{x_2}\\0&\frac{x_2^2w_{42}}{x_4}&x_4w_{44}&\frac{w_{43}}{x_4}\\0&\frac{x_2^2w_{32}}{x_3}&\frac{w_{34}}{x_3}&x_3w_{33}\end{pmatrix}$
$\begin{pmatrix}1&0&1&1\\0&0&w_{23}&w_{24}\\0&w_{32}&0&w_{34}\\0&w_{42}&w_{43}&w_{44}\end{pmatrix}$	$\begin{pmatrix}1&0&0&0\\0&x_2&0&0\\0&0&x_3&0\\0&0&0&x_4\end{pmatrix}$	$\begin{pmatrix}1&0&1&1\\0&0&\frac{w_{23}}{x_2}&\frac{w_{24}}{x_2}\\0&\frac{x_2^2w_{32}}{x_3}&0&\frac{w_{34}}{x_3}\\0&\frac{x_2^2w_{42}}{x_4}&\frac{w_{43}}{x_4}&x_4w_{44}\end{pmatrix}$	$\begin{pmatrix}1&0&1&1\\0&w_{22}&w_{23}&w_{24}\\0&w_{32}&w_{33}&w_{34}\\0&w_{42}&w_{43}&w_{44}\end{pmatrix}$	$\begin{pmatrix}1&0&0&0\\0&x_2&0&0\\0&0&x_3&0\\0&0&0&x_4\end{pmatrix}$	$\begin{pmatrix}1&0&1&1\\0&x_2w_{22}&\frac{w_{23}}{x_2}&\frac{w_{24}}{x_2}\\0&\frac{x_2^2w_{32}}{x_3}&x_3w_{33}&\frac{w_{34}}{x_3}\\0&\frac{x_2^2w_{42}}{x_4}&\frac{w_{43}}{x_4}&x_4w_{44}\end{pmatrix}$
$\begin{pmatrix}1&0&1&1\\0&w_{22}&w_{23}&w_{24}\\0&w_{32}&0&w_{34}\\0&w_{42}&w_{43}&0\end{pmatrix}$	$\begin{pmatrix}1&0&0&0\\0&x_2&0&0\\0&0&x_3&0\\0&0&0&x_4\end{pmatrix}$	$\begin{pmatrix}1&0&1&1\\0&x_2w_{22}&\frac{w_{23}}{x_2}&\frac{w_{24}}{x_2}\\0&\frac{x_2^2w_{32}}{x_3}&0&\frac{w_{34}}{x_3}\\0&\frac{x_2^2w_{42}}{x_4}&\frac{w_{43}}{x_4}&0\end{pmatrix}$	$\begin{pmatrix}1&0&1&1\\0&w_{22}&w_{23}&w_{24}\\0&w_{32}&w_{33}&w_{34}\\0&w_{42}&w_{43}&w_{44}\end{pmatrix}$	$\begin{pmatrix}1&0&0&0\\0&x_2&0&0\\0&0&0&x_3\\0&0&x_4&0\end{pmatrix}$	$\begin{pmatrix}1&0&1&1\\0&x_2w_{22}&\frac{w_{24}}{x_2}&\frac{w_{23}}{x_2}\\0&\frac{x_2^2w_{42}}{x_4}&x_4w_{44}&\frac{w_{43}}{x_4}\\0&\frac{x_2^2w_{32}}{x_3}&\frac{w_{34}}{x_3}&x_3w_{33}\end{pmatrix}$

2.3 The Matrix U Has No Nonzero Entries

M_B	$P_{B'B}$	M'_B	M_B	$P_{B'B}$	M'_B	M_B	$P_{B'B}$	M'_B
$\begin{pmatrix}1&1&1&1\\0&0&0&w_{24}\\0&w_{32}&0&0\\0&0&w_{43}&0\end{pmatrix}$	$\begin{pmatrix}1&0&0&0\\0&x_3x_4z&0&0\\0&0&x_3&0\\0&0&0&x_4\end{pmatrix}$	$\begin{pmatrix}1&1&1&1\\0&0&0&\frac{w_{24}}{x_3x_4z}\\0&\frac{w_{32}}{x_3}&0&0\\0&0&\frac{w_{43}}{x_4}&0\end{pmatrix}$	$\begin{pmatrix}1&1&1&1\\0&w_{22}&w_{23}&0\\0&0&w_{33}&w_{34}\\0&w_{42}&0&w_{44}\end{pmatrix}$	$\begin{pmatrix}1&0&0&0\\0&x_3x_4z&0&0\\0&0&x_3&0\\0&0&0&x_4\end{pmatrix}$	$\begin{pmatrix}1&1&1&1\\0&x_3x_4zw_{22}&\frac{w_{23}}{x_3x_4z}&0\\0&0&x_3w_{33}&\frac{w_{34}}{x_3}\\0&\frac{w_{42}}{x_4}&0&x_4w_{44}\end{pmatrix}$	$\begin{pmatrix}1&1&1&1\\0&w_{22}&w_{23}&w_{24}\\0&w_{32}&w_{33}&0\\0&0&w_{43}&0\end{pmatrix}$	$\begin{pmatrix}1&0&0&0\\0&x_3x_4z&0&0\\0&0&x_3&0\\0&0&0&x_4\end{pmatrix}$	$\begin{pmatrix}1&1&1&1\\0&x_3x_4zw_{22}&\frac{w_{23}}{x_3x_4z}&\frac{w_{24}}{x_3x_4z}\\0&\frac{w_{32}}{x_3}&x_3w_{33}&0\\0&0&\frac{w_{43}}{x_4}&0\end{pmatrix}$
$\begin{pmatrix}1&1&1&1\\0&0&0&w_{24}\\0&w_{32}&0&0\\0&0&w_{43}&0\end{pmatrix}$	$\begin{pmatrix}1&0&0&0\\0&0&x_3x_4z&0\\0&0&0&x_3\\0&x_4&0&0\end{pmatrix}$	$\begin{pmatrix}1&1&1&1\\0&0&0&\frac{w_{43}}{x_4}\\0&\frac{w_{24}}{x_3x_4z}&0&0\\0&0&\frac{w_{32}}{x_3}&0\end{pmatrix}$	$\begin{pmatrix}1&1&1&1\\0&w_{22}&w_{23}&0\\0&0&w_{33}&w_{34}\\0&w_{42}&0&w_{44}\end{pmatrix}$	$\begin{pmatrix}1&0&0&0\\0&0&x_3x_4z&0\\0&0&0&x_3\\0&x_4&0&0\end{pmatrix}$	$\begin{pmatrix}1&1&1&1\\0&x_4w_{44}&\frac{w_{42}}{x_4}&0\\0&0&x_3x_4zw_{22}&\frac{w_{23}}{x_3x_4z}\\0&\frac{w_{34}}{x_3}&0&x_3w_{33}\end{pmatrix}$	$\begin{pmatrix}1&1&1&1\\0&w_{22}&w_{23}&w_{24}\\0&w_{32}&0&0\\0&w_{42}&w_{43}&0\end{pmatrix}$	$\begin{pmatrix}1&0&0&0\\0&x_3x_4z&0&0\\0&0&x_3&0\\0&0&0&x_4\end{pmatrix}$	$\begin{pmatrix}1&1&1&1\\0&x_3x_4zw_{22}&\frac{w_{23}}{x_3x_4z}&\frac{w_{24}}{x_3x_4z}\\0&\frac{w_{32}}{x_3}&0&0\\0&\frac{w_{42}}{x_4}&\frac{w_{43}}{x_4}&0\end{pmatrix}$
$\begin{pmatrix}1&1&1&1\\0&0&0&w_{24}\\0&w_{32}&0&0\\0&0&w_{43}&0\end{pmatrix}$	$\begin{pmatrix}1&0&0&0\\0&0&0&x_3x_4z\\0&x_3&0&0\\0&0&x_4&0\end{pmatrix}$	$\begin{pmatrix}1&1&1&1\\0&0&0&\frac{w_{32}}{x_3}\\0&\frac{w_{43}}{x_4}&0&0\\0&0&\frac{w_{24}}{x_3x_4z}&0\end{pmatrix}$	$\begin{pmatrix}1&1&1&1\\0&w_{22}&w_{23}&0\\0&0&w_{33}&w_{34}\\0&w_{42}&0&w_{44}\end{pmatrix}$	$\begin{pmatrix}1&0&0&0\\0&0&0&x_3x_4z\\0&x_3&0&0\\0&0&x_4&0\end{pmatrix}$	$\begin{pmatrix}1&1&1&1\\0&x_3w_{33}&\frac{w_{34}}{x_3}&0\\0&0&x_4w_{44}&\frac{w_{42}}{x_4}\\0&\frac{w_{23}}{x_3x_4z}&0&x_3x_4zw_{22}\end{pmatrix}$	$\begin{pmatrix}1&1&1&1\\0&0&w_{23}&w_{24}\\0&w_{32}&0&w_{34}\\0&w_{42}&w_{43}&0\end{pmatrix}$	$\begin{pmatrix}1&0&0&0\\0&x_3x_4z&0&0\\0&0&x_3&0\\0&0&0&x_4\end{pmatrix}$	$\begin{pmatrix}1&1&1&1\\0&0&\frac{w_{23}}{x_3x_4z}&\frac{w_{24}}{x_3x_4z}\\0&\frac{w_{32}}{x_3}&0&\frac{w_{34}}{x_3}\\0&\frac{w_{42}}{x_4}&\frac{w_{43}}{x_4}&0\end{pmatrix}$
$\begin{pmatrix}1&1&1&1\\0&w_{22}&0&w_{24}\\0&w_{32}&0&0\\0&0&w_{43}&0\end{pmatrix}$	$\begin{pmatrix}1&0&0&0\\0&x_3x_4z&0&0\\0&0&x_3&0\\0&0&0&x_4\end{pmatrix}$	$\begin{pmatrix}1&1&1&1\\0&x_3x_4zw_{22}&0&\frac{w_{24}}{x_3x_4z}\\0&\frac{w_{32}}{x_3}&0&0\\0&0&\frac{w_{43}}{x_4}&0\end{pmatrix}$	$\begin{pmatrix}1&1&1&1\\0&w_{22}&w_{23}&w_{24}\\0&w_{32}&0&0\\0&w_{42}&0&w_{44}\end{pmatrix}$	$\begin{pmatrix}1&0&0&0\\0&x_3x_4z&0&0\\0&0&x_3&0\\0&0&0&x_4\end{pmatrix}$	$\begin{pmatrix}1&1&1&1\\0&x_3x_4zw_{22}&\frac{w_{23}}{x_3x_4z}&\frac{w_{24}}{x_3x_4z}\\0&\frac{w_{32}}{x_3}&0&0\\0&\frac{w_{42}}{x_4}&0&x_4w_{44}\end{pmatrix}$	$\begin{pmatrix}1&1&1&1\\0&0&w_{23}&w_{24}\\0&w_{32}&0&w_{34}\\0&w_{42}&w_{43}&0\end{pmatrix}$	$\begin{pmatrix}1&0&0&0\\0&0&x_3x_4z&0\\0&0&0&x_3\\0&x_4&0&0\end{pmatrix}$	$\begin{pmatrix}1&1&1&1\\0&0&\frac{w_{42}}{x_4}&\frac{w_{43}}{x_4}\\0&\frac{w_{24}}{x_3x_4z}&0&\frac{w_{23}}{x_3x_4z}\\0&\frac{w_{34}}{x_3}&\frac{w_{32}}{x_3}&0\end{pmatrix}$
$\begin{pmatrix}1&1&1&1\\0&0&w_{23}&0\\0&w_{32}&0&w_{34}\\0&w_{42}&0&w_{44}\end{pmatrix}$	$\begin{pmatrix}1&0&0&0\\0&x_3x_4z&0&0\\0&0&x_3&0\\0&0&0&x_4\end{pmatrix}$	$\begin{pmatrix}1&1&1&1\\0&0&\frac{w_{23}}{x_3x_4z}&0\\0&\frac{w_{32}}{x_3}&0&\frac{w_{34}}{x_3}\\0&\frac{w_{42}}{x_4}&0&x_4w_{44}\end{pmatrix}$	$\begin{pmatrix}1&1&1&1\\0&w_{22}&w_{23}&w_{24}\\0&w_{32}&0&0\\0&0&w_{43}&w_{44}\end{pmatrix}$	$\begin{pmatrix}1&0&0&0\\0&x_3x_4z&0&0\\0&0&x_3&0\\0&0&0&x_4\end{pmatrix}$	$\begin{pmatrix}1&1&1&1\\0&x_3x_4zw_{22}&\frac{w_{23}}{x_3x_4z}&\frac{w_{24}}{x_3x_4z}\\0&\frac{w_{32}}{x_3}&0&0\\0&0&\frac{w_{43}}{x_4}&x_4w_{44}\end{pmatrix}$	$\begin{pmatrix}1&1&1&1\\0&0&w_{23}&w_{24}\\0&w_{32}&0&w_{34}\\0&w_{42}&w_{43}&0\end{pmatrix}$	$\begin{pmatrix}1&0&0&0\\0&0&0&x_3x_4z\\0&x_3&0&0\\0&0&x_4&0\end{pmatrix}$	$\begin{pmatrix}1&1&1&1\\0&0&\frac{w_{34}}{x_3}&\frac{w_{32}}{x_3}\\0&\frac{w_{43}}{x_4}&0&\frac{w_{42}}{x_4}\\0&\frac{w_{23}}{x_3x_4z}&\frac{w_{24}}{x_3x_4z}&0\end{pmatrix}$
$\begin{pmatrix}1&1&1&1\\0&0&w_{23}&0\\0&w_{32}&0&w_{34}\\0&0&w_{43}&w_{44}\end{pmatrix}$	$\begin{pmatrix}1&0&0&0\\0&x_3x_4z&0&0\\0&0&x_3&0\\0&0&0&x_4\end{pmatrix}$	$\begin{pmatrix}1&1&1&1\\0&0&\frac{w_{23}}{x_3x_4z}&0\\0&\frac{w_{32}}{x_3}&0&\frac{w_{34}}{x_3}\\0&0&\frac{w_{43}}{x_4}&x_4w_{44}\end{pmatrix}$	$\begin{pmatrix}1&1&1&1\\0&0&w_{23}&w_{24}\\0&w_{32}&w_{33}&0\\0&w_{42}&0&w_{44}\end{pmatrix}$	$\begin{pmatrix}1&0&0&0\\0&x_3x_4z&0&0\\0&0&x_3&0\\0&0&0&x_4\end{pmatrix}$	$\begin{pmatrix}1&1&1&1\\0&0&\frac{w_{23}}{x_3x_4z}&\frac{w_{24}}{x_3x_4z}\\0&\frac{w_{32}}{x_3}&x_3w_{33}&0\\0&\frac{w_{42}}{x_4}&0&x_4w_{44}\end{pmatrix}$	$\begin{pmatrix}1&1&1&1\\0&0&w_{23}&w_{24}\\0&w_{32}&0&w_{34}\\0&w_{42}&w_{43}&0\end{pmatrix}$	$\begin{pmatrix}1&0&0&0\\0&x_3x_4z&0&0\\0&0&0&x_3\\0&0&x_4&0\end{pmatrix}$	$\begin{pmatrix}1&1&1&1\\0&0&\frac{w_{24}}{x_3x_4z}&\frac{w_{23}}{x_3x_4z}\\0&\frac{w_{42}}{x_4}&0&\frac{w_{43}}{x_4}\\0&\frac{w_{32}}{x_3}&\frac{w_{34}}{x_3}&0\end{pmatrix}$
$\begin{pmatrix}1&1&1&1\\0&w_{22}&w_{23}&w_{24}\\0&w_{32}&0&0\\0&0&w_{43}&0\end{pmatrix}$	$\begin{pmatrix}1&0&0&0\\0&x_3x_4z&0&0\\0&0&x_3&0\\0&0&0&x_4\end{pmatrix}$	$\begin{pmatrix}1&1&1&1\\0&x_3x_4zw_{22}&\frac{w_{23}}{x_3x_4z}&\frac{w_{24}}{x_3x_4z}\\0&\frac{w_{32}}{x_3}&0&0\\0&0&\frac{w_{43}}{x_4}&0\end{pmatrix}$	$\begin{pmatrix}1&1&1&1\\0&0&w_{23}&w_{24}\\0&w_{32}&w_{33}&0\\0&w_{42}&0&w_{44}\end{pmatrix}$	$\begin{pmatrix}1&0&0&0\\0&x_3x_4z&0&0\\0&0&0&x_3\\0&0&x_4&0\end{pmatrix}$	$\begin{pmatrix}1&1&1&1\\0&0&\frac{w_{24}}{x_3x_4z}&\frac{w_{23}}{x_3x_4z}\\0&\frac{w_{42}}{x_4}&x_4w_{44}&0\\0&\frac{w_{32}}{x_3}&0&x_3w_{33}\end{pmatrix}$	$\begin{pmatrix}1&1&1&1\\0&0&w_{23}&w_{24}\\0&w_{32}&0&w_{34}\\0&w_{42}&w_{43}&0\end{pmatrix}$	$\begin{pmatrix}1&0&0&0\\0&0&0&x_3x_4z\\0&0&x_3&0\\0&x_4&0&0\end{pmatrix}$	$\begin{pmatrix}1&1&1&1\\0&0&\frac{w_{43}}{x_4}&\frac{w_{42}}{x_4}\\0&\frac{w_{34}}{x_3}&0&\frac{w_{32}}{x_3}\\0&\frac{w_{24}}{x_3x_4z}&\frac{w_{23}}{x_3x_4z}&0\end{pmatrix}$
$\begin{pmatrix}1&1&1&1\\0&w_{22}&0&w_{24}\\0&w_{32}&w_{33}&0\\0&0&w_{43}&0\end{pmatrix}$	$\begin{pmatrix}1&0&0&0\\0&x_3x_4z&0&0\\0&0&x_3&0\\0&0&0&x_4\end{pmatrix}$	$\begin{pmatrix}1&1&1&1\\0&x_3x_4zw_{22}&0&\frac{w_{24}}{x_3x_4z}\\0&\frac{w_{32}}{x_3}&x_3w_{33}&0\\0&0&\frac{w_{43}}{x_4}&0\end{pmatrix}$	$\begin{pmatrix}1&1&1&1\\0&0&w_{23}&w_{24}\\0&w_{32}&w_{33}&0\\0&0&w_{43}&w_{44}\end{pmatrix}$	$\begin{pmatrix}1&0&0&0\\0&x_3x_4z&0&0\\0&0&x_3&0\\0&0&0&x_4\end{pmatrix}$	$\begin{pmatrix}1&1&1&1\\0&0&\frac{w_{23}}{x_3x_4z}&\frac{w_{24}}{x_3x_4z}\\0&\frac{w_{32}}{x_3}&x_3w_{33}&0\\0&0&\frac{w_{43}}{x_4}&x_4w_{44}\end{pmatrix}$	$\begin{pmatrix}1&1&1&1\\0&0&w_{23}&w_{24}\\0&w_{32}&0&w_{34}\\0&w_{42}&w_{43}&0\end{pmatrix}$	$\begin{pmatrix}1&0&0&0\\0&0&x_3x_4z&0\\0&x_3&0&0\\0&0&0&x_4\end{pmatrix}$	$\begin{pmatrix}1&1&1&1\\0&0&\frac{w_{32}}{x_3}&\frac{w_{34}}{x_3}\\0&\frac{w_{23}}{x_3x_4z}&0&\frac{w_{24}}{x_3x_4z}\\0&\frac{w_{42}}{x_4}&\frac{w_{43}}{x_4}&0\end{pmatrix}$
$\begin{pmatrix}1&1&1&1\\0&w_{22}&0&w_{24}\\0&w_{32}&0&0\\0&w_{42}&w_{43}&0\end{pmatrix}$	$\begin{pmatrix}1&0&0&0\\0&x_3x_4z&0&0\\0&0&x_3&0\\0&0&0&x_4\end{pmatrix}$	$\begin{pmatrix}1&1&1&1\\0&x_3x_4zw_{22}&0&\frac{w_{24}}{x_3x_4z}\\0&\frac{w_{32}}{x_3}&0&0\\0&\frac{w_{42}}{x_4}&\frac{w_{43}}{x_4}&0\end{pmatrix}$	$\begin{pmatrix}1&1&1&1\\0&0&w_{23}&w_{24}\\0&w_{32}&0&w_{34}\\0&w_{42}&0&w_{44}\end{pmatrix}$	$\begin{pmatrix}1&0&0&0\\0&x_3x_4z&0&0\\0&0&x_3&0\\0&0&0&x_4\end{pmatrix}$	$\begin{pmatrix}1&1&1&1\\0&0&\frac{w_{23}}{x_3x_4z}&\frac{w_{24}}{x_3x_4z}\\0&\frac{w_{32}}{x_3}&0&\frac{w_{34}}{x_3}\\0&\frac{w_{42}}{x_4}&0&x_4w_{44}\end{pmatrix}$	$\begin{pmatrix}1&1&1&1\\0&w_{22}&w_{23}&w_{24}\\0&w_{32}&w_{33}&0\\0&w_{42}&0&w_{44}\end{pmatrix}$	$\begin{pmatrix}1&0&0&0\\0&x_3x_4z&0&0\\0&0&x_3&0\\0&0&0&x_4\end{pmatrix}$	$\begin{pmatrix}1&1&1&1\\0&x_3x_4zw_{22}&\frac{w_{23}}{x_3x_4z}&\frac{w_{24}}{x_3x_4z}\\0&\frac{w_{32}}{x_3}&x_3w_{33}&0\\0&\frac{w_{42}}{x_4}&0&x_4w_{44}\end{pmatrix}$
$\begin{pmatrix}1&1&1&1\\0&0&w_{23}&w_{24}\\0&w_{32}&0&w_{34}\\0&0&w_{43}&0\end{pmatrix}$	$\begin{pmatrix}1&0&0&0\\0&x_3x_4z&0&0\\0&0&x_3&0\\0&0&0&x_4\end{pmatrix}$	$\begin{pmatrix}1&1&1&1\\0&0&\frac{w_{23}}{x_3x_4z}&\frac{w_{24}}{x_3x_4z}\\0&\frac{w_{32}}{x_3}&0&\frac{w_{34}}{x_3}\\0&0&\frac{w_{43}}{x_4}&0\end{pmatrix}$	$\begin{pmatrix}1&1&1&1\\0&0&w_{23}&w_{24}\\0&w_{32}&0&0\\0&w_{42}&w_{43}&w_{44}\end{pmatrix}$	$\begin{pmatrix}1&0&0&0\\0&x_3x_4z&0&0\\0&0&x_3&0\\0&0&0&x_4\end{pmatrix}$	$\begin{pmatrix}1&1&1&1\\0&0&\frac{w_{23}}{x_3x_4z}&\frac{w_{24}}{x_3x_4z}\\0&\frac{w_{32}}{x_3}&0&0\\0&\frac{w_{42}}{x_4}&\frac{w_{43}}{x_4}&x_4w_{44}\end{pmatrix}$	$\begin{pmatrix}1&1&1&1\\0&w_{22}&w_{23}&w_{24}\\0&w_{32}&w_{33}&0\\0&w_{42}&0&w_{44}\end{pmatrix}$	$\begin{pmatrix}1&0&0&0\\0&x_3x_4z&0&0\\0&0&0&x_3\\0&0&x_4&0\end{pmatrix}$	$\begin{pmatrix}1&1&1&1\\0&x_3x_4zw_{22}&\frac{w_{24}}{x_3x_4z}&\frac{w_{23}}{x_3x_4z}\\0&\frac{w_{42}}{x_4}&x_4w_{44}&0\\0&\frac{w_{32}}{x_3}&0&x_3w_{33}\end{pmatrix}$

M_B	$P_{B'B}$	M'_B
$\begin{pmatrix}1&1&1&1\\0&w_{22}&w_{23}&w_{24}\\0&w_{32}&w_{33}&0\\0&0&w_{43}&w_{44}\end{pmatrix}$	$\begin{pmatrix}1&0&0&0\\0&x_3x_4z&0&0\\0&0&x_3&0\\0&0&0&x_4\end{pmatrix}$	$\begin{pmatrix}1&1&1&1\\0&x_3x_4zw_{22}&\frac{w_{23}}{x_3x_4z}&\frac{w_{24}}{x_3x_4z}\\0&\frac{w_{32}}{x_3}&x_3w_{33}&0\\0&0&\frac{w_{43}}{x_4}&x_4w_{44}\end{pmatrix}$
$\begin{pmatrix}1&1&1&1\\0&w_{22}&w_{23}&w_{24}\\0&w_{32}&0&0\\0&w_{42}&w_{43}&w_{44}\end{pmatrix}$	$\begin{pmatrix}1&0&0&0\\0&x_3x_4z&0&0\\0&0&x_3&0\\0&0&0&x_4\end{pmatrix}$	$\begin{pmatrix}1&1&1&1\\0&x_3x_4zw_{22}&\frac{w_{23}}{x_3x_4z}&\frac{w_{24}}{x_3x_4z}\\0&\frac{w_{32}}{x_3}&0&0\\0&\frac{w_{42}}{x_4}&\frac{w_{43}}{x_4}&x_4w_{44}\end{pmatrix}$
$\begin{pmatrix}1&1&1&1\\0&w_{22}&w_{23}&w_{24}\\0&w_{32}&0&w_{34}\\0&w_{42}&0&w_{44}\end{pmatrix}$	$\begin{pmatrix}1&0&0&0\\0&x_3x_4z&0&0\\0&0&x_3&0\\0&0&0&x_4\end{pmatrix}$	$\begin{pmatrix}1&1&1&1\\0&x_3x_4zw_{22}&\frac{w_{23}}{x_3x_4z}&\frac{w_{24}}{x_3x_4z}\\0&\frac{w_{32}}{x_3}&0&\frac{w_{34}}{x_3}\\0&\frac{w_{42}}{x_4}&0&x_4w_{44}\end{pmatrix}$
$\begin{pmatrix}1&1&1&1\\0&0&w_{23}&w_{24}\\0&w_{32}&w_{33}&w_{34}\\0&w_{42}&0&w_{44}\end{pmatrix}$	$\begin{pmatrix}1&0&0&0\\0&x_3x_4z&0&0\\0&0&x_3&0\\0&0&0&x_4\end{pmatrix}$	$\begin{pmatrix}1&1&1&1\\0&0&\frac{w_{23}}{x_3x_4z}&\frac{w_{24}}{x_3x_4z}\\0&\frac{w_{32}}{x_3}&x_3w_{33}&\frac{w_{34}}{x_3}\\0&\frac{w_{42}}{x_4}&0&x_4w_{44}\end{pmatrix}$
$\begin{pmatrix}1&1&1&1\\0&0&w_{23}&w_{24}\\0&w_{32}&0&w_{34}\\0&w_{42}&w_{43}&w_{44}\end{pmatrix}$	$\begin{pmatrix}1&0&0&0\\0&x_3x_4z&0&0\\0&0&x_3&0\\0&0&0&x_4\end{pmatrix}$	$\begin{pmatrix}1&1&1&1\\0&0&\frac{w_{23}}{x_3x_4z}&\frac{w_{24}}{x_3x_4z}\\0&\frac{w_{32}}{x_3}&0&\frac{w_{34}}{x_3}\\0&\frac{w_{42}}{x_4}&\frac{w_{43}}{x_4}&x_4w_{44}\end{pmatrix}$
$\begin{pmatrix}1&1&1&1\\0&0&w_{23}&w_{24}\\0&w_{32}&0&w_{34}\\0&w_{42}&w_{43}&w_{44}\end{pmatrix}$	$\begin{pmatrix}1&0&0&0\\0&0&x_3x_4z&0\\0&x_3&0&0\\0&0&0&x_4\end{pmatrix}$	$\begin{pmatrix}1&1&1&1\\0&0&\frac{w_{32}}{x_3}&\frac{w_{34}}{x_3}\\0&\frac{w_{23}}{x_3x_4z}&0&\frac{w_{24}}{x_3x_4z}\\0&\frac{w_{43}}{x_4}&\frac{w_{42}}{x_4}&x_4w_{44}\end{pmatrix}$
$\begin{pmatrix}1&1&1&1\\0&w_{22}&w_{23}&w_{24}\\0&w_{32}&w_{33}&w_{34}\\0&w_{42}&0&w_{44}\end{pmatrix}$	$\begin{pmatrix}1&0&0&0\\0&x_3x_4z&0&0\\0&0&x_3&0\\0&0&0&x_4\end{pmatrix}$	$\begin{pmatrix}1&1&1&1\\0&x_3x_4zw_{22}&\frac{w_{23}}{x_3x_4z}&\frac{w_{24}}{x_3x_4z}\\0&\frac{w_{32}}{x_3}&x_3w_{33}&\frac{w_{34}}{x_3}\\0&\frac{w_{42}}{x_4}&0&x_4w_{44}\end{pmatrix}$
$\begin{pmatrix}1&1&1&1\\0&w_{22}&w_{23}&w_{24}\\0&w_{32}&w_{33}&w_{34}\\0&w_{42}&w_{43}&0\end{pmatrix}$	$\begin{pmatrix}1&0&0&0\\0&x_3x_4z&0&0\\0&0&x_3&0\\0&0&0&x_4\end{pmatrix}$	$\begin{pmatrix}1&1&1&1\\0&x_3x_4zw_{22}&\frac{w_{23}}{x_3x_4z}&\frac{w_{24}}{x_3x_4z}\\0&\frac{w_{32}}{x_3}&x_3w_{33}&\frac{w_{34}}{x_3}\\0&\frac{w_{42}}{x_4}&\frac{w_{43}}{x_4}&0\end{pmatrix}$
$\begin{pmatrix}1&1&1&1\\0&w_{22}&w_{23}&w_{24}\\0&w_{32}&w_{33}&w_{34}\\0&w_{42}&w_{43}&0\end{pmatrix}$	$\begin{pmatrix}1&0&0&0\\0&0&x_3x_4z&0\\0&x_3&0&0\\0&0&0&x_4\end{pmatrix}$	$\begin{pmatrix}1&1&1&1\\0&x_3w_{33}&\frac{w_{32}}{x_3}&\frac{w_{34}}{x_3}\\0&\frac{w_{23}}{x_3x_4z}&x_3x_4zw_{22}&\frac{w_{24}}{x_3x_4z}\\0&\frac{w_{43}}{x_4}&\frac{w_{42}}{x_4}&0\end{pmatrix}$
$\begin{pmatrix}1&1&1&1\\0&w_{22}&w_{23}&w_{24}\\0&w_{32}&w_{33}&w_{34}\\0&w_{42}&w_{43}&w_{44}\end{pmatrix}$	$\begin{pmatrix}1&0&0&0\\0&x_3x_4z&0&0\\0&0&x_3&0\\0&0&0&x_4\end{pmatrix}$	$\begin{pmatrix}1&1&1&1\\0&x_3x_4zw_{22}&\frac{w_{23}}{x_3x_4z}&\frac{w_{24}}{x_3x_4z}\\0&\frac{w_{32}}{x_3}&x_3w_{33}&\frac{w_{34}}{x_3}\\0&\frac{w_{42}}{x_4}&\frac{w_{43}}{x_4}&x_4w_{44}\end{pmatrix}$

M_B	$P_{B'B}$	M'_B
$\begin{pmatrix}1&1&1&1\\0&w_{22}&w_{23}&w_{24}\\0&w_{32}&w_{33}&w_{34}\\0&w_{42}&w_{43}&w_{44}\end{pmatrix}$	$\begin{pmatrix}1&0&0&0\\0&0&x_3x_4z&0\\0&0&0&x_3\\0&x_4&0&0\end{pmatrix}$	$\begin{pmatrix}1&1&1&1\\0&x_4w_{44}&\frac{w_{42}}{x_4}&\frac{w_{43}}{x_4}\\0&\frac{w_{24}}{x_3x_4z}&x_3x_4zw_{22}&\frac{w_{23}}{x_3x_4z}\\0&\frac{w_{34}}{x_3}&\frac{w_{32}}{x_3}&x_3w_{33}\end{pmatrix}$
$\begin{pmatrix}1&1&1&1\\0&w_{22}&w_{23}&w_{24}\\0&w_{32}&w_{33}&w_{34}\\0&w_{42}&w_{43}&w_{44}\end{pmatrix}$	$\begin{pmatrix}1&0&0&0\\0&0&0&x_3x_4z\\0&x_3&0&0\\0&0&x_4&0\end{pmatrix}$	$\begin{pmatrix}1&1&1&1\\0&x_3w_{33}&\frac{w_{34}}{x_3}&\frac{w_{32}}{x_3}\\0&\frac{w_{43}}{x_4}&x_4w_{44}&\frac{w_{42}}{x_4}\\0&\frac{w_{23}}{x_3x_4z}&\frac{w_{24}}{x_3x_4z}&x_3x_4zw_{22}\end{pmatrix}$
$\begin{pmatrix}1&1&1&1\\0&w_{22}&w_{23}&w_{24}\\0&w_{32}&w_{33}&w_{34}\\0&w_{42}&w_{43}&w_{44}\end{pmatrix}$	$\begin{pmatrix}1&0&0&0\\0&x_3x_4z&0&0\\0&0&0&x_3\\0&0&x_4&0\end{pmatrix}$	$\begin{pmatrix}1&1&1&1\\0&x_3x_4zw_{22}&\frac{w_{24}}{x_3x_4z}&\frac{w_{23}}{x_3x_4z}\\0&\frac{w_{42}}{x_4}&x_4w_{44}&\frac{w_{43}}{x_4}\\0&\frac{w_{32}}{x_3}&\frac{w_{34}}{x_3}&x_3w_{33}\end{pmatrix}$
$\begin{pmatrix}1&1&1&1\\0&w_{22}&w_{23}&w_{24}\\0&w_{32}&w_{33}&w_{34}\\0&w_{42}&w_{43}&w_{44}\end{pmatrix}$	$\begin{pmatrix}1&0&0&0\\0&0&0&x_3x_4z\\0&0&x_3&0\\0&x_4&0&0\end{pmatrix}$	$\begin{pmatrix}1&1&1&1\\0&x_4w_{44}&\frac{w_{43}}{x_4}&\frac{w_{42}}{x_4}\\0&\frac{w_{34}}{x_3}&x_3w_{33}&\frac{w_{32}}{x_3}\\0&\frac{w_{24}}{x_3x_4z}&\frac{w_{23}}{x_3x_4z}&x_3x_4zw_{22}\end{pmatrix}$
$\begin{pmatrix}1&1&1&1\\0&w_{22}&w_{23}&w_{24}\\0&w_{32}&w_{33}&w_{34}\\0&w_{42}&w_{43}&w_{44}\end{pmatrix}$	$\begin{pmatrix}1&0&0&0\\0&0&x_3x_4z&0\\0&x_3&0&0\\0&0&0&x_4\end{pmatrix}$	$\begin{pmatrix}1&1&1&1\\0&x_3w_{33}&\frac{w_{32}}{x_3}&\frac{w_{34}}{x_3}\\0&\frac{w_{23}}{x_3x_4z}&x_3x_4zw_{22}&\frac{w_{24}}{x_3x_4z}\\0&\frac{w_{43}}{x_4}&\frac{w_{42}}{x_4}&x_4w_{44}\end{pmatrix}$

3 The Maximal Basic Ideal Is Two-Dimensional

Now we consider that he structure matrix M_B is $\left(\begin{array}{c|c} W & U \\ \hline \begin{matrix}0&0\\0&0\end{matrix} & Y \end{array}\right)$. Observe that the matrix U cannot be the zero because otherwise the algebra will be reducible.

The classification into families of non-isomorphic evolution algebras given in [1, Sect. 4] depends on the matrix W. Concretely, W must be the structure matrix of a perfect two dimensional evolution algebra. The subsections that follow corresponds to each of these matrices, which can be found in [5, Theorem 3.3 (III)]. In order to not enlarge the paper, we have not distinguished the subsections in the Sects. 3.1–3.4. Each table in this section corresponds to an election of W jointly with an election of the number of nonzero entries in the matrix U, according to [1, Sect. 4].

3.1 Take $W = \begin{pmatrix}1&0\\0&1\end{pmatrix}$

M_B	$P_{B'B}$	M'_B
$\begin{pmatrix}1&0&0&0\\0&1&0&w_{24}\\0&0&0&1\\0&0&1&0\end{pmatrix}$	$\begin{pmatrix}1&0&0&0\\0&1&0&0\\0&0&x_4^2&0\\0&0&0&x_4\end{pmatrix}$	$\begin{pmatrix}1&0&0&0\\0&1&0&x_4^2w_{24}\\0&0&0&1\\0&0&1&0\end{pmatrix}$
$\begin{pmatrix}1&0&0&0\\0&1&0&w_{24}\\0&0&0&1\\0&0&1&w_{44}\end{pmatrix}$	$\begin{pmatrix}1&0&0&0\\0&1&0&0\\0&0&x_4^2&0\\0&0&0&x_4\end{pmatrix}$	$\begin{pmatrix}1&0&0&0\\0&1&0&x_4^2w_{24}\\0&0&0&1\\0&0&1&x_4w_{44}\end{pmatrix}$
$\begin{pmatrix}1&0&0&0\\0&1&0&w_{24}\\0&0&w_{33}&1\\0&0&1&0\end{pmatrix}$	$\begin{pmatrix}1&0&0&0\\0&1&0&0\\0&0&x_4^2&0\\0&0&0&x_4\end{pmatrix}$	$\begin{pmatrix}1&0&0&0\\0&1&0&x_4^2w_{24}\\0&0&x_4^2w_{33}&1\\0&0&1&0\end{pmatrix}$
$\begin{pmatrix}1&0&0&0\\0&1&w_{23}&w_{24}\\0&0&1&w_{34}\\0&0&w_{43}&1\end{pmatrix}$	$\begin{pmatrix}1&0&0&0\\0&1&0&0\\0&0&0&1\\0&0&1&0\end{pmatrix}$	$\begin{pmatrix}1&0&0&0\\0&1&w_{24}&w_{23}\\0&0&1&w_{43}\\0&0&w_{34}&1\end{pmatrix}$
$\begin{pmatrix}1&0&0&0\\0&1&w_{23}&w_{24}\\0&0&1&w_{34}\\0&0&w_{43}&1\end{pmatrix}$	$\begin{pmatrix}1&0&0&0\\0&1&0&0\\0&0&0&1\\0&0&1&0\end{pmatrix}$	$\begin{pmatrix}1&0&0&0\\0&1&w_{24}&w_{23}\\0&0&1&w_{43}\\0&0&w_{34}&1\end{pmatrix}$
$\begin{pmatrix}1&0&0&0\\0&1&w_{23}&w_{24}\\0&0&0&1\\0&0&1&0\end{pmatrix}$	$\begin{pmatrix}1&0&0&0\\0&1&0&0\\0&0&x_4^2&0\\0&0&0&x_4\end{pmatrix}$	$\begin{pmatrix}1&0&0&0\\0&1&x_4w_{23}&x_4^2w_{24}\\0&0&0&1\\0&0&1&0\end{pmatrix}$
$\begin{pmatrix}1&0&0&0\\0&1&w_{23}&w_{24}\\0&0&0&1\\0&0&1&0\end{pmatrix}$	$\begin{pmatrix}1&0&0&0\\0&1&0&0\\0&0&0&x_4^2\\0&0&x_4&0\end{pmatrix}$	$\begin{pmatrix}1&0&0&0\\0&1&x_4^2w_{24}&x_4w_{23}\\0&0&0&1\\0&0&1&0\end{pmatrix}$
$\begin{pmatrix}1&0&0&0\\0&1&w_{23}&w_{24}\\0&0&0&1\\0&0&1&w_{44}\end{pmatrix}$	$\begin{pmatrix}1&0&0&0\\0&1&0&0\\0&0&x_4^2&0\\0&0&0&x_4\end{pmatrix}$	$\begin{pmatrix}1&0&0&0\\0&1&x_4w_{23}&x_4^2w_{24}\\0&0&0&1\\0&0&1&x_4w_{44}\end{pmatrix}$
$\begin{pmatrix}1&0&0&w_{14}\\0&1&0&w_{24}\\0&0&1&w_{34}\\0&0&w_{43}&1\end{pmatrix}$	$\begin{pmatrix}0&1&0&0\\1&0&0&0\\0&0&1&0\\0&0&0&1\end{pmatrix}$	$\begin{pmatrix}1&0&0&w_{24}\\0&1&0&w_{14}\\0&0&1&w_{34}\\0&0&w_{43}&1\end{pmatrix}$
$\begin{pmatrix}1&0&0&w_{14}\\0&1&0&w_{24}\\0&0&0&1\\0&0&1&0\end{pmatrix}$	$\begin{pmatrix}1&0&0&0\\0&1&0&0\\0&0&x_4^2&0\\0&0&0&x_4\end{pmatrix}$	$\begin{pmatrix}1&0&0&x_4^2w_{14}\\0&1&0&x_4^2w_{24}\\0&0&0&1\\0&0&1&0\end{pmatrix}$

M_B	$P_{B'B}$	M'_B
$\begin{pmatrix}1&0&0&w_{14}\\0&1&0&w_{24}\\0&0&0&1\\0&0&1&0\end{pmatrix}$	$\begin{pmatrix}0&1&0&0\\1&0&0&0\\0&0&x_4^2&0\\0&0&0&x_4\end{pmatrix}$	$\begin{pmatrix}1&0&0&x_4^2w_{24}\\0&1&0&x_4^2w_{14}\\0&0&0&1\\0&0&1&0\end{pmatrix}$
$\begin{pmatrix}1&0&0&w_{14}\\0&1&0&w_{24}\\0&0&0&1\\0&0&1&w_{44}\end{pmatrix}$	$\begin{pmatrix}1&0&0&0\\0&1&0&0\\0&0&x_4^2&0\\0&0&0&x_4\end{pmatrix}$	$\begin{pmatrix}1&0&0&x_4^2w_{14}\\0&1&0&x_4^2w_{24}\\0&0&0&1\\0&0&1&x_4w_{44}\end{pmatrix}$
$\begin{pmatrix}1&0&0&w_{14}\\0&1&0&w_{24}\\0&0&0&1\\0&0&1&w_{44}\end{pmatrix}$	$\begin{pmatrix}0&1&0&0\\1&0&0&0\\0&0&x_4^2&0\\0&0&0&x_4\end{pmatrix}$	$\begin{pmatrix}1&0&0&x_4^2w_{24}\\0&1&0&x_4^2w_{14}\\0&0&0&1\\0&0&1&x_4w_{44}\end{pmatrix}$
$\begin{pmatrix}1&0&0&w_{14}\\0&1&0&w_{24}\\0&0&w_{33}&1\\0&0&1&0\end{pmatrix}$	$\begin{pmatrix}1&0&0&0\\0&1&0&0\\0&0&x_4^2&0\\0&0&0&x_4\end{pmatrix}$	$\begin{pmatrix}1&0&0&x_4^2w_{14}\\0&1&0&x_4^2w_{24}\\0&0&x_4^2w_{33}&1\\0&0&1&0\end{pmatrix}$
$\begin{pmatrix}1&0&0&w_{14}\\0&1&0&w_{24}\\0&0&w_{33}&1\\0&0&1&0\end{pmatrix}$	$\begin{pmatrix}0&1&0&0\\1&0&0&0\\0&0&x_4^2&0\\0&0&0&x_4\end{pmatrix}$	$\begin{pmatrix}1&0&0&x_4^2w_{24}\\0&1&0&x_4^2w_{14}\\0&0&x_4^2w_{33}&1\\0&0&1&0\end{pmatrix}$
$\begin{pmatrix}1&0&0&w_{14}\\0&1&w_{23}&0\\0&0&1&w_{34}\\0&0&w_{43}&1\end{pmatrix}$	$\begin{pmatrix}0&1&0&0\\1&0&0&0\\0&0&0&1\\0&0&1&0\end{pmatrix}$	$\begin{pmatrix}1&0&0&w_{23}\\0&1&w_{14}&0\\0&0&1&w_{43}\\0&0&w_{34}&1\end{pmatrix}$
$\begin{pmatrix}1&0&0&w_{14}\\0&1&w_{23}&0\\0&0&0&1\\0&0&1&0\end{pmatrix}$	$\begin{pmatrix}1&0&0&0\\0&1&0&0\\0&0&x_4^2&0\\0&0&0&x_4\end{pmatrix}$	$\begin{pmatrix}1&0&0&x_4^2w_{14}\\0&1&x_4w_{23}&0\\0&0&0&1\\0&0&1&0\end{pmatrix}$
$\begin{pmatrix}1&0&0&w_{14}\\0&1&w_{23}&0\\0&0&0&1\\0&0&1&0\end{pmatrix}$	$\begin{pmatrix}0&1&0&0\\1&0&0&0\\0&0&0&x_4^2\\0&0&x_4&0\end{pmatrix}$	$\begin{pmatrix}1&0&0&x_4w_{23}\\0&1&x_4^2w_{14}&0\\0&0&0&1\\0&0&1&0\end{pmatrix}$
$\begin{pmatrix}1&0&0&w_{14}\\0&1&w_{23}&0\\0&0&0&1\\0&0&1&w_{44}\end{pmatrix}$	$\begin{pmatrix}1&0&0&0\\0&1&0&0\\0&0&x_4^2&0\\0&0&0&x_4\end{pmatrix}$	$\begin{pmatrix}1&0&0&x_4^2w_{14}\\0&1&x_4w_{23}&0\\0&0&0&1\\0&0&1&x_4w_{44}\end{pmatrix}$
$\begin{pmatrix}1&0&w_{13}&w_{14}\\0&1&w_{23}&0\\0&0&0&1\\0&0&1&0\end{pmatrix}$	$\begin{pmatrix}1&0&0&0\\0&1&0&0\\0&0&x_4^2&0\\0&0&0&x_4\end{pmatrix}$	$\begin{pmatrix}1&0&x_4w_{13}&x_4^2w_{14}\\0&1&x_4w_{23}&0\\0&0&0&1\\0&0&1&0\end{pmatrix}$

M_B	$P_{B'B}$	M'_B
$\begin{pmatrix}1&0&w_{13}&w_{14}\\0&1&w_{23}&0\\0&0&0&1\\0&0&1&w_{44}\end{pmatrix}$	$\begin{pmatrix}1&0&0&0\\0&1&0&0\\0&0&x_4^2&0\\0&0&0&x_4\end{pmatrix}$	$\begin{pmatrix}1&0&x_4w_{13}&x_4^2w_{14}\\0&1&x_4w_{23}&0\\0&0&0&1\\0&0&1&x_4w_{44}\end{pmatrix}$
$\begin{pmatrix}1&0&w_{13}&w_{14}\\0&1&w_{23}&0\\0&0&w_{33}&1\\0&0&1&0\end{pmatrix}$	$\begin{pmatrix}1&0&0&0\\0&1&0&0\\0&0&x_4^2&0\\0&0&0&x_4\end{pmatrix}$	$\begin{pmatrix}1&0&x_4w_{13}&x_4^2w_{14}\\0&1&x_4w_{23}&0\\0&0&x_4^2w_{33}&1\\0&0&1&0\end{pmatrix}$
$\begin{pmatrix}1&0&w_{13}&w_{14}\\0&1&w_{23}&w_{24}\\0&0&1&w_{34}\\0&0&w_{43}&1\end{pmatrix}$	$\begin{pmatrix}0&1&0&0\\1&0&0&0\\0&0&1&0\\0&0&0&1\end{pmatrix}$	$\begin{pmatrix}1&0&w_{23}&w_{24}\\0&1&w_{13}&w_{14}\\0&0&1&w_{34}\\0&0&w_{43}&1\end{pmatrix}$
$\begin{pmatrix}1&0&w_{13}&w_{14}\\0&1&w_{23}&w_{24}\\0&0&1&w_{34}\\0&0&w_{43}&1\end{pmatrix}$	$\begin{pmatrix}1&0&0&0\\0&1&0&0\\0&0&0&1\\0&0&1&0\end{pmatrix}$	$\begin{pmatrix}1&0&w_{14}&w_{13}\\0&1&w_{24}&w_{23}\\0&0&1&w_{43}\\0&0&w_{34}&1\end{pmatrix}$
$\begin{pmatrix}1&0&w_{13}&w_{14}\\0&1&w_{23}&w_{24}\\0&0&1&w_{34}\\0&0&w_{43}&1\end{pmatrix}$	$\begin{pmatrix}0&1&0&0\\1&0&0&0\\0&0&0&1\\0&0&1&0\end{pmatrix}$	$\begin{pmatrix}1&0&w_{24}&w_{23}\\0&1&w_{14}&w_{13}\\0&0&1&w_{43}\\0&0&w_{34}&1\end{pmatrix}$
$\begin{pmatrix}1&0&w_{13}&w_{14}\\0&1&w_{23}&w_{24}\\0&0&0&1\\0&0&1&0\end{pmatrix}$	$\begin{pmatrix}1&0&0&0\\0&1&0&0\\0&0&x_4^2&0\\0&0&0&x_4\end{pmatrix}$	$\begin{pmatrix}1&0&x_4w_{13}&x_4^2w_{14}\\0&1&x_4w_{23}&x_4^2w_{24}\\0&0&0&1\\0&0&1&0\end{pmatrix}$
$\begin{pmatrix}1&0&w_{13}&w_{14}\\0&1&w_{23}&w_{24}\\0&0&0&1\\0&0&1&0\end{pmatrix}$	$\begin{pmatrix}0&1&0&0\\1&0&0&0\\0&0&x_4^2&0\\0&0&0&x_4\end{pmatrix}$	$\begin{pmatrix}1&0&x_4w_{23}&x_4^2w_{24}\\0&1&x_4w_{13}&x_4^2w_{14}\\0&0&0&1\\0&0&1&0\end{pmatrix}$
$\begin{pmatrix}1&0&w_{13}&w_{14}\\0&1&w_{23}&w_{24}\\0&0&0&1\\0&0&1&0\end{pmatrix}$	$\begin{pmatrix}1&0&0&0\\0&1&0&0\\0&0&0&x_4^2\\0&0&x_4&0\end{pmatrix}$	$\begin{pmatrix}1&0&x_4^2w_{14}&x_4w_{13}\\0&1&x_4^2w_{24}&x_4w_{23}\\0&0&0&1\\0&0&1&0\end{pmatrix}$
$\begin{pmatrix}1&0&w_{13}&w_{14}\\0&1&w_{23}&w_{24}\\0&0&0&1\\0&0&1&0\end{pmatrix}$	$\begin{pmatrix}0&1&0&0\\1&0&0&0\\0&0&0&x_4^2\\0&0&x_4&0\end{pmatrix}$	$\begin{pmatrix}1&0&x_4^2w_{24}&x_4w_{23}\\0&1&x_4^2w_{14}&x_4w_{13}\\0&0&0&1\\0&0&1&0\end{pmatrix}$
$\begin{pmatrix}1&0&w_{13}&w_{14}\\0&1&w_{23}&w_{24}\\0&0&0&1\\0&0&1&w_{44}\end{pmatrix}$	$\begin{pmatrix}1&0&0&0\\0&1&0&0\\0&0&x_4^2&0\\0&0&0&x_4\end{pmatrix}$	$\begin{pmatrix}1&0&x_4w_{13}&x_4^2w_{14}\\0&1&x_4w_{23}&x_4^2w_{24}\\0&0&0&1\\0&0&1&x_4w_{44}\end{pmatrix}$
$\begin{pmatrix}1&0&w_{13}&w_{14}\\0&1&w_{23}&w_{24}\\0&0&0&1\\0&0&1&w_{44}\end{pmatrix}$	$\begin{pmatrix}0&1&0&0\\1&0&0&0\\0&0&x_4^2&0\\0&0&0&x_4\end{pmatrix}$	$\begin{pmatrix}1&0&x_4w_{23}&x_4^2w_{24}\\0&1&x_4w_{13}&x_4^2w_{14}\\0&0&0&1\\0&0&1&x_4w_{44}\end{pmatrix}$

3.2 Take $W = \begin{pmatrix}1&0\\\omega_{21}&1\end{pmatrix}$

M_B	$P_{B'B}$	M'_B
$\begin{pmatrix}1&0&0&0\\w_{21}&1&0&w_{24}\\0&0&0&1\\0&0&1&0\end{pmatrix}$	$\begin{pmatrix}1&0&0&0\\0&1&0&0\\0&0&x_4^2&0\\0&0&0&x_4\end{pmatrix}$	$\begin{pmatrix}1&0&0&0\\w_{21}&1&0&x_4^2w_{24}\\0&0&0&1\\0&0&1&0\end{pmatrix}$
$\begin{pmatrix}1&0&0&0\\w_{21}&1&0&w_{24}\\0&0&0&1\\0&0&1&w_{44}\end{pmatrix}$	$\begin{pmatrix}1&0&0&0\\0&1&0&0\\0&0&x_4^2&0\\0&0&0&x_4\end{pmatrix}$	$\begin{pmatrix}1&0&0&0\\w_{21}&1&0&x_4^2w_{24}\\0&0&0&1\\0&0&1&x_4w_{44}\end{pmatrix}$
$\begin{pmatrix}1&0&0&0\\w_{21}&1&0&w_{24}\\0&0&w_{33}&1\\0&0&1&0\end{pmatrix}$	$\begin{pmatrix}1&0&0&0\\0&1&0&0\\0&0&x_4^2&0\\0&0&0&x_4\end{pmatrix}$	$\begin{pmatrix}1&0&0&0\\w_{21}&1&0&x_4^2w_{24}\\0&0&x_4^2w_{33}&1\\0&0&1&0\end{pmatrix}$
$\begin{pmatrix}1&0&0&w_{14}\\w_{21}&1&0&0\\0&0&0&1\\0&0&1&0\end{pmatrix}$	$\begin{pmatrix}1&0&0&0\\0&1&0&0\\0&0&x_4^2&0\\0&0&0&x_4\end{pmatrix}$	$\begin{pmatrix}1&0&0&x_4^2w_{14}\\w_{21}&1&0&0\\0&0&0&1\\0&0&1&0\end{pmatrix}$
$\begin{pmatrix}1&0&0&w_{14}\\w_{21}&1&0&0\\0&0&0&1\\0&0&1&w_{44}\end{pmatrix}$	$\begin{pmatrix}1&0&0&0\\0&1&0&0\\0&0&x_4^2&0\\0&0&0&x_4\end{pmatrix}$	$\begin{pmatrix}1&0&0&x_4^2w_{14}\\w_{21}&1&0&0\\0&0&0&1\\0&0&1&x_4w_{44}\end{pmatrix}$
$\begin{pmatrix}1&0&0&w_{14}\\w_{21}&1&0&0\\0&0&w_{33}&1\\0&0&1&0\end{pmatrix}$	$\begin{pmatrix}1&0&0&0\\0&1&0&0\\0&0&x_4^2&0\\0&0&0&x_4\end{pmatrix}$	$\begin{pmatrix}1&0&0&x_4^2w_{14}\\w_{21}&1&0&0\\0&0&x_4^2w_{33}&1\\0&0&1&0\end{pmatrix}$
$\begin{pmatrix}1&0&0&0\\w_{21}&1&w_{23}&w_{24}\\0&0&1&w_{34}\\0&0&w_{43}&1\end{pmatrix}$	$\begin{pmatrix}1&0&0&0\\0&1&0&0\\0&0&0&1\\0&0&1&0\end{pmatrix}$	$\begin{pmatrix}1&0&0&0\\w_{21}&1&w_{24}&w_{23}\\0&0&1&w_{43}\\0&0&w_{34}&1\end{pmatrix}$
$\begin{pmatrix}1&0&0&0\\w_{21}&1&w_{23}&w_{24}\\0&0&0&1\\0&0&1&0\end{pmatrix}$	$\begin{pmatrix}1&0&0&0\\0&1&0&0\\0&0&x_4^2&0\\0&0&0&x_4\end{pmatrix}$	$\begin{pmatrix}1&0&0&0\\w_{21}&1&x_4w_{23}&x_4^2w_{24}\\0&0&0&1\\0&0&1&0\end{pmatrix}$
$\begin{pmatrix}1&0&0&0\\w_{21}&1&w_{23}&w_{24}\\0&0&0&1\\0&0&1&0\end{pmatrix}$	$\begin{pmatrix}1&0&0&0\\0&1&0&0\\0&0&0&x_4^2\\0&0&x_4&0\end{pmatrix}$	$\begin{pmatrix}1&0&0&0\\w_{21}&1&x_4^2w_{24}&x_4w_{23}\\0&0&0&1\\0&0&1&0\end{pmatrix}$

M_B	$P_{B'B}$	M'_B
$\begin{pmatrix}1&0&0&0\\w_{21}&1&w_{23}&w_{24}\\0&0&0&1\\0&0&1&w_{44}\end{pmatrix}$	$\begin{pmatrix}1&0&0&0\\0&1&0&0\\0&0&x_4^2&0\\0&0&0&x_4\end{pmatrix}$	$\begin{pmatrix}1&0&0&0\\w_{21}&1&x_4w_{23}&x_4^2w_{24}\\0&0&0&1\\0&0&1&x_4w_{44}\end{pmatrix}$
$\begin{pmatrix}1&0&w_{13}&w_{14}\\w_{21}&1&0&0\\0&0&0&1\\0&0&1&0\end{pmatrix}$	$\begin{pmatrix}1&0&0&0\\0&1&0&0\\0&0&x_4^2&0\\0&0&0&x_4\end{pmatrix}$	$\begin{pmatrix}1&0&x_4w_{13}&x_4^2w_{14}\\w_{21}&1&0&0\\0&0&0&1\\0&0&1&0\end{pmatrix}$
$\begin{pmatrix}1&0&w_{13}&w_{14}\\w_{21}&1&0&0\\0&0&0&1\\0&0&1&0\end{pmatrix}$	$\begin{pmatrix}1&0&0&0\\0&1&0&0\\0&0&0&x_4^2\\0&0&x_4&0\end{pmatrix}$	$\begin{pmatrix}1&0&x_4^2w_{14}&x_4w_{13}\\w_{21}&1&0&0\\0&0&0&1\\0&0&1&0\end{pmatrix}$
$\begin{pmatrix}1&0&w_{13}&w_{14}\\w_{21}&1&0&0\\0&0&0&1\\0&0&1&w_{44}\end{pmatrix}$	$\begin{pmatrix}1&0&0&0\\0&1&0&0\\0&0&x_4^2&0\\0&0&0&x_4\end{pmatrix}$	$\begin{pmatrix}1&0&x_4w_{13}&x_4^2w_{14}\\w_{21}&1&0&0\\0&0&0&1\\0&0&1&x_4w_{44}\end{pmatrix}$
$\begin{pmatrix}1&0&0&w_{14}\\w_{21}&1&0&w_{24}\\0&0&0&1\\0&0&1&0\end{pmatrix}$	$\begin{pmatrix}1&0&0&0\\0&1&0&0\\0&0&x_4^2&0\\0&0&0&x_4\end{pmatrix}$	$\begin{pmatrix}1&0&0&x_4^2w_{14}\\w_{21}&1&0&x_4^2w_{24}\\0&0&0&1\\0&0&1&0\end{pmatrix}$
$\begin{pmatrix}1&0&0&w_{14}\\w_{21}&1&0&w_{24}\\0&0&0&1\\0&0&1&w_{44}\end{pmatrix}$	$\begin{pmatrix}1&0&0&0\\0&1&0&0\\0&0&x_4^2&0\\0&0&0&x_4\end{pmatrix}$	$\begin{pmatrix}1&0&0&x_4^2w_{14}\\w_{21}&1&0&x_4^2w_{24}\\0&0&0&1\\0&0&1&x_4w_{44}\end{pmatrix}$
$\begin{pmatrix}1&0&0&w_{14}\\w_{21}&1&0&w_{24}\\0&0&w_{33}&1\\0&0&1&0\end{pmatrix}$	$\begin{pmatrix}1&0&0&0\\0&1&0&0\\0&0&x_4^2&0\\0&0&0&x_4\end{pmatrix}$	$\begin{pmatrix}1&0&0&x_4^2w_{14}\\w_{21}&1&0&x_4^2w_{24}\\0&0&x_4^2w_{33}&1\\0&0&1&0\end{pmatrix}$
$\begin{pmatrix}1&0&0&w_{14}\\w_{21}&1&w_{23}&0\\0&0&0&1\\0&0&1&0\end{pmatrix}$	$\begin{pmatrix}1&0&0&0\\0&1&0&0\\0&0&x_4^2&0\\0&0&0&x_4\end{pmatrix}$	$\begin{pmatrix}1&0&0&x_4^2w_{14}\\w_{21}&1&x_4w_{23}&0\\0&0&0&1\\0&0&1&0\end{pmatrix}$
$\begin{pmatrix}1&0&0&w_{14}\\w_{21}&1&w_{23}&0\\0&0&0&1\\0&0&1&w_{44}\end{pmatrix}$	$\begin{pmatrix}1&0&0&0\\0&1&0&0\\0&0&x_4^2&0\\0&0&0&x_4\end{pmatrix}$	$\begin{pmatrix}1&0&0&x_4^2w_{14}\\w_{21}&1&x_4w_{23}&0\\0&0&0&1\\0&0&1&x_4w_{44}\end{pmatrix}$
$\begin{pmatrix}1&0&0&w_{14}\\w_{21}&1&w_{23}&0\\0&0&w_{33}&1\\0&0&1&0\end{pmatrix}$	$\begin{pmatrix}1&0&0&0\\0&1&0&0\\0&0&x_4^2&0\\0&0&0&x_4\end{pmatrix}$	$\begin{pmatrix}1&0&0&x_4^2w_{14}\\w_{21}&1&x_4w_{23}&0\\0&0&x_4^2w_{33}&1\\0&0&1&0\end{pmatrix}$

M_B	$P_{B'B}$	M'_B
$\begin{pmatrix}1&0&w_{13}&w_{14}\\w_{21}&1&w_{23}&0\\0&0&0&1\\0&0&1&0\end{pmatrix}$	$\begin{pmatrix}1&0&0&0\\0&1&0&0\\0&0&x_4^2&0\\0&0&0&x_4\end{pmatrix}$	$\begin{pmatrix}1&0&x_4w_{13}&x_4^2w_{14}\\w_{21}&1&x_4w_{23}&0\\0&0&0&1\\0&0&1&0\end{pmatrix}$
$\begin{pmatrix}1&0&w_{13}&w_{14}\\w_{21}&1&w_{23}&0\\0&0&0&1\\0&0&1&w_{44}\end{pmatrix}$	$\begin{pmatrix}1&0&0&0\\0&1&0&0\\0&0&x_4^2&0\\0&0&0&x_4\end{pmatrix}$	$\begin{pmatrix}1&0&x_4w_{13}&x_4^2w_{14}\\w_{21}&1&x_4w_{23}&0\\0&0&0&1\\0&0&1&x_4w_{44}\end{pmatrix}$
$\begin{pmatrix}1&0&w_{13}&w_{14}\\w_{21}&1&w_{23}&0\\0&0&w_{33}&1\\0&0&1&0\end{pmatrix}$	$\begin{pmatrix}1&0&0&0\\0&1&0&0\\0&0&x_4^2&0\\0&0&0&x_4\end{pmatrix}$	$\begin{pmatrix}1&0&x_4w_{13}&x_4^2w_{14}\\w_{21}&1&x_4w_{23}&0\\0&0&x_4^2w_{33}&1\\0&0&1&0\end{pmatrix}$
$\begin{pmatrix}1&0&w_{13}&0\\w_{21}&1&w_{23}&w_{24}\\0&0&0&1\\0&0&1&0\end{pmatrix}$	$\begin{pmatrix}1&0&0&0\\0&1&0&0\\0&0&x_4^2&0\\0&0&0&x_4\end{pmatrix}$	$\begin{pmatrix}1&0&x_4w_{13}&0\\w_{21}&1&x_4w_{23}&x_4^2w_{24}\\0&0&0&1\\0&0&1&0\end{pmatrix}$
$\begin{pmatrix}1&0&w_{13}&0\\w_{21}&1&w_{23}&w_{24}\\0&0&0&1\\0&0&1&w_{44}\end{pmatrix}$	$\begin{pmatrix}1&0&0&0\\0&1&0&0\\0&0&x_4^2&0\\0&0&0&x_4\end{pmatrix}$	$\begin{pmatrix}1&0&x_4w_{13}&0\\w_{21}&1&x_4w_{23}&x_4^2w_{24}\\0&0&0&1\\0&0&1&x_4w_{44}\end{pmatrix}$
$\begin{pmatrix}1&0&w_{13}&0\\w_{21}&1&w_{23}&w_{24}\\0&0&w_{33}&1\\0&0&1&0\end{pmatrix}$	$\begin{pmatrix}1&0&0&0\\0&1&0&0\\0&0&x_4^2&0\\0&0&0&x_4\end{pmatrix}$	$\begin{pmatrix}1&0&x_4w_{13}&0\\w_{21}&1&x_4w_{23}&x_4^2w_{24}\\0&0&x_4^2w_{33}&1\\0&0&1&0\end{pmatrix}$
$\begin{pmatrix}1&0&w_{13}&w_{14}\\w_{21}&1&w_{23}&w_{24}\\0&0&1&w_{34}\\0&0&w_{43}&1\end{pmatrix}$	$\begin{pmatrix}1&0&0&0\\0&1&0&0\\0&0&0&1\\0&0&1&0\end{pmatrix}$	$\begin{pmatrix}1&0&w_{14}&w_{13}\\w_{21}&1&w_{24}&w_{23}\\0&0&1&w_{43}\\0&0&w_{34}&1\end{pmatrix}$
$\begin{pmatrix}1&0&w_{13}&w_{14}\\w_{21}&1&w_{23}&w_{24}\\0&0&0&1\\0&0&1&0\end{pmatrix}$	$\begin{pmatrix}1&0&0&0\\0&1&0&0\\0&0&x_4^2&0\\0&0&0&x_4\end{pmatrix}$	$\begin{pmatrix}1&0&x_4w_{13}&x_4^2w_{14}\\w_{21}&1&x_4w_{23}&x_4^2w_{24}\\0&0&0&1\\0&0&1&0\end{pmatrix}$
$\begin{pmatrix}1&0&w_{13}&w_{14}\\w_{21}&1&w_{23}&w_{24}\\0&0&0&1\\0&0&1&0\end{pmatrix}$	$\begin{pmatrix}1&0&0&0\\0&1&0&0\\0&0&0&x_4^2\\0&0&x_4&0\end{pmatrix}$	$\begin{pmatrix}1&0&x_4^2w_{14}&x_4w_{13}\\w_{21}&1&x_4^2w_{24}&x_4w_{23}\\0&0&0&1\\0&0&1&0\end{pmatrix}$
$\begin{pmatrix}1&0&w_{13}&w_{14}\\w_{21}&1&w_{23}&w_{24}\\0&0&0&1\\0&0&1&w_{44}\end{pmatrix}$	$\begin{pmatrix}1&0&0&0\\0&1&0&0\\0&0&x_4^2&0\\0&0&0&x_4\end{pmatrix}$	$\begin{pmatrix}1&0&x_4w_{13}&x_4^2w_{14}\\w_{21}&1&x_4w_{23}&x_4^2w_{24}\\0&0&0&1\\0&0&1&x_4w_{44}\end{pmatrix}$

3.3 Take $W = \begin{pmatrix} 1 & \omega_{12} \\ \omega_{21} & 1 \end{pmatrix}$

M_B	$P_{B'B}$	M'_B
$\begin{pmatrix} 1 & w_{12} & 0 & 0 \\ w_{21} & 1 & 0 & w_{24} \\ 0 & 0 & 0 & 1 \\ 0 & 0 & 1 & 0 \end{pmatrix}$	$\begin{pmatrix} 1 & 0 & 0 & 0 \\ 0 & 1 & 0 & 0 \\ 0 & 0 & x_4^2 & 0 \\ 0 & 0 & 0 & x_4 \end{pmatrix}$	$\begin{pmatrix} 1 & w_{12} & 0 & 0 \\ w_{21} & 1 & 0 & x_4^2 w_{24} \\ 0 & 0 & 0 & 1 \\ 0 & 0 & 1 & 0 \end{pmatrix}$
$\begin{pmatrix} 1 & w_{12} & 0 & 0 \\ w_{21} & 1 & 0 & w_{24} \\ 0 & 0 & 0 & 1 \\ 0 & 0 & 1 & w_{44} \end{pmatrix}$	$\begin{pmatrix} 1 & 0 & 0 & 0 \\ 0 & 1 & 0 & 0 \\ 0 & 0 & x_4^2 & 0 \\ 0 & 0 & 0 & x_4 \end{pmatrix}$	$\begin{pmatrix} 1 & w_{12} & 0 & 0 \\ w_{21} & 1 & 0 & x_4^2 w_{24} \\ 0 & 0 & 0 & 1 \\ 0 & 0 & 1 & x_4 w_{44} \end{pmatrix}$
$\begin{pmatrix} 1 & w_{12} & 0 & 0 \\ w_{21} & 1 & 0 & w_{24} \\ 0 & 0 & w_{33} & 1 \\ 0 & 0 & 1 & 0 \end{pmatrix}$	$\begin{pmatrix} 1 & 0 & 0 & 0 \\ 0 & 1 & 0 & 0 \\ 0 & 0 & x_4^2 & 0 \\ 0 & 0 & 0 & x_4 \end{pmatrix}$	$\begin{pmatrix} 1 & w_{12} & 0 & 0 \\ w_{21} & 1 & 0 & x_4^2 w_{24} \\ 0 & 0 & x_4^2 w_{33} & 1 \\ 0 & 0 & 1 & 0 \end{pmatrix}$
$\begin{pmatrix} 1 & w_{12} & 0 & 0 \\ w_{21} & 1 & w_{23} & w_{24} \\ 0 & 0 & 1 & w_{34} \\ 0 & 0 & w_{43} & 1 \end{pmatrix}$	$\begin{pmatrix} 1 & 0 & 0 & 0 \\ 0 & 1 & 0 & 0 \\ 0 & 0 & 0 & 1 \\ 0 & 0 & 1 & 0 \end{pmatrix}$	$\begin{pmatrix} 1 & w_{12} & 0 & 0 \\ w_{21} & 1 & w_{24} & w_{23} \\ 0 & 0 & 1 & w_{43} \\ 0 & 0 & w_{34} & 1 \end{pmatrix}$
$\begin{pmatrix} 1 & w_{12} & 0 & 0 \\ w_{21} & 1 & w_{23} & w_{24} \\ 0 & 0 & 0 & 1 \\ 0 & 0 & 1 & 0 \end{pmatrix}$	$\begin{pmatrix} 1 & 0 & 0 & 0 \\ 0 & 1 & 0 & 0 \\ 0 & 0 & x_4^2 & 0 \\ 0 & 0 & 0 & x_4 \end{pmatrix}$	$\begin{pmatrix} 1 & w_{12} & 0 & 0 \\ w_{21} & 1 & x_4 w_{23} & x_4^2 w_{24} \\ 0 & 0 & 0 & 1 \\ 0 & 0 & 1 & 0 \end{pmatrix}$
$\begin{pmatrix} 1 & w_{12} & 0 & 0 \\ w_{21} & 1 & w_{23} & w_{24} \\ 0 & 0 & 0 & 1 \\ 0 & 0 & 1 & 0 \end{pmatrix}$	$\begin{pmatrix} 1 & 0 & 0 & 0 \\ 0 & 1 & 0 & 0 \\ 0 & 0 & 0 & x_4^2 \\ 0 & 0 & x_4 & 0 \end{pmatrix}$	$\begin{pmatrix} 1 & w_{12} & 0 & 0 \\ w_{21} & 1 & x_4^2 w_{24} & x_4 w_{23} \\ 0 & 0 & 0 & 1 \\ 0 & 0 & 1 & 0 \end{pmatrix}$
$\begin{pmatrix} 1 & w_{12} & 0 & 0 \\ w_{21} & 1 & w_{23} & w_{24} \\ 0 & 0 & 0 & 1 \\ 0 & 0 & 1 & w_{44} \end{pmatrix}$	$\begin{pmatrix} 1 & 0 & 0 & 0 \\ 0 & 1 & 0 & 0 \\ 0 & 0 & x_4^2 & 0 \\ 0 & 0 & 0 & x_4 \end{pmatrix}$	$\begin{pmatrix} 1 & w_{12} & 0 & 0 \\ w_{21} & 1 & x_4 w_{23} & x_4^2 w_{24} \\ 0 & 0 & 0 & 1 \\ 0 & 0 & 1 & x_4 w_{44} \end{pmatrix}$
$\begin{pmatrix} 1 & w_{12} & 0 & w_{14} \\ w_{21} & 1 & 0 & w_{24} \\ 0 & 0 & 1 & w_{34} \\ 0 & 0 & w_{43} & 1 \end{pmatrix}$	$\begin{pmatrix} 0 & 1 & 0 & 0 \\ 1 & 0 & 0 & 0 \\ 0 & 0 & 1 & 0 \\ 0 & 0 & 0 & 1 \end{pmatrix}$	$\begin{pmatrix} 1 & w_{21} & 0 & w_{24} \\ w_{12} & 1 & 0 & w_{14} \\ 0 & 0 & 1 & w_{34} \\ 0 & 0 & w_{43} & 1 \end{pmatrix}$
$\begin{pmatrix} 1 & w_{12} & 0 & w_{14} \\ w_{21} & 1 & 0 & w_{24} \\ 0 & 0 & 0 & 1 \\ 0 & 0 & 1 & 0 \end{pmatrix}$	$\begin{pmatrix} 1 & 0 & 0 & 0 \\ 0 & 1 & 0 & 0 \\ 0 & 0 & x_4^2 & 0 \\ 0 & 0 & 0 & x_4 \end{pmatrix}$	$\begin{pmatrix} 1 & w_{12} & 0 & x_4^2 w_{14} \\ w_{21} & 1 & 0 & x_4^2 w_{24} \\ 0 & 0 & 0 & 1 \\ 0 & 0 & 1 & 0 \end{pmatrix}$
$\begin{pmatrix} 1 & w_{12} & 0 & w_{14} \\ w_{21} & 1 & 0 & w_{24} \\ 0 & 0 & 0 & 1 \\ 0 & 0 & 1 & 0 \end{pmatrix}$	$\begin{pmatrix} 0 & 1 & 0 & 0 \\ 1 & 0 & 0 & 0 \\ 0 & 0 & x_4^2 & 0 \\ 0 & 0 & 0 & x_4 \end{pmatrix}$	$\begin{pmatrix} 1 & w_{21} & 0 & x_4^2 w_{24} \\ w_{12} & 1 & 0 & x_4^2 w_{14} \\ 0 & 0 & 0 & 1 \\ 0 & 0 & 1 & 0 \end{pmatrix}$

M_B	$P_{B'B}$	M'_B
$\begin{pmatrix} 1 & w_{12} & 0 & w_{14} \\ w_{21} & 1 & 0 & w_{24} \\ 0 & 0 & 0 & 1 \\ 0 & 0 & 1 & w_{44} \end{pmatrix}$	$\begin{pmatrix} 1 & 0 & 0 & 0 \\ 0 & 1 & 0 & 0 \\ 0 & 0 & x_4^2 & 0 \\ 0 & 0 & 0 & x_4 \end{pmatrix}$	$\begin{pmatrix} 1 & w_{12} & 0 & x_4^2 w_{14} \\ w_{21} & 1 & 0 & x_4^2 w_{24} \\ 0 & 0 & 0 & 1 \\ 0 & 0 & 1 & x_4 w_{44} \end{pmatrix}$
$\begin{pmatrix} 1 & w_{12} & 0 & w_{14} \\ w_{21} & 1 & 0 & w_{24} \\ 0 & 0 & 0 & 1 \\ 0 & 0 & 1 & w_{44} \end{pmatrix}$	$\begin{pmatrix} 0 & 1 & 0 & 0 \\ 1 & 0 & 0 & 0 \\ 0 & 0 & x_4^2 & 0 \\ 0 & 0 & 0 & x_4 \end{pmatrix}$	$\begin{pmatrix} 1 & w_{21} & 0 & x_4^2 w_{24} \\ w_{12} & 1 & 0 & x_4^2 w_{14} \\ 0 & 0 & 0 & 1 \\ 0 & 0 & 1 & x_4 w_{44} \end{pmatrix}$
$\begin{pmatrix} 1 & w_{12} & 0 & w_{14} \\ w_{21} & 1 & 0 & w_{24} \\ 0 & 0 & w_{33} & 1 \\ 0 & 0 & 1 & 0 \end{pmatrix}$	$\begin{pmatrix} 1 & 0 & 0 & 0 \\ 0 & 1 & 0 & 0 \\ 0 & 0 & x_4^2 & 0 \\ 0 & 0 & 0 & x_4 \end{pmatrix}$	$\begin{pmatrix} 1 & w_{12} & 0 & x_4^2 w_{14} \\ w_{21} & 1 & 0 & x_4^2 w_{24} \\ 0 & 0 & x_4^2 w_{33} & 1 \\ 0 & 0 & 1 & 0 \end{pmatrix}$
$\begin{pmatrix} 1 & w_{12} & 0 & w_{14} \\ w_{21} & 1 & 0 & w_{24} \\ 0 & 0 & w_{33} & 1 \\ 0 & 0 & 1 & 0 \end{pmatrix}$	$\begin{pmatrix} 0 & 1 & 0 & 0 \\ 1 & 0 & 0 & 0 \\ 0 & 0 & x_4^2 & 0 \\ 0 & 0 & 0 & x_4 \end{pmatrix}$	$\begin{pmatrix} 1 & w_{21} & 0 & x_4^2 w_{24} \\ w_{12} & 1 & 0 & x_4^2 w_{14} \\ 0 & 0 & x_4^2 w_{33} & 1 \\ 0 & 0 & 1 & 0 \end{pmatrix}$
$\begin{pmatrix} 1 & w_{12} & 0 & w_{14} \\ w_{21} & 1 & w_{23} & 0 \\ 0 & 0 & 1 & w_{34} \\ 0 & 0 & w_{43} & 1 \end{pmatrix}$	$\begin{pmatrix} 0 & 1 & 0 & 0 \\ 1 & 0 & 0 & 0 \\ 0 & 0 & 0 & 1 \\ 0 & 0 & 1 & 0 \end{pmatrix}$	$\begin{pmatrix} 1 & w_{21} & 0 & w_{23} \\ w_{12} & 1 & w_{14} & 0 \\ 0 & 0 & 1 & w_{43} \\ 0 & 0 & w_{34} & 1 \end{pmatrix}$
$\begin{pmatrix} 1 & w_{12} & 0 & w_{14} \\ w_{21} & 1 & w_{23} & 0 \\ 0 & 0 & 0 & 1 \\ 0 & 0 & 1 & 0 \end{pmatrix}$	$\begin{pmatrix} 1 & 0 & 0 & 0 \\ 0 & 1 & 0 & 0 \\ 0 & 0 & x_4^2 & 0 \\ 0 & 0 & 0 & x_4 \end{pmatrix}$	$\begin{pmatrix} 1 & w_{12} & 0 & x_4^2 w_{14} \\ w_{21} & 1 & x_4 w_{23} & 0 \\ 0 & 0 & 0 & 1 \\ 0 & 0 & 1 & 0 \end{pmatrix}$
$\begin{pmatrix} 1 & w_{12} & 0 & w_{14} \\ w_{21} & 1 & w_{23} & 0 \\ 0 & 0 & 0 & 1 \\ 0 & 0 & 1 & 0 \end{pmatrix}$	$\begin{pmatrix} 0 & 1 & 0 & 0 \\ 1 & 0 & 0 & 0 \\ 0 & 0 & 0 & x_4^2 \\ 0 & 0 & x_4 & 0 \end{pmatrix}$	$\begin{pmatrix} 1 & w_{21} & 0 & x_4 w_{23} \\ w_{12} & 1 & x_4^2 w_{14} & 0 \\ 0 & 0 & 0 & 1 \\ 0 & 0 & 1 & 0 \end{pmatrix}$
$\begin{pmatrix} 1 & w_{12} & 0 & w_{14} \\ w_{21} & 1 & w_{23} & 0 \\ 0 & 0 & 0 & 1 \\ 0 & 0 & 1 & w_{44} \end{pmatrix}$	$\begin{pmatrix} 1 & 0 & 0 & 0 \\ 0 & 1 & 0 & 0 \\ 0 & 0 & x_4^2 & 0 \\ 0 & 0 & 0 & x_4 \end{pmatrix}$	$\begin{pmatrix} 1 & w_{12} & 0 & x_4^2 w_{14} \\ w_{21} & 1 & x_4 w_{23} & 0 \\ 0 & 0 & 0 & 1 \\ 0 & 0 & 1 & x_4 w_{44} \end{pmatrix}$
$\begin{pmatrix} 1 & w_{12} & w_{13} & w_{14} \\ w_{21} & 1 & w_{23} & 0 \\ 0 & 0 & 0 & 1 \\ 0 & 0 & 1 & 0 \end{pmatrix}$	$\begin{pmatrix} 1 & 0 & 0 & 0 \\ 0 & 1 & 0 & 0 \\ 0 & 0 & x_4^2 & 0 \\ 0 & 0 & 0 & x_4 \end{pmatrix}$	$\begin{pmatrix} 1 & w_{12} & x_4 w_{13} & x_4^2 w_{14} \\ w_{21} & 1 & x_4 w_{23} & 0 \\ 0 & 0 & 0 & 1 \\ 0 & 0 & 1 & 0 \end{pmatrix}$
$\begin{pmatrix} 1 & w_{12} & w_{13} & w_{14} \\ w_{21} & 1 & w_{23} & 0 \\ 0 & 0 & 0 & 1 \\ 0 & 0 & 1 & w_{44} \end{pmatrix}$	$\begin{pmatrix} 1 & 0 & 0 & 0 \\ 0 & 1 & 0 & 0 \\ 0 & 0 & x_4^2 & 0 \\ 0 & 0 & 0 & x_4 \end{pmatrix}$	$\begin{pmatrix} 1 & w_{12} & x_4 w_{13} & x_4^2 w_{14} \\ w_{21} & 1 & x_4 w_{23} & 0 \\ 0 & 0 & 0 & 1 \\ 0 & 0 & 1 & x_4 w_{44} \end{pmatrix}$

M_B	$P_{B'B}$	M'_B
$\begin{pmatrix} 1 & w_{12} & w_{13} & w_{14} \\ w_{21} & 1 & w_{23} & 0 \\ 0 & 0 & w_{33} & 1 \\ 0 & 0 & 1 & 0 \end{pmatrix}$	$\begin{pmatrix} 1 & 0 & 0 & 0 \\ 0 & 1 & 0 & 0 \\ 0 & 0 & x_4^2 & 0 \\ 0 & 0 & 0 & x_4 \end{pmatrix}$	$\begin{pmatrix} 1 & w_{12} & x_4 w_{13} & x_4^2 w_{14} \\ w_{21} & 1 & x_4 w_{23} & 0 \\ 0 & 0 & x_4^2 w_{33} & 1 \\ 0 & 0 & 1 & 0 \end{pmatrix}$
$\begin{pmatrix} 1 & w_{12} & w_{13} & w_{14} \\ w_{21} & 1 & w_{23} & w_{24} \\ 0 & 0 & 1 & w_{34} \\ 0 & 0 & w_{43} & 1 \end{pmatrix}$	$\begin{pmatrix} 0 & 1 & 0 & 0 \\ 1 & 0 & 0 & 0 \\ 0 & 0 & 1 & 0 \\ 0 & 0 & 0 & 1 \end{pmatrix}$	$\begin{pmatrix} 1 & w_{21} & w_{23} & w_{24} \\ w_{12} & 1 & w_{13} & w_{14} \\ 0 & 0 & 1 & w_{34} \\ 0 & 0 & w_{43} & 1 \end{pmatrix}$
$\begin{pmatrix} 1 & w_{12} & w_{13} & w_{14} \\ w_{21} & 1 & w_{23} & w_{24} \\ 0 & 0 & 1 & w_{34} \\ 0 & 0 & w_{43} & 1 \end{pmatrix}$	$\begin{pmatrix} 1 & 0 & 0 & 0 \\ 0 & 1 & 0 & 0 \\ 0 & 0 & 0 & 1 \\ 0 & 0 & 1 & 0 \end{pmatrix}$	$\begin{pmatrix} 1 & w_{12} & w_{14} & w_{13} \\ w_{21} & 1 & w_{24} & w_{23} \\ 0 & 0 & 1 & w_{43} \\ 0 & 0 & w_{34} & 1 \end{pmatrix}$
$\begin{pmatrix} 1 & w_{12} & w_{13} & w_{14} \\ w_{21} & 1 & w_{23} & w_{24} \\ 0 & 0 & 1 & w_{34} \\ 0 & 0 & w_{43} & 1 \end{pmatrix}$	$\begin{pmatrix} 0 & 1 & 0 & 0 \\ 1 & 0 & 0 & 0 \\ 0 & 0 & 0 & 1 \\ 0 & 0 & 1 & 0 \end{pmatrix}$	$\begin{pmatrix} 1 & w_{21} & w_{24} & w_{23} \\ w_{12} & 1 & w_{14} & w_{13} \\ 0 & 0 & 1 & w_{43} \\ 0 & 0 & w_{34} & 1 \end{pmatrix}$
$\begin{pmatrix} 1 & w_{12} & w_{13} & w_{14} \\ w_{21} & 1 & w_{23} & w_{24} \\ 0 & 0 & 0 & 1 \\ 0 & 0 & 1 & 0 \end{pmatrix}$	$\begin{pmatrix} 1 & 0 & 0 & 0 \\ 0 & 1 & 0 & 0 \\ 0 & 0 & x_4^2 & 0 \\ 0 & 0 & 0 & x_4 \end{pmatrix}$	$\begin{pmatrix} 1 & w_{12} & x_4 w_{13} & x_4^2 w_{14} \\ w_{21} & 1 & x_4 w_{23} & x_4^2 w_{24} \\ 0 & 0 & 0 & 1 \\ 0 & 0 & 1 & 0 \end{pmatrix}$
$\begin{pmatrix} 1 & w_{12} & w_{13} & w_{14} \\ w_{21} & 1 & w_{23} & w_{24} \\ 0 & 0 & 0 & 1 \\ 0 & 0 & 1 & 0 \end{pmatrix}$	$\begin{pmatrix} 0 & 1 & 0 & 0 \\ 1 & 0 & 0 & 0 \\ 0 & 0 & x_4^2 & 0 \\ 0 & 0 & 0 & x_4 \end{pmatrix}$	$\begin{pmatrix} 1 & w_{21} & x_4 w_{23} & x_4^2 w_{24} \\ w_{12} & 1 & x_4 w_{13} & x_4^2 w_{14} \\ 0 & 0 & 0 & 1 \\ 0 & 0 & 1 & 0 \end{pmatrix}$
$\begin{pmatrix} 1 & w_{12} & w_{13} & w_{14} \\ w_{21} & 1 & w_{23} & w_{24} \\ 0 & 0 & 0 & 1 \\ 0 & 0 & 1 & 0 \end{pmatrix}$	$\begin{pmatrix} 1 & 0 & 0 & 0 \\ 0 & 1 & 0 & 0 \\ 0 & 0 & 0 & x_4^2 \\ 0 & 0 & x_4 & 0 \end{pmatrix}$	$\begin{pmatrix} 1 & w_{12} & x_4^2 w_{14} & x_4 w_{13} \\ w_{21} & 1 & x_4^2 w_{24} & x_4 w_{23} \\ 0 & 0 & 0 & 1 \\ 0 & 0 & 1 & 0 \end{pmatrix}$
$\begin{pmatrix} 1 & w_{12} & w_{13} & w_{14} \\ w_{21} & 1 & w_{23} & w_{24} \\ 0 & 0 & 0 & 1 \\ 0 & 0 & 1 & 0 \end{pmatrix}$	$\begin{pmatrix} 0 & 1 & 0 & 0 \\ 1 & 0 & 0 & 0 \\ 0 & 0 & 0 & x_4^2 \\ 0 & 0 & x_4 & 0 \end{pmatrix}$	$\begin{pmatrix} 1 & w_{21} & x_4^2 w_{24} & x_4 w_{23} \\ w_{12} & 1 & x_4^2 w_{14} & x_4 w_{13} \\ 0 & 0 & 0 & 1 \\ 0 & 0 & 1 & 0 \end{pmatrix}$
$\begin{pmatrix} 1 & w_{12} & w_{13} & w_{14} \\ w_{21} & 1 & w_{23} & w_{24} \\ 0 & 0 & 0 & 1 \\ 0 & 0 & 1 & w_{44} \end{pmatrix}$	$\begin{pmatrix} 1 & 0 & 0 & 0 \\ 0 & 1 & 0 & 0 \\ 0 & 0 & x_4^2 & 0 \\ 0 & 0 & 0 & x_4 \end{pmatrix}$	$\begin{pmatrix} 1 & w_{12} & x_4 w_{13} & x_4^2 w_{14} \\ w_{21} & 1 & x_4 w_{23} & x_4^2 w_{24} \\ 0 & 0 & 0 & 1 \\ 0 & 0 & 1 & x_4 w_{44} \end{pmatrix}$
$\begin{pmatrix} 1 & w_{12} & w_{13} & w_{14} \\ w_{21} & 1 & w_{23} & w_{24} \\ 0 & 0 & 0 & 1 \\ 0 & 0 & 1 & w_{44} \end{pmatrix}$	$\begin{pmatrix} 0 & 1 & 0 & 0 \\ 1 & 0 & 0 & 0 \\ 0 & 0 & x_4^2 & 0 \\ 0 & 0 & 0 & x_4 \end{pmatrix}$	$\begin{pmatrix} 1 & w_{21} & x_4 w_{23} & x_4^2 w_{24} \\ w_{12} & 1 & x_4 w_{13} & x_4^2 w_{14} \\ 0 & 0 & 0 & 1 \\ 0 & 0 & 1 & x_4 w_{44} \end{pmatrix}$

3.4 Take $W = \begin{pmatrix} 0 & 1 \\ 1 & 0 \end{pmatrix}$

M_B	$P_{B'B}$	M'_B
$\begin{pmatrix} 0 & 1 & 0 & 0 \\ 1 & 0 & 0 & w_{24} \\ 0 & 0 & 1 & w_{34} \\ 0 & 0 & w_{43} & 1 \end{pmatrix}$	$\begin{pmatrix} x_2^2 & 0 & 0 & 0 \\ 0 & x_2 & 0 & 0 \\ 0 & 0 & 1 & 0 \\ 0 & 0 & 0 & 1 \end{pmatrix}$	$\begin{pmatrix} 0 & 1 & 0 & 0 \\ 1 & 0 & 0 & \frac{w_{24}}{x_2} \\ 0 & 0 & 1 & w_{34} \\ 0 & 0 & w_{43} & 1 \end{pmatrix}$
$\begin{pmatrix} 0 & 1 & 0 & 0 \\ 1 & 0 & 0 & w_{24} \\ 0 & 0 & 0 & 1 \\ 0 & 0 & 1 & 0 \end{pmatrix}$	$\begin{pmatrix} x_2^2 & 0 & 0 & 0 \\ 0 & x_2 & 0 & 0 \\ 0 & 0 & x_4^2 & 0 \\ 0 & 0 & 0 & x_4 \end{pmatrix}$	$\begin{pmatrix} 0 & 1 & 0 & 0 \\ 1 & 0 & 0 & \frac{x_4^2 w_{24}}{x_2} \\ 0 & 0 & 0 & 1 \\ 0 & 0 & 1 & 0 \end{pmatrix}$
$\begin{pmatrix} 0 & 1 & 0 & 0 \\ 1 & 0 & 0 & w_{24} \\ 0 & 0 & 0 & 1 \\ 0 & 0 & 1 & w_{44} \end{pmatrix}$	$\begin{pmatrix} x_2^2 & 0 & 0 & 0 \\ 0 & x_2 & 0 & 0 \\ 0 & 0 & x_4^2 & 0 \\ 0 & 0 & 0 & x_4 \end{pmatrix}$	$\begin{pmatrix} 0 & 1 & 0 & 0 \\ 1 & 0 & 0 & \frac{x_4^2 w_{24}}{x_2} \\ 0 & 0 & 0 & 1 \\ 0 & 0 & 1 & x_4 w_{44} \end{pmatrix}$
$\begin{pmatrix} 0 & 1 & 0 & 0 \\ 1 & 0 & 0 & w_{24} \\ 0 & 0 & w_{33} & 1 \\ 0 & 0 & 1 & 0 \end{pmatrix}$	$\begin{pmatrix} x_2^2 & 0 & 0 & 0 \\ 0 & x_2 & 0 & 0 \\ 0 & 0 & x_4^2 & 0 \\ 0 & 0 & 0 & x_4 \end{pmatrix}$	$\begin{pmatrix} 0 & 1 & 0 & 0 \\ 1 & 0 & 0 & \frac{x_4^2 w_{24}}{x_2} \\ 0 & 0 & x_4^2 w_{33} & 1 \\ 0 & 0 & 1 & 0 \end{pmatrix}$
$\begin{pmatrix} 0 & 1 & 0 & 0 \\ 1 & 0 & w_{23} & w_{24} \\ 0 & 0 & 1 & w_{34} \\ 0 & 0 & w_{43} & 1 \end{pmatrix}$	$\begin{pmatrix} x_2^2 & 0 & 0 & 0 \\ 0 & x_2 & 0 & 0 \\ 0 & 0 & 1 & 0 \\ 0 & 0 & 0 & 1 \end{pmatrix}$	$\begin{pmatrix} 0 & 1 & 0 & 0 \\ 1 & 0 & \frac{w_{23}}{x_2} & \frac{w_{24}}{x_2} \\ 0 & 0 & 1 & w_{34} \\ 0 & 0 & w_{43} & 1 \end{pmatrix}$
$\begin{pmatrix} 0 & 1 & 0 & 0 \\ 1 & 0 & w_{23} & w_{24} \\ 0 & 0 & 1 & w_{34} \\ 0 & 0 & w_{43} & 1 \end{pmatrix}$	$\begin{pmatrix} x_2^2 & 0 & 0 & 0 \\ 0 & x_2 & 0 & 0 \\ 0 & 0 & 0 & 1 \\ 0 & 0 & 1 & 0 \end{pmatrix}$	$\begin{pmatrix} 0 & 1 & 0 & 0 \\ 1 & 0 & \frac{w_{24}}{x_2} & \frac{w_{23}}{x_2} \\ 0 & 0 & 1 & w_{43} \\ 0 & 0 & w_{34} & 1 \end{pmatrix}$
$\begin{pmatrix} 0 & 1 & 0 & 0 \\ 1 & 0 & w_{23} & w_{24} \\ 0 & 0 & 0 & 1 \\ 0 & 0 & 1 & 0 \end{pmatrix}$	$\begin{pmatrix} x_2^2 & 0 & 0 & 0 \\ 0 & x_2 & 0 & 0 \\ 0 & 0 & x_4^2 & 0 \\ 0 & 0 & 0 & x_4 \end{pmatrix}$	$\begin{pmatrix} 0 & 1 & 0 & 0 \\ 1 & 0 & \frac{x_4 w_{23}}{x_2} & \frac{x_4^2 w_{24}}{x_2} \\ 0 & 0 & 0 & 1 \\ 0 & 0 & 1 & 0 \end{pmatrix}$
$\begin{pmatrix} 0 & 1 & 0 & 0 \\ 1 & 0 & w_{23} & w_{24} \\ 0 & 0 & 0 & 1 \\ 0 & 0 & 1 & 0 \end{pmatrix}$	$\begin{pmatrix} x_2^2 & 0 & 0 & 0 \\ 0 & x_2 & 0 & 0 \\ 0 & 0 & 0 & x_4^2 \\ 0 & 0 & x_4 & 0 \end{pmatrix}$	$\begin{pmatrix} 0 & 1 & 0 & 0 \\ 1 & 0 & \frac{x_4^2 w_{24}}{x_2} & \frac{x_4 w_{23}}{x_2} \\ 0 & 0 & 0 & 1 \\ 0 & 0 & 1 & 0 \end{pmatrix}$
$\begin{pmatrix} 0 & 1 & 0 & 0 \\ 1 & 0 & w_{23} & w_{24} \\ 0 & 0 & 0 & 1 \\ 0 & 0 & 1 & w_{44} \end{pmatrix}$	$\begin{pmatrix} x_2^2 & 0 & 0 & 0 \\ 0 & x_2 & 0 & 0 \\ 0 & 0 & x_4^2 & 0 \\ 0 & 0 & 0 & x_4 \end{pmatrix}$	$\begin{pmatrix} 0 & 1 & 0 & 0 \\ 1 & 0 & \frac{x_4 w_{23}}{x_2} & \frac{x_4^2 w_{24}}{x_2} \\ 0 & 0 & 0 & 1 \\ 0 & 0 & 1 & x_4 w_{44} \end{pmatrix}$
$\begin{pmatrix} 0 & 1 & 0 & w_{14} \\ 1 & 0 & 0 & w_{24} \\ 0 & 0 & 1 & w_{34} \\ 0 & 0 & w_{43} & 1 \end{pmatrix}$	$\begin{pmatrix} x_2^2 & 0 & 0 & 0 \\ 0 & x_2 & 0 & 0 \\ 0 & 0 & 1 & 0 \\ 0 & 0 & 0 & 1 \end{pmatrix}$	$\begin{pmatrix} 0 & 1 & 0 & \frac{w_{14}}{x_2^2} \\ 1 & 0 & 0 & \frac{w_{24}}{x_2} \\ 0 & 0 & 1 & w_{34} \\ 0 & 0 & w_{43} & 1 \end{pmatrix}$

M_B	$P_{B'B}$	M'_B
$\begin{pmatrix} 0 & 1 & 0 & w_{14} \\ 1 & 0 & 0 & w_{24} \\ 0 & 0 & 1 & w_{34} \\ 0 & 0 & w_{43} & 1 \end{pmatrix}$	$\begin{pmatrix} 0 & x_2^2 & 0 & 0 \\ x_2 & 0 & 0 & 0 \\ 0 & 0 & 1 & 0 \\ 0 & 0 & 0 & 1 \end{pmatrix}$	$\begin{pmatrix} 0 & 1 & 0 & \frac{w_{24}}{x_2} \\ 1 & 0 & 0 & \frac{w_{14}}{x_2^2} \\ 0 & 0 & 1 & w_{34} \\ 0 & 0 & w_{43} & 1 \end{pmatrix}$
$\begin{pmatrix} 0 & 1 & 0 & w_{14} \\ 1 & 0 & 0 & w_{24} \\ 0 & 0 & 0 & 1 \\ 0 & 0 & 1 & 0 \end{pmatrix}$	$\begin{pmatrix} x_2^2 & 0 & 0 & 0 \\ 0 & x_2 & 0 & 0 \\ 0 & 0 & x_4^2 & 0 \\ 0 & 0 & 0 & x_4 \end{pmatrix}$	$\begin{pmatrix} 0 & 1 & 0 & \frac{x_4^2 w_{14}}{x_2^2} \\ 1 & 0 & 0 & \frac{x_4^2 w_{24}}{x_2} \\ 0 & 0 & 0 & 1 \\ 0 & 0 & 1 & 0 \end{pmatrix}$
$\begin{pmatrix} 0 & 1 & 0 & w_{14} \\ 1 & 0 & 0 & w_{24} \\ 0 & 0 & 0 & 1 \\ 0 & 0 & 1 & 0 \end{pmatrix}$	$\begin{pmatrix} 0 & x_2^2 & 0 & 0 \\ x_2 & 0 & 0 & 0 \\ 0 & 0 & x_4^2 & 0 \\ 0 & 0 & 0 & x_4 \end{pmatrix}$	$\begin{pmatrix} 0 & 1 & 0 & \frac{x_4^2 w_{24}}{x_2} \\ 1 & 0 & 0 & \frac{x_4^2 w_{14}}{x_2^2} \\ 0 & 0 & 0 & 1 \\ 0 & 0 & 1 & 0 \end{pmatrix}$
$\begin{pmatrix} 0 & 1 & 0 & w_{14} \\ 1 & 0 & 0 & w_{24} \\ 0 & 0 & 0 & 1 \\ 0 & 0 & 1 & w_{44} \end{pmatrix}$	$\begin{pmatrix} x_2^2 & 0 & 0 & 0 \\ 0 & x_2 & 0 & 0 \\ 0 & 0 & x_4^2 & 0 \\ 0 & 0 & 0 & x_4 \end{pmatrix}$	$\begin{pmatrix} 0 & 1 & 0 & \frac{x_4^2 w_{14}}{x_2^2} \\ 1 & 0 & 0 & \frac{x_4^2 w_{24}}{x_2} \\ 0 & 0 & 0 & 1 \\ 0 & 0 & 1 & x_4 w_{44} \end{pmatrix}$
$\begin{pmatrix} 0 & 1 & 0 & w_{14} \\ 1 & 0 & 0 & w_{24} \\ 0 & 0 & 0 & 1 \\ 0 & 0 & 1 & w_{44} \end{pmatrix}$	$\begin{pmatrix} 0 & x_2^2 & 0 & 0 \\ x_2 & 0 & 0 & 0 \\ 0 & 0 & x_4^2 & 0 \\ 0 & 0 & 0 & x_4 \end{pmatrix}$	$\begin{pmatrix} 0 & 1 & 0 & \frac{x_4^2 w_{24}}{x_2} \\ 1 & 0 & 0 & \frac{x_4^2 w_{14}}{x_2^2} \\ 0 & 0 & 0 & 1 \\ 0 & 0 & 1 & x_4 w_{44} \end{pmatrix}$
$\begin{pmatrix} 0 & 1 & 0 & w_{14} \\ 1 & 0 & 0 & w_{24} \\ 0 & 0 & w_{33} & 1 \\ 0 & 0 & 1 & 0 \end{pmatrix}$	$\begin{pmatrix} x_2^2 & 0 & 0 & 0 \\ 0 & x_2 & 0 & 0 \\ 0 & 0 & x_4^2 & 0 \\ 0 & 0 & 0 & x_4 \end{pmatrix}$	$\begin{pmatrix} 0 & 1 & 0 & \frac{x_4^2 w_{14}}{x_2^2} \\ 1 & 0 & 0 & \frac{x_4^2 w_{24}}{x_2} \\ 0 & 0 & x_4^2 w_{33} & 1 \\ 0 & 0 & 1 & 0 \end{pmatrix}$
$\begin{pmatrix} 0 & 1 & 0 & w_{14} \\ 1 & 0 & 0 & w_{24} \\ 0 & 0 & w_{33} & 1 \\ 0 & 0 & 1 & 0 \end{pmatrix}$	$\begin{pmatrix} 0 & x_2^2 & 0 & 0 \\ x_2 & 0 & 0 & 0 \\ 0 & 0 & x_4^2 & 0 \\ 0 & 0 & 0 & x_4 \end{pmatrix}$	$\begin{pmatrix} 0 & 1 & 0 & \frac{x_4^2 w_{24}}{x_2} \\ 1 & 0 & 0 & \frac{x_4^2 w_{14}}{x_2^2} \\ 0 & 0 & x_4^2 w_{33} & 1 \\ 0 & 0 & 1 & 0 \end{pmatrix}$
$\begin{pmatrix} 0 & 1 & 0 & w_{14} \\ 1 & 0 & w_{23} & 0 \\ 0 & 0 & 1 & w_{34} \\ 0 & 0 & w_{43} & 1 \end{pmatrix}$	$\begin{pmatrix} x_2^2 & 0 & 0 & 0 \\ 0 & x_2 & 0 & 0 \\ 0 & 0 & 1 & 0 \\ 0 & 0 & 0 & 1 \end{pmatrix}$	$\begin{pmatrix} 0 & 1 & 0 & \frac{w_{14}}{x_2^2} \\ 1 & 0 & \frac{w_{23}}{x_2} & 0 \\ 0 & 0 & 1 & w_{34} \\ 0 & 0 & w_{43} & 1 \end{pmatrix}$
$\begin{pmatrix} 0 & 1 & 0 & w_{14} \\ 1 & 0 & w_{23} & 0 \\ 0 & 0 & 1 & w_{34} \\ 0 & 0 & w_{43} & 1 \end{pmatrix}$	$\begin{pmatrix} 0 & x_2^2 & 0 & 0 \\ x_2 & 0 & 0 & 0 \\ 0 & 0 & 0 & 1 \\ 0 & 0 & 1 & 0 \end{pmatrix}$	$\begin{pmatrix} 0 & 1 & 0 & \frac{w_{23}}{x_2} \\ 1 & 0 & \frac{w_{14}}{x_2^2} & 0 \\ 0 & 0 & 1 & w_{43} \\ 0 & 0 & w_{34} & 1 \end{pmatrix}$
$\begin{pmatrix} 0 & 1 & 0 & w_{14} \\ 1 & 0 & w_{23} & 0 \\ 0 & 0 & 0 & 1 \\ 0 & 0 & 1 & 0 \end{pmatrix}$	$\begin{pmatrix} x_2^2 & 0 & 0 & 0 \\ 0 & x_2 & 0 & 0 \\ 0 & 0 & x_4^2 & 0 \\ 0 & 0 & 0 & x_4 \end{pmatrix}$	$\begin{pmatrix} 0 & 1 & 0 & \frac{x_4^2 w_{14}}{x_2^2} \\ 1 & 0 & \frac{x_4 w_{23}}{x_2} & 0 \\ 0 & 0 & 0 & 1 \\ 0 & 0 & 1 & 0 \end{pmatrix}$

M_B	$P_{B'B}$	M'_B
$\begin{pmatrix} 0 & 1 & 0 & w_{14} \\ 1 & 0 & w_{23} & 0 \\ 0 & 0 & 0 & 1 \\ 0 & 0 & 1 & 0 \end{pmatrix}$	$\begin{pmatrix} 0 & x_2^2 & 0 & 0 \\ x_2 & 0 & 0 & 0 \\ 0 & 0 & 0 & x_4^2 \\ 0 & 0 & x_4 & 0 \end{pmatrix}$	$\begin{pmatrix} 0 & 1 & 0 & \frac{x_4 w_{23}}{x_2} \\ 1 & 0 & \frac{x_4^2 w_{14}}{x_2^2} & 0 \\ 0 & 0 & 0 & 1 \\ 0 & 0 & 1 & 0 \end{pmatrix}$
$\begin{pmatrix} 0 & 1 & 0 & w_{14} \\ 1 & 0 & w_{23} & 0 \\ 0 & 0 & 0 & 1 \\ 0 & 0 & 1 & w_{44} \end{pmatrix}$	$\begin{pmatrix} x_2^2 & 0 & 0 & 0 \\ 0 & x_2 & 0 & 0 \\ 0 & 0 & x_4^2 & 0 \\ 0 & 0 & 0 & x_4 \end{pmatrix}$	$\begin{pmatrix} 0 & 1 & 0 & \frac{x_4^2 w_{14}}{x_2^2} \\ 1 & 0 & \frac{x_4 w_{23}}{x_2} & 0 \\ 0 & 0 & 0 & 1 \\ 0 & 0 & 1 & x_4 w_{44} \end{pmatrix}$
$\begin{pmatrix} 0 & 1 & w_{13} & w_{14} \\ 1 & 0 & w_{23} & 0 \\ 0 & 0 & 1 & w_{34} \\ 0 & 0 & w_{43} & 1 \end{pmatrix}$	$\begin{pmatrix} x_2^2 & 0 & 0 & 0 \\ 0 & x_2 & 0 & 0 \\ 0 & 0 & 1 & 0 \\ 0 & 0 & 0 & 1 \end{pmatrix}$	$\begin{pmatrix} 0 & 1 & \frac{w_{13}}{x_2^2} & \frac{w_{14}}{x_2^2} \\ 1 & 0 & \frac{w_{23}}{x_2} & 0 \\ 0 & 0 & 1 & w_{34} \\ 0 & 0 & w_{43} & 1 \end{pmatrix}$
$\begin{pmatrix} 0 & 1 & w_{13} & w_{14} \\ 1 & 0 & w_{23} & 0 \\ 0 & 0 & 0 & 1 \\ 0 & 0 & 1 & 0 \end{pmatrix}$	$\begin{pmatrix} x_2^2 & 0 & 0 & 0 \\ 0 & x_2 & 0 & 0 \\ 0 & 0 & x_4^2 & 0 \\ 0 & 0 & 0 & x_4 \end{pmatrix}$	$\begin{pmatrix} 0 & 1 & \frac{x_4 w_{13}}{x_2^2} & \frac{x_4^2 w_{14}}{x_2^2} \\ 1 & 0 & \frac{x_4 w_{23}}{x_2} & 0 \\ 0 & 0 & 0 & 1 \\ 0 & 0 & 1 & 0 \end{pmatrix}$
$\begin{pmatrix} 0 & 1 & w_{13} & w_{14} \\ 1 & 0 & w_{23} & 0 \\ 0 & 0 & 0 & 1 \\ 0 & 0 & 1 & w_{44} \end{pmatrix}$	$\begin{pmatrix} x_2^2 & 0 & 0 & 0 \\ 0 & x_2 & 0 & 0 \\ 0 & 0 & x_4^2 & 0 \\ 0 & 0 & 0 & x_4 \end{pmatrix}$	$\begin{pmatrix} 0 & 1 & \frac{x_4 w_{13}}{x_2^2} & \frac{x_4^2 w_{14}}{x_2^2} \\ 1 & 0 & \frac{x_4 w_{23}}{x_2} & 0 \\ 0 & 0 & 0 & 1 \\ 0 & 0 & 1 & x_4 w_{44} \end{pmatrix}$
$\begin{pmatrix} 0 & 1 & w_{13} & w_{14} \\ 1 & 0 & w_{23} & 0 \\ 0 & 0 & w_{33} & 1 \\ 0 & 0 & 1 & 0 \end{pmatrix}$	$\begin{pmatrix} x_2^2 & 0 & 0 & 0 \\ 0 & x_2 & 0 & 0 \\ 0 & 0 & x_4^2 & 0 \\ 0 & 0 & 0 & x_4 \end{pmatrix}$	$\begin{pmatrix} 0 & 1 & \frac{x_4 w_{13}}{x_2^2} & \frac{x_4^2 w_{14}}{x_2^2} \\ 1 & 0 & \frac{x_4 w_{23}}{x_2} & 0 \\ 0 & 0 & x_4^2 w_{33} & 1 \\ 0 & 0 & 1 & 0 \end{pmatrix}$
$\begin{pmatrix} 0 & 1 & w_{13} & w_{14} \\ 1 & 0 & w_{23} & w_{24} \\ 0 & 0 & 1 & w_{34} \\ 0 & 0 & w_{43} & 1 \end{pmatrix}$	$\begin{pmatrix} x_2^2 & 0 & 0 & 0 \\ 0 & x_2 & 0 & 0 \\ 0 & 0 & 1 & 0 \\ 0 & 0 & 0 & 1 \end{pmatrix}$	$\begin{pmatrix} 0 & 1 & \frac{w_{13}}{x_2^2} & \frac{w_{14}}{x_2^2} \\ 1 & 0 & \frac{w_{23}}{x_2} & \frac{w_{24}}{x_2} \\ 0 & 0 & 1 & w_{34} \\ 0 & 0 & w_{43} & 1 \end{pmatrix}$
$\begin{pmatrix} 0 & 1 & w_{13} & w_{14} \\ 1 & 0 & w_{23} & w_{24} \\ 0 & 0 & 1 & w_{34} \\ 0 & 0 & w_{43} & 1 \end{pmatrix}$	$\begin{pmatrix} 0 & x_2^2 & 0 & 0 \\ x_2 & 0 & 0 & 0 \\ 0 & 0 & 1 & 0 \\ 0 & 0 & 0 & 1 \end{pmatrix}$	$\begin{pmatrix} 0 & 1 & \frac{w_{23}}{x_2} & \frac{w_{24}}{x_2} \\ 1 & 0 & \frac{w_{13}}{x_2^2} & \frac{w_{14}}{x_2^2} \\ 0 & 0 & 1 & w_{34} \\ 0 & 0 & w_{43} & 1 \end{pmatrix}$
$\begin{pmatrix} 0 & 1 & w_{13} & w_{14} \\ 1 & 0 & w_{23} & w_{24} \\ 0 & 0 & 1 & w_{34} \\ 0 & 0 & w_{43} & 1 \end{pmatrix}$	$\begin{pmatrix} x_2^2 & 0 & 0 & 0 \\ 0 & x_2 & 0 & 0 \\ 0 & 0 & 0 & 1 \\ 0 & 0 & 1 & 0 \end{pmatrix}$	$\begin{pmatrix} 0 & 1 & \frac{w_{14}}{x_2^2} & \frac{w_{13}}{x_2^2} \\ 1 & 0 & \frac{w_{24}}{x_2} & \frac{w_{23}}{x_2} \\ 0 & 0 & 1 & w_{43} \\ 0 & 0 & w_{34} & 1 \end{pmatrix}$
$\begin{pmatrix} 0 & 1 & w_{13} & w_{14} \\ 1 & 0 & w_{23} & w_{24} \\ 0 & 0 & 1 & w_{34} \\ 0 & 0 & w_{43} & 1 \end{pmatrix}$	$\begin{pmatrix} 0 & x_2^2 & 0 & 0 \\ x_2 & 0 & 0 & 0 \\ 0 & 0 & 0 & 1 \\ 0 & 0 & 1 & 0 \end{pmatrix}$	$\begin{pmatrix} 0 & 1 & \frac{w_{24}}{x_2} & \frac{w_{23}}{x_2} \\ 1 & 0 & \frac{w_{14}}{x_2^2} & \frac{w_{13}}{x_2^2} \\ 0 & 0 & 1 & w_{43} \\ 0 & 0 & w_{34} & 1 \end{pmatrix}$

M_B	$P_{B'B}$	M'_B
$\begin{pmatrix} 0&1&w_{13}&w_{14}\\ 1&0&w_{23}&w_{24}\\ 0&0&0&1\\ 0&0&1&0 \end{pmatrix}$	$\begin{pmatrix} x_2^2&0&0&0\\ 0&x_2&0&0\\ 0&0&x_4^2&0\\ 0&0&0&x_4 \end{pmatrix}$	$\begin{pmatrix} 0&1&\frac{x_4w_{13}}{x_2^2}&\frac{x_4^2w_{14}}{x_2^2}\\ 1&0&\frac{x_4w_{23}}{x_2}&\frac{x_4^2w_{24}}{x_2}\\ 0&0&0&1\\ 0&0&1&0 \end{pmatrix}$
$\begin{pmatrix} 0&1&w_{13}&w_{14}\\ 1&0&w_{23}&w_{24}\\ 0&0&0&1\\ 0&0&1&0 \end{pmatrix}$	$\begin{pmatrix} 0&x_2^2&0&0\\ x_2&0&0&0\\ 0&0&x_4^2&0\\ 0&0&0&x_4 \end{pmatrix}$	$\begin{pmatrix} 0&1&\frac{x_4w_{23}}{x_2}&\frac{x_4^2w_{24}}{x_2}\\ 1&0&\frac{x_4w_{13}}{x_2^2}&\frac{x_4^2w_{14}}{x_2^2}\\ 0&0&0&1\\ 0&0&1&0 \end{pmatrix}$
$\begin{pmatrix} 0&1&w_{13}&w_{14}\\ 1&0&w_{23}&w_{24}\\ 0&0&0&1\\ 0&0&1&0 \end{pmatrix}$	$\begin{pmatrix} x_2^2&0&0&0\\ 0&x_2&0&0\\ 0&0&0&x_4^2\\ 0&0&x_4&0 \end{pmatrix}$	$\begin{pmatrix} 0&1&\frac{x_4^2w_{14}}{x_2^2}&\frac{x_4w_{13}}{x_2^2}\\ 1&0&\frac{x_4^2w_{24}}{x_2}&\frac{x_4w_{23}}{x_2}\\ 0&0&0&1\\ 0&0&1&0 \end{pmatrix}$
$\begin{pmatrix} 0&1&w_{13}&w_{14}\\ 1&0&w_{23}&w_{24}\\ 0&0&0&1\\ 0&0&1&0 \end{pmatrix}$	$\begin{pmatrix} 0&x_2^2&0&0\\ x_2&0&0&0\\ 0&0&0&x_4^2\\ 0&0&x_4&0 \end{pmatrix}$	$\begin{pmatrix} 0&1&\frac{x_4^2w_{24}}{x_2}&\frac{x_4w_{23}}{x_2}\\ 1&0&\frac{x_4^2w_{14}}{x_2^2}&\frac{x_4w_{13}}{x_2^2}\\ 0&0&0&1\\ 0&0&1&0 \end{pmatrix}$
$\begin{pmatrix} 0&1&w_{13}&w_{14}\\ 1&0&w_{23}&w_{24}\\ 0&0&0&1\\ 0&0&1&w_{44} \end{pmatrix}$	$\begin{pmatrix} x_2^2&0&0&0\\ 0&x_2&0&0\\ 0&0&x_4^2&0\\ 0&0&0&x_4 \end{pmatrix}$	$\begin{pmatrix} 0&1&\frac{x_4w_{13}}{x_2^2}&\frac{x_4^2w_{14}}{x_2^2}\\ 1&0&\frac{x_4w_{23}}{x_2}&\frac{x_4^2w_{24}}{x_2}\\ 0&0&0&1\\ 0&0&1&x_4w_{44} \end{pmatrix}$
$\begin{pmatrix} 0&1&w_{13}&w_{14}\\ 1&0&w_{23}&w_{24}\\ 0&0&0&1\\ 0&0&1&w_{44} \end{pmatrix}$	$\begin{pmatrix} 0&x_2^2&0&0\\ x_2&0&0&0\\ 0&0&x_4^2&0\\ 0&0&0&x_4 \end{pmatrix}$	$\begin{pmatrix} 0&1&\frac{x_4w_{23}}{x_2}&\frac{x_4^2w_{24}}{x_2}\\ 1&0&\frac{x_4w_{13}}{x_2^2}&\frac{x_4^2w_{14}}{x_2^2}\\ 0&0&0&1\\ 0&0&1&x_4w_{44} \end{pmatrix}$

M_B	$P_{B'B}$	M'_B
$\begin{pmatrix} 0&1&0&0\\ 1&w_{22}&0&w_{24}\\ 0&0&1&w_{34}\\ 0&0&w_{43}&1 \end{pmatrix}$	$\begin{pmatrix} x_2^2&0&0&0\\ 0&x_2&0&0\\ 0&0&1&0\\ 0&0&0&1 \end{pmatrix}$	$\begin{pmatrix} 0&1&0&0\\ 1&x_2w_{22}&0&\frac{w_{24}}{x_2}\\ 0&0&1&w_{34}\\ 0&0&w_{43}&1 \end{pmatrix}$
$\begin{pmatrix} 0&1&0&0\\ 1&w_{22}&0&w_{24}\\ 0&0&0&1\\ 0&0&1&0 \end{pmatrix}$	$\begin{pmatrix} x_2^2&0&0&0\\ 0&x_2&0&0\\ 0&0&x_4^2&0\\ 0&0&0&x_4 \end{pmatrix}$	$\begin{pmatrix} 0&1&0&0\\ 1&x_2w_{22}&0&\frac{x_4^2w_{24}}{x_2}\\ 0&0&0&1\\ 0&0&1&0 \end{pmatrix}$
$\begin{pmatrix} 0&1&0&0\\ 1&w_{22}&0&w_{24}\\ 0&0&0&1\\ 0&0&1&w_{44} \end{pmatrix}$	$\begin{pmatrix} x_2^2&0&0&0\\ 0&x_2&0&0\\ 0&0&x_4^2&0\\ 0&0&0&x_4 \end{pmatrix}$	$\begin{pmatrix} 0&1&0&0\\ 1&x_2w_{22}&0&\frac{x_4^2w_{24}}{x_2}\\ 0&0&0&1\\ 0&0&1&x_4w_{44} \end{pmatrix}$
$\begin{pmatrix} 0&1&0&0\\ 1&w_{22}&0&w_{24}\\ 0&0&w_{33}&1\\ 0&0&1&0 \end{pmatrix}$	$\begin{pmatrix} x_2^2&0&0&0\\ 0&x_2&0&0\\ 0&0&x_4^2&0\\ 0&0&0&x_4 \end{pmatrix}$	$\begin{pmatrix} 0&1&0&0\\ 1&x_2w_{22}&0&\frac{x_4^2w_{24}}{x_2}\\ 0&0&x_4^2w_{33}&1\\ 0&0&1&0 \end{pmatrix}$
$\begin{pmatrix} 0&1&0&w_{14}\\ 1&w_{22}&0&0\\ 0&0&1&w_{34}\\ 0&0&w_{43}&1 \end{pmatrix}$	$\begin{pmatrix} x_2^2&0&0&0\\ 0&x_2&0&0\\ 0&0&1&0\\ 0&0&0&1 \end{pmatrix}$	$\begin{pmatrix} 0&1&0&\frac{w_{14}}{x_2^2}\\ 1&x_2w_{22}&0&0\\ 0&0&1&w_{34}\\ 0&0&w_{43}&1 \end{pmatrix}$
$\begin{pmatrix} 0&1&0&w_{14}\\ 1&w_{22}&0&0\\ 0&0&0&1\\ 0&0&1&0 \end{pmatrix}$	$\begin{pmatrix} x_2^2&0&0&0\\ 0&x_2&0&0\\ 0&0&x_4^2&0\\ 0&0&0&x_4 \end{pmatrix}$	$\begin{pmatrix} 0&1&0&\frac{x_4^2w_{14}}{x_2^2}\\ 1&x_2w_{22}&0&0\\ 0&0&0&1\\ 0&0&1&0 \end{pmatrix}$
$\begin{pmatrix} 0&1&0&w_{14}\\ 1&w_{22}&0&0\\ 0&0&0&1\\ 0&0&1&w_{44} \end{pmatrix}$	$\begin{pmatrix} x_2^2&0&0&0\\ 0&x_2&0&0\\ 0&0&x_4^2&0\\ 0&0&0&x_4 \end{pmatrix}$	$\begin{pmatrix} 0&1&0&\frac{x_4^2w_{14}}{x_2^2}\\ 1&x_2w_{22}&0&0\\ 0&0&0&1\\ 0&0&1&x_4w_{44} \end{pmatrix}$
$\begin{pmatrix} 0&1&0&w_{14}\\ 1&w_{22}&0&0\\ 0&0&w_{33}&1\\ 0&0&1&0 \end{pmatrix}$	$\begin{pmatrix} x_2^2&0&0&0\\ 0&x_2&0&0\\ 0&0&x_4^2&0\\ 0&0&0&x_4 \end{pmatrix}$	$\begin{pmatrix} 0&1&0&\frac{x_4^2w_{14}}{x_2^2}\\ 1&x_2w_{22}&0&0\\ 0&0&x_4^2w_{33}&1\\ 0&0&1&0 \end{pmatrix}$
$\begin{pmatrix} 0&1&0&0\\ 1&w_{22}&w_{23}&w_{24}\\ 0&0&1&w_{34}\\ 0&0&w_{43}&1 \end{pmatrix}$	$\begin{pmatrix} x_2^2&0&0&0\\ 0&x_2&0&0\\ 0&0&1&0\\ 0&0&0&1 \end{pmatrix}$	$\begin{pmatrix} 0&1&0&0\\ 1&x_2w_{22}&\frac{w_{23}}{x_2}&\frac{w_{24}}{x_2}\\ 0&0&1&w_{34}\\ 0&0&w_{43}&1 \end{pmatrix}$
$\begin{pmatrix} 0&1&0&0\\ 1&w_{22}&w_{23}&w_{24}\\ 0&0&1&w_{34}\\ 0&0&w_{43}&1 \end{pmatrix}$	$\begin{pmatrix} x_2^2&0&0&0\\ 0&x_2&0&0\\ 0&0&0&1\\ 0&0&1&0 \end{pmatrix}$	$\begin{pmatrix} 0&1&0&0\\ 1&x_2w_{22}&\frac{w_{24}}{x_2}&\frac{w_{23}}{x_2}\\ 0&0&1&w_{43}\\ 0&0&w_{34}&1 \end{pmatrix}$

M_B	$P_{B'B}$	M'_B
$\begin{pmatrix} 0&1&0&0\\ 1&w_{22}&w_{23}&w_{24}\\ 0&0&0&1\\ 0&0&1&0 \end{pmatrix}$	$\begin{pmatrix} x_2^2&0&0&0\\ 0&x_2&0&0\\ 0&0&x_4^2&0\\ 0&0&0&x_4 \end{pmatrix}$	$\begin{pmatrix} 0&1&0&0\\ 1&x_2w_{22}&\frac{x_4w_{23}}{x_2}&\frac{x_4^2w_{24}}{x_2}\\ 0&0&0&1\\ 0&0&1&0 \end{pmatrix}$
$\begin{pmatrix} 0&1&0&0\\ 1&w_{22}&w_{23}&w_{24}\\ 0&0&0&1\\ 0&0&1&0 \end{pmatrix}$	$\begin{pmatrix} x_2^2&0&0&0\\ 0&x_2&0&0\\ 0&0&0&x_4^2\\ 0&0&x_4&0 \end{pmatrix}$	$\begin{pmatrix} 0&1&0&0\\ 1&x_2w_{22}&\frac{x_4^2w_{24}}{x_2}&\frac{x_4w_{23}}{x_2}\\ 0&0&0&1\\ 0&0&1&0 \end{pmatrix}$
$\begin{pmatrix} 0&1&0&0\\ 1&w_{22}&w_{23}&w_{24}\\ 0&0&0&1\\ 0&0&1&w_{44} \end{pmatrix}$	$\begin{pmatrix} x_2^2&0&0&0\\ 0&x_2&0&0\\ 0&0&x_4^2&0\\ 0&0&0&x_4 \end{pmatrix}$	$\begin{pmatrix} 0&1&0&0\\ 1&x_2w_{22}&\frac{x_4w_{23}}{x_2}&\frac{x_4^2w_{24}}{x_2}\\ 0&0&0&1\\ 0&0&1&x_4w_{44} \end{pmatrix}$
$\begin{pmatrix} 0&1&w_{13}&w_{14}\\ 1&w_{22}&0&0\\ 0&0&1&w_{34}\\ 0&0&w_{43}&1 \end{pmatrix}$	$\begin{pmatrix} x_2^2&0&0&0\\ 0&x_2&0&0\\ 0&0&1&0\\ 0&0&0&1 \end{pmatrix}$	$\begin{pmatrix} 0&1&\frac{w_{13}}{x_2^2}&\frac{w_{14}}{x_2^2}\\ 1&x_2w_{22}&0&0\\ 0&0&1&w_{34}\\ 0&0&w_{43}&1 \end{pmatrix}$
$\begin{pmatrix} 0&1&w_{13}&w_{14}\\ 1&w_{22}&0&0\\ 0&0&0&1\\ 0&0&1&0 \end{pmatrix}$	$\begin{pmatrix} x_2^2&0&0&0\\ 0&x_2&0&0\\ 0&0&x_4^2&0\\ 0&0&0&x_4 \end{pmatrix}$	$\begin{pmatrix} 0&1&\frac{x_4w_{13}}{x_2^2}&\frac{x_4^2w_{14}}{x_2^2}\\ 1&x_2w_{22}&0&0\\ 0&0&0&1\\ 0&0&1&0 \end{pmatrix}$
$\begin{pmatrix} 0&1&w_{13}&w_{14}\\ 1&w_{22}&0&0\\ 0&0&0&1\\ 0&0&1&0 \end{pmatrix}$	$\begin{pmatrix} x_2^2&0&0&0\\ 0&x_2&0&0\\ 0&0&0&x_4^2\\ 0&0&x_4&0 \end{pmatrix}$	$\begin{pmatrix} 0&1&\frac{x_4^2w_{14}}{x_2^2}&\frac{x_4w_{13}}{x_2^2}\\ 1&x_2w_{22}&0&0\\ 0&0&0&1\\ 0&0&1&0 \end{pmatrix}$
$\begin{pmatrix} 0&1&w_{13}&w_{14}\\ 1&w_{22}&0&0\\ 0&0&0&1\\ 0&0&1&w_{44} \end{pmatrix}$	$\begin{pmatrix} x_2^2&0&0&0\\ 0&x_2&0&0\\ 0&0&x_4^2&0\\ 0&0&0&x_4 \end{pmatrix}$	$\begin{pmatrix} 0&1&\frac{x_4w_{13}}{x_2^2}&\frac{x_4^2w_{14}}{x_2^2}\\ 1&x_2w_{22}&0&0\\ 0&0&0&1\\ 0&0&1&x_4w_{44} \end{pmatrix}$
$\begin{pmatrix} 0&1&0&w_{14}\\ 1&w_{22}&0&w_{24}\\ 0&0&1&w_{34}\\ 0&0&w_{43}&1 \end{pmatrix}$	$\begin{pmatrix} x_2^2&0&0&0\\ 0&x_2&0&0\\ 0&0&1&0\\ 0&0&0&1 \end{pmatrix}$	$\begin{pmatrix} 0&1&0&\frac{w_{14}}{x_2^2}\\ 1&x_2w_{22}&0&\frac{w_{24}}{x_2}\\ 0&0&1&w_{34}\\ 0&0&w_{43}&1 \end{pmatrix}$
$\begin{pmatrix} 0&1&0&w_{14}\\ 1&w_{22}&0&w_{24}\\ 0&0&0&1\\ 0&0&1&0 \end{pmatrix}$	$\begin{pmatrix} x_2^2&0&0&0\\ 0&x_2&0&0\\ 0&0&x_4^2&0\\ 0&0&0&x_4 \end{pmatrix}$	$\begin{pmatrix} 0&1&0&\frac{x_4^2w_{14}}{x_2^2}\\ 1&x_2w_{22}&0&\frac{x_4^2w_{24}}{x_2}\\ 0&0&0&1\\ 0&0&1&0 \end{pmatrix}$
$\begin{pmatrix} 0&1&0&w_{14}\\ 1&w_{22}&0&w_{24}\\ 0&0&0&1\\ 0&0&1&w_{44} \end{pmatrix}$	$\begin{pmatrix} x_2^2&0&0&0\\ 0&x_2&0&0\\ 0&0&x_4^2&0\\ 0&0&0&x_4 \end{pmatrix}$	$\begin{pmatrix} 0&1&0&\frac{x_4^2w_{14}}{x_2^2}\\ 1&x_2w_{22}&0&\frac{x_4^2w_{24}}{x_2}\\ 0&0&0&1\\ 0&0&1&x_4w_{44} \end{pmatrix}$

M_B	$P_{B'B}$	M'_B
$\begin{pmatrix} 0&1&0&w_{14}\\ 1&w_{22}&0&w_{24}\\ 0&0&w_{33}&1\\ 0&0&1&0 \end{pmatrix}$	$\begin{pmatrix} x_2^2&0&0&0\\ 0&x_2&0&0\\ 0&0&x_4^2&0\\ 0&0&0&x_4 \end{pmatrix}$	$\begin{pmatrix} 0&1&0&\frac{x_4^2w_{14}}{x_2^2}\\ 1&x_2w_{22}&0&\frac{x_4^2w_{24}}{x_2}\\ 0&0&x_4^2w_{33}&1\\ 0&0&1&0 \end{pmatrix}$
$\begin{pmatrix} 0&1&0&w_{14}\\ 1&w_{22}&w_{23}&0\\ 0&0&1&w_{34}\\ 0&0&w_{43}&1 \end{pmatrix}$	$\begin{pmatrix} x_2^2&0&0&0\\ 0&x_2&0&0\\ 0&0&1&0\\ 0&0&0&1 \end{pmatrix}$	$\begin{pmatrix} 0&1&0&\frac{w_{14}}{x_2^2}\\ 1&x_2w_{22}&\frac{w_{23}}{x_2}&0\\ 0&0&1&w_{34}\\ 0&0&w_{43}&1 \end{pmatrix}$
$\begin{pmatrix} 0&1&0&w_{14}\\ 1&w_{22}&w_{23}&0\\ 0&0&0&1\\ 0&0&1&0 \end{pmatrix}$	$\begin{pmatrix} x_2^2&0&0&0\\ 0&x_2&0&0\\ 0&0&x_4^2&0\\ 0&0&0&x_4 \end{pmatrix}$	$\begin{pmatrix} 0&1&0&\frac{x_4^2w_{14}}{x_2^2}\\ 1&x_2w_{22}&\frac{x_4w_{23}}{x_2}&0\\ 0&0&0&1\\ 0&0&1&0 \end{pmatrix}$
$\begin{pmatrix} 0&1&0&w_{14}\\ 1&w_{22}&w_{23}&0\\ 0&0&0&1\\ 0&0&1&w_{44} \end{pmatrix}$	$\begin{pmatrix} x_2^2&0&0&0\\ 0&x_2&0&0\\ 0&0&x_4^2&0\\ 0&0&0&x_4 \end{pmatrix}$	$\begin{pmatrix} 0&1&0&\frac{x_4^2w_{14}}{x_2^2}\\ 1&x_2w_{22}&\frac{x_4w_{23}}{x_2}&0\\ 0&0&0&1\\ 0&0&1&x_4w_{44} \end{pmatrix}$
$\begin{pmatrix} 0&1&0&w_{14}\\ 1&w_{22}&w_{23}&0\\ 0&0&w_{33}&1\\ 0&0&1&0 \end{pmatrix}$	$\begin{pmatrix} x_2^2&0&0&0\\ 0&x_2&0&0\\ 0&0&x_4^2&0\\ 0&0&0&x_4 \end{pmatrix}$	$\begin{pmatrix} 0&1&0&\frac{x_4^2w_{14}}{x_2^2}\\ 1&x_2w_{22}&\frac{x_4w_{23}}{x_2}&0\\ 0&0&x_4^2w_{33}&1\\ 0&0&1&0 \end{pmatrix}$
$\begin{pmatrix} 0&1&w_{13}&w_{14}\\ 1&w_{22}&w_{23}&0\\ 0&0&1&w_{34}\\ 0&0&w_{43}&1 \end{pmatrix}$	$\begin{pmatrix} x_2^2&0&0&0\\ 0&x_2&0&0\\ 0&0&1&0\\ 0&0&0&1 \end{pmatrix}$	$\begin{pmatrix} 0&1&\frac{w_{13}}{x_2^2}&\frac{w_{14}}{x_2^2}\\ 1&x_2w_{22}&\frac{w_{23}}{x_2}&0\\ 0&0&1&w_{34}\\ 0&0&w_{43}&1 \end{pmatrix}$
$\begin{pmatrix} 0&1&w_{13}&w_{14}\\ 1&w_{22}&w_{23}&0\\ 0&0&0&1\\ 0&0&1&0 \end{pmatrix}$	$\begin{pmatrix} x_2^2&0&0&0\\ 0&x_2&0&0\\ 0&0&x_4^2&0\\ 0&0&0&x_4 \end{pmatrix}$	$\begin{pmatrix} 0&1&\frac{x_4w_{13}}{x_2^2}&\frac{x_4^2w_{14}}{x_2^2}\\ 1&x_2w_{22}&\frac{x_4w_{23}}{x_2}&0\\ 0&0&0&1\\ 0&0&1&0 \end{pmatrix}$
$\begin{pmatrix} 0&1&w_{13}&w_{14}\\ 1&w_{22}&w_{23}&0\\ 0&0&0&1\\ 0&0&1&w_{44} \end{pmatrix}$	$\begin{pmatrix} x_2^2&0&0&0\\ 0&x_2&0&0\\ 0&0&x_4^2&0\\ 0&0&0&x_4 \end{pmatrix}$	$\begin{pmatrix} 0&1&\frac{x_4w_{13}}{x_2^2}&\frac{x_4^2w_{14}}{x_2^2}\\ 1&x_2w_{22}&\frac{x_4w_{23}}{x_2}&0\\ 0&0&0&1\\ 0&0&1&x_4w_{44} \end{pmatrix}$
$\begin{pmatrix} 0&1&w_{13}&w_{14}\\ 1&w_{22}&w_{23}&0\\ 0&0&w_{33}&1\\ 0&0&1&0 \end{pmatrix}$	$\begin{pmatrix} x_2^2&0&0&0\\ 0&x_2&0&0\\ 0&0&x_4^2&0\\ 0&0&0&x_4 \end{pmatrix}$	$\begin{pmatrix} 0&1&\frac{x_4w_{13}}{x_2^2}&\frac{x_4^2w_{14}}{x_2^2}\\ 1&x_2w_{22}&\frac{x_4w_{23}}{x_2}&0\\ 0&0&x_4^2w_{33}&1\\ 0&0&1&0 \end{pmatrix}$
$\begin{pmatrix} 0&1&w_{13}&0\\ 1&w_{22}&w_{23}&w_{24}\\ 0&0&1&w_{34}\\ 0&0&w_{43}&1 \end{pmatrix}$	$\begin{pmatrix} x_2^2&0&0&0\\ 0&x_2&0&0\\ 0&0&1&0\\ 0&0&0&1 \end{pmatrix}$	$\begin{pmatrix} 0&1&\frac{w_{13}}{x_2^2}&0\\ 1&x_2w_{22}&\frac{w_{23}}{x_2}&\frac{w_{24}}{x_2}\\ 0&0&1&w_{34}\\ 0&0&w_{43}&1 \end{pmatrix}$

M_B	$P_{B'B}$	M'_B
$\begin{pmatrix} 0&1&w_{13}&0\\ 1&w_{22}&w_{23}&w_{24}\\ 0&0&0&1\\ 0&0&1&0 \end{pmatrix}$	$\begin{pmatrix} x_2^2&0&0&0\\ 0&x_2&0&0\\ 0&0&x_4^2&0\\ 0&0&0&x_4 \end{pmatrix}$	$\begin{pmatrix} 0&1&\frac{x_4w_{13}}{x_2^2}&0\\ 1&x_2w_{22}&\frac{x_4w_{23}}{x_2}&\frac{x_4^2w_{24}}{x_2}\\ 0&0&0&1\\ 0&0&1&0 \end{pmatrix}$
$\begin{pmatrix} 0&1&w_{13}&0\\ 1&w_{22}&w_{23}&w_{24}\\ 0&0&0&1\\ 0&0&1&w_{44} \end{pmatrix}$	$\begin{pmatrix} x_2^2&0&0&0\\ 0&x_2&0&0\\ 0&0&x_4^2&0\\ 0&0&0&x_4 \end{pmatrix}$	$\begin{pmatrix} 0&1&\frac{x_4w_{13}}{x_2^2}&0\\ 1&x_2w_{22}&\frac{x_4w_{23}}{x_2}&\frac{x_4^2w_{24}}{x_2}\\ 0&0&0&1\\ 0&0&1&x_4w_{44} \end{pmatrix}$
$\begin{pmatrix} 0&1&w_{13}&0\\ 1&w_{22}&w_{23}&w_{24}\\ 0&0&w_{33}&1\\ 0&0&1&0 \end{pmatrix}$	$\begin{pmatrix} x_2^2&0&0&0\\ 0&x_2&0&0\\ 0&0&x_4^2&0\\ 0&0&0&x_4 \end{pmatrix}$	$\begin{pmatrix} 0&1&\frac{x_4w_{13}}{x_2^2}&0\\ 1&x_2w_{22}&\frac{x_4w_{23}}{x_2}&\frac{x_4^2w_{24}}{x_2}\\ 0&0&x_4^2w_{33}&1\\ 0&0&1&0 \end{pmatrix}$
$\begin{pmatrix} 0&1&w_{13}&w_{14}\\ 1&w_{22}&w_{23}&w_{24}\\ 0&0&1&w_{34}\\ 0&0&w_{43}&1 \end{pmatrix}$	$\begin{pmatrix} x_2^2&0&0&0\\ 0&x_2&0&0\\ 0&0&1&0\\ 0&0&0&1 \end{pmatrix}$	$\begin{pmatrix} 0&1&\frac{w_{13}}{x_2^2}&\frac{w_{14}}{x_2^2}\\ 1&x_2w_{22}&\frac{w_{23}}{x_2}&\frac{w_{24}}{x_2}\\ 0&0&1&w_{34}\\ 0&0&w_{43}&1 \end{pmatrix}$
$\begin{pmatrix} 0&1&w_{13}&w_{14}\\ 1&w_{22}&w_{23}&w_{24}\\ 0&0&0&1\\ 0&0&1&0 \end{pmatrix}$	$\begin{pmatrix} x_2^2&0&0&0\\ 0&x_2&0&0\\ 0&0&x_4^2&0\\ 0&0&0&x_4 \end{pmatrix}$	$\begin{pmatrix} 0&1&\frac{x_4w_{13}}{x_2^2}&\frac{x_4^2w_{14}}{x_2^2}\\ 1&x_2w_{22}&\frac{x_4w_{23}}{x_2}&\frac{x_4^2w_{24}}{x_2}\\ 0&0&0&1\\ 0&0&1&0 \end{pmatrix}$
$\begin{pmatrix} 0&1&w_{13}&w_{14}\\ 1&w_{22}&w_{23}&w_{24}\\ 0&0&0&1\\ 0&0&1&0 \end{pmatrix}$	$\begin{pmatrix} x_2^2&0&0&0\\ 0&x_2&0&0\\ 0&0&0&x_4^2\\ 0&0&x_4&0 \end{pmatrix}$	$\begin{pmatrix} 0&1&\frac{x_4^2w_{14}}{x_2^2}&\frac{x_4w_{13}}{x_2^2}\\ 1&x_2w_{22}&\frac{x_4^2w_{24}}{x_2}&\frac{x_4w_{23}}{x_2}\\ 0&0&0&1\\ 0&0&1&0 \end{pmatrix}$
$\begin{pmatrix} 0&1&w_{13}&w_{14}\\ 1&w_{22}&w_{23}&w_{24}\\ 0&0&0&1\\ 0&0&1&w_{44} \end{pmatrix}$	$\begin{pmatrix} x_2^2&0&0&0\\ 0&x_2&0&0\\ 0&0&x_4^2&0\\ 0&0&0&x_4 \end{pmatrix}$	$\begin{pmatrix} 0&1&\frac{x_4w_{13}}{x_2^2}&\frac{x_4^2w_{14}}{x_2^2}\\ 1&x_2w_{22}&\frac{x_4w_{23}}{x_2}&\frac{x_4^2w_{24}}{x_2}\\ 0&0&0&1\\ 0&0&1&x_4w_{44} \end{pmatrix}$

4 The Evolution Algebra Has a 3-Basic Ideal

In this last section we will consider that the evolution algebra A has a maximal basic ideal I of dimension 3. Assume that the structure matrix is $M_B = \left(\begin{array}{c|c} W & U \\ \hline 0\,0\,0 & 1 \end{array}\right)$. The classification given in [1, Sect. 5] depends on considering whether or not the ideal I has a 2-basic ideal. In the affirmative case, the authors distinguished whether or not this 2-basic ideal is contained in another 3-basic ideal. When this happens, it is said that the evolution algebra satisfies Condition (3, 2, 3).

4.1 A Has a 3-Basic Ideal Which Has a 2-Basic Ideal

A Satisfies Condition (3, 2, 3)

M_B	$P_{B'B}$	M'_B
$\begin{pmatrix} w_{11} & w_{12} & 0 & 1 \\ 0 & w_{22} & 0 & 0 \\ 0 & 0 & w_{33} & 1 \\ 0 & 0 & 0 & 1 \end{pmatrix}$	$\begin{pmatrix} 1 & 0 & 0 & 0 \\ 0 & x_2 & 0 & 0 \\ 0 & 0 & 1 & 0 \\ 0 & 0 & 0 & 1 \end{pmatrix}$	$\begin{pmatrix} w_{11} & x_2^2 w_{12} & 0 & 1 \\ 0 & x_2 w_{22} & 0 & 0 \\ 0 & 0 & w_{33} & 1 \\ 0 & 0 & 0 & 1 \end{pmatrix}$
$\begin{pmatrix} 0 & w_{12} & w_{13} & 1 \\ w_{21} & 0 & 0 & 0 \\ 0 & 0 & w_{33} & 0 \\ 0 & 0 & 0 & 1 \end{pmatrix}$	$\begin{pmatrix} 1 & 0 & 0 & 0 \\ 0 & x_2 & 0 & 0 \\ 0 & 0 & x_3 & 0 \\ 0 & 0 & 0 & 1 \end{pmatrix}$	$\begin{pmatrix} 0 & x_2^2 w_{12} & x_3^2 w_{13} & 1 \\ \frac{w_{21}}{x_2} & 0 & 0 & 0 \\ 0 & 0 & x_3 w_{33} & 0 \\ 0 & 0 & 0 & 1 \end{pmatrix}$
$\begin{pmatrix} 0 & w_{12} & w_{13} & 1 \\ w_{21} & 0 & 0 & 0 \\ 0 & 0 & w_{33} & 0 \\ 0 & 0 & 0 & 1 \end{pmatrix}$	$\begin{pmatrix} x_3^2 w_{13} & 0 & 0 & 0 \\ 0 & x_2 & 0 & 0 \\ 0 & 0 & 0 & x_3 \\ 0 & 0 & x_4 & 0 \end{pmatrix}$	$\begin{pmatrix} 0 & \frac{x_2^2 w_{12}}{x_3^2 w_{13}} & \frac{x_4^2}{x_3^2 w_{13}} & 1 \\ \frac{x_3^4 w_{13}^2 w_{21}}{x_2} & 0 & 0 & 0 \\ 0 & 0 & x_4 & 0 \\ 0 & 0 & 0 & 1 \end{pmatrix}$
$\begin{pmatrix} 0 & w_{12} & w_{13} & 0 \\ w_{21} & 0 & 0 & 1 \\ 0 & 0 & w_{33} & 0 \\ 0 & 0 & 0 & 1 \end{pmatrix}$	$\begin{pmatrix} x_1 & 0 & 0 & 0 \\ 0 & 1 & 0 & 0 \\ 0 & 0 & x_3 & 0 \\ 0 & 0 & 0 & 1 \end{pmatrix}$	$\begin{pmatrix} 0 & \frac{w_{12}}{x_1} & \frac{x_3^2 w_{13}}{x_1} & 0 \\ x_1^2 w_{21} & 0 & 0 & 1 \\ 0 & 0 & x_3 w_{33} & 0 \\ 0 & 0 & 0 & 1 \end{pmatrix}$
$\begin{pmatrix} 0 & w_{12} & w_{13} & 0 \\ w_{21} & 0 & 0 & 1 \\ 0 & 0 & w_{33} & 0 \\ 0 & 0 & 0 & 1 \end{pmatrix}$	$\begin{pmatrix} 0 & x_3^2 w_{13} & 0 & 0 \\ x_2 & 0 & 0 & 0 \\ 0 & 0 & 0 & x_3 \\ 0 & 0 & x_4 & 0 \end{pmatrix}$	$\begin{pmatrix} 0 & \frac{x_3^4 w_{13}^2 w_{21}}{x_2} & \frac{x_4^2}{x_2} & 0 \\ \frac{x_2^2 w_{12}}{x_3^2 w_{13}} & 0 & 0 & 1 \\ 0 & 0 & x_4 & 0 \\ 0 & 0 & 0 & 1 \end{pmatrix}$
$\begin{pmatrix} 0 & w_{12} & w_{13} & 1 \\ w_{21} & 0 & 0 & 1 \\ 0 & 0 & w_{33} & 0 \\ 0 & 0 & 0 & 1 \end{pmatrix}$	$\begin{pmatrix} 1 & 0 & 0 & 0 \\ 0 & 1 & 0 & 0 \\ 0 & 0 & x_3 & 0 \\ 0 & 0 & 0 & 1 \end{pmatrix}$	$\begin{pmatrix} 0 & w_{12} & x_3^2 w_{13} & 1 \\ w_{21} & 0 & 0 & 1 \\ 0 & 0 & x_3 w_{33} & 0 \\ 0 & 0 & 0 & 1 \end{pmatrix}$
$\begin{pmatrix} w_{11} & w_{12} & w_{13} & 1 \\ 0 & w_{22} & 0 & 0 \\ 0 & 0 & w_{33} & 0 \\ 0 & 0 & 0 & 1 \end{pmatrix}$	$\begin{pmatrix} 1 & 0 & 0 & 0 \\ 0 & x_2 & 0 & 0 \\ 0 & 0 & x_3 & 0 \\ 0 & 0 & 0 & 1 \end{pmatrix}$	$\begin{pmatrix} w_{11} & x_2^2 w_{12} & x_3^2 w_{13} & 1 \\ 0 & x_2 w_{22} & 0 & 0 \\ 0 & 0 & x_3 w_{33} & 0 \\ 0 & 0 & 0 & 1 \end{pmatrix}$
$\begin{pmatrix} w_{11} & w_{12} & w_{13} & 1 \\ 0 & w_{22} & 0 & 0 \\ 0 & 0 & w_{33} & 0 \\ 0 & 0 & 0 & 1 \end{pmatrix}$	$\begin{pmatrix} x_3^2 w_{13} & 0 & 0 & 0 \\ 0 & 0 & x_2 & 0 \\ 0 & 0 & 0 & x_3 \\ 0 & x_4 & 0 & 0 \end{pmatrix}$	$\begin{pmatrix} x_3^2 w_{11} w_{13} & \frac{x_4^2}{x_3^2 w_{13}} & \frac{x_2^2 w_{12}}{x_3^2 w_{13}} & 1 \\ 0 & x_4 & 0 & 0 \\ 0 & 0 & x_2 w_{22} & 0 \\ 0 & 0 & 0 & 1 \end{pmatrix}$
$\begin{pmatrix} w_{11} & w_{12} & w_{13} & 1 \\ 0 & w_{22} & 0 & 0 \\ 0 & 0 & w_{33} & 0 \\ 0 & 0 & 0 & 1 \end{pmatrix}$	$\begin{pmatrix} x_2^2 w_{12} & 0 & 0 & 0 \\ 0 & 0 & 0 & x_2 \\ 0 & x_3 & 0 & 0 \\ 0 & 0 & x_4 & 0 \end{pmatrix}$	$\begin{pmatrix} x_2^2 w_{11} w_{12} & \frac{x_3^2 w_{13}}{x_2^2 w_{12}} & \frac{x_4^2}{x_2^2 w_{12}} & 1 \\ 0 & x_3 w_{33} & 0 & 0 \\ 0 & 0 & x_4 & 0 \\ 0 & 0 & 0 & 1 \end{pmatrix}$
$\begin{pmatrix} w_{11} & w_{12} & w_{13} & 1 \\ 0 & w_{22} & 0 & 0 \\ 0 & 0 & w_{33} & 0 \\ 0 & 0 & 0 & 1 \end{pmatrix}$	$\begin{pmatrix} x_3^2 w_{13} & 0 & 0 & 0 \\ 0 & x_2 & 0 & 0 \\ 0 & 0 & 0 & x_3 \\ 0 & 0 & x_4 & 0 \end{pmatrix}$	$\begin{pmatrix} x_3^2 w_{11} w_{13} & \frac{x_2^2 w_{12}}{x_3^2 w_{13}} & \frac{x_4^2}{x_3^2 w_{13}} & 1 \\ 0 & x_2 w_{22} & 0 & 0 \\ 0 & 0 & x_4 & 0 \\ 0 & 0 & 0 & 1 \end{pmatrix}$
$\begin{pmatrix} w_{11} & w_{12} & w_{13} & 1 \\ 0 & w_{22} & 0 & 0 \\ 0 & 0 & w_{33} & 0 \\ 0 & 0 & 0 & 1 \end{pmatrix}$	$\begin{pmatrix} x_2^2 w_{12} & 0 & 0 & 0 \\ 0 & 0 & 0 & x_2 \\ 0 & 0 & x_3 & 0 \\ 0 & x_4 & 0 & 0 \end{pmatrix}$	$\begin{pmatrix} x_2^2 w_{11} w_{12} & \frac{x_4^2}{x_2^2 w_{12}} & \frac{x_3^2 w_{13}}{x_2^2 w_{12}} & 1 \\ 0 & x_4 & 0 & 0 \\ 0 & 0 & x_3 w_{33} & 0 \\ 0 & 0 & 0 & 1 \end{pmatrix}$
$\begin{pmatrix} w_{11} & w_{12} & w_{13} & 1 \\ 0 & w_{22} & 0 & 0 \\ 0 & 0 & w_{33} & 0 \\ 0 & 0 & 0 & 1 \end{pmatrix}$	$\begin{pmatrix} 1 & 0 & 0 & 0 \\ 0 & 0 & x_2 & 0 \\ 0 & x_3 & 0 & 0 \\ 0 & 0 & 0 & 1 \end{pmatrix}$	$\begin{pmatrix} w_{11} & x_3^2 w_{13} & x_2^2 w_{12} & 1 \\ 0 & x_3 w_{33} & 0 & 0 \\ 0 & 0 & x_2 w_{22} & 0 \\ 0 & 0 & 0 & 1 \end{pmatrix}$
$\begin{pmatrix} w_{11} & w_{12} & w_{13} & 0 \\ 0 & w_{22} & 0 & 1 \\ 0 & 0 & w_{33} & 0 \\ 0 & 0 & 0 & 1 \end{pmatrix}$	$\begin{pmatrix} x_1 & 0 & 0 & 0 \\ 0 & 1 & 0 & 0 \\ 0 & 0 & x_3 & 0 \\ 0 & 0 & 0 & 1 \end{pmatrix}$	$\begin{pmatrix} x_1 w_{11} & \frac{w_{12}}{x_1} & \frac{x_3^2 w_{13}}{x_1} & 0 \\ 0 & w_{22} & 0 & 1 \\ 0 & 0 & x_3 w_{33} & 0 \\ 0 & 0 & 0 & 1 \end{pmatrix}$
$\begin{pmatrix} w_{11} & w_{12} & w_{13} & 1 \\ 0 & w_{22} & 0 & 1 \\ 0 & 0 & w_{33} & 0 \\ 0 & 0 & 0 & 1 \end{pmatrix}$	$\begin{pmatrix} 1 & 0 & 0 & 0 \\ 0 & 1 & 0 & 0 \\ 0 & 0 & x_3 & 0 \\ 0 & 0 & 0 & 1 \end{pmatrix}$	$\begin{pmatrix} w_{11} & w_{12} & x_3^2 w_{13} & 1 \\ 0 & w_{22} & 0 & 1 \\ 0 & 0 & x_3 w_{33} & 0 \\ 0 & 0 & 0 & 1 \end{pmatrix}$
$\begin{pmatrix} w_{11} & w_{12} & 0 & 0 \\ 0 & w_{22} & w_{23} & 1 \\ 0 & 0 & w_{33} & 0 \\ 0 & 0 & 0 & 1 \end{pmatrix}$	$\begin{pmatrix} x_1 & 0 & 0 & 0 \\ 0 & 1 & 0 & 0 \\ 0 & 0 & x_3 & 0 \\ 0 & 0 & 0 & 1 \end{pmatrix}$	$\begin{pmatrix} x_1 w_{11} & \frac{w_{12}}{x_1} & 0 & 0 \\ 0 & w_{22} & x_3^2 w_{23} & 1 \\ 0 & 0 & x_3 w_{33} & 0 \\ 0 & 0 & 0 & 1 \end{pmatrix}$
$\begin{pmatrix} w_{11} & w_{12} & 0 & 0 \\ 0 & w_{22} & w_{23} & 1 \\ 0 & 0 & w_{33} & 0 \\ 0 & 0 & 0 & 1 \end{pmatrix}$	$\begin{pmatrix} x_1 & 0 & 0 & 0 \\ 0 & x_3^2 w_{23} & 0 & 0 \\ 0 & 0 & 0 & x_3 \\ 0 & 0 & x_4 & 0 \end{pmatrix}$	$\begin{pmatrix} x_1 w_{11} & \frac{x_3^4 w_{12} w_{23}^2}{x_1} & 0 & 0 \\ 0 & x_3^2 w_{22} w_{23} & \frac{x_4^2}{x_3^2 w_{23}} & 1 \\ 0 & 0 & x_4 & 0 \\ 0 & 0 & 0 & 1 \end{pmatrix}$
$\begin{pmatrix} w_{11} & w_{12} & 0 & 1 \\ 0 & w_{22} & w_{23} & 1 \\ 0 & 0 & w_{33} & 0 \\ 0 & 0 & 0 & 1 \end{pmatrix}$	$\begin{pmatrix} 1 & 0 & 0 & 0 \\ 0 & 1 & 0 & 0 \\ 0 & 0 & x_3 & 0 \\ 0 & 0 & 0 & 1 \end{pmatrix}$	$\begin{pmatrix} w_{11} & w_{12} & 0 & 1 \\ 0 & w_{22} & x_3^2 w_{23} & 1 \\ 0 & 0 & x_3 w_{33} & 0 \\ 0 & 0 & 0 & 1 \end{pmatrix}$
$\begin{pmatrix} w_{11} & 0 & w_{13} & 1 \\ 0 & w_{22} & w_{23} & 1 \\ 0 & 0 & w_{33} & 0 \\ 0 & 0 & 0 & 1 \end{pmatrix}$	$\begin{pmatrix} 1 & 0 & 0 & 0 \\ 0 & 1 & 0 & 0 \\ 0 & 0 & x_3 & 0 \\ 0 & 0 & 0 & 1 \end{pmatrix}$	$\begin{pmatrix} w_{11} & 0 & x_3^2 w_{13} & 1 \\ 0 & w_{22} & x_3^2 w_{23} & 1 \\ 0 & 0 & x_3 w_{33} & 0 \\ 0 & 0 & 0 & 1 \end{pmatrix}$
$\begin{pmatrix} w_{11} & 0 & w_{13} & 1 \\ 0 & w_{22} & w_{23} & 1 \\ 0 & 0 & w_{33} & 0 \\ 0 & 0 & 0 & 1 \end{pmatrix}$	$\begin{pmatrix} 0 & 1 & 0 & 0 \\ 1 & 0 & 0 & 0 \\ 0 & 0 & x_3 & 0 \\ 0 & 0 & 0 & 1 \end{pmatrix}$	$\begin{pmatrix} w_{22} & 0 & x_3^2 w_{23} & 1 \\ 0 & w_{11} & x_3^2 w_{13} & 1 \\ 0 & 0 & x_3 w_{33} & 0 \\ 0 & 0 & 0 & 1 \end{pmatrix}$
$\begin{pmatrix} w_{11} & 0 & w_{13} & 1 \\ 0 & w_{22} & w_{23} & 1 \\ 0 & 0 & w_{33} & 0 \\ 0 & 0 & 0 & 1 \end{pmatrix}$	$\begin{pmatrix} x_3^2 w_{13} & 0 & 0 & 0 \\ 0 & x_3^2 w_{23} & 0 & 0 \\ 0 & 0 & 0 & x_3 \\ 0 & 0 & x_4 & 0 \end{pmatrix}$	$\begin{pmatrix} x_3^2 w_{11} w_{13} & 0 & \frac{x_4^2}{x_3^2 w_{13}} & 1 \\ 0 & x_3^2 w_{22} w_{23} & \frac{x_4^2}{x_3^2 w_{23}} & 1 \\ 0 & 0 & x_4 & 0 \\ 0 & 0 & 0 & 1 \end{pmatrix}$
$\begin{pmatrix} w_{11} & 0 & w_{13} & 1 \\ 0 & w_{22} & w_{23} & 1 \\ 0 & 0 & w_{33} & 0 \\ 0 & 0 & 0 & 1 \end{pmatrix}$	$\begin{pmatrix} 0 & x_3^2 w_{13} & 0 & 0 \\ x_3^2 w_{23} & 0 & 0 & 0 \\ 0 & 0 & 0 & x_3 \\ 0 & 0 & x_4 & 0 \end{pmatrix}$	$\begin{pmatrix} x_3^2 w_{22} w_{23} & 0 & \frac{x_4^2}{x_3^2 w_{23}} & 1 \\ 0 & x_3^2 w_{11} w_{13} & \frac{x_4^2}{x_3^2 w_{13}} & 1 \\ 0 & 0 & x_4 & 0 \\ 0 & 0 & 0 & 1 \end{pmatrix}$
$\begin{pmatrix} w_{11} & w_{12} & w_{13} & 1 \\ w_{21} & 0 & 0 & 0 \\ 0 & 0 & w_{33} & 0 \\ 0 & 0 & 0 & 1 \end{pmatrix}$	$\begin{pmatrix} 1 & 0 & 0 & 0 \\ 0 & x_2 & 0 & 0 \\ 0 & 0 & x_3 & 0 \\ 0 & 0 & 0 & 1 \end{pmatrix}$	$\begin{pmatrix} w_{11} & x_2^2 w_{12} & x_3^2 w_{13} & 1 \\ \frac{w_{21}}{x_2} & 0 & 0 & 0 \\ 0 & 0 & x_3 w_{33} & 0 \\ 0 & 0 & 0 & 1 \end{pmatrix}$
$\begin{pmatrix} w_{11} & w_{12} & w_{13} & 1 \\ w_{21} & 0 & 0 & 0 \\ 0 & 0 & w_{33} & 0 \\ 0 & 0 & 0 & 1 \end{pmatrix}$	$\begin{pmatrix} x_3^2 w_{13} & 0 & 0 & 0 \\ 0 & x_2 & 0 & 0 \\ 0 & 0 & 0 & x_3 \\ 0 & 0 & x_4 & 0 \end{pmatrix}$	$\begin{pmatrix} x_3^2 w_{11} w_{13} & \frac{x_2^2 w_{12}}{x_3^2 w_{13}} & \frac{x_4^2}{x_3^2 w_{13}} & 1 \\ \frac{x_3^4 w_{13}^2 w_{21}}{x_2} & 0 & 0 & 0 \\ 0 & 0 & x_4 & 0 \\ 0 & 0 & 0 & 1 \end{pmatrix}$
$\begin{pmatrix} w_{11} & w_{12} & w_{13} & 0 \\ w_{21} & 0 & 0 & 1 \\ 0 & 0 & w_{33} & 0 \\ 0 & 0 & 0 & 1 \end{pmatrix}$	$\begin{pmatrix} x_1 & 0 & 0 & 0 \\ 0 & 1 & 0 & 0 \\ 0 & 0 & x_3 & 0 \\ 0 & 0 & 0 & 1 \end{pmatrix}$	$\begin{pmatrix} x_1 w_{11} & \frac{w_{12}}{x_1} & \frac{x_3^2 w_{13}}{x_1} & 0 \\ x_1^2 w_{21} & 0 & 0 & 1 \\ 0 & 0 & x_3 w_{33} & 0 \\ 0 & 0 & 0 & 1 \end{pmatrix}$
$\begin{pmatrix} w_{11} & w_{12} & w_{13} & 1 \\ w_{21} & 0 & 0 & 1 \\ 0 & 0 & w_{33} & 0 \\ 0 & 0 & 0 & 1 \end{pmatrix}$	$\begin{pmatrix} 1 & 0 & 0 & 0 \\ 0 & 1 & 0 & 0 \\ 0 & 0 & x_3 & 0 \\ 0 & 0 & 0 & 1 \end{pmatrix}$	$\begin{pmatrix} w_{11} & w_{12} & x_3^2 w_{13} & 1 \\ w_{21} & 0 & 0 & 1 \\ 0 & 0 & x_3 w_{33} & 0 \\ 0 & 0 & 0 & 1 \end{pmatrix}$
$\begin{pmatrix} 0 & w_{12} & 0 & 1 \\ w_{21} & w_{22} & w_{23} & 0 \\ 0 & 0 & w_{33} & 0 \\ 0 & 0 & 0 & 1 \end{pmatrix}$	$\begin{pmatrix} 1 & 0 & 0 & 0 \\ 0 & x_2 & 0 & 0 \\ 0 & 0 & x_3 & 0 \\ 0 & 0 & 0 & 1 \end{pmatrix}$	$\begin{pmatrix} 0 & x_2^2 w_{12} & 0 & 1 \\ \frac{w_{21}}{x_2} & x_2 w_{22} & \frac{x_3^2 w_{23}}{x_2} & 0 \\ 0 & 0 & x_3 w_{33} & 0 \\ 0 & 0 & 0 & 1 \end{pmatrix}$
$\begin{pmatrix} 0 & w_{12} & 0 & 1 \\ w_{21} & w_{22} & w_{23} & 0 \\ 0 & 0 & w_{33} & 0 \\ 0 & 0 & 0 & 1 \end{pmatrix}$	$\begin{pmatrix} 1 & 0 & 0 & 0 \\ 0 & x_2 & 0 & 0 \\ 0 & 0 & x_3 & 0 \\ 0 & 0 & 0 & 1 \end{pmatrix}$	$\begin{pmatrix} 0 & x_2^2 w_{12} & 0 & 1 \\ \frac{w_{21}}{x_2} & x_2 w_{22} & \frac{x_3^2 w_{23}}{x_2} & 0 \\ 0 & 0 & x_3 w_{33} & 0 \\ 0 & 0 & 0 & 1 \end{pmatrix}$
$\begin{pmatrix} 0 & w_{12} & 0 & 0 \\ w_{21} & w_{22} & w_{23} & 1 \\ 0 & 0 & w_{33} & 0 \\ 0 & 0 & 0 & 1 \end{pmatrix}$	$\begin{pmatrix} x_1 & 0 & 0 & 0 \\ 0 & 1 & 0 & 0 \\ 0 & 0 & x_3 & 0 \\ 0 & 0 & 0 & 1 \end{pmatrix}$	$\begin{pmatrix} 0 & \frac{w_{12}}{x_1} & 0 & 0 \\ x_1^2 w_{21} & w_{22} & x_3^2 w_{23} & 1 \\ 0 & 0 & x_3 w_{33} & 0 \\ 0 & 0 & 0 & 1 \end{pmatrix}$
$\begin{pmatrix} 0 & w_{12} & 0 & 0 \\ w_{21} & w_{22} & w_{23} & 1 \\ 0 & 0 & w_{33} & 0 \\ 0 & 0 & 0 & 1 \end{pmatrix}$	$\begin{pmatrix} x_1 & 0 & 0 & 0 \\ 0 & x_3^2 w_{23} & 0 & 0 \\ 0 & 0 & 0 & x_3 \\ 0 & 0 & x_4 & 0 \end{pmatrix}$	$\begin{pmatrix} 0 & \frac{x_3^4 w_{12} w_{23}^2}{x_1} & 0 & 0 \\ \frac{x_1^2 w_{21}}{x_3^2 w_{23}} & x_3^2 w_{22} w_{23} & \frac{x_4^2}{x_3^2 w_{23}} & 1 \\ 0 & 0 & x_4 & 0 \\ 0 & 0 & 0 & 1 \end{pmatrix}$
$\begin{pmatrix} 0 & w_{12} & 0 & 1 \\ w_{21} & w_{22} & w_{23} & 1 \\ 0 & 0 & w_{33} & 0 \\ 0 & 0 & 0 & 1 \end{pmatrix}$	$\begin{pmatrix} 1 & 0 & 0 & 0 \\ 0 & 1 & 0 & 0 \\ 0 & 0 & x_3 & 0 \\ 0 & 0 & 0 & 1 \end{pmatrix}$	$\begin{pmatrix} 0 & w_{12} & 0 & 1 \\ w_{21} & w_{22} & x_3^2 w_{23} & 1 \\ 0 & 0 & x_3 w_{33} & 0 \\ 0 & 0 & 0 & 1 \end{pmatrix}$

M_B	$P_{B'B}$	M'_B
$\begin{pmatrix} w_{11} & w_{12} & 0 & 1 \\ w_{21} & 0 & w_{23} & 0 \\ 0 & 0 & w_{33} & 0 \\ 0 & 0 & 0 & 1 \end{pmatrix}$	$\begin{pmatrix} 1 & 0 & 0 & 0 \\ 0 & x_2 & 0 & 0 \\ 0 & 0 & x_3 & 0 \\ 0 & 0 & 0 & 1 \end{pmatrix}$	$\begin{pmatrix} w_{11} & x_2^2w_{12} & 0 & 1 \\ \frac{w_{21}}{x_2} & 0 & \frac{x_3^2w_{23}}{x_2} & 0 \\ 0 & 0 & x_3w_{33} & 0 \\ 0 & 0 & 0 & 1 \end{pmatrix}$
$\begin{pmatrix} w_{11} & w_{12} & 0 & 0 \\ w_{21} & 0 & w_{23} & 1 \\ 0 & 0 & w_{33} & 0 \\ 0 & 0 & 0 & 1 \end{pmatrix}$	$\begin{pmatrix} x_1 & 0 & 0 & 0 \\ 0 & 1 & 0 & 0 \\ 0 & 0 & x_3 & 0 \\ 0 & 0 & 0 & 1 \end{pmatrix}$	$\begin{pmatrix} x_1w_{11} & \frac{w_{12}}{x_1} & 0 & 0 \\ x_1^2w_{21} & 0 & x_3^2w_{23} & 1 \\ 0 & 0 & x_3w_{33} & 0 \\ 0 & 0 & 0 & 1 \end{pmatrix}$
$\begin{pmatrix} w_{11} & w_{12} & 0 & 0 \\ w_{21} & 0 & w_{23} & 1 \\ 0 & 0 & w_{33} & 0 \\ 0 & 0 & 0 & 1 \end{pmatrix}$	$\begin{pmatrix} x_1 & 0 & 0 & 0 \\ 0 & x_3^2w_{23} & 0 & 0 \\ 0 & 0 & 0 & x_3 \\ 0 & 0 & x_4 & 0 \end{pmatrix}$	$\begin{pmatrix} x_1w_{11} & \frac{x_3^2w_{12}w_{23}}{x_1} & 0 & 0 \\ \frac{x_1^2w_{21}}{x_3^2w_{23}} & 0 & \frac{x_4^2}{x_3^2w_{23}} & 1 \\ 0 & 0 & x_4 & 0 \\ 0 & 0 & 0 & 1 \end{pmatrix}$
$\begin{pmatrix} w_{11} & w_{12} & 0 & 1 \\ w_{21} & 0 & w_{23} & 1 \\ 0 & 0 & w_{33} & 0 \\ 0 & 0 & 0 & 1 \end{pmatrix}$	$\begin{pmatrix} 1 & 0 & 0 & 0 \\ 0 & 1 & 0 & 0 \\ 0 & 0 & x_3 & 0 \\ 0 & 0 & 0 & 1 \end{pmatrix}$	$\begin{pmatrix} w_{11} & w_{12} & 0 & 1 \\ w_{21} & 0 & x_3^2w_{23} & 1 \\ 0 & 0 & x_3w_{33} & 0 \\ 0 & 0 & 0 & 1 \end{pmatrix}$
$\begin{pmatrix} 0 & w_{12} & w_{13} & 0 \\ w_{21} & 0 & w_{23} & 1 \\ 0 & 0 & w_{33} & 0 \\ 0 & 0 & 0 & 1 \end{pmatrix}$	$\begin{pmatrix} x_1 & 0 & 0 & 0 \\ 0 & 1 & 0 & 0 \\ 0 & 0 & x_3 & 0 \\ 0 & 0 & 0 & 1 \end{pmatrix}$	$\begin{pmatrix} 0 & \frac{w_{12}}{x_1} & \frac{x_3^2w_{13}}{x_1} & 0 \\ x_1^2w_{21} & 0 & x_3^2w_{23} & 1 \\ 0 & 0 & x_3w_{33} & 0 \\ 0 & 0 & 0 & 1 \end{pmatrix}$
$\begin{pmatrix} 0 & w_{12} & w_{13} & 1 \\ w_{21} & 0 & w_{23} & 1 \\ 0 & 0 & w_{33} & 0 \\ 0 & 0 & 0 & 1 \end{pmatrix}$	$\begin{pmatrix} 1 & 0 & 0 & 0 \\ 0 & 1 & 0 & 0 \\ 0 & 0 & x_3 & 0 \\ 0 & 0 & 0 & 1 \end{pmatrix}$	$\begin{pmatrix} 0 & w_{12} & x_3^2w_{13} & 1 \\ w_{21} & 0 & x_3^2w_{23} & 1 \\ 0 & 0 & x_3w_{33} & 0 \\ 0 & 0 & 0 & 1 \end{pmatrix}$
$\begin{pmatrix} 0 & w_{12} & w_{13} & 1 \\ w_{21} & 0 & w_{23} & 1 \\ 0 & 0 & w_{33} & 0 \\ 0 & 0 & 0 & 1 \end{pmatrix}$	$\begin{pmatrix} 0 & 1 & 0 & 0 \\ 1 & 0 & 0 & 0 \\ 0 & 0 & x_3 & 0 \\ 0 & 0 & 0 & 1 \end{pmatrix}$	$\begin{pmatrix} 0 & w_{21} & x_3^2w_{23} & 1 \\ w_{12} & 0 & x_3^2w_{13} & 1 \\ 0 & 0 & x_3w_{33} & 0 \\ 0 & 0 & 0 & 1 \end{pmatrix}$
$\begin{pmatrix} 0 & w_{12} & w_{13} & 1 \\ w_{21} & 0 & w_{23} & 1 \\ 0 & 0 & w_{33} & 0 \\ 0 & 0 & 0 & 1 \end{pmatrix}$	$\begin{pmatrix} x_3^2w_{13} & 0 & 0 & 0 \\ 0 & x_3^2w_{23} & 0 & 0 \\ 0 & 0 & 0 & x_3 \\ 0 & 0 & x_4 & 0 \end{pmatrix}$	$\begin{pmatrix} 0 & \frac{x_3^2w_{12}w_{23}^2}{w_{13}} & \frac{x_4^2}{x_3^2w_{13}} & 1 \\ \frac{x_3^2w_{13}^2w_{21}}{w_{23}} & 0 & \frac{x_4^2}{x_3^2w_{23}} & 1 \\ 0 & 0 & x_4 & 0 \\ 0 & 0 & 0 & 1 \end{pmatrix}$
$\begin{pmatrix} 0 & w_{12} & w_{13} & 1 \\ w_{21} & 0 & w_{23} & 1 \\ 0 & 0 & w_{33} & 0 \\ 0 & 0 & 0 & 1 \end{pmatrix}$	$\begin{pmatrix} 0 & x_3^2w_{13} & 0 & 0 \\ x_3^2w_{23} & 0 & 0 & 0 \\ 0 & 0 & 0 & x_3 \\ 0 & 0 & x_4 & 0 \end{pmatrix}$	$\begin{pmatrix} 0 & \frac{x_3^2w_{13}^2w_{21}}{w_{23}} & \frac{x_4^2}{x_3^2w_{23}} & 1 \\ \frac{x_3^2w_{12}w_{23}^2}{w_{13}} & 0 & \frac{x_4^2}{x_3^2w_{13}} & 1 \\ 0 & 0 & x_4 & 0 \\ 0 & 0 & 0 & 1 \end{pmatrix}$
$\begin{pmatrix} w_{11} & w_{12} & w_{13} & 1 \\ w_{21} & w_{22} & 0 & 0 \\ 0 & 0 & w_{33} & 0 \\ 0 & 0 & 0 & 1 \end{pmatrix}$	$\begin{pmatrix} 1 & 0 & 0 & 0 \\ 0 & x_2 & 0 & 0 \\ 0 & 0 & x_3 & 0 \\ 0 & 0 & 0 & 1 \end{pmatrix}$	$\begin{pmatrix} w_{11} & x_2^2w_{12} & x_3^2w_{13} & 1 \\ \frac{w_{21}}{x_2} & x_2w_{22} & 0 & 0 \\ 0 & 0 & x_3w_{33} & 0 \\ 0 & 0 & 0 & 1 \end{pmatrix}$

M_B	$P_{B'B}$	M'_B
$\begin{pmatrix} w_{11} & w_{12} & w_{13} & 1 \\ w_{21} & w_{22} & 0 & 0 \\ 0 & 0 & w_{33} & 0 \\ 0 & 0 & 0 & 1 \end{pmatrix}$	$\begin{pmatrix} x_3^2w_{13} & 0 & 0 & 0 \\ 0 & x_2 & 0 & 0 \\ 0 & 0 & 0 & x_3 \\ 0 & 0 & x_4 & 0 \end{pmatrix}$	$\begin{pmatrix} x_3^2w_{11}w_{13} & \frac{x_2^2w_{12}}{x_3^2w_{13}} & \frac{x_4^2}{x_3^2w_{13}} & 1 \\ \frac{x_3^4w_{13}^2w_{21}}{x_2} & x_2w_{22} & 0 & 0 \\ 0 & 0 & x_4 & 0 \\ 0 & 0 & 0 & 1 \end{pmatrix}$
$\begin{pmatrix} w_{11} & w_{12} & w_{13} & 0 \\ w_{21} & w_{22} & 0 & 1 \\ 0 & 0 & w_{33} & 0 \\ 0 & 0 & 0 & 1 \end{pmatrix}$	$\begin{pmatrix} x_1 & 0 & 0 & 0 \\ 0 & 1 & 0 & 0 \\ 0 & 0 & x_3 & 0 \\ 0 & 0 & 0 & 1 \end{pmatrix}$	$\begin{pmatrix} x_1w_{11} & \frac{w_{12}}{x_1} & \frac{x_3^2w_{13}}{x_1} & 0 \\ x_1^2w_{21} & w_{22} & 0 & 1 \\ 0 & 0 & x_3w_{33} & 0 \\ 0 & 0 & 0 & 1 \end{pmatrix}$
$\begin{pmatrix} w_{11} & w_{12} & w_{13} & 0 \\ w_{21} & w_{22} & 0 & 1 \\ 0 & 0 & w_{33} & 0 \\ 0 & 0 & 0 & 1 \end{pmatrix}$	$\begin{pmatrix} 0 & x_3^2w_{13} & 0 & 0 \\ x_2 & 0 & 0 & 0 \\ 0 & 0 & 0 & x_3 \\ 0 & 0 & x_4 & 0 \end{pmatrix}$	$\begin{pmatrix} x_2w_{22} & \frac{x_3^4w_{13}^2w_{21}}{x_2} & \frac{x_4^2}{x_2} & 0 \\ \frac{x_2^2w_{12}}{x_3^2w_{13}} & x_3^2w_{11}w_{13} & 0 & 1 \\ 0 & 0 & x_4 & 0 \\ 0 & 0 & 0 & 1 \end{pmatrix}$
$\begin{pmatrix} w_{11} & w_{12} & w_{13} & 1 \\ w_{21} & w_{22} & 0 & 1 \\ 0 & 0 & w_{33} & 0 \\ 0 & 0 & 0 & 1 \end{pmatrix}$	$\begin{pmatrix} 1 & 0 & 0 & 0 \\ 0 & 1 & 0 & 0 \\ 0 & 0 & x_3 & 0 \\ 0 & 0 & 0 & 1 \end{pmatrix}$	$\begin{pmatrix} w_{11} & w_{12} & x_3^2w_{13} & 1 \\ w_{21} & w_{22} & 0 & 1 \\ 0 & 0 & x_3w_{33} & 0 \\ 0 & 0 & 0 & 1 \end{pmatrix}$
$\begin{pmatrix} w_{11} & 0 & w_{13} & 1 \\ w_{21} & w_{22} & w_{23} & 1 \\ 0 & 0 & w_{33} & 0 \\ 0 & 0 & 0 & 1 \end{pmatrix}$	$\begin{pmatrix} 1 & 0 & 0 & 0 \\ 0 & 1 & 0 & 0 \\ 0 & 0 & x_3 & 0 \\ 0 & 0 & 0 & 1 \end{pmatrix}$	$\begin{pmatrix} w_{11} & 0 & x_3^2w_{13} & 1 \\ w_{21} & w_{22} & x_3^2w_{23} & 1 \\ 0 & 0 & x_3w_{33} & 0 \\ 0 & 0 & 0 & 1 \end{pmatrix}$
$\begin{pmatrix} w_{11} & 0 & w_{13} & 1 \\ w_{21} & w_{22} & w_{23} & 1 \\ 0 & 0 & w_{33} & 0 \\ 0 & 0 & 0 & 1 \end{pmatrix}$	$\begin{pmatrix} x_3^2w_{13} & 0 & 0 & 0 \\ 0 & x_3^2w_{23} & 0 & 0 \\ 0 & 0 & 0 & x_3 \\ 0 & 0 & x_4 & 0 \end{pmatrix}$	$\begin{pmatrix} x_3^2w_{11}w_{13} & 0 & \frac{x_4^2}{x_3^2w_{13}} & 1 \\ \frac{x_3^2w_{13}^2w_{21}}{w_{23}} & x_3^2w_{22}w_{23} & \frac{x_4^2}{x_3^2w_{23}} & 1 \\ 0 & 0 & x_4 & 0 \\ 0 & 0 & 0 & 1 \end{pmatrix}$
$\begin{pmatrix} w_{11} & w_{12} & w_{13} & 1 \\ w_{21} & 0 & w_{23} & 1 \\ 0 & 0 & w_{33} & 0 \\ 0 & 0 & 0 & 1 \end{pmatrix}$	$\begin{pmatrix} 1 & 0 & 0 & 0 \\ 0 & 1 & 0 & 0 \\ 0 & 0 & x_3 & 0 \\ 0 & 0 & 0 & 1 \end{pmatrix}$	$\begin{pmatrix} w_{11} & w_{12} & x_3^2w_{13} & 1 \\ w_{21} & 0 & x_3^2w_{23} & 1 \\ 0 & 0 & x_3w_{33} & 0 \\ 0 & 0 & 0 & 1 \end{pmatrix}$
$\begin{pmatrix} w_{11} & w_{12} & w_{13} & 1 \\ w_{21} & 0 & w_{23} & 1 \\ 0 & 0 & w_{33} & 0 \\ 0 & 0 & 0 & 1 \end{pmatrix}$	$\begin{pmatrix} x_3^2w_{13} & 0 & 0 & 0 \\ 0 & x_3^2w_{23} & 0 & 0 \\ 0 & 0 & 0 & x_3 \\ 0 & 0 & x_4 & 0 \end{pmatrix}$	$\begin{pmatrix} x_3^2w_{11}w_{13} & \frac{x_3^2w_{12}w_{23}^2}{w_{13}} & \frac{x_4^2}{x_3^2w_{13}} & 1 \\ \frac{x_3^2w_{13}^2w_{21}}{w_{23}} & 0 & \frac{x_4^2}{x_3^2w_{23}} & 1 \\ 0 & 0 & x_4 & 0 \\ 0 & 0 & 0 & 1 \end{pmatrix}$
$\begin{pmatrix} w_{11} & w_{12} & w_{13} & 1 \\ w_{21} & w_{22} & w_{23} & 1 \\ 0 & 0 & w_{33} & 0 \\ 0 & 0 & 0 & 1 \end{pmatrix}$	$\begin{pmatrix} 1 & 0 & 0 & 0 \\ 0 & 1 & 0 & 0 \\ 0 & 0 & x_3 & 0 \\ 0 & 0 & 0 & 1 \end{pmatrix}$	$\begin{pmatrix} w_{11} & w_{12} & x_3^2w_{13} & 1 \\ w_{21} & w_{22} & x_3^2w_{23} & 1 \\ 0 & 0 & x_3w_{33} & 0 \\ 0 & 0 & 0 & 1 \end{pmatrix}$
$\begin{pmatrix} w_{11} & w_{12} & w_{13} & 1 \\ w_{21} & w_{22} & w_{23} & 1 \\ 0 & 0 & w_{33} & 0 \\ 0 & 0 & 0 & 1 \end{pmatrix}$	$\begin{pmatrix} 0 & 1 & 0 & 0 \\ 1 & 0 & 0 & 0 \\ 0 & 0 & x_3 & 0 \\ 0 & 0 & 0 & 1 \end{pmatrix}$	$\begin{pmatrix} w_{22} & w_{21} & x_3^2w_{23} & 1 \\ w_{12} & w_{11} & x_3^2w_{13} & 1 \\ 0 & 0 & x_3w_{33} & 0 \\ 0 & 0 & 0 & 1 \end{pmatrix}$

M_B	$P_{B'B}$	M'_B
$\begin{pmatrix} w_{11} & w_{12} & w_{13} & 1 \\ w_{21} & w_{22} & w_{23} & 1 \\ 0 & 0 & w_{33} & 0 \\ 0 & 0 & 0 & 1 \end{pmatrix}$	$\begin{pmatrix} x_3^2w_{13} & 0 & 0 & 0 \\ 0 & x_3^2w_{23} & 0 & 0 \\ 0 & 0 & 0 & x_3 \\ 0 & 0 & x_4 & 0 \end{pmatrix}$	$\begin{pmatrix} x_3^2w_{11}w_{13} & \frac{x_3^2w_{12}w_{23}^2}{w_{13}} & \frac{x_4^2}{x_3^2w_{13}} & 1 \\ \frac{x_3^2w_{13}^2w_{21}}{w_{23}} & x_3^2w_{22}w_{23} & \frac{x_4^2}{x_3^2w_{23}} & 1 \\ 0 & 0 & x_4 & 0 \\ 0 & 0 & 0 & 1 \end{pmatrix}$
$\begin{pmatrix} w_{11} & w_{12} & w_{13} & 1 \\ w_{21} & w_{22} & w_{23} & 1 \\ 0 & 0 & w_{33} & 0 \\ 0 & 0 & 0 & 1 \end{pmatrix}$	$\begin{pmatrix} 0 & x_3^2w_{13} & 0 & 0 \\ x_3^2w_{23} & 0 & 0 & 0 \\ 0 & 0 & 0 & x_3 \\ 0 & 0 & x_4 & 0 \end{pmatrix}$	$\begin{pmatrix} x_3^2w_{22}w_{23} & \frac{x_3^2w_{13}^2w_{21}}{w_{23}} & \frac{x_4^2}{x_3^2w_{23}} & 1 \\ \frac{x_3^2w_{12}w_{23}^2}{w_{13}} & x_3^2w_{11}w_{13} & \frac{x_4^2}{x_3^2w_{13}} & 1 \\ 0 & 0 & x_4 & 0 \\ 0 & 0 & 0 & 1 \end{pmatrix}$

The Algebra A Does Not Satisfy Condition (3, 2, 3)

M_B	$P_{B'B}$	M'_B
$\begin{pmatrix} w_{11} & 0 & 0 & 1 \\ 0 & w_{22} & 0 & 1 \\ 0 & 0 & w_{33} & 1 \\ 0 & 0 & 0 & 1 \end{pmatrix}$	$\begin{pmatrix} 0 & 1 & 0 & 0 \\ 1 & 0 & 0 & 0 \\ 0 & 0 & 1 & 0 \\ 0 & 0 & 0 & 1 \end{pmatrix}$	$\begin{pmatrix} w_{22} & 0 & 0 & 1 \\ 0 & w_{11} & 0 & 1 \\ 0 & 0 & w_{33} & 1 \\ 0 & 0 & 0 & 1 \end{pmatrix}$
$\begin{pmatrix} w_{11} & 0 & 0 & 1 \\ 0 & w_{22} & 0 & 1 \\ 0 & 0 & w_{33} & 1 \\ 0 & 0 & 0 & 1 \end{pmatrix}$	$\begin{pmatrix} 0 & 0 & 1 & 0 \\ 1 & 0 & 0 & 0 \\ 0 & 1 & 0 & 0 \\ 0 & 0 & 0 & 1 \end{pmatrix}$	$\begin{pmatrix} w_{22} & 0 & 0 & 1 \\ 0 & w_{33} & 0 & 1 \\ 0 & 0 & w_{11} & 1 \\ 0 & 0 & 0 & 1 \end{pmatrix}$
$\begin{pmatrix} w_{11} & 0 & 0 & 1 \\ 0 & w_{22} & 0 & 1 \\ 0 & 0 & w_{33} & 1 \\ 0 & 0 & 0 & 1 \end{pmatrix}$	$\begin{pmatrix} 0 & 0 & 1 & 0 \\ 0 & 1 & 0 & 0 \\ 1 & 0 & 0 & 0 \\ 0 & 0 & 0 & 1 \end{pmatrix}$	$\begin{pmatrix} w_{33} & 0 & 0 & 1 \\ 0 & w_{22} & 0 & 1 \\ 0 & 0 & w_{11} & 1 \\ 0 & 0 & 0 & 1 \end{pmatrix}$
$\begin{pmatrix} w_{11} & 0 & 0 & 1 \\ 0 & w_{22} & 0 & 1 \\ 0 & 0 & w_{33} & 1 \\ 0 & 0 & 0 & 1 \end{pmatrix}$	$\begin{pmatrix} 1 & 0 & 0 & 0 \\ 0 & 0 & 1 & 0 \\ 0 & 1 & 0 & 0 \\ 0 & 0 & 0 & 1 \end{pmatrix}$	$\begin{pmatrix} w_{11} & 0 & 0 & 1 \\ 0 & w_{33} & 0 & 1 \\ 0 & 0 & w_{22} & 1 \\ 0 & 0 & 0 & 1 \end{pmatrix}$
$\begin{pmatrix} w_{11} & 0 & 0 & 1 \\ 0 & w_{22} & 0 & 1 \\ 0 & 0 & w_{33} & 1 \\ 0 & 0 & 0 & 1 \end{pmatrix}$	$\begin{pmatrix} 0 & 1 & 0 & 0 \\ 0 & 0 & 1 & 0 \\ 1 & 0 & 0 & 0 \\ 0 & 0 & 0 & 1 \end{pmatrix}$	$\begin{pmatrix} w_{33} & 0 & 0 & 1 \\ 0 & w_{11} & 0 & 1 \\ 0 & 0 & w_{22} & 1 \\ 0 & 0 & 0 & 1 \end{pmatrix}$
$\begin{pmatrix} 0 & w_{12} & 0 & 1 \\ w_{21} & 0 & 0 & 1 \\ 0 & 0 & w_{33} & 1 \\ 0 & 0 & 0 & 1 \end{pmatrix}$	$\begin{pmatrix} 0 & 1 & 0 & 0 \\ 1 & 0 & 0 & 0 \\ 0 & 0 & 1 & 0 \\ 0 & 0 & 0 & 1 \end{pmatrix}$	$\begin{pmatrix} 0 & w_{21} & 0 & 1 \\ w_{12} & 0 & 0 & 1 \\ 0 & 0 & w_{33} & 1 \\ 0 & 0 & 0 & 1 \end{pmatrix}$
$\begin{pmatrix} w_{11} & w_{12} & w_{13} & 1 \\ 0 & w_{22} & 0 & 1 \\ 0 & 0 & w_{33} & 1 \\ 0 & 0 & 0 & 1 \end{pmatrix}$	$\begin{pmatrix} 1 & 0 & 0 & 0 \\ 0 & 0 & 1 & 0 \\ 0 & 1 & 0 & 0 \\ 0 & 0 & 0 & 1 \end{pmatrix}$	$\begin{pmatrix} w_{11} & w_{13} & w_{12} & 1 \\ 0 & w_{33} & 0 & 1 \\ 0 & 0 & w_{22} & 1 \\ 0 & 0 & 0 & 1 \end{pmatrix}$
$\begin{pmatrix} w_{11} & w_{12} & 0 & 1 \\ w_{21} & w_{22} & 0 & 1 \\ 0 & 0 & w_{33} & 1 \\ 0 & 0 & 0 & 1 \end{pmatrix}$	$\begin{pmatrix} 0 & 1 & 0 & 0 \\ 1 & 0 & 0 & 0 \\ 0 & 0 & 1 & 0 \\ 0 & 0 & 0 & 1 \end{pmatrix}$	$\begin{pmatrix} w_{22} & w_{21} & 0 & 1 \\ w_{12} & w_{11} & 0 & 1 \\ 0 & 0 & w_{33} & 1 \\ 0 & 0 & 0 & 1 \end{pmatrix}$
$\begin{pmatrix} w_{11} & 0 & w_{13} & 1 \\ 0 & w_{22} & w_{23} & 1 \\ 0 & 0 & w_{33} & 1 \\ 0 & 0 & 0 & 1 \end{pmatrix}$	$\begin{pmatrix} 0 & 1 & 0 & 0 \\ 1 & 0 & 0 & 0 \\ 0 & 0 & 1 & 0 \\ 0 & 0 & 0 & 1 \end{pmatrix}$	$\begin{pmatrix} w_{22} & 0 & w_{23} & 1 \\ 0 & w_{11} & w_{13} & 1 \\ 0 & 0 & w_{33} & 1 \\ 0 & 0 & 0 & 1 \end{pmatrix}$
$\begin{pmatrix} 0 & w_{12} & w_{13} & 1 \\ w_{21} & 0 & w_{23} & 1 \\ 0 & 0 & w_{33} & 1 \\ 0 & 0 & 0 & 1 \end{pmatrix}$	$\begin{pmatrix} 0 & 1 & 0 & 0 \\ 1 & 0 & 0 & 0 \\ 0 & 0 & 1 & 0 \\ 0 & 0 & 0 & 1 \end{pmatrix}$	$\begin{pmatrix} 0 & w_{21} & w_{23} & 1 \\ w_{12} & 0 & w_{13} & 1 \\ 0 & 0 & w_{33} & 1 \\ 0 & 0 & 0 & 1 \end{pmatrix}$

M_B	$P_{B'B}$	M'_B
$\begin{pmatrix} w_{11} & w_{12} & w_{13} & 1 \\ w_{21} & w_{22} & w_{23} & 1 \\ 0 & 0 & w_{33} & 1 \\ 0 & 0 & 0 & 1 \end{pmatrix}$	$\begin{pmatrix} 0 & 1 & 0 & 0 \\ 1 & 0 & 0 & 0 \\ 0 & 0 & 1 & 0 \\ 0 & 0 & 0 & 1 \end{pmatrix}$	$\begin{pmatrix} w_{22} & w_{21} & w_{23} & 1 \\ w_{12} & w_{11} & w_{13} & 1 \\ 0 & 0 & w_{33} & 1 \\ 0 & 0 & 0 & 1 \end{pmatrix}$
$\begin{pmatrix} 0 & w_{12} & w_{13} & 0 \\ w_{21} & 0 & 0 & 0 \\ 0 & 0 & w_{33} & 1 \\ 0 & 0 & 0 & 1 \end{pmatrix}$	$\begin{pmatrix} x_1 & 0 & 0 & 0 \\ 0 & x_2 & 0 & 0 \\ 0 & 0 & 1 & 0 \\ 0 & 0 & 0 & 1 \end{pmatrix}$	$\begin{pmatrix} 0 & \frac{x_2^2w_{12}}{x_1} & \frac{w_{13}}{x_1} & 0 \\ \frac{x_1^2w_{21}}{x_2} & 0 & 0 & 0 \\ 0 & 0 & w_{33} & 1 \\ 0 & 0 & 0 & 1 \end{pmatrix}$
$\begin{pmatrix} w_{11} & w_{12} & 0 & 0 \\ 0 & w_{22} & w_{23} & 0 \\ 0 & 0 & w_{33} & 1 \\ 0 & 0 & 0 & 1 \end{pmatrix}$	$\begin{pmatrix} x_1 & 0 & 0 & 0 \\ 0 & x_2 & 0 & 0 \\ 0 & 0 & 1 & 0 \\ 0 & 0 & 0 & 1 \end{pmatrix}$	$\begin{pmatrix} x_1w_{11} & \frac{x_2^2w_{12}}{x_1} & 0 & 0 \\ 0 & x_2w_{22} & \frac{w_{23}}{x_2} & 0 \\ 0 & 0 & w_{33} & 1 \\ 0 & 0 & 0 & 1 \end{pmatrix}$
$\begin{pmatrix} w_{11} & 0 & w_{13} & 0 \\ 0 & w_{22} & w_{23} & 0 \\ 0 & 0 & w_{33} & 1 \\ 0 & 0 & 0 & 1 \end{pmatrix}$	$\begin{pmatrix} x_1 & 0 & 0 & 0 \\ 0 & x_2 & 0 & 0 \\ 0 & 0 & 1 & 0 \\ 0 & 0 & 0 & 1 \end{pmatrix}$	$\begin{pmatrix} x_1w_{11} & 0 & \frac{w_{13}}{x_1} & 0 \\ 0 & x_2w_{22} & \frac{w_{23}}{x_2} & 0 \\ 0 & 0 & w_{33} & 1 \\ 0 & 0 & 0 & 1 \end{pmatrix}$
$\begin{pmatrix} w_{11} & 0 & w_{13} & 0 \\ 0 & w_{22} & w_{23} & 0 \\ 0 & 0 & w_{33} & 1 \\ 0 & 0 & 0 & 1 \end{pmatrix}$	$\begin{pmatrix} 0 & x_1 & 0 & 0 \\ x_2 & 0 & 0 & 0 \\ 0 & 0 & 1 & 0 \\ 0 & 0 & 0 & 1 \end{pmatrix}$	$\begin{pmatrix} x_2w_{22} & 0 & \frac{w_{23}}{x_2} & 0 \\ 0 & x_1w_{11} & \frac{w_{13}}{x_1} & 0 \\ 0 & 0 & w_{33} & 1 \\ 0 & 0 & 0 & 1 \end{pmatrix}$
$\begin{pmatrix} w_{11} & w_{12} & w_{13} & 0 \\ w_{21} & 0 & 0 & 0 \\ 0 & 0 & w_{33} & 1 \\ 0 & 0 & 0 & 1 \end{pmatrix}$	$\begin{pmatrix} x_1 & 0 & 0 & 0 \\ 0 & x_2 & 0 & 0 \\ 0 & 0 & 1 & 0 \\ 0 & 0 & 0 & 1 \end{pmatrix}$	$\begin{pmatrix} x_1w_{11} & \frac{x_2^2w_{12}}{x_1} & \frac{w_{13}}{x_1} & 0 \\ \frac{x_1^2w_{21}}{x_2} & 0 & 0 & 0 \\ 0 & 0 & w_{33} & 1 \\ 0 & 0 & 0 & 1 \end{pmatrix}$
$\begin{pmatrix} w_{11} & w_{12} & 0 & 0 \\ w_{21} & 0 & w_{23} & 0 \\ 0 & 0 & w_{33} & 1 \\ 0 & 0 & 0 & 1 \end{pmatrix}$	$\begin{pmatrix} x_1 & 0 & 0 & 0 \\ 0 & x_2 & 0 & 0 \\ 0 & 0 & 1 & 0 \\ 0 & 0 & 0 & 1 \end{pmatrix}$	$\begin{pmatrix} x_1w_{11} & \frac{x_2^2w_{12}}{x_1} & 0 & 0 \\ \frac{x_1^2w_{21}}{x_2} & 0 & \frac{w_{23}}{x_2} & 0 \\ 0 & 0 & w_{33} & 1 \\ 0 & 0 & 0 & 1 \end{pmatrix}$
$\begin{pmatrix} 0 & w_{12} & w_{13} & 0 \\ w_{21} & 0 & w_{23} & 0 \\ 0 & 0 & w_{33} & 1 \\ 0 & 0 & 0 & 1 \end{pmatrix}$	$\begin{pmatrix} x_1 & 0 & 0 & 0 \\ 0 & x_2 & 0 & 0 \\ 0 & 0 & 1 & 0 \\ 0 & 0 & 0 & 1 \end{pmatrix}$	$\begin{pmatrix} 0 & \frac{x_2^2w_{12}}{x_1} & \frac{w_{13}}{x_1} & 0 \\ \frac{x_1^2w_{21}}{x_2} & 0 & \frac{w_{23}}{x_2} & 0 \\ 0 & 0 & w_{33} & 1 \\ 0 & 0 & 0 & 1 \end{pmatrix}$
$\begin{pmatrix} 0 & w_{12} & w_{13} & 0 \\ w_{21} & 0 & w_{23} & 0 \\ 0 & 0 & w_{33} & 1 \\ 0 & 0 & 0 & 1 \end{pmatrix}$	$\begin{pmatrix} 0 & x_1 & 0 & 0 \\ x_2 & 0 & 0 & 0 \\ 0 & 0 & 1 & 0 \\ 0 & 0 & 0 & 1 \end{pmatrix}$	$\begin{pmatrix} 0 & \frac{x_1^2w_{21}}{x_2} & \frac{w_{23}}{x_2} & 0 \\ \frac{x_2^2w_{12}}{x_1} & 0 & \frac{w_{13}}{x_1} & 0 \\ 0 & 0 & w_{33} & 1 \\ 0 & 0 & 0 & 1 \end{pmatrix}$
$\begin{pmatrix} w_{11} & w_{12} & w_{13} & 0 \\ w_{21} & w_{22} & 0 & 0 \\ 0 & 0 & w_{33} & 1 \\ 0 & 0 & 0 & 1 \end{pmatrix}$	$\begin{pmatrix} x_1 & 0 & 0 & 0 \\ 0 & x_2 & 0 & 0 \\ 0 & 0 & 1 & 0 \\ 0 & 0 & 0 & 1 \end{pmatrix}$	$\begin{pmatrix} x_1w_{11} & \frac{x_2^2w_{12}}{x_1} & \frac{w_{13}}{x_1} & 0 \\ \frac{x_1^2w_{21}}{x_2} & x_2w_{22} & 0 & 0 \\ 0 & 0 & w_{33} & 1 \\ 0 & 0 & 0 & 1 \end{pmatrix}$

M_B	$P_{B'B}$	M'_B
$\begin{pmatrix} w_{11} & w_{12} & w_{13} & 0 \\ w_{21} & w_{22} & 0 & 0 \\ 0 & 0 & w_{33} & 1 \\ 0 & 0 & 0 & 1 \end{pmatrix}$	$\begin{pmatrix} x_1 & 0 & 0 & 0 \\ 0 & x_2 & 0 & 0 \\ 0 & 0 & 1 & 0 \\ 0 & 0 & 0 & 1 \end{pmatrix}$	$\begin{pmatrix} x_1w_{11} & \frac{x_2^2w_{12}}{x_1} & \frac{w_{13}}{x_1} & 0 \\ \frac{x_1^2w_{21}}{x_2} & x_2w_{22} & 0 & 0 \\ 0 & 0 & w_{33} & 1 \\ 0 & 0 & 0 & 1 \end{pmatrix}$
$\begin{pmatrix} w_{11} & w_{12} & w_{13} & 0 \\ w_{21} & 0 & w_{23} & 0 \\ 0 & 0 & w_{33} & 1 \\ 0 & 0 & 0 & 1 \end{pmatrix}$	$\begin{pmatrix} x_1 & 0 & 0 & 0 \\ 0 & x_2 & 0 & 0 \\ 0 & 0 & 1 & 0 \\ 0 & 0 & 0 & 1 \end{pmatrix}$	$\begin{pmatrix} x_1w_{11} & \frac{x_2^2w_{12}}{x_1} & \frac{w_{13}}{x_1} & 0 \\ \frac{x_1^2w_{21}}{x_2} & 0 & \frac{w_{23}}{x_2} & 0 \\ 0 & 0 & w_{33} & 1 \\ 0 & 0 & 0 & 1 \end{pmatrix}$
$\begin{pmatrix} w_{11} & w_{12} & w_{13} & 0 \\ w_{21} & w_{22} & w_{23} & 0 \\ 0 & 0 & w_{33} & 1 \\ 0 & 0 & 0 & 1 \end{pmatrix}$	$\begin{pmatrix} x_1 & 0 & 0 & 0 \\ 0 & x_2 & 0 & 0 \\ 0 & 0 & 1 & 0 \\ 0 & 0 & 0 & 1 \end{pmatrix}$	$\begin{pmatrix} x_1w_{11} & \frac{x_2^2w_{12}}{x_1} & \frac{w_{13}}{x_1} & 0 \\ \frac{x_1^2w_{21}}{x_2} & x_2w_{22} & \frac{w_{23}}{x_2} & 0 \\ 0 & 0 & w_{33} & 1 \\ 0 & 0 & 0 & 1 \end{pmatrix}$
$\begin{pmatrix} w_{11} & w_{12} & w_{13} & 0 \\ w_{21} & w_{22} & w_{23} & 0 \\ 0 & 0 & w_{33} & 1 \\ 0 & 0 & 0 & 1 \end{pmatrix}$	$\begin{pmatrix} 0 & x_1 & 0 & 0 \\ x_2 & 0 & 0 & 0 \\ 0 & 0 & 1 & 0 \\ 0 & 0 & 0 & 1 \end{pmatrix}$	$\begin{pmatrix} x_2w_{22} & \frac{x_1^2w_{21}}{x_2} & \frac{w_{23}}{x_2} & 0 \\ \frac{x_2^2w_{12}}{x_1} & x_1w_{11} & \frac{w_{13}}{x_1} & 0 \\ 0 & 0 & w_{33} & 1 \\ 0 & 0 & 0 & 1 \end{pmatrix}$
$\begin{pmatrix} 0 & w_{12} & 0 & 1 \\ w_{21} & 0 & 0 & 0 \\ 0 & 0 & w_{33} & 1 \\ 0 & 0 & 0 & 1 \end{pmatrix}$	$\begin{pmatrix} 1 & 0 & 0 & 0 \\ 0 & x_2 & 0 & 0 \\ 0 & 0 & 1 & 0 \\ 0 & 0 & 0 & 1 \end{pmatrix}$	$\begin{pmatrix} 0 & x_2^2w_{12} & 0 & 1 \\ \frac{w_{21}}{x_2} & 0 & 0 & 0 \\ 0 & 0 & w_{33} & 1 \\ 0 & 0 & 0 & 1 \end{pmatrix}$
$\begin{pmatrix} w_{11} & w_{12} & 0 & 1 \\ w_{21} & 0 & 0 & 0 \\ 0 & 0 & w_{33} & 1 \\ 0 & 0 & 0 & 1 \end{pmatrix}$	$\begin{pmatrix} 1 & 0 & 0 & 0 \\ 0 & x_2 & 0 & 0 \\ 0 & 0 & 1 & 0 \\ 0 & 0 & 0 & 1 \end{pmatrix}$	$\begin{pmatrix} w_{11} & x_2^2w_{12} & 0 & 1 \\ \frac{w_{21}}{x_2} & 0 & 0 & 0 \\ 0 & 0 & w_{33} & 1 \\ 0 & 0 & 0 & 1 \end{pmatrix}$
$\begin{pmatrix} 0 & w_{12} & w_{13} & 1 \\ w_{21} & 0 & 0 & 0 \\ 0 & 0 & w_{33} & 1 \\ 0 & 0 & 0 & 1 \end{pmatrix}$	$\begin{pmatrix} 1 & 0 & 0 & 0 \\ 0 & x_2 & 0 & 0 \\ 0 & 0 & 1 & 0 \\ 0 & 0 & 0 & 1 \end{pmatrix}$	$\begin{pmatrix} 0 & x_2^2w_{12} & w_{13} & 1 \\ \frac{w_{21}}{x_2} & 0 & 0 & 0 \\ 0 & 0 & w_{33} & 1 \\ 0 & 0 & 0 & 1 \end{pmatrix}$
$\begin{pmatrix} w_{11} & w_{12} & 0 & 1 \\ w_{21} & w_{22} & 0 & 0 \\ 0 & 0 & w_{33} & 1 \\ 0 & 0 & 0 & 1 \end{pmatrix}$	$\begin{pmatrix} 1 & 0 & 0 & 0 \\ 0 & x_2 & 0 & 0 \\ 0 & 0 & 1 & 0 \\ 0 & 0 & 0 & 1 \end{pmatrix}$	$\begin{pmatrix} w_{11} & x_2^2w_{12} & 0 & 1 \\ \frac{w_{21}}{x_2} & x_2w_{22} & 0 & 0 \\ 0 & 0 & w_{33} & 1 \\ 0 & 0 & 0 & 1 \end{pmatrix}$
$\begin{pmatrix} w_{11} & 0 & w_{13} & 1 \\ 0 & w_{22} & w_{23} & 0 \\ 0 & 0 & w_{33} & 1 \\ 0 & 0 & 0 & 1 \end{pmatrix}$	$\begin{pmatrix} 1 & 0 & 0 & 0 \\ 0 & x_2 & 0 & 0 \\ 0 & 0 & 1 & 0 \\ 0 & 0 & 0 & 1 \end{pmatrix}$	$\begin{pmatrix} w_{11} & 0 & w_{13} & 1 \\ 0 & x_2w_{22} & \frac{w_{23}}{x_2} & 0 \\ 0 & 0 & w_{33} & 1 \\ 0 & 0 & 0 & 1 \end{pmatrix}$
$\begin{pmatrix} w_{11} & w_{12} & 0 & 1 \\ 0 & w_{22} & w_{23} & 0 \\ 0 & 0 & w_{33} & 1 \\ 0 & 0 & 0 & 1 \end{pmatrix}$	$\begin{pmatrix} 1 & 0 & 0 & 0 \\ 0 & x_2 & 0 & 0 \\ 0 & 0 & 1 & 0 \\ 0 & 0 & 0 & 1 \end{pmatrix}$	$\begin{pmatrix} w_{11} & x_2^2w_{12} & 0 & 1 \\ 0 & x_2w_{22} & \frac{w_{23}}{x_2} & 0 \\ 0 & 0 & w_{33} & 1 \\ 0 & 0 & 0 & 1 \end{pmatrix}$

M_B	$P_{B'B}$	M'_B	M_B	$P_{B'B}$	M'_B
$\begin{pmatrix}w_{11}&w_{12}&w_{13}&1\\w_{21}&0&0&0\\0&0&w_{33}&1\\0&0&0&1\end{pmatrix}$	$\begin{pmatrix}1&0&0&0\\0&x_2&0&0\\0&0&1&0\\0&0&0&1\end{pmatrix}$	$\begin{pmatrix}w_{11}&x_2^2w_{12}&w_{13}&1\\\frac{w_{21}}{x_2}&0&0&0\\0&0&w_{33}&1\\0&0&0&1\end{pmatrix}$	$\begin{pmatrix}w_{11}&w_{12}&w_{13}&0\\0&w_{22}&0&1\\0&0&w_{33}&1\\0&0&0&1\end{pmatrix}$	$\begin{pmatrix}x_1&0&0&0\\0&1&0&0\\0&0&1&0\\0&0&0&1\end{pmatrix}$	$\begin{pmatrix}x_1w_{11}&\frac{w_{12}}{x_1}&\frac{w_{13}}{x_1}&0\\0&w_{22}&0&1\\0&0&w_{33}&1\\0&0&0&1\end{pmatrix}$
$\begin{pmatrix}w_{11}&w_{12}&0&1\\w_{21}&0&w_{23}&0\\0&0&w_{33}&1\\0&0&0&1\end{pmatrix}$	$\begin{pmatrix}1&0&0&0\\0&x_2&0&0\\0&0&1&0\\0&0&0&1\end{pmatrix}$	$\begin{pmatrix}w_{11}&x_2^2w_{12}&0&1\\\frac{w_{21}}{x_2}&0&\frac{w_{23}}{x_2}&0\\0&0&w_{33}&1\\0&0&0&1\end{pmatrix}$	$\begin{pmatrix}w_{11}&w_{12}&w_{13}&0\\0&w_{22}&0&1\\0&0&w_{33}&1\\0&0&0&1\end{pmatrix}$	$\begin{pmatrix}x_1&0&0&0\\0&0&1&0\\0&1&0&0\\0&0&0&1\end{pmatrix}$	$\begin{pmatrix}x_1w_{11}&\frac{w_{13}}{x_1}&\frac{w_{12}}{x_1}&0\\0&w_{33}&0&1\\0&0&w_{22}&1\\0&0&0&1\end{pmatrix}$
$\begin{pmatrix}0&w_{12}&w_{13}&1\\w_{21}&0&w_{23}&0\\0&0&w_{33}&1\\0&0&0&1\end{pmatrix}$	$\begin{pmatrix}1&0&0&0\\0&x_2&0&0\\0&0&1&0\\0&0&0&1\end{pmatrix}$	$\begin{pmatrix}0&x_2^2w_{12}&w_{13}&1\\\frac{w_{21}}{x_2}&0&\frac{w_{23}}{x_2}&0\\0&0&w_{33}&1\\0&0&0&1\end{pmatrix}$	$\begin{pmatrix}w_{11}&w_{12}&0&0\\0&w_{22}&w_{23}&1\\0&0&w_{33}&1\\0&0&0&1\end{pmatrix}$	$\begin{pmatrix}x_1&0&0&0\\0&1&0&0\\0&0&1&0\\0&0&0&1\end{pmatrix}$	$\begin{pmatrix}x_1w_{11}&\frac{w_{12}}{x_1}&0&0\\0&w_{22}&w_{23}&1\\0&0&w_{33}&1\\0&0&0&1\end{pmatrix}$
$\begin{pmatrix}w_{11}&w_{12}&w_{13}&1\\w_{21}&w_{22}&0&0\\0&0&w_{33}&1\\0&0&0&1\end{pmatrix}$	$\begin{pmatrix}1&0&0&0\\0&x_2&0&0\\0&0&1&0\\0&0&0&1\end{pmatrix}$	$\begin{pmatrix}w_{11}&x_2^2w_{12}&w_{13}&1\\\frac{w_{21}}{x_2}&x_2w_{22}&0&0\\0&0&w_{33}&1\\0&0&0&1\end{pmatrix}$	$\begin{pmatrix}w_{11}&w_{12}&w_{13}&0\\w_{21}&0&0&1\\0&0&w_{33}&1\\0&0&0&1\end{pmatrix}$	$\begin{pmatrix}x_1&0&0&0\\0&1&0&0\\0&0&1&0\\0&0&0&1\end{pmatrix}$	$\begin{pmatrix}x_1w_{11}&\frac{w_{12}}{x_1}&\frac{w_{13}}{x_1}&0\\x_1^2w_{21}&0&0&1\\0&0&w_{33}&1\\0&0&0&1\end{pmatrix}$
$\begin{pmatrix}w_{11}&w_{12}&w_{13}&1\\0&w_{22}&w_{23}&0\\0&0&w_{33}&1\\0&0&0&1\end{pmatrix}$	$\begin{pmatrix}1&0&0&0\\0&x_2&0&0\\0&0&1&0\\0&0&0&1\end{pmatrix}$	$\begin{pmatrix}w_{11}&x_2^2w_{12}&w_{13}&1\\0&x_2w_{22}&\frac{w_{23}}{x_2}&0\\0&0&w_{33}&1\\0&0&0&1\end{pmatrix}$	$\begin{pmatrix}w_{11}&w_{12}&w_{13}&0\\0&w_{22}&0&1\\0&0&w_{33}&1\\0&0&0&1\end{pmatrix}$	$\begin{pmatrix}x_1&0&0&0\\0&1&0&0\\0&0&1&0\\0&0&0&1\end{pmatrix}$	$\begin{pmatrix}x_1w_{11}&\frac{w_{12}}{x_1}&\frac{w_{13}}{x_1}&0\\0&w_{22}&0&1\\0&0&w_{33}&1\\0&0&0&1\end{pmatrix}$
$\begin{pmatrix}w_{11}&w_{12}&w_{13}&1\\w_{21}&0&w_{23}&0\\0&0&w_{33}&1\\0&0&0&1\end{pmatrix}$	$\begin{pmatrix}1&0&0&0\\0&x_2&0&0\\0&0&1&0\\0&0&0&1\end{pmatrix}$	$\begin{pmatrix}w_{11}&x_2^2w_{12}&w_{13}&1\\\frac{w_{21}}{x_2}&0&\frac{w_{23}}{x_2}&0\\0&0&w_{33}&1\\0&0&0&1\end{pmatrix}$	$\begin{pmatrix}w_{11}&w_{12}&w_{13}&0\\0&w_{22}&0&1\\0&0&w_{33}&1\\0&0&0&1\end{pmatrix}$	$\begin{pmatrix}x_1&0&0&0\\0&0&1&0\\0&1&0&0\\0&0&0&1\end{pmatrix}$	$\begin{pmatrix}x_1w_{11}&\frac{w_{13}}{x_1}&\frac{w_{12}}{x_1}&0\\0&w_{33}&0&1\\0&0&w_{22}&1\\0&0&0&1\end{pmatrix}$
$\begin{pmatrix}w_{11}&w_{12}&w_{13}&1\\w_{21}&w_{22}&w_{23}&0\\0&0&w_{33}&1\\0&0&0&1\end{pmatrix}$	$\begin{pmatrix}1&0&0&0\\0&x_2&0&0\\0&0&1&0\\0&0&0&1\end{pmatrix}$	$\begin{pmatrix}w_{11}&x_2^2w_{12}&w_{13}&1\\\frac{w_{21}}{x_2}&x_2w_{22}&\frac{w_{23}}{x_2}&0\\0&0&w_{33}&1\\0&0&0&1\end{pmatrix}$	$\begin{pmatrix}w_{11}&w_{12}&0&0\\w_{21}&0&w_{23}&1\\0&0&w_{33}&1\\0&0&0&1\end{pmatrix}$	$\begin{pmatrix}x_1&0&0&0\\0&1&0&0\\0&0&1&0\\0&0&0&1\end{pmatrix}$	$\begin{pmatrix}x_1w_{11}&\frac{w_{12}}{x_1}&0&0\\x_1^2w_{21}&0&w_{23}&1\\0&0&w_{33}&1\\0&0&0&1\end{pmatrix}$
$\begin{pmatrix}w_{11}&w_{12}&0&0\\0&w_{22}&0&1\\0&0&w_{33}&1\\0&0&0&1\end{pmatrix}$	$\begin{pmatrix}x_1&0&0&0\\0&1&0&0\\0&0&1&0\\0&0&0&1\end{pmatrix}$	$\begin{pmatrix}x_1w_{11}&\frac{w_{12}}{x_1}&0&0\\0&w_{22}&0&1\\0&0&w_{33}&1\\0&0&0&1\end{pmatrix}$	$\begin{pmatrix}w_{11}&w_{12}&w_{13}&0\\w_{21}&w_{22}&0&1\\0&0&w_{33}&1\\0&0&0&1\end{pmatrix}$	$\begin{pmatrix}x_1&0&0&0\\0&1&0&0\\0&0&1&0\\0&0&0&1\end{pmatrix}$	$\begin{pmatrix}x_1w_{11}&\frac{w_{12}}{x_1}&\frac{w_{13}}{x_1}&0\\x_1^2w_{21}&w_{22}&0&1\\0&0&w_{33}&1\\0&0&0&1\end{pmatrix}$
$\begin{pmatrix}w_{11}&w_{12}&0&0\\w_{21}&0&0&1\\0&0&w_{33}&1\\0&0&0&1\end{pmatrix}$	$\begin{pmatrix}x_1&0&0&0\\0&1&0&0\\0&0&1&0\\0&0&0&1\end{pmatrix}$	$\begin{pmatrix}x_1w_{11}&\frac{w_{12}}{x_1}&0&0\\x_1^2w_{21}&0&0&1\\0&0&w_{33}&1\\0&0&0&1\end{pmatrix}$	$\begin{pmatrix}w_{11}&w_{12}&w_{13}&0\\0&w_{22}&w_{23}&1\\0&0&w_{33}&1\\0&0&0&1\end{pmatrix}$	$\begin{pmatrix}x_1&0&0&0\\0&1&0&0\\0&0&1&0\\0&0&0&1\end{pmatrix}$	$\begin{pmatrix}x_1w_{11}&\frac{w_{12}}{x_1}&\frac{w_{13}}{x_1}&0\\0&w_{22}&w_{23}&1\\0&0&w_{33}&1\\0&0&0&1\end{pmatrix}$
$\begin{pmatrix}0&w_{12}&w_{13}&0\\w_{21}&0&0&1\\0&0&w_{33}&1\\0&0&0&1\end{pmatrix}$	$\begin{pmatrix}x_1&0&0&0\\0&1&0&0\\0&0&1&0\\0&0&0&1\end{pmatrix}$	$\begin{pmatrix}0&\frac{w_{12}}{x_1}&\frac{w_{13}}{x_1}&0\\x_1^2w_{21}&0&0&1\\0&0&w_{33}&1\\0&0&0&1\end{pmatrix}$	$\begin{pmatrix}w_{11}&w_{12}&w_{13}&0\\w_{21}&0&w_{23}&1\\0&0&w_{33}&1\\0&0&0&1\end{pmatrix}$	$\begin{pmatrix}x_1&0&0&0\\0&1&0&0\\0&0&1&0\\0&0&0&1\end{pmatrix}$	$\begin{pmatrix}x_1w_{11}&\frac{w_{12}}{x_1}&\frac{w_{13}}{x_1}&0\\x_1^2w_{21}&0&w_{23}&1\\0&0&w_{33}&1\\0&0&0&1\end{pmatrix}$

4.2 No 3-Basic Ideal of A Has a 2-Basic Ideal

We divide this subsection into different items depending on the number of zero entries in the matrix U given at the beginning of the section.

The Matrix U Has One Nonzero Entry

M_B	$P_{B'B}$	M'_B
$\begin{pmatrix}0&0&w_{13}&0\\w_{21}&0&0&0\\0&w_{32}&0&1\\0&0&0&1\end{pmatrix}$	$\begin{pmatrix}x_1&0&0&0\\0&x_2&0&0\\0&0&1&0\\0&0&0&1\end{pmatrix}$	$\begin{pmatrix}0&0&\frac{w_{13}}{x_1}&0\\\frac{x_1^2w_{21}}{x_2}&0&0&0\\0&x_2^2w_{32}&0&1\\0&0&0&1\end{pmatrix}$
$\begin{pmatrix}0&w_{12}&0&0\\w_{21}&0&0&0\\w_{31}&0&w_{33}&1\\0&0&0&1\end{pmatrix}$	$\begin{pmatrix}x_1&0&0&0\\0&x_2&0&0\\0&0&1&0\\0&0&0&1\end{pmatrix}$	$\begin{pmatrix}0&\frac{x_2^2w_{12}}{x_1}&0&0\\\frac{x_1^2w_{21}}{x_2}&0&0&0\\x_1^2w_{31}&0&w_{33}&1\\0&0&0&1\end{pmatrix}$
$\begin{pmatrix}w_{11}&0&w_{13}&0\\0&0&w_{23}&0\\0&w_{32}&0&1\\0&0&0&1\end{pmatrix}$	$\begin{pmatrix}x_1&0&0&0\\0&x_2&0&0\\0&0&1&0\\0&0&0&1\end{pmatrix}$	$\begin{pmatrix}x_1w_{11}&0&\frac{w_{13}}{x_1}&0\\0&0&\frac{w_{23}}{x_2}&0\\0&x_2^2w_{32}&0&1\\0&0&0&1\end{pmatrix}$
$\begin{pmatrix}0&0&w_{13}&0\\w_{21}&w_{22}&0&0\\w_{31}&0&0&1\\0&0&0&1\end{pmatrix}$	$\begin{pmatrix}x_1&0&0&0\\0&x_2&0&0\\0&0&1&0\\0&0&0&1\end{pmatrix}$	$\begin{pmatrix}0&0&\frac{w_{13}}{x_1}&0\\\frac{x_1^2w_{21}}{x_2}&x_2w_{22}&0&0\\x_1^2w_{31}&0&0&1\\0&0&0&1\end{pmatrix}$
$\begin{pmatrix}w_{11}&0&w_{13}&0\\w_{21}&0&0&0\\0&w_{32}&0&1\\0&0&0&1\end{pmatrix}$	$\begin{pmatrix}x_1&0&0&0\\0&x_2&0&0\\0&0&1&0\\0&0&0&1\end{pmatrix}$	$\begin{pmatrix}x_1w_{11}&0&\frac{w_{13}}{x_1}&0\\\frac{x_1^2w_{21}}{x_2}&0&0&0\\0&x_2^2w_{32}&0&1\\0&0&0&1\end{pmatrix}$
$\begin{pmatrix}0&w_{12}&0&0\\0&0&w_{23}&0\\w_{31}&0&w_{33}&1\\0&0&0&1\end{pmatrix}$	$\begin{pmatrix}x_1&0&0&0\\0&x_2&0&0\\0&0&1&0\\0&0&0&1\end{pmatrix}$	$\begin{pmatrix}0&\frac{x_2^2w_{12}}{x_1}&0&0\\0&0&\frac{w_{23}}{x_2}&0\\x_1^2w_{31}&0&w_{33}&1\\0&0&0&1\end{pmatrix}$
$\begin{pmatrix}w_{11}&w_{12}&0&0\\0&0&w_{23}&0\\w_{31}&0&0&1\\0&0&0&1\end{pmatrix}$	$\begin{pmatrix}x_1&0&0&0\\0&x_2&0&0\\0&0&1&0\\0&0&0&1\end{pmatrix}$	$\begin{pmatrix}x_1w_{11}&\frac{x_2^2w_{12}}{x_1}&0&0\\0&0&\frac{w_{23}}{x_2}&0\\x_1^2w_{31}&0&0&1\\0&0&0&1\end{pmatrix}$
$\begin{pmatrix}0&w_{12}&w_{13}&0\\w_{21}&0&0&0\\0&w_{32}&0&1\\0&0&0&1\end{pmatrix}$	$\begin{pmatrix}x_1&0&0&0\\0&x_2&0&0\\0&0&1&0\\0&0&0&1\end{pmatrix}$	$\begin{pmatrix}0&\frac{x_2^2w_{12}}{x_1}&\frac{w_{13}}{x_1}&0\\\frac{x_1^2w_{21}}{x_2}&0&0&0\\0&x_2^2w_{32}&0&1\\0&0&0&1\end{pmatrix}$
$\begin{pmatrix}0&w_{12}&0&0\\0&0&w_{23}&0\\w_{31}&w_{32}&0&1\\0&0&0&1\end{pmatrix}$	$\begin{pmatrix}x_1&0&0&0\\0&x_2&0&0\\0&0&1&0\\0&0&0&1\end{pmatrix}$	$\begin{pmatrix}0&\frac{x_2^2w_{12}}{x_1}&0&0\\0&0&\frac{w_{23}}{x_2}&0\\x_1^2w_{31}&x_2^2w_{32}&0&1\\0&0&0&1\end{pmatrix}$
$\begin{pmatrix}0&w_{12}&w_{13}&0\\0&0&w_{23}&0\\w_{31}&0&0&1\\0&0&0&1\end{pmatrix}$	$\begin{pmatrix}x_1&0&0&0\\0&x_2&0&0\\0&0&1&0\\0&0&0&1\end{pmatrix}$	$\begin{pmatrix}0&\frac{x_2^2w_{12}}{x_1}&\frac{w_{13}}{x_1}&0\\0&0&\frac{w_{23}}{x_2}&0\\x_1^2w_{31}&0&0&1\\0&0&0&1\end{pmatrix}$

M_B	$P_{B'B}$	M'_B
$\begin{pmatrix}w_{11}&w_{12}&0&0\\w_{21}&0&0&0\\w_{31}&0&w_{33}&1\\0&0&0&1\end{pmatrix}$	$\begin{pmatrix}x_1&0&0&0\\0&x_2&0&0\\0&0&1&0\\0&0&0&1\end{pmatrix}$	$\begin{pmatrix}x_1w_{11}&\frac{x_2^2w_{12}}{x_1}&0&0\\\frac{x_1^2w_{21}}{x_2}&0&0&0\\x_1^2w_{31}&0&w_{33}&1\\0&0&0&1\end{pmatrix}$
$\begin{pmatrix}w_{11}&0&w_{13}&0\\0&0&w_{23}&0\\0&w_{32}&w_{33}&1\\0&0&0&1\end{pmatrix}$	$\begin{pmatrix}x_1&0&0&0\\0&x_2&0&0\\0&0&1&0\\0&0&0&1\end{pmatrix}$	$\begin{pmatrix}x_1w_{11}&0&\frac{w_{13}}{x_1}&0\\0&0&\frac{w_{23}}{x_2}&0\\0&x_2^2w_{32}&w_{33}&1\\0&0&0&1\end{pmatrix}$
$\begin{pmatrix}w_{11}&0&w_{13}&0\\w_{21}&w_{22}&0&0\\w_{31}&0&0&1\\0&0&0&1\end{pmatrix}$	$\begin{pmatrix}x_1&0&0&0\\0&x_2&0&0\\0&0&1&0\\0&0&0&1\end{pmatrix}$	$\begin{pmatrix}x_1w_{11}&0&\frac{w_{13}}{x_1}&0\\\frac{x_1^2w_{21}}{x_2}&x_2w_{22}&0&0\\x_1^2w_{31}&0&0&1\\0&0&0&1\end{pmatrix}$
$\begin{pmatrix}w_{11}&w_{12}&0&0\\w_{21}&0&0&0\\0&w_{32}&w_{33}&1\\0&0&0&1\end{pmatrix}$	$\begin{pmatrix}x_1&0&0&0\\0&x_2&0&0\\0&0&1&0\\0&0&0&1\end{pmatrix}$	$\begin{pmatrix}x_1w_{11}&\frac{x_2^2w_{12}}{x_1}&0&0\\\frac{x_1^2w_{21}}{x_2}&0&0&0\\0&x_2^2w_{32}&w_{33}&1\\0&0&0&1\end{pmatrix}$
$\begin{pmatrix}w_{11}&w_{12}&0&0\\0&0&w_{23}&0\\0&w_{32}&w_{33}&1\\0&0&0&1\end{pmatrix}$	$\begin{pmatrix}x_1&0&0&0\\0&x_2&0&0\\0&0&1&0\\0&0&0&1\end{pmatrix}$	$\begin{pmatrix}x_1w_{11}&\frac{x_2^2w_{12}}{x_1}&0&0\\0&0&\frac{w_{23}}{x_2}&0\\0&x_2^2w_{32}&w_{33}&1\\0&0&0&1\end{pmatrix}$
$\begin{pmatrix}w_{11}&0&w_{13}&0\\0&w_{22}&w_{23}&0\\w_{31}&0&0&1\\0&0&0&1\end{pmatrix}$	$\begin{pmatrix}x_1&0&0&0\\0&x_2&0&0\\0&0&1&0\\0&0&0&1\end{pmatrix}$	$\begin{pmatrix}x_1w_{11}&0&\frac{w_{13}}{x_1}&0\\0&x_2w_{22}&\frac{w_{23}}{x_2}&0\\x_1^2w_{31}&0&0&1\\0&0&0&1\end{pmatrix}$
$\begin{pmatrix}0&w_{12}&0&0\\w_{21}&0&w_{23}&0\\w_{31}&0&w_{33}&1\\0&0&0&1\end{pmatrix}$	$\begin{pmatrix}x_1&0&0&0\\0&x_2&0&0\\0&0&1&0\\0&0&0&1\end{pmatrix}$	$\begin{pmatrix}0&\frac{x_2^2w_{12}}{x_1}&0&0\\\frac{x_1^2w_{21}}{x_2}&0&\frac{w_{23}}{x_2}&0\\x_1^2w_{31}&0&w_{33}&1\\0&0&0&1\end{pmatrix}$
$\begin{pmatrix}w_{11}&0&w_{13}&0\\w_{21}&0&w_{23}&0\\0&w_{32}&0&1\\0&0&0&1\end{pmatrix}$	$\begin{pmatrix}x_1&0&0&0\\0&x_2&0&0\\0&0&1&0\\0&0&0&1\end{pmatrix}$	$\begin{pmatrix}x_1w_{11}&0&\frac{w_{13}}{x_1}&0\\\frac{x_1^2w_{21}}{x_2}&0&\frac{w_{23}}{x_2}&0\\0&x_2^2w_{32}&0&1\\0&0&0&1\end{pmatrix}$
$\begin{pmatrix}0&0&w_{13}&0\\w_{21}&w_{22}&0&0\\w_{31}&w_{32}&0&1\\0&0&0&1\end{pmatrix}$	$\begin{pmatrix}x_1&0&0&0\\0&x_2&0&0\\0&0&1&0\\0&0&0&1\end{pmatrix}$	$\begin{pmatrix}0&0&\frac{w_{13}}{x_1}&0\\\frac{x_1^2w_{21}}{x_2}&x_2w_{22}&0&0\\x_1^2w_{31}&x_2^2w_{32}&0&1\\0&0&0&1\end{pmatrix}$
$\begin{pmatrix}0&w_{12}&0&0\\w_{21}&0&w_{23}&0\\0&w_{32}&w_{33}&1\\0&0&0&1\end{pmatrix}$	$\begin{pmatrix}x_1&0&0&0\\0&x_2&0&0\\0&0&1&0\\0&0&0&1\end{pmatrix}$	$\begin{pmatrix}0&\frac{x_2^2w_{12}}{x_1}&0&0\\\frac{x_1^2w_{21}}{x_2}&0&\frac{w_{23}}{x_2}&0\\0&x_2^2w_{32}&w_{33}&1\\0&0&0&1\end{pmatrix}$

M_B	$P_{B'B}$	M'_B
$\begin{pmatrix}w_{11}&w_{12}&0&0\\w_{21}&0&w_{23}&0\\0&w_{32}&0&1\\0&0&0&1\end{pmatrix}$	$\begin{pmatrix}x_1&0&0&0\\0&x_2&0&0\\0&0&1&0\\0&0&0&1\end{pmatrix}$	$\begin{pmatrix}x_1w_{11}&\frac{x_2^2w_{12}}{x_1}&0&0\\\frac{x_1^2w_{21}}{x_2}&0&\frac{w_{23}}{x_2}&0\\0&x_2^2w_{32}&0&1\\0&0&0&1\end{pmatrix}$
$\begin{pmatrix}0&0&w_{13}&0\\0&w_{22}&w_{23}&0\\w_{31}&w_{32}&0&1\\0&0&0&1\end{pmatrix}$	$\begin{pmatrix}x_1&0&0&0\\0&x_2&0&0\\0&0&1&0\\0&0&0&1\end{pmatrix}$	$\begin{pmatrix}0&0&\frac{w_{13}}{x_1}&0\\0&x_2w_{22}&\frac{w_{23}}{x_2}&0\\x_1^2w_{31}&x_2^2w_{32}&0&1\\0&0&0&1\end{pmatrix}$
$\begin{pmatrix}0&w_{12}&0&0\\w_{21}&0&0&0\\w_{31}&w_{32}&w_{33}&1\\0&0&0&1\end{pmatrix}$	$\begin{pmatrix}x_1&0&0&0\\0&x_2&0&0\\0&0&1&0\\0&0&0&1\end{pmatrix}$	$\begin{pmatrix}0&\frac{x_2^2w_{12}}{x_1}&0&0\\\frac{x_1^2w_{21}}{x_2}&0&0&0\\x_1^2w_{31}&x_2^2w_{32}&w_{33}&1\\0&0&0&1\end{pmatrix}$
$\begin{pmatrix}0&w_{12}&0&0\\w_{21}&0&0&0\\w_{31}&w_{32}&w_{33}&1\\0&0&0&1\end{pmatrix}$	$\begin{pmatrix}0&x_1&0&0\\x_2&0&0&0\\0&0&1&0\\0&0&0&1\end{pmatrix}$	$\begin{pmatrix}0&\frac{x_1^2w_{21}}{x_2}&0&0\\\frac{x_2^2w_{12}}{x_1}&0&0&0\\x_2^2w_{32}&x_1^2w_{31}&w_{33}&1\\0&0&0&1\end{pmatrix}$
$\begin{pmatrix}w_{11}&w_{12}&w_{13}&0\\0&0&w_{23}&0\\0&w_{32}&0&1\\0&0&0&1\end{pmatrix}$	$\begin{pmatrix}x_1&0&0&0\\0&x_2&0&0\\0&0&1&0\\0&0&0&1\end{pmatrix}$	$\begin{pmatrix}x_1w_{11}&\frac{x_2^2w_{12}}{x_1}&\frac{w_{13}}{x_1}&0\\0&0&\frac{w_{23}}{x_2}&0\\0&x_2^2w_{32}&0&1\\0&0&0&1\end{pmatrix}$
$\begin{pmatrix}w_{11}&w_{12}&w_{13}&0\\w_{21}&0&0&0\\0&w_{32}&0&1\\0&0&0&1\end{pmatrix}$	$\begin{pmatrix}x_1&0&0&0\\0&x_2&0&0\\0&0&1&0\\0&0&0&1\end{pmatrix}$	$\begin{pmatrix}x_1w_{11}&\frac{x_2^2w_{12}}{x_1}&\frac{w_{13}}{x_1}&0\\\frac{x_1^2w_{21}}{x_2}&0&0&0\\0&x_2^2w_{32}&0&1\\0&0&0&1\end{pmatrix}$
$\begin{pmatrix}0&w_{12}&0&0\\0&0&w_{23}&0\\w_{31}&w_{32}&w_{33}&1\\0&0&0&1\end{pmatrix}$	$\begin{pmatrix}x_1&0&0&0\\0&x_2&0&0\\0&0&1&0\\0&0&0&1\end{pmatrix}$	$\begin{pmatrix}0&\frac{x_2^2w_{12}}{x_1}&0&0\\0&0&\frac{w_{23}}{x_2}&0\\x_1^2w_{31}&x_2^2w_{32}&w_{33}&1\\0&0&0&1\end{pmatrix}$
$\begin{pmatrix}w_{11}&w_{12}&w_{13}&0\\0&0&w_{23}&0\\w_{31}&0&0&1\\0&0&0&1\end{pmatrix}$	$\begin{pmatrix}x_1&0&0&0\\0&x_2&0&0\\0&0&1&0\\0&0&0&1\end{pmatrix}$	$\begin{pmatrix}x_1w_{11}&\frac{x_2^2w_{12}}{x_1}&\frac{w_{13}}{x_1}&0\\0&0&\frac{w_{23}}{x_2}&0\\x_1^2w_{31}&0&0&1\\0&0&0&1\end{pmatrix}$
$\begin{pmatrix}w_{11}&0&w_{13}&0\\w_{21}&w_{22}&0&0\\0&w_{32}&0&1\\0&0&0&1\end{pmatrix}$	$\begin{pmatrix}x_1&0&0&0\\0&x_2&0&0\\0&0&1&0\\0&0&0&1\end{pmatrix}$	$\begin{pmatrix}x_1w_{11}&0&\frac{w_{13}}{x_1}&0\\\frac{x_1^2w_{21}}{x_2}&x_2w_{22}&0&0\\0&x_2^2w_{32}&0&1\\0&0&0&1\end{pmatrix}$
$\begin{pmatrix}0&w_{12}&0&0\\0&w_{22}&w_{23}&0\\w_{31}&0&w_{33}&1\\0&0&0&1\end{pmatrix}$	$\begin{pmatrix}x_1&0&0&0\\0&x_2&0&0\\0&0&1&0\\0&0&0&1\end{pmatrix}$	$\begin{pmatrix}0&\frac{x_2^2w_{12}}{x_1}&0&0\\0&x_2w_{22}&\frac{w_{23}}{x_2}&0\\x_1^2w_{31}&0&w_{33}&1\\0&0&0&1\end{pmatrix}$

M_B	$P_{B'B}$	M'_B
$\begin{pmatrix}w_{11}&w_{12}&0&0\\0&0&w_{23}&0\\w_{31}&0&w_{33}&1\\0&0&0&1\end{pmatrix}$	$\begin{pmatrix}x_1&0&0&0\\0&x_2&0&0\\0&0&1&0\\0&0&0&1\end{pmatrix}$	$\begin{pmatrix}x_1w_{11}&\frac{x_2^2w_{12}}{x_1}&0&0\\0&0&\frac{w_{23}}{x_2}&0\\x_1^2w_{31}&0&w_{33}&1\\0&0&0&1\end{pmatrix}$
$\begin{pmatrix}w_{11}&0&w_{13}&0\\w_{21}&0&0&0\\w_{31}&w_{32}&0&1\\0&0&0&1\end{pmatrix}$	$\begin{pmatrix}x_1&0&0&0\\0&x_2&0&0\\0&0&1&0\\0&0&0&1\end{pmatrix}$	$\begin{pmatrix}x_1w_{11}&0&\frac{w_{13}}{x_1}&0\\\frac{x_1^2w_{21}}{x_2}&0&0&0\\x_1^2w_{31}&x_2^2w_{32}&0&1\\0&0&0&1\end{pmatrix}$
$\begin{pmatrix}0&w_{12}&w_{13}&0\\0&0&w_{23}&0\\w_{31}&0&w_{33}&1\\0&0&0&1\end{pmatrix}$	$\begin{pmatrix}x_1&0&0&0\\0&x_2&0&0\\0&0&1&0\\0&0&0&1\end{pmatrix}$	$\begin{pmatrix}0&\frac{x_2^2w_{12}}{x_1}&\frac{w_{13}}{x_1}&0\\0&0&\frac{w_{23}}{x_2}&0\\x_1^2w_{31}&0&w_{33}&1\\0&0&0&1\end{pmatrix}$
$\begin{pmatrix}w_{11}&w_{12}&0&0\\w_{21}&0&w_{23}&0\\w_{31}&0&0&1\\0&0&0&1\end{pmatrix}$	$\begin{pmatrix}x_1&0&0&0\\0&x_2&0&0\\0&0&1&0\\0&0&0&1\end{pmatrix}$	$\begin{pmatrix}x_1w_{11}&\frac{x_2^2w_{12}}{x_1}&0&0\\\frac{x_1^2w_{21}}{x_2}&0&\frac{w_{23}}{x_2}&0\\x_1^2w_{31}&0&0&1\\0&0&0&1\end{pmatrix}$
$\begin{pmatrix}0&w_{12}&w_{13}&0\\w_{21}&0&w_{23}&0\\0&w_{32}&0&1\\0&0&0&1\end{pmatrix}$	$\begin{pmatrix}x_1&0&0&0\\0&x_2&0&0\\0&0&1&0\\0&0&0&1\end{pmatrix}$	$\begin{pmatrix}0&\frac{x_2^2w_{12}}{x_1}&\frac{w_{13}}{x_1}&0\\\frac{x_1^2w_{21}}{x_2}&0&\frac{w_{23}}{x_2}&0\\0&x_2^2w_{32}&0&1\\0&0&0&1\end{pmatrix}$
$\begin{pmatrix}0&w_{12}&0&0\\w_{21}&0&w_{23}&0\\w_{31}&w_{32}&0&1\\0&0&0&1\end{pmatrix}$	$\begin{pmatrix}x_1&0&0&0\\0&x_2&0&0\\0&0&1&0\\0&0&0&1\end{pmatrix}$	$\begin{pmatrix}0&\frac{x_2^2w_{12}}{x_1}&0&0\\\frac{x_1^2w_{21}}{x_2}&0&\frac{w_{23}}{x_2}&0\\x_1^2w_{31}&x_2^2w_{32}&0&1\\0&0&0&1\end{pmatrix}$
$\begin{pmatrix}0&w_{12}&w_{13}&0\\0&0&w_{23}&0\\w_{31}&w_{32}&0&1\\0&0&0&1\end{pmatrix}$	$\begin{pmatrix}x_1&0&0&0\\0&x_2&0&0\\0&0&1&0\\0&0&0&1\end{pmatrix}$	$\begin{pmatrix}0&\frac{x_2^2w_{12}}{x_1}&\frac{w_{13}}{x_1}&0\\0&0&\frac{w_{23}}{x_2}&0\\x_1^2w_{31}&x_2^2w_{32}&0&1\\0&0&0&1\end{pmatrix}$
$\begin{pmatrix}w_{11}&w_{12}&0&0\\0&w_{22}&w_{23}&0\\w_{31}&0&w_{33}&1\\0&0&0&1\end{pmatrix}$	$\begin{pmatrix}x_1&0&0&0\\0&x_2&0&0\\0&0&1&0\\0&0&0&1\end{pmatrix}$	$\begin{pmatrix}x_1w_{11}&\frac{x_2^2w_{12}}{x_1}&0&0\\0&x_2w_{22}&\frac{w_{23}}{x_2}&0\\x_1^2w_{31}&0&w_{33}&1\\0&0&0&1\end{pmatrix}$
$\begin{pmatrix}w_{11}&w_{12}&0&0\\w_{21}&w_{22}&0&0\\w_{31}&0&w_{33}&1\\0&0&0&1\end{pmatrix}$	$\begin{pmatrix}x_1&0&0&0\\0&x_2&0&0\\0&0&1&0\\0&0&0&1\end{pmatrix}$	$\begin{pmatrix}x_1w_{11}&\frac{x_2^2w_{12}}{x_1}&0&0\\\frac{x_1^2w_{21}}{x_2}&x_2w_{22}&0&0\\x_1^2w_{31}&0&w_{33}&1\\0&0&0&1\end{pmatrix}$
$\begin{pmatrix}w_{11}&0&w_{13}&0\\0&w_{22}&w_{23}&0\\0&w_{32}&w_{33}&1\\0&0&0&1\end{pmatrix}$	$\begin{pmatrix}x_1&0&0&0\\0&x_2&0&0\\0&0&1&0\\0&0&0&1\end{pmatrix}$	$\begin{pmatrix}x_1w_{11}&0&\frac{w_{13}}{x_1}&0\\0&x_2w_{22}&\frac{w_{23}}{x_2}&0\\0&x_2^2w_{32}&w_{33}&1\\0&0&0&1\end{pmatrix}$

M_B	$P_{B'B}$	M'_B
$\begin{pmatrix}w_{11}&0&w_{13}&0\\w_{21}&w_{22}&0&0\\w_{31}&0&w_{33}&1\\0&0&0&1\end{pmatrix}$	$\begin{pmatrix}x_1&0&0&0\\0&x_2&0&0\\0&0&1&0\\0&0&0&1\end{pmatrix}$	$\begin{pmatrix}x_1w_{11}&0&\frac{w_{13}}{x_1}&0\\\frac{x_1^2w_{21}}{x_2}&x_2w_{22}&0&0\\x_1^2w_{31}&0&w_{33}&1\\0&0&0&1\end{pmatrix}$
$\begin{pmatrix}w_{11}&w_{12}&w_{13}&0\\w_{21}&0&0&0\\w_{31}&0&w_{33}&1\\0&0&0&1\end{pmatrix}$	$\begin{pmatrix}x_1&0&0&0\\0&x_2&0&0\\0&0&1&0\\0&0&0&1\end{pmatrix}$	$\begin{pmatrix}x_1w_{11}&\frac{x_2^2w_{12}}{x_1}&\frac{w_{13}}{x_1}&0\\\frac{x_1^2w_{21}}{x_2}&0&0&0\\x_1^2w_{31}&0&w_{33}&1\\0&0&0&1\end{pmatrix}$
$\begin{pmatrix}w_{11}&0&w_{13}&0\\0&0&w_{23}&0\\w_{31}&w_{32}&w_{33}&1\\0&0&0&1\end{pmatrix}$	$\begin{pmatrix}x_1&0&0&0\\0&x_2&0&0\\0&0&1&0\\0&0&0&1\end{pmatrix}$	$\begin{pmatrix}x_1w_{11}&0&\frac{w_{13}}{x_1}&0\\0&0&\frac{w_{23}}{x_2}&0\\x_1^2w_{31}&x_2^2w_{32}&w_{33}&1\\0&0&0&1\end{pmatrix}$
$\begin{pmatrix}w_{11}&w_{12}&w_{13}&0\\w_{21}&w_{22}&0&0\\w_{31}&0&0&1\\0&0&0&1\end{pmatrix}$	$\begin{pmatrix}x_1&0&0&0\\0&x_2&0&0\\0&0&1&0\\0&0&0&1\end{pmatrix}$	$\begin{pmatrix}x_1w_{11}&\frac{x_2^2w_{12}}{x_1}&\frac{w_{13}}{x_1}&0\\\frac{x_1^2w_{21}}{x_2}&x_2w_{22}&0&0\\x_1^2w_{31}&0&0&1\\0&0&0&1\end{pmatrix}$
$\begin{pmatrix}w_{11}&w_{12}&w_{13}&0\\w_{21}&0&0&0\\0&w_{32}&w_{33}&1\\0&0&0&1\end{pmatrix}$	$\begin{pmatrix}x_1&0&0&0\\0&x_2&0&0\\0&0&1&0\\0&0&0&1\end{pmatrix}$	$\begin{pmatrix}x_1w_{11}&\frac{x_2^2w_{12}}{x_1}&\frac{w_{13}}{x_1}&0\\\frac{x_1^2w_{21}}{x_2}&0&0&0\\0&x_2^2w_{32}&w_{33}&1\\0&0&0&1\end{pmatrix}$
$\begin{pmatrix}w_{11}&w_{12}&0&0\\0&0&w_{23}&0\\w_{31}&w_{32}&w_{33}&1\\0&0&0&1\end{pmatrix}$	$\begin{pmatrix}x_1&0&0&0\\0&x_2&0&0\\0&0&1&0\\0&0&0&1\end{pmatrix}$	$\begin{pmatrix}x_1w_{11}&\frac{x_2^2w_{12}}{x_1}&0&0\\0&0&\frac{w_{23}}{x_2}&0\\x_1^2w_{31}&x_2^2w_{32}&w_{33}&1\\0&0&0&1\end{pmatrix}$
$\begin{pmatrix}w_{11}&w_{12}&w_{13}&0\\0&w_{22}&w_{23}&0\\w_{31}&0&0&1\\0&0&0&1\end{pmatrix}$	$\begin{pmatrix}x_1&0&0&0\\0&x_2&0&0\\0&0&1&0\\0&0&0&1\end{pmatrix}$	$\begin{pmatrix}x_1w_{11}&\frac{x_2^2w_{12}}{x_1}&\frac{w_{13}}{x_1}&0\\0&x_2w_{22}&\frac{w_{23}}{x_2}&0\\x_1^2w_{31}&0&0&1\\0&0&0&1\end{pmatrix}$
$\begin{pmatrix}0&w_{12}&w_{13}&0\\w_{21}&w_{22}&0&0\\w_{31}&0&w_{33}&1\\0&0&0&1\end{pmatrix}$	$\begin{pmatrix}x_1&0&0&0\\0&x_2&0&0\\0&0&1&0\\0&0&0&1\end{pmatrix}$	$\begin{pmatrix}0&\frac{x_2^2w_{12}}{x_1}&\frac{w_{13}}{x_1}&0\\\frac{x_1^2w_{21}}{x_2}&x_2w_{22}&0&0\\x_1^2w_{31}&0&w_{33}&1\\0&0&0&1\end{pmatrix}$
$\begin{pmatrix}w_{11}&0&w_{13}&0\\0&w_{22}&w_{23}&0\\w_{31}&w_{32}&0&1\\0&0&0&1\end{pmatrix}$	$\begin{pmatrix}x_1&0&0&0\\0&x_2&0&0\\0&0&1&0\\0&0&0&1\end{pmatrix}$	$\begin{pmatrix}x_1w_{11}&0&\frac{w_{13}}{x_1}&0\\0&x_2w_{22}&\frac{w_{23}}{x_2}&0\\x_1^2w_{31}&x_2^2w_{32}&0&1\\0&0&0&1\end{pmatrix}$
$\begin{pmatrix}w_{11}&0&w_{13}&0\\0&w_{22}&w_{23}&0\\w_{31}&w_{32}&0&1\\0&0&0&1\end{pmatrix}$	$\begin{pmatrix}0&x_1&0&0\\x_2&0&0&0\\0&0&1&0\\0&0&0&1\end{pmatrix}$	$\begin{pmatrix}x_2w_{22}&0&\frac{w_{23}}{x_2}&0\\0&x_1w_{11}&\frac{w_{13}}{x_1}&0\\x_2^2w_{32}&x_1^2w_{31}&0&1\\0&0&0&1\end{pmatrix}$

M_B	$P_{B'B}$	M'_B
$\begin{pmatrix}0&w_{12}&w_{13}&0\\w_{21}&w_{22}&0&0\\0&w_{32}&w_{33}&1\\0&0&0&1\end{pmatrix}$	$\begin{pmatrix}x_1&0&0&0\\0&x_2&0&0\\0&0&1&0\\0&0&0&1\end{pmatrix}$	$\begin{pmatrix}0&\frac{x_2^2w_{12}}{x_1}&\frac{w_{13}}{x_1}&0\\\frac{x_1^2w_{21}}{x_2}&x_2w_{22}&0&0\\0&x_2^2w_{32}&w_{33}&1\\0&0&0&1\end{pmatrix}$
$\begin{pmatrix}w_{11}&w_{12}&0&0\\0&w_{22}&w_{23}&0\\w_{31}&w_{32}&0&1\\0&0&0&1\end{pmatrix}$	$\begin{pmatrix}x_1&0&0&0\\0&x_2&0&0\\0&0&1&0\\0&0&0&1\end{pmatrix}$	$\begin{pmatrix}x_1w_{11}&\frac{x_2^2w_{12}}{x_1}&0&0\\0&x_2w_{22}&\frac{w_{23}}{x_2}&0\\x_1^2w_{31}&x_2^2w_{32}&0&1\\0&0&0&1\end{pmatrix}$
$\begin{pmatrix}0&w_{12}&w_{13}&0\\0&w_{22}&w_{23}&0\\w_{31}&0&w_{33}&1\\0&0&0&1\end{pmatrix}$	$\begin{pmatrix}x_1&0&0&0\\0&x_2&0&0\\0&0&1&0\\0&0&0&1\end{pmatrix}$	$\begin{pmatrix}0&\frac{x_2^2w_{12}}{x_1}&\frac{w_{13}}{x_1}&0\\0&x_2w_{22}&\frac{w_{23}}{x_2}&0\\x_1^2w_{31}&0&w_{33}&1\\0&0&0&1\end{pmatrix}$
$\begin{pmatrix}0&w_{12}&w_{13}&0\\w_{21}&0&w_{23}&0\\w_{31}&0&w_{33}&1\\0&0&0&1\end{pmatrix}$	$\begin{pmatrix}x_1&0&0&0\\0&x_2&0&0\\0&0&1&0\\0&0&0&1\end{pmatrix}$	$\begin{pmatrix}0&\frac{x_2^2w_{12}}{x_1}&\frac{w_{13}}{x_1}&0\\\frac{x_1^2w_{21}}{x_2}&0&\frac{w_{23}}{x_2}&0\\x_1^2w_{31}&0&w_{33}&1\\0&0&0&1\end{pmatrix}$
$\begin{pmatrix}w_{11}&0&w_{13}&0\\w_{21}&0&w_{23}&0\\w_{31}&w_{32}&0&1\\0&0&0&1\end{pmatrix}$	$\begin{pmatrix}x_1&0&0&0\\0&x_2&0&0\\0&0&1&0\\0&0&0&1\end{pmatrix}$	$\begin{pmatrix}x_1w_{11}&0&\frac{w_{13}}{x_1}&0\\\frac{x_1^2w_{21}}{x_2}&0&\frac{w_{23}}{x_2}&0\\x_1^2w_{31}&x_2^2w_{32}&0&1\\0&0&0&1\end{pmatrix}$
$\begin{pmatrix}0&w_{12}&w_{13}&0\\w_{21}&w_{22}&0&0\\w_{31}&w_{32}&0&1\\0&0&0&1\end{pmatrix}$	$\begin{pmatrix}x_1&0&0&0\\0&x_2&0&0\\0&0&1&0\\0&0&0&1\end{pmatrix}$	$\begin{pmatrix}0&\frac{x_2^2w_{12}}{x_1}&\frac{w_{13}}{x_1}&0\\\frac{x_1^2w_{21}}{x_2}&x_2w_{22}&0&0\\x_1^2w_{31}&x_2^2w_{32}&0&1\\0&0&0&1\end{pmatrix}$
$\begin{pmatrix}0&w_{12}&w_{13}&0\\w_{21}&0&0&0\\w_{31}&w_{32}&w_{33}&1\\0&0&0&1\end{pmatrix}$	$\begin{pmatrix}x_1&0&0&0\\0&x_2&0&0\\0&0&1&0\\0&0&0&1\end{pmatrix}$	$\begin{pmatrix}0&\frac{x_2^2w_{12}}{x_1}&\frac{w_{13}}{x_1}&0\\\frac{x_1^2w_{21}}{x_2}&0&0&0\\x_1^2w_{31}&x_2^2w_{32}&w_{33}&1\\0&0&0&1\end{pmatrix}$
$\begin{pmatrix}w_{11}&w_{12}&w_{13}&0\\0&0&w_{23}&0\\w_{31}&w_{32}&0&1\\0&0&0&1\end{pmatrix}$	$\begin{pmatrix}x_1&0&0&0\\0&x_2&0&0\\0&0&1&0\\0&0&0&1\end{pmatrix}$	$\begin{pmatrix}x_1w_{11}&\frac{x_2^2w_{12}}{x_1}&\frac{w_{13}}{x_1}&0\\0&0&\frac{w_{23}}{x_2}&0\\x_1^2w_{31}&x_2^2w_{32}&0&1\\0&0&0&1\end{pmatrix}$
$\begin{pmatrix}0&w_{12}&w_{13}&0\\w_{21}&w_{22}&w_{23}&0\\w_{31}&0&0&1\\0&0&0&1\end{pmatrix}$	$\begin{pmatrix}x_1&0&0&0\\0&x_2&0&0\\0&0&1&0\\0&0&0&1\end{pmatrix}$	$\begin{pmatrix}0&\frac{x_2^2w_{12}}{x_1}&\frac{w_{13}}{x_1}&0\\\frac{x_1^2w_{21}}{x_2}&x_2w_{22}&\frac{w_{23}}{x_2}&0\\x_1^2w_{31}&0&0&1\\0&0&0&1\end{pmatrix}$
$\begin{pmatrix}0&w_{12}&0&0\\w_{21}&w_{22}&0&0\\w_{31}&w_{32}&w_{33}&1\\0&0&0&1\end{pmatrix}$	$\begin{pmatrix}x_1&0&0&0\\0&x_2&0&0\\0&0&1&0\\0&0&0&1\end{pmatrix}$	$\begin{pmatrix}0&\frac{x_2^2w_{12}}{x_1}&0&0\\\frac{x_1^2w_{21}}{x_2}&x_2w_{22}&0&0\\x_1^2w_{31}&x_2^2w_{32}&w_{33}&1\\0&0&0&1\end{pmatrix}$

M_B	$P_{B'B}$	M'_B
$\begin{pmatrix}w_{11}&w_{12}&w_{13}&0\\0&w_{22}&w_{23}&0\\0&w_{32}&0&1\\0&0&0&1\end{pmatrix}$	$\begin{pmatrix}x_1&0&0&0\\0&x_2&0&0\\0&0&1&0\\0&0&0&1\end{pmatrix}$	$\begin{pmatrix}x_1w_{11}&\frac{x_2^2w_{12}}{x_1}&\frac{w_{13}}{x_1}&0\\0&x_2w_{22}&\frac{w_{23}}{x_2}&0\\0&x_2^2w_{32}&0&1\\0&0&0&1\end{pmatrix}$
$\begin{pmatrix}0&0&w_{13}&0\\w_{21}&w_{22}&w_{23}&0\\w_{31}&0&w_{33}&1\\0&0&0&1\end{pmatrix}$	$\begin{pmatrix}x_1&0&0&0\\0&x_2&0&0\\0&0&1&0\\0&0&0&1\end{pmatrix}$	$\begin{pmatrix}0&0&\frac{w_{13}}{x_1}&0\\\frac{x_1^2w_{21}}{x_2}&x_2w_{22}&\frac{w_{23}}{x_2}&0\\x_1^2w_{31}&0&w_{33}&1\\0&0&0&1\end{pmatrix}$
$\begin{pmatrix}w_{11}&w_{12}&w_{13}&0\\w_{21}&w_{22}&0&0\\0&w_{32}&0&1\\0&0&0&1\end{pmatrix}$	$\begin{pmatrix}x_1&0&0&0\\0&x_2&0&0\\0&0&1&0\\0&0&0&1\end{pmatrix}$	$\begin{pmatrix}x_1w_{11}&\frac{x_2^2w_{12}}{x_1}&\frac{w_{13}}{x_1}&0\\\frac{x_1^2w_{21}}{x_2}&x_2w_{22}&0&0\\0&x_2^2w_{32}&0&1\\0&0&0&1\end{pmatrix}$
$\begin{pmatrix}0&w_{12}&0&0\\0&w_{22}&w_{23}&0\\w_{31}&w_{32}&w_{33}&1\\0&0&0&1\end{pmatrix}$	$\begin{pmatrix}x_1&0&0&0\\0&x_2&0&0\\0&0&1&0\\0&0&0&1\end{pmatrix}$	$\begin{pmatrix}0&\frac{x_2^2w_{12}}{x_1}&0&0\\0&x_2w_{22}&\frac{w_{23}}{x_2}&0\\x_1^2w_{31}&x_2^2w_{32}&w_{33}&1\\0&0&0&1\end{pmatrix}$
$\begin{pmatrix}w_{11}&w_{12}&w_{13}&0\\0&0&w_{23}&0\\w_{31}&0&w_{33}&1\\0&0&0&1\end{pmatrix}$	$\begin{pmatrix}x_1&0&0&0\\0&x_2&0&0\\0&0&1&0\\0&0&0&1\end{pmatrix}$	$\begin{pmatrix}x_1w_{11}&\frac{x_2^2w_{12}}{x_1}&\frac{w_{13}}{x_1}&0\\0&0&\frac{w_{23}}{x_2}&0\\x_1^2w_{31}&0&w_{33}&1\\0&0&0&1\end{pmatrix}$
$\begin{pmatrix}w_{11}&w_{12}&w_{13}&0\\w_{21}&0&0&0\\w_{31}&w_{32}&0&1\\0&0&0&1\end{pmatrix}$	$\begin{pmatrix}x_1&0&0&0\\0&x_2&0&0\\0&0&1&0\\0&0&0&1\end{pmatrix}$	$\begin{pmatrix}x_1w_{11}&\frac{x_2^2w_{12}}{x_1}&\frac{w_{13}}{x_1}&0\\\frac{x_1^2w_{21}}{x_2}&0&0&0\\x_1^2w_{31}&x_2^2w_{32}&0&1\\0&0&0&1\end{pmatrix}$
$\begin{pmatrix}0&w_{12}&w_{13}&0\\0&0&w_{23}&0\\w_{31}&w_{32}&w_{33}&1\\0&0&0&1\end{pmatrix}$	$\begin{pmatrix}x_1&0&0&0\\0&x_2&0&0\\0&0&1&0\\0&0&0&1\end{pmatrix}$	$\begin{pmatrix}0&\frac{x_2^2w_{12}}{x_1}&\frac{w_{13}}{x_1}&0\\0&0&\frac{w_{23}}{x_2}&0\\x_1^2w_{31}&x_2^2w_{32}&w_{33}&1\\0&0&0&1\end{pmatrix}$
$\begin{pmatrix}w_{11}&w_{12}&w_{13}&0\\w_{21}&0&w_{23}&0\\w_{31}&0&0&1\\0&0&0&1\end{pmatrix}$	$\begin{pmatrix}x_1&0&0&0\\0&x_2&0&0\\0&0&1&0\\0&0&0&1\end{pmatrix}$	$\begin{pmatrix}x_1w_{11}&\frac{x_2^2w_{12}}{x_1}&\frac{w_{13}}{x_1}&0\\\frac{x_1^2w_{21}}{x_2}&0&\frac{w_{23}}{x_2}&0\\x_1^2w_{31}&0&0&1\\0&0&0&1\end{pmatrix}$
$\begin{pmatrix}0&w_{12}&w_{13}&0\\w_{21}&0&w_{23}&0\\w_{31}&w_{32}&0&1\\0&0&0&1\end{pmatrix}$	$\begin{pmatrix}x_1&0&0&0\\0&x_2&0&0\\0&0&1&0\\0&0&0&1\end{pmatrix}$	$\begin{pmatrix}0&\frac{x_2^2w_{12}}{x_1}&\frac{w_{13}}{x_1}&0\\\frac{x_1^2w_{21}}{x_2}&0&\frac{w_{23}}{x_2}&0\\x_1^2w_{31}&x_2^2w_{32}&0&1\\0&0&0&1\end{pmatrix}$
$\begin{pmatrix}0&w_{12}&w_{13}&0\\w_{21}&0&w_{23}&0\\w_{31}&w_{32}&0&1\\0&0&0&1\end{pmatrix}$	$\begin{pmatrix}0&x_1&0&0\\x_2&0&0&0\\0&0&1&0\\0&0&0&1\end{pmatrix}$	$\begin{pmatrix}0&\frac{x_1^2w_{21}}{x_2}&\frac{w_{23}}{x_2}&0\\\frac{x_2^2w_{12}}{x_1}&0&\frac{w_{13}}{x_1}&0\\x_2^2w_{32}&x_1^2w_{31}&0&1\\0&0&0&1\end{pmatrix}$

M_B	$P_{B'B}$	M'_B
$\begin{pmatrix}w_{11}&w_{12}&w_{13}&0\\w_{21}&w_{22}&0&0\\w_{31}&0&w_{33}&1\\0&0&0&1\end{pmatrix}$	$\begin{pmatrix}x_1&0&0&0\\0&x_2&0&0\\0&0&1&0\\0&0&0&1\end{pmatrix}$	$\begin{pmatrix}x_1w_{11}&\frac{x_2^2w_{12}}{x_1}&\frac{w_{13}}{x_1}&0\\\frac{x_1^2w_{21}}{x_2}&x_2w_{22}&0&0\\x_1^2w_{31}&0&w_{33}&1\\0&0&0&1\end{pmatrix}$
$\begin{pmatrix}w_{11}&0&w_{13}&0\\0&w_{22}&w_{23}&0\\w_{31}&w_{32}&w_{33}&1\\0&0&0&1\end{pmatrix}$	$\begin{pmatrix}x_1&0&0&0\\0&x_2&0&0\\0&0&1&0\\0&0&0&1\end{pmatrix}$	$\begin{pmatrix}x_1w_{11}&0&\frac{w_{13}}{x_1}&0\\0&x_2w_{22}&\frac{w_{23}}{x_2}&0\\x_1^2w_{31}&x_2^2w_{32}&w_{33}&1\\0&0&0&1\end{pmatrix}$
$\begin{pmatrix}w_{11}&0&w_{13}&0\\0&w_{22}&w_{23}&0\\w_{31}&w_{32}&w_{33}&1\\0&0&0&1\end{pmatrix}$	$\begin{pmatrix}0&x_1&0&0\\x_2&0&0&0\\0&0&1&0\\0&0&0&1\end{pmatrix}$	$\begin{pmatrix}x_2w_{22}&0&\frac{w_{23}}{x_2}&0\\0&x_1w_{11}&\frac{w_{13}}{x_1}&0\\x_2^2w_{32}&x_1^2w_{31}&w_{33}&1\\0&0&0&1\end{pmatrix}$
$\begin{pmatrix}w_{11}&w_{12}&w_{13}&0\\w_{21}&w_{22}&0&0\\0&w_{32}&w_{33}&1\\0&0&0&1\end{pmatrix}$	$\begin{pmatrix}x_1&0&0&0\\0&x_2&0&0\\0&0&1&0\\0&0&0&1\end{pmatrix}$	$\begin{pmatrix}x_1w_{11}&\frac{x_2^2w_{12}}{x_1}&\frac{w_{13}}{x_1}&0\\\frac{x_1^2w_{21}}{x_2}&x_2w_{22}&0&0\\0&x_2^2w_{32}&w_{33}&1\\0&0&0&1\end{pmatrix}$
$\begin{pmatrix}w_{11}&w_{12}&0&0\\0&w_{22}&w_{23}&0\\w_{31}&w_{32}&w_{33}&1\\0&0&0&1\end{pmatrix}$	$\begin{pmatrix}x_1&0&0&0\\0&x_2&0&0\\0&0&1&0\\0&0&0&1\end{pmatrix}$	$\begin{pmatrix}x_1w_{11}&\frac{x_2^2w_{12}}{x_1}&0&0\\0&x_2w_{22}&\frac{w_{23}}{x_2}&0\\x_1^2w_{31}&x_2^2w_{32}&w_{33}&1\\0&0&0&1\end{pmatrix}$
$\begin{pmatrix}w_{11}&w_{12}&w_{13}&0\\0&w_{22}&w_{23}&0\\w_{31}&0&w_{33}&1\\0&0&0&1\end{pmatrix}$	$\begin{pmatrix}x_1&0&0&0\\0&x_2&0&0\\0&0&1&0\\0&0&0&1\end{pmatrix}$	$\begin{pmatrix}x_1w_{11}&\frac{x_2^2w_{12}}{x_1}&\frac{w_{13}}{x_1}&0\\0&x_2w_{22}&\frac{w_{23}}{x_2}&0\\x_1^2w_{31}&0&w_{33}&1\\0&0&0&1\end{pmatrix}$
$\begin{pmatrix}w_{11}&0&w_{13}&0\\w_{21}&w_{22}&w_{23}&0\\w_{31}&0&w_{33}&1\\0&0&0&1\end{pmatrix}$	$\begin{pmatrix}x_1&0&0&0\\0&x_2&0&0\\0&0&1&0\\0&0&0&1\end{pmatrix}$	$\begin{pmatrix}x_1w_{11}&0&\frac{w_{13}}{x_1}&0\\\frac{x_1^2w_{21}}{x_2}&x_2w_{22}&\frac{w_{23}}{x_2}&0\\x_1^2w_{31}&0&w_{33}&1\\0&0&0&1\end{pmatrix}$
$\begin{pmatrix}w_{11}&w_{12}&0&0\\w_{21}&w_{22}&0&0\\w_{31}&w_{32}&w_{33}&1\\0&0&0&1\end{pmatrix}$	$\begin{pmatrix}x_1&0&0&0\\0&x_2&0&0\\0&0&1&0\\0&0&0&1\end{pmatrix}$	$\begin{pmatrix}x_1w_{11}&\frac{x_2^2w_{12}}{x_1}&0&0\\\frac{x_1^2w_{21}}{x_2}&x_2w_{22}&0&0\\x_1^2w_{31}&x_2^2w_{32}&w_{33}&1\\0&0&0&1\end{pmatrix}$
$\begin{pmatrix}w_{11}&w_{12}&0&0\\w_{21}&w_{22}&0&0\\w_{31}&w_{32}&w_{33}&1\\0&0&0&1\end{pmatrix}$	$\begin{pmatrix}0&x_1&0&0\\x_2&0&0&0\\0&0&1&0\\0&0&0&1\end{pmatrix}$	$\begin{pmatrix}x_2w_{22}&\frac{x_1^2w_{21}}{x_2}&0&0\\\frac{x_2^2w_{12}}{x_1}&x_1w_{11}&0&0\\x_2^2w_{32}&x_1^2w_{31}&w_{33}&1\\0&0&0&1\end{pmatrix}$
$\begin{pmatrix}w_{11}&w_{12}&w_{13}&0\\w_{21}&0&0&0\\w_{31}&w_{32}&w_{33}&1\\0&0&0&1\end{pmatrix}$	$\begin{pmatrix}x_1&0&0&0\\0&x_2&0&0\\0&0&1&0\\0&0&0&1\end{pmatrix}$	$\begin{pmatrix}x_1w_{11}&\frac{x_2^2w_{12}}{x_1}&\frac{w_{13}}{x_1}&0\\\frac{x_1^2w_{21}}{x_2}&0&0&0\\x_1^2w_{31}&x_2^2w_{32}&w_{33}&1\\0&0&0&1\end{pmatrix}$

M_B	$P_{B'B}$	M'_B
$\begin{pmatrix} w_{11} & w_{12} & w_{13} & 0 \\ 0 & 0 & w_{23} & 0 \\ w_{31} & w_{32} & w_{33} & 1 \\ 0 & 0 & 0 & 1 \end{pmatrix}$	$\begin{pmatrix} x_1 & 0 & 0 & 0 \\ 0 & x_2 & 0 & 0 \\ 0 & 0 & 1 & 0 \\ 0 & 0 & 0 & 1 \end{pmatrix}$	$\begin{pmatrix} x_1 w_{11} & \frac{x_2^2 w_{12}}{x_1} & \frac{w_{13}}{x_1} & 0 \\ 0 & 0 & \frac{w_{23}}{x_2} & 0 \\ x_1^2 w_{31} & x_2^2 w_{32} & w_{33} & 1 \\ 0 & 0 & 0 & 1 \end{pmatrix}$
$\begin{pmatrix} w_{11} & w_{12} & w_{13} & 0 \\ w_{21} & w_{22} & w_{23} & 0 \\ w_{31} & 0 & 0 & 1 \\ 0 & 0 & 0 & 1 \end{pmatrix}$	$\begin{pmatrix} x_1 & 0 & 0 & 0 \\ 0 & x_2 & 0 & 0 \\ 0 & 0 & 1 & 0 \\ 0 & 0 & 0 & 1 \end{pmatrix}$	$\begin{pmatrix} x_1 w_{11} & \frac{x_2^2 w_{12}}{x_1} & \frac{w_{13}}{x_1} & 0 \\ \frac{x_1^2 w_{21}}{x_2} & x_2 w_{22} & \frac{w_{23}}{x_2} & 0 \\ x_1^2 w_{31} & 0 & 0 & 1 \\ 0 & 0 & 0 & 1 \end{pmatrix}$
$\begin{pmatrix} w_{11} & w_{12} & w_{13} & 0 \\ w_{21} & 0 & w_{23} & 0 \\ w_{31} & 0 & w_{33} & 1 \\ 0 & 0 & 0 & 1 \end{pmatrix}$	$\begin{pmatrix} x_1 & 0 & 0 & 0 \\ 0 & x_2 & 0 & 0 \\ 0 & 0 & 1 & 0 \\ 0 & 0 & 0 & 1 \end{pmatrix}$	$\begin{pmatrix} x_1 w_{11} & \frac{x_2^2 w_{12}}{x_1} & \frac{w_{13}}{x_1} & 0 \\ \frac{x_1^2 w_{21}}{x_2} & 0 & \frac{w_{23}}{x_2} & 0 \\ x_1^2 w_{31} & 0 & w_{33} & 1 \\ 0 & 0 & 0 & 1 \end{pmatrix}$
$\begin{pmatrix} w_{11} & 0 & w_{13} & 0 \\ w_{21} & 0 & w_{23} & 0 \\ w_{31} & w_{32} & w_{33} & 1 \\ 0 & 0 & 0 & 1 \end{pmatrix}$	$\begin{pmatrix} x_1 & 0 & 0 & 0 \\ 0 & x_2 & 0 & 0 \\ 0 & 0 & 1 & 0 \\ 0 & 0 & 0 & 1 \end{pmatrix}$	$\begin{pmatrix} x_1 w_{11} & 0 & \frac{w_{13}}{x_1} & 0 \\ \frac{x_1^2 w_{21}}{x_2} & 0 & \frac{w_{23}}{x_2} & 0 \\ x_1^2 w_{31} & x_2^2 w_{32} & w_{33} & 1 \\ 0 & 0 & 0 & 1 \end{pmatrix}$
$\begin{pmatrix} w_{11} & w_{12} & w_{13} & 0 \\ w_{21} & w_{22} & 0 & 0 \\ w_{31} & w_{32} & 0 & 1 \\ 0 & 0 & 0 & 1 \end{pmatrix}$	$\begin{pmatrix} x_1 & 0 & 0 & 0 \\ 0 & x_2 & 0 & 0 \\ 0 & 0 & 1 & 0 \\ 0 & 0 & 0 & 1 \end{pmatrix}$	$\begin{pmatrix} x_1 w_{11} & \frac{x_2^2 w_{12}}{x_1} & \frac{w_{13}}{x_1} & 0 \\ \frac{x_1^2 w_{21}}{x_2} & x_2 w_{22} & 0 & 0 \\ x_1^2 w_{31} & x_2^2 w_{32} & 0 & 1 \\ 0 & 0 & 0 & 1 \end{pmatrix}$
$\begin{pmatrix} 0 & w_{12} & w_{13} & 0 \\ w_{21} & w_{22} & w_{23} & 0 \\ w_{31} & 0 & w_{33} & 1 \\ 0 & 0 & 0 & 1 \end{pmatrix}$	$\begin{pmatrix} x_1 & 0 & 0 & 0 \\ 0 & x_2 & 0 & 0 \\ 0 & 0 & 1 & 0 \\ 0 & 0 & 0 & 1 \end{pmatrix}$	$\begin{pmatrix} 0 & \frac{x_2^2 w_{12}}{x_1} & \frac{w_{13}}{x_1} & 0 \\ \frac{x_1^2 w_{21}}{x_2} & x_2 w_{22} & \frac{w_{23}}{x_2} & 0 \\ x_1^2 w_{31} & 0 & w_{33} & 1 \\ 0 & 0 & 0 & 1 \end{pmatrix}$
$\begin{pmatrix} w_{11} & 0 & w_{13} & 0 \\ w_{21} & w_{22} & w_{23} & 0 \\ w_{31} & w_{32} & 0 & 1 \\ 0 & 0 & 0 & 1 \end{pmatrix}$	$\begin{pmatrix} x_1 & 0 & 0 & 0 \\ 0 & x_2 & 0 & 0 \\ 0 & 0 & 1 & 0 \\ 0 & 0 & 0 & 1 \end{pmatrix}$	$\begin{pmatrix} x_1 w_{11} & 0 & \frac{w_{13}}{x_1} & 0 \\ \frac{x_1^2 w_{21}}{x_2} & x_2 w_{22} & \frac{w_{23}}{x_2} & 0 \\ x_1^2 w_{31} & x_2^2 w_{32} & 0 & 1 \\ 0 & 0 & 0 & 1 \end{pmatrix}$
$\begin{pmatrix} 0 & w_{12} & w_{13} & 0 \\ w_{21} & w_{22} & 0 & 0 \\ w_{31} & w_{32} & w_{33} & 1 \\ 0 & 0 & 0 & 1 \end{pmatrix}$	$\begin{pmatrix} x_1 & 0 & 0 & 0 \\ 0 & x_2 & 0 & 0 \\ 0 & 0 & 1 & 0 \\ 0 & 0 & 0 & 1 \end{pmatrix}$	$\begin{pmatrix} 0 & \frac{x_2^2 w_{12}}{x_1} & \frac{w_{13}}{x_1} & 0 \\ \frac{x_1^2 w_{21}}{x_2} & x_2 w_{22} & 0 & 0 \\ x_1^2 w_{31} & x_2^2 w_{32} & w_{33} & 1 \\ 0 & 0 & 0 & 1 \end{pmatrix}$
$\begin{pmatrix} 0 & w_{12} & w_{13} & 0 \\ w_{21} & 0 & w_{23} & 0 \\ w_{31} & w_{32} & w_{33} & 1 \\ 0 & 0 & 0 & 1 \end{pmatrix}$	$\begin{pmatrix} x_1 & 0 & 0 & 0 \\ 0 & x_2 & 0 & 0 \\ 0 & 0 & 1 & 0 \\ 0 & 0 & 0 & 1 \end{pmatrix}$	$\begin{pmatrix} 0 & \frac{x_2^2 w_{12}}{x_1} & \frac{w_{13}}{x_1} & 0 \\ \frac{x_1^2 w_{21}}{x_2} & 0 & \frac{w_{23}}{x_2} & 0 \\ x_1^2 w_{31} & x_2^2 w_{32} & w_{33} & 1 \\ 0 & 0 & 0 & 1 \end{pmatrix}$
$\begin{pmatrix} 0 & w_{12} & w_{13} & 0 \\ w_{21} & 0 & w_{23} & 0 \\ w_{31} & w_{32} & w_{33} & 1 \\ 0 & 0 & 0 & 1 \end{pmatrix}$	$\begin{pmatrix} 0 & x_1 & 0 & 0 \\ x_2 & 0 & 0 & 0 \\ 0 & 0 & 1 & 0 \\ 0 & 0 & 0 & 1 \end{pmatrix}$	$\begin{pmatrix} 0 & \frac{x_1^2 w_{21}}{x_2} & \frac{w_{23}}{x_2} & 0 \\ \frac{x_2^2 w_{12}}{x_1} & 0 & \frac{w_{13}}{x_1} & 0 \\ x_2^2 w_{32} & x_1^2 w_{31} & w_{33} & 1 \\ 0 & 0 & 0 & 1 \end{pmatrix}$

M_B	$P_{B'B}$	M'_B
$\begin{pmatrix} w_{11} & w_{12} & w_{13} & 0 \\ w_{21} & 0 & w_{23} & 0 \\ w_{31} & w_{32} & 0 & 1 \\ 0 & 0 & 0 & 1 \end{pmatrix}$	$\begin{pmatrix} x_1 & 0 & 0 & 0 \\ 0 & x_2 & 0 & 0 \\ 0 & 0 & 1 & 0 \\ 0 & 0 & 0 & 1 \end{pmatrix}$	$\begin{pmatrix} x_1 w_{11} & \frac{x_2^2 w_{12}}{x_1} & \frac{w_{13}}{x_1} & 0 \\ \frac{x_1^2 w_{21}}{x_2} & 0 & \frac{w_{23}}{x_2} & 0 \\ x_1^2 w_{31} & x_2^2 w_{32} & 0 & 1 \\ 0 & 0 & 0 & 1 \end{pmatrix}$
$\begin{pmatrix} w_{11} & w_{12} & w_{13} & 0 \\ w_{21} & w_{22} & w_{23} & 0 \\ w_{31} & 0 & w_{33} & 1 \\ 0 & 0 & 0 & 1 \end{pmatrix}$	$\begin{pmatrix} x_1 & 0 & 0 & 0 \\ 0 & x_2 & 0 & 0 \\ 0 & 0 & 1 & 0 \\ 0 & 0 & 0 & 1 \end{pmatrix}$	$\begin{pmatrix} x_1 w_{11} & \frac{x_2^2 w_{12}}{x_1} & \frac{w_{13}}{x_1} & 0 \\ \frac{x_1^2 w_{21}}{x_2} & x_2 w_{22} & \frac{w_{23}}{x_2} & 0 \\ x_1^2 w_{31} & 0 & w_{33} & 1 \\ 0 & 0 & 0 & 1 \end{pmatrix}$
$\begin{pmatrix} w_{11} & 0 & w_{13} & 0 \\ w_{21} & w_{22} & w_{23} & 0 \\ w_{31} & w_{32} & w_{33} & 1 \\ 0 & 0 & 0 & 1 \end{pmatrix}$	$\begin{pmatrix} x_1 & 0 & 0 & 0 \\ 0 & x_2 & 0 & 0 \\ 0 & 0 & 1 & 0 \\ 0 & 0 & 0 & 1 \end{pmatrix}$	$\begin{pmatrix} x_1 w_{11} & 0 & \frac{w_{13}}{x_1} & 0 \\ \frac{x_1^2 w_{21}}{x_2} & x_2 w_{22} & \frac{w_{23}}{x_2} & 0 \\ x_1^2 w_{31} & x_2^2 w_{32} & w_{33} & 1 \\ 0 & 0 & 0 & 1 \end{pmatrix}$
$\begin{pmatrix} w_{11} & w_{12} & w_{13} & 0 \\ w_{21} & w_{22} & 0 & 0 \\ w_{31} & w_{32} & w_{33} & 1 \\ 0 & 0 & 0 & 1 \end{pmatrix}$	$\begin{pmatrix} x_1 & 0 & 0 & 0 \\ 0 & x_2 & 0 & 0 \\ 0 & 0 & 1 & 0 \\ 0 & 0 & 0 & 1 \end{pmatrix}$	$\begin{pmatrix} x_1 w_{11} & \frac{x_2^2 w_{12}}{x_1} & \frac{w_{13}}{x_1} & 0 \\ \frac{x_1^2 w_{21}}{x_2} & x_2 w_{22} & 0 & 0 \\ x_1^2 w_{31} & x_2^2 w_{32} & w_{33} & 1 \\ 0 & 0 & 0 & 1 \end{pmatrix}$
$\begin{pmatrix} w_{11} & w_{12} & w_{13} & 0 \\ w_{21} & w_{22} & w_{23} & 0 \\ w_{31} & w_{32} & 0 & 1 \\ 0 & 0 & 0 & 1 \end{pmatrix}$	$\begin{pmatrix} x_1 & 0 & 0 & 0 \\ 0 & x_2 & 0 & 0 \\ 0 & 0 & 1 & 0 \\ 0 & 0 & 0 & 1 \end{pmatrix}$	$\begin{pmatrix} x_1 w_{11} & \frac{x_2^2 w_{12}}{x_1} & \frac{w_{13}}{x_1} & 0 \\ \frac{x_1^2 w_{21}}{x_2} & x_2 w_{22} & \frac{w_{23}}{x_2} & 0 \\ x_1^2 w_{31} & x_2^2 w_{32} & 0 & 1 \\ 0 & 0 & 0 & 1 \end{pmatrix}$
$\begin{pmatrix} w_{11} & w_{12} & w_{13} & 0 \\ w_{21} & w_{22} & w_{23} & 0 \\ w_{31} & w_{32} & 0 & 1 \\ 0 & 0 & 0 & 1 \end{pmatrix}$	$\begin{pmatrix} 0 & x_1 & 0 & 0 \\ x_2 & 0 & 0 & 0 \\ 0 & 0 & 1 & 0 \\ 0 & 0 & 0 & 1 \end{pmatrix}$	$\begin{pmatrix} x_2 w_{22} & \frac{x_1^2 w_{21}}{x_2} & \frac{w_{23}}{x_2} & 0 \\ \frac{x_2^2 w_{12}}{x_1} & x_1 w_{11} & \frac{w_{13}}{x_1} & 0 \\ x_2^2 w_{32} & x_1^2 w_{31} & 0 & 1 \\ 0 & 0 & 0 & 1 \end{pmatrix}$
$\begin{pmatrix} 0 & w_{12} & w_{13} & 0 \\ w_{21} & w_{22} & w_{23} & 0 \\ w_{31} & w_{32} & w_{33} & 1 \\ 0 & 0 & 0 & 1 \end{pmatrix}$	$\begin{pmatrix} x_1 & 0 & 0 & 0 \\ 0 & x_2 & 0 & 0 \\ 0 & 0 & 1 & 0 \\ 0 & 0 & 0 & 1 \end{pmatrix}$	$\begin{pmatrix} 0 & \frac{x_2^2 w_{12}}{x_1} & \frac{w_{13}}{x_1} & 0 \\ \frac{x_1^2 w_{21}}{x_2} & x_2 w_{22} & \frac{w_{23}}{x_2} & 0 \\ x_1^2 w_{31} & x_2^2 w_{32} & w_{33} & 1 \\ 0 & 0 & 0 & 1 \end{pmatrix}$
$\begin{pmatrix} w_{11} & w_{12} & w_{13} & 0 \\ w_{21} & w_{22} & w_{23} & 0 \\ w_{31} & w_{32} & w_{33} & 1 \\ 0 & 0 & 0 & 1 \end{pmatrix}$	$\begin{pmatrix} x_1 & 0 & 0 & 0 \\ 0 & x_2 & 0 & 0 \\ 0 & 0 & 1 & 0 \\ 0 & 0 & 0 & 1 \end{pmatrix}$	$\begin{pmatrix} x_1 w_{11} & \frac{x_2^2 w_{12}}{x_1} & \frac{w_{13}}{x_1} & 0 \\ \frac{x_1^2 w_{21}}{x_2} & x_2 w_{22} & \frac{w_{23}}{x_2} & 0 \\ x_1^2 w_{31} & x_2^2 w_{32} & w_{33} & 1 \\ 0 & 0 & 0 & 1 \end{pmatrix}$
$\begin{pmatrix} w_{11} & w_{12} & w_{13} & 0 \\ w_{21} & w_{22} & w_{23} & 0 \\ w_{31} & w_{32} & w_{33} & 1 \\ 0 & 0 & 0 & 1 \end{pmatrix}$	$\begin{pmatrix} 0 & x_1 & 0 & 0 \\ x_2 & 0 & 0 & 0 \\ 0 & 0 & 1 & 0 \\ 0 & 0 & 0 & 1 \end{pmatrix}$	$\begin{pmatrix} x_2 w_{22} & \frac{x_1^2 w_{21}}{x_2} & \frac{w_{23}}{x_2} & 0 \\ \frac{x_2^2 w_{12}}{x_1} & x_1 w_{11} & \frac{w_{13}}{x_1} & 0 \\ x_2^2 w_{32} & x_1^2 w_{31} & w_{33} & 1 \\ 0 & 0 & 0 & 1 \end{pmatrix}$

The Matrix U Has Two Nonzero Entries

M_B	$P_{B'B}$	M'_B
$\begin{pmatrix} 0 & 0 & w_{13} & 1 \\ w_{21} & 0 & 0 & 1 \\ 0 & w_{32} & 0 & 0 \\ 0 & 0 & 0 & 1 \end{pmatrix}$	$\begin{pmatrix} 1 & 0 & 0 & 0 \\ 0 & 1 & 0 & 0 \\ 0 & 0 & x_3 & 0 \\ 0 & 0 & 0 & 1 \end{pmatrix}$	$\begin{pmatrix} 0 & 0 & x_3^2 w_{13} & 1 \\ w_{21} & 0 & 0 & 1 \\ 0 & \frac{w_{32}}{x_3} & 0 & 0 \\ 0 & 0 & 0 & 1 \end{pmatrix}$
$\begin{pmatrix} 0 & w_{12} & 0 & 1 \\ w_{21} & 0 & 0 & 1 \\ w_{31} & 0 & w_{33} & 0 \\ 0 & 0 & 0 & 1 \end{pmatrix}$	$\begin{pmatrix} 1 & 0 & 0 & 0 \\ 0 & 1 & 0 & 0 \\ 0 & 0 & x_3 & 0 \\ 0 & 0 & 0 & 1 \end{pmatrix}$	$\begin{pmatrix} 0 & w_{12} & 0 & 1 \\ w_{21} & 0 & 0 & 1 \\ \frac{w_{31}}{x_3} & 0 & x_3 w_{33} & 0 \\ 0 & 0 & 0 & 1 \end{pmatrix}$
$\begin{pmatrix} w_{11} & 0 & w_{13} & 1 \\ 0 & 0 & w_{23} & 1 \\ 0 & w_{32} & 0 & 0 \\ 0 & 0 & 0 & 1 \end{pmatrix}$	$\begin{pmatrix} 1 & 0 & 0 & 0 \\ 0 & 1 & 0 & 0 \\ 0 & 0 & x_3 & 0 \\ 0 & 0 & 0 & 1 \end{pmatrix}$	$\begin{pmatrix} w_{11} & 0 & x_3^2 w_{13} & 1 \\ 0 & 0 & x_3^2 w_{23} & 1 \\ 0 & \frac{w_{32}}{x_3} & 0 & 0 \\ 0 & 0 & 0 & 1 \end{pmatrix}$
$\begin{pmatrix} 0 & 0 & w_{13} & 1 \\ w_{21} & w_{22} & 0 & 1 \\ w_{31} & 0 & 0 & 0 \\ 0 & 0 & 0 & 1 \end{pmatrix}$	$\begin{pmatrix} 1 & 0 & 0 & 0 \\ 0 & 1 & 0 & 0 \\ 0 & 0 & x_3 & 0 \\ 0 & 0 & 0 & 1 \end{pmatrix}$	$\begin{pmatrix} 0 & 0 & x_3^2 w_{13} & 1 \\ w_{21} & w_{22} & 0 & 1 \\ \frac{w_{31}}{x_3} & 0 & 0 & 0 \\ 0 & 0 & 0 & 1 \end{pmatrix}$
$\begin{pmatrix} w_{11} & 0 & w_{13} & 1 \\ w_{21} & 0 & 0 & 1 \\ 0 & w_{32} & 0 & 0 \\ 0 & 0 & 0 & 1 \end{pmatrix}$	$\begin{pmatrix} 1 & 0 & 0 & 0 \\ 0 & 1 & 0 & 0 \\ 0 & 0 & x_3 & 0 \\ 0 & 0 & 0 & 1 \end{pmatrix}$	$\begin{pmatrix} w_{11} & 0 & x_3^2 w_{13} & 1 \\ w_{21} & 0 & 0 & 1 \\ 0 & \frac{w_{32}}{x_3} & 0 & 0 \\ 0 & 0 & 0 & 1 \end{pmatrix}$
$\begin{pmatrix} 0 & w_{12} & 0 & 1 \\ 0 & 0 & w_{23} & 1 \\ w_{31} & 0 & w_{33} & 0 \\ 0 & 0 & 0 & 1 \end{pmatrix}$	$\begin{pmatrix} 1 & 0 & 0 & 0 \\ 0 & 1 & 0 & 0 \\ 0 & 0 & x_3 & 0 \\ 0 & 0 & 0 & 1 \end{pmatrix}$	$\begin{pmatrix} 0 & w_{12} & 0 & 1 \\ 0 & 0 & x_3^2 w_{23} & 1 \\ \frac{w_{31}}{x_3} & 0 & x_3 w_{33} & 0 \\ 0 & 0 & 0 & 1 \end{pmatrix}$
$\begin{pmatrix} w_{11} & w_{12} & 0 & 1 \\ 0 & 0 & w_{23} & 1 \\ w_{31} & 0 & 0 & 0 \\ 0 & 0 & 0 & 1 \end{pmatrix}$	$\begin{pmatrix} 1 & 0 & 0 & 0 \\ 0 & 1 & 0 & 0 \\ 0 & 0 & x_3 & 0 \\ 0 & 0 & 0 & 1 \end{pmatrix}$	$\begin{pmatrix} w_{11} & w_{12} & 0 & 1 \\ 0 & 0 & x_3^2 w_{23} & 1 \\ \frac{w_{31}}{x_3} & 0 & 0 & 0 \\ 0 & 0 & 0 & 1 \end{pmatrix}$
$\begin{pmatrix} 0 & w_{12} & w_{13} & 1 \\ w_{21} & 0 & 0 & 1 \\ 0 & w_{32} & 0 & 0 \\ 0 & 0 & 0 & 1 \end{pmatrix}$	$\begin{pmatrix} 1 & 0 & 0 & 0 \\ 0 & 1 & 0 & 0 \\ 0 & 0 & x_3 & 0 \\ 0 & 0 & 0 & 1 \end{pmatrix}$	$\begin{pmatrix} 0 & w_{12} & x_3^2 w_{13} & 1 \\ w_{21} & 0 & 0 & 1 \\ 0 & \frac{w_{32}}{x_3} & 0 & 0 \\ 0 & 0 & 0 & 1 \end{pmatrix}$
$\begin{pmatrix} 0 & w_{12} & 0 & 1 \\ 0 & 0 & w_{23} & 1 \\ w_{31} & w_{32} & 0 & 0 \\ 0 & 0 & 0 & 1 \end{pmatrix}$	$\begin{pmatrix} 1 & 0 & 0 & 0 \\ 0 & 1 & 0 & 0 \\ 0 & 0 & x_3 & 0 \\ 0 & 0 & 0 & 1 \end{pmatrix}$	$\begin{pmatrix} 0 & w_{12} & 0 & 1 \\ 0 & 0 & x_3^2 w_{23} & 1 \\ \frac{w_{31}}{x_3} & \frac{w_{32}}{x_3} & 0 & 0 \\ 0 & 0 & 0 & 1 \end{pmatrix}$
$\begin{pmatrix} 0 & w_{12} & w_{13} & 1 \\ 0 & 0 & w_{23} & 1 \\ w_{31} & 0 & 0 & 0 \\ 0 & 0 & 0 & 1 \end{pmatrix}$	$\begin{pmatrix} 1 & 0 & 0 & 0 \\ 0 & 1 & 0 & 0 \\ 0 & 0 & x_3 & 0 \\ 0 & 0 & 0 & 1 \end{pmatrix}$	$\begin{pmatrix} 0 & w_{12} & x_3^2 w_{13} & 1 \\ 0 & 0 & x_3^2 w_{23} & 1 \\ \frac{w_{31}}{x_3} & 0 & 0 & 0 \\ 0 & 0 & 0 & 1 \end{pmatrix}$

M_B	$P_{B'B}$	M'_B
$\begin{pmatrix} w_{11} & w_{12} & 0 & 1 \\ w_{21} & 0 & 0 & 1 \\ w_{31} & 0 & w_{33} & 0 \\ 0 & 0 & 0 & 1 \end{pmatrix}$	$\begin{pmatrix} 1 & 0 & 0 & 0 \\ 0 & 1 & 0 & 0 \\ 0 & 0 & x_3 & 0 \\ 0 & 0 & 0 & 1 \end{pmatrix}$	$\begin{pmatrix} w_{11} & w_{12} & 0 & 1 \\ w_{21} & 0 & 0 & 1 \\ \frac{w_{31}}{x_3} & 0 & x_3 w_{33} & 0 \\ 0 & 0 & 0 & 1 \end{pmatrix}$
$\begin{pmatrix} w_{11} & 0 & w_{13} & 1 \\ 0 & 0 & w_{23} & 1 \\ 0 & w_{32} & w_{33} & 0 \\ 0 & 0 & 0 & 1 \end{pmatrix}$	$\begin{pmatrix} 1 & 0 & 0 & 0 \\ 0 & 1 & 0 & 0 \\ 0 & 0 & x_3 & 0 \\ 0 & 0 & 0 & 1 \end{pmatrix}$	$\begin{pmatrix} w_{11} & 0 & x_3^2 w_{13} & 1 \\ 0 & 0 & x_3^2 w_{23} & 1 \\ 0 & \frac{w_{32}}{x_3} & x_3 w_{33} & 0 \\ 0 & 0 & 0 & 1 \end{pmatrix}$
$\begin{pmatrix} w_{11} & 0 & w_{13} & 1 \\ w_{21} & w_{22} & 0 & 1 \\ w_{31} & 0 & 0 & 0 \\ 0 & 0 & 0 & 1 \end{pmatrix}$	$\begin{pmatrix} 1 & 0 & 0 & 0 \\ 0 & 1 & 0 & 0 \\ 0 & 0 & x_3 & 0 \\ 0 & 0 & 0 & 1 \end{pmatrix}$	$\begin{pmatrix} w_{11} & 0 & x_3^2 w_{13} & 1 \\ w_{21} & w_{22} & 0 & 1 \\ \frac{w_{31}}{x_3} & 0 & 0 & 0 \\ 0 & 0 & 0 & 1 \end{pmatrix}$
$\begin{pmatrix} w_{11} & w_{12} & 0 & 1 \\ w_{21} & 0 & 0 & 1 \\ 0 & w_{32} & w_{33} & 0 \\ 0 & 0 & 0 & 1 \end{pmatrix}$	$\begin{pmatrix} 1 & 0 & 0 & 0 \\ 0 & 1 & 0 & 0 \\ 0 & 0 & x_3 & 0 \\ 0 & 0 & 0 & 1 \end{pmatrix}$	$\begin{pmatrix} w_{11} & w_{12} & 0 & 1 \\ w_{21} & 0 & 0 & 1 \\ 0 & \frac{w_{32}}{x_3} & x_3 w_{33} & 0 \\ 0 & 0 & 0 & 1 \end{pmatrix}$
$\begin{pmatrix} w_{11} & w_{12} & 0 & 1 \\ 0 & 0 & w_{23} & 1 \\ 0 & w_{32} & w_{33} & 0 \\ 0 & 0 & 0 & 1 \end{pmatrix}$	$\begin{pmatrix} 1 & 0 & 0 & 0 \\ 0 & 1 & 0 & 0 \\ 0 & 0 & x_3 & 0 \\ 0 & 0 & 0 & 1 \end{pmatrix}$	$\begin{pmatrix} w_{11} & w_{12} & 0 & 1 \\ 0 & 0 & x_3^2 w_{23} & 1 \\ 0 & \frac{w_{32}}{x_3} & x_3 w_{33} & 0 \\ 0 & 0 & 0 & 1 \end{pmatrix}$
$\begin{pmatrix} w_{11} & 0 & w_{13} & 1 \\ 0 & w_{22} & w_{23} & 1 \\ w_{31} & 0 & 0 & 0 \\ 0 & 0 & 0 & 1 \end{pmatrix}$	$\begin{pmatrix} 1 & 0 & 0 & 0 \\ 0 & 1 & 0 & 0 \\ 0 & 0 & x_3 & 0 \\ 0 & 0 & 0 & 1 \end{pmatrix}$	$\begin{pmatrix} w_{11} & 0 & x_3^2 w_{13} & 1 \\ 0 & w_{22} & x_3^2 w_{23} & 1 \\ \frac{w_{31}}{x_3} & 0 & 0 & 0 \\ 0 & 0 & 0 & 1 \end{pmatrix}$
$\begin{pmatrix} 0 & w_{12} & 0 & 1 \\ w_{21} & 0 & w_{23} & 1 \\ w_{31} & 0 & w_{33} & 0 \\ 0 & 0 & 0 & 1 \end{pmatrix}$	$\begin{pmatrix} 1 & 0 & 0 & 0 \\ 0 & 1 & 0 & 0 \\ 0 & 0 & x_3 & 0 \\ 0 & 0 & 0 & 1 \end{pmatrix}$	$\begin{pmatrix} 0 & w_{12} & 0 & 1 \\ w_{21} & 0 & x_3^2 w_{23} & 1 \\ \frac{w_{31}}{x_3} & 0 & x_3 w_{33} & 0 \\ 0 & 0 & 0 & 1 \end{pmatrix}$
$\begin{pmatrix} w_{11} & 0 & w_{13} & 1 \\ w_{21} & 0 & w_{23} & 1 \\ 0 & w_{32} & 0 & 0 \\ 0 & 0 & 0 & 1 \end{pmatrix}$	$\begin{pmatrix} 1 & 0 & 0 & 0 \\ 0 & 1 & 0 & 0 \\ 0 & 0 & x_3 & 0 \\ 0 & 0 & 0 & 1 \end{pmatrix}$	$\begin{pmatrix} w_{11} & 0 & x_3^2 w_{13} & 1 \\ w_{21} & 0 & x_3^2 w_{23} & 1 \\ 0 & \frac{w_{32}}{x_3} & 0 & 0 \\ 0 & 0 & 0 & 1 \end{pmatrix}$
$\begin{pmatrix} 0 & 0 & w_{13} & 1 \\ w_{21} & w_{22} & 0 & 1 \\ w_{31} & w_{32} & 0 & 0 \\ 0 & 0 & 0 & 1 \end{pmatrix}$	$\begin{pmatrix} 1 & 0 & 0 & 0 \\ 0 & 1 & 0 & 0 \\ 0 & 0 & x_3 & 0 \\ 0 & 0 & 0 & 1 \end{pmatrix}$	$\begin{pmatrix} 0 & 0 & x_3^2 w_{13} & 1 \\ w_{21} & w_{22} & 0 & 1 \\ \frac{w_{31}}{x_3} & \frac{w_{32}}{x_3} & 0 & 0 \\ 0 & 0 & 0 & 1 \end{pmatrix}$
$\begin{pmatrix} 0 & w_{12} & 0 & 1 \\ w_{21} & 0 & w_{23} & 1 \\ 0 & w_{32} & w_{33} & 0 \\ 0 & 0 & 0 & 1 \end{pmatrix}$	$\begin{pmatrix} 1 & 0 & 0 & 0 \\ 0 & 1 & 0 & 0 \\ 0 & 0 & x_3 & 0 \\ 0 & 0 & 0 & 1 \end{pmatrix}$	$\begin{pmatrix} 0 & w_{12} & 0 & 1 \\ w_{21} & 0 & x_3^2 w_{23} & 1 \\ 0 & \frac{w_{32}}{x_3} & x_3 w_{33} & 0 \\ 0 & 0 & 0 & 1 \end{pmatrix}$

M_B	$P_{B'B}$	M'_B
$\begin{pmatrix} w_{11} & w_{12} & 0 & 1 \\ w_{21} & 0 & w_{23} & 1 \\ 0 & w_{32} & 0 & 0 \\ 0 & 0 & 0 & 1 \end{pmatrix}$	$\begin{pmatrix} 1 & 0 & 0 & 0 \\ 0 & 1 & 0 & 0 \\ 0 & 0 & x_3 & 0 \\ 0 & 0 & 0 & 1 \end{pmatrix}$	$\begin{pmatrix} w_{11} & w_{12} & 0 & 1 \\ w_{21} & 0 & x_3^2 w_{23} & 1 \\ 0 & \frac{w_{32}}{x_3} & 0 & 0 \\ 0 & 0 & 0 & 1 \end{pmatrix}$
$\begin{pmatrix} 0 & 0 & w_{13} & 1 \\ 0 & w_{22} & w_{23} & 1 \\ w_{31} & w_{32} & 0 & 0 \\ 0 & 0 & 0 & 1 \end{pmatrix}$	$\begin{pmatrix} 1 & 0 & 0 & 0 \\ 0 & 1 & 0 & 0 \\ 0 & 0 & x_3 & 0 \\ 0 & 0 & 0 & 1 \end{pmatrix}$	$\begin{pmatrix} 0 & 0 & x_3^2 w_{13} & 1 \\ 0 & w_{22} & x_3^2 w_{23} & 1 \\ \frac{w_{31}}{x_3} & \frac{w_{32}}{x_3} & 0 & 0 \\ 0 & 0 & 0 & 1 \end{pmatrix}$
$\begin{pmatrix} 0 & w_{12} & 0 & 1 \\ w_{21} & 0 & 0 & 1 \\ w_{31} & w_{32} & w_{33} & 0 \\ 0 & 0 & 0 & 1 \end{pmatrix}$	$\begin{pmatrix} 1 & 0 & 0 & 0 \\ 0 & 1 & 0 & 0 \\ 0 & 0 & x_3 & 0 \\ 0 & 0 & 0 & 1 \end{pmatrix}$	$\begin{pmatrix} 0 & w_{12} & 0 & 1 \\ w_{21} & 0 & 0 & 1 \\ \frac{w_{31}}{x_3} & \frac{w_{32}}{x_3} & x_3 w_{33} & 0 \\ 0 & 0 & 0 & 1 \end{pmatrix}$
$\begin{pmatrix} 0 & w_{12} & 0 & 1 \\ w_{21} & 0 & 0 & 1 \\ w_{31} & w_{32} & w_{33} & 0 \\ 0 & 0 & 0 & 1 \end{pmatrix}$	$\begin{pmatrix} 0 & 1 & 0 & 0 \\ 1 & 0 & 0 & 0 \\ 0 & 0 & x_3 & 0 \\ 0 & 0 & 0 & 1 \end{pmatrix}$	$\begin{pmatrix} 0 & w_{21} & 0 & 1 \\ w_{12} & 0 & 0 & 1 \\ \frac{w_{32}}{x_3} & \frac{w_{31}}{x_3} & x_3 w_{33} & 0 \\ 0 & 0 & 0 & 1 \end{pmatrix}$
$\begin{pmatrix} w_{11} & w_{12} & w_{13} & 1 \\ 0 & 0 & w_{23} & 1 \\ 0 & w_{32} & 0 & 0 \\ 0 & 0 & 0 & 1 \end{pmatrix}$	$\begin{pmatrix} 1 & 0 & 0 & 0 \\ 0 & 1 & 0 & 0 \\ 0 & 0 & x_3 & 0 \\ 0 & 0 & 0 & 1 \end{pmatrix}$	$\begin{pmatrix} w_{11} & w_{12} & x_3^2 w_{13} & 1 \\ 0 & 0 & x_3^2 w_{23} & 1 \\ 0 & \frac{w_{32}}{x_3} & 0 & 0 \\ 0 & 0 & 0 & 1 \end{pmatrix}$
$\begin{pmatrix} w_{11} & w_{12} & w_{13} & 1 \\ w_{21} & 0 & 0 & 1 \\ 0 & w_{32} & 0 & 0 \\ 0 & 0 & 0 & 1 \end{pmatrix}$	$\begin{pmatrix} 1 & 0 & 0 & 0 \\ 0 & 1 & 0 & 0 \\ 0 & 0 & x_3 & 0 \\ 0 & 0 & 0 & 1 \end{pmatrix}$	$\begin{pmatrix} w_{11} & w_{12} & x_3^2 w_{13} & 1 \\ w_{21} & 0 & 0 & 1 \\ 0 & \frac{w_{32}}{x_3} & 0 & 0 \\ 0 & 0 & 0 & 1 \end{pmatrix}$
$\begin{pmatrix} 0 & w_{12} & 0 & 1 \\ 0 & 0 & w_{23} & 1 \\ w_{31} & w_{32} & w_{33} & 0 \\ 0 & 0 & 0 & 1 \end{pmatrix}$	$\begin{pmatrix} 1 & 0 & 0 & 0 \\ 0 & 1 & 0 & 0 \\ 0 & 0 & x_3 & 0 \\ 0 & 0 & 0 & 1 \end{pmatrix}$	$\begin{pmatrix} 0 & w_{12} & 0 & 1 \\ 0 & 0 & x_3^2 w_{23} & 1 \\ \frac{w_{31}}{x_3} & \frac{w_{32}}{x_3} & x_3 w_{33} & 0 \\ 0 & 0 & 0 & 1 \end{pmatrix}$
$\begin{pmatrix} w_{11} & w_{12} & w_{13} & 0 \\ 0 & 0 & w_{23} & 0 \\ w_{31} & 0 & 0 & 1 \\ 0 & 0 & 0 & 1 \end{pmatrix}$	$\begin{pmatrix} x_1 & 0 & 0 & 0 \\ 0 & x_2 & 0 & 0 \\ 0 & 0 & 1 & 0 \\ 0 & 0 & 0 & 1 \end{pmatrix}$	$\begin{pmatrix} x_1 w_{11} & \frac{x_2^2 w_{12}}{x_1} & \frac{w_{13}}{x_1} & 0 \\ 0 & 0 & \frac{w_{23}}{x_2} & 0 \\ x_1^2 w_{31} & 0 & 0 & 1 \\ 0 & 0 & 0 & 1 \end{pmatrix}$
$\begin{pmatrix} w_{11} & 0 & w_{13} & 1 \\ w_{21} & w_{22} & 0 & 1 \\ 0 & w_{32} & 0 & 0 \\ 0 & 0 & 0 & 1 \end{pmatrix}$	$\begin{pmatrix} 1 & 0 & 0 & 0 \\ 0 & 1 & 0 & 0 \\ 0 & 0 & x_3 & 0 \\ 0 & 0 & 0 & 1 \end{pmatrix}$	$\begin{pmatrix} w_{11} & 0 & x_3^2 w_{13} & 1 \\ w_{21} & w_{22} & 0 & 1 \\ 0 & \frac{w_{32}}{x_3} & 0 & 0 \\ 0 & 0 & 0 & 1 \end{pmatrix}$
$\begin{pmatrix} 0 & w_{12} & 0 & 1 \\ 0 & w_{22} & w_{23} & 1 \\ w_{31} & 0 & w_{33} & 0 \\ 0 & 0 & 0 & 1 \end{pmatrix}$	$\begin{pmatrix} 1 & 0 & 0 & 0 \\ 0 & 1 & 0 & 0 \\ 0 & 0 & x_3 & 0 \\ 0 & 0 & 0 & 1 \end{pmatrix}$	$\begin{pmatrix} 0 & w_{12} & 0 & 1 \\ 0 & w_{22} & x_3^2 w_{23} & 1 \\ \frac{w_{31}}{x_3} & 0 & x_3 w_{33} & 0 \\ 0 & 0 & 0 & 1 \end{pmatrix}$

M_B	$P_{B'B}$	M'_B
$\begin{pmatrix}w_{11}&w_{12}&0&1\\0&0&w_{23}&1\\w_{31}&0&w_{33}&0\\0&0&0&1\end{pmatrix}$	$\begin{pmatrix}1&0&0&0\\0&1&0&0\\0&0&x_3&0\\0&0&0&1\end{pmatrix}$	$\begin{pmatrix}w_{11}&w_{12}&0&1\\0&0&x_3^2w_{23}&1\\\frac{w_{31}}{x_3}&0&x_3w_{33}&0\\0&0&0&1\end{pmatrix}$
$\begin{pmatrix}w_{11}&0&w_{13}&1\\w_{21}&0&0&1\\w_{31}&w_{32}&0&0\\0&0&0&1\end{pmatrix}$	$\begin{pmatrix}1&0&0&0\\0&1&0&0\\0&0&x_3&0\\0&0&0&1\end{pmatrix}$	$\begin{pmatrix}w_{11}&0&x_3^2w_{13}&1\\w_{21}&0&0&1\\\frac{w_{31}}{x_3}&\frac{w_{32}}{x_3}&0&0\\0&0&0&1\end{pmatrix}$
$\begin{pmatrix}0&w_{12}&w_{13}&1\\0&0&w_{23}&1\\w_{31}&0&w_{33}&0\\0&0&0&1\end{pmatrix}$	$\begin{pmatrix}1&0&0&0\\0&1&0&0\\0&0&x_3&0\\0&0&0&1\end{pmatrix}$	$\begin{pmatrix}0&w_{12}&x_3^2w_{13}&1\\0&0&x_3^2w_{23}&1\\\frac{w_{31}}{x_3}&0&x_3w_{33}&0\\0&0&0&1\end{pmatrix}$
$\begin{pmatrix}w_{11}&w_{12}&0&1\\w_{21}&0&w_{23}&1\\w_{31}&0&0&0\\0&0&0&1\end{pmatrix}$	$\begin{pmatrix}1&0&0&0\\0&1&0&0\\0&0&x_3&0\\0&0&0&1\end{pmatrix}$	$\begin{pmatrix}w_{11}&w_{12}&0&1\\w_{21}&0&x_3^2w_{23}&1\\\frac{w_{31}}{x_3}&0&0&0\\0&0&0&1\end{pmatrix}$
$\begin{pmatrix}0&w_{12}&w_{13}&1\\w_{21}&0&w_{23}&1\\0&w_{32}&0&0\\0&0&0&1\end{pmatrix}$	$\begin{pmatrix}1&0&0&0\\0&1&0&0\\0&0&x_3&0\\0&0&0&1\end{pmatrix}$	$\begin{pmatrix}0&w_{12}&x_3^2w_{13}&1\\w_{21}&0&x_3^2w_{23}&1\\0&\frac{w_{32}}{x_3}&0&0\\0&0&0&1\end{pmatrix}$
$\begin{pmatrix}0&w_{12}&0&1\\w_{21}&0&w_{23}&1\\w_{31}&w_{32}&0&0\\0&0&0&1\end{pmatrix}$	$\begin{pmatrix}1&0&0&0\\0&1&0&0\\0&0&x_3&0\\0&0&0&1\end{pmatrix}$	$\begin{pmatrix}0&w_{12}&0&1\\w_{21}&0&x_3^2w_{23}&1\\\frac{w_{31}}{x_3}&\frac{w_{32}}{x_3}&0&0\\0&0&0&1\end{pmatrix}$
$\begin{pmatrix}0&w_{12}&w_{13}&1\\0&0&w_{23}&1\\w_{31}&w_{32}&0&0\\0&0&0&1\end{pmatrix}$	$\begin{pmatrix}1&0&0&0\\0&1&0&0\\0&0&x_3&0\\0&0&0&1\end{pmatrix}$	$\begin{pmatrix}0&w_{12}&x_3^2w_{13}&1\\0&0&x_3^2w_{23}&1\\\frac{w_{31}}{x_3}&\frac{w_{32}}{x_3}&0&0\\0&0&0&1\end{pmatrix}$
$\begin{pmatrix}w_{11}&w_{12}&0&1\\0&w_{22}&w_{23}&1\\w_{31}&0&w_{33}&0\\0&0&0&1\end{pmatrix}$	$\begin{pmatrix}1&0&0&0\\0&1&0&0\\0&0&x_3&0\\0&0&0&1\end{pmatrix}$	$\begin{pmatrix}w_{11}&w_{12}&0&1\\0&w_{22}&x_3^2w_{23}&1\\\frac{w_{31}}{x_3}&0&x_3w_{33}&0\\0&0&0&1\end{pmatrix}$
$\begin{pmatrix}w_{11}&w_{12}&0&1\\w_{21}&w_{22}&0&1\\w_{31}&0&w_{33}&0\\0&0&0&1\end{pmatrix}$	$\begin{pmatrix}1&0&0&0\\0&1&0&0\\0&0&x_3&0\\0&0&0&1\end{pmatrix}$	$\begin{pmatrix}w_{11}&w_{12}&0&1\\w_{21}&w_{22}&0&1\\\frac{w_{31}}{x_3}&0&x_3w_{33}&0\\0&0&0&1\end{pmatrix}$
$\begin{pmatrix}w_{11}&0&w_{13}&1\\0&w_{22}&w_{23}&1\\0&w_{32}&w_{33}&0\\0&0&0&1\end{pmatrix}$	$\begin{pmatrix}1&0&0&0\\0&1&0&0\\0&0&x_3&0\\0&0&0&1\end{pmatrix}$	$\begin{pmatrix}w_{11}&0&x_3^2w_{13}&1\\0&w_{22}&x_3^2w_{23}&1\\0&\frac{w_{32}}{x_3}&x_3w_{33}&0\\0&0&0&1\end{pmatrix}$

M_B	$P_{B'B}$	M'_B
$\begin{pmatrix}w_{11}&0&w_{13}&1\\w_{21}&w_{22}&0&1\\w_{31}&0&w_{33}&0\\0&0&0&1\end{pmatrix}$	$\begin{pmatrix}1&0&0&0\\0&1&0&0\\0&0&x_3&0\\0&0&0&1\end{pmatrix}$	$\begin{pmatrix}w_{11}&0&x_3^2w_{13}&1\\w_{21}&w_{22}&0&1\\\frac{w_{31}}{x_3}&0&x_3w_{33}&0\\0&0&0&1\end{pmatrix}$
$\begin{pmatrix}w_{11}&w_{12}&w_{13}&1\\w_{21}&0&0&1\\w_{31}&0&w_{33}&0\\0&0&0&1\end{pmatrix}$	$\begin{pmatrix}1&0&0&0\\0&1&0&0\\0&0&x_3&0\\0&0&0&1\end{pmatrix}$	$\begin{pmatrix}w_{11}&w_{12}&x_3^2w_{13}&1\\w_{21}&0&0&1\\\frac{w_{31}}{x_3}&0&x_3w_{33}&0\\0&0&0&1\end{pmatrix}$
$\begin{pmatrix}w_{11}&0&w_{13}&1\\0&0&w_{23}&1\\w_{31}&w_{32}&w_{33}&0\\0&0&0&1\end{pmatrix}$	$\begin{pmatrix}1&0&0&0\\0&1&0&0\\0&0&x_3&0\\0&0&0&1\end{pmatrix}$	$\begin{pmatrix}w_{11}&0&x_3^2w_{13}&1\\0&0&x_3^2w_{23}&1\\\frac{w_{31}}{x_3}&\frac{w_{32}}{x_3}&x_3w_{33}&0\\0&0&0&1\end{pmatrix}$
$\begin{pmatrix}w_{11}&w_{12}&w_{13}&1\\w_{21}&w_{22}&0&1\\w_{31}&0&0&0\\0&0&0&1\end{pmatrix}$	$\begin{pmatrix}1&0&0&0\\0&1&0&0\\0&0&x_3&0\\0&0&0&1\end{pmatrix}$	$\begin{pmatrix}w_{11}&w_{12}&x_3^2w_{13}&1\\w_{21}&w_{22}&0&1\\\frac{w_{31}}{x_3}&0&0&0\\0&0&0&1\end{pmatrix}$
$\begin{pmatrix}w_{11}&w_{12}&w_{13}&1\\w_{21}&0&0&1\\0&w_{32}&w_{33}&0\\0&0&0&1\end{pmatrix}$	$\begin{pmatrix}1&0&0&0\\0&1&0&0\\0&0&x_3&0\\0&0&0&1\end{pmatrix}$	$\begin{pmatrix}w_{11}&w_{12}&x_3^2w_{13}&1\\w_{21}&0&0&1\\0&\frac{w_{32}}{x_3}&x_3w_{33}&0\\0&0&0&1\end{pmatrix}$
$\begin{pmatrix}w_{11}&w_{12}&0&1\\0&0&w_{23}&1\\w_{31}&w_{32}&w_{33}&0\\0&0&0&1\end{pmatrix}$	$\begin{pmatrix}1&0&0&0\\0&1&0&0\\0&0&x_3&0\\0&0&0&1\end{pmatrix}$	$\begin{pmatrix}w_{11}&w_{12}&0&1\\0&0&x_3^2w_{23}&1\\\frac{w_{31}}{x_3}&\frac{w_{32}}{x_3}&x_3w_{33}&0\\0&0&0&1\end{pmatrix}$
$\begin{pmatrix}w_{11}&w_{12}&w_{13}&1\\0&w_{22}&w_{23}&1\\w_{31}&0&0&0\\0&0&0&1\end{pmatrix}$	$\begin{pmatrix}1&0&0&0\\0&1&0&0\\0&0&x_3&0\\0&0&0&1\end{pmatrix}$	$\begin{pmatrix}w_{11}&w_{12}&x_3^2w_{13}&1\\0&w_{22}&x_3^2w_{23}&1\\\frac{w_{31}}{x_3}&0&0&0\\0&0&0&1\end{pmatrix}$
$\begin{pmatrix}0&w_{12}&w_{13}&1\\w_{21}&w_{22}&0&1\\w_{31}&0&w_{33}&0\\0&0&0&1\end{pmatrix}$	$\begin{pmatrix}1&0&0&0\\0&1&0&0\\0&0&x_3&0\\0&0&0&1\end{pmatrix}$	$\begin{pmatrix}0&w_{12}&x_3^2w_{13}&1\\w_{21}&w_{22}&0&1\\\frac{w_{31}}{x_3}&0&x_3w_{33}&0\\0&0&0&1\end{pmatrix}$
$\begin{pmatrix}w_{11}&0&w_{13}&1\\0&w_{22}&w_{23}&1\\w_{31}&w_{32}&0&0\\0&0&0&1\end{pmatrix}$	$\begin{pmatrix}1&0&0&0\\0&1&0&0\\0&0&x_3&0\\0&0&0&1\end{pmatrix}$	$\begin{pmatrix}w_{11}&0&x_3^2w_{13}&1\\0&w_{22}&x_3^2w_{23}&1\\\frac{w_{31}}{x_3}&\frac{w_{32}}{x_3}&0&0\\0&0&0&1\end{pmatrix}$
$\begin{pmatrix}w_{11}&0&w_{13}&1\\0&w_{22}&w_{23}&1\\w_{31}&w_{32}&0&0\\0&0&0&1\end{pmatrix}$	$\begin{pmatrix}0&1&0&0\\1&0&0&0\\0&0&x_3&0\\0&0&0&1\end{pmatrix}$	$\begin{pmatrix}w_{22}&0&x_3^2w_{23}&1\\0&w_{11}&x_3^2w_{13}&1\\\frac{w_{32}}{x_3}&\frac{w_{31}}{x_3}&0&0\\0&0&0&1\end{pmatrix}$

M_B	$P_{B'B}$	M'_B
$\begin{pmatrix}0&w_{12}&w_{13}&1\\w_{21}&w_{22}&0&1\\0&w_{32}&w_{33}&0\\0&0&0&1\end{pmatrix}$	$\begin{pmatrix}1&0&0&0\\0&1&0&0\\0&0&x_3&0\\0&0&0&1\end{pmatrix}$	$\begin{pmatrix}0&w_{12}&x_3^2w_{13}&1\\w_{21}&w_{22}&0&1\\0&\frac{w_{32}}{x_3}&x_3w_{33}&0\\0&0&0&1\end{pmatrix}$
$\begin{pmatrix}w_{11}&w_{12}&0&1\\0&w_{22}&w_{23}&1\\w_{31}&w_{32}&0&0\\0&0&0&1\end{pmatrix}$	$\begin{pmatrix}1&0&0&0\\0&1&0&0\\0&0&x_3&0\\0&0&0&1\end{pmatrix}$	$\begin{pmatrix}w_{11}&w_{12}&0&1\\0&w_{22}&x_3^2w_{23}&1\\\frac{w_{31}}{x_3}&\frac{w_{32}}{x_3}&0&0\\0&0&0&1\end{pmatrix}$
$\begin{pmatrix}0&w_{12}&w_{13}&1\\0&w_{22}&w_{23}&1\\w_{31}&0&w_{33}&0\\0&0&0&1\end{pmatrix}$	$\begin{pmatrix}1&0&0&0\\0&1&0&0\\0&0&x_3&0\\0&0&0&1\end{pmatrix}$	$\begin{pmatrix}0&w_{12}&x_3^2w_{13}&1\\0&w_{22}&x_3^2w_{23}&1\\\frac{w_{31}}{x_3}&0&x_3w_{33}&0\\0&0&0&1\end{pmatrix}$
$\begin{pmatrix}0&w_{12}&w_{13}&1\\w_{21}&0&w_{23}&1\\w_{31}&0&w_{33}&0\\0&0&0&1\end{pmatrix}$	$\begin{pmatrix}1&0&0&0\\0&1&0&0\\0&0&x_3&0\\0&0&0&1\end{pmatrix}$	$\begin{pmatrix}0&w_{12}&x_3^2w_{13}&1\\w_{21}&0&x_3^2w_{23}&1\\\frac{w_{31}}{x_3}&0&x_3w_{33}&0\\0&0&0&1\end{pmatrix}$
$\begin{pmatrix}w_{11}&0&w_{13}&1\\w_{21}&0&w_{23}&1\\w_{31}&w_{32}&0&0\\0&0&0&1\end{pmatrix}$	$\begin{pmatrix}1&0&0&0\\0&1&0&0\\0&0&x_3&0\\0&0&0&1\end{pmatrix}$	$\begin{pmatrix}w_{11}&0&x_3^2w_{13}&1\\w_{21}&0&x_3^2w_{23}&1\\\frac{w_{31}}{x_3}&\frac{w_{32}}{x_3}&0&0\\0&0&0&1\end{pmatrix}$
$\begin{pmatrix}0&w_{12}&w_{13}&1\\w_{21}&w_{22}&0&1\\w_{31}&w_{32}&0&0\\0&0&0&1\end{pmatrix}$	$\begin{pmatrix}1&0&0&0\\0&1&0&0\\0&0&x_3&0\\0&0&0&1\end{pmatrix}$	$\begin{pmatrix}0&w_{12}&x_3^2w_{13}&1\\w_{21}&w_{22}&0&1\\\frac{w_{31}}{x_3}&\frac{w_{32}}{x_3}&0&0\\0&0&0&1\end{pmatrix}$
$\begin{pmatrix}0&w_{12}&w_{13}&1\\w_{21}&0&0&1\\w_{31}&w_{32}&w_{33}&0\\0&0&0&1\end{pmatrix}$	$\begin{pmatrix}1&0&0&0\\0&1&0&0\\0&0&x_3&0\\0&0&0&1\end{pmatrix}$	$\begin{pmatrix}0&w_{12}&x_3^2w_{13}&1\\w_{21}&0&0&1\\\frac{w_{31}}{x_3}&\frac{w_{32}}{x_3}&x_3w_{33}&0\\0&0&0&1\end{pmatrix}$
$\begin{pmatrix}w_{11}&w_{12}&w_{13}&1\\0&0&w_{23}&1\\w_{31}&w_{32}&0&0\\0&0&0&1\end{pmatrix}$	$\begin{pmatrix}1&0&0&0\\0&1&0&0\\0&0&x_3&0\\0&0&0&1\end{pmatrix}$	$\begin{pmatrix}w_{11}&w_{12}&x_3^2w_{13}&1\\0&0&x_3^2w_{23}&1\\\frac{w_{31}}{x_3}&\frac{w_{32}}{x_3}&0&0\\0&0&0&1\end{pmatrix}$
$\begin{pmatrix}0&w_{12}&w_{13}&1\\w_{21}&w_{22}&w_{23}&1\\w_{31}&0&0&0\\0&0&0&1\end{pmatrix}$	$\begin{pmatrix}1&0&0&0\\0&1&0&0\\0&0&x_3&0\\0&0&0&1\end{pmatrix}$	$\begin{pmatrix}0&w_{12}&x_3^2w_{13}&1\\w_{21}&w_{22}&x_3^2w_{23}&1\\\frac{w_{31}}{x_3}&0&0&0\\0&0&0&1\end{pmatrix}$
$\begin{pmatrix}0&w_{12}&0&1\\w_{21}&w_{22}&0&1\\w_{31}&w_{32}&w_{33}&0\\0&0&0&1\end{pmatrix}$	$\begin{pmatrix}1&0&0&0\\0&1&0&0\\0&0&x_3&0\\0&0&0&1\end{pmatrix}$	$\begin{pmatrix}0&w_{12}&0&1\\w_{21}&w_{22}&0&1\\\frac{w_{31}}{x_3}&\frac{w_{32}}{x_3}&x_3w_{33}&0\\0&0&0&1\end{pmatrix}$

M_B	$P_{B'B}$	M'_B
$\begin{pmatrix}w_{11}&w_{12}&w_{13}&1\\0&w_{22}&w_{23}&1\\0&w_{32}&0&0\\0&0&0&1\end{pmatrix}$	$\begin{pmatrix}1&0&0&0\\0&1&0&0\\0&0&x_3&0\\0&0&0&1\end{pmatrix}$	$\begin{pmatrix}w_{11}&w_{12}&x_3^2w_{13}&1\\0&w_{22}&x_3^2w_{23}&1\\0&\frac{w_{32}}{x_3}&0&0\\0&0&0&1\end{pmatrix}$
$\begin{pmatrix}0&0&w_{13}&1\\w_{21}&w_{22}&w_{23}&1\\w_{31}&0&w_{33}&0\\0&0&0&1\end{pmatrix}$	$\begin{pmatrix}1&0&0&0\\0&1&0&0\\0&0&x_3&0\\0&0&0&1\end{pmatrix}$	$\begin{pmatrix}0&0&x_3^2w_{13}&1\\w_{21}&w_{22}&x_3^2w_{23}&1\\\frac{w_{31}}{x_3}&0&x_3w_{33}&0\\0&0&0&1\end{pmatrix}$
$\begin{pmatrix}w_{11}&w_{12}&w_{13}&1\\w_{21}&w_{22}&0&1\\0&w_{32}&0&0\\0&0&0&1\end{pmatrix}$	$\begin{pmatrix}1&0&0&0\\0&1&0&0\\0&0&x_3&0\\0&0&0&1\end{pmatrix}$	$\begin{pmatrix}w_{11}&w_{12}&x_3^2w_{13}&1\\w_{21}&w_{22}&0&1\\0&\frac{w_{32}}{x_3}&0&0\\0&0&0&1\end{pmatrix}$
$\begin{pmatrix}0&w_{12}&0&1\\0&w_{22}&w_{23}&1\\w_{31}&w_{32}&w_{33}&0\\0&0&0&1\end{pmatrix}$	$\begin{pmatrix}1&0&0&0\\0&1&0&0\\0&0&x_3&0\\0&0&0&1\end{pmatrix}$	$\begin{pmatrix}0&w_{12}&0&1\\0&w_{22}&x_3^2w_{23}&1\\\frac{w_{31}}{x_3}&\frac{w_{32}}{x_3}&x_3w_{33}&0\\0&0&0&1\end{pmatrix}$
$\begin{pmatrix}w_{11}&w_{12}&w_{13}&1\\0&0&w_{23}&1\\w_{31}&0&w_{33}&0\\0&0&0&1\end{pmatrix}$	$\begin{pmatrix}1&0&0&0\\0&1&0&0\\0&0&x_3&0\\0&0&0&1\end{pmatrix}$	$\begin{pmatrix}w_{11}&w_{12}&x_3^2w_{13}&1\\0&0&x_3^2w_{23}&1\\\frac{w_{31}}{x_3}&0&x_3w_{33}&0\\0&0&0&1\end{pmatrix}$
$\begin{pmatrix}w_{11}&w_{12}&w_{13}&1\\w_{21}&0&0&1\\w_{31}&w_{32}&0&0\\0&0&0&1\end{pmatrix}$	$\begin{pmatrix}1&0&0&0\\0&1&0&0\\0&0&x_3&0\\0&0&0&1\end{pmatrix}$	$\begin{pmatrix}w_{11}&w_{12}&x_3^2w_{13}&1\\w_{21}&0&0&1\\\frac{w_{31}}{x_3}&\frac{w_{32}}{x_3}&0&0\\0&0&0&1\end{pmatrix}$
$\begin{pmatrix}0&w_{12}&w_{13}&1\\0&0&w_{23}&1\\w_{31}&w_{32}&w_{33}&0\\0&0&0&1\end{pmatrix}$	$\begin{pmatrix}1&0&0&0\\0&1&0&0\\0&0&x_3&0\\0&0&0&1\end{pmatrix}$	$\begin{pmatrix}0&w_{12}&x_3^2w_{13}&1\\0&0&x_3^2w_{23}&1\\\frac{w_{31}}{x_3}&\frac{w_{32}}{x_3}&x_3w_{33}&0\\0&0&0&1\end{pmatrix}$
$\begin{pmatrix}w_{11}&w_{12}&w_{13}&1\\w_{21}&0&w_{23}&1\\w_{31}&0&0&0\\0&0&0&1\end{pmatrix}$	$\begin{pmatrix}1&0&0&0\\0&1&0&0\\0&0&x_3&0\\0&0&0&1\end{pmatrix}$	$\begin{pmatrix}w_{11}&w_{12}&x_3^2w_{13}&1\\w_{21}&0&x_3^2w_{23}&1\\\frac{w_{31}}{x_3}&0&0&0\\0&0&0&1\end{pmatrix}$
$\begin{pmatrix}0&w_{12}&w_{13}&1\\w_{21}&0&w_{23}&1\\w_{31}&w_{32}&0&0\\0&0&0&1\end{pmatrix}$	$\begin{pmatrix}1&0&0&0\\0&1&0&0\\0&0&x_3&0\\0&0&0&1\end{pmatrix}$	$\begin{pmatrix}0&w_{12}&x_3^2w_{13}&1\\w_{21}&0&x_3^2w_{23}&1\\\frac{w_{31}}{x_3}&\frac{w_{32}}{x_3}&0&0\\0&0&0&1\end{pmatrix}$
$\begin{pmatrix}0&w_{12}&w_{13}&1\\w_{21}&0&w_{23}&1\\w_{31}&w_{32}&0&0\\0&0&0&1\end{pmatrix}$	$\begin{pmatrix}0&1&0&0\\1&0&0&0\\0&0&x_3&0\\0&0&0&1\end{pmatrix}$	$\begin{pmatrix}0&w_{21}&x_3^2w_{23}&1\\w_{12}&0&x_3^2w_{13}&1\\\frac{w_{32}}{x_3}&\frac{w_{31}}{x_3}&0&0\\0&0&0&1\end{pmatrix}$

M_B	$P_{B'B}$	M'_B
$\begin{pmatrix}w_{11}&w_{12}&w_{13}&1\\w_{21}&w_{22}&0&1\\w_{31}&0&w_{33}&0\\0&0&0&1\end{pmatrix}$	$\begin{pmatrix}1&0&0&0\\0&1&0&0\\0&0&x_3&0\\0&0&0&1\end{pmatrix}$	$\begin{pmatrix}w_{11}&w_{12}&x_3^2w_{13}&1\\w_{21}&w_{22}&0&1\\\frac{w_{31}}{x_3}&0&x_3w_{33}&0\\0&0&0&1\end{pmatrix}$
$\begin{pmatrix}w_{11}&0&w_{13}&1\\0&w_{22}&w_{23}&1\\w_{31}&w_{32}&w_{33}&0\\0&0&0&1\end{pmatrix}$	$\begin{pmatrix}1&0&0&0\\0&1&0&0\\0&0&x_3&0\\0&0&0&1\end{pmatrix}$	$\begin{pmatrix}w_{11}&0&x_3^2w_{13}&1\\0&w_{22}&x_3^2w_{23}&1\\\frac{w_{31}}{x_3}&\frac{w_{32}}{x_3}&x_3w_{33}&0\\0&0&0&1\end{pmatrix}$
$\begin{pmatrix}w_{11}&0&w_{13}&1\\0&w_{22}&w_{23}&1\\w_{31}&w_{32}&w_{33}&0\\0&0&0&1\end{pmatrix}$	$\begin{pmatrix}0&1&0&0\\1&0&0&0\\0&0&x_3&0\\0&0&0&1\end{pmatrix}$	$\begin{pmatrix}w_{22}&0&x_3^2w_{23}&1\\0&w_{11}&x_3^2w_{13}&1\\\frac{w_{32}}{x_3}&\frac{w_{31}}{x_3}&x_3w_{33}&0\\0&0&0&1\end{pmatrix}$
$\begin{pmatrix}w_{11}&w_{12}&w_{13}&1\\w_{21}&w_{22}&0&1\\0&w_{32}&w_{33}&0\\0&0&0&1\end{pmatrix}$	$\begin{pmatrix}1&0&0&0\\0&1&0&0\\0&0&x_3&0\\0&0&0&1\end{pmatrix}$	$\begin{pmatrix}w_{11}&w_{12}&x_3^2w_{13}&1\\w_{21}&w_{22}&0&1\\0&\frac{w_{32}}{x_3}&x_3w_{33}&0\\0&0&0&1\end{pmatrix}$
$\begin{pmatrix}w_{11}&w_{12}&0&1\\0&w_{22}&w_{23}&1\\w_{31}&w_{32}&w_{33}&0\\0&0&0&1\end{pmatrix}$	$\begin{pmatrix}1&0&0&0\\0&1&0&0\\0&0&x_3&0\\0&0&0&1\end{pmatrix}$	$\begin{pmatrix}w_{11}&w_{12}&0&1\\0&w_{22}&x_3^2w_{23}&1\\\frac{w_{31}}{x_3}&\frac{w_{32}}{x_3}&x_3w_{33}&0\\0&0&0&1\end{pmatrix}$
$\begin{pmatrix}w_{11}&w_{12}&w_{13}&0\\0&w_{22}&w_{23}&0\\w_{31}&0&w_{33}&1\\0&0&0&1\end{pmatrix}$	$\begin{pmatrix}x_1&0&0&0\\0&x_2&0&0\\0&0&1&0\\0&0&0&1\end{pmatrix}$	$\begin{pmatrix}x_1w_{11}&\frac{x_2^2w_{12}}{x_1}&\frac{w_{13}}{x_1}&0\\0&x_2w_{22}&\frac{w_{23}}{x_2}&0\\x_1^2w_{31}&0&w_{33}&1\\0&0&0&1\end{pmatrix}$
$\begin{pmatrix}w_{11}&0&w_{13}&1\\w_{21}&w_{22}&w_{23}&1\\w_{31}&0&w_{33}&0\\0&0&0&1\end{pmatrix}$	$\begin{pmatrix}1&0&0&0\\0&1&0&0\\0&0&x_3&0\\0&0&0&1\end{pmatrix}$	$\begin{pmatrix}w_{11}&0&x_3^2w_{13}&1\\w_{21}&w_{22}&x_3^2w_{23}&1\\\frac{w_{31}}{x_3}&0&x_3w_{33}&0\\0&0&0&1\end{pmatrix}$
$\begin{pmatrix}w_{11}&w_{12}&0&1\\w_{21}&w_{22}&0&1\\w_{31}&w_{32}&w_{33}&0\\0&0&0&1\end{pmatrix}$	$\begin{pmatrix}1&0&0&0\\0&1&0&0\\0&0&x_3&0\\0&0&0&1\end{pmatrix}$	$\begin{pmatrix}w_{11}&w_{12}&0&1\\w_{21}&w_{22}&0&1\\\frac{w_{31}}{x_3}&\frac{w_{32}}{x_3}&x_3w_{33}&0\\0&0&0&1\end{pmatrix}$
$\begin{pmatrix}w_{11}&w_{12}&0&1\\w_{21}&w_{22}&0&1\\w_{31}&w_{32}&w_{33}&0\\0&0&0&1\end{pmatrix}$	$\begin{pmatrix}0&1&0&0\\1&0&0&0\\0&0&x_3&0\\0&0&0&1\end{pmatrix}$	$\begin{pmatrix}w_{22}&w_{21}&0&1\\w_{12}&w_{11}&0&1\\\frac{w_{32}}{x_3}&\frac{w_{31}}{x_3}&x_3w_{33}&0\\0&0&0&1\end{pmatrix}$
$\begin{pmatrix}w_{11}&w_{12}&w_{13}&1\\w_{21}&0&0&1\\w_{31}&w_{32}&w_{33}&0\\0&0&0&1\end{pmatrix}$	$\begin{pmatrix}1&0&0&0\\0&1&0&0\\0&0&x_3&0\\0&0&0&1\end{pmatrix}$	$\begin{pmatrix}w_{11}&w_{12}&x_3^2w_{13}&1\\w_{21}&0&0&1\\\frac{w_{31}}{x_3}&\frac{w_{32}}{x_3}&x_3w_{33}&0\\0&0&0&1\end{pmatrix}$

M_B	$P_{B'B}$	M'_B
$\begin{pmatrix}w_{11}&w_{12}&w_{13}&1\\0&0&w_{23}&1\\w_{31}&w_{32}&w_{33}&0\\0&0&0&1\end{pmatrix}$	$\begin{pmatrix}1&0&0&0\\0&1&0&0\\0&0&x_3&0\\0&0&0&1\end{pmatrix}$	$\begin{pmatrix}w_{11}&w_{12}&x_3^2w_{13}&1\\0&0&x_3^2w_{23}&1\\\frac{w_{31}}{x_3}&\frac{w_{32}}{x_3}&x_3w_{33}&0\\0&0&0&1\end{pmatrix}$
$\begin{pmatrix}w_{11}&w_{12}&w_{13}&1\\w_{21}&w_{22}&w_{23}&1\\w_{31}&0&0&0\\0&0&0&1\end{pmatrix}$	$\begin{pmatrix}1&0&0&0\\0&1&0&0\\0&0&x_3&0\\0&0&0&1\end{pmatrix}$	$\begin{pmatrix}w_{11}&w_{12}&x_3^2w_{13}&1\\w_{21}&w_{22}&x_3^2w_{23}&1\\\frac{w_{31}}{x_3}&0&0&0\\0&0&0&1\end{pmatrix}$
$\begin{pmatrix}w_{11}&w_{12}&w_{13}&1\\w_{21}&0&w_{23}&1\\w_{31}&0&w_{33}&0\\0&0&0&1\end{pmatrix}$	$\begin{pmatrix}1&0&0&0\\0&1&0&0\\0&0&x_3&0\\0&0&0&1\end{pmatrix}$	$\begin{pmatrix}w_{11}&w_{12}&x_3^2w_{13}&1\\w_{21}&0&x_3^2w_{23}&1\\\frac{w_{31}}{x_3}&0&x_3w_{33}&0\\0&0&0&1\end{pmatrix}$
$\begin{pmatrix}w_{11}&0&w_{13}&1\\w_{21}&0&w_{23}&1\\w_{31}&w_{32}&w_{33}&0\\0&0&0&1\end{pmatrix}$	$\begin{pmatrix}1&0&0&0\\0&1&0&0\\0&0&x_3&0\\0&0&0&1\end{pmatrix}$	$\begin{pmatrix}w_{11}&0&x_3^2w_{13}&1\\w_{21}&0&x_3^2w_{23}&1\\\frac{w_{31}}{x_3}&\frac{w_{32}}{x_3}&x_3w_{33}&0\\0&0&0&1\end{pmatrix}$
$\begin{pmatrix}w_{11}&w_{12}&w_{13}&1\\w_{21}&w_{22}&0&1\\w_{31}&w_{32}&0&0\\0&0&0&1\end{pmatrix}$	$\begin{pmatrix}1&0&0&0\\0&1&0&0\\0&0&x_3&0\\0&0&0&1\end{pmatrix}$	$\begin{pmatrix}w_{11}&w_{12}&x_3^2w_{13}&1\\w_{21}&w_{22}&0&1\\\frac{w_{31}}{x_3}&\frac{w_{32}}{x_3}&0&0\\0&0&0&1\end{pmatrix}$
$\begin{pmatrix}0&w_{12}&w_{13}&1\\w_{21}&w_{22}&w_{23}&1\\w_{31}&0&w_{33}&0\\0&0&0&1\end{pmatrix}$	$\begin{pmatrix}1&0&0&0\\0&1&0&0\\0&0&x_3&0\\0&0&0&1\end{pmatrix}$	$\begin{pmatrix}0&w_{12}&x_3^2w_{13}&1\\w_{21}&w_{22}&x_3^2w_{23}&1\\\frac{w_{31}}{x_3}&0&x_3w_{33}&0\\0&0&0&1\end{pmatrix}$
$\begin{pmatrix}w_{11}&0&w_{13}&1\\w_{21}&w_{22}&w_{23}&1\\w_{31}&w_{32}&0&0\\0&0&0&1\end{pmatrix}$	$\begin{pmatrix}1&0&0&0\\0&1&0&0\\0&0&x_3&0\\0&0&0&1\end{pmatrix}$	$\begin{pmatrix}w_{11}&0&x_3^2w_{13}&1\\w_{21}&w_{22}&x_3^2w_{23}&1\\\frac{w_{31}}{x_3}&\frac{w_{32}}{x_3}&0&0\\0&0&0&1\end{pmatrix}$
$\begin{pmatrix}0&w_{12}&w_{13}&1\\w_{21}&w_{22}&0&1\\w_{31}&w_{32}&w_{33}&0\\0&0&0&1\end{pmatrix}$	$\begin{pmatrix}1&0&0&0\\0&1&0&0\\0&0&x_3&0\\0&0&0&1\end{pmatrix}$	$\begin{pmatrix}0&w_{12}&x_3^2w_{13}&1\\w_{21}&w_{22}&0&1\\\frac{w_{31}}{x_3}&\frac{w_{32}}{x_3}&x_3w_{33}&0\\0&0&0&1\end{pmatrix}$
$\begin{pmatrix}0&w_{12}&w_{13}&1\\w_{21}&0&w_{23}&1\\w_{31}&w_{32}&w_{33}&0\\0&0&0&1\end{pmatrix}$	$\begin{pmatrix}1&0&0&0\\0&1&0&0\\0&0&x_3&0\\0&0&0&1\end{pmatrix}$	$\begin{pmatrix}0&w_{12}&x_3^2w_{13}&1\\w_{21}&0&x_3^2w_{23}&1\\\frac{w_{31}}{x_3}&\frac{w_{32}}{x_3}&x_3w_{33}&0\\0&0&0&1\end{pmatrix}$
$\begin{pmatrix}0&w_{12}&w_{13}&1\\w_{21}&0&w_{23}&1\\w_{31}&w_{32}&w_{33}&0\\0&0&0&1\end{pmatrix}$	$\begin{pmatrix}0&1&0&0\\1&0&0&0\\0&0&x_3&0\\0&0&0&1\end{pmatrix}$	$\begin{pmatrix}0&w_{21}&x_3^2w_{23}&1\\w_{12}&0&x_3^2w_{13}&1\\\frac{w_{32}}{x_3}&\frac{w_{31}}{x_3}&x_3w_{33}&0\\0&0&0&1\end{pmatrix}$

M_B	$P_{B'B}$	M'_B
$\begin{pmatrix} w_{11} & w_{12} & w_{13} & 1 \\ w_{21} & 0 & w_{23} & 1 \\ w_{31} & w_{32} & 0 & 0 \\ 0 & 0 & 0 & 1 \end{pmatrix}$	$\begin{pmatrix} 1 & 0 & 0 & 0 \\ 0 & 1 & 0 & 0 \\ 0 & 0 & x_3 & 0 \\ 0 & 0 & 0 & 1 \end{pmatrix}$	$\begin{pmatrix} w_{11} & w_{12} & x_3^2 w_{13} & 1 \\ w_{21} & 0 & x_3^2 w_{23} & 1 \\ \frac{w_{31}}{x_3} & \frac{w_{32}}{x_3} & 0 & 0 \\ 0 & 0 & 0 & 1 \end{pmatrix}$
$\begin{pmatrix} w_{11} & w_{12} & w_{13} & 1 \\ w_{21} & w_{22} & w_{23} & 1 \\ w_{31} & 0 & w_{33} & 0 \\ 0 & 0 & 0 & 1 \end{pmatrix}$	$\begin{pmatrix} 1 & 0 & 0 & 0 \\ 0 & 1 & 0 & 0 \\ 0 & 0 & x_3 & 0 \\ 0 & 0 & 0 & 1 \end{pmatrix}$	$\begin{pmatrix} w_{11} & w_{12} & x_3^2 w_{13} & 1 \\ w_{21} & w_{22} & x_3^2 w_{23} & 1 \\ \frac{w_{31}}{x_3} & 0 & x_3 w_{33} & 0 \\ 0 & 0 & 0 & 1 \end{pmatrix}$
$\begin{pmatrix} w_{11} & 0 & w_{13} & 1 \\ w_{21} & w_{22} & w_{23} & 1 \\ w_{31} & w_{32} & w_{33} & 0 \\ 0 & 0 & 0 & 1 \end{pmatrix}$	$\begin{pmatrix} 1 & 0 & 0 & 0 \\ 0 & 1 & 0 & 0 \\ 0 & 0 & x_3 & 0 \\ 0 & 0 & 0 & 1 \end{pmatrix}$	$\begin{pmatrix} w_{11} & 0 & x_3^2 w_{13} & 1 \\ w_{21} & w_{22} & x_3^2 w_{23} & 1 \\ \frac{w_{31}}{x_3} & \frac{w_{32}}{x_3} & x_3 w_{33} & 0 \\ 0 & 0 & 0 & 1 \end{pmatrix}$
$\begin{pmatrix} w_{11} & w_{12} & w_{13} & 1 \\ w_{21} & w_{22} & 0 & 1 \\ w_{31} & w_{32} & w_{33} & 0 \\ 0 & 0 & 0 & 1 \end{pmatrix}$	$\begin{pmatrix} 1 & 0 & 0 & 0 \\ 0 & 1 & 0 & 0 \\ 0 & 0 & x_3 & 0 \\ 0 & 0 & 0 & 1 \end{pmatrix}$	$\begin{pmatrix} w_{11} & w_{12} & x_3^2 w_{13} & 1 \\ w_{21} & w_{22} & 0 & 1 \\ \frac{w_{31}}{x_3} & \frac{w_{32}}{x_3} & x_3 w_{33} & 0 \\ 0 & 0 & 0 & 1 \end{pmatrix}$
$\begin{pmatrix} w_{11} & w_{12} & w_{13} & 1 \\ w_{21} & w_{22} & w_{23} & 1 \\ w_{31} & w_{32} & 0 & 0 \\ 0 & 0 & 0 & 1 \end{pmatrix}$	$\begin{pmatrix} 1 & 0 & 0 & 0 \\ 0 & 1 & 0 & 0 \\ 0 & 0 & x_3 & 0 \\ 0 & 0 & 0 & 1 \end{pmatrix}$	$\begin{pmatrix} w_{11} & w_{12} & x_3^2 w_{13} & 1 \\ w_{21} & w_{22} & x_3^2 w_{23} & 1 \\ \frac{w_{31}}{x_3} & \frac{w_{32}}{x_3} & 0 & 0 \\ 0 & 0 & 0 & 1 \end{pmatrix}$
$\begin{pmatrix} w_{11} & w_{12} & w_{13} & 1 \\ w_{21} & w_{22} & w_{23} & 1 \\ w_{31} & w_{32} & 0 & 0 \\ 0 & 0 & 0 & 1 \end{pmatrix}$	$\begin{pmatrix} 0 & 1 & 0 & 0 \\ 1 & 0 & 0 & 0 \\ 0 & 0 & x_3 & 0 \\ 0 & 0 & 0 & 1 \end{pmatrix}$	$\begin{pmatrix} w_{22} & w_{21} & x_3^2 w_{23} & 1 \\ w_{12} & w_{11} & x_3^2 w_{13} & 1 \\ \frac{w_{32}}{x_3} & \frac{w_{31}}{x_3} & 0 & 0 \\ 0 & 0 & 0 & 1 \end{pmatrix}$
$\begin{pmatrix} 0 & w_{12} & w_{13} & 1 \\ w_{21} & w_{22} & w_{23} & 1 \\ w_{31} & w_{32} & w_{33} & 0 \\ 0 & 0 & 0 & 1 \end{pmatrix}$	$\begin{pmatrix} 1 & 0 & 0 & 0 \\ 0 & 1 & 0 & 0 \\ 0 & 0 & x_3 & 0 \\ 0 & 0 & 0 & 1 \end{pmatrix}$	$\begin{pmatrix} 0 & w_{12} & x_3^2 w_{13} & 1 \\ w_{21} & w_{22} & x_3^2 w_{23} & 1 \\ \frac{w_{31}}{x_3} & \frac{w_{32}}{x_3} & x_3 w_{33} & 0 \\ 0 & 0 & 0 & 1 \end{pmatrix}$
$\begin{pmatrix} w_{11} & w_{12} & w_{13} & 1 \\ w_{21} & w_{22} & w_{23} & 1 \\ w_{31} & w_{32} & w_{33} & 0 \\ 0 & 0 & 0 & 1 \end{pmatrix}$	$\begin{pmatrix} 1 & 0 & 0 & 0 \\ 0 & 1 & 0 & 0 \\ 0 & 0 & x_3 & 0 \\ 0 & 0 & 0 & 1 \end{pmatrix}$	$\begin{pmatrix} w_{11} & w_{12} & x_3^2 w_{13} & 1 \\ w_{21} & w_{22} & x_3^2 w_{23} & 1 \\ \frac{w_{31}}{x_3} & \frac{w_{32}}{x_3} & x_3 w_{33} & 0 \\ 0 & 0 & 0 & 1 \end{pmatrix}$
$\begin{pmatrix} w_{11} & w_{12} & w_{13} & 1 \\ w_{21} & w_{22} & w_{23} & 1 \\ w_{31} & w_{32} & w_{33} & 0 \\ 0 & 0 & 0 & 1 \end{pmatrix}$	$\begin{pmatrix} 0 & 1 & 0 & 0 \\ 1 & 0 & 0 & 0 \\ 0 & 0 & x_3 & 0 \\ 0 & 0 & 0 & 1 \end{pmatrix}$	$\begin{pmatrix} w_{22} & w_{21} & x_3^2 w_{23} & 1 \\ w_{12} & w_{11} & x_3^2 w_{13} & 1 \\ \frac{w_{32}}{x_3} & \frac{w_{31}}{x_3} & x_3 w_{33} & 0 \\ 0 & 0 & 0 & 1 \end{pmatrix}$

The Matrix U Has No Nonzero Entries

M_B	$P_{B'B}$	M'_B
$\begin{pmatrix} 0 & 0 & w_{13} & 1 \\ w_{21} & 0 & 0 & 1 \\ 0 & w_{32} & 0 & 1 \\ 0 & 0 & 0 & 1 \end{pmatrix}$	$\begin{pmatrix} 0&0&1&0 \\ 1&0&0&0 \\ 0&1&0&0 \\ 0&0&0&1 \end{pmatrix}$	$\begin{pmatrix} 0 & 0 & w_{21} & 1 \\ w_{32} & 0 & 0 & 1 \\ 0 & w_{13} & 0 & 1 \\ 0 & 0 & 0 & 1 \end{pmatrix}$
$\begin{pmatrix} 0 & 0 & w_{13} & 1 \\ w_{21} & 0 & 0 & 1 \\ 0 & w_{32} & 0 & 1 \\ 0 & 0 & 0 & 1 \end{pmatrix}$	$\begin{pmatrix} 0&1&0&0 \\ 0&0&1&0 \\ 1&0&0&0 \\ 0&0&0&1 \end{pmatrix}$	$\begin{pmatrix} 0 & 0 & w_{32} & 1 \\ w_{13} & 0 & 0 & 1 \\ 0 & w_{21} & 0 & 1 \\ 0 & 0 & 0 & 1 \end{pmatrix}$
$\begin{pmatrix} 0 & w_{12} & 0 & 1 \\ w_{21} & 0 & 0 & 1 \\ w_{31} & w_{32} & w_{33} & 1 \\ 0 & 0 & 0 & 1 \end{pmatrix}$	$\begin{pmatrix} 0&1&0&0 \\ 1&0&0&0 \\ 0&0&1&0 \\ 0&0&0&1 \end{pmatrix}$	$\begin{pmatrix} 0 & w_{21} & 0 & 1 \\ w_{12} & 0 & 0 & 1 \\ w_{32} & w_{31} & w_{33} & 1 \\ 0 & 0 & 0 & 1 \end{pmatrix}$
$\begin{pmatrix} w_{11} & w_{12} & 0 & 1 \\ 0 & w_{22} & w_{23} & 1 \\ w_{31} & 0 & w_{33} & 1 \\ 0 & 0 & 0 & 1 \end{pmatrix}$	$\begin{pmatrix} 0&0&1&0 \\ 1&0&0&0 \\ 0&1&0&0 \\ 0&0&0&1 \end{pmatrix}$	$\begin{pmatrix} w_{22} & w_{23} & 0 & 1 \\ 0 & w_{33} & w_{31} & 1 \\ w_{12} & 0 & w_{11} & 1 \\ 0 & 0 & 0 & 1 \end{pmatrix}$
$\begin{pmatrix} w_{11} & w_{12} & 0 & 1 \\ 0 & w_{22} & w_{23} & 1 \\ w_{31} & 0 & w_{33} & 1 \\ 0 & 0 & 0 & 1 \end{pmatrix}$	$\begin{pmatrix} 0&1&0&0 \\ 0&0&1&0 \\ 1&0&0&0 \\ 0&0&0&1 \end{pmatrix}$	$\begin{pmatrix} w_{33} & w_{31} & 0 & 1 \\ 0 & w_{11} & w_{12} & 1 \\ w_{23} & 0 & w_{22} & 1 \\ 0 & 0 & 0 & 1 \end{pmatrix}$
$\begin{pmatrix} 0 & w_{12} & w_{13} & 1 \\ w_{21} & w_{22} & 0 & 1 \\ w_{31} & 0 & w_{33} & 1 \\ 0 & 0 & 0 & 1 \end{pmatrix}$	$\begin{pmatrix} 1&0&0&0 \\ 0&0&1&0 \\ 0&1&0&0 \\ 0&0&0&1 \end{pmatrix}$	$\begin{pmatrix} 0 & w_{13} & w_{12} & 1 \\ w_{31} & w_{33} & 0 & 1 \\ w_{21} & 0 & w_{22} & 1 \\ 0 & 0 & 0 & 1 \end{pmatrix}$
$\begin{pmatrix} 0 & w_{12} & w_{13} & 1 \\ w_{21} & 0 & w_{23} & 1 \\ w_{31} & w_{32} & 0 & 1 \\ 0 & 0 & 0 & 1 \end{pmatrix}$	$\begin{pmatrix} 0&1&0&0 \\ 1&0&0&0 \\ 0&0&1&0 \\ 0&0&0&1 \end{pmatrix}$	$\begin{pmatrix} 0 & w_{21} & w_{23} & 1 \\ w_{12} & 0 & w_{13} & 1 \\ w_{32} & w_{31} & 0 & 1 \\ 0 & 0 & 0 & 1 \end{pmatrix}$
$\begin{pmatrix} 0 & w_{12} & w_{13} & 1 \\ w_{21} & 0 & w_{23} & 1 \\ w_{31} & w_{32} & 0 & 1 \\ 0 & 0 & 0 & 1 \end{pmatrix}$	$\begin{pmatrix} 0&0&1&0 \\ 1&0&0&0 \\ 0&1&0&0 \\ 0&0&0&1 \end{pmatrix}$	$\begin{pmatrix} 0 & w_{23} & w_{21} & 1 \\ w_{32} & 0 & w_{31} & 1 \\ w_{12} & w_{13} & 0 & 1 \\ 0 & 0 & 0 & 1 \end{pmatrix}$
$\begin{pmatrix} 0 & w_{12} & w_{13} & 1 \\ w_{21} & 0 & w_{23} & 1 \\ w_{31} & w_{32} & 0 & 1 \\ 0 & 0 & 0 & 1 \end{pmatrix}$	$\begin{pmatrix} 0&0&1&0 \\ 0&1&0&0 \\ 1&0&0&0 \\ 0&0&0&1 \end{pmatrix}$	$\begin{pmatrix} 0 & w_{32} & w_{31} & 1 \\ w_{23} & 0 & w_{21} & 1 \\ w_{13} & w_{12} & 0 & 1 \\ 0 & 0 & 0 & 1 \end{pmatrix}$
$\begin{pmatrix} 0 & w_{12} & w_{13} & 1 \\ w_{21} & 0 & w_{23} & 1 \\ w_{31} & w_{32} & 0 & 1 \\ 0 & 0 & 0 & 1 \end{pmatrix}$	$\begin{pmatrix} 1&0&0&0 \\ 0&0&1&0 \\ 0&1&0&0 \\ 0&0&0&1 \end{pmatrix}$	$\begin{pmatrix} 0 & w_{13} & w_{12} & 1 \\ w_{31} & 0 & w_{32} & 1 \\ w_{21} & w_{23} & 0 & 1 \\ 0 & 0 & 0 & 1 \end{pmatrix}$

M_B	$P_{B'B}$	M'_B
$\begin{pmatrix} 0 & w_{12} & w_{13} & 1 \\ w_{21} & 0 & w_{23} & 1 \\ w_{31} & w_{32} & 0 & 1 \\ 0 & 0 & 0 & 1 \end{pmatrix}$	$\begin{pmatrix} 0&1&0&0 \\ 0&0&1&0 \\ 1&0&0&0 \\ 0&0&0&1 \end{pmatrix}$	$\begin{pmatrix} 0 & w_{31} & w_{32} & 1 \\ w_{13} & 0 & w_{12} & 1 \\ w_{23} & w_{21} & 0 & 1 \\ 0 & 0 & 0 & 1 \end{pmatrix}$
$\begin{pmatrix} w_{11} & w_{12} & w_{13} & 1 \\ w_{21} & w_{22} & 0 & 1 \\ w_{31} & 0 & w_{33} & 1 \\ 0 & 0 & 0 & 1 \end{pmatrix}$	$\begin{pmatrix} 1&0&0&0 \\ 0&0&1&0 \\ 0&1&0&0 \\ 0&0&0&1 \end{pmatrix}$	$\begin{pmatrix} w_{11} & w_{13} & w_{12} & 1 \\ w_{31} & w_{33} & 0 & 1 \\ w_{21} & 0 & w_{22} & 1 \\ 0 & 0 & 0 & 1 \end{pmatrix}$
$\begin{pmatrix} w_{11} & 0 & w_{13} & 1 \\ w_{21} & w_{22} & w_{23} & 1 \\ w_{31} & 0 & w_{33} & 1 \\ 0 & 0 & 0 & 1 \end{pmatrix}$	$\begin{pmatrix} 0&0&1&0 \\ 0&1&0&0 \\ 1&0&0&0 \\ 0&0&0&1 \end{pmatrix}$	$\begin{pmatrix} w_{33} & 0 & w_{31} & 1 \\ w_{23} & w_{22} & w_{21} & 1 \\ w_{13} & 0 & w_{11} & 1 \\ 0 & 0 & 0 & 1 \end{pmatrix}$
$\begin{pmatrix} 0 & w_{12} & w_{13} & 1 \\ w_{21} & 0 & w_{23} & 1 \\ w_{31} & w_{32} & w_{33} & 1 \\ 0 & 0 & 0 & 1 \end{pmatrix}$	$\begin{pmatrix} 0&1&0&0 \\ 1&0&0&0 \\ 0&0&1&0 \\ 0&0&0&1 \end{pmatrix}$	$\begin{pmatrix} 0 & w_{21} & w_{23} & 1 \\ w_{12} & 0 & w_{13} & 1 \\ w_{32} & w_{31} & w_{33} & 1 \\ 0 & 0 & 0 & 1 \end{pmatrix}$
$\begin{pmatrix} w_{11} & w_{12} & w_{13} & 1 \\ w_{21} & w_{22} & w_{23} & 1 \\ w_{31} & w_{32} & 0 & 1 \\ 0 & 0 & 0 & 1 \end{pmatrix}$	$\begin{pmatrix} 0&1&0&0 \\ 1&0&0&0 \\ 0&0&1&0 \\ 0&0&0&1 \end{pmatrix}$	$\begin{pmatrix} w_{22} & w_{21} & w_{23} & 1 \\ w_{12} & w_{11} & w_{13} & 1 \\ w_{32} & w_{31} & 0 & 1 \\ 0 & 0 & 0 & 1 \end{pmatrix}$
$\begin{pmatrix} w_{11} & w_{12} & w_{13} & 1 \\ w_{21} & w_{22} & w_{23} & 1 \\ w_{31} & w_{32} & w_{33} & 1 \\ 0 & 0 & 0 & 1 \end{pmatrix}$	$\begin{pmatrix} 0&1&0&0 \\ 1&0&0&0 \\ 0&0&1&0 \\ 0&0&0&1 \end{pmatrix}$	$\begin{pmatrix} w_{22} & w_{21} & w_{23} & 1 \\ w_{12} & w_{11} & w_{13} & 1 \\ w_{32} & w_{31} & w_{33} & 1 \\ 0 & 0 & 0 & 1 \end{pmatrix}$
$\begin{pmatrix} w_{11} & w_{12} & w_{13} & 1 \\ w_{21} & w_{22} & w_{23} & 1 \\ w_{31} & w_{32} & w_{33} & 1 \\ 0 & 0 & 0 & 1 \end{pmatrix}$	$\begin{pmatrix} 0&0&1&0 \\ 1&0&0&0 \\ 0&1&0&0 \\ 0&0&0&1 \end{pmatrix}$	$\begin{pmatrix} w_{22} & w_{23} & w_{21} & 1 \\ w_{32} & w_{33} & w_{31} & 1 \\ w_{12} & w_{13} & w_{11} & 1 \\ 0 & 0 & 0 & 1 \end{pmatrix}$
$\begin{pmatrix} w_{11} & w_{12} & w_{13} & 1 \\ w_{21} & w_{22} & w_{23} & 1 \\ w_{31} & w_{32} & w_{33} & 1 \\ 0 & 0 & 0 & 1 \end{pmatrix}$	$\begin{pmatrix} 0&0&1&0 \\ 0&1&0&0 \\ 1&0&0&0 \\ 0&0&0&1 \end{pmatrix}$	$\begin{pmatrix} w_{33} & w_{32} & w_{31} & 1 \\ w_{23} & w_{22} & w_{21} & 1 \\ w_{13} & w_{12} & w_{11} & 1 \\ 0 & 0 & 0 & 1 \end{pmatrix}$
$\begin{pmatrix} w_{11} & w_{12} & w_{13} & 1 \\ w_{21} & w_{22} & w_{23} & 1 \\ w_{31} & w_{32} & w_{33} & 1 \\ 0 & 0 & 0 & 1 \end{pmatrix}$	$\begin{pmatrix} 1&0&0&0 \\ 0&0&1&0 \\ 0&1&0&0 \\ 0&0&0&1 \end{pmatrix}$	$\begin{pmatrix} w_{11} & w_{13} & w_{12} & 1 \\ w_{31} & w_{33} & w_{32} & 1 \\ w_{21} & w_{23} & w_{22} & 1 \\ 0 & 0 & 0 & 1 \end{pmatrix}$
$\begin{pmatrix} w_{11} & w_{12} & w_{13} & 1 \\ w_{21} & w_{22} & w_{23} & 1 \\ w_{31} & w_{32} & w_{33} & 1 \\ 0 & 0 & 0 & 1 \end{pmatrix}$	$\begin{pmatrix} 0&1&0&0 \\ 0&0&1&0 \\ 1&0&0&0 \\ 0&0&0&1 \end{pmatrix}$	$\begin{pmatrix} w_{33} & w_{31} & w_{32} & 1 \\ w_{13} & w_{11} & w_{12} & 1 \\ w_{23} & w_{21} & w_{22} & 1 \\ 0 & 0 & 0 & 1 \end{pmatrix}$

References

1. Cabrera Casado, Y., Kanuni, M., Siles Molina, M.: Basic ideals in evolution algebras. Linear Algebra Appl. (to appear)
2. Cabrera Casado, Y., Siles Molina, M., Velasco, M.V.: Classification of three-dimensional evolution algebras. Linear Algebra Appl. **524**, 68–108 (2017)
3. Cabrera Casado, Y., Siles Molina, M., Velasco, M.V.: Evolution algebras of arbitrary dimension and their decompositions. Linear Algebra Appl. **495**, 122–162 (2016)
4. Cabrera Casado, Y., Siles Molina, M.: Classification of four dimensional simple evolution algebras, Preprint
5. Cabrera Casado, Y.: Evolution algebras. Doctoral dissertation, Universidad de Málaga. http://hdl.handle.net/10630/14175 (2016)
6. Casas, José M., Ladra, Manuel, Omirov, Bakhrom A., Rozikov, Utkir A.: On evolution algebras. Algebra Colloq. **21**, 331–342 (2014)
7. Tian, J.P., Vojtechovsky, P.: Mathematical concepts of evolution algebras in non-Mendelian genetics. Quasigroups Relat. Syst. **14**(1), 111–122 (2006)
8. Tian, J.P.: Evolution Algebras and Their Applications. Lecture Notes in Mathematics, vol. 1921. Springer, Berlin (2008)

Power-Associative Evolution Algebras

Moussa Ouattara and Souleymane Savadogo

Abstract The paper is devoted to the study of evolution algebras that are power-associative algebras. We give the Wedderburn decomposition of evolution algebras that are power-associative algebras and we prove that these algebras are Jordan algebras. Finally, we use this decomposition to classify these algebras up to dimension six.

Keywords Evolution algebras · Power associativity · Wedderburn decomposition · Nilalgebras

1 Introduction

The notion of evolution algebras was introduced in 2006 by J.P. Tian and P. Vojtěchovský [13]. In 2008, J.P. Tian laid the foundations of this algebra in his monograph [14].

In ([13], Theorem 1.5), the authors show that evolution algebras are commutative (hence flexible) not necessarily power-associative (nor associative).

Let $\mathcal{E}$ be n-dimensional algebra over a commutative field F. We say that $\mathcal{E}$ is an *n-dimensional evolution algebra* if it admits a basis $B = \{e_i; 1 \leq i \leq n\}$ such that:

$$e_i^2 = \sum_{k=1}^{n} a_{ik} e_k \text{ for all } 1 \leq i \leq n \text{ and } e_i e_j = 0 \text{ for } i \neq j. \tag{1}$$

M. Ouattara (✉)
Département de Mathématiques et Informatique, Université Joseph KI-ZERBO, 03 BP 7021, Ouagadougou 03, Burkina Faso
e-mail: ouatt_ken@yahoo.fr

S. Savadogo
Université Norbert Zongo, BP 376, Koudougou, Burkina Faso
e-mail: sara01souley@yahoo.fr

M. Siles Molina et al. (eds.), *Associative and Non-Associative Algebras and Applications*, Springer Proceedings in Mathematics & Statistics 311, https://doi.org/10.1007/978-3-030-35256-1_2

The basis B is called *natural basis* of $\mathcal{E}$ and the matrix (a_{ij}) is the matrix of structural constants of $\mathcal{E}$, relative to the natural basis B.

In ([2], p.127), the authors exhibit a necessary condition for an evolution algebra to be power-associative. They show in particular that the only power-associative evolution algebras are such that $a_{ii}^2 = a_{ii}$ $(i = 1, \ldots, n)$. However, this condition is not sufficient. It is verified for all nil evolution algebras of the fact that for any $i = 1, \ldots, n$, $a_{ii} = 0$ ([5, Proof of Theorem 2.2]).

In [4, Example 4.8], for $k = n = 4$, we find a power-associative algebra which is not associative. Indeed, it is isomorphic to $N_{4,6}$ (see Table 1). Except the type [1, 1, 1], all algebras in [7, Table 1] are nil associative evolution algebras. In [7, Table 2], the first line is associative, while the third is power-associative. In [3, Table 1], the fifth, sixth and seventh algebras, of dimension 3, are associative. They are respectively isomorphic to $\mathcal{E}_{3,4}$, $N_{3,3}(1)$ and $N_{3,2}$ (see Table 1). In [9, Theorem 6.1], the algebras $\mathcal{E}_{4,1}$, $\mathcal{E}_{4,2}$, $\mathcal{E}_{4,3}$, $\mathcal{E}_{4,5}$ and $\mathcal{E}_{4,10}$ are associative.

The evolution algebras are not defined by identities and thus do not form a variety of non-associative algebras, such as that of Lie algebras, alternative algebras or Jordan algebras. Therefore, research on these algebras follows different paths [2–7, 9, 11, 13].

In Sect. 2, we give an example of an algebra which is power-associative but which is not an evolution algebra. We also recall some definitions and known results about nil evolution algebras.

In Sect. 3, we give Wedderburn decomposition of a power-associative evolution algebra. Since the semisimple component is associative, we deduce that an evolution algebra is power-associative if and only if its nil radical is power-associative. We show that the nil power-associative evolution algebras are Jordan algebras. Thus, power-associative evolution algebras are identified with those which are Jordan algebras.

In Sect. 4, we determine power-associative evolution algebras up to dimension six.

2 Preliminaries

2.1 *Example of Jordan Algebra That Is Not Evolution Algebra*

An algebra $\mathcal{E}$ is *baric* if there is a non trivial homomorphism of algebra $\omega : \mathcal{E} \longrightarrow F$.

Example 1 In [2, Example 2.2], the authors show that zygotic algebra for simple Mendelian inheritance $Z(n, 2)$ (case $n = 2$) is not an evolution algebra. This algebra is a Jordan algebra. In fact, the algebra $Z(n, 2)$ is the commutative duplicate of the gametic algebra for simple Mendelian $G(n, 2)$.

Example 2 The gametic algebra for simple Mendelian inheritance $G(n, 2)$ is not an evolution algebra.

Proof The multiplication table of $G(n,2)$ in the basis $B_1 = \{a_i; 1 \leq i \leq n\}$ is given by $a_ia_j = \frac{1}{2}(a_i + a_j)$ for all $i, j \in \{1, \dots, n\}$ and the linear mapping $\omega : a_i \longmapsto 1$ for all $i \in \{1, \dots, n\}$ is a weight function of $G(n,2)$. For all $x, y \in G(n,2)$, we have $xy = \frac{1}{2}(\omega(x)y + \omega(y)x)$ (see [15]). Suppose that $G(n,2)$ is an evolution algebra in the basis $B_2 = \{e_i; 1 \leq i \leq n\}$.

Since ω is a weight function, there is $i_0 \in \{1, \dots, n\}$ such that $\omega(e_{i_0}) \neq 0$. For $j \neq i_0$, $e_{i_0}e_j = 0$ leads to $\omega(e_{i_0})\omega(e_j) = 0$, so $\omega(e_j) = 0$. Then $0 = e_{i_0}e_j = \frac{1}{2}\omega(e_{i_0})e_j$ gives $e_j = 0$, which is impossible. Such a basis B_2 does not exist.

2.2 Results About Nil Evolution Algebras

For an element $a \in \mathcal{E}$, we define principal powers as follows $a^1 = a$ and $a^{k+1} = a^ka$ while that of $\mathcal{E}$ are defined by

$$\mathcal{E}^1 = \mathcal{E}, \quad \mathcal{E}^{k+1} = \mathcal{E}^k\mathcal{E} \quad (k \geq 1).$$

Definition 1 We will say that algebra $\mathcal{E}$ is:

(i) *right nilpotent* if there is a nonzero integer n such that $\mathcal{E}^n = 0$ and the minimal such number is called the *index* of right nilpotency;

(ii) *nil* if there is a nonzero integer $n(a)$ such that $a^{n(a)} = 0$, for all $a \in \mathcal{E}$ and the minimal such number is called the *nil-index* of $\mathcal{E}$.

Theorem 1 ([5, Theorem 2.7]) *The following statements are equivalent for an n-dimensional evolution algebra $\mathcal{E}$:*

(a) The matrix corresponding to $\mathcal{E}$ can be written as

$$\widehat{A} = \begin{pmatrix} 0 & a_{12} & a_{13} & \cdots & a_{1n} \\ 0 & 0 & a_{23} & \cdots & a_{2n} \\ 0 & 0 & 0 & \cdots & a_{3n} \\ \vdots & \vdots & \vdots & \cdots & \vdots \\ 0 & 0 & 0 & \cdots & 0 \end{pmatrix}.$$

(b) $\mathcal{E}$ is a right nilpotent algebra.

(c) $\mathcal{E}$ is a nil algebra.

We define the following subspace of $\mathcal{E}$:

- The *annihilator* $\mathrm{ann}^1(\mathcal{E}) = \mathrm{ann}(\mathcal{E}) = \{x \in \mathcal{E} : x\mathcal{E} = 0\}$;
- $\mathrm{ann}^i(\mathcal{E})$ is defined by $\mathrm{ann}^i(\mathcal{E})/\mathrm{ann}^{i-1}(\mathcal{E}) = \mathrm{ann}\big(\mathcal{E}/\mathrm{ann}^{i-1}(\mathcal{E})\big)$.
- $\mathcal{U}_i \oplus \mathcal{U}_1 = \{x \in \mathrm{ann}^i(\mathcal{E}) \mid x\mathrm{ann}^{i-1}(\mathcal{E}) = 0\}$.

In ([6, Lemma 2.7]), the authors show that $\mathrm{ann}(\mathcal{E}) = span\{e_i \in B \mid e_i^2 = 0\}$.

In [7], the authors show that $\mathrm{ann}^i(\mathcal{E}) = span\{e \in B \mid e^2 \in \mathrm{ann}^{i-1}(\mathcal{E})\}$ and that the basis $B = B_1 \cup \cdots \cup B_r$, where $B_i = \{e \in B \mid e^2 \in \mathrm{ann}^{i-1}(\mathcal{E}), e \notin \mathrm{ann}^{i-1}(\mathcal{E})\}$.

Then, for $\mathcal{U}_i := span\{B_i\}$, $i = 1, \ldots, r$, we have $\mathcal{U}_1 \oplus \cdots \oplus \mathcal{U}_i = \text{ann}^i(\mathcal{E})$ $(i = 1, \ldots, r)$. They prove then that $\mathcal{U}_i \oplus \mathcal{U}_1$ is an invariant for nil evolution algebras.

The type of a nil evolution algebra $\mathcal{E}$ is the sequence $[n_1, n_2, \ldots, n_r]$ where r and n_i are integers defined by

$$\text{ann}^r(\mathcal{E}) = \mathcal{E};\ n_i = \dim_F(\text{ann}^i(\mathcal{E})) - \dim_F(\text{ann}^{i-1}(\mathcal{E}))$$

and $n_1 + \cdots + n_i = dim_F(\text{ann}^i(\mathcal{E}))$ for all $i \in \{1, \ldots, r\}$.

3 Power Associativity

3.1 Some Definitions

Definition 2 An algebra $\mathcal{E}$ is called:

(i) *associative* if for all $x,\ y,\ z \in \mathcal{E}$, $(x, y, z) = 0$ where $(x, y, z) = (xy)z - x(yz)$ is the associator of the elements $x,\ y,\ z$ of $\mathcal{E}$;

(ii) *Jordan* if it is commutative and $(x^2, y, x) = 0$, for all $x, y \in \mathcal{E}$;

(iii) *power-associative* if the subalgebra generated by x is associative. In other words, for all $x \in \mathcal{E}$, $x^i x^j = x^{i+j}$ for all integers $i,\ j \geq 1$.

Definition 3 Algebra $\mathcal{E}$ is said to be *third power-associative* if for all $x \in \mathcal{E}$, $x^2x = xx^2$.

Remark 1 Since any commutative algebra is third power-associative, we deduce that any evolution algebra is third power-associative.

Definition 4 An algebra $\mathcal{E}$ is said to be *fourth power-associative* if $x^2x^2 = x^4$ for all $x \in \mathcal{E}$.

Theorem 2 ([1]) *Let F be a field of characteristic $\neq 2, 3, 5$. An algebra $\mathcal{E}$ is power-associative if and only if $x^2x^2 = x^4$, for all $x \in \mathcal{E}$.*

3.2 Characterization of Associative Algebras

If $\mathcal{E}$ is a n-dimensional algebra with basis $\{e_i;\ 1 \leq i \leq n\}$, then $\mathcal{E}$ is associative if and only if $(e_ie_j)e_k = e_i(e_je_k)$ for all $1 \leq i, j, k \leq n$ [10, Proposition 1]. We then deduce:

Lemma 1 *An evolution algebra with natural basis $B = \{e_i;\ 1 \leq i \leq n\}$ is associative if and only if $e_i^2e_j = e_i(e_ie_j) = 0$ for all $1 \leq i \neq j \leq n$.*

Theorem 3 ([8, Theorem 1]) *A commutative power-associative nilalgebra A of nilindex 3 and of dimension 4 over a field F of characteristic $\neq 2$ is associative, and $A^3 = 0$.*

For the evolution algebra, this theorem has the following generalization:

Theorem 4 *Let $\mathcal{E}$ be a finite-dimensional evolution algebra. Then $\mathcal{E}$ is an associative nilalgebra if and only if $x^3 = 0$, for all $x \in \mathcal{E}$. In this case $\mathcal{E}^3 = 0$.*

Proof Let $B = \{e_i;\ 1 \leq i \leq n\}$ be a natural basis of $\mathcal{E}$.

Let's suppose that $\mathcal{E}$ is an associative nilalgebra. Then for all integer i we have $e_i^3 = 0$. Set $x = \sum_{i=1}^n x_i e_i$. We have $x^3 = \sum_{i,j=1}^n x_i^2 x_j e_i^2 e_j = 0$ because $e_i^2 e_j = 0$ since $\mathcal{E}$ is nil and associative.

Conversely the partial linearization of the identity $x^3 = 0$ give

$$2(xy)x + x^2 y = 0. \tag{2}$$

For i distinct from j we have $e_i^2 e_j = 0$ and $\mathcal{E}$ is associative by Lemma 1. We have $\mathcal{E}^3 = \langle e_i^2 e_j;\ 1 \leq i, j \leq n \rangle = 0$. □

3.3 Characterization of Power-Associative Algebras

Let $\mathcal{E}$ be an n-dimensional evolution algebra over the field F of characteristic $\neq 2$ and with a natural basis $B = \{e_i;\ 1 \leq i \leq n\}$ whose multiplication table is given by (1).

Lemma 2 *An evolution algebra $\mathcal{E}$ is a fourth power-associative if and only if the following conditions are satisfied:*

(1) $e_i^4 = e_i^2 e_i^2$ for all $1 \leq i \leq n$;

(2) $2e_i^2 e_j^2 = (e_i^2 e_j)e_j + (e_j^2 e_i)e_i$ for all $1 \leq i < j \leq n$;

(3) $e_i^3 e_j + (e_i^2 e_j)e_i = 0$ for all $1 \leq i \neq j \leq n$;

(4) $(e_i^2 e_j)e_k + (e_i^2 e_k)e_j = 0$ for all $1 \leq i \leq n$ and $1 \leq j < k \leq n$ with $i \neq j$ and $i \neq k$.

Proof Suppose that $\mathcal{E}$ is a fourth power-associative evolution algebra and $x, y, z \in \mathcal{E}$. We have

$$(x, x, x^2) = 0. \tag{3}$$

The partial linearization of (3) gives:

$$(y, x, x^2) + (x, y, x^2) + 2(x, x, xy) = 0 \tag{4}$$

$$\begin{aligned}(y, z, x^2) &+ 2(y, x, xz) + (z, y, x^2) + 2(x, y, xz) + 2(z, x, xy) + 2(x, z, xy)\\ &+ 2(x, x, zy) = 0\end{aligned} \tag{5}$$

Since the basis vectors $\{e_i;\ 1 \leq i \leq n\}$ check identities (3), (4) and (5), we have
(3) $\Longrightarrow$ (1) taking $x = e_i$.
(4) $\Longrightarrow$ (3) taking $x = e_i$ and $y = e_j$ with $i \neq j$.
(5) $\Longrightarrow$ (2) taking $x = e_i$ and $y = z = e_j$ with $i \neq j$.
(5) $\Longrightarrow$ (4) taking $x = e_i$, $y = e_j$ and $z = e_k$; i, j, k are pairwise distinct.

Conversely, suppose that conditions (1), (2), (3) and (4) are satisfied. Let $x = \sum_{i=1}^{n} x_i e_i$ be an element of $\mathcal{E}$. We have the following equalities:

$$
\begin{aligned}
x^2 &= \sum_{i=1}^{n} x_i^2 e_i^2, \\
x^3 &= \sum_{i,j=1}^{n} x_i^2 x_j e_i^2 e_j = \sum_{i=1}^{n} x_i^3 e_i^3 + \sum_{i=1}^{n} \sum_{j=1,\ j\neq i}^{n} x_i^2 x_j e_i^2 e_j, \\
x^2 x^2 &= \sum_{i,j=1}^{n} x_i^2 x_j^2 e_i^2 e_j^2 = \sum_{i=1}^{n} x_i^4 e_i^2 e_i^2 + \sum_{i=1}^{n} \sum_{j=1,\ j\neq i}^{n} x_i^2 x_j^2 e_i^2 e_j^2 \\
&= \sum_{i=1}^{n} x_i^4 e_i^2 e_i^2 + 2 \sum_{1\leq i<j\leq n} x_i^2 x_j^2 e_i^2 e_j^2, \\
x^4 &= \sum_{i,j=1}^{n} x_i^3 x_j e_i^3 e_j + \sum_{i=1}^{n} \sum_{\substack{j=1 \\ j\neq i}}^{n} \sum_{k=1}^{n} x_i^2 x_j x_k (e_i^2 e_j) e_k \\
&= \sum_{i=1}^{n} x_i^4 e_i^4 + \sum_{i=1}^{n} \sum_{j=1,\ j\neq i}^{n} x_i^3 x_j \big(e_i^3 e_j + (e_i^2 e_j) e_i\big) + \sum_{i=1}^{n} \sum_{j=1,\ j\neq i}^{n} x_i^2 x_j^2 (e_i^2 e_j) e_j \\
&\quad + \sum_{i=1}^{n} \sum_{j=1,\ j\neq i}^{n} \sum_{\substack{k=1 \\ k\neq i \\ k\neq j}}^{n} x_i^2 x_j x_k (e_i^2 e_j) e_k.
\end{aligned}
$$

We have

$$
\sum_{i=1}^{n} \sum_{\substack{j=1 \\ j\neq i}}^{n} x_i^2 x_j^2 (e_i^2 e_j) e_j = \sum_{1\leq i<j\leq n} x_i^2 x_j^2 \big((e_i^2 e_j) e_j + (e_j^2 e_i) e_i\big),
$$

and

$$
\sum_{i=1}^{n} \sum_{\substack{j=1 \\ j\neq i}}^{n} \sum_{\substack{k=1 \\ k\neq i \\ k\neq j}}^{n} x_i^2 x_j x_k (e_i^2 e_j) e_k = \sum_{i=1}^{n} \sum_{\substack{1\leq j<k\leq n \\ j\neq i \\ k\neq i}} x_i^2 x_j x_k \big((e_i^2 e_j) e_k + (e_i^2 e_k) e_j\big) = 0.
$$

So

$$x^4 = \sum_{i=1}^{n} x_i^4 e_i^4 + \sum_{1 \leq i < j \leq n} x_i^2 x_j^2 \big((e_i^2 e_j)e_j + (e_j^2 e_i)e_i\big) = x^2 x^2.$$

Corollary 1 *Let $\mathcal{E}$ be a nil evolution algebra. Then $\mathcal{E}$ is fourth power-associative algebra if and only if the following conditions are satisfied:*

(1) $e_i^2 e_j^2 = 0$ *for all* $1 \leq i \leq j \leq n$;
(2) $(e_i^2 e_j)e_k = 0$ *for all* $1 \leq i, j, k \leq n$.

Proof Since $\mathcal{E}$ is a nil evolution algebra, then for all $1 \leq i, j, k \leq n$, we have $e_i^3 = 0$ and $a_{jk}a_{kj} = 0$ [5, proof of Theorem 2.2]. According to (2) of Lemma 2, we have $2e_i^2 e_j^2 = (e_i^2 e_j)e_j + (e_j^2 e_i)e_i = a_{ij}e_j^3 + a_{ji}e_i^3 = 0$, hence (1).

Let's show that $(e_i^2 e_j)e_k = 0$ for i, j, k pairwise different and consider (4) of Lemma 2, i.e.,

$$0 = (e_i^2 e_j)e_k + (e_i^2 e_k)e_j = a_{ij}a_{jk}e_k^2 + a_{ik}a_{kj}e_j^2. \tag{6}$$

Since $a_{jk}a_{kj} = 0$, we consider two cases:

If $a_{jk} = 0$, then $(e_i^2 e_j)e_k = 0$ and (6) leads to $(e_i^2 e_k)e_j = 0$;

If $a_{kj} = 0$, then $(e_i^2 e_k)e_j = 0$ and (6) gives $(e_i^2 e_k)e_j = 0$.

Thus (4) of Lemma 2 is equivalent to $(e_i^2 e_j)e_k = 0$, for all i, j, k pairwise distinct. □

3.4 Characterization of Jordan Algebras

Assume that $\mathcal{E}$ is an n-dimensional evolution algebra over the field F of characteristic $\neq 2$ and with a natural basis $B = \{e_i; 1 \leq i \leq n\}$ whose multiplication table is given by (1).

Lemma 3 *The algebra $\mathcal{E}$ is a Jordan evolution algebra if and only if the following conditions are satisfied:*

(1) $e_i^2 e_i^2 = e_i^4$ *for all* $1 \leq i \leq n$;
(2) $e_i^3 e_j = 0$ *for all* $1 \leq i \neq j \leq n$;
(3) $(e_i^2 e_j)e_i = 0$ *for all* $1 \leq i \neq j \leq n$;
(4) $e_i^2 e_j^2 = (e_i^2 e_j)e_j = (e_j^2 e_i)e_i$ *for all* $1 \leq i \neq j \leq n$;
(5) $(e_i^2 e_j)e_k = 0$ *for all* $1 \leq i, j, k \leq n$ *with* i, j, k *pairwise distinct.*

Proof Suppose that $\mathcal{E}$ is a Jordan algebra and $x, y, z \in \mathcal{E}$. We have

$$(x^2, y, x) = 0. \tag{7}$$

By linearizing (7), we obtain

$$2(xz, y, x) + (x^2, y, z) = 0 \tag{8}$$

The vectors $\{e_i; 1 \le i \le n\}$ in the natural basis verify identities (7), (8).

(7) $\Longrightarrow$ (1) taking $x = y = e_i$.

(7) $\Longrightarrow$ (3) taking $x = e_i$ and $y = e_j$ with $i \neq j$.

(8) $\Longrightarrow$ (2) taking $x = y = e_i$ and $z = e_j$ with $i \neq j$.

(8) $\Longrightarrow$ (4) taking $x = e_i$ and $y = z = e_j$; we also take $x = e_j$ and $y = z = e_i$ with $i \neq j$.

(8) $\Longrightarrow$ (5) taking $x = e_i$, $y = e_j$ and $z = e_k$ with i, j, k pairwise distinct.

Conversely, suppose that conditions (1), (2), (3), (4) and (5) are satisfied. Let $\sum_{i=1}^{n} x_i e_i$ and $y = \sum_{i=1}^{n} y_i e_i$ be elements of $\mathcal{E}$. We have the following equalities:

$$x^2 = \sum_{i=1}^{n} x_i^2 e_i^2, \quad yx = \sum_{i=1}^{n} y_i x_i e_i^2,$$

$$x^2 y = \sum_{i,j=1}^{n} x_i^2 y_j e_i^2 e_j = \sum_{i=1}^{n} x_i^2 y_i e_i^3 + \sum_{i=1}^{n} \sum_{\substack{j=1 \\ j\neq i}}^{n} x_i^2 y_j e_i^2 e_j,$$

$$(x^2 y)x = \sum_{i,j=1}^{n} x_i^2 y_i x_j e_i^3 e_j + \sum_{i=1}^{n} \sum_{\substack{j=1 \\ j\neq i}}^{n} \sum_{k=1}^{n} x_i^2 y_j x_k (e_i^2 e_j) e_k$$

$$= \sum_{i=1}^{n} x_i^3 y_i e_i^4 + \sum_{i=1}^{n} \sum_{\substack{j=1 \\ j\neq i}}^{n} x_i^2 y_i x_j e_i^3 e_j + \sum_{i=1}^{n} \sum_{\substack{j=1 \\ j\neq i}}^{n} x_i^3 y_j (e_i^2 e_j) e_i$$

$$+ \sum_{i=1}^{n} \sum_{\substack{j=1 \\ j\neq i}}^{n} x_i^2 y_j x_j (e_i^2 e_j) e_j + \sum_{i=1}^{n} \sum_{\substack{j=1 \\ j\neq i}}^{n} \sum_{\substack{k=1 \\ k\neq i \\ k\neq j}}^{n} x_i^2 y_j x_k (e_i^2 e_j) e_k,$$

$$(x^2 y)x = \sum_{i=1}^{n} x_i^3 y_i e_i^4 + \sum_{i=1}^{n} \sum_{\substack{j=1 \\ j\neq i}}^{n} x_i^2 y_j x_j (e_i^2 e_j) e_j,$$

$$x^2(yx) = \sum_{i,j=1}^{n} x_i^2 y_j x_j e_i^2 e_j^2 = \sum_{i=1}^{n} x_i^3 y_i e_i^2 e_i^2 + \sum_{i=1}^{n} \sum_{\substack{j=1 \\ j\neq i}}^{n} x_i^2 y_j x_j e_i^2 e_j^2$$

$$= \sum_{i=1}^{n} x_i^3 y_i e_i^4 + \sum_{i=1}^{n} \sum_{\substack{j=1 \\ j\neq i}}^{n} x_i^2 y_j x_j (e_i^2 e_j) e_j,$$

$$x^2(yx) = (x^2 y)x.$$

Corollary 2 *Let $\mathcal{E}$ be a nil evolution algebra. Then $\mathcal{E}$ is a Jordan algebra if and only if the following conditions are satisfied:*

(1) $e_i^2 e_j^2 = 0$ *for all* $1 \le i \le j \le n$;

(2) $(e_i^2 e_j) e_k = 0$ *for all* $1 \le i, j, k \le n$.

Proof Since (2) is exactly (5) of Lemma 3, it remains to show (1). Since $\mathcal{E}$ is a nilalgebra, then for all $1 \leq i \leq n$, we have $e_i^3 = 0$. According to (4) of Lemma 3, we have $e_i^2 e_j^2 = (e_i^2 e_j)e_j = a_{ij} e_j^3 = 0$, hence (1). □

Proposition 1 *Let $\mathcal{E}$ be a nil finite-dimensional evolution algebra. The following statements are equivalent:*
(1) *$\mathcal{E}$ is a fourth power-associative algebra.*
(2) *$\mathcal{E}$ is a Jordan algebra.*

Proof This proposition is a consequence of Corollaries 1 and 2. □

3.5 Wedderburn Decomposition

Any finite-dimensional commutative power-associative algebra which is not a nilalgebra contains at least one nonzero idempotent ([12, Proposition 3.3, p. 33]).

Moreover, in [1] the author shows that any commutative power-associative algebra $\mathcal{E}$, which has a nonzero idempotent, admits the following Peirce decomposition: $\mathcal{E} = \mathcal{E}_1(e) \oplus \mathcal{E}_0(e) \oplus \mathcal{E}_{\frac{1}{2}}(e)$ where $\mathcal{E}_\lambda(e) = \{x \in \mathcal{E};\ ex = \lambda x\}$.

Definition 5 ([2, Definition 2.4]) An *evolution subalgebra* of an evolution algebra $\mathcal{E}$ is a subalgebra $\mathcal{E}' \subseteq \mathcal{E}$ such that $\mathcal{E}'$ is an evolution algebra i.e. $\mathcal{E}'$ has a natural basis. We say that $\mathcal{E}'$ has the *extension property* if there exists a natural basis B' of $\mathcal{E}'$ which can be extended to a natural basis B of $\mathcal{E}$.

Let $\mathcal{E}$ be an n-dimensional evolution algebra over a field F of characteristic $\neq 0$ and with a natural basis $B = \{e_i;\ 1 \leq i \leq n\}$ whose multiplication table is given by (1).

Lemma 4 *If $\mathcal{E}$ is fourth power-associative evolution algebra, then for all $i \neq j$, $e_i^3 e_j = (e_i^2 e_j)e_i = 0$.*

Proof Suppose that there exists $i_0 \neq j_0 \in \{1, \ldots, n\}$ such that $e_{i_0}^3 e_{j_0} \neq 0$. Then $a_{i_0 i_0} a_{i_0 j_0} e_{j_0}^2 \neq 0$. Since $\mathcal{E}$ is power-associative, $0 = e_{i_0}^3 e_{j_0} + (e_{i_0}^2 e_{j_0}) e_{i_0} = a_{i_0 i_0} a_{i_0 j_0} e_{j_0}^2 + a_{i_0 j_0} a_{j_0 i_0} e_{i_0}^2$. And since $a_{i_0 j_0} \neq 0$, then

$$a_{i_0 i_0} e_{j_0}^2 + a_{j_0 i_0} e_{i_0}^2 = 0. \tag{9}$$

By multiplying (9) by e_{i_0}, we get $0 = a_{i_0 i_0} e_{j_0}^2 e_{i_0} + a_{j_0 i_0} e_{i_0}^2 e_{i_0} = a_{i_0 i_0} a_{j_0 i_0} e_{i_0}^2 + a_{j_0 i_0} a_{i_0 i_0} e_{i_0}^2 = 2 a_{i_0 i_0} a_{j_0 i_0} e_{i_0}^2$ then $a_{j_0 i_0} = 0$.
(9) leads to $a_{i_0 i_0} e_{j_0}^2 = 0$, which is impossible. We deduce that for all $i \neq j$, $e_i^3 e_j = 0$ and $(e_i^2 e_j)e_i = 0$. □

Lemma 5 *Let $\mathcal{E}$ be a fourth power-associative evolution algebra and not nilalgebra. Then $\mathcal{E}$ admits a nonzero idempotent e such that the evolution subalgebra of $\mathcal{E}$ generated by e has the extension property.*

Proof Since $\mathcal{E}$ is not nilalgebra, there exists $i_0 \in \{1, \ldots, n\}$ such that $e_{i_0}^3 \neq 0$. Without loss of generality, we can assume that $e_1^3 \neq 0$. Thus $e = a_{11}^{-2} e_1^2$ is an idempotent. Indeed, $e^2 = a_{11}^{-4} e_1^2 e_1^2 = a_{11}^{-4} e_1^4 = a_{11}^{-2} e_1^2 = e$. Let's show that the family $\{e, e_2, e_3, \ldots, e_n\}$ is a natural basis of $\mathcal{E}$. Let $\alpha_1, \ldots, \alpha_n$ be scalars such that

$$0 = \alpha_1 e + \alpha_2 e_2 + \cdots + \alpha_n e_n. \tag{10}$$

By multiplying (10) by e, we get

$$0 = \alpha_1 e^2 + a_{11}^{-2}(\alpha_2 e_1^2 e_2 + \cdots + \alpha_n e_1^2 e_n) = \alpha_1 e + a_{11}^{-2}(\alpha_2 e_1^2 e_2 + \cdots + \alpha_n e_1^2 e_n) \tag{11}$$

For $i \neq 1$, $0 = e_1^3 e_i = a_{11} e_1^2 e_i$ leads to $e_1^2 e_i = 0$ because $a_{11} \neq 0$. Thus (11) leads to $\alpha_1 e = 0$, then $\alpha_1 = 0$ and (10) leads to $\alpha_2 = \alpha_3 = \cdots = \alpha_n = 0$ because $\{e_2, e_3, \ldots, e_n\}$ is linearly independent. It has been shown in passing that $ee_i = 0$ for $i = 2, \ldots, n$. So the family $\{e, e_2, e_3, \ldots, e_n\}$ is a natural basis of $\mathcal{E}$. □

Theorem 5 (Wedderburn) *Let $\mathcal{E}$ be a not nil power-associative evolution algebra. Then $\mathcal{E}$ admits s nonzero pairwise orthogonal idempotents $u_1, u_2, \ldots, u_s$, such that*

$$\mathcal{E} = Fu_1 \oplus Fu_2 \oplus \cdots \oplus Fu_s \oplus N, \tag{12}$$

direct sum of algebras, where $s \geq 1$ is an integer and N is either zero or a nil power-associative evolution algebra.

Furthermore $\mathcal{E}_{ss} = Fu_1 \oplus Fu_2 \oplus \cdots \oplus Fu_s$ is the semisimple component of $\mathcal{E}$ and $N = Rad(\mathcal{E})$ is the nil radical of $\mathcal{E}$.

Proof Suppose that $\mathcal{E}$ is a not nilalgebra. Then $\mathcal{E}$ admits nonzero idempotent u_1 such that the evolution subalgebra Fu_1 generated by u_1 has the extension property and we denote $\mathcal{E} = \mathcal{E}_1(u_1) \oplus \mathcal{E}_0(u_1) \oplus \mathcal{E}_{\frac{1}{2}}(u_1)$ the Peirce decomposition of $\mathcal{E}$ relative to u_1. Then $\{u_1, e_2, e_3, \ldots, e_n\}$ is a natural basis of $\mathcal{E}$.

Take $v = \alpha_1 u_1 + \alpha_2 e_2 + \alpha_3 e_3 + \cdots + \alpha_n e_n \in \mathcal{E}_{\frac{1}{2}}(u_1)$. We have $u_1 v = \alpha_1 u_1^2 = \alpha_1 u_1 = \frac{1}{2} v = \frac{1}{2}(\alpha_1 u_1 + \alpha_2 e_2 + \alpha_3 e_3 + \cdots + \alpha_n e_n)$. Then $\alpha_1 = \frac{1}{2}\alpha_1$ and $\alpha_i = 0$, for $i \neq 1$. We also have $\alpha_1 = 0$ and $v = 0$, i.e. $\mathcal{E}_{\frac{1}{2}}(u_1) = 0$.

Now let $v = \alpha_1 u_1 + \alpha_2 e_2 + \alpha_3 e_3 + \cdots + \alpha_n e_n \in \mathcal{E}_1(u_1)$. We have $u_1 v = \alpha_1 u_1 = v = \alpha_1 u_1 + \alpha_2 e_2 + \alpha_3 e_3 + \cdots + \alpha_n e_n$ leads to $\alpha_i = 0$ for $i \neq 1$. Thus, $\mathcal{E}_1(u_1) = Fu_1$, so $\mathcal{E} = Fu_1 \oplus \mathcal{E}_0(u_1)$, which is a direct sum of algebras.

If $\mathcal{E}_0(u_1)$ is a nilalgebra, $s = 1$ and we are done. Otherwise we repeat the reasoning with the evolution subalgebra $\mathcal{E}_0(u_1)$ of dimension $n - 1$. Continuing in this form, we get the result because $\dim_F(\mathcal{E})$ is finite. □

Theorem 6 *The following statements are equivalent:*

(1) *$\mathcal{E}$ is power-associative evolution algebra.*

(2) *$\mathcal{E}$ is Jordan evolution algebra.*

Proof (2) $\Longrightarrow$ (1) because any Jordan algebra is power-associative.

(1) $\Longrightarrow$ (2) Suppose that $\mathcal{E}$ is a power-associative evolution algebra. Let $\mathcal{E} = \mathcal{E}_{ss} \oplus N$ be the Wedderburn decomposition of $\mathcal{E}$ where $\mathcal{E}_{ss}$ is the semisimple component of $\mathcal{E}$ and N is the nil radical of $\mathcal{E}$.

Since $\mathcal{E}_{ss}$ is associative and that (12) is a direct sum of algebras, then $\mathcal{E}$ is a Jordan algebra if and only if N is a Jordan algebra. And since N is a nil fourth power-associative evolution algebra, then it is a Jordan algebra by Proposition 1. □

4 Classification

According to Theorems 5 and 6, the classification problem of power-associative evolution algebras is reduced to that of nil power-associative evolution algebras.

Definition 6 An algebra $\mathcal{E}$ is *decomposable* if there are nonzero ideals $\mathcal{I}$ and $\mathcal{J}$ such that $\mathcal{E} = \mathcal{I} \oplus \mathcal{J}$. Otherwise, it is *indecomposable*.

Lemma 6 ([7, Corollary 2.6]) *Let $\mathcal{E}$ be a finite-dimensional evolution algebra such that* $\dim_F(\text{ann}(\mathcal{E})) \geq \frac{1}{2}\dim_F(\mathcal{E}) \geq 1$. *Then $\mathcal{E}$ is decomposable.*

Moreover for classification, it is sufficient to determine the indecomposable evolution algebras. We are interested in the classification of indecomposable evolution algebras up to dimension 6. Then $\dim_F(\text{ann}(\mathcal{E})) = 1$ or 2.

4.1 Nil Indecomposable Associative Evolution Algebra

Suppose that N is an associative evolution algebra and $B = \{e_i \mid 1 \leq i \leq n\}$ is a natural basis. We have $e_i^2 e_j = 0$, for all $i, j = 1, \ldots, n$. Then for all $i \in \{1, \ldots, n\}$ such that $e_i^2 \neq 0$, we have $e_i^2 \in \text{ann}(N)$.

(1) $\dim(\text{ann}(N)) = 1$.

Without loss of generality, we can suppose that $\text{ann}(N) = Fe_n$. Thus, for all $i \neq n$, we have $e_i^2 = \alpha_i e_n$ with $\alpha_i \neq 0$. We set $v_i = e_i$ and $v_n = \alpha_1 e_n$ with $1 \leq i \leq n-1$. We have:
$v_1^2 = v_n$, $v_i^2 = \alpha_i \alpha_1^{-1} v_n = \beta_i v_n$, $v_n^2 = 0$ with $\beta_i \neq 0$ and $2 \leq i \leq n-1$.

(2) $\dim(\text{ann}(N)) = 2$.

Without loss of generality, we can suppose that $\text{ann}(N) = Fe_{n-1} \oplus Fe_n$. Thus, for all $i \neq n-1, n$, we have $e_i^2 = \alpha_{i,n-1}e_{n-1} + \alpha_{i,n}e_n$ with $(\alpha_{i,n-1}, \alpha_{in}) \neq (0, 0)$. We can suppose that $N^2 = Fe_1^2 \oplus Fe_{n-2}^2$. Otherwise we swap the vectors $e_1, e_2, \ldots, e_{n-2}$. We set $v_i = e_i$ $(1 \leq i \leq n-2)$, $v_{n-1} = v_1^2$, $v_n = v_{n-2}^2$, $v_i^2 = \alpha_{i,n-1}v_{n-1} + \alpha_{i,n}v_n$. Also, there is $i_0 \in \{2, \ldots, n-3\}$ such that $\alpha_{i_0,n-1}\alpha_{i_0 n} \neq 0$. Otherwise N would be decomposable.

One-Dimensional Classification

Theorem 7 *The only* 1*-dimensional nil evolution algebra indecomposable and associative, is* $N_{1,1}$*, which satisfies* $e_1^2 = 0$ *and has type* [1].

Two-Dimensional Classification

Theorem 8 *The only* 2*-dimensional nil evolution algebra indecomposable and associative is* $N_{2,2}$*, which satisfies* $e_1^2 = e_2$*,* $e_2^2 = 0$ *and has type* [1, 1].

Proof $\dim(N) = 2$ leads to $\dim(\text{ann}(N)) = 1$, otherwise N would be decomposable. So, N is isomorphic to $N_{2,2}$, which satisfies $e_1^2 = e_2$ and $e_2^2 = 0$. □

Three-Dimensional Classification

Theorem 9 *The only* 3*-dimensional nil evolution algebra indecomposable and associative is* $N_{3,3}(\alpha)$*, which satisfies* $e_1^2 = e_3$*,* $e_2^2 = \alpha e_3$*,* $e_3^2 = 0$ *and has type* [1, 2]*, where* $\alpha \in F^*$.

Proof $\dim(N) = 3$ leads to $\dim(\text{ann}(N)) < \frac{1}{2}\dim(N) = 1.5$, so $\dim(\text{ann}(N)) = 1$. Thus, N is isomorphic to $N_{3,3}(\alpha)$: $e_1^2 = e_3$, $e_2^2 = \alpha e_3$ and $e_3^2 = 0$ with $\alpha \in F^*$. □

Four-Dimensional Classification

Theorem 10 *The only* 4*-dimensional nil evolution algebra indecomposable and associative is* $N_{4,5}(\alpha, \beta)$*, which satisfies* $e_1^2 = e_4$*,* $e_2^2 = \alpha e_4$*,* $e_3^2 = \beta e_4$*,* $e_4^2 = 0$ *and has type* [1, 3]*, where* $\alpha, \beta \in F^*$.

Proof $\dim(N) = 4$ leads to $\dim(\text{ann}(N)) < \frac{1}{2}\dim(N) = 2$, so $\dim(\text{ann}(N)) = 1$. Thus, N is isomorphic to $N_{4,5}(\alpha, \beta)$: $e_1^2 = e_4$, $e_2^2 = \alpha e_4$, $e_3^2 = \beta e_4$, $e_4^2 = 0$ where $\alpha, \beta \in F^*$. □

Five-Dimensional Classification

Theorem 11 *Let N be a nil indecomposable associative evolution algebra of dimension five. Then N is isomorphic to one and only one of the algebras* $N_{5,8}(\alpha, \beta, \gamma)$ *and* $N_{5,9}(\alpha, \beta)$ *in Table* 2.

Proof $\dim(N) = 5$ leads to $\dim(\text{ann}(N)) < \frac{1}{2}\dim(N) = 2.5$, so $\dim(\text{ann}(N)) = 1, 2$.

(1) $\dim(\text{ann}(N)) = 1$; N is isomorphic to $N_{5,8}(\alpha, \beta, \gamma)$, which satisfies $e_1^2 = e_5$, $e_2^2 = \alpha e_5$, $e_3^2 = \beta e_5$, $e_4^2 = \gamma e_5$, $e_5^2 = 0$ with $\alpha, \beta, \gamma \in F^*$.

(2) $\dim(\text{ann}(N)) = 2$; N is isomorphic to $N_{5,9}(\alpha, \beta)$, which satisfies $e_1^2 = e_4$, $e_2^2 = \alpha e_4 + \beta e_5$, $e_3^2 = e_5$, $e_4^2 = e_5^2 = 0$ with $\alpha, \beta \in F^*$. □

Six-Dimensional Classification

Theorem 12 *Let N be a nil indecomposable associative evolution algebra of dimension six. Then N is isomorphic to one and only one of the algebras* $N_{6,16}(\alpha, \beta, \gamma, \delta)$, $N_{6,17}(\alpha, \beta, \gamma)$ *and* $N_{6,18}(\alpha, \beta, \gamma, \delta)$ *in Table 3.*

Proof $\dim(N) = 6$ leads to $\dim(\text{ann}(N)) < \frac{1}{2}\dim(N) = 3$ so $\dim(\text{ann}(N)) = 1, 2$.

(1) $\dim(\text{ann}(N)) = 1$; N is isomorphic to $N_{6,16}(\alpha, \beta, \gamma, \delta)$, which satisfies $e_1^2 = e_6$, $e_2^2 = \alpha e_6$, $e_3^2 = \beta e_6$, $e_4^2 = \gamma e_6$, $e_5^2 = \delta e_6$, $e_6^2 = 0$ with $\alpha, \beta, \gamma, \delta \in F^*$.

(2) $\dim(\text{ann}(N)) = 2$; the multiplication table of N is of the form: $e_1^2 = e_5$, $e_2^2 = \alpha e_5 + \beta e_6$, $e_3^2 = \gamma e_5 + \delta e_6$, $e_4^2 = e_6$, $e_5^2 = e_6^2 = 0$ with $\alpha\beta \neq 0$ and $(\gamma, \delta) \neq 0$.

(a) If $\alpha\delta - \beta\gamma \neq 0$ and $\gamma\delta \neq 0$, then the algebra is isomorphic to $N_{6,18}(\alpha, \beta, \gamma, \delta)$.
(b) $\gamma\delta = 0$ or $\alpha\delta - \gamma\beta = 0$. We consider three cases:

(i) If $\gamma = 0$ ($\delta \neq 0$), then N is isomorphic to $N_{6,17}(\alpha, \beta, \delta)$, which satisfies $e_1^2 = e_5$, $e_2^2 = \alpha e_5 + \beta e_6$, $e_3^2 = \delta e_6$, $e_4^2 = e_6$, $e_5^2 = e_6^2 = 0$.
(ii) If $\delta = 0$ ($\gamma \neq 0$), by swapping e_5 and e_6 we find again $N_{6,17}(\beta, \alpha, \gamma)$.
(iii) $\alpha\delta - \beta\gamma = 0$ and $\gamma\delta \neq 0$. We have $\delta = \alpha^{-1}\beta\gamma$ and by setting $w_1 = e_1$, $w_2 = e_4$, $w_3 = e_3$, $w_4 = e_2$, $w_5 = e_5$ and $w_6 = \alpha e_5 + \beta e_6$, we get algebra $N_{6,17}(-\alpha\beta^{-1}, \beta^{-1}, \alpha^{-1}\gamma)$. □

4.2 Nil Indecomposable Evolution Algebra Which Is Not Associative

Suppose that N is a nil indecomposable power-associative evolution algebra which is not associative. There are $i_0, j_0 \in \{1, \ldots, n\}$ distinct such that $e_{i_0}^2 e_{j_0} \neq 0$. We have $0 \neq e_{i_0}^2 e_{j_0} = a_{i_0 j_0} e_{j_0}^2$, which leads to $a_{i_0 j_0} \neq 0$, $e_{i_0}^2 \neq 0$ and $e_{j_0}^2 \neq 0$.

Since N is power-associative, we have $0 = (e_{i_0}^2 e_{j_0})e_k = a_{i_0 j_0} e_{j_0}^2 e_k$ then $e_{j_0}^2 e_k = 0$, for all integers $k \neq j_0$ and since $e_{j_0}^3 = 0$, then $e_{j_0}^2 e_k = 0$, for all integers k. We deduce that $e_{j_0}^2 \in \text{ann}(N)$.

Since $e_{i_0}^2 \notin \text{ann}(N)$, necessarily $a_{i i_0} = 0$ for all $1 \leq i \leq n$ if not, assuming that $a_{i_1 i_0} \neq 0$ for some i_1, we have, for all $1 \leq k \leq n$, $0 = (e_{i_1}^2 e_{i_0})e_k = a_{i_1 i_0} e_{i_0}^2 e_k$, so $e_{i_0}^2 e_k = 0$ i.e. $e_{i_0}^2 \in \text{ann}(N)$: contradiction.

Suppose that for all $i \neq i_0, j_0$, $e_{i_0}^2 e_i = 0$. Since

$$e_{i_0}^2 = a_{i_0 j_0} e_{j_0} + \sum_{i \neq i_0, j_0} a_{i_0 i} e_i,$$

We have

$$0 = e_{i_0}^2 e_{i_0}^2 = a_{i_0 j_0}^2 e_{j_0}^2 + \sum_{i \neq i_0, j_0} a_{i_0 i}^2 e_i^2 = a_{i_0 j_0}^2 e_{j_0}^2 \text{ because } a_{i_0 i}^2 e_i^2 = a_{i_0 i} e_{i_0}^2 e_i = 0.$$

This is impossible because $a_{i_0 j_0} e_{j_0}^2 \neq 0$. Then there is j_1 distinct from i_0 and j_0 such that $e_{i_0}^2 e_{j_1} \neq 0$. The argument at the beginning of the paragraph tells us that $e_{j_1}^2 \in \text{ann}(N)$. Since N is nil, then $\text{ann}(N) \neq 0$, $\dim(N) > 3$ and without loss of generality, we will set $i_0 = 1$, $j_0 = 2$ and $j_1 = 3$.

Theorem 13 ([8, Theorem 2]) *Let A be a commutative power-associative nilalgebra of nil-index* 4 *and dimension* 4 *over a field F of characteristic* $\neq 2$*, then* $A^4 = 0$ *and there is* $y \notin A^2$ *such that* $yA^2 = 0$.

For the evolution algebras, this theorem has the following generalization:

Theorem 14 *Let* $\mathcal{E}$ *be a finite-dimensional nil power-associative evolution algebra which is not associative. Then* $\mathcal{E}$ *is of nil-index* 4, $\mathcal{E}^4 = 0$ *and there is* $y \notin \mathcal{E}^2$ *such that* $y\mathcal{E}^2 = 0$.

Proof Let us suppose that $\mathcal{E}$ is a power-associative evolution nilalgebra which is not associative and $x = \sum_{i=1}^n x_i e_i$. We have $x^3 = \sum_{i,j=1}^n x_i^2 x_j e_i^2 e_j$ and $x^4 = \sum_{i,j,k=1} x_i^2 x_j x_k (e_i^2 e_j) e_k = 0$ by Corollary 1. Since $\mathcal{E}$ is not associative, there are integers $i_0 \neq j_0$ such that $e_{i_0}^2 e_{j_0} \neq 0$. In this case $e_{j_0}^2 e_{i_0} = 0$ and then the element $a = e_{i_0} + e_{j_0}$ satisfies $a^3 = e_{i_0}^2 e_{j_0} + e_{j_0}^2 e_{i_0} = e_{i_0}^2 e_{j_0} \neq 0$. Thus the nilindex of $\mathcal{E}$ is 4. We have $\mathcal{E}^3 = \langle e_i^2 e_j;\ 1 \leq i, j \leq n \rangle \neq 0$ because $\mathcal{E}$ is not associative and $\mathcal{E}^4 = \langle (e_i^2 e_j) e_k;\ 1 \leq i, j, k \leq n \rangle = 0$ by Corollary 1. Since $e_{i_0}^2 \notin \text{ann}(\mathcal{E})$ necessarily $a_{k i_0} = 0$ and $e_k^2 e_{i_0} = a_{k i_0} e_{i_0}^2 = 0$ for all integer k, hence $e_{i_0} \mathcal{E}^2 = 0$. In fact $e_{i_0} \notin \mathcal{E}^2$ otherwise we would have $e_{i_0}^2 \in \mathcal{E}^{[3]} = \mathcal{E}^2 \mathcal{E}^2 = \langle e_i^2 e_j^2;\ 1 \leq i, j \leq n \rangle = 0$, which is absurd because $e_{i_0}^2 \neq 0$, hence the theorem is proved. □

Four-Dimensional Classification

Theorem 15 *The only* 4*-dimensional nil indecomposable power-associative evolution algebra, which is not associative, is* $N_{4,6}$ *in Table 1.*

Proof Since $\dim(N) = 4$, we have $\dim(\text{ann}(N)) = 1$. From the above

$$e_1^2 = a_{12} e_2 + a_{13} e_3 + a_{14} e_4,\ e_2^2 = a_{24} e_4,\ e_3^2 = a_{34} e_4,\ e_4^2 = 0$$

with $a_{12} a_{13} a_{24} a_{34} \neq 0$.

Set $v_2 = a_{12} e_2 + a_{14} e_4$, $v_3 = a_{13} e_3$. We have $e_1^2 = v_2 + v_3$ and $0 = e_1^2 e_1^2 = v_2^2 + v_3^2$ leads to $v_3^2 = -v_2^2$. Then we set $v_4 = v_2^2 = a_{12}^2 a_{24} e_4$; so N is isomorphic to $N_{4,6}$, which satisfies:

$$e_1^2 = e_2 + e_3, \quad e_2^2 = e_4, \quad e_3^2 = -e_4, \quad e_4^2 = 0$$

and has type $[1, 2, 1]$. □

Five-Dimensional Classification

Theorem 16 *Let N be a nil indecomposable power-associative which is not associative evolution algebra, of dimension five. Then N is isomorphic to one and only one of the algebras $N_{5,10}(\alpha)$, $N_{5,11}(\alpha)$ and $N_{5,12}(\alpha, \beta)$ in Table 2.*

Proof We have $\dim(N) = 5$ leads to $\dim(\mathrm{ann}(N)) < \frac{1}{2}\dim(N) = 2.5$, so $\dim(\mathrm{ann}(N)) \in \{1, 2\}$.

(1) $\dim(\mathrm{ann}(N)) = 1$.

$$e_1^2 = a_{12}e_2 + a_{13}e_3 + a_{14}e_4 + a_{15}e_5, e_2^2 = a_{25}e_5, e_3^2 = a_{35}e_5$$
$$e_4^2 = a_{42}e_2 + a_{43}e_3 + a_{45}e_5, e_5^2 = 0 \text{ with } a_{12}a_{13}a_{25}a_{35} \neq 0 \text{ and } e_4^2 \neq 0.$$

Since $0 = (e_1^2 e_4)e_j = a_{14}a_{4j}e_j^2$, we have $a_{14}a_{4j} = 0$ (with $j = 2, 3$). We then distinguish two cases.

(1.1) $a_{14} = 0$ i.e. $e_1^2 = a_{12}e_2 + a_{13}e_3 + a_{15}e_5$ with $a_{12}a_{13} \neq 0$. Set $v_2 = a_{12}e_2 + a_{15}e_5$ and $v_3 = a_{13}e_3$. We have $e_1^2 = v_2 + v_3$ and $0 = e_1^2 e_1^2 = v_2^2 + v_3^2$, so $v_3^2 = -v_2^2$. We set $v_5 = v_2^2 = a_{12}^2 a_{25}e_5$, so $v_3^2 = -v_5$ and there are scalars α_{42}, α_{43} and α_{45} not all zero such that $e_4^2 = \alpha_{42}v_2 + \alpha_{43}v_3 + \alpha_{45}v_5$. We have $0 = e_1^2 e_4^2 = \alpha_{42}v_2^2 + \alpha_{43}v_3^2 = (\alpha_{42} - \alpha_{43})v_5$, so $\alpha_{42} = \alpha_{43}$; this relation ensures the power associativity of N. The family $\{e_1, v_2, v_3, e_4, v_5\}$ is a natural basis of N and its multiplication table is defined by:

$$e_1^2 = v_2 + v_3, v_2^2 = v_5, v_3^2 = -v_5, e_4^2 = \alpha_{42}(v_2 + v_3) + \alpha_{45}v_5, v_5^2 = 0$$

with $(\alpha_{42}, \alpha_{45}) \neq 0$.

We have $\mathrm{ann}(N) = \langle v_5 \rangle$, $v_2, v_3 \in \mathrm{ann}^2(N)$, $e_1 \in \mathrm{ann}^3(N)$ and since $e_4^2 \in \langle v_2, v_3, v_5 \rangle$ then the possible types of N are $[1, 3, 1]$ or $[1, 2, 2]$.

(1.1.1) The type of N is $[1, 3, 1]$, then $\alpha_{42} = 0$. We have $\mathrm{ann}(N) = \langle v_5 \rangle$, $\mathrm{ann}^2(N) = \langle v_2, v_3, e_4, v_5 \rangle$ and $\mathrm{ann}^3(N) = N$. So

$$\mathcal{U}_3 \oplus \mathcal{U}_1 = \{x \in \mathrm{ann}^3(N);\ x[\mathrm{ann}^2(N)] = 0\} = \langle e_1, v_5 \rangle$$

and $(\mathcal{U}_3 \oplus \mathcal{U}_1)^2 = \langle v_2 + v_3 \rangle$. It happens that N is isomorphic to $N_{5,10}(\alpha)$, which satisfies $e_1^2 = e_2 + e_3, e_2^2 = e_5, e_3^2 = -e_5, e_4^2 = \alpha e_5, e_5^2 = 0$ with $\alpha \neq 0$.

(1.1.2) The type of N is $[1, 2, 2]$; then $\alpha_{42} \neq 0$.

We have $\mathrm{ann}(N) = \langle v_5 \rangle$, $\mathrm{ann}^2(N) = \langle v_2, v_3, v_5 \rangle$ and $\mathrm{ann}^3(N) = N$. So $\mathcal{U}_3 \oplus \mathcal{U}_1 = \langle e_1, e_4, v_5 \rangle$ and $(\mathcal{U}_3 \oplus \mathcal{U}_1)^2 = \langle v_2 + v_3, e_4^2 \rangle$. Thus, $\dim\left((\mathcal{U}_3 \oplus \mathcal{U}_1)^2\right) = 1, 2$.

(*a*) $\dim\left((\mathcal{U}_3 \oplus \mathcal{U}_1)^2\right) = 1$ implies $\alpha_{45} = 0$. Then N is isomorphic to $N_{5,11}(\alpha)$, which satisfies $e_1^2 = e_2 + e_3, e_2^2 = e_5, e_3^2 = -e_5, e_4^2 = \alpha(e_2 + e_3), e_5^2 = 0$ with $\alpha \neq 0$.

(*b*) $\dim\left((\mathcal{U}_3 \oplus \mathcal{U}_1)^2\right) = 2$ implies $\alpha_{45} \neq 0$. Then N is isomorphic to $N_{5,12}(\alpha, \beta)$, which satisfies $e_1^2 = e_2 + e_3, e_2^2 = e_5, e_3^2 = -e_5, e_4^2 = \alpha(e_2 + e_3) + \beta e_5, e_5^2 = 0$ with $\alpha\beta \neq 0$.

(1.2) $a_{14} \neq 0$, i.e., $e_1^2 = a_{12}e_2 + a_{13}e_3 + a_{14}e_4 + a_{15}e_5$ and $e_4^2 = a_{45}e_5$ with $a_{12}a_{13}a_{14}a_{45} \neq 0$. Set $v_2 = a_{12}e_2 + a_{15}e_5$, $v_3 = a_{13}e_3$ and $v_4 = a_{14}e_4$; we have $e_1^2 = v_2 + v_3 + v_4$ and $0 = e_1^2e_1^2 = v_2^2 + v_3^2 + v_4^2$ implies $v_4^2 = -(v_2^2 + v_3^2)$. We set $v_5 = v_2^2 = a_{12}^2a_{25}e_5$. There is $\alpha \in F^*$ such that $v_3^2 = \alpha v_2^2 = \alpha v_5$.

The family $\{e_1, v_2, v_3, v_4, v_5\}$ is a natural basis of N and its multiplication table is defined by: $e_1^2 = v_2 + v_3 + v_4, v_2^2 = v_5, v_3^2 = \alpha v_5, v_4^2 = -(1+\alpha)v_5$, and $v_5^2 = 0$ with $\alpha(1+\alpha) \neq 0$.

Let us show that N is isomorphic to $N_{5,10}(\beta)$ for some nonzero β. We set $w_3 = v_3 + v_4$ and $w_4 = av_3 + bv_4$; we have that $0 = w_3w_4 = av_3^2 + bv_4^2 = [\alpha a - (1+\alpha)b]v_5$, then $b = \frac{\alpha a}{1+\alpha}$. For $a = 1$, $b = \frac{\alpha}{1+\alpha}$ and $w_4 = v_3 + \frac{\alpha}{1+\alpha}v_4$. The family $\{e_1, v_2, w_3, w_4, v_5\}$ is a natural basis of N and its multiplication table is defined by: $e_1^2 = v_2 + w_3, v_2^2 = v_5, w_3^2 = -v_5, w_4^2 = [\alpha - \frac{\alpha^2}{1+\alpha}]v_5 = \frac{\alpha}{1+\alpha}v_5, v_5^2 = 0$ with $\frac{\alpha}{1+\alpha} \neq 0$. We deduce that $N \simeq N_{5,10}(\beta)$ where $\beta = \frac{\alpha}{1+\alpha}$.

(2) $\dim(\text{ann}(N)) = 2$. We have $e_1^2 = a_{12}e_2 + a_{13}e_3 + a_{14}e_4 + a_{15}e_5$, $e_2^2 = a_{24}e_4 + a_{25}e_5$, $e_3^2 = a_{34}e_4 + a_{35}e_5$, $e_4^2 = e_5^2 = 0$ with $a_{12}a_{13} \neq 0$, $(a_{24}, a_{25}) \neq 0$ and $(a_{34}, a_{35}) \neq 0$.

We set $v_2 = a_{12}e_2 + a_{15}e_5$ and $v_3 = a_{13}e_3 + a_{14}e_4$. We have: $e_1^2 = v_2 + v_3$ and $0 = e_1^2e_1^2 = v_2^2 + v_3^2$ implies $v_3^2 = -v_2^2$. Set $v_4 = v_2^2 = a_{12}^2e_2^2$, so $v_3^2 = -v_4$. Then $N = \langle e_1, v_2, v_3, v_4\rangle \oplus \langle v_5\rangle$ direct sum of evolution algebras where $\text{ann}(N) = Fv_4 \oplus Fv_5$. □

Six-Dimensional Classification

Theorem 17 *Let N be a nil indecomposable power-associative evolution algebra, which is not associative of dimension six. Then N is isomorphic to one and only one of the seven algebras $N_{6,19}(\alpha, \beta)$ to $N_{6,26}$ in Table 3.*

Proof Since $\dim(N) = 6$, we have $\dim(\text{ann}(N)) < \frac{1}{2}\dim(N) = 3$ and $\dim(\text{ann}(N)) \in \{1, 2\}$.

(1) $\dim(\text{ann}(N)) = 1$.

$$\begin{aligned}
&e_1^2 = a_{12}e_2 + a_{13}e_3 + a_{14}e_4 + a_{15}e_5 + a_{16}e_6,\ e_2^2 = a_{26}e_6,\ e_3^2 = a_{36}e_6\\
&e_4^2 = a_{42}e_2 + a_{43}e_3 + a_{45}e_5 + a_{46}e_6,\ e_5^2 = a_{52}e_2 + a_{53}e_3 + a_{54}e_4 + a_{56}e_6\\
&e_6^2 = 0 \text{ with } a_{12}a_{13}a_{26}a_{36} \neq 0,\ e_4^2 \neq 0 \text{ and } e_5^2 \neq 0.
\end{aligned}$$

Since $0 = (e_1^2e_4)e_i = a_{14}a_{4i}e_i^2$, then $a_{14}a_{4i} = 0$ (with $i = 2, 3, 5$) and $0 = (e_1^2e_5)e_j = a_{15}a_{5j}e_j^2$ implies $a_{15}a_{5j} = 0$ (with $j = 2, 3, 4$), we then distinguish four cases:

(1.1) $a_{14} = a_{15} = 0$ i.e. $e_1^2 = a_{12}e_2 + a_{13}e_3 + a_{16}e_6$. We set $v_2 = a_{12}e_2 + a_{16}e_6$ and $v_3 = a_{13}e_3$. We have $e_1^2 = v_2 + v_3$ and $0 = e_1^2 e_1^2 = v_2^2 + v_3^2$ implies $v_3^2 = -v_2^2$. Set $v_6 = v_2^2 = a_{12}^2 a_{26} e_6$; so $v_3^2 = -v_6$ and there are scalars $\alpha_{42}, \alpha_{43}, \alpha_{45}, \alpha_{46}$ not all zero such that $e_4^2 = \alpha_{42}v_2 + \alpha_{43}v_3 + \alpha_{45}e_5 + \alpha_{46}v_6$. Similarly, there are scalars $\alpha_{52}, \alpha_{53}, \alpha_{54}, \alpha_{56}$ not all zero such that $e_5^2 = \alpha_{52}v_2 + \alpha_{53}v_3 + \alpha_{54}e_4 + \alpha_{56}v_6$. Thus, we have

$$0 = e_1^2 e_4^2 = \alpha_{42}v_2^2 + \alpha_{43}v_3^2 = (\alpha_{42} - \alpha_{43})v_6 \Longrightarrow \alpha_{42} = \alpha_{43}.$$
$$0 = e_1^2 e_5^2 = \alpha_{52}v_2^2 + \alpha_{53}v_3^2 = (\alpha_{52} - \alpha_{53})v_6 \Longrightarrow \alpha_{52} = \alpha_{53}.$$
$$0 = e_4^2 e_4^2 = (\alpha_{42}^2 - \alpha_{43}^2)v_6 + \alpha_{45}e_5^2 = \alpha_{45}e_5^2 \Longrightarrow \alpha_{45} = 0.$$
$$0 = e_5^2 e_5^2 = (\alpha_{52}^2 - \alpha_{53}^2)v_6 + \alpha_{54}e_4^2 = \alpha_{54}e_4^2 \Longrightarrow \alpha_{54} = 0.$$

These relations ensure power associativity of N. The family $\{e_1, v_2, v_3, e_4, e_5, v_6\}$ is a natural basis of N and its multiplication table is given by: $e_1^2 = v_2 + v_3$, $v_2^2 = v_6$, $v_3^2 = -v_6$, $e_4^2 = \alpha_{42}(v_2 + v_3) + \alpha_{46}v_6$, $e_5^2 = \alpha_{52}(v_2 + v_3) + \alpha_{56}v_6$, $v_6^2 = 0$ with $(\alpha_{42}, \alpha_{46}) \neq 0$ and $(\alpha_{52}, \alpha_{56}) \neq 0$.

We have: $\operatorname{ann}(N) = \langle v_6 \rangle$, $v_2, v_3 \in \operatorname{ann}^2(N)$ and $e_1 \in \operatorname{ann}^3(N)$. Since $e_4^2, e_5^2 \in \langle v_2, v_3, v_6 \rangle$ then the possible types of N are: $[1, 4, 1]$, $[1, 3, 2]$ or $[1, 2, 3]$.

(1.1.1) Type $[1, 4, 1]$ leads to $\alpha_{42} = \alpha_{52} = 0$. we have $\operatorname{ann}^2(N) = \langle v_2, v_3, e_4, e_5, v_6 \rangle$ and $\operatorname{ann}^3(N) = N$. So $\mathcal{U}_3 \oplus \mathcal{U}_1 = \langle e_1, v_6 \rangle$ and $(\mathcal{U}_3 \oplus \mathcal{U}_1)^2 = \langle v_2 + v_3 \rangle$.

$N \simeq N_{6,19}(\alpha, \beta) : e_1^2 = e_2 + e_3, e_2^2 = e_6, e_3^2 = -e_6, e_4^2 = \alpha e_6, e_5^2 = \beta e_6, e_6^2 = 0$ with $\alpha, \beta \in F^*$.

(1.1.2) Type $[1, 3, 2]$ leads to $\alpha_{42} \neq 0$ and $\alpha_{52} = 0$ or $\alpha_{42} = 0$ and $\alpha_{52} \neq 0$.

(*a*) For $\alpha_{42} \neq 0$ and $\alpha_{52} = 0$, we have: $\operatorname{ann}^2(N) = \langle v_2, v_3, e_5, v_6 \rangle$ and $\operatorname{ann}^3(N) = N$. So $\mathcal{U}_3 \oplus \mathcal{U}_1 = \langle e_1, e_4, v_6 \rangle$ and $\left(\mathcal{U}_3 \oplus \mathcal{U}_1\right)^2 = \langle v_2 + v_3, e_4^2 \rangle$. Thus $\dim\left((\mathcal{U}_3 \oplus \mathcal{U}_1)^2\right) = 1, 2$.

- $\dim\left((\mathcal{U}_3 \oplus \mathcal{U}_1)^2\right) = 1 \Longrightarrow \alpha_{46} = 0$.
  $N \simeq N_{6,20}(\alpha, \beta) : e_1^2 = e_2 + e_3, e_2^2 = e_6, e_3^2 = -e_6, e_4^2 = \alpha(e_2 + e_3), e_5^2 = \beta e_6, e_6^2 = 0$ with $\alpha, \beta \in F^*$.
- $\dim\left((\mathcal{U}_3 \oplus \mathcal{U}_1)^2\right) = 2 \Longrightarrow \alpha_{46} \neq 0$.
  $N \simeq N_{6,21}(\alpha, \beta, \gamma) : e_1^2 = e_2 + e_3, e_2^2 = e_6, e_3^2 = -e_6, e_4^2 = \alpha(e_2 + e_3) + \beta e_6, e_5^2 = \gamma e_6, e_6^2 = 0$ with $\alpha, \beta, \gamma \in F^*$.

(*b*) For $\alpha_{42} = 0$ and $\alpha_{52} \neq 0$, we have again case (*a*) by permuting vectors e_4 and e_5.

(1.1.3) Type $[1, 2, 3]$ leads to $\alpha_{42}\alpha_{52} \neq 0$. We have $\operatorname{ann}^2(N) = \langle v_2, v_3, v_6 \rangle$ and $\operatorname{ann}^3(N) = N$. So $\mathcal{U}_3 \oplus \mathcal{U}_1 = \langle e_1, e_4, e_5, v_6 \rangle$ and $\left(\mathcal{U}_3 \oplus \mathcal{U}_1\right)^2 = \langle v_2 + v_3, e_4^2, e_5^2 \rangle$. Thus, $\dim\left((\mathcal{U}_3 \oplus \mathcal{U}_1)^2\right) = 1, 2$.

- $\dim\left((\mathcal{U}_3 \oplus \mathcal{U}_1)^2\right) = 1 \Longrightarrow \alpha_{46} = \alpha_{56} = 0$. Then $N \simeq N_{6,22}(\alpha, \beta)$, which satisfies $e_1^2 = e_2 + e_3, e_2^2 = e_6, e_3^2 = -e_6, e_4^2 = \alpha(e_2 + e_3), e_5^2 = \beta(e_2 + e_3), e_6^2 = 0$ with $\alpha, \beta \in F^*$.

- $\dim\big((\mathcal{U}_3\oplus\mathcal{U}_1)^2\big)=2 \Longrightarrow (\alpha_{46},\alpha_{56})\neq 0$. We have $N(\alpha,\beta,\gamma,\delta): e_1^2=e_2+e_3, e_2^2=e_6, e_3^2=-e_6, e_4^2=\alpha(e_2+e_3)+\beta e_6, e_5^2=\gamma(e_2+e_3)+\delta e_6, e_6^2=0$ with $\alpha\gamma\neq 0$ and $(\beta,\delta)\neq 0$.
  We distinguish four cases:

  (i) $\beta\delta\neq 0$ and $\alpha\delta-\beta\gamma\neq 0$ gives the algebra $N_{6,23}(\alpha,\beta,\gamma,\delta)$.
  (ii) $\delta=0$ and $\beta\neq 0$ gives $N\simeq N_{6,24}(\alpha,\beta,\gamma)$.
  (iii) $\beta=0$ and $\delta\neq 0$. By permuting e_4 and e_5 we come back to case (ii).
  (iv) $\beta\delta\neq 0$ and $\alpha\delta-\beta\gamma=0$. Since $\delta=\alpha^{-1}\beta\gamma$, we have $e_4^2=\alpha(e_2+e_3+\alpha^{-1}\beta e_6)$ and $e_5^2=\gamma(e_2+e_3+\alpha^{-1}\beta e_6)$. We do the change of basis: $w_1=e_4, w_2=\alpha e_2, w_3=\alpha e_3+\beta e_6, w_4=e_1, w_5=e_5$ and $w_6=\alpha^2 e_6$. In the natural basis $\{w_1,w_2,w_3,w_4,w_5,w_6\}$ the multiplication table leads to $N_{6,24}(\alpha^{-1},-\alpha^{-3}\beta,\alpha^{-1}\gamma)$.

(1.2) $a_{14}\neq 0$ and $a_{15}=0$ i.e. $e_1^2=a_{12}e_2+a_{13}e_3+a_{14}e_4+a_{16}e_6$ and $e_4^2=a_{46}e_6$ with $a_{12}a_{13}a_{14}a_{46}\neq 0$. Set $v_2=a_{12}e_2+a_{16}e_6$; $v_3=a_{13}e_3$ and $v_4=a_{14}e_4$; we have: $e_1^2=v_2+v_3+v_4$ and $0=e_1^2e_1^2=v_2^2+v_3^2+v_4^2$ then $v_4^2=-(v_2^2+v_3^2)$. Set $v_6=v_2^2=a_{12}^2a_{26}e_6$. There is $\alpha\in F^*$ such that $v_3^2=\alpha v_2^2=\alpha v_6$. Thus, $v_4^2=-(1+\alpha)v_6$ with $1+\alpha\neq 0$. There are scalars $\alpha_{52},\alpha_{53},\alpha_{54},\alpha_{56}$ not all zero such that $e_5^2=\alpha_{52}v_2+\alpha_{53}v_3+\alpha_{54}v_4+\alpha_{56}v_6$. We have $0=e_1^2e_5^2=[\alpha_{52}+\alpha\alpha_{53}-(1+\alpha)\alpha_{54}]v_6$, which implies $\alpha_{52}=-\alpha\alpha_{53}+(1+\alpha)\alpha_{54}$. We also have that $0=e_5^2e_5^2=[\alpha_{52}^2+\alpha\alpha_{53}^2-(1+\alpha)\alpha_{54}^2]v_6$, which leads to $0=\alpha_{52}^2+\alpha\alpha_{53}^2-(1+\alpha)\alpha_{54}^2$. We have: $\alpha_{52}^2+\alpha\alpha_{53}^2-(1+\alpha)\alpha_{54}^2=(1+\alpha)^2\alpha_{54}^2+\alpha^2\alpha_{53}^2-2\alpha(1+\alpha)\alpha_{53}\alpha_{54}+\alpha\alpha_{53}^2-(1+\alpha)\alpha_{54}^2=\alpha(1+\alpha)(\alpha_{54}-\alpha_{53})^2$. So, $0=\alpha_{52}^2+\alpha\alpha_{53}^2-(1+\alpha)\alpha_{54}^2=\alpha(1+\alpha)(\alpha_{54}-\alpha_{53})^2\Longrightarrow \alpha_{54}=\alpha_{53}$. Moreover $\alpha_{52}=(1+\alpha)\alpha_{54}-\alpha\alpha_{53}=\alpha_{54}$. Thus,

$$e_1^2=v_2+v_3+v_4,\quad v_2^2=v_6,\quad v_3^2=\alpha v_6,\quad v_4^2=-(1+\alpha)v_6$$
$$e_5^2=\alpha_{52}(v_2+v_3+v_4)+\alpha_{56}v_6,\ v_6^2=0 \text{ with } \alpha,\ 1+\alpha\in F^* \text{ and } (\alpha_{52},\alpha_{56})\neq 0.$$

Let us show that N is isomorphic to one of the following algebras: $N_{6,18}(\alpha',\beta')$, $N_{6,20}(\alpha',\beta')$, $N_{6,21}(\alpha',\beta',\gamma')$ for some (α',β',γ') with $\alpha',\beta',\gamma'\in F^*$.

Set $w_3=v_3+v_4$, $w_4=e_5$ and $w_5=av_3+bv_4$; we have: $0=w_3w_5=\big(\alpha a-(1+\alpha)b\big)v_6$ implies $b=\frac{\alpha a}{1+\alpha}$. For $a=1$, $b=\frac{\alpha}{1+\alpha}$ and $w_5=v_3+\frac{\alpha}{1+\alpha}v_4$.

The family $\{e_1,v_2,w_3,w_4,w_5,v_6\}$ is a natural basis of N and its multiplication table is defined by: $e_1^2=v_2+w_3, v_2^2=v_6, w_3^2=-v_6, w_4^2=\alpha_{52}(v_2+w_3)+\alpha_{56}v_6, w_5^2=[\alpha-\frac{\alpha^2}{1+\alpha}]v_6=[\frac{\alpha}{1+\alpha}]v_6, v_6^2=0$ with $\frac{\alpha}{1+\alpha}\neq 0$ and $(\alpha_{52},\alpha_{56})\neq 0$.

For $\alpha_{52} = 0$, then $N \simeq N_{6,19}(\alpha_{56}, \frac{\alpha}{1+\alpha})$ with $\alpha_{56}, \frac{\alpha}{1+\alpha} \in F^*$.
For $\alpha_{56} = 0$, then $N \simeq N_{6,20}(\alpha_{52}, \frac{\alpha}{1+\alpha})$ with $\alpha_{56}, \frac{\alpha}{1+\alpha} \in F^*$.
For $\alpha_{52}\alpha_{56} \neq 0$, then $N \simeq N_{6,21}(\alpha_{52}, \alpha_{56}, \frac{\alpha}{1+\alpha})$ with $\alpha_{52}, \alpha_{56}, \frac{\alpha}{1+\alpha} \in F^*$.

(1.3) $a_{14} = 0$ and $a_{15} \neq 0$, i.e., $e_1^2 = a_{12}e_2 + a_{13}e_3 + a_{15}e_5 + a_{16}e_6$ and $e_5^2 = a_{56}e_6$ with $a_{12}a_{13}a_{15}a_{56} \neq 0$. By permuting the vectors e_4 and e_5 of natural basis, we reach again Case 1.2

(1.4) $a_{14}a_{15} \neq 0$, i.e., $e_1^2 = a_{12}e_2 + a_{13}e_3 + a_{14}e_4 + a_{15}e_5 + a_{16}e_6$, $e_4^2 = a_{46}e_6$ and $e_5^2 = a_{56}e_6$ with $a_{12}a_{13}a_{14}a_{15}a_{46}a_{56} \neq 0$. Set $v_2 = a_{12}e_2 + a_{16}e_6$, $v_3 = a_{13}e_3$, $v_4 = a_{14}e_4$ and $v_5 = a_{15}e_5$; we have: $e_1^2 = v_2 + v_3 + v_4 + v_5$ and $0 = e_1^2e_1^2 = v_2^2 + v_3^2 + v_4^2 + v_5^2$ implies $v_5^2 = -(v_2^2 + v_3^2 + v_4^2)$.
Set $v_6 = v_2^2 = a_{12}^2a_{26}e_6$. There are $\alpha, \beta \in F^*$ such that $v_3^2 = \alpha v_6$ and $v_4^2 = \beta v_6$. Thus, $v_5^2 = -(1 + \alpha + \beta)v_6$ with $1 + \alpha + \beta \neq 0$.
The family $\{e_1, v_2, v_3, v_4, v_5, v_6\}$ is a natural basis of N and its multiplication table is defined by: $e_1^2 = v_2 + v_3 + v_4 + v_5$, $v_2^2 = v_6$, $v_3^2 = \alpha v_6$, $v_4^2 = \beta v_6$, $v_5^2 = -(1 + \alpha + \beta)v_6$, $v_6^2 = 0$ with $\alpha\beta(1 + \alpha + \beta) \neq 0$.

Let us show that there are vectors $x = x_3v_3 + x_4v_4 + x_5v_5$ and $y = y_3v_3 + y_4v_4 + y_5v_5$ such that the family $\{e_1, v_2, v_3 + v_4 + v_5, x, y, v_6\}$ is a natural basis of N. We set $w_3 = v_3 + v_4 + v_5$. The family $\{x, y\}$ is orthogonal to w_3 if and only if

$$x_5 = (1 + \alpha + \beta)^{-1}(x_3\alpha + x_4\beta), \tag{13}$$
$$y_5 = (1 + \alpha + \beta)^{-1}(y_3\alpha + y_4\beta). \tag{14}$$

Then x and y are written

$$x = x_3[v_3 + \alpha(1 + \alpha + \beta)^{-1}v_5] + x_4[v_4 + \beta(1 + \alpha + \beta)^{-1}v_5],$$
$$y = y_3[v_3 + \alpha(1 + \alpha + \beta)^{-1}v_5] + y_4[v_4 + \beta(1 + \alpha + \beta)^{-1}v_5].$$

The orthogonality of x and y leads to $\alpha(1 + \beta)x_3y_3 + \beta(1 + \alpha)x_4y_4 - \alpha\beta(x_3y_4 + x_4y_3) = 0$, so

$$x_3\alpha[(1 + \beta)y_3 - \beta y_4] + x_4\beta[(1 + \alpha)y_4 - \alpha y_3] = 0. \tag{15}$$

We distinguish three cases:

(i) If $\alpha + 1 \neq 0$, by taking $x_3 = 0$ and $y_3 = \alpha^{-1}(1 + \alpha)y_4$, it comes that

$$x = x_4[v_4 + \beta(1 + \alpha + \beta)^{-1}v_5],$$
$$y = y_4[\alpha^{-1}(1 + \alpha)v_3 + v_4 + v_5].$$

(ii) If $\beta + 1 \neq 0$, by taking $x_4 = 0$ and $y_4 = \beta^{-1}(1 + \beta)y_3$, it comes that

$$x = x_3[v_3 + \alpha(1 + \alpha + \beta)^{-1}v_5],$$
$$y = y_3[v_3 + \beta^{-1}(1 + \beta)v_4 + v_5].$$

(iii) $\alpha = \beta = -1$. Identity (15) leads to $x_3y_4 + x_4y_3 = 0$ and we must have $x_3y_4 - x_4y_3 = 2x_3y_4 \neq 0$. Then

$$x = x_3[v_3 + v_5] + x_4[v_4 + v_5],$$
$$y = y_3[v_3 + v_5] + y_4[v_4 + v_5].$$

In each of the three cases, there are $\alpha', \beta' \in F^*$ such that $x^2 = \alpha' v_6$ and $y^2 = \beta' v_6$. Thus, the multiplication table of the natural basis $\{e_1, v_2, v_3 + v_4 + v_5, x, y, v_6\}$ of N is given by: $e_1^2 = v_2 + w_3, v_2^2 = v_6, w_3^2 = -v_6, x^2 = \alpha' v_6, y^2 = \beta' v_6, v_6^2 = 0$ with $\alpha'\beta' \neq 0$. So N is isomorphic to $N_{6,19}(\alpha', \beta')$.

(2) $\dim(\text{ann}(N)) = 2$. We have:

$$e_1^2 = a_{12}e_2 + a_{13}e_3 + a_{14}e_4 + a_{15}e_5 + a_{16}e_6, \quad e_2^2 = a_{25}e_5 + a_{26}e_6,$$
$$e_3^2 = a_{35}e_5 + a_{36}e_6, \quad e_4^2 = a_{42}e_2 + a_{43}e_3 + a_{45}e_5 + a_{46}e_6,$$
$$e_5^2 = 0, \quad e_6^2 = 0 \text{ with } a_{12}a_{13} \neq 0\,;\, (a_{25}, a_{26}) \neq 0\,;\, (a_{35}, a_{36}) \text{ and } e_4^2 \neq 0.$$

We have $0 = (e_1^2 e_4)e_2 = a_{14}a_{42}e_2^2$ and $0 = (e_1^2 e_4)e_3 = a_{14}a_{43}e_3^2$, so $a_{14}a_{42} = a_{14}a_{43} = 0$.

We then distinguish two cases

(2.1) $a_{14} = 0$ i.e. $e_1^2 = a_{12}e_2 + a_{13}e_3 + a_{15}e_5 + a_{16}e_6$ with $a_{12}a_{13} \neq 0$. Set $v_2 = a_{12}e_2 + a_{15}e_5 + a_{16}e_6$ and $v_3 = a_{13}e_3$, we have $e_1^2 = v_2 + v_3$ and $0 = e_1^2 e_1^2 = v_2^2 + v_3^2$ leads to $v_3^2 = -v_2^2$.
Set $v_5 = v_2^2 = a_{12}^2 e_2^2$. So $v_3^2 = -v_5$ and there are scalars α, β, γ and δ not all zero such that $e_4^2 = \alpha v_2 + \beta v_3 + \gamma v_5 + \delta e_6$. By setting $w = \gamma v_5 + \delta e_6$, we have $e_4^2 = \alpha v_2 + \beta v_3 + w$.

Moreover $0 = e_1^2 e_4^2 = \alpha v_2^2 + \beta v_3^2 = (\alpha - \beta)v_5$ leads to $\alpha = \beta$. This relation ensure the power associativity of N.

(2.1.1) The family $\{v_5, w\}$ is linearly independent. In that case, the family $\{e_1, v_2, v_3, e_4, v_5, w\}$ is a natural basis of N and its multiplication table is defined by

$$e_1^2 = v_2 + v_3, v_2^2 = v_5, v_3^2 = -v_5,$$
$$e_4^2 = \alpha(v_2 + v_3) + w, v_5^2 = 0, w^2 = 0 \text{ with } \alpha \in F.$$

If $\alpha = 0$, then $N = \langle e_1, v_2, v_3, v_5 \rangle \oplus \langle e_4, w \rangle$, which is a direct sum of evolution algebras.

We deduce that $\alpha \neq 0$ and N is isomorphic to $N_{6,25}(\alpha)$, which satisfies $e_1^2 = e_2 + e_3, e_2^2 = e_5\; e_3^2 = -e_5, e_4^2 = \alpha(e_2 + e_3) + e_6, e_5^2 = 0, e_6^2 = 0$ with $\alpha \in F^*$ and its type is $[2, 2, 2]$. We have $\text{ann}(N) = \langle e_5, e_6 \rangle$; $\text{ann}^2(N) = \langle e_2, e_3, e_5, e_6 \rangle$ and $\text{ann}^3(N) = N$. So, $\mathcal{U}_3 \oplus U_1 = \langle e_1, e_4, e_6 \rangle$ and $(U_3 \oplus U_1)^2 = \langle e_2 + e_3, e_6 \rangle$.

(2.1.2) The family $\{v_5, w\}$ is linearly dependent. Then, there is $\gamma \in F$ such that $w = \gamma v_5$. In this case $N = \langle e_1, v_2, v_3, v_5, e_4\rangle \oplus \langle v_6\rangle$ is a direct sum of evolution algebras, where $\mathrm{ann}(N) = Fv_5 \oplus Fv_6$.

(2.2) $a_{14} \neq 0$, i.e., $e_1^2 = a_{12}e_2 + a_{13}e_3 + a_{14}e_4 + a_{15}e_5 + a_{16}e_6$ and $e_4^2 = a_{45}e_5 + a_{46}e_6$ with $a_{12}a_{13}a_{14} \neq 0$ and $(a_{45}, a_{46}) \neq 0$. We set $v_2 = a_{12}e_2 + a_{15}e_5 + a_{16}e_6$, $v_3 = a_{13}e_3$ and $v_4 = a_{14}e_4$. We have $e_1^2 = v_2 + v_3 + v_4$ and $0 = e_1^2e_1^2 = v_2^2 + v_3^2 + v_4^2$ leads to $v_4^2 = -(v_2^2 + v_3^2)$ and we set $v_5 = v_2^2 = a_{12}^2e_2^2$.

(2.2.1) The family $\{v_5, v_3^2\}$ is linearly independent. By setting $v_6 = v_3^2 = a_{13}^2e_3^2$, we have $N \simeq N_{6,26} : e_1^2 = e_2 + e_3 + e_4, e_2^2 = e_5, e_3^2 = e_6, e_4^2 = -(e_5 + e_6), e_5^2 = 0$, $e_6^2 = 0$ and its type is $[2, 3, 1]$. We have: $\mathrm{ann}(N) = \langle e_5, e_6\rangle$; $\mathrm{ann}^2(N) = \langle e_2, e_3, e_4, e_5, e_6\rangle$ and $\mathrm{ann}^3(N) = N$. So, $\mathcal{U}_3 \oplus \mathcal{U}_1 = \langle e_1, e_6\rangle$ and $(\mathcal{U}_3 \oplus \mathcal{U}_1)^2 = \langle e_2 + e_3 + e_4\rangle$.

(2.2.2) The family $\{v_5, v_3^2\}$ is linearly dependent. There is $\alpha \in F$ such that $v_3^2 = \alpha v_5$. We have $v_4^2 = -(1 + \alpha)v_5$ and $N = \langle e_1, v_2, v_3, v_4, v_5\rangle \oplus \langle v_6\rangle$, which is a direct sum of evolution algebras, where $\mathrm{ann}(N) = Fv_5 \oplus Fv_6$. □

4.3 General Classification

Let $\mathcal{E}_{i,j}$ be the jth power-associative evolution algebra of dimension i over a field F. If $\mathcal{E}_{i,j}$ is a nilalgebra, we denote it by $N_{i,j}$. According to Definition 4 and Theorem 5, we determine $\mathcal{E}_{i,j}$ for $1 \leq i \leq 6$. Moreover, $\mathcal{E}_{ss}$ denotes the semisimple component in the Wedderburn decomposition of $\mathcal{E}_{i,j}$ where s is the number of idempotents pairwise orthogonal.

Proposition 2 *Let $\mathcal{E}$ be a power-associative evolution algebra of dimension ≤ 4. Then $\mathcal{E}$ is isomorphic to one and only one of the algebras in Table 1, where $\mathcal{E}^\star$ is not associative.*

Proposition 3 *Let $\mathcal{E}$ be a power-associative evolution algebra of dimension 5. Then $\mathcal{E}$ is isomorphic to one and only one of the algebras in Table 2 where $\mathcal{E}^\star$ is not associative.*

Proposition 4 *Let $\mathcal{E}$ be a power-associative evolution algebra of dimension 6. Then $\mathcal{E}$ is isomorphic to one and only one of the algebras in Table 3, where $\mathcal{E}^\star$ is not associative.*

Table 1 $\dim(\mathcal{E}) \leq 4$

dim	$\mathcal{E}$	Multiplication	Type
1	$N_{1,1}$	$e_1^2 = 0$	[1]
	$\mathcal{E}_{1,2} = \mathcal{E}_{11}$	$e_1^2 = e_1$	–
2	$N_{2,1} = N_{1,1} \oplus N_{1,1}$	$e_1^2 = e_2^2 = 0$	[2]
	$N_{2,2}$	$e_1^2 = e_2,\ e_2^2 = 0$	[1, 1]
	$\mathcal{E}_{2,3} = \mathcal{E}_{11} \oplus N_{1,1}$	$e_1^2 = e_1,\ e_2^2 = 0$	–
	$\mathcal{E}_{2,4} = \mathcal{E}_{22}$	$e_1^2 = e_1,\ e_2^2 = e_2$	–
3	$N_{3,1} = N_{2,1} \oplus N_{1,1}$	$e_1^2 = e_2^2 = e_3^2 = 0$	[3]
	$N_{3,2} = N_{2,2} \oplus N_{1,1}$	$e_1^2 = e_2,\ e_2^2 = e_3^2 = 0$	[2, 1]
	$N_{3,3}(\alpha)$	$e_1^2 = e_3,\ e_2^2 = \alpha e_3,\ e_3^2 = 0$ with $\alpha \in F^*$	[1, 2]
	$\mathcal{E}_{3,4} = \mathcal{E}_{11} \oplus N_{2,1}$	$e_1^2 = e_1,\ e_2^2 = e_3^2 = 0$	–
	$\mathcal{E}_{3,5} = \mathcal{E}_{11} \oplus N_{2,2}$	$e_1^2 = e_1,\ e_2^2 = e_3,\ e_3^2 = 0$	–
	$\mathcal{E}_{3,6} = \mathcal{E}_{22} \oplus N_{1,1}$	$e_1^2 = e_1,\ e_2^2 = e_2,\ e_3^2 = 0$	–
	$\mathcal{E}_{3,7} = \mathcal{E}_{33}$	$e_1^2 = e_1,\ e_2^2 = e_2,\ e_3^2 = e_3$	–
4	$N_{4,1} = N_{3,1} \oplus N_{1,1}$	$e_1^2 = e_2^2 = e_3^2 = e_4^2 = 0$	[4]
	$N_{4,2} = N_{3,2} \oplus N_{1,1}$	$e_1^2 = e_2,\ e_2^2 = e_3^2 = e_4^2 = 0$	[3, 1]
	$N_{4,3}(\alpha) = N_{3,3}(\alpha) \oplus N_{1,1}$	$e_1^2 = e_3,\ e_2^2 = \alpha e_3,\ e_3^2 = e_4^2 = 0$ with $\alpha \in F^*$	[2, 2]
	$N_{4,4} = N_{2,2} \oplus N_{2,2}$	$e_1^2 = e_2,\ e_2^2 = 0,\ e_3^2 = e_4,\ e_4^2 = 0$	[2, 2]
	$N_{4,5}(\alpha, \beta)$	$e_1^2 = e_4,\ e_2^2 = \alpha e_4,\ e_3^2 = \beta e_4,\ e_4^2 = 0$ with $\alpha, \beta \in F^*$	[1, 3]
	$N_{4,6}^{\star}$	$e_1^2 = e_2 + e_3,\ e_2^2 = e_4,\ e_3^2 = -e_4,\ e_4^2 = 0$	[1, 2, 1]
	$\mathcal{E}_{4,7} = \mathcal{E}_{11} \oplus N_{3,1}$	$e_1^2 = e_1,\ e_2^2 = e_3^2 = e_4^2 = 0$	–
	$\mathcal{E}_{4,8} = \mathcal{E}_{11} \oplus N_{3,2}$	$e_1^2 = e_1,\ e_2^2 = e_3,\ e_3^2 = e_4^2 = 0$	–
	$\mathcal{E}_{4,9}(\alpha) = \mathcal{E}_{11} \oplus N_{3,3}(\alpha)$	$e_1^2 = e_1,\ e_2^2 = e_4,\ e_3^2 = \alpha e_4,\ e_4^2 = 0$ with $\alpha \in F^*$	–
	$\mathcal{E}_{4,10} = \mathcal{E}_{22} \oplus N_{2,1}$	$e_1^2 = e_1,\ e_2^2 = e_2,\ e_3^2 = e_4^2 = 0$	–
	$\mathcal{E}_{4,11} = \mathcal{E}_{22} \oplus N_{2,2}$	$e_1^2 = e_1,\ e_2^2 = e_2,\ e_3^2 = e_4,\ e_4^2 = 0$	–
	$\mathcal{E}_{4,12} = \mathcal{E}_{33} \oplus N_{1,1}$	$e_1^2 = e_1,\ e_2^2 = e_2,\ e_3^2 = e_3,\ e_4^2 = 0$	–
	$\mathcal{E}_{4,13} = \mathcal{E}_{44}$	$e_1^2 = e_1,\ e_2^2 = e_2,\ e_3^2 = e_3,\ e_4^2 = e_4$	–

Table 2 $\dim(\mathcal{E}) = 5$

$\mathcal{E}$	Multiplication	Type
$N_{5,1} = N_{4,1} \oplus N_{1,1}$	$e_1^2 = e_2^2 = e_3^2 = e_4^2 = e_5^2 = 0$	[5]
$N_{5,2} = N_{4,2} \oplus N_{1,1}$	$e_1^2 = e_2,\ e_2^2 = e_3^2 = e_4^2 = e_5^2 = 0$	[4, 1]
$N_{5,3}(\alpha) = N_{4,3}(\alpha) \oplus N_{1,1}$	$e_1^2 = e_3,\ e_2^2 = \alpha e_3,\ e_3^2 = e_4^2 = e_5^2 = 0$ with $\alpha \in F^*$	[3, 2]
$N_{5,4} = N_{4,4} \oplus N_{1,1}$	$e_1^2 = e_2,\ e_2^2 = 0,\ e_3^2 = e_4,\ e_4^2 = e_5^2 = 0$	[3, 2]
$N_{5,5}(\alpha, \beta) = N_{4,5}(\alpha, \beta) \oplus N_{1,1}$	$e_1^2 = e_4,\ e_2^2 = \alpha e_4,\ e_3^2 = \beta e_4,\ e_4^2 = e_5^2 = 0$ with $\alpha, \beta \in F^*$	[2, 3]
$N_{5,6}(\alpha) = N_{3,3}(\alpha) \oplus N_{2,2}$	$e_1^2 = e_3,\ e_2^2 = \alpha e_3,\ e_3^2 = 0,\ e_4^2 = e_5,\ e_5^2 = 0$ with $\alpha \in F^*$	[2, 3]
$N_{5,7}^{\star} = N_{4,6} \oplus N_{1,1}$	$e_1^2 = e_2 + e_3,\ e_2^2 = e_4,\ e_3^2 = -e_4,\ e_4^2 = e_5^2 = 0$	[2, 2, 1]
$N_{5,8}(\alpha, \beta, \gamma)$	$e_1^2 = e_5,\ e_2^2 = \alpha e_5,\ e_3^2 = \beta e_5,\ e_4^2 = \gamma e_5,\ e_5^2 = 0$ with $\alpha, \beta, \gamma \in F^*$	[1, 4]
$N_{5,9}(\alpha, \beta)$	$e_1^2 = e_4,\ e_2^2 = \alpha e_4 + \beta e_5,\ e_3^2 = e_5,\ e_4^2 = e_5^2 = 0$ with $\alpha, \beta \in F^*$	[2, 3]
$N_{5,10}^{\star}(\alpha)$	$e_1^2 = e_2 + e_3,\ e_2^2 = e_5,\ e_3^2 = -e_5,\ e_4^2 = \alpha e_5,\ e_5^2 = 0$ with $\alpha \in F^*$	[1, 3, 1]
$N_{5,11}^{\star}(\alpha)$	$e_1^2 = e_2 + e_3,\ e_2^2 = e_5,\ e_3^2 = -e_5,\ e_4^2 = \alpha(e_2 + e_3),\ e_5^2 = 0$ with $\alpha \in F^*$	[1, 2, 2]
$N_{5,12}^{\star}(\alpha, \beta)$	$e_1^2 = e_2 + e_3,\ e_2^2 = e_5,\ e_3^2 = -e_5,\ e_4^2 = \alpha(e_2 + e_3) + \beta e_5,$ $e_5^2 = 0$ with $\alpha, \beta \in F^*$	[1, 2, 2]
$\mathcal{E}_{5,13} = \mathcal{E}_{11} \oplus N_{4,1}$	$e_1^2 = e_1,\ e_2^2 = e_3^2 = e_4^2 = e_5^2 = 0$	–
$\mathcal{E}_{5,14} = \mathcal{E}_{11} \oplus N_{4,2}$	$e_1^2 = e_1,\ e_2^2 = e_3,\ e_3^2 = e_4^2 = e_5^2 = 0$	–
$\mathcal{E}_{5,15}(\alpha) = \mathcal{E}_{11} \oplus N_{4,3}(\alpha)$	$e_1^2 = e_1,\ e_2^2 = e_4,\ e_3^2 = \alpha e_4,\ e_4^2 = e_5^2 = 0$ with $\alpha \in F^*$	–
$\mathcal{E}_{5,16} = \mathcal{E}_{11} \oplus N_{4,4}$	$e_1^2 = e_1,\ e_2^2 = e_3,\ e_3^2 = 0,\ e_4^2 = e_5,\ e_5^2 = 0$	–
$\mathcal{E}_{5,17}(\alpha, \beta) = \mathcal{E}_{11} \oplus N_{4,5}(\alpha, \beta)$	$e_1^2 = e_1,\ e_2^2 = e_5,\ e_3^2 = \alpha e_5,\ e_4^2 = \beta e_5,\ e_5^2 = 0$ with $\alpha, \beta \in F^*$	–
$\mathcal{E}_{5,18}^{\star} = \mathcal{E}_{11} \oplus N_{4,6}$	$e_1^2 = e_1,\ e_2^2 = e_3 + e_4,\ e_3^2 = e_5,\ e_4^2 = -e_5,\ e_5^2 = 0$	–
$\mathcal{E}_{5,19} = \mathcal{E}_{22} \oplus N_{3,1}$	$e_1^2 = e_1,\ e_2^2 = e_2,\ e_3^2 = e_4^2 = e_5^2 = 0$	–
$\mathcal{E}_{5,20} = \mathcal{E}_{22} \oplus N_{3,2}$	$e_1^2 = e_1,\ e_2^2 = e_2,\ e_3^2 = e_4,\ e_4^2 = e_5^2 = 0$	–
$\mathcal{E}_{5,21}(\alpha) = \mathcal{E}_{22} \oplus N_{3,3}(\alpha)$	$e_1^2 = e_1,\ e_2^2 = e_2,\ e_3^2 = e_5,\ e_4^2 = \alpha e_5,\ e_5^2 = 0$ with $\alpha \in F^*$	–
$\mathcal{E}_{5,22} = \mathcal{E}_{33} \oplus N_{2,1}$	$e_1^2 = e_1,\ e_2^2 = e_2,\ e_3^2 = e_3,\ e_4^2 = e_5^2 = 0$	–
$\mathcal{E}_{5,23} = \mathcal{E}_{33} \oplus N_{2,2}$	$e_1^2 = e_1,\ e_2^2 = e_2,\ e_3^2 = e_3,\ e_4^2 = e_5,\ e_5^2 = 0$	–
$\mathcal{E}_{5,24} = \mathcal{E}_{44} \oplus N_{1,1}$	$e_1^2 = e_1,\ e_2^2 = e_2,\ e_3^2 = e_3,\ e_4^2 = e_4,\ e_5^2 = 0$	–
$\mathcal{E}_{5,25} = \mathcal{E}_{55}$	$e_1^2 = e_1,\ e_2^2 = e_2,\ e_3^2 = e_3,\ e_4^2 = e_4,\ e_5^2 = e_5$	–

Table 3 $\dim(\mathcal{E}) = 6$

$\mathcal{E}$	Multiplication	Type
$N_{6,1} = N_{5,1} \oplus N_{1,1}$	$e_1^2 = e_2^2 = e_3^2 = e_4^2 = e_5^2 = e_6^2 = 0$	[6]
$N_{6,2} = N_{5,2} \oplus N_{1,1}$	$e_1^2 = e_2$, $e_2^2 = e_3^2 = e_4^2 = e_5^2 = e_6^2 = 0$	[5, 1]
$N_{6,3}(\alpha) = N_{5,3}(\alpha) \oplus N_{1,1}$	$e_1^2 = e_3$, $e_2^2 = \alpha e_3$, $e_3^2 = e_4^2 = e_5^2 = e_6^2 = 0$ with $\alpha \in F^*$	[4, 2]
$N_{6,4} = N_{5,4} \oplus N_{1,1}$	$e_1^2 = e_2$, $e_2^2 = 0$, $e_3^2 = e_4$, $e_4^2 = e_5^2 = e_6^2 = 0$	[4, 2]
$N_{6,5}(\alpha, \beta) = N_{5,5}(\alpha, \beta) \oplus N_{1,1}$	$e_1^2 = e_4$, $e_2^2 = \alpha e_4$, $e_3^2 = \beta e_4$, $e_4^2 = e_5^2 = e_6^2 = 0$ with $\alpha, \beta \in F^*$	[3, 3]
$N_{6,6}(\alpha) = N_{5,6}(\alpha) \oplus N_{1,1}$	$e_1^2 = e_3$, $e_2^2 = \alpha e_3$, $e_3^2 = 0$, $e_4^2 = e_5$, $e_5^2 = e_6^2 = 0$ with $\alpha \in F^*$	[3, 3]
$N_{6,7}(\alpha, \beta, \gamma) = N_{5,8}(\alpha, \beta, \gamma) \oplus N_{1,1}$	$e_1^2 = e_5$, $e_2^2 = \alpha e_5$, $e_3^2 = \beta e_5$, $e_4^2 = \gamma e_5$, $e_5^2 = e_6^2 = 0$ with $\alpha, \beta, \gamma \in F^*$	[2, 4]
$N_{6,8}(\alpha, \beta) = N_{5,9}(\alpha, \beta) \oplus N_{1,1}$	$e_1^2 = e_4$, $e_2^2 = \alpha e_4 + \beta e_5$, $e_3^2 = e_5$, $e_4^2 = e_5^2 = e_6^2 = 0$ with $\alpha, \beta \in F^*$	[3, 3]
$N_{6,9}(\alpha, \beta) = N_{4,5}(\alpha, \beta) \oplus N_{2,2}$	$e_1^2 = e_4$, $e_2^2 = \alpha e_4$, $e_3^2 = \beta e_4$, $e_4^2 = 0$, $e_5^2 = e_6$, $e_6^2 = 0$ with $\alpha, \beta \in F^*$	[2, 4]
$N_{6,10}(\alpha, \beta) = N_{3,3}(\alpha) \oplus N_{3,3}(\beta)$	$e_1^2 = e_3$, $e_2^2 = \alpha e_3$, $e_3^2 = 0$, $e_4^2 = e_6$, $e_5^2 = \beta e_6$, $e_6^2 = 0$ with $\alpha, \beta \in F^*$	[2, 4]
$N_{6,11}^\star = N_{5,7} \oplus N_{1,1}$	$e_1^2 = e_2 + e_3$, $e_2^2 = e_4$, $e_3^2 = -e_4$, $e_4^2 = e_5^2 = e_6^2 = 0$	[3, 2, 1]
$N_{6,12}^\star(\alpha) = N_{5,10}(\alpha) \oplus N_{1,1}$	$e_1^2 = e_2 + e_3$, $e_2^2 = e_5$, $e_3^2 = -e_5$, $e_4^2 = \alpha e_5$, $e_5^2 = e_6^2 = 0$ with $\alpha \in F^*$	[2, 3, 1]
$N_{6,13}^\star(\alpha) = N_{5,11}(\alpha) \oplus N_{1,1}$	$e_1^2 = e_2 + e_3$, $e_2^2 = e_5$, $e_3^2 = -e_5$, $e_4^2 = \alpha(e_2 + e_3)$, $e_5^2 = e_6^2 = 0$ with $\alpha \in F^*$	[2, 2, 2]
$N_{6,14}^\star(\alpha, \beta) = N_{5,12}(\alpha, \beta) \oplus N_{1,1}$	$e_1^2 = e_2 + e_3$, $e_2^2 = e_5$, $e_3^2 = -e_5$, $e_4^2 = \alpha(e_2 + e_3) + \beta e_5$, $e_5^2 = e_6^2 = 0$ with $\alpha, \beta \in F^*$	[2, 2, 2]
$N_{6,15}^\star = N_{4,6} \oplus N_{2,2}$	$e_1^2 = e_2 + e_3$, $e_2^2 = e_4$, $e_3^2 = -e_4$, $e_4^2 = 0$, $e_5^2 = e_6$, $e_6^2 = 0$	[2, 3, 1]
$N_{6,16}(\alpha, \beta, \gamma, \delta)$	$e_1^2 = e_6$, $e_2^2 = \alpha e_6$, $e_3^2 = \beta e_6$, $e_4^2 = \gamma e_6$, $e_5^2 = \delta e_6$, $e_6^2 = 0$ with $\alpha, \beta, \gamma, \delta \in F^*$	[1, 5]
$N_{6,17}(\alpha, \beta, \gamma)$	$e_1^2 = e_5$, $e_2^2 = \alpha e_5 + \beta e_6$, $e_3^2 = \gamma e_6$, $e_4^2 = e_6$, $e_5^2 = e_6^2 = 0$ with $\alpha\beta\gamma \neq 0$	[2, 4]
$N_{6,18}(\alpha, \beta, \gamma, \delta)$	$e_1^2 = e_5$, $e_2^2 = \alpha e_5 + \beta e_6$, $e_3^2 = \gamma e_5 + \delta e_6$, $e_4^2 = e_6$, $e_5^2 = e_6^2 = 0$ with $\alpha\beta \neq 0$, $\gamma\delta \neq 0$ and $\alpha\delta - \beta\gamma \neq 0$	[2, 4]
$N_{6,19}^\star(\alpha, \beta)$	$e_1^2 = e_2 + e_3$, $e_2^2 = e_6$, $e_3^2 = -e_6$, $e_4^2 = \alpha e_6$, $e_5^2 = \beta e_6$, $e_6^2 = 0$ with $\alpha, \beta \in F^*$	[1, 4, 1]

(continued)

Table 3 (continued)

$\mathcal{E}$	Multiplication	Type
$N^{\star}_{6,20}(\alpha, \beta)$	$e_1^2 = e_2 + e_3$, $e_2^2 = e_6$, $e_3^2 = -e_6$, $e_4^2 = \alpha(e_2 + e_3)$, $e_5^2 = \beta e_6$, $e_6^2 = 0$ with $\alpha, \beta \in F^*$	[1, 3, 2]
$N^{\star}_{6,21}(\alpha, \beta, \gamma)$	$e_1^2 = e_2 + e_3$, $e_2^2 = e_6$, $e_3^2 = -e_6$, $e_4^2 = \alpha(e_2 + e_3) + \beta e_6$, $e_5^2 = \gamma e_6$, $e_6^2 = 0$ with $\alpha, \beta, \gamma \in F^*$	[1, 3, 2]
$N^{\star}_{6,22}(\alpha, \beta)$	$e_1^2 = e_2 + e_3$, $e_2^2 = e_6$, $e_3^2 = -e_6$, $e_4^2 = \alpha(e_2 + e_3)$, $e_5^2 = \beta(e_2 + e_3)$, $e_6^2 = 0$ with $\alpha, \beta \in F^*$	[1, 2, 3]
$N^{\star}_{6,23}(\alpha, \beta, \gamma, \delta)$	$e_1^2 = e_2 + e_3$, $e_2^2 = e_6$, $e_3^2 = -e_6$, $e_4^2 = \alpha(e_2 + e_3) + \beta e_6$, $e_5^2 = \gamma(e_2 + e_3) + \delta e_6$, $e_6^2 = 0$ with $\alpha\delta - \beta\gamma \neq 0$, $\alpha\gamma \neq 0$ and $\beta\delta \neq 0$	[1, 2, 3]
$N^{\star}_{6,24}(\alpha, \beta, \gamma)$	$e_1^2 = e_2 + e_3$, $e_2^2 = e_6$, $e_3^2 = -e_6$, $e_4^2 = \alpha(e_2 + e_3) + \beta e_6$, $e_5^2 = \gamma(e_2 + e_3)$, $e_6^2 = 0$ with $\alpha\beta\gamma \neq 0$	[1, 2, 3]
$N^{\star}_{6,25}(\alpha)$	$e_1^2 = e_2 + e_3$, $e_2^2 = e_5$, $e_3^2 = -e_5$, $e_4^2 = \alpha(e_2 + e_3) + e_6$, $e_5^2 = e_6^2 = 0$ with $\alpha \in F^*$	[2, 2, 2]
$N^{\star}_{6,26}$	$e_1^2 = e_2 + e_3 + e_4$, $e_2^2 = e_5$, $e_3^2 = e_6$, $e_4^2 = -(e_5 + e_6)$, $e_5^2 = e_6^2 = 0$	[2, 3, 1]
$\mathcal{E}_{6,27} = \mathcal{E}_{11} \oplus N_{5,1}$	$e_1^2 = e_1$, $e_2^2 = e_3^2 = e_4^2 = e_5^2 = e_6^2 = 0$	–
$\mathcal{E}_{6,28} = \mathcal{E}_{11} \oplus N_{5,2}$	$e_1^2 = e_1$, $e_2^2 = e_3$, $e_3^2 = e_4^2 = e_5^2 = e_6^2 = 0$	–
$\mathcal{E}_{6,29}(\alpha) = \mathcal{E}_{11} \oplus N_{5,3}(\alpha)$	$e_1^2 = e_1$, $e_2^2 = e_4$, $e_3^2 = \alpha e_4$, $e_4^2 = e_5^2 = e_6^2 = 0$ with $\alpha \in F^*$	–
$\mathcal{E}_{6,30} = \mathcal{E}_{11} \oplus N_{5,4}$	$e_1^2 = e_1$, $e_2^2 = e_3$, $e_3^2 = 0$, $e_4^2 = e_5$, $e_5^2 = e_6^2 = 0$	–
$\mathcal{E}_{6,31}(\alpha, \beta) = \mathcal{E}_{11} \oplus N_{5,5}(\alpha, \beta)$	$e_1^2 = e_1$, $e_2^2 = e_5$, $e_3^2 = \alpha e_5$, $e_4^2 = \beta e_5$, $e_5^2 = e_6^2 = 0$ with $\alpha, \beta \in F^*$	–
$\mathcal{E}_{6,32}(\alpha) = \mathcal{E}_{11} \oplus N_{5,6}(\alpha)$	$e_1^2 = e_1$, $e_2^2 = e_4$, $e_3^2 = \alpha e_4$, $e_4^2 = 0$, $e_5^2 = e_6$, $e_6^2 = 0$ with $\alpha \neq 0$.	–
$\mathcal{E}_{6,33}(\alpha, \beta, \gamma) = \mathcal{E}_{11} \oplus N_{5,8}(\alpha, \beta, \gamma)$	$e_1^2 = e_1$, $e_2^2 = e_6$, $e_3^2 = \alpha e_6$, $e_4^2 = \beta e_6$, $e_5^2 = \gamma e_6$, $e_6^2 = 0$ with $\alpha, \beta, \gamma \in F^*$	–
$\mathcal{E}_{6,34}(\alpha, \beta) = \mathcal{E}_{11} \oplus N_{5,9}(\alpha, \beta)$	$e_1^2 = e_1$, $e_2^2 = e_5$, $e_3^2 = \alpha e_5 + \beta e_6$, $e_4^2 = e_6$, $e_5^2 = e_6^2 = 0$ with $\alpha, \beta \in F^*$	–
$\mathcal{E}^{\star}_{6,35} = \mathcal{E}_{11} \oplus N_{5,7}$	$e_1^2 = e_1$, $e_2^2 = e_3 + e_4$, $e_3^2 = e_5$, $e_4^2 = -e_5$, $e_5^2 = e_6^2 = 0$	–
$\mathcal{E}^{\star}_{6,36}(\alpha) = \mathcal{E}_{11} \oplus N_{5,10}(\alpha)$	$e_1^2 = e_1$, $e_2^2 = e_3 + e_4$, $e_3^2 = e_6$, $e_4^2 = -e_6$, $e_5^2 = \alpha e_6$, $e_6^2 = 0$ with $\alpha \in F^*$	–

(continued)

Table 3 (continued)

$\mathcal{E}$	Multiplication	Type
$\mathcal{E}^{\star}_{6,37}(\alpha) = \mathcal{E}_{11} \oplus N_{5,11}(\alpha)$	$e_1^2 = e_1$, $e_2^2 = e_3 + e_4$, $e_3^2 = e_6$, $e_4^2 = -e_6$, $e_5^2 = \alpha(e_3 + e_4)$, $e_6^2 = 0$ with $\alpha \in F^*$	–
$\mathcal{E}^{\star}_{6,38}(\alpha, \beta) = \mathcal{E}_{11} \oplus N_{5,12}(\alpha, \beta)$	$e_1^2 = e_1$, $e_2^2 = e_3 + e_4$, $e_3^2 = e_6$, $e_4^2 = -e_6$, $e_5^2 = \alpha(e_3 + e_4) + \beta e_6$, $e_6^2 = 0$ with $\alpha, \beta \in F^*$	–
$\mathcal{E}_{6,39} = \mathcal{E}_{22} \oplus N_{4,1}$	$e_1^2 = e_1$, $e_2^2 = e_2$, $e_3^2 = e_4^2 = e_5^2 = e_6^2 = 0$	–
$\mathcal{E}_{6,40} = \mathcal{E}_{22} \oplus N_{4,2}$	$e_1^2 = e_1$, $e_2^2 = e_2$, $e_3^2 = e_4$, $e_4^2 = e_5^2 = e_6^2 = 0$	–
$\mathcal{E}_{6,41}(\alpha) = \mathcal{E}_{22} \oplus N_{4,3}(\alpha)$	$e_1^2 = e_1$, $e_2^2 = e_2$, $e_3^2 = e_5$, $e_4^2 = \alpha e_5$, $e_5^2 = e_6^2 = 0$ with $\alpha \in F^*$	–
$\mathcal{E}_{6,42} = \mathcal{E}_{22} \oplus N_{4,4}$	$e_1^2 = e_1$, $e_2^2 = e_2$, $e_3^2 = e_4$, $e_4^2 = 0$, $e_5^2 = e_6$, $e_6^2 = 0$	–
$\mathcal{E}_{6,43}(\alpha, \beta) = \mathcal{E}_{22} \oplus N_{4,5}(\alpha, \beta)$	$e_1^2 = e_1$, $e_2^2 = e_2$, $e_3^2 = e_6$, $e_4^2 = \alpha e_6$, $e_5^2 = \beta e_6$, $e_6^2 = 0$ $\alpha, \beta \in F^*$	–
$\mathcal{E}^{\star}_{6,44} = \mathcal{E}_{22} \oplus N_{4,6}$	$e_1^2 = e_1$, $e_2^2 = e_2$, $e_3^2 = e_4 + e_5$, $e_4^2 = e_6$, $e_5^2 = -e_6$, $e_6^2 = 0$	–
$\mathcal{E}_{6,45} = \mathcal{E}_{33} \oplus N_{3,1}$	$e_1^2 = e_1$, $e_2^2 = e_2$, $e_3^2 = e_3$, $e_4^2 = e_5^2 = e_6^2 = 0$	–
$\mathcal{E}_{6,46} = \mathcal{E}_{33} \oplus N_{3,2}$	$e_1^2 = e_1$, $e_2^2 = e_2$, $e_3^2 = e_3$, $e_4^2 = e_5$, $e_5^2 = e_6^2 = 0$	–
$\mathcal{E}_{6,47}(\alpha) = \mathcal{E}_{33} \oplus N_{3,3}(\alpha)$	$e_1^2 = e_1$, $e_2^2 = e_2$, $e_3^2 = e_3$, $e_4^2 = e_6$, $e_5^2 = \alpha e_6$, $e_6^2 = 0$ with $\alpha \in F^*$	–
$\mathcal{E}_{6,48} = \mathcal{E}_{44} \oplus N_{2,1}$	$e_1^2 = e_1$, $e_2^2 = e_2$, $e_3^2 = e_3$, $e_4^2 = e_4$, $e_5^2 = e_6^2 = 0$	–
$\mathcal{E}_{6,49} = \mathcal{E}_{44} \oplus N_{2,2}$	$e_1^2 = e_1$, $e_2^2 = e_2$, $e_3^2 = e_3$, $e_4^2 = e_4$, $e_5^2 = e_6$, $e_6^2 = 0$	–
$\mathcal{E}_{6,50} = \mathcal{E}_{55} \oplus N_{1,1}$	$e_1^2 = e_1$, $e_2^2 = e_2$, $e_3^2 = e_3$, $e_4^2 = e_4$, $e_5^2 = e_5$, $e_6^2 = 0$	–
$\mathcal{E}_{6,51} = \mathcal{E}_{66}$	$e_1^2 = e_1$, $e_2^2 = e_2$, $e_3^2 = e_3$, $e_4^2 = e_4$, $e_5^2 = e_5$, $e_6^2 = e_6$	–

Acknowledgements The authors are grateful to the referee for her/his useful comments.

References

1. Albert, A.A.: A theory of power-associative commutative algebras. Trans. Am. Math. Soc. **69**, 503–527 (1950). https://doi.org/10.2307/1990496
2. Casado, Y.C., Molina, M.S., Velasco, M.V.: Evolution algebras of arbitrary dimension and their decompositions. Linear Algebra Appl. **495**, 122–162 (2016). https://doi.org/10.1016/j.laa.2016.01.007

3. Casado, Y.C., Molina, M.S., Velasco, M.V.: Classification of three-dimensional evolution algebras. Linear Algebra Appl. **524**, 68–108 (2017). https://doi.org/10.1016/j.laa.2017.02.015
4. Camacho, L.M., Gómez, J.R., Omirov, B.A., Turdibaev, R.M.: Some properties of evolution algebras. Bull. Korean Math. Soc. **50**(5), 1481–1494 (2013). https://doi.org/10.4134/BKMS.2013.50.5.1481
5. Casas, J.M., Ladra, M., Omirov, B.A., Rozikov, U.A.: On evolution algebras. Algebra Colloq. **21**(2), 331–342 (2014). https://doi.org/10.1142/S1005386714000285
6. Elduque, A., Labra, A.: Evolution algebras and graphs. J. Algebra Appl. **14**(7), 10p (2015). https://doi.org/10.1142/S0219498815501030
7. Elduque, A., Labra, A.: On nilpotent evolution algebras. Linear Algebra Appl. **505**, 11–31 (2016). https://doi.org/10.1016/j.laa.2016.04.025
8. Gerstenhaber, M., Myung, H.C.: On commutative power-associative nilalgebras of low dimension. Proc. Am. Math. Soc. **48**, 29–32 (1975). https://doi.org/10.2307/2040687
9. Hegazi, A.S., Hani, A.: Nilpotent evolution algebras over arbitrary fields. Linear Algebra Appl. **486**, 345–360 (2015). https://doi.org/10.1016/j.laa.2015.07.041
10. Jacobson, N.: Lie Algebras. Dover Publications, New York (1979)
11. Labra, A., Ladra, M., Rozikov, U.A.: An evolution algebra in population genetics. Linear Algebra Appl. **457**, 348–362 (2014). https://doi.org/10.1016/j.laa.2014.05.036
12. Schafer, R.D.: An introduction to nonassociative algebras. Academic, New York (1966)
13. Tian, J.P., Vojtěchovský, P.: Mathematical concepts of evolution algebras in non-Mendelian genetics. Quasigroups Relat. Syst. **14**(1), 111–122 (2006)
14. Tian, J.P.: Evolution Algebras and Their Applications. Lecture Note in Mathematics, vol. 1921. Springer, Berlin (2008). https://doi.org/10.1007/978-3-540-74284-5
15. Wörz-Busekros, A.: Algebras in Genetics. Lecture Notes in Biomathematics, vol. 36. Springer, Berlin (1980)

A Survey on Isometries Between Lipschitz Spaces

M. G. Cabrera-Padilla, A. Jiménez-Vargas and Moisés Villegas-Vallecillos

Abstract The famous Banach–Stone theorem, which characterizes surjective linear isometries between $C(X)$ spaces as certain weighted composition operators, has motivated the study of isometries defined on different function spaces (see [33, 34]). The research on surjective linear isometries between spaces of Lipschitz functions is a subject of long tradition which goes back to the sixties with the works of de Leeuw [61] and Roy [81], and followed by those by Mayer-Wolf [67], Weaver [97], Araujo and Dubarbie [3], and Botelho, Fleming and Jamison [8]. This topic continues to attract the attention of some authors (see [44, 52, 62]). In the setting of Lipschitz spaces, we present a survey on non-necessarily surjective linear isometries and codimension 1 linear isometries [55], vector-valued linear isometries [56], local isometries and generalized bi-circular projections [54], 2-local isometries [52, 57], projections and averages of isometries [12] and hermitian operators [13, 14]. We also raise some open problems on bilinear isometries and approximate isometries in the same context.

Keywords Lipschitz function · Linear isometry · Weighted composition operator · Banach–Stone type theorem

1 Introduction

Given two metric spaces (X, d_X) and (Y, d_Y), let us recall that a map $f: X \to Y$ is said to be Lipschitz if there exists a constant $k \geq 0$ such that $d_Y(f(x), f(w)) \leq k \cdot d_X(x, w)$ for all $x, w \in X$. In this case, the number

M. G. Cabrera-Padilla · A. Jiménez-Vargas (✉)
Departamento de Matemáticas, Universidad de Almería, 04120 Almería, Spain
e-mail: ajimenez@ual.es

M. G. Cabrera-Padilla
e-mail: m_gador@hotmail.com

M. Villegas-Vallecillos
Departamento de Matemáticas, Universidad de Cádiz, 11510 Puerto Real, Spain
e-mail: moises.villegas@uca.es

M. Siles Molina et al. (eds.), *Associative and Non-Associative Algebras and Applications*, Springer Proceedings in Mathematics & Statistics 311,
https://doi.org/10.1007/978-3-030-35256-1_3

$$\mathrm{Lip}(f) = \sup\left\{\frac{d_Y(f(x), f(w))}{d_X(x, w)} : x, w \in X,\ x \neq w\right\}$$

is called the Lipschitz constant of f. If f is bijective and both f and f^{-1} are Lipschitz, it is said that f is a Lipschitz homeomorphism.

For a metric space X and a Banach space E over $\mathbb{K}$ (the field of real or complex numbers), the Lipschitz space $\mathrm{Lip}(X, E)$ is the Banach space of all bounded Lipschitz maps from X to E, with the maximum norm defined by

$$\|f\| = \max\{\|f\|_\infty, \mathrm{Lip}(f)\},$$

where

$$\|f\|_\infty = \sup\{\|f(x)\| : x \in X\}.$$

In some results of this survey, we will consider the vector space of all bounded Lipschitz maps from X to E equipped with another classical norm, the so-called sum norm:

$$\|f\|_s = \|f\|_\infty + \mathrm{Lip}(f).$$

An important subset of $\mathrm{Lip}(X, E)$ is the closed subspace $\mathrm{lip}(X, E)$ consisting of all functions $f \in \mathrm{Lip}(X, E)$ verifying the following condition:

$$\forall \varepsilon > 0,\ \exists \delta > 0 \colon x, w \in X,\ 0 < d_X(x, w) < \delta \quad \Rightarrow \quad \|f(x) - f(w)\| < \varepsilon\, d_X(x, w).$$

The elements of $\mathrm{lip}(X, E)$ are called little Lipschitz functions. Given a metric space (X, d) and a number $\alpha \in]0, 1[$, the set X with the distance d^α is named Hölder metric space and it is usual to denote it by X^α. The maps in $\mathrm{Lip}(X^\alpha, E)$ are known as Hölder functions. The space $\mathrm{lip}(X^\alpha, E)$ contains $\mathrm{Lip}(X, E)$, but there exist $\mathrm{lip}(X, E)$ spaces containing only constant maps.

The spaces $\mathrm{Lip}(X) := \mathrm{Lip}(X, \mathbb{K})$ and $\mathrm{lip}(X) := \mathrm{lip}(X, \mathbb{K})$, endowed with the pointwise product, are unital algebras whose unit is $\mathbf{1}_X$, the map which is constantly equal to 1 on X. For a complete information on Lipschitz spaces, we refer the reader to the monograph [99] by Nik Weaver.

This paper essentially gathers our contribution to the study about isometries between Lipschitz spaces, published over the past years in a series of papers [12–14, 52, 54–57].

In order to describe in a precise way its contents, we need to introduce a bit of nomenclature and some notation.

Throughout this paper and, unless otherwise stated, E and F will be nonzero normed spaces, and the symbols B_E, S_E and $\mathrm{Ext}(B_E)$ will stand for the closed unit ball, the unit sphere of E and the set of extreme points of B_E, respectively. The space of bounded linear operators from E to F will be represented by $\mathcal{L}(E, F)$ and we will write $\mathcal{L}(E) = \mathcal{L}(E, E)$ and $E^* = \mathcal{L}(E, \mathbb{K})$ for short. For a bounded metric space X, we denote its diameter by $\mathrm{diam}(X)$.

Given a topological space X, we will use $C(X, E)$, $C_b(X, E)$ and $C_0(X, E)$ to denote the spaces of continuous functions, bounded and continuous functions and continuous functions vanishing at the infinity from X to E, respectively. When $E = \mathbb{K}$, we will usually write $C(X)$, $C_b(X)$ and $C_0(X)$.

For two sets X and Y and two linear spaces E and F, let $\mathcal{M}$ and $\mathcal{N}$ be two families of functions from X to E and from Y into F, respectively. A map $T: \mathcal{M} \to \mathcal{N}$ is said to be a weighted composition operator on a subset $Y_0 \subset Y$ if there exist two maps φ from Y_0 to X and $\widehat{T}$ from Y_0 to the space of all linear operators from E to F such that

$$T(f)(y) = \widehat{T}(y)(f(\varphi(y))), \quad \forall y \in Y_0, \ \forall f \in \mathcal{M}.$$

The maps φ and $\widehat{T}$ are said to be the symbol and the weight of T, respectively. In the case $E = F = \mathbb{K}$, a weighted composition operator on Y_0 is of the form

$$T(f)(y) = \tau(y) f(\varphi(y)), \quad \forall y \in Y_0, \ \forall f \in \mathcal{M},$$

for a symbol $\varphi: Y_0 \to X$ and a weight $\tau: Y_0 \to \mathbb{K}$.

We have divided this survey into a series of sections which collect our main results about different types of isometries on Lipschitz spaces and some operators related to those isometries.

The first section is devoted to linear isometries between spaces of scalar-valued Lipschitz functions. The study of these maps has its origin in the 1960s, when Roy [81] described the surjective linear isometries T on $\mathrm{Lip}(X)$ when X is a compact, connected space whose diameter is less than or equal to 1. Namely, he proved that T is a weighted composition operator whose symbol is a surjective isometry on X and whose weight is a unimodular constant function.

In 1969, Vasavada [92] extended this result by establishing that if X and Y are compact r-connected metric spaces for some $r \in]0, 1[$ with their diameters being less than or equal to 2, then any surjective linear isometry from $\mathrm{Lip}(X)$ to $\mathrm{Lip}(Y)$ comes from a surjective isometry from Y into X as in the aforementioned description.

Let us recall that a metric space X is r-connected for some $r \in \mathbb{R}^+$ if it is not possible to decompose it into two nonempty subsets A and B such that $d(A, B) \geq r$.

The condition that metric spaces have diameters less than or equal to 2 is not too restrictive, since, if (X, d) is a metric space and X' is the set X endowed with the distance $d'(x, y) = \min\{2, d(x, y)\}$, then $\mathrm{diam}(X') \leq 2$ and $\mathrm{Lip}(X')$ is isometrically isomorphic to $\mathrm{Lip}(X)$ [98, Proposition 1.7.1].

We must also mention the works by Novinger [75] and Mayer-Wolf [67] in which surjective linear isometries between $\mathrm{Lip}(X^\alpha)$ spaces, with X being compact, were classified. Jarosz and Pathak [48] also analyzed surjective linear isometries on $\mathrm{Lip}(X)$ and $\mathrm{lip}(X^\alpha)$, even though their most relevant contribution was made with respect to the sum norm. Recently, Hatori and Oi [44] have continued this study on the isometry group of $\mathrm{Lip}(X)$ with the sum norm.

On the other hand, Weaver [97] developed a technique to eliminate the compactness hypothesis on metric spaces and obtained the same description as Vasavada for surjective linear isometries $T: \mathrm{Lip}(X) \to \mathrm{Lip}(Y)$ when X and Y are complete, 1-connected and whose diameters are less than or equal to 2.

Onto linear isometries on Lip(X) have been considerably studied in contrast with the case of into linear isometries. This fact is surprising if we compare it with the formidable literature on nonsurjective isometries in the context of $C(X)$ spaces. In this setting, the most important result is the famous Holsztyński theorem [45]. He proved that if T is a linear isometry from $C(X)$ into $C(Y)$, then there exist a closed subset Y_0 of Y, a surjective continuous map φ from Y_0 into X and a function $\tau \in C(Y)$ with $\|\tau\|_\infty = 1$ and $|\tau(y)| = 1$ for all $y \in Y_0$ such that T is a weighted composition operator on Y_0 with symbol φ and weight τ. Holsztyński theorem was extended to $C_0(X)$ spaces by Jeang and Wong [49], and to subspaces of continuous functions by Araujo and Font [4]. It has been also extended to the setting of C*-algebras, JB*-triples, and real C*-algebras and JB*-triples by Chu and Wong [20], Chu and Mackey [19], and Apazoglou and Peralta [2], respectively.

We show in Sect. 2 that Holsztyński theorem has a natural formulation in the context of Lip(X) spaces. Given two compact metric spaces X and Y, our main result (Theorem 1) states that every linear isometry T from Lip(X) to Lip(Y) such that $T(\mathbf{1}_X)$ is a contraction ($\mathrm{Lip}(T(\mathbf{1}_X)) < 1$) admits a representation in terms of a weighted composition operator on a closed subset Y_0 of Y, whose symbol $\varphi\colon Y_0 \to X$ is a surjective Lipschitz map with $\mathrm{Lip}(\varphi) \le \max\{1, \mathrm{diam}(X)/2\}$ and whose weight $\tau : Y \to \mathbb{K}$ is a Lipschitz map with $|\tau| = 1$ on Y_0. Moreover, we evidence that Y_0 is the largest subset of Y in which T admits this representation as a weighted composition operator (Corollary 1). In general, our result does not hold when $T(\mathbf{1}_X)$ is not a contraction. A simple counterexample can be found in [97].

In the proof of our theorem, we use techniques of extreme points as it was done in [24, 75, 81] and, because of it, we start Sect. 2 with a description of the extreme points of the unit ball of Lip(X)*.

Our Lipschitz version of Holsztyński theorem will be used along the rest of sections in order to study surjective isometries, codimension 1 isometries and local isometries.

In the surjective case, we show that every linear isometry T from Lip(X) onto Lip(Y) such that $T(\mathbf{1}_X)$ is a contraction, is a weighted composition operator in which the symbol $\varphi\colon Y \to X$ is a Lipschitz homeomorphism and the weight $\tau : Y \to S_{\mathbb{K}}$ is a Lipschitz function verifying that $d(\varphi(y), \varphi(z)) = d(y, z)$ and $\tau(y) = \tau(z)$ for any $y, z \in Y$ with $d(y, z) < 2$ (Theorem 2).

In Sect. 3, we use Theorem 1 to investigate about codimension 1 linear isometries between Lip(X) spaces.

In the context of spaces of continuous functions, codimension 1 linear isometries and, in particular, isometric shift operators were studied by several authors by using Holsztyński theorem. We cite, for instance, the works by Gutek, Hart, Jamison and Rajagopalan [40] and Farid and Varadarajan [27].

Making use of a similar technique, we classify codimension 1 linear isometries $T\colon \mathrm{Lip}(X) \to \mathrm{Lip}(Y)$ such that $\mathrm{Lip}(T(\mathbf{1}_X)) < 1$ into two types (Definition 4). We say that T is of type I if there exists an isolated point $p \in Y$ such that T is a weighted composition operator on $Y \backslash \{p\}$ whose symbol function is a surjective Lipschitz map and whose weight function is a unimodular Lipschitz map. We say that T is of type II if it admits the same representation in all Y. These two types are not self-excluding

(Proposition 6). There are also codimension 1 linear isometries of type I which are not of type II and vice versa (Propositions 7 and 8). Moreover, we prove that if T is of type I, then its symbol is a Lipschitz homeomorphism.

Section 4 is focused on studying the algebraic reflexivity of the isometry group of $\text{Lip}(X)$. In the last decades several works on this topic have appeared in the context of operator algebras. The main question consists of investigating when the local action of some set of transformations (derivations, automorphisms or isometries) completely determines the mentioned set. Consult, for instance, the paper by Brešar and Šemrl [15] on local automorphisms on $\mathcal{L}(H)$ where H is a Hilbert space. Nevertheless, the algebraic reflexivity can be defined by considering any normed space instead of an operator algebra. Given a normed space E and a nonempty subset $\mathcal{S}$ of $\mathcal{L}(E)$, denote

$$\text{ref}_{\text{al}}(\mathcal{S}) = \{\Phi \in \mathcal{L}(E)\colon \forall e \in E \;\; \exists \Phi_e \in \mathcal{S} \text{ with } \Phi(e) = \Phi_e(e)\}.$$

We say that $\mathcal{S}$ is algebraically reflexive if $\text{ref}_{\text{al}}(\mathcal{S}) = \mathcal{S}$. In particular, if $\mathcal{G}(E)$ is the set of all surjective linear isometries on E, the elements of $\text{ref}_{\text{al}}(\mathcal{G}(E))$ are called local linear isometries, and E is said to be iso-reflexive if $\mathcal{G}(E)$ is algebraically reflexive.

Molnár [69] proved that $\mathcal{L}(H)$ is iso-reflexive when H is an infinite-dimensional separable Hilbert space and, in a joint work with Zalar [73], they investigated the algebraic reflexivity of the isometry group of other important Banach spaces. For example, they showed that $C(X, \mathbb{C})$ is iso-reflexive if X is a compact Hausdorff space verifying the first axiom of countability. Their theorem is not true without this condition [16, Example 2 and Theorem 9].

Motivated by these results, we prove in Sect. 4 that $\text{Lip}(X)$ is iso-reflexive (Theorem 4). Furthermore, we use this property of $\text{Lip}(X)$ to get the algebraic reflexivity of some sets of isometries and projections on $\text{Lip}(X)$.

Let us recall that, given a metric space X, it is said that an isometry $\varphi\colon X \to X$ is involutive if $\varphi^2 = \text{Id}$, where Id is the identity map on X. In particular, if E is a normed space, involutive linear isometries of E are usually given the name of isometric reflections of E.

On the other hand, a (linear) projection $P\colon E \to E$ is bi-circular if $P + \lambda(\text{Id} - P)$ is an isometry for all $\lambda \in S_{\mathbb{K}}$, and it is generalized bi-circular if $P + \lambda(\text{Id} - P)$ is an isometry for some $\lambda \in S_{\mathbb{K}} \setminus \{1\}$. Bi-circular projections were first studied by Stachó and Zalar [84] in 2004. Shortly after, Fosner, Ilisevic and Li [37] extended this notion by introducing the generalized bi-circular projections.

We prove that the sets of isometric reflections and generalized bi-circular projections on $\text{Lip}(X)$ are algebraically reflexive (Corollaries 6 and 7). To this end, we first give a description of such operators. We show that every isometric reflection of $\text{Lip}(X)$ is, either a composition operator whose symbol is an involutive isometry on X, or its inverse (Corollary 5), and that every generalized bi-circular projection on $\text{Lip}(X)$ is the average of the identity map and an isometric reflection (Proposition 9).

Generalized bi-circular projections were studied by Botelho and Jamison [9, 10] for Lipschitz spaces. We must point out that Dutta and Rao [25] investigated the algebraic reflexivity of the sets of isometric reflections and generalized bi-circular projections on $C(X)$ where X is a compact Hausdorff space verifying the first axiom of countability.

Section 5 is devoted to the study of projections on $\text{Lip}(X)$ which can be expressed as the average of two or three surjective linear isometries, when X is a compact 1-connected metric space with diameter at most 2. We show that generalized bi-circular projections on $\text{Lip}(X)$ are the only projections on $\text{Lip}(X)$ which can be represented as the average of two surjective linear isometries (Theorem 5). In order to achieve this objective, we first characterize when the average of two isometries is a projection on $\text{Lip}(X)$ (Proposition 11). In [7, 11], similar results were obtained for such projections on the spaces $C(X, \mathbb{C})$ and $C(X, E)$ where E is a strictly convex space.

Our method will be extended to investigate when the average of three isometries is a projection on $\text{Lip}(X)$ (Theorem 6). The concept of n-circular projection allows us to establish that the average P of two (three) surjective linear isometries on $\text{Lip}(X)$ is a projection if and only if P is either a trivial projection or a 2-circular (3-circular, respectively) projection (Theorem 7).

All the results that we mentioned above on the algebraic reflexivity correspond to linear structures. Taking into account the important achievements obtained, it is natural to want to extend this research line to more general structures where linearity is not a premise. However, the locality condition satisfied by the elements of $\text{ref}_{\text{al}}(\mathcal{S})$ is too weak to get reasonable results without considering linearity.

In 1997, Šemrl [82] introduced the concept of 2-locality for automorphisms and derivations. Given an algebra $\mathcal{A}$, a map $\Phi\colon \mathcal{A} \to \mathcal{A}$ (which is not assumed to be linear) is a 2-local automorphism (respectively, a 2-local derivation) if for any $a, b \in \mathcal{A}$ there exists an automorphism (respectively, a derivation) $\Phi_{a,b}\colon \mathcal{A} \to \mathcal{A}$ such that $\Phi(a) = \Phi_{a,b}(a)$ and $\Phi(b) = \Phi_{a,b}(b)$. Šemrl proved in this paper that if H is an infinite-dimensional separable Hilbert space, then every 2-local automorphism on $\mathcal{L}(H)$ is an automorphism, and he established a similar statement for 2-local derivations.

Molnár [70] extended it to isometries in 2002. Given a Banach space E, a map $T\colon E \to E$ is said to be a 2-local isometry if for every $x, y \in E$ there exists a surjective linear isometry $\Phi_{x,y}\colon E \to E$, depending on x and y, such that $T(x) = \Phi_{x,y}(x)$ and $T(y) = \Phi_{x,y}(y)$ (a priori, it is not assumed that T is linear or surjective). Molnár [70] showed that every 2-local isometry on $\mathcal{L}(H)$ is a surjective linear isometry and posed the problem of studying 2-local isometries on other Banach spaces.

In recent years numerous papers on 2-locality in the setting of spaces of operators have been published (see for example [63, 71, 72]). In function algebras, the first result in this research line was presented by Györy [41] in 2001. He showed that if X is a σ-compact first countable Hausdorff space, then every 2-local isometry on $C_0(X, \mathbb{C})$ is a surjective linear isometry. Furthermore, Hatori, Miura, Oka and Takagi [42] considered 2-local isometries and 2-local automorphisms on uniform algebras, including certain algebras of holomorphic functions of one and two variables.

We devote Sect. 6 to the study of 2-local isometries on $\text{Lip}(X)$. We consider the space $\text{Lip}(X)$ equipped indistinctly with the maximum norm and the sum norm. It is said that the isometry group of $\text{Lip}(X)$ is canonical if every surjective linear isometry on $\text{Lip}(X)$ can be expressed as a weighted composition operator, where the weight is a unimodular constant and the symbol is a surjective isometry on X. We prove that if X is a separable bounded metric space and the isometry group of $\mathrm{Lip}(X)$ is canonical, then each 2-local isometry on $\mathrm{Lip}(X)$ is a surjective linear isometry (Theorem 9). Moreover, we give a complete description of all 2-local isometries of $\mathrm{Lip}(X)$ when X is bounded (Theorem 8).

In the final part of the section, we present 2-locality for the spaces of vector-valued Lipschitz maps. Under convenient conditions on a compact metric space X and a Banach space E, we provide a description of (standard) 2-local isometries on the Banach space $\mathrm{Lip}(X, E)$ in terms of a generalized composition operator (Theorems 12 and 13) and study in which cases every (standard) 2-local isometry on $\mathrm{Lip}(X, E)$ is linear and surjective (Theorems 14 and 15).

Another approach in the study of 2-local isometries on uniform algebras and Lipschitz algebras was introduced by Li, Peralta, Wang and Wang in [62]. They focused on the notion of weak-2-local isometry, established spherical variants of the Gleason–Kahane–Żelazko [39, 58] and Kowalski–Słodkowski theorems [60], and applied them to prove that every weak-2-local isometry between two uniform algebras is a linear map. In this way, they resolved a couple of problems raised in [42]. Some contributions also are given for weak-2-local isometries on Lipschitz algebras.

Linear isometries between spaces of vector-valued Lipschitz maps are dealed in Sect. 7. In 1950, Jerison [50] extended the Banach–Stone theorem to $C(X, E)$ spaces and Cambern [17] generalized in 1978 Jerison's theorem considering non necessarily surjective linear isometries $T: C(X, E) \to C(Y, F)$, where X, Y are compact Hausdorff spaces and E, F are normed spaces with F strictly convex.

Our purpose in this section is to state Cambern and Jerison theorems in the context of Lipschitz spaces. If X, Y are compact metric spaces and E is a strictly convex normed space, then every linear isometry T from $\mathrm{Lip}(X, E)$ to $\mathrm{Lip}(Y, E)$ fixing some constant function with modulus one is a weighted composition operator on a nonempty closed subset $Y_0 \subset Y$, whose symbol is a surjective Lipschitz map $\varphi: Y_0 \to X$ with $\mathrm{Lip}(\varphi) \leq \max\{1, \mathrm{diam}(X)/2\}$ and whose weight is a Lipschitz map $\widehat{T}: Y \to \mathcal{L}(E)$ with $\|\widehat{T}(y)\| = 1$ for all $y \in Y$ (Theorem 17).

If, in addition, T is surjective and X and Y have diameters less than or equal to 2, then $Y_0 = Y$, φ is a surjective isometry and $\widehat{T}(y)$ is a surjective linear isometry for all $y \in Y$ (Corollary 9).

In the surjective case, Araujo and Dubarbie [3] generalized our results by considering (non-necessarily compact) complete metric spaces, and weakening the additional condition imposed to the isometry T. Furthermore, Botelho, Fleming and Jamison [8] also tackled this problem by removing the hypothesis of strict convexity on E and F, but they added other constraints.

Sections 8 and 9 focus on hermitian operators between Lipschitz spaces. The notion of an hermitian operator on a Banach space goes back to the early work of Lumer [65] and also Vidav [94]. Hermitian operators have played an important role in the characterization of surjective isometries of various Banach spaces, but they are also interesting in themselves. For certain Banach spaces, hermitian operators are trivial, which means that they are real scalar multiples of the identity. See [33, 34] for a complete information on hermitian operators.

In Sect. 8, we show that the spaces Lip(X) and lip(X^α) $(0 < \alpha < 1)$, equipped with the sum norm, support only trivial hermitian operators when X is a compact metric space (Theorem 20). This generalizes some known results [6] on hermitian operators between Lipschitz function spaces on $[0, 1]$. We also deal with the natural connection between hermitian operators and bi-circular projections and our result implies that the only such projections on Lip(X) and lip(X^α) with $0 < \alpha < 1$ are the trivial projections, 0 and Id.

In Sect. 9, we investigate the class of hermitian operators on Lipschitz spaces Lip(X, E) on a compact and 2-connected metric space and with values on a complex Banach space, following a scheme employed by Fleming and Jamison in [32] in the characterization of the hermitian operators on $C(X, E)$. This approach relies on the existence of semi-inner products on such spaces which are compatible with the norm. To be more precise, we start by embedding Lip(X, E) isometrically into a space of vector valued continuous functions defined on a compact space. Then we construct a semi-inner product on Lip(X, E) compatible with the norm. This approach allows us to describe the hermitian operators as multiplication operators via a hermitian operator on E. In particular we conclude that the space of all scalar valued Lipschitz functions only supports trivial hermitian operators (Theorem 21). These results yield the form for the normal and the adjoint abelian operators on this setting (Theorem 22).

Recently, Hatori and Oi [43] studied hermitian operators on a Banach algebra of Lipschitz maps Lip(X, A), where X is a compact metric space which is not necessarily 2-connected, with values in a uniform algebra A. It is proved that if A is a uniform algebra on a compact Hausdorff space Y, then T is a hermitian operator on Lip(X, A) if and only if there is a real-valued function $f \in A$ such that T is the corresponding left multiplication operator $M_{\mathbf{1}_X \otimes f}$.

In Sect. 10, some open problems will be posed. Just as a very succinct presentation of them, we would like to comment that we intend to give a Banach–Stone type representation for two types of isometries that have not been studied until now in the context of Lipschitz spaces: linear isometries of finite codimension (Problem 1) and bilinear isometries (Problem 2). Concerning to approximate isometries, we also propose the study of problems of Hyers–Ulam type stability on the approximation, perhaps in the uniform norm, of an ε-nonexpansive map or an ε-approximate isometry on Lip(X) by a nonexpansive map or an isometry, respectively (Problem 3).

Problem 4 is a particularization of the so-called Tingley's problem in the setting of Lipschitz spaces. This problem asks on the possibility of extending a surjective isometry between the unit spheres of two normed spaces, and is related to the Mazur–Ulam's property. It is a subject with a certain tradition, the first work of Tingley dates back to 1987, and a renewed interest has attracted a widen audience since the first contribution by Ding [22] in 2002. More recently, Tan [85–87] in 2011–12, Peralta and Tanaka [79] in 2016, Fernández and Peralta [29–31] in 2017, and Tanaka [88–90] in 2014–17, have given it a new and vigorous impulse.

To finish the exposition of open questions, we propose in Problem 6 the study of contractive and bicontractive projections on Lip(X) to determine under what conditions on the metric space X, the space Lip(X) belongs to the class of Banach spaces in which every bicontractive projection is the mean of the identity mapping and an involutive isometry. Our interest in this issue is justified by our previous study of generalized bi-circular projections on Lip(X). In fact, we ask in Problem 5 if the average of $n > 4$ surjective linear isometries on Lip(X) (where X is a compact 1-connected metric space with diam$(X) \leq 2$) is a projection if and only if it is either a trivial projection or a n-circular projection.

2 Scalar-Valued Linear Isometries

The main objective of this section is to present a version for Lip(X)-spaces of the renowned Holztyński theorem [45] on linear isometries (not necessarily surjective) between $C(X)$-spaces. This aim is achieved at Theorem 1.

We begin by describing the set of extreme points of the unit ball of Lip$(X)^*$. Let us recall the notion of extreme point.

Definition 1 Let E be a normed space and let C be a subset of E. A point $e \in C$ is said to be an extreme point of C if given $t \in]0, 1[$ and $x, y \in C$ such that $e = tx + (1-t)y$, then $e = x = y$. The set of all extreme points of C will be denoted by Ext(C). It is said that E is strictly convex if Ext$(B_E) = S_E$.

A simple application of the Hahn–Banach theorem, the Banach–Alaoglu theorem and the Krein–Milman theorem yields the following well-known fact.

Proposition 1 *Let E be a normed space and let M be a vector subspace of E. For each $\eta \in$ Ext(B_{M^*}) there exists $\varphi \in$ Ext(B_{E^*}) such that $\varphi|_M = \eta$.*

Let us recall that a Hausdorff topological space X is completely regular if given a closed subset $C \subset X$ and $x_0 \in X \backslash C$, there exists a continuous function $f : X \to [0, 1]$ such that $f(x_0) = 0$ and $f(x) = 1$ for all $x \in C$.

For instance, every metric space X is completely regular. Note that the Lipschitz map $f : X \to [0, 1]$, defined by $f(x) = \max\{0, 1 - d(x, C)/d(x_0, C)\}$, satisfies the required conditions.

The next result shows that it is possible to compactify a completely regular topological space so that bounded continuous functions can be extended and their norms can be preserved.

Proposition 2 ([21, 6.2. Stone–Čech compactification]) *Let X be a completely regular topological space. Then there exist a compact Hausdorff space βX (the Stone–Čech compactification of X) and a continuous map $\Delta : X \to \beta X$ such that*

1. *$\Delta : X \to \Delta(X)$ is a homeomorphism.*
2. *$\Delta(X)$ is dense in βX.*

3. *If* $f \in C_b(X)$*, then there exists a unique continuous map* $\beta f : \beta X \to \mathbb{K}$ *such that* $\beta f \circ \Delta = f$.

Moreover, if Ω *is a compact Hausdorff space and* $\pi : X \to \Omega$ *is a continuous map for which the aforementioned properties hold, then* βX *is homeomorphic to* Ω *and* π *is the composition of* Δ *with the corresponding homeomorphism.*

Our next step is to define the so-called de Leeuw's map.

Definition 2 Let X be a metric space and consider the set

$$\widetilde{X} = \left\{(x, y) \in X^2 : x \neq y\right\}.$$

Then, for each map $f \in \mathrm{Lip}(X)$, we define $\widetilde{f} : \widetilde{X} \to \mathbb{K}$ as follows

$$\widetilde{f}(x, y) = \frac{f(x) - f(y)}{d(x, y)}, \quad \forall (x, y) \in \widetilde{X}.$$

The map $f \mapsto \widetilde{f}$ from $\mathrm{Lip}(X)$ into $C_b(\widetilde{X})$ is called the de Leeuw's map.

Since $\widetilde{X}$ is a metric space with the distance given by

$$((x, y), (x_0, y_0)) \mapsto d(x, x_0) + d(y, y_0), \quad \forall (x, y), (x_0, y_0) \in \widetilde{X},$$

then $\widetilde{X}$ is completely regular. Thus, we can consider the Stone–Čech compactification of $\widetilde{X}$, $\beta\widetilde{X}$. So, for any map $f \in \mathrm{Lip}(X)$ there exists a unique extension of $\widetilde{f}$, $\beta\widetilde{f} \in C(\beta\widetilde{X})$, such that $\mathrm{Lip}(f) = ||\widetilde{f}||_\infty = ||\beta\widetilde{f}||_\infty$. Note that if X is a compact space, then the disjoint union $X \cup \beta\widetilde{X}$ is a compact Hausdorff topological space endowed with the topology

$$\mathcal{T} = \left\{U \cup V : U \text{ is open in } X \text{ and } V \text{ is open in } \beta\widetilde{X}\right\}.$$

This construction identifies $\mathrm{Lip}(X)$ with a closed subspace of continuous maps:

Lemma 1 *Let* X *be a compact metric space. Then* $\Phi : \mathrm{Lip}(X) \to C\left(X \cup \beta\widetilde{X}\right)$ *given by*

$$\Phi(f)(w) = \begin{cases} f(w) & \text{if } w \in X, \\ \left(\beta\widetilde{f}\right)(w) & \text{if } w \in \beta\widetilde{X}, \end{cases}$$

for all $f \in \mathrm{Lip}(X)$*, is a linear isometry.*

For each $w \in X \cup \beta\widetilde{X}$, we define $\delta_w \in C\left(X \cup \beta\widetilde{X}\right)^*$ and $\widetilde{\delta}_w \in \mathrm{Lip}(X)^*$ as

$$\delta_w(g) = g(w), \quad \forall g \in C\left(X \cup \beta\widetilde{X}\right), \quad \widetilde{\delta}_w(f) = \Phi(f)(w), \quad \forall f \in \mathrm{Lip}(X).$$

It is plain that $|\widetilde{\delta}_w(f)| \leq \|f\|$ for all $f \in \mathrm{Lip}(X)$, and so $\widetilde{\delta}_w \in B_{\mathrm{Lip}(X)^*}$.

It is well-known [24, p. 441] that the extreme points of $B_{\Phi(\mathrm{Lip}(X))^*}$ are functionals of the form $\tau \cdot \delta_w|_{\Phi(\mathrm{Lip}(X))}$ where $\tau \in S_{\mathbb{K}}$ and $w \in X \cup \beta\widetilde{X}$. From this, it is possible to deduce the following

Proposition 3 *If X is a compact metric space, then every extreme point of $B_{\mathrm{Lip}(X)^*}$ is of the form $\tau \cdot \widetilde{\delta}_w$ where $\tau \in S_{\mathbb{K}}$ and $w \in X \cup \beta\widetilde{X}$.*

For our purposes we need to present part of the converse statement: $\widetilde{\delta}_w$ is an extreme point of $B_{\mathrm{Lip}(X)^*}$ when $w \in X$. This fact was proved by Roy [81, Lemma 1.2] by using a result due to de Leeuw [61, Lemma 3.2].

Proposition 4 *Let X be a metric space and $x \in X$. Then the functional $\widetilde{\delta}_x \in \mathrm{Lip}(X)^*$ given by $\widetilde{\delta}_x(f) = f(x)$ for all $f \in \mathrm{Lip}(X)$ is an extreme point of $B_{\mathrm{Lip}(X)^*}$.*

Before formulating our Lipschitz version of Holztyński theorem, we introduce an auxiliary result which makes the proof easier.

Lemma 2 ([95, Lema 2.8]) *Let X and Y be compact metric spaces and let $T\colon \mathrm{Lip}(X) \to \mathrm{Lip}(Y)$ be a linear isometry. Suppose that there exist a nonempty subset $Y_0 \subset Y$ and a map $\varphi\colon Y_0 \to X$ such that*

$$|T(\mathbf{1}_X)(y)| = 1, \quad T(f)(y) = T(\mathbf{1}_X)(y)\, f(\varphi(y)), \quad \forall f \in \mathrm{Lip}(X),\ \forall y \in Y_0.$$

Then φ is a Lipschitz map and $\mathrm{Lip}(\varphi) \leq \max\{1, \mathrm{diam}(X)/2\}$. Furthermore, if $y, z \in Y_0$ and $d(y, z) < 2$, then $d(\varphi(y), \varphi(z)) \leq d(y, z)$.

Let us recall that a contraction is a Lipschitz map whose Lipschitz constant is less than 1. After this preparation, we can formulate the announced result.

Theorem 1 ([55, Theorem 2.4]) *Let X and Y be compact metric spaces and let $T\colon \mathrm{Lip}(X) \to \mathrm{Lip}(Y)$ be a linear isometry. Suppose that $T(\mathbf{1}_X)$ is a contraction. Then there exist a nonempty closed subset $Y_0 \subset Y$, a surjective Lipschitz map $\varphi\colon Y_0 \to X$ with $\mathrm{Lip}(\varphi) \leq \max\{1, \mathrm{diam}(X)/2\}$ and a function $\tau \in \mathrm{Lip}(Y)$ with $\|\tau\| = 1$ and $|\tau(y)| = 1$ for all $y \in Y_0$ such that*

$$T(f)(y) = \tau(y) f(\varphi(y)), \quad \forall y \in Y_0,\ \forall f \in \mathrm{Lip}(X).$$

In Theorem 1, we cannot remove the condition $\mathrm{Lip}(T(\mathbf{1}_X)) < 1$. Weaver [97] gives an example of this fact by considering the metric space $X = \{p, q\}$, where $d(p, q) = 1$, and defining $T\colon \mathrm{Lip}(X) \to \mathrm{Lip}(X)$ as $T(f)(p) = f(p)$ and $T(f)(q) = f(p) - f(q)$ for all $f \in \mathrm{Lip}(X)$.

The next result shows that the triple (Y_0, τ, φ) associated to the isometry T in Theorem 1 possesses a universal property.

Corollary 1 ([55, Corollary 2.5]) *Let X and Y be compact metric spaces and $T\colon \mathrm{Lip}(X) \to \mathrm{Lip}(Y)$ a linear isometry. Suppose that $T(\mathbf{1}_X)$ is a contraction and that Y_0, τ and φ are defined as in Theorem 1. If $Y_0' \subset Y$, and $\tau'\colon Y_0' \to \mathbb{K}$ and $\varphi'\colon Y_0' \to X$ are two Lipschitz maps such that*

$$T(f)(y) = \tau'(y) f(\varphi'(y)), \quad \forall y \in Y_0', \ \forall f \in \mathrm{Lip}(X);$$

then $Y_0' \subset Y_0$, $\tau' = \tau|_{Y_0'}$ *and* $\varphi' = \varphi|_{Y_0'}$.

Now, we study the onto case.

Lemma 3 ([95, Lema 2.11]) *Let X and Y be compact metric spaces, $T: \mathrm{Lip}(X) \to \mathrm{Lip}(Y)$ a surjective linear isometry and $Y_1 = \{y \in Y : T(\mathbf{1}_X)(y) \neq 0\}$. Then $Y_1 \neq \emptyset$, $|T(\mathbf{1}_X)(y)| = 1$ for all $y \in Y_1$ and there exists an injective Lipschitz map $\varphi: Y_1 \to X$ such that*

$$T(f)(y) = T(\mathbf{1}_X)(y) f(\varphi(y)), \quad \forall y \in Y_1, \ \forall f \in \mathrm{Lip}(X).$$

Next, we improve this lemma by using Theorem 1 and the following notion.

Definition 3 Let X be a metric space, $r > 0$ and let C be a subset of X. C is said to be r-connected if it is not possible to find nonempty sets $A, B \subset C$ such that $C = A \cup B$ and $d(A, B) \geq r$.

Based on this concept, we define on X the following equivalence relation: $x \sim z$ if and only if there exist $x_1, \ldots, x_k, x_{k+1} \in X$ such that $x_1 = x$, $x_{k+1} = z$ and $d(x_j, x_{j+1}) < r$ for all $j \in \{1, \ldots, k\}$. It can easily be checked that the equivalence classes of this relation are maximal r-connected subsets on X. This equivalence classes of the relation $\sim$ are called r-connected components of X, and it is straightforward to prove that they are (pairwise disjoint) open and closed subsets of X.

We get the mentioned improvement of Lemma 3:

Theorem 2 ([95, Teorema 2.14], compare to [55, Theorem 3.1]) *Let X and Y be compact metric spaces and let T be a linear isometry from* $\mathrm{Lip}(X)$ *onto* $\mathrm{Lip}(Y)$ *such that $T(\mathbf{1}_X)$ is a contraction. Then there exist a Lipschitz map $\tau: Y \to S_{\mathbb{K}}$ and a Lipschitz homeomorphism $\varphi: Y \to X$ with* $\mathrm{Lip}(\varphi) \leq \max\{1, \mathrm{diam}(X)/2\}$ *and* $\mathrm{Lip}(\varphi^{-1}) \leq \max\{1, \mathrm{diam}(Y)/2\}$ *such that*

$$T(f)(y) = \tau(y) f(\varphi(y)), \quad \forall y \in Y, \ \forall f \in \mathrm{Lip}(X).$$

Moreover, if $y, z \in Y$ and $d(y, z) < 2$, then $d(\varphi(y), \varphi(z)) = d(y, z)$ and $\tau(y) = \tau(z)$.

Under the conditions given in Theorem 2, if we also suppose that T is unital, then we get that T is an algebra isomorphism. More specifically:

Corollary 2 ([95, Corolario 2.15], see also [55, Corollary 3.2]) *Let X and Y be compact metric spaces and T:* $\mathrm{Lip}(X) \to \mathrm{Lip}(Y)$ *a surjective linear isometry such that $T(\mathbf{1}_X) = \mathbf{1}_Y$. Then there exist a Lipschitz homeomorphism φ from Y to X such that $d(\varphi(y), \varphi(z)) = d(y, z)$ for any $y, z \in Y$ with $d(y, z) < 2$,* $\mathrm{Lip}(\varphi) \leq \max\{1, \mathrm{diam}(X)/2\}$, $\mathrm{Lip}(\varphi^{-1}) \leq \max\{1, \mathrm{diam}(Y)/2\}$ *and $T(f) = f \circ \varphi$ for all $f \in$* $\mathrm{Lip}(X)$.

If $\alpha \in S_{\mathbb{K}}$ and $\varphi\colon Y \to X$ is a surjective isometry, it is easy to prove that $T\colon \mathrm{Lip}(X) \to \mathrm{Lip}(Y)$ defined as $T(f) = \alpha \cdot (f \circ \varphi)$ for all $f \in \mathrm{Lip}(X)$ is a surjective linear isometry. Conversely, we have the following consequence of Lemma 3.

Corollary 3 ([95, Corolario 2.16], see also [55, Theorem 3.3]) *Let X and Y be compact metric spaces whose diameters are less than 2 and let $T\colon \mathrm{Lip}(X) \to \mathrm{Lip}(Y)$ be a surjective linear isometry such that $T(\mathbf{1}_X)$ is a contraction. Then there exist $\alpha \in S_{\mathbb{K}}$ and a surjective isometry $\varphi\colon Y \to X$ such that $T(f) = \alpha \cdot (f \circ \varphi)$ for all $f \in \mathrm{Lip}(X)$.*

If we replace the hypothesis $\mathrm{Lip}(T(\mathbf{1}_X)) < 1$ in Theorem 2 by the condition that the metric spaces are 1-connected, we can get, as a consequence, the following result by Weaver [98, Theorem 2.6.7]. Note that conditions $\mathrm{diam}(X) \leq 2$ and $\mathrm{diam}(Y) \leq 2$ are not restrictive, since we can apply Vasavada's reduction [98, Proposition 1.7.1].

Corollary 4 ([95, Corolario 2.17]) *Let X and Y be compact 1-connected metric spaces whose diameters are less than or equal to 2 and let $T\colon \mathrm{Lip}(X) \to \mathrm{Lip}(Y)$ be a surjective linear isometry. Then there exist $\alpha \in S_{\mathbb{K}}$ and a surjective isometry $\varphi\colon Y \to X$ such that $T(f) = \alpha \cdot (f \circ \varphi)$ for all $f \in \mathrm{Lip}(X)$.*

3 Codimension 1 Linear Isometries

Let us recall that a linear map $T\colon \mathrm{Lip}(X) \to \mathrm{Lip}(Y)$ has codimension 1 if $T(\mathrm{Lip}(X))$ is a maximal vector subspace of $\mathrm{Lip}(Y)$.

This section is devoted to describe codimension 1 linear isometries T from $\mathrm{Lip}(X)$ to $\mathrm{Lip}(Y)$ such that $T(\mathbf{1}_X)$ is a contraction. With this aim, we make use of Theorem 1. First of all, we see that these maps can be of two types.

Proposition 5 ([55, Theorem 4.1]) *Let X and Y be compact metric spaces and let $T\colon \mathrm{Lip}(X) \to \mathrm{Lip}(Y)$ be a codimension 1 linear isometry such that $T(\mathbf{1}_X)$ is a contraction. Then there exist a subset $Y_0 \subset Y$, a surjective Lipschitz map $\varphi\colon Y_0 \to X$ and a Lipschitz function $\tau\colon Y_0 \to S_{\mathbb{K}}$ such that*

$$T(f)(y) = \tau(y) f(\varphi(y)), \quad \forall y \in Y_0, \ \forall f \in \mathrm{Lip}(X).$$

Furthermore, either $Y_0 = Y$ or $Y_0 = Y \backslash \{p\}$ for some isolated point $p \in Y$.

Proposition 5 allows us to make the following classification.

Definition 4 Let X and Y be compact metric spaces and let $T\colon \mathrm{Lip}(X) \to \mathrm{Lip}(Y)$ be a codimension 1 linear isometry such that $T(\mathbf{1}_X)$ is a contraction. We say that:

1. T is of type I if there exists an isolated point $p \in Y$ such that T is a weighted composition operator on $Y \backslash \{p\}$ whose symbol function is a surjective Lipschitz map $\varphi\colon Y \backslash \{p\} \to X$ and its weight function is a Lipschitz map $\tau\colon Y \backslash \{p\} \to S_{\mathbb{K}}$.

2. T is of type II if it is a weighted composition operator on Y whose symbol function is a surjective Lipschitz map $\varphi\colon Y \to X$ and whose weight map is a Lipschitz function $\tau\colon Y \to S_{\mathbb{K}}$.

These two types are not disjoint as the next result shows:

Proposition 6 ([55, Proposition 4.2]) *Let Y be a compact metric space and suppose that there exist two points $y_0, p \in Y$ such that $y_0 \neq p$ and $d(y, y_0) \leq d(y, p)$ for all $y \in Y\backslash\{p\}$ (in particular, this happens when $\operatorname{diam}(Y\backslash\{p\}) \leq d(p, Y\backslash\{p\})$). Then p is an isolated point of Y and the map $T\colon \operatorname{Lip}(Y\backslash\{p\}) \to \operatorname{Lip}(Y)$ defined by*

$$T(f)(y) = f(y), \quad \forall y \in Y\backslash\{p\}, \quad T(f)(p) = f(y_0), \quad \forall f \in \operatorname{Lip}(Y\backslash\{p\}),$$

is a codimension 1 linear isometry with $\operatorname{Lip}(T(\mathbf{1}_X)) < 1$ which is simultaneously of type I and type II.

Next we give a method for constructing codimension 1 linear isometries of type I which are not of type II.

Proposition 7 ([55, Proposition 4.3]) *Let X and Y be compact metric spaces and let $\alpha \in S_{\mathbb{K}}$. Suppose that there exist $p \in Y$ such that $1 < d(p, Y\backslash\{p\})$ and a surjective isometry $\varphi\colon Y\backslash\{p\} \to X$. Then the map $T\colon \operatorname{Lip}(X) \to \operatorname{Lip}(Y)$ defined by*

$$T(f)(y) = \alpha f(\varphi(y)), \quad \forall y \in Y\backslash\{p\}, \quad T(f)(p) = 0, \quad \forall f \in \operatorname{Lip}(X),$$

is a codimension 1 linear isometry with $\operatorname{Lip}(T(\mathbf{1}_X)) < 1$ of type I but it is not of type II.

There are also codimension 1 linear isometries of type II which are not of type I:

Proposition 8 *Consider the metric spaces $X = [0, 2]$ and $Y = [0, 1] \cup [4, 5]$ equipped with their usual distances. Let $\varphi\colon Y \to X$ and $\tau\colon Y \to S_{\mathbb{K}}$ be the maps given by*

$$\varphi(y) = \begin{cases} y & \text{if } y \in [0, 1], \\ y - 3 & \text{if } y \in [4, 5]; \end{cases} \qquad \tau(y) = \begin{cases} 1 & \text{if } y \in [0, 1], \\ -1 & \text{if } y \in [4, 5]. \end{cases}$$

The map $T\colon \operatorname{Lip}(X) \to \operatorname{Lip}(Y)$ defined by

$$T(f)(y) = \tau(y) f(\varphi(y)), \quad \forall y \in Y, \ \forall f \in \operatorname{Lip}(X),$$

is a codimension 1 linear isometry. Observe that T is of type II, but it is not of type I because Y has no isolated points.

Next, we study some properties of the map φ. In particular, we show that φ is a Lipschitz homeomorphism when T is of type I.

Theorem 3 ([55, Proposition 4.6]) *Let X and Y be compact metric spaces and let $T\colon \mathrm{Lip}(X) \to \mathrm{Lip}(Y)$ be a codimension 1 linear isometry with $\mathrm{Lip}(T(\mathbf{1}_X)) < 1$. Take Y_0, φ and τ as in Proposition 5. The following statements hold:*

1. *Given $x \in X$, $\varphi^{-1}(\{x\})$ has at most two elements.*
2. *If there exists a point $x_0 \in X$ such that $\varphi^{-1}(\{x_0\})$ has two distinct points, then, for all $x \in X\backslash\{x_0\}$, $\varphi^{-1}(\{x\})$ consists of an only element.*
3. *If T is of type I, then, for the corresponding isolated point $p \in Y$, we have that $\varphi\colon Y\backslash\{p\} \to X$ is a Lipschitz homeomorphism.*

4 Local Isometries

This section is a contribution to an interesting current topic, the study of normed spaces whose isometry group is determined by the local behavior of its elements.

Definition 5 Let E be a normed space and let $\mathcal{S}$ be a nonempty subset of $\mathcal{L}(E)$. We denote by $\mathrm{ref}_{\mathrm{al}}(\mathcal{S})$ the set of operators $T \in \mathcal{L}(E)$ verifying the following property:

$$\forall e \in E \;\; \exists\, \Phi_e \in \mathcal{S}\colon \; T(e) = \Phi_e(e).$$

It is said that $\mathcal{S}$ is algebraically reflexive if $\mathrm{ref}_{\mathrm{al}}(\mathcal{S}) = \mathcal{S}$.

We say that E is iso-reflexive if $\mathcal{G}(E)$ is algebraically reflexive, where $\mathcal{G}(E)$ stands for the set of all surjective linear isometries from E to E.

The elements of $\mathrm{ref}_{\mathrm{al}}(\mathcal{G}(E))$ are called locally surjective linear isometries or local isometries. Thus, E is iso-reflexive if and only if every locally surjective linear isometry is surjective.

Using those descriptions of into linear isometries (Theorem 1) and onto linear isometries (Corollary 4) between $\mathrm{Lip}(X)$-spaces, we can state the following result on the algebraic reflexivity of $\mathrm{Lip}(X)$.

Theorem 4 ([95, Teorema 2.26]) *Let X be a compact 1-connected metric space. Then $\mathrm{Lip}(X)$ is iso-reflexive.*

Now, we use the iso-reflexivity of $\mathrm{Lip}(X)$ in order to study the algebraic reflexivity of some subsets of isometries and projections on $\mathrm{Lip}(X)$. Let us introduce these sets. We use the symbol Id_A to denote the identity map on a nonempty set A, or Id if there is no confusion.

Definition 6 Let X be a metric space and let E be a normed space.

1. An isometry $\varphi\colon X \to X$ is said to be involutive if φ^2 is the identity on X.
2. An isometric reflection of E is an involutive linear isometry on E.
3. A linear map $P\colon E \to E$ is a projection if $P^2 = P$.

4. It is said that a projection $P\colon E \to E$ is bi-circular if $\alpha P + \gamma(\mathrm{Id} - P)$ is an isometry for any $\alpha, \gamma \in S_{\mathbb{K}}$. It is clear that P is bi-circular if and only if $P + \lambda \cdot (\mathrm{Id} - P)$ is an isometry for all $\lambda \in S_{\mathbb{K}}$.
5. A projection $P\colon E \to E$ is called generalized bi-circular if there exists $\lambda \in S_{\mathbb{K}} \backslash \{1\}$ such that $P + \lambda \cdot (\mathrm{Id} - P)$ is an isometry.
6. A projection $P\colon E \to E$ with $\|P\| \leq 1$ is said to be contractive. If $\|P\| \leq 1$ and $\|\mathrm{Id} - P\| \leq 1$, we say that P is bi-contractive.

From Corollary 4, we can easily get the description of the isometric reflections of $\mathrm{Lip}(X)$ as weighted composition operators:

Corollary 5 ([95, Corolario 2.28]) *Let X be a compact 1-connected metric space whose diameter is less than or equal to 2. A map $T : \mathrm{Lip}(X) \to \mathrm{Lip}(X)$ is an isometric reflection if and only if there exist a constant $\tau \in \{-1, 1\}$ and an involutive isometry φ on X such that*

$$T(f)(x) = \tau \cdot f(\varphi(x)), \quad \forall x \in X, \ \forall f \in \mathrm{Lip}(X).$$

Corollary 6 ([95, Corolario 2.29]) *Let X be a compact 1-connected metric space. Then the set of isometric reflections of $\mathrm{Lip}(X)$ is algebraically reflexive.*

Now, we proceed in the same way with generalized bi-circular projections. Firstly, we give a description of them:

Proposition 9 ([95, Proposición 2.30]) *Let X be a compact 1-connected metric space whose diameter is less than or equal to 2. A map $P\colon \mathrm{Lip}(X) \to \mathrm{Lip}(X)$ is a generalized bi-circular projection if and only if there exist a number $\tau \in \{-1, 1\}$ and an involutive isometry $\varphi\colon X \to X$ such that*

$$P(f)(x) = \frac{1}{2}\left[f(x) + \tau \cdot f(\varphi(x))\right], \quad \forall x \in X, \ \forall f \in \mathrm{Lip}(X).$$

Secondly, we show the algebraic reflexivity of the set of generalized bi-circular projections:

Corollary 7 ([95, Corolario 2.31]) *Let X be a compact 1-connected metric space. Then the set of all generalized bi-circular projections on $\mathrm{Lip}(X)$ is algebraically reflexive.*

The following result is an immediate consequence of Proposition 9.

Corollary 8 ([95, Corolario 2.32]) *Let X be a compact 1-connected metric space. Every generalized bi-circular projection on $\mathrm{Lip}(X)$ is a bi-contractive projection.*

All the results of this section were stated with similar proofs in [54] for $\mathrm{Lip}(X)$ endowed with the sum norm, when X is a compact metric space.

5 Projections and Averages of Linear Isometries

We have just shown that every generalized bi-circular projection on $\mathrm{Lip}(X)$ can be expressed as the average of two isometries. Now, we prove that such projections are the only projections on $\mathrm{Lip}(X)$ having this property. In order to achieve these aim, we first characterize when the average of two isometries is a projection on $\mathrm{Lip}(X)$.

First of all, we study the projections on $\mathrm{Lip}(X)$ which can be expressed as convex combinations of two surjective linear isometries. Let I_1 and I_2 be two surjective linear isometries on $\mathrm{Lip}(X)$ given by

$$I_k(f)(x) = \tau_k f(\varphi_k(x)), \quad \forall f \in \mathrm{Lip}(X),\ \forall x \in X \qquad (k = 1, 2),$$

where $\tau_k \in \mathbb{K}$ with $|\tau_k| = 1$ and $\varphi_k \colon X \to X$ is a surjective isometry.

Our initial objective is to find conditions on the constants τ_k, the functions φ_k and the parameter $0 < \lambda < 1$ in order to $\lambda I_1 + (1 - \lambda) I_2$ is a projection on $\mathrm{Lip}(X)$:

Proposition 10 ([12, Proposition 2.1]) *Let P be a projection on* $\mathrm{Lip}(X)$ *and let* $0 < \lambda < 1$. *If $P = \lambda I_1 + (1 - \lambda) I_2$, then*

(i) $\tau_1 = \tau_2 = 1$, *or* $\tau_1 = -\tau_2$ *and* $\lambda = 1/2$.
(ii) *If* $\varphi_1(x) \neq \varphi_2(x)$, *then either* $\varphi_1(x) = x$ *or* $\varphi_2(x) = x$.
(iii) *If* $x = \varphi_1(x) \neq \varphi_2(x)$, *then* $\varphi_1(\varphi_2(x)) = \varphi_2(x)$, $\varphi_2^2(x) = x$, $\lambda = 1/2$, $\tau_1 = 1$ *and* $\tau_2^2 = 1$.
(iv) *If* $x = \varphi_2(x) \neq \varphi_1(x)$, *then* $\varphi_2(\varphi_1(x)) = \varphi_1(x)$, $\varphi_1^2(x) = x$, $\lambda = 1/2$, $\tau_2 = 1$ *and* $\tau_1^2 = 1$.

Now, we characterize the operators $(I_1 + I_2)/2$ which are projections on $\mathrm{Lip}(X)$.

Proposition 11 ([12, Proposition 2.2]) *The operator $(I_1 + I_2)/2$ is a projection on* $\mathrm{Lip}(X)$ *if and only if one of the following statements holds:*

1. $\tau_1 = \tau_2 = 1$ *and for each* $x \in X$ *it is verified that:*

   a. $x = \varphi_1(x) = \varphi_2(x)$, *or*
   b. $x = \varphi_1(x) \neq \varphi_2(x)$, $\varphi_1(\varphi_2(x)) = \varphi_2(x)$ *and* $\varphi_2^2(x) = x$, *or*
   c. $x = \varphi_2(x) \neq \varphi_1(x)$, $\varphi_2(\varphi_1(x)) = \varphi_1(x)$ *and* $\varphi_1^2(x) = x$.

2. $\tau_1 = -\tau_2$ *and* $\varphi_1(x) = \varphi_2(x)$ *for all* $x \in X$, *namely* $((I_1 + I_2)/2)(f)(x) = 0$, *for all* $f \in \mathrm{Lip}(X)$.
3. $\tau_1 = 1$, $\tau_2 = -1$ *and each* $x \in X$ *satisfies:*

   a. $\varphi_1(x) = \varphi_2(x)$, *or*
   b. $x = \varphi_1(x) \neq \varphi_2(x)$, $\varphi_1(\varphi_2(x)) = \varphi_2(x)$ *and* $\varphi_2^2(x) = x$.

4. $\tau_1 = -1$, $\tau_2 = 1$ *and each* $x \in X$ *verifies:*

   a. $\varphi_1(x) = \varphi_2(x)$, *or*
   b. $x = \varphi_2(x) \neq \varphi_1(x)$, $\varphi_2(\varphi_1(x)) = \varphi_1(x)$ *and* $\varphi_1^2(x) = x$.

We know that a generalized bi-circular projection on $\mathrm{Lip}(X)$ is the average of the identity map and an involutive isometry. By applying Proposition 11 yields:

Theorem 5 ([12, Theorem 2.3]) *A linear projection on* $\mathrm{Lip}(X)$ *is the average of two surjective linear isometries if and only if it is a generalized bi-circular projection.*

Following a process similar to the one above, we next characterize the projections on $\mathrm{Lip}(X)$ which can be expressed as the average of three surjective linear isometries.

Theorem 6 ([12, Theorem 3.8]) *For* $k = 1, 2, 3$, *let* I_k *be a surjective linear isometry on* $\mathrm{Lip}(X)$ *given by*

$$I_k(f)(x) = \tau_k f(\varphi_k(x)) \quad (f \in \mathrm{Lip}(X),\ x \in X) \quad (k = 1, 2, 3),$$

where τ_k *is a unimodular scalar and* φ_k *is a surjective isometry on* X, *and let* Q *be the average of* I_1, I_2 *and* I_3. *Then* Q *is a projection on* $\mathrm{Lip}(X)$ *if and only if there exist a* $\tau \in \mathbb{K}$ *with* $\tau^3 = 1$ *and a surjective isometry* φ *on* X *with* $\varphi^3 = \mathrm{Id}$ *such that*

$$Q(f)(x) = \frac{f(x) + \tau f(\varphi(x)) + \tau^2 f(\varphi^2(x))}{3}$$

for all $f \in \mathrm{Lip}(X)$ *and* $x \in X$.

The previous theorem motivates the following concept.

Definition 7 Let $n \in \mathbb{N}$ with $n \geq 2$. It is said that a continuous linear operator Q on $\mathrm{Lip}(X)$ is a n-circular projection if there exist a scalar $\tau \in \mathbb{K}$ such that $\tau^n = 1$ and a surjective isometry φ on X such that $\varphi^n = \mathrm{Id}$ and $\varphi^k \neq \mathrm{Id}$ for $k = 1, \ldots, n-1$ satisfying

$$Q(f)(x) = \frac{\sum_{k=0}^{n-1} \tau^k f(\varphi^k(x))}{n},$$

for every $f \in \mathrm{Lip}(X)$ and $x \in X$. We take $\varphi^0 = \mathrm{Id}$.

Theorems 5 and 6 can be restated as in the following theorem. We refer to a projection as being trivial if it is equal to either the zero or the identity operators.

Theorem 7 ([12, Theorem 4.2]) *Let* X *be a compact* 1*-connected metric space with* $\mathrm{diam}(X) \leq 2$.

1. *The average of two surjective isometries on* $\mathrm{Lip}(X)$ *is a projection if and only if it is either a trivial projection or a* 2*-circular projection.*
2. *The average of three surjective isometries on* $\mathrm{Lip}(X)$ *is a projection if and only if it is either a trivial projection or a* 3*-circular projection.*

We end this section with two remarks motivated by the previous results.

Remark 1 It can be checked that if X is a compact 1-connected metric space with $\text{diam}(X) \leq 2$, then 3-circular projections on $\text{Lip}(X)$ cannot be expressed as the average of two surjective linear isometries on $\text{Lip}(X)$.

Remark 2 It is well-known that generalized bi-circular projections are bicontractive [37]. Observe that 3-circular projections are not necessarily bicontractive. In fact, let $X = \{a, b, c\}$ equipped with the metric $d(a, b) = d(b, c) = d(a, c) = 2$. Consider $P = (\text{Id} + R + R^2)/3$ with $R(f) = f \circ \varphi$ and φ a period 3 isometry on X ($\varphi(a) = b$, $\varphi(b) = c$ and $\varphi(c) = a$). Then $\text{Id} - P = (2\text{Id} - R - R^2)/3$. Consider $f \in \text{Lip}(X)$ such that $f(\varphi(a)) = f(\varphi^2(a)) = -1$ and $f(a) = 1$. Note that $\|f\| = 6/5$ and $\|(\text{Id} - P)(f)\| = 23/15$, hence $\|\text{Id} - P\| > 1$.

6 2-Local Isometries

The notion of locally surjective linear isometry can be generalized as follows: on the one hand, linearity can be suppressed; on the other hand, we can consider several points (more than one) in which the map coincides with a surjective linear isometry. Taking this into account, it is not surprising to introduce the following definition.

Definition 8 Let E and F be normed spaces. We say that a map $\Phi : E \to F$ is a 2-local isometry if for any $x, y \in E$, there exists a surjective linear isometry $\Phi_{x,y} : E \to F$, depending, in general, on x and y, such that $\Phi(x) = \Phi_{x,y}(x)$ and $\Phi(y) = \Phi_{x,y}(y)$.

It is said that a Banach space E is 2-iso-reflexive if every 2-local isometry from E to E is linear and surjective.

This concept was introduced by Molnár in [70], where he showed that the space of linear bounded operators on a Hilbert space is 2-iso-reflexive. Numerous papers on 2-iso-reflexivity of Banach spaces have appeared since then (cf. [1, 41, 42, 51, 62]).

In the following results, we consider the space $\text{Lip}(X)$ endowed with the maximum norm $\max\{\|f\|_\infty, \text{Lip}(f)\}$ or with the sum norm $\|f\|_\infty + \text{Lip}(f)$. It is easy to check that if τ is a unimodular scalar and φ is a surjective isometry on X, then the weighted composition operator given by

$$\Phi(f) = \tau \cdot (f \circ \varphi), \quad \forall f \in \text{Lip}(X),$$

is a surjective linear isometry on $\text{Lip}(X)$. If every surjective linear isometry on $\text{Lip}(X)$ can be expressed in this way, it is said that the isometry group of $\text{Lip}(X)$ is canonical. The isometry group of $\text{Lip}(X)$ is not canonical in general (see an example in [97]). However, Hatori and Oi [44] showed that the isometry group of $\text{Lip}(X)$ with the sum norm is canonical when X is compact (see also [48]). On the other hand, the same conclusion was obtained for the maximum norm of $\text{Lip}(X)$ by Roy [81] when X is compact and connected with diameter at most 1; and, independently, by Vasavada [92] when X is compact and satisfies certain separation conditions.

First of all, we give a description of 2-local isometries on $\mathrm{Lip}(X)$ in terms of a weighted composition operator.

Theorem 8 ([57, Theorem 2.1]) *Let X be a bounded metric space and let Φ be a 2-local isometry on* $\mathrm{Lip}(X)$ *whose isometry group is canonical. Then there exist a subset X_0 of X, a unimodular scalar τ and a Lipschitz bijection $\varphi\colon X_0 \to X$ such that*

$$\Phi(f)|_{X_0} = \tau \cdot (f \circ \varphi), \quad \forall f \in \mathrm{Lip}(X).$$

Clearly, every 2-local isometry on a Banach space is an isometry, so the main problem is to study if 2-local isometries are linear and surjective. We can give a positive answer for 2-local isometries on $\mathrm{Lip}(X)$. More precisely, when X is, in addition, separable, we show that $X_0 = X$ and φ is a Lipschitz homeomorphism, and thus Φ is a surjective linear isometry on $\mathrm{Lip}(X)$. For this aim, we use the following result which is interesting in itself. It is a Lipschitz version of a result due to Győry [41] for continuous functions. A detailed reading of its proof shows that its adaptation to Lipschitz functions is far from being simple.

Proposition 12 ([57, Proposition 3.2]) *Let X be a metric space and let $R = \{r_n \colon n \in \mathbb{N}\}$ be a countable set of distinct points of X. Then there exist two Lipschitz maps $f, g\colon X \to [0, 1]$ satisfying that f has a strict local maximum in each point of R and*

$$\{z \in X \colon (f(z), g(z)) = (f(r_n), g(r_n))\} = \{r_n\}, \quad \forall n \in \mathbb{N}.$$

We have all the ingredients to prove the most important result on 2-locality on $\mathrm{Lip}(X)$.

Theorem 9 ([57, Theorem 3.3]) *Let X be a separable bounded metric space. If the isometry group of* $\mathrm{Lip}(X)$ *is canonical, then* $\mathrm{Lip}(X)$ *is 2-iso-reflexive.*

Now, let us recall the following concept.

Definition 9 If E and F are normed spaces, it is said that a map $\Phi\colon E \to F$ is a weak 2-local isometry if for any $x, y \in E$ and $\varphi \in F^*$, there exists a surjective linear isometry $\Phi_{x,y,\varphi}\colon E \to F$, depending in general on x, y, φ, such that $\varphi\Phi(x) = \varphi\Phi_{x,y,\varphi}(x)$ and $\varphi\Phi(y) = \Phi_{x,y,\varphi}(y)$.

In [62], the authors stated a spherical version of the Kowalski–Slodkowski theorem [60], and applied this result to prove the following fact:

Theorem 10 ([62, Theorem 3.5 and Corollaries 3.8 and 3.9]) *Let X be a metric space and assume that the isometry group of* $\mathrm{Lip}(X)$ *is canonical. Then every weak 2-local isometry on* $\mathrm{Lip}(X)$ *is linear. As a consequence, if X is in addition compact, then* $\mathrm{Lip}(X)$ *is 2-iso-reflexive.*

The final part of this section is devoted to the study of 2-locality on the space of vector-valued Lipschitz maps $\mathrm{Lip}(X, E)$ equipped with the maximum norm. To achieve our goals it is necessary to get a description of the surjective linear isometries of such spaces. In our case, we make use of the following result by Botelho, Fleming and Jamison.

Theorem 11 ([8, Theorem 10]) *Let X and Y be pathwise-connected compact metric spaces, let E and F be smooth reflexive Banach spaces, and let* $T\colon \mathrm{Lip}(X, E) \to \mathrm{Lip}(Y, F)$ *be a surjective linear isometry. Then, there exist a Lipschitz homeomorphism* $\varphi\colon Y \to X$ *with* $\mathrm{Lip}(\varphi) \leq \max\{1, \mathrm{diam}(X)\}$ *and* $\mathrm{Lip}(\varphi^{-1}) \leq \max\{1, \mathrm{diam}(Y)\}$, *and a Lipschitz map* $V\colon Y \to \mathrm{Iso}(E, F)$ *such that*

$$T(f)(y) = V(y)(f(\varphi(y))), \quad \forall y \in Y,\ f \in \mathrm{Lip}(X, E).$$

Araujo and Dubarbie [3] introduced a subclass of surjective linear isometries admitting an expression similar to the one given in Theorem 11, which is also adequate for our purposes.

Given X and Y two metric spaces, a bijection $\varphi\colon Y \to X$ is said to preserve distances less than 2 if $d(\varphi(x), \varphi(y)) = d(x, y)$ when $d(x, y) < 2$. We denote by $\mathrm{Iso}_{<2}(Y, X)$ the set of all bijections $\varphi\colon Y \to X$ such that φ and φ^{-1} preserve distances which are less than 2. Note that every element $\varphi \in \mathrm{Iso}_{<2}(Y, X)$ is a Lipschitz homeomorphism when X and Y are bounded.

Definition 10 ([3, Definition 2.3]) Let X and Y be metric spaces and let E and F be normed spaces. We say that a map $T\colon \mathrm{Lip}(X, E) \to \mathrm{Lip}(Y, F)$ is a standard isometry if there exist a bijection $\varphi \in \mathrm{Iso}_{<2}(Y, X)$ and a map $J\colon Y \to \mathrm{Iso}(E, F)$ which is constant in each 2-component of Y such that

$$T(f)(y) = J(y)(f(\varphi(y))), \quad \forall y \in Y,\ f \in \mathrm{Lip}(X, E).$$

In fact, J is a Lipschitz map.

Note that every standard isometry is a surjective linear isometry. Theorem 3.1 in [3] provides us a condition under which both kinds of isometries coincide.

Definition 11 Let X and Y be metric spaces and let E and F be normed spaces. A map $\Delta\colon \mathrm{Lip}(X, E) \to \mathrm{Lip}(Y, F)$ is said to be a 2-local standard isometry if for any $f, g \in \mathrm{Lip}(X, E)$, there exists a standard isometry $T_{f,g}\colon \mathrm{Lip}(X, E) \to \mathrm{Lip}(Y, F)$, depending in general on f and g, such that $\Delta(f) = T_{f,g}(f)$ and $\Delta(g) = T_{f,g}(g)$.

We say that $\mathrm{Lip}(X, E)$ is 2-standard-iso-reflexive if every 2-local standard isometry on $\mathrm{Lip}(X, E)$ is linear and surjective.

Under certain conditions on a compact metric space X and a normed space E, we can give a description of all 2-local isometries and all 2-local standard isometries on $\mathrm{Lip}(X, E)$ as a generalized composition operator, and study when the space $\mathrm{Lip}(X, E)$ is 2-iso-reflexive or 2-standard-iso-reflexive.

Let us recall that a normed space E is smooth if for every point $e \in S_E$, there exists a unique functional $e^* \in S_{E^*}$ such that $e^*(e) = 1$. It is known that a space E is smooth if and only if its norm is Gâteaux-differentiable at every point of S_E.

We establish a Lipschitz version of [1, Theorem 6].

Theorem 12 ([52, Theorem 2.3]) *Let X and Y be pathwise-connected compact metric spaces, let E be a smooth reflexive Banch space which is 2-iso-reflexive, and let $\Delta\colon \mathrm{Lip}(X, E) \to \mathrm{Lip}(Y, E)$ be a 2-local isometry. Then there exist a subset Y_0 of Y, a Lipschitz bijection ψ from Y_0 into X, and a continuous map $y \in Y_0 \mapsto V(y) \in \mathrm{Iso}(E)$ from Y_0 to $\mathrm{Iso}(E)$ endowed with the strong operator topology (SOT for short) such that*

$$\Delta(f)(y) = V(y)(f(\psi(y))), \quad \forall y \in Y_0,\ f \in \mathrm{Lip}(X, E).$$

Similar reasoning given in the aforementioned theorem can be used to describe 2-local standard isometries.

Theorem 13 ([52, Theorem 2.4]) *Let X and Y be compact metric spaces, let E be a smooth and reflexive Banach space which is also 2-iso-reflexive, and let $\Delta\colon \mathrm{Lip}(X, E) \to \mathrm{Lip}(Y, E)$ be a 2-local standard isometry. Then there exist a subset Y_0 of Y, a Lipschitz bijection ψ from Y_0 into X, and a continuous map V from Y_0 to $(\mathrm{Iso}(E), \mathrm{SOT})$ such that*

$$\Delta(f)(y) = V(y)(f(\psi(y))), \quad \forall y \in Y_0,\ f \in \mathrm{Lip}(X, E).$$

Theorem 12 and Proposition 12 provide us the tools to apply the proof method introduced by Győry in [41].

Theorem 14 ([52, Theorem 2.6]) *Let X and Y be pathwise-connected compact metric spaces, let E be a smooth and reflexive Banach space which is 2-iso-reflexive, and let $\Delta\colon \mathrm{Lip}(X, E) \to \mathrm{Lip}(Y, E)$ a 2-local isometry. Then there exist a Lipschitz homeomorphism ψ from Y into X and a continuous map V from Y to $(\mathrm{Iso}(E), \mathrm{SOT})$ such that*

$$\Delta f(y) = V(y)(f(\psi(y))), \quad \forall y \in Y,\ f \in \mathrm{Lip}(X, E).$$

Furthermore, Δ is a surjective linear isometry.

Thanks to Theorem 13, the arguments used at the proof of the previous theorem also work in the case of 2-local standard isometries.

Theorem 15 ([52, Theorem 2.7]) *Let X and Y be compact metric spaces, let E be a smooth and reflexive which is 2-iso-reflexive, and let $\Delta\colon \mathrm{Lip}(X, E) \to \mathrm{Lip}(Y, E)$ be a 2-local standard isometry. Then there exist a Lipschitz homeomorphism ψ from Y to X and a continuous map V from Y into $(\mathrm{Iso}(E), \mathrm{SOT})$ such that*

$$\Delta f(y) = V(y)(f(\psi(y))), \quad \forall y \in Y,\ f \in \mathrm{Lip}(X, E).$$

Moreover, Δ is a surjective linear isometry.

In Theorems 14 and 15, it was shown that, under certain hypotheses on X and E, the space $\mathrm{Lip}(X, E)$ is 2-iso-reflexive and 2-standard-iso-reflexive, respectively. One of these requirements is that E is 2-iso-reflexive. In fact, this condition is necessary to get the conclusion of these theorems.

Theorem 16 ([52, Theorem 2.8]) *Let X be a metric space and let E be a Banach space. If* $\mathrm{Lip}(X, E)$ *is 2-iso-reflexive or 2-standard-iso-reflexive, then E is 2-iso-reflexive.*

Finally, it is interesting to point out that our results also hold if, in their hypotheses, we consider a 2-local standard isometry Δ from $\mathrm{Lip}(X, E)$ into $\mathrm{Lip}(Y, F)$, where E and F are smooth and reflexive Banach spaces for which each 2-local isometry from E into F is linear and surjective. In that case, the map V in the conclusion takes its values on $\mathrm{Iso}(E, F)$.

7 Vector-Valued Linear Isometries

We give now a complete description of linear isometries between $\mathrm{Lip}(X, E)$-spaces satisfying some mild conditions. So we establish a Lipschitz version of a known theorem by Cambern [17] on linear isometries between $C(X, E)$-spaces.

Theorem 17 ([56, Theorem 2.1]) *Let X and Y be compact metric spaces, let E be a strictly convex normed space and let $T\colon \mathrm{Lip}(X, E) \to \mathrm{Lip}(Y, E)$ be a linear isometry. Suppose that $T(\mathbf{1}_X \otimes e) = \mathbf{1}_Y \otimes e$ for some $e \in S_E$. Then there exist a closed nonempty subset Y_0 of Y, a surjective Lipschitz map $\varphi\colon Y_0 \to X$ with $\mathrm{Lip}(\varphi) \leq \max\{1, \mathrm{diam}(X)/2\}$ and a Lipschitz map $\widehat{T}\colon Y \to \mathcal{L}(E)$ with $\|\widehat{T}(y)\| = 1$ for all $y \in Y$ such that*

$$T(f)(y) = \widehat{T}(y)(f(\varphi(y))), \quad \forall y \in Y_0, \ \forall f \in \mathrm{Lip}(X, E).$$

The above condition, $T(\mathbf{1}_X \otimes e) = \mathbf{1}_Y \otimes e$ for some $e \in S_E$, is not too restrictive if one studies some known results in the scalar case. Our condition in this case means $T(\mathbf{1}_X) = \mathbf{1}_Y$ and notice that connectedness assumptions on the metric spaces in [81, Lemma 1.5] and [97, Lemma 6] imply that the function $T(\mathbf{1}_X)$ is constant.

Jerison's theorem [50] on surjective linear isometries between $C(X, E)$-spaces also has a natural formulation for Lipschitz maps:

Theorem 18 ([56, Theorem 3.1]) *Let X and Y be compact metric spaces, E a strictly convex normed space and $T\colon \mathrm{Lip}(X, E) \to \mathrm{Lip}(Y, E)$ a surjective linear isometry such that $T(\mathbf{1}_X \otimes e) = \mathbf{1}_Y \otimes e$ for some $e \in S_E$. Then there exist a Lipschitz homeomorphism $\varphi\colon Y \to X$ with $\mathrm{Lip}(\varphi) \leq \max\{1, \mathrm{diam}(X)/2\}$ and $\mathrm{Lip}(\varphi^{-1}) \leq \max\{1, \mathrm{diam}(Y)/2\}$, and a Lipschitz map $\widehat{T}\colon Y \to \mathcal{L}(E)$ such that $\widehat{T}(y)$ is an isometry from E onto itself for all $y \in Y$ and*

$$T(f)(y) = \widehat{T}(y)(f(\varphi(y))), \quad \forall y \in Y, \ \forall f \in \mathrm{Lip}(X, E).$$

As a direct consequence of Theorem 18, we obtain the following:

Corollary 9 ([56, Corollary 3.2]) *Let X and Y be two compact metric spaces whose diameters are less than or equal to 2, let E be a strictly convex normed space and $T: \mathrm{Lip}(X, E) \to \mathrm{Lip}(Y, E)$ a surjective linear isometry such that $T(\mathbf{1}_X \otimes e) = \mathbf{1}_Y \otimes e$ for some $e \in S_E$. Then there exist a surjective isometry $\varphi: Y \to X$ and a Lipschitz map $\widehat{T}: Y \to \mathcal{L}(E)$ such that $\widehat{T}(y)$ is a surjective isometry from E to E for all $y \in Y$ and we have that*

$$T(f)(y) = \widehat{T}(y)(f(\varphi(y))), \quad \forall y \in Y, \ \forall f \in \mathrm{Lip}(X, E).$$

When E is a Hilbert space, Theorems 17 and 18 can be improved:

Corollary 10 ([56, Corollary 3.3]) *Let X and Y be compact metric spaces and let E be a Hilbert space. Let T be a linear isometry from $\mathrm{Lip}(X, E)$ into $\mathrm{Lip}(Y, E)$ such that $T(1_X \otimes e)$ is a constant function for some $e \in S_E$. Then there exists a Lipschitz map φ from a closed subset Y_0 of Y to X with $\mathrm{Lip}(\varphi) \leq \max\{1, \mathrm{diam}(X)/2\}$, and a Lipschitz map $\widehat{T}$ from Y into $\mathcal{L}(E)$ with $||\widehat{T}(y)|| = 1$ for all $y \in Y$, such that*

$$T(f)(y) = \widehat{T}(y)(f(\varphi(y))), \quad \forall y \in Y_0, \ \forall f \in \mathrm{Lip}(X, E).$$

If, in addition, T is surjective, then $Y_0 = Y$, φ is a Lipschitz homeomorphism with $\mathrm{Lip}(\varphi^{-1}) \leq \max\{1, \mathrm{diam}(Y)/2\}$ and, for each $y \in Y$, $\widehat{T}(y)$ is a surjective isometry.

Some versions of the preceding results for isometries from $\mathrm{Lip}(X, E)$ to $\mathrm{Lip}(Y, F)$ with $E \neq F$ can be consulted in [95].

8 Scalar-Valued Hermitian Operators

Let A be a complex Banach algebra with unity I and let A^* be its dual space. Given $a \in A$, recall that the algebraic numerical range $V(a)$ is given by

$$V(a) = \left\{F(a) \colon F \in A^*, \ \|F\| = F(I) = 1\right\}.$$

An element $a \in A$ is said to be hermitian if $V(a) \subset \mathbb{R}$. It is known that $a \in A$ is hermitian if and only if $\|\exp(ita)\| = 1$ for all $t \in \mathbb{R}$.

Let E be a complex Banach space and $\mathcal{L}(E)$ the Banach algebra of all bounded linear operators on E equipped with the operator norm. It is well-known that an operator $T \in \mathcal{L}(E)$ is hermitian if and only if $\exp(itT)$ is an isometry for each $t \in \mathbb{R}$. The set of hermitian operators on E is a real subspace of $\mathcal{L}(E)$ which contains all operators of the form $\lambda\mathrm{Id}$, where λ is a real number. A hermitian operator is said to be trivial if it is a real multiple of the identity operator.

We study in this section the hermitian operators defined on spaces $\mathrm{Lip}(X)$ and $\mathrm{lip}(X^\alpha)$ with $\alpha \in]0, 1[$, equipped with the sum norm. In [6], it was shown that hermitian operators on $\text{Lip}([0,1])$ and $\text{lip}([0,1]^\alpha)$ are trivial. We generalize here this result by showing that the same property holds for the spaces $\text{Lip}(X)$ and $\text{lip}(X^\alpha)$ with $\alpha \in]0,1[$, whenever X is a compact metric space.

The proof of our main theorem requires some preliminary lemmas. Firstly, we need to have a good representation for surjective linear isometries on such spaces.

Theorem 19 ([14, Theorem 2.1]) *Let X be a compact metric space. Then $T\colon \text{Lip}(X) \to \text{Lip}(X)$ is a surjective linear isometry such that $T(\mathbf{1}_X) = \mathbf{1}_X$ if and only if there exists a surjective isometry $\varphi\colon X \to X$ such that T is of the form $T(f) = f \circ \varphi$ for all $f \in \text{Lip}(X)$. The same characterization holds for a surjective linear isometry T on $\text{lip}(X^\alpha)$ $(0 < \alpha < 1)$ such that $T(\mathbf{1}_X) = \mathbf{1}_X$.*

Secondly, we characterize the hermitian elements of such spaces.

Lemma 4 ([14, Lemma 2.2]) *Let X be a compact metric space and $h \in \text{Lip}(X)$ (or $\text{lip}(X^\alpha)$, $0 < \alpha < 1$). Then h is a hermitian element in $\text{Lip}(X)$ (or $\text{lip}(X^\alpha)$) if and only if h is a real constant function.*

Following an idea of de Leeuw [61], we embed the Banach spaces $\text{Lip}(X)$ and $\text{lip}(X^\alpha)$ $(0 < \alpha < 1)$ isometrically into some suitable spaces of complex-valued continuous functions.

Let X be a compact metric space. Since $\widetilde{X}$ is completely regular, we can consider $\beta\widetilde{X}$, the Stone-Čech compactification of $\widetilde{X}$. Let $C(X \cup \beta\widetilde{X})$ denote the Banach space of all complex-valued continuous functions on $X \cup \beta\widetilde{X}$, under the norm

$$\|f\| = \left\| f|_X \right\|_\infty + \|f|_{\beta\widetilde{X}}\|_\infty \qquad (f \in C(X \cup \beta\widetilde{X})),$$

and let $C_0(X \cup \widetilde{X})$ denote the Banach space of all complex-valued continuous functions on $X \cup \widetilde{X}$ vanishing at the infinity, endowed with the norm

$$\|f\| = \left\| f|_X \right\|_\infty + \left\| f|_{\widetilde{X}} \right\|_\infty \qquad (f \in C_0(X \cup \widetilde{X})).$$

Let us recall that the Riesz representation theorem states that the map $\mu \mapsto F_\mu$, given by

$$F_\mu(f) = \int_{X \cup \beta\widetilde{X}} f\, d\mu \qquad (f \in C(X \cup \beta\widetilde{X})),$$

defines an isometric isomorphism from the Banach space $\mathcal{M}(X \cup \beta\widetilde{X})$ of all complex-valued regular Borel measures on $X \cup \beta\widetilde{X}$ equipped with the norm of total variation:

$$\|\mu\| = |\mu|\,(X \cup \beta\widetilde{X}) \qquad (\mu \in \mathcal{M}(X \cup \beta\widetilde{X}))$$

onto the dual space of $(C(X \cup \beta\widetilde{X}), \|\cdot\|_\infty)$. Similarly, the map $\nu \mapsto G_\nu$ defined by

$$G_\nu(f) = \int_{X \cup \widetilde{X}} f\, d\nu \qquad (f \in C_0(X \cup \widetilde{X}))$$

is an isometric isomorphism from the Banach space $\mathcal{M}(X \cup \widetilde{X})$ with the norm:

$$\|\nu\| = |\nu|\,(X \cup \widetilde{X}) \qquad (\nu \in \mathcal{M}(X \cup \widetilde{X}))$$

onto the dual space of $(C_0(X \cup \widetilde{X}), \|\cdot\|_\infty)$.

For each $f \in \mathrm{Lip}(X)$ or $f \in \mathrm{lip}(X^\alpha)$, $0 < \alpha < 1$, we set $\widetilde{f} : \widetilde{X} \to \mathbb{C}$ to be the map given by

$$\widetilde{f}(x, y) = \frac{f(x) - f(y)}{d(x, y)^\alpha}, \quad \forall (x, y) \in \widetilde{X},$$

where $\alpha = 1$ when $f \in \mathrm{Lip}(X)$. It is easy to show that $\widetilde{f}$ is continuous on $\widetilde{X}$ and $||\widetilde{f}||_\infty = p_\alpha(f)$ $(0 < \alpha \leq 1)$. Hence there exists a unique continuous function $\beta\widetilde{f}$ on $\beta\widetilde{X}$ such that $(\beta\widetilde{f})|_{\widetilde{X}} = \widetilde{f}$ and $||\beta\widetilde{f}||_\infty = ||\widetilde{f}||_\infty$. Furthermore, if $f \in \mathrm{lip}(X^\alpha)$, then $\widetilde{f}$ vanishes at infinity on $\widetilde{X}$. The maps $\Phi : \mathrm{Lip}(X) \to C(X \cup \beta\widetilde{X})$ and $\Psi : \mathrm{lip}(X^\alpha) \to C_0(X \cup \widetilde{X})$, defined by

$$\Phi(f)(w) = \begin{cases} f(w) & \text{if } w \in X, \\ \beta\widetilde{f}(w) & \text{if } w \in \beta\widetilde{X}, \end{cases}$$

and

$$\Psi(f)(w) = \begin{cases} f(w) & \text{if } w \in X, \\ \widetilde{f}(w) & \text{if } w \in \widetilde{X}, \end{cases}$$

are isometric linear embeddings from $\mathrm{Lip}(X)$ with the norm $\|\cdot\|_s$ into $C(X \cup \beta\widetilde{X})$, and from $\mathrm{lip}(X^\alpha)$ with the norm $\|\cdot\|_s$ into $C_0(X \cup \widetilde{X})$, respectively.

Using these identifications, the Hahn–Banach theorem and the Riesz representation theorem provide the following descriptions of the dual spaces of $\mathrm{Lip}(X)$ and $\mathrm{lip}(X^\alpha)$.

Lemma 5 ([14, Lemma 2.3]) *Let X be a compact metric space.*

1. *For each $F \in \mathrm{Lip}(X)^*$, there exists $\mu \in \mathcal{M}(X \cup \beta\widetilde{X})$ with $\|F\| \leq \|\mu\|$ satisfying*

$$F(f) = \int_{X \cup \beta\widetilde{X}} \Phi(f)(w)\, d\mu(w), \quad \forall f \in \mathrm{Lip}(X).$$

2. *Let $\alpha \in (0, 1)$. For each $G \in \mathrm{lip}(X^\alpha)^*$, there exists $\nu \in \mathcal{M}(X \cup \widetilde{X})$ with $\|G\| \leq \|\nu\|$ such that*

$$G(f) = \int_{X \cup \widetilde{X}} \Psi(f)(w)\, d\nu(w), \quad \forall f \in \mathrm{lip}(X^\alpha).$$

Finally, we need some properties of hermitian operators on our Lipschitz spaces.

Lemma 6 ([14, Lemmas 3.2 and 3.3]) *Let* $A(X)$ *denote either* $\mathrm{Lip}(X)$ *or* $\mathrm{lip}(X^\alpha)$, $0 < \alpha < 1$. *Recall that* $\alpha = 1$ *in the case* $A(X) = \mathrm{Lip}(X)$. *If* $T: A(X) \to A(X)$ *is a hermitian bounded linear operator, then the following statements hold:*

(i) *There exists* $\lambda \in \mathbb{R}$ *such that* $T(\mathbf{1}_X) = \lambda \mathbf{1}_X$.
(ii) *For each* $t \in \mathbb{R}$, $\exp(it(T - \lambda \mathrm{Id}))$ *is a surjective linear isometry on* $A(X)$ *fixing* $\mathbf{1}_X$.
(iii) *For each* $t \in \mathbb{R}$, *there exists a surjective isometry* φ_t *on* X *such that*

$$\exp(it(T - \lambda \mathrm{Id}))(f)(x) = f(\varphi_t(x)), \qquad \forall f \in A(X),\ \forall x \in X.$$

(iv) $\{\varphi_t\}_{t \in \mathbb{R}}$ *is a one-parameter group of surjective isometries on* X *such that, for each* $x \in X$, *the map* $t \mapsto \varphi_t(x)$ *from* $\mathbb{R}$ *to* X *is continuous.*
(v) *For every* $f \in A(X)$,

$$\lim_{t \to 0} (f \circ \varphi_t - f)(x) = 0, \qquad \forall x \in X,$$

and

$$\lim_{t \to 0} \frac{(f \circ \varphi_t - f)(x) - (f \circ \varphi_t - f)(y)}{d(x, y)^\alpha} = 0, \qquad \forall (x, y) \in \widetilde{X}.$$

(vi) *For every* $f \in A(X)$,

$$\lim_{t \to 0} \beta \widetilde{f_t}(w) = 0, \qquad \forall w \in \beta \widetilde{X},$$

where, for each $t \in \mathbb{R}$, f_t *denotes the function* $f \circ \varphi_t - f$.

Using the previous results, we may prove our main theorem which describes all the hermitian operators on $\mathrm{Lip}(X)$ or $\mathrm{lip}(X^\alpha)$ with $0 < \alpha < 1$.

Theorem 20 ([14, Theorem 3.1]) *Let* X *be a compact metric space. A bounded linear operator* $T: \mathrm{Lip}(X) \to \mathrm{Lip}(X)$ *is hermitian if and only if* T *is a real multiple of the identity operator on* $\mathrm{Lip}(X)$. *An analogous assertion holds for* $T: \mathrm{lip}(X^\alpha) \to \mathrm{lip}(X^\alpha)$ *with* $0 < \alpha < 1$.

We also mention the natural connection between hermitian operators and the class of bi-circular projections. Jamison [47] showed that these projections are exactly the hermitian projections. Our result implies that the only bi-circular projections on $\mathrm{Lip}(X)$ and $\mathrm{lip}(X^\alpha)$ with $0 < \alpha < 1$ are the trivial projections, 0 and Id.

9 Vector-Valued Hermitian Operators

Our objective in this section is to characterize the hermitian bounded operators on $\mathrm{Lip}(X, E)$, equipped with the maximum norm, with X a compact and 2-connected metric space and E a complex Banach space with norm $\|\cdot\|_E$.

We now review the definition of semi-inner product on a complex Banach space, as presented in [65, 66]. Given a complex Banach space E, a function $[\cdot,\cdot]_E\colon E\times E\to\mathbb{C}$ is called a semi-inner product if, for every $x, y, z\in E$ and $\lambda\in\mathbb{C}$, the following properties hold:

1. $[x+y, z]_E = [x, z]_E + [y, z]_E$,
2. $[\lambda x, y]_E = \lambda[x, y]_E$,
3. $[x, x]_E > 0$ for $x\neq 0$,
4. $|[x, y]_E|^2 \le [x, x]_E[y, y]_E$.

A semi-inner product $[\cdot,\cdot]_E$ is said to be compatible with the norm $\|\cdot\|_E$ if $[x, x]_E = \|x\|_E^2$ for every $x\in E$. The existence of semi-inner products compatible with the norm follows from the Hahn–Banach theorem which guarantees the existence of duality maps $u\mapsto\varphi_u$ from E into E^* which satisfy $\|\varphi_u\| = 1$ and $\varphi_u(u) = \|u\|_E$. Such a duality map yields a semi-inner product by defining $[u, v]_E = \varphi_v(u)$. Such maps are not unique and so there are several semi-inner products compatible with the existing norm unless the unit ball of E is smooth. We denote the sets of hermitian bounded operators on E by $H(E)$, respectively. See [33] for these results.

A bounded operator T on E is hermitian if and only if there exists a semi-inner product $[\cdot,\cdot]_E$ compatible with the norm such that $[Tx, x]_E\in\mathbb{R}$ for every $x\in E$. It is important to mention that if T is hermitian, then for every semi-inner product $[\cdot,\cdot]$ on E compatible with the norm, $[Tx, x]\in\mathbb{R}$ for every $x\in E$, cf. [33].

Note that the Stone–Čech compactification $\beta(\widetilde{X}\times B(E^*))$ of $\widetilde{X}\times B(E^*)$ is a compact space containing $\widetilde{X}\times B(E^*)$ as a dense subspace. For each $f\in \mathrm{Lip}(X, E)$, the bounded continuous mapping $\widetilde{f}\colon\widetilde{X}\times B(E^*)\to\mathbb{C}$, given by

$$\widetilde{f}((x, y), \varphi) = \varphi\left(\frac{f(x) - f(y)}{d(x, y)}\right),$$

has a unique continuous extension $\beta(\widetilde{f})\colon\beta(\widetilde{X}\times B(E^*))\to\mathbb{C}$ such that $||\beta(\widetilde{f})||_\infty = ||\widetilde{f}||_\infty$.

We now consider the isometric embedding

$$\Gamma\colon \mathrm{Lip}(X, E)\to C(X\cup\beta(\widetilde{X}\times B(E^*)), E\oplus_\infty\mathbb{C})$$

given by

$$\begin{aligned} f\to\Gamma(f)\colon X\cup\beta(\widetilde{X}\times B(E^*)) &\to E\oplus_\infty\mathbb{C}\\ x\in X &\to (f(x), 0),\\ \xi\in\beta(\widetilde{X}\times B(E^*)) &\to (0, \beta(\widetilde{f})(\xi)). \end{aligned}$$

Standard techniques show that Γ is a linear isometry. For each $g\in\mathrm{Lip}(X, E)$, we define

$$P_g = \left\{t\in X\cup\beta(\widetilde{X}\times B(E^*))\colon \|\Gamma(g)(t)\|_{E\oplus_\infty\mathbb{C}} = \|g\|\right\}.$$

We first choose a semi-inner product on E, $[\cdot, \cdot]_E$, compatible with the norm. Then we define the following semi-inner product $[\cdot, \cdot]_{E\oplus_\infty \mathbb{C}}$ by

$$[(u_0, \lambda_0), (u_1, \lambda_1)]_{E\oplus_\infty \mathbb{C}} = \begin{cases} [u_0, u_1]_E \text{ if } \|(u_1, \lambda_1)\|_{E\oplus_\infty \mathbb{C}} = \|u_1\|_E, \\ \\ \lambda_0 \overline{\lambda_1} \qquad \text{ if } \|(u_1, \lambda_1)\|_{E\oplus_\infty \mathbb{C}} \neq \|u_1\|_E. \end{cases}$$

This semi-inner product is compatible with the norm on $E \oplus_\infty \mathbb{C}$. It is easy to see that this semi-inner product induces the following semi-inner products:

$$[(u, 0), (v, 0)]_{E\oplus_\infty \mathbb{C}} = [u, v]_E,$$
$$[(0, \lambda_0), (0, \lambda_1)]_{E\oplus_\infty \mathbb{C}} = \lambda_0 \overline{\lambda_1},$$

on the respective component spaces $\{(u, 0) : u \in E\}$ and $\{(0, \lambda) : \lambda \in \mathbb{C}\}$, compatible with the existing norms.

Let $\psi : \mathrm{Lip}(X, E) \to \cup_{g\in \mathrm{Lip}(X,E)} P_g$ be a mapping such that $\psi(g) \in P_g$ for each $g \in \mathrm{Lip}(X, E)$. We now define

$$[f, g]_\psi = [\Gamma(f)(\psi(g)), \Gamma(g)(\psi(g))]_{E\oplus_\infty \mathbb{C}} \qquad (f, g \in \mathrm{Lip}(X, E)).$$

This is a semi-inner product in $\mathrm{Lip}(X, E)$ compatible with the norm on $\mathrm{Lip}(X, E)$, since

$$[f, f]_\psi = [\Gamma(f)(\psi(f)), \Gamma(f)(\psi(f))]_{E\oplus_\infty \mathbb{C}} = \|\Gamma(f)(\psi(f))\|^2_{E\oplus_\infty \mathbb{C}} = \|f\|^2$$

for all $f \in \mathrm{Lip}(X, E)$. Given $v \in E$, the symbol $\mathbf{v}$ represents the constant function on X everywhere equal to v.

Lemma 7 ([13, Lemma 2.1]) *Let X be a compact metric space, E a complex Banach space and T a hermitian bounded operator on* $\mathrm{Lip}(X, E)$. *Then the function $A : X \to \mathcal{L}(E)$, given by $A(x)(v) = T(\mathbf{v})(x)$ for all $x \in X$ and $v \in E$, is Lipschitz on X and with values in $H(E)$.*

Proposition 13 ([13, Proposition 2.2]) *Let X be a compact metric space, E a complex Banach space and T a hermitian bounded operator on* $\mathrm{Lip}(X, E)$. *If $f \in \mathrm{Lip}(X, E)$ and $x_0 \in X$ are such that $f(x_0) = 0$, then $T(f)(x_0) = 0$.*

Using the two preceding results, we can give the following description of the bounded hermitian operators on $\mathrm{Lip}(X, E)$.

Proposition 14 ([13, Proposition 2.3]) *Let X be a compact metric space, E a complex Banach space and T a bounded operator on* $\mathrm{Lip}(X, E)$. *If T is hermitian, then there exists a mapping $A \in \mathrm{Lip}(X, H(E))$ such that $T(f)(x) = A(x)(f(x))$ for every $f \in \mathrm{Lip}(X, E)$ and $x \in X$.*

We now are ready to characterize such operators.

Theorem 21 ([13, Theorem 2.4]) *Let X be a compact and 2-connected metric space, E a complex Banach space and $T\colon \mathrm{Lip}(X, E) \to \mathrm{Lip}(X, E)$ a bounded operator. Then T is hermitian if and only if there exists a hermitian bounded operator $A\colon E \to E$ such that $T(f)(x) = A(f(x))$ for every $f \in \mathrm{Lip}(X, E)$ and $x \in X$.*

Taking into account that the metric space X is compact and that the 2-connected components of X are open sets in X, the next corollary follows straightforwardly from Theorem 21 and Proposition 14.

Corollary 11 ([13, Corollary 2.6]) *Let X be a compact metric space, E a complex Banach space, $T\colon \mathrm{Lip}(X, E) \to \mathrm{Lip}(X, E)$ a hermitian bounded operator and $X_1, \ldots, X_m$ the 2-connected components of X. Then there exist m hermitian bounded operators $A_1, \ldots, A_m\colon E \to E$ such that*

$$T(f)(x) = \sum_{j=1}^{m} A_j(\chi_j(f)(x)), \qquad \forall x \in X,\ \forall f \in \mathrm{Lip}(X, E),$$

where, for each $j \in \{1, \ldots, m\}$, $\chi_j(f)(x) = f(x)$ if $x \in X_j$ and $\chi_j(f)(x) = 0$ otherwise.

We now state the result for the scalar case which follows as a particular case of Theorem 21.

Corollary 12 ([13, Corollary 2.7]) *Let X be a compact and 2-connected metric space and T a bounded operator on $\mathrm{Lip}(X)$. Then T is hermitian if and only if T is a real multiple of the identity operator on $\mathrm{Lip}(X)$.*

We make now some remarks on adjoint abelian and normal operators on $\mathrm{Lip}(X, E)$. We start with the definitions of adjoint abelian and normal operators as presented in [83] and in [59].

Definition 12 Let E be a complex Banach space and let $T\colon E \to E$ be a bounded operator.

1. T is adjoint abelian if and only if there exists a semi-inner product $[\cdot, \cdot]$ compatible with the norm of E such that $[Tx, y] = [x, Ty]$ for all $x, y \in E$.
2. T is normal if and only if there exist two hermitian and commuting operators T_0 and T_1 on E such that $T = T_0 + iT_1$.

The results presented before imply the following.

Theorem 22 ([13, Theorem 3.2]) *Let X be a compact and 2-connected metric space, E a complex Banach space and T a bounded operator on $\mathrm{Lip}(X, E)$.*

1. *If T is an adjoint abelian hermitian operator, then there exist an adjoint abelian hermitian operator A on E such that $T(f) = Af$ for every $f \in \mathrm{Lip}(X, E)$.*
2. *T is normal if and only if there exist commuting hermitian operators A and B on E such that $T(f) = Af + iBf$ for all $f \in \mathrm{Lip}(X, E)$.*

10 Open Problems

According to the antecedents and the current state of the research lines previously showed in this paper, we next present some concrete problems which are pretended to be studied in the near future.

Our previous study of linear isometries on Lipschitz spaces suggests the immediate approach of a couple of problems. We studied codimension 1 linear isometries on $\mathrm{Lip}(X)$ in [55]. Araujo and Font tackled in [5] a related issue for finite codimension linear isometries between $C(X, E)$-spaces. On the other hand, the study of isometries of spaces of Lipschitz maps is restricted to the case in which these maps are defined between the spaces $\mathrm{Lip}(X, E)$ but, apparently, very few is known when these maps are defined between subspaces of $\mathrm{Lip}(X, E)$.

Problem 1 Making use of our Lipschitz version of Cambern theorem (17) for the first question and the extreme point method for the second one, we pose:

1. To give a complete description of linear isometries from $\mathrm{Lip}(X, E)$ into $\mathrm{Lip}(Y, F)$, whose ranges have finite codimension.
2. To establish Banach–Stone type theorems for linear isometries defined between subspaces of vector-valued Lipschitz maps.

A problem which has not been addressed yet is the study of bilinear isometries between Lipschitz spaces. Let us recall that a bilinear map T from $C(X, E) \times C(Y, E)$ to $C(Z, E)$ is an isometry if

$$\|T(f, g)\|_\infty = \|f\|_\infty \|g\|_\infty , \quad \forall (f, g) \in C(X, E) \times C(Y, E).$$

Some examples of these bilinear isometries can be found in reference [80].

In [74], Moreno and Rodríguez proved the following bilinear version of the well-known Holsztyński theorem on (not necessarily surjective) linear isometries of the spaces $C(X)$ (see [45] and also [4]): if $T : C(X) \times C(Y) \to C(Z)$ is a bilinear isometry, then there exist a closed subset Z_0 of Z, a surjective continuous map h from Z_0 to $X \times Y$ and a norm one map $a \in C(Z)$ such that

$$T(f, g)(z) = a(z)(f(\pi_X(h(z))), g(\pi_Y(h(z)))), \quad \forall z \in Z_0, \ \forall (f, g) \in C(X) \times C(Y).$$

The proof of this fact is essentially based on Stone–Weierstrass theorem.

In [35], Font and Sanchís extended these results to certain subspaces of scalar-valued continuous functions, where this theorem cannot been applied. Furthermore, they studied in [36] conditions under which a Banach–Stone type representation for bilinear isometries on $C(X, E)$ can be obtained, which we can see below.

Given X, Y, Z compact Hausdorff spaces and E_1, E_2, E_3 Banach spaces, we say that a map $T : C(X, E_1) \times C(Y, E_2) \to C(Z, E_3)$ is stable on constants if, for any $(f, g) \in C(X, E_1) \times C(Y, E_2)$ and $z \in Z$, it is satisfied that

$$\|T(f, \mathbf{1}_X \otimes e_2)(z)\| = \|T(f, \mathbf{1}_X \otimes u_2)(z)\|$$

for all $e_2, u_2 \in S_{E_2}$, and

$$\|T(\mathbf{1}_X \otimes e_1, g)(z)\| = \|T(\mathbf{1}_X \otimes u_1, g)(z)\|$$

for every $e_1, u_1 \in S_{E_1}$. We denote by $\mathrm{Bil}(E_1 \times E_2, E_3)$ the space of all continuous bilinear maps from $E_1 \times E_2$ into E_3, equipped with the strong operator topology.

Font and Sanchís proved that if $T: C(X, E_1) \times C(Y, E_2) \to C(Z, E_3)$ is a stable-on-constants isometry and E_3 is strictly convex, then there exist a set $Z_0 \subset Z$, a continuous function $\omega: Z_0 \to \mathrm{Bil}(E_1 \times E_2, E_3)$ and a surjective continuous map $h: Z_0 \to X \times Y$ such that

$$T(f, g)(z) = \omega(z)(f(\pi_X(h(z))), g(\pi_Y(h(z)))), \ \forall z \in Z_0, \ \forall (f, g) \in C(X, E_1) \times C(Y, E_2).$$

In some sense, they showed that the stability of T on constants is a necessary condition. The aforementioned result extends, on the one hand, the theorem by Moreno and Rodríguez [74] taking $E_1 = E_2 = E_3 = \mathbb{K}$, and, on the other hand, the vector-valued version of Holsztyński theorem proved by Cambern [17] supposing that Y is a singleton and E_2 is the field $\mathbb{K}$.

Naturally, the notion of bilinear isometry can be defined between other distinguished subspaces of continuous functions and, this way, it is possible to give rise to

Problem 2 Study the Banach–Stone type representation of bilinear isometries defined between $\mathrm{Lip}(X, E)$ spaces.

Other kind of isometries which can be interesting to study is the class of approximate isometries. In [93], Vestfrid introduced the concept of ε-nonexpansive map as follows. Given two metric spaces (X, d_X) and (Y, d_Y) and $\varepsilon > 0$, it is said that $f: X \to Y$ is ε-nonexpansive if

$$d_Y(f(x), f(y)) \leq d_X(x, y) + \varepsilon, \quad \forall x, y \in X.$$

We say that f is a ε-isometric map or a ε-isometry if

$$|d_Y(f(x), f(y)) - d_X(x, y)| \leq \varepsilon, \quad \forall x, y \in X.$$

An ε-isometry defined between normed spaces is called standard if $f(0) = 0$.

A notorious result, proved by Mazur and Ulam [68] in 1932, states that if $f: X \to Y$ is a surjective isometry between two real Banach spaces, then f is affine. Hyers and Ulam [46] showed in 1945 that for every standard surjective ε-isometry $f: X \to Y$ between two real Hilbert spaces, there exists a surjective linear isometry $g: X \to Y$ such that $\|f(x) - g(x)\| \leq 10\varepsilon$ for all $x \in X$. The reader is referred to the introduction of [93] for a wide discussion on some extensions of these results and others related topics.

In his work, Vestfrid obtained Hyers–Ulam type results for maps from $C(X)$ into $C(Y)$. His main theorem assures that if X is a compact Hausdorff space, Y is a metrizable compact space and $T: C(X) \to C(Y)$ is a standard ε-isometry, then there exists an isometry (which is not necessarily linear) $S: C(X) \to C(Y)$ such that $\|T(f) - S(f)\|_\infty \leq 5\varepsilon$ for all $f \in C(X)$. If, in addition, for each proper closed subset S of Y, there exists $f \in C(X)$ such that $|T(f)(z)| < \|T(f)\|_\infty - 3.5\varepsilon$ for all $z \in S$, then S can be chosen so that it is linear. This assertion does not hold, for instance, for the ℓ_p-norm with $1 < p < \infty$. The proof is based on some results concerning the approximation of ε-nonexpansive maps from a metric space M to $\ell_\infty(S)$ where S is an arbitrary set, or to $C_b(S)$ where S is a topological space, by means of nonexpansive maps.

The previous results motivate the approach of the following problems in the setting of spaces of Lipschitz functions:

Problem 3 Study the Hyers–Ulam type stability with respect to, either the uniform norm, or the maximum or the sum Lipschitz norms, of both (not necessarily surjective) ε-nonexpansive maps and standard ε-isometries from $\mathrm{Lip}(X, E)$ into $\mathrm{Lip}(Y, F)$.

Now we present some questions which are also open for $C(X)$-spaces: given compact metric spaces X and Y,

1. Is there any constant c such that for each (continuous) ε-nonexpansive map T from $\mathrm{Lip}(X)$ to $\mathrm{Lip}(Y)$, there exists a nonexpansive map S from $\mathrm{Lip}(X)$ into $\mathrm{Lip}(Y)$ whose distance to T is less than or equal to $c\varepsilon$?
2. Is there any constant c such that for each ε-isometry $T: \mathrm{Lip}(X) \to \mathrm{Lip}(Y)$, there exists an isometry S from $\mathrm{Lip}(X)$ to $\mathrm{Lip}(Y)$ whose distance to T is less than or equal to $c\varepsilon$?
3. If the answer to any of the previous questions is positive, what is the best possible constant c?

The Mazur–Ulam theorem [68] has a wide amount of applications in different areas of Mathematics. In 1972, Mankiewicz proved a local version of this theorem showing that every bijective isometry between convex sets in normed linear spaces with nonempty interiors admits a unique extension to a bijective affine isometry between the corresponding spaces. Thus this result states that it is not necessary to have a surjective isometry defined between the totality of the spaces X and Y to identify them through an affine transformation, it is enough to make an isometric identification of their respective closed unit balls.

One of the most modern variants of the Mazur–Ulam theorem is due to a result by Tingley [91] in 1987. The result, known as Tingley's problem, asks when a surjective isometry $T: S_X \to S_Y$ can be extended to a real linear isometry from X into Y. It is known that this problem has a positive answer for certain classical Banach spaces such as sequence spaces, continuous function algebras and spaces of measurable functions [85, 87, 96]. The work [100] can give a general vision on the advances in this problem.

The list of spaces for which Tingley's question has a positive answer has been increased by the recent works of Fernández, Peralta and Tanaka and it includes the space of compact operators, compact C*-algebras and weakly compact JB*-triples [29, 79], von Neumann factors of type I, atomic von Neumann algebras and atomic JBW*-triples [30, 31], spaces of trace class operators [28] and positive operator algebras [77]. See [78] for a compilation of recent results on Tingley's problem in operator algebras.

The Mazur–Ulam property is intrinsically related to Tingley's problem. According to [18], it is said that a Banach space Z satisfies the Mazur–Ulam property if every surjective isometry from S_Z into S_Y, where Y is any Banach space, admits an (unique) extension to a real surjective linear isometry from Z to Y.

A pioneer contribution due to Ding [23] shows that $c_0(\mathbb{N}, \mathbb{R})$ satisfies the Mazur–Ulam property. The list of real Banach spaces having this property includes $c(\Gamma, \mathbb{R})$, $c_0(\Gamma, \mathbb{R})$ and $\ell_\infty(\Gamma, \mathbb{R})$ where Γ is an infinite set endowed with the discrete topology, and $C(K, \mathbb{R})$ where K is a compact metric space (see [26, 64] and their references). In [85–87], it was shown that the space $L^p((\Omega, \Sigma, \mu), \mathbb{R})$ of all real measurable functions on a σ-finite measure space (Ω, Σ, μ) satisfies the Mazur–Ulam property for all $1 \leq p \leq \infty$. In the complex case, the problem for the spaces $c_0(\Gamma, \mathbb{C})$ and $\ell_\infty(\Gamma, \mathbb{C})$ was closed in [53] and [76], respectively. All this literature justifies our interest in tackling the following:

Problem 4 Study Tingley's problem and the Mazur–Ulam property for the spaces of Lipschitz functions.

Our previous study on projections motivates the last two problems of this section. Theorem 7 claims that, under certain constraints, the average of two (three) isometries on $\mathrm{Lip}(X)$ is a nontrivial projection if and only if it is a 2-circular (respectively, 3-circular) projection, and, therefore, it is natural to arise the following question:

Problem 5 Let X be a compact 1-connected metric space with $\mathrm{diam}(X) \leq 2$ and $n \geq 2$. Is the average of n pairwise distinct isometries on $\mathrm{Lip}(X)$ a projection if and only if it is a trivial projection or a n-circular projection?

It is well-known by a celebrated work by Friedman and Russo [38] that if X is a compact Hausdorff metric space, then a linear projection $P\colon C(X) \to C(X)$ is bicontractive if and only if $P = (1/2)(\mathrm{Id} + T)$, where T is an involutive isometry on $C(X)$.

Problem 6 Study contractive and bicontractive projections on $\mathrm{Lip}(X)$ in order to determine under which conditions on the metric space X, the space $\mathrm{Lip}(X)$ belongs to the class of Banach spaces E in which every bicontractive projection $P\colon E \to E$ is of the form $P = (1/2)(\mathrm{Id} + T)$, where $T\colon E \to E$ is an involutive isometry.

Acknowledgements The authors wish to thank the referee for making several suggestions which improved this paper.

The first author's research was partially supported by Predoctoral contract for the Personnel Research Training 2016 of University of Almería, and the second author's one by the Spanish Ministry of Economy and Competitiveness project no. MTM2014-58984-P and the European Regional Development Fund (ERDF), and Junta of Andalucía grant FQM-194.

Some of the results of this survey were presented in the talks "Advances about isometries on Lipschitz spaces" by A. Jiménez–Vargas and "Bilinear isometries on algebras of Lipschitz functions" by Moisés Villegas-Vallecillos at *The 3rd Moroccan Andalusian Meeting on Algebras and their Applications (Chefchaouen, Morocco, April 12–14, 2018)*. They would like to thank the organizers for the hospitality during their stay.

References

1. Al-Halees, H., Fleming, R.J.: On 2-local isometries on continuous vector-valued function spaces. J. Math. Anal. Appl. **354**, 70–77 (2009)
2. Apazoglou, M., Peralta, A.M.: Linear isometries between real JB*-triples and C*-algebras. Quart. J. Math. (Oxford) **65**, 485–503 (2014)
3. Araujo, J., Dubarbie, L.: Noncompactness and noncompleteness in isometries of Lipschitz spaces. J. Math. Anal. Appl. **377**, 15–29 (2011)
4. Araujo, J., Font, J.J.: Linear isometries between subspaces of continuous functions. Trans. Amer. Math. Soc. **349**, 413–442 (1997)
5. Araujo, J., Font, J.J.: Finite codimensional isometries on spaces of vector-valued continuous functions. J. Math. Anal. Appl. **421**, 186–205 (2015)
6. Berkson, E., Sourour, A.: The hermitian operators on some Banach spaces. Studia Math. **52**, 33–41 (1974)
7. Botelho, F.: Projections as convex combinations of surjective isometries on $C(\Omega)$. J. Math. Anal. Appl. **341**, 1163–1169 (2008)
8. Botelho, F., Fleming, R.J., Jamison, J.E.: Extreme points and isometries on vector-valued Lipschitz spaces. J. Math. Anal. Appl. **381**, 821–832 (2011)
9. Botelho, F., Jamison, J.E.: Generalized bi-circular projections on Lipschitz spaces. Acta Sci. Math. **75**, 103–112 (2009)
10. Botelho, F., Jamison, J.E.: Algebraic reflexivity of sets of bounded operators on vector valued Lipschitz functions. Linear Algebra Appl. **432**, 3337–3342 (2010)
11. Botelho, F., Jamison, J.E.: Projections in the convex hull of surjective isometries. Can. Math. Bull. **53**, 398–403 (2010)
12. Botelho, F., Jamison, J.E., Jiménez-Vargas, A.: Projections and averages of isometries on Lipschitz spaces. J. Math. Anal. Appl. **386**, 910–920 (2012)
13. Botelho, F., Jamison, J., Jiménez-Vargas, A., Villegas-Vallecillos, M.: Hermitian operators on Lipschitz function spaces. Studia Math. **215**(2), 127–137 (2013)
14. Botelho, F., Jamison, J., Jiménez-Vargas, A., Villegas-Vallecillos, M.: Hermitian operators on Banach algebras of Lipschitz functions. Proc. Amer. Math. Soc. **142**(10), 3469–3481 (2014)
15. Brešar, M., Šemrl, P.: On local automorphisms and mappings that preserve idempotents. Studia Math. **113**, 101–108 (1995)
16. Cabello Sánchez, F., Molnár, L.: Reflexivity of the isometry group of some classical spaces. Rev. Mat. Iberoam. **18**, 409–430 (2002)
17. Cambern, M.: A Holsztyński theorem for spaces of continuous vector-valued functions. Studia Math. **63**, 213–217 (1978)
18. Cheng, L., Dong, Y.: On a generalized Mazur-Ulam question: extension of isometries between unit spheres of Banach spaces. J. Math. Anal. Appl. **377**, 464–470 (2011)
19. Chu, C.-H., Mackey, M.: Isometries between JB*-triples. Math. Z. **251**, 615–633 (2005)

20. Chu, C.-H., Wong, N.-C.: Isometries between C*-algebras. Rev. Mat. Iberoam. **20**, 87–105 (2004)
21. Conway, J.B.: A Course in Functional Analysis. Graduate Texts in Mathematics, vol. 96. Springer, New York (1985)
22. Ding, G.G.: The 1-Lipschitz mapping between the unit spheres of two Hilbert spaces can be extended to a real linear isometry of the whole space. Sci. China Ser. A **45**(4), 479–483 (2002)
23. Ding, G.G.: The isometric extension of the into mapping from a $\mathcal{L}^{\infty}(\Gamma)$-type space to some Banach space. Illinois J. Math. **51**(2), 445–453 (2007)
24. Dunford, N., Schwartz, J.T.: Linear Operators, Part I: General Theory. Wiley, New York (1958)
25. Dutta, S., Rao, T.S.S.R.K.: Algebraic reflexivity of some subsets of the isometry group. Linear Algebra Appl. **429**, 1522–1527 (2008)
26. Fang, X.N., Wang, J.H.: Extension of isometries between the unit spheres of normed space E and $C(\Omega)$. Acta Math. Sinica (Engl. Ser.) **22**, 1819–1824 (2006)
27. Farid, F.O., Varadajan, K.: Isometric shift operators on $C(X)$. Can. J. Math. **46**, 532–542 (1994)
28. Fernández-Polo, F.J., Garcés, J.J., Peralta, A.M., Villanueva, I.: Tingley's problem for spaces of trace class operators. Linear Algebra Appl. **529**, 294–323 (2017)
29. Fernández-Polo, F.J., Peralta, A.M.: Low rank compact operators and Tingley's problem. Adv. Math. **338**, 1–40 (2018)
30. Fernández-Polo, F.J., Peralta, A.M.: Tingley's problem through the facial structure of an atomic JBW*-triple. J. Math. Anal. Appl. **455**(1), 750–760 (2017)
31. Fernández-Polo, F.J., Peralta, A.M.: On the extension of isometries between the unit spheres of a C*-algebra and $B(H)$. Trans. Amer. Math. Soc. Ser. B **5**, 63–80 (2018)
32. Fleming, R., Jamison, J.E.: Hermitian Operators on $C(X, E)$ and the Banach-Stone theorem. Math. Z. **170**, 77–84 (1980)
33. Fleming, R., Jamison, J.E.: Isometries on Banach Spaces: Function Spaces. Monographs and Surveys in Pure and Applied Mathematics, vol. 129:1. Chapman & Hall, London (2003)
34. Fleming, R., Jamison, J.E.: Isometries on Banach Spaces: Vector-Valued Function Spaces. Monographs and Surveys in Pure and Applied Mathematics, vol. 129:2. Chapman & Hall, London (2008)
35. Font, J.J., Sanchís, M.: Bilinear isometries on subspaces of continuous functions. Math. Nachr. **283**(4), 568–572 (2010)
36. Font, J.J., Sanchís, M.: Bilinear isometries on spaces of vector-valued continuous functions. J. Math. Anal. Appl. **385**, 340–344 (2012)
37. Fosner, M., Ilisevic, D., Li, C.: G-invariant norms and bicircular projections. Linear Algebra Appl. **420**, 596–608 (2007)
38. Friedman, Y., Russo, B.: Contractive projections on $C_0(K)$. Trans. Amer. Math. Soc. **273**(1), 57–73 (1982)
39. Gleason, A.M.: A characterization of maximal ideals. J. Analyse Math. **19**(1), 171–172 (1967)
40. Gutek, A., Hart, D., Jamison, J., Rajagopalan, M.: Shift operators on Banach spaces. J. Funct. Anal. **101**, 97–119 (1991)
41. Győry, M.: 2-local isometries of $C_0(X)$. Acta Sci. Math. (Szeged) **67**, 735–746 (2001)
42. Hatori, O., Miura, T., Oka, H., Takagi, H.: 2-local isometries and 2-local automorphisms on uniform algebras. Int. Math. Forum **50**, 2491–2502 (2007)
43. Hatori, O., Oi, S.: Hermitian operators on Banach algebras of vector-valued Lipschitz maps. J. Math. Anal. Appl. **452**(1), 378–387 (2017)
44. Hatori, O., Oi, S.: Isometries on Banach algebras of vector-valued maps. Acta Sci. Math. Szeged **84**(1–2), 151–183 (2018)
45. Holsztyński, W.: Continuous mappings induced by isometries of spaces of continuous functions. Studia Math. **26**, 133–136 (1966)
46. Hyers, D.H., Ulam, S.M.: On approximate isometries. Bull. Amer. Math. Soc. **51**, 288–292 (1945)
47. Jamison, J.E.: Bicircular projections on some Banach spaces. Linear Algebra Appl. **420**, 29–33 (2007)

48. Jarosz, K., Pathak, V.: Isometries between function spaces. Trans. Amer. Math. Soc. **305**, 193–206 (1988)
49. Jeang, J.-S., Wong, N.-C.: Weighted composition operators of $C_0(X)'s$. J. Math. Anal. Appl. **201**, 981–993 (1996)
50. Jerison, M.: The space of bounded maps into a Banach space. Ann. of Math. **52**, 309–327 (1950)
51. Ji, P.S., Wei, C.P.: Isometries and 2-local isometries on Cartan bimodule algebras in hyperfinite factors of type II_1. Acta Math. Sinica (Chin. Ser.) **49**, 51–58 (2006)
52. Jiménez-Vargas, A., Li, L., Peralta, A.M., Wang, L., Wang, Y.-S.: 2-local standard isometries on vector-valued Lipschitz function spaces. J. Math. Anal. Appl. **461**, 1287–1298 (2018)
53. Jiménez-Vargas, A., Morales, A., Peralta, A.M., Ramírez, M.I.: The Mazur-Ulam property for the space of complex null sequences. Linear Multilinear Algebra **67**(4), 799–816 (2019)
54. Jiménez-Vargas, A., Morales Campoy, A., Villegas-Vallecillos, M.: Algebraic reflexivity of the isometry group of some spaces of Lipschitz functions. J. Math. Anal. Appl. **366**, 195–201 (2010)
55. Jiménez-Vargas, A., Villegas-Vallecillos, M.: Into linear isometries between spaces of Lipschitz functions. Houston J. Math. **34**, 1165–1184 (2008)
56. Jiménez-Vargas, A., Villegas-Vallecillos, M.: Linear isometries between spaces of vector-valued Lipschitz functions. Proc. Amer. Math. Soc. **137**, 1381–1388 (2009)
57. Jiménez-Vargas, A., Villegas-Vallecillos, M.: 2-local isometries on spaces of Lipschitz functions. Can. Math. Bull. **54**(4), 680–692 (2011)
58. Kahane, J.-P., Żelazko, W.: A characterization of maximal ideals in commutative Banach algebras. Studia Math. **29**, 339–343 (1968)
59. Kirsti, M.: On normal and compact operators on Banach spaces. Israel J. Math. **37**(1–2), 164–170 (1980)
60. Kowalski, S., Slodkowski, Z.: A characterization of multiplicative linear functionals in Banach algebras. Studia Math. **67**(3), 215–223 (1980)
61. de Leeuw, K.: Banach spaces of Lipschitz functions. Studia Math. **21**, 55–66 (1961/62)
62. Li, L., Peralta, A.M., Wang, L., Wang, Y.-S.: Weak-2-local isometries on uniform algebras and Lipschitz algebras. Publ. Mat. **63**(1), 241–264 (2019)
63. Liu, J.H., Wong, N.C.: 2-local automorphisms of operators algebras. J. Math. Anal. Appl. **321**, 741–750 (2006)
64. Liu, R.: On extension of isometries between unit spheres of $\mathcal{L}^\infty(\Gamma)$-type space and a Banach space E. J. Math. Anal. Appl. **333**, 959–970 (2007)
65. Lumer, G.: Semi-inner product spaces. Trans. Amer. Math. Soc. **100**, 29–43 (1961)
66. Lumer, G.: Isometries of Reflexive Orlicz spaces. Ann. Inst. Fourier (Grenoble) **13**, 99–109 (1963)
67. Mayer-Wolf, E.: Isometries between Banach spaces of Lipschitz functions. Israel J. Math. **38**, 58–74 (1981)
68. Mazur, S., Ulam, S.: Sur les transformations isométriques d'espaces vectoriels normés. Comp. Rend. Paris **194**, 946–948 (1932)
69. Molnár, L.: The set of automorphisms of $B(H)$ is topologically reflexive in $B(B(H))$. Studia Math. **122**, 183–193 (1997)
70. Molnár, L.: 2-local isometries of some operators algebras. Proc. Edinb. Math. Soc. **45**, 349–352 (2002)
71. Molnár, L.: Local automorphisms of operator algebras on Banach spaces. Proc. Amer. Math. Soc. **131**, 1867–1874 (2003)
72. Molnár, L., Šemrl, P.: Local automorphisms of the unitary group and the general linear group on a Hilbert space. Exp. Math. **18**, 231–238 (2000)
73. Molnár, L., Zalar, B.: Reflexivity of the group of surjective isometries of some Banach spaces. Proc. Edinb. Math. Soc. **42**, 17–36 (1999)
74. Moreno, A., Rodríguez, A.: A bilinear version of Holsztyński theorem on isometries of C(X)-spaces. Studia Math. **166**, 83–91 (2005)

75. Novinger, W.: Linear isometries of subspaces of continuous functions. Studia Math. **53**, 273–276 (1975)
76. Peralta, A.M.: Extending surjective isometries defined on the unit sphere of $\ell_\infty(\Gamma)$. Rev. Mat. Complut. **32**(1), 99–114 (2019)
77. Peralta, A.M.: On the unit sphere of positive operators. Banach J. Math. Anal. **13**(1), 91–112 (2019)
78. Peralta, A.M.: A survey on Tingley's problem for operator algebras. Acta Sci. Math. (Szeged) **84**, 81–123 (2018)
79. Peralta, A.M., Tanaka, R.: A solution to Tingley's problem for isometries between the unit spheres of compact C*-algebras and JB*-triples. Sci. China Math. **62**(3), 553–568 (2019)
80. Rodríguez, A.: Absolute valued algebras and absolute-valuable Banach spaces. In: Advanced Courses of Mathematical Analysis I: Proceedings of the First International School, Cádiz, Spain, 2002, World Scientific Publishing, 2004, pp. 99–155
81. Roy, A.K.: Extreme points and linear isometries of the Banach space of Lipschitz functions. Can. J. Math. **20**, 1150–1164 (1968)
82. Šemrl, P.: Local automorphisms and derivations on $B(H)$. Proc. Amer. Math. Soc. **125**, 2677–2680 (1997)
83. Stampfli, J.G.: Adjoint abelian operators on Banach spaces. Can. J. Math. **21**, 233–246 (1969)
84. Stachó, L.L., Zalar, B.: Bicircular projections on some matrix and operator spaces. Linear Algebra Appl. **384**, 9–20 (2004)
85. Tan, D.: Extension of isometries on unit sphere of L^∞. Taiwan. J. Math. **15**, 819–827 (2011)
86. Tan, D.: On extension of isometries on the unit spheres of L^p-spaces for $0<p\leq 1$. Nonlinear Anal. **74**, 6981–6987 (2011)
87. Tan, D.: Extension of isometries on the unit sphere of L^p-spaces. Acta Math. Sin. (Engl. Ser.) **28**, 1197–1208 (2012)
88. Tanaka, R.: A further property of spherical isometries. Bull. Aust. Math. Soc. **90**, 304–310 (2014)
89. Tanaka, R.: The solution of Tingley's problem for the operator norm unit sphere of complex $n\times n$ matrices. Linear Algebra Appl. **494**, 274–285 (2016)
90. Tanaka, R.: Spherical isometries of finite dimensional C*-algebras. J. Math. Anal. Appl. **445**(1), 337–341 (2017)
91. Tingley, D.: Isometries of the unit sphere. Geom. Dedicata **22**, 371–378 (1987)
92. Vasavada, M.H.: Closed ideals and linear isometries of certain function spaces. Ph.D. thesis, University of Wisconsin (1969)
93. Vestfrid, I.A.: Hyers-Ulam stability of isometries and non-expansive maps between spaces of continuous functions. Proc. Amer. Math. Soc. **145**, 2481–2494 (2017)
94. Vidav, I.: Eine metrische Kennzeichnung der selbstadjungierten Operatoren. Math. Z. **66**, 121–128 (1956)
95. Villegas-Vallecillos, M.: Teoremas de tipo Banach-Stone en espacios de funciones lipschitzianas. Tesis doctoral, Universidad de Almería (2011)
96. Wang, R.S.: Isometries between the unit spheres of $C_0(\Omega)$ type spaces. Acta Math. Sci. (English Ed.) **14**(1), 82–89 (1994)
97. Weaver, N.: Isometries of noncompact Lipschitz spaces. Can. Math. Bull. **38**, 242–249 (1995)
98. Weaver, N.: Lipschitz Algebras. World Scientific Publishing Co. Inc, River Edge (1999)
99. Weaver, N.: Lipschitz Algebras, 2nd edn. World Scientific Publishing Co. Pte. Ltd., Hackensack (2018)
100. Yang, X., Zhao, X.: On the extension problems of isometric and nonexpansive mappings. In: Rassias, Themistocles M., Pardalos, Panos M. (eds.) Mathematics Without Boundaries, pp. 725–748. Springer, New York (2014)

The Principal Eigenvalue for a Class of Singular Quasilinear Elliptic Operators and Applications

José Carmona, Salvador López-Martínez and Pedro J. Martínez-Aparicio

Dedicado a Amin Kaidi por su 70º cumpleaños.

Abstract We characterize the principal eigenvalue associated to the singular quasilinear elliptic operator $-\Delta u - \mu(x)\frac{|\nabla u|^q}{u^{q-1}}$ in a bounded smooth domain $\Omega \subset \mathbb{R}^N$ with zero Dirichlet boundary conditions. Here, $1 < q \leq 2$ and $0 \leq \mu \in L^\infty(\Omega)$. As applications we derive some existence of solutions results (as well as uniqueness, nonexistence and homogenization results) to a problem whose model is

$$\begin{cases} -\Delta u = \lambda u + \mu(x)\dfrac{|\nabla u|^q}{|u|^{q-1}} + f(x) & \text{in } \Omega, \\ u = 0 & \text{on } \partial\Omega, \end{cases}$$

where $\lambda \in \mathbb{R}$ and $f \in L^p(\Omega)$ for some $p > \frac{N}{2}$.

Research supported by PGC2018-096422-B-I00 (MCIU/AEI/FEDER, UE), Junta de Andalucía FQM-194 (first author) and FQM-116, Programa de Contratos Predoctorales del Plan Propio de la Universidad de Granada (second author). First author also thanks the support from CDTIME.

J. Carmona (✉) · P. J. Martínez-Aparicio
Departamento de Matemáticas, Universidad de Almería, Ctra. Sacramento s/n, La Cañada de San Urbano, 04120 Almería, Spain
e-mail: jcarmona@ual.es

P. J. Martínez-Aparicio
e-mail: pedroj.ma@ual.es

S. López-Martínez (✉)
Departamento de Análisis Matemático, Universidad de Granada, Facultad de Ciencias, Avenida Fuentenueva s/n, 18071 Granada, Spain
e-mail: salvadorlopez@ugr.es

M. Siles Molina et al. (eds.), *Associative and Non-Associative Algebras and Applications*, Springer Proceedings in Mathematics & Statistics 311,
https://doi.org/10.1007/978-3-030-35256-1_4

Keywords Quasilinear elliptic equations · Singular problems

1 Introduction

We consider a bounded domain $\Omega \subset \mathbb{R}^N$ $(N \geq 3)$ with $C^{1,1}$ boundary and study the quasilinear elliptic problem:

$$\begin{cases} -\mathrm{div}(m(x)\nabla u) = \lambda u + \mu(x)\dfrac{|\nabla u|^q}{|u|^{q-1}} + f(x) & \text{in } \Omega, \\ u = 0 & \text{on } \partial\Omega, \end{cases} \qquad (P_\lambda)$$

where $\lambda \in \mathbb{R}$, $0 \leq \mu \in L^\infty(\Omega)$, $f \in L^p(\Omega)$ with $p > \frac{N}{2}$, $1 < q \leq 2$ and, for some constant η, $0 < \eta \leq m \in L^\infty(\Omega) \cap W^{1,\infty}_{\mathrm{loc}}(\Omega)$. We say that a solution to problem (P_λ) is a function $u \in H_0^1(\Omega) \cap L^\infty(\Omega)$ such that $\mu(x)\frac{|\nabla u|^q}{|u|^{q-1}} \in L^1(\{|u| > 0\})$ and

$$\int_\Omega m(x)\nabla u \nabla \phi = \lambda \int_\Omega u\phi + \int_{\{|u|>0\}} \mu(x)\frac{|\nabla u|^q}{|u|^{q-1}}\phi + \int_\Omega f(x)\phi,$$

for every $\phi \in H_0^1(\Omega) \cap L^\infty(\Omega)$.

The aim of this note is to summarize the known results, obtained in [6] and [7], concerning the existence, uniqueness, homogenization and nonexistence of solution to problem (P_λ) (which improve, in some sense, those contained in [2] for $q = 2$). In these mentioned papers, it is shown that the validity of such results depends on the existence of a principal eigenvalue for the eigenvalue problem

$$\begin{cases} -\mathrm{div}(m(x)\nabla u) = \lambda u + \mu(x)\dfrac{|\nabla u|^q}{|u|^{q-1}} & \text{in } \Omega, \\ u = 0 & \text{on } \partial\Omega. \end{cases} \qquad (E_\lambda)$$

Inspired by [3], the principal eigenvalue can be characterized by

$$\lambda^* = \sup\left\{\lambda \in \mathbb{R} \;\middle|\; \begin{array}{l} \text{there exists a supersolution } v \text{ to } (E_\lambda) \\ \text{such that } v \geq c \text{ in } \Omega \text{ for some } c > 0 \end{array}\right\}, \qquad (1)$$

where the precise meaning of supersolution used in (1) is specified in the next section.

2 Principal Eigenvalue

We say that $v \in H^1(\Omega) \cap L^\infty(\Omega)$ is a *supersolution* to (E_λ) if $v > 0$ a.e. in Ω, $\frac{|\nabla v|^q}{v^{q-1}} \in L^1_{\mathrm{loc}}(\Omega)$ and the following inequality holds

$$\int_\Omega m(x)\nabla v \nabla\phi \geq \lambda \int_\Omega v\phi + \int_\Omega \mu(x)\frac{|\nabla v|^q}{v^{q-1}}\phi, \forall \phi \in H_0^1(\Omega)\cap L^\infty(\Omega), \phi \geq 0. \tag{2}$$

Analogously it is defined the concept of supersolution for (P_λ) and, with the reverse inequality, the concept of subsolution. Moreover, we say that

- v satisfies condition (v_1) if $v \geq c$ in Ω for some $c > 0$.
- v satisfies condition (v_2) if $v - c \in H_0^1(\Omega)$ for some $c > 0$.
- v satisfies condition (v_3) if, for some $\gamma_0 < 1$, $v^\gamma \in H^1(\Omega)$ for every $\gamma > \gamma_0$.

Thus, in order to summarize the main properties and characterizations of λ^*, we define for $i = 1, 2, 3$,

$$I_i = \left\{\lambda \in \mathbb{R} \,\middle|\, \begin{array}{c}\text{there exists a supersolution } v \text{ to } (E_\lambda)\\ \text{such that } v \text{ satisfies } (v_i)\end{array}\right\}.$$

Proposition 1 *Assume that* $1 < q \leq 2$, $0 \leq \mu \in L^\infty(\Omega)$ *and* $0 < \eta \leq m \in L^\infty(\Omega)$. *Then the sets* I_1, I_2 *and* I_3 *are nonempty intervals which are unbounded from below, so* $\lambda^* = \sup I_1$ *is well defined. Moreover,* $I_1 = I_2$ *and* $\lambda^* = \sup I_3$. *In addition,* $0 < \lambda^* \leq \lambda_1(m) \equiv \inf_{w\in H_0^1(\Omega)\setminus\{0\}} \frac{\int_\Omega m(x)|\nabla w|^2}{\int_\Omega w^2}$.

Proof We include here the main steps in the proof, further detail may be found in [6] in the case $m(x) = 1$.

First we observe that, from the concept of supersolution it is easily deduced that $(-\infty, \lambda] \subset I_i$ whenever $\lambda \in I_i$. Moreover, taking $v = 1$ as a supersolution to (E_0) we derive that $(-\infty, 0] \subset I_i$. In particular, I_i is an interval unbounded from below.

Step 1. $I_1 = I_2$. Observe that, since $(-\infty, 0] \subset I_1 \cap I_2$ then it is enough to prove that $I_1 \cap (0, +\infty) = I_2 \cap (0, +\infty)$. Assume that $0 < \lambda \in I_2 \cap (0, +\infty)$. Hence, there exist $v \in H^1(\Omega) \cap L^\infty(\Omega)$ and $c > 0$ with $v > 0$ in Ω, $v - c \in H_0^1(\Omega)$, and

$$-\mathrm{div}(m(x)\nabla(v-c)) = -\mathrm{div}(m(x)\nabla v) \geq \lambda v + \mu(x)\frac{|\nabla v|^q}{v^{q-1}} \geq 0 \text{ in } \Omega.$$

Therefore, the maximum principle yields to $v \geq c$ in Ω, and so $\lambda \in I_1 \cap (0, +\infty)$.

Conversely if $0 < \lambda \in I_1$ and $-\mathrm{div}(m(x)\nabla v) \geq \lambda v + \mu(x)\frac{|\nabla v|^q}{v^{q-1}}$ in Ω for some $v \in H^1(\Omega) \cap L^\infty(\Omega)$ with $v \geq c > 0$ then $v - c$ is a non-negative supersolution to the problem (without singularity)

$$\begin{cases} -\mathrm{div}(m(x)\nabla u) = \lambda u + \mu(x)\dfrac{|\nabla u|^q}{(|u| + c)^{q-1}} + \lambda c & \text{in } \Omega, \\ u = 0 & \text{on } \partial\Omega. \end{cases} \tag{3}$$

Since the trivial function it is a subsolution there exists a solution $u \in H_0^1(\Omega) \cap L^\infty(\Omega)$ to (3) (see [5, Théorème 3.1]) with $0 \leq u \leq v - c$ in Ω. Thus, $u + c$ is a supersolution to (E_λ) that satisfies (v_2) and therefore $\lambda \in I_2$.

Step 2. $\lambda^* = \sup I_3$. First we observe the trivial inclusion $I_1 \subset I_3$. Thus in order to prove Step 2 we are going to show that $I_3 - \epsilon \subset I_1$ for every $\epsilon > 0$ small enough. Indeed, assume that $\lambda \in I_3$, i.e. there exist $u \in H^1(\Omega) \cap L^\infty(\Omega)$ and $\widetilde{\gamma} \in (0, 1)$ satisfying

$$u > 0 \text{ in } \Omega, \quad -\text{div}(m(x)\nabla u) \geq \lambda u + \mu(x)\frac{|\nabla u|^q}{u^{q-1}} \text{ in } \Omega, \quad u^\gamma \in H^1(\Omega) \quad \forall \gamma > \widetilde{\gamma}.$$

Then we show that $\lambda - \epsilon \in I_1$. In fact we prove that a supersolution to $(E_{\lambda-\epsilon})$ is $v = \epsilon(\varphi_1^\gamma + 1) + u^\gamma$ where ϵ is a small enough positive constant, $\gamma \in \left(\max\left\{\frac{1}{2}, \widetilde{\gamma}, \frac{\lambda-\epsilon}{\lambda}\right\}, 1\right)$ and $\varphi_1 > 0$ is the the principal positive and normalized eigenfunction associated to $\lambda_1(m)$, that is,

$$\begin{cases} -\text{div}(m(x)\nabla\varphi_1) = \lambda_1(m)\varphi_1 & \text{in } \Omega, \\ \varphi_1 = 0 & \text{on } \partial\Omega. \end{cases}$$

Observe that, since $\gamma > \frac{1}{2}$ it is easy to deduce that $\varphi_1^\gamma \in H_0^1(\Omega)$. Indeed, take $(\varphi_1 + \delta)^{2\gamma-1} - \delta^{2\gamma-1}$ as test function in the equation satisfied by φ_1 and use Fatou lemma as $\delta \to 0$.

Thus, since $\gamma > \widetilde{\gamma}$, we have $v \in H^1(\Omega) \cap L^\infty(\Omega)$ and, clearly, $v \geq \epsilon$ in Ω and only remains to prove that v is a supersolution to $(E_{\lambda-\epsilon})$.

Let $\phi \in H_0^1(\Omega) \cap L^\infty(\Omega)$ be such that $\phi \geq 0$ in Ω and has compact support. Direct computations yield to

$$\begin{aligned} \int_\Omega \left(-m(x)\nabla v\nabla\phi + (\lambda - \epsilon)v\phi + \mu(x)\frac{|\nabla v|^q}{v^{q-1}}\phi\right) &\leq -(\gamma\lambda - (\lambda - \epsilon)) \int_\Omega u^\gamma\phi \\ +\varepsilon \int_\Omega \left(-\gamma(1-\gamma)m(x)\frac{|\nabla\varphi_1|^2}{\varphi_1^{2-\gamma}} + ((\lambda - \epsilon) - \gamma\lambda_1(m))\varphi_1^\gamma\right)\phi& \\ + \int_\Omega \left((\lambda - \epsilon) + \|\mu\|_{L^\infty(\Omega)}C_1\frac{|\nabla\varphi_1|^q}{\varphi_1^{q(1-\gamma)}}\right)\phi.& \end{aligned} \tag{4}$$

Using Hopf lemma we can assure that $|\nabla\varphi_1|$ is bounded away from zero in a small neighborhood Ω_δ of the boundary. Using also that $\gamma > \frac{\lambda-\epsilon}{\lambda}$ and $q(1-\gamma) < 2-\gamma$, we choose δ sufficiently small and independent of ϵ, such that, in Ω_δ

$$\begin{aligned} \Psi(x) \equiv -\gamma(1-\gamma)m(x)\frac{|\nabla\varphi_1|^2}{\varphi_1^{2-\gamma}} + ((\lambda - \epsilon) - \gamma\lambda_1(m))\varphi_1^\gamma \\ +(\lambda - \epsilon) + \|\mu\|_{L^\infty(\Omega)}C_1\frac{|\nabla\varphi_1|^q}{\varphi_1^{q(1-\gamma)}} \leq 0. \end{aligned} \tag{5}$$

Consequently, we take ϵ small enough in order to have in $\Omega \setminus \Omega_\delta$

$$\epsilon\Psi(x) \le \epsilon C_2 \le (\gamma\lambda - (\lambda - \epsilon)) \inf_{\Omega\setminus\Omega_\delta} (u^\gamma) \le (\gamma\lambda - (\lambda - \epsilon))u^\gamma. \quad (6)$$

Gathering (4), (5) and (6) together we conclude that

$$\int_\Omega m(x)\nabla v \nabla\phi \ge (\lambda - \epsilon)\int_\Omega v\phi + \int_\Omega \mu(x)\frac{|\nabla v|^q}{v^{q-1}}\phi.$$

Step 3. $\lambda^* > 0$. First we choose $c > 0$ large enough and $\delta > 0$ small enough in order to assure (using [9, Theorem 3.4]) the existence of solution $0 \le u \in H_0^1(\Omega) \cap L^\infty(\Omega)$ to

$$\begin{cases} -\mathrm{div}(m(x)\nabla u) = \dfrac{\mu(x)}{c^{q-1}}|\nabla u|^q + \delta & \text{in } \Omega, \\ u = 0 & \text{on } \partial\Omega. \end{cases} \quad (7)$$

Then, for some $\lambda > 0$ small, $v = u + c \in H^1(\Omega) \cap L^\infty(\Omega)$ is a supersolution to (E_λ) which satisfies (v_1). Indeed,

$$-\mathrm{div}(m(x)\nabla v) = -\mathrm{div}(m(x)\nabla u) \ge \mu(x)\frac{|\nabla v|^q}{v^{q-1}} + \lambda v + (\delta - \lambda\|v\|_{L^\infty(\Omega)}).$$

Step 4. $\lambda^* \le \lambda_1(m)$. Assume that $0 < \lambda \in I_2$. From Step 1, there exists $\psi \ge 0$ solution to (3) for some $c > 0$. Taking φ_1 as test function in (3) we have

$$\lambda_1(m)\int_\Omega \varphi_1\psi = \int_\Omega m(x)\nabla\varphi_1\nabla\psi = \lambda\int_\Omega \psi\varphi_1 + \int_\Omega \mu(x)\frac{|\nabla\psi|^q}{(\psi + c)^{q-1}}\varphi_1 + \lambda c\int_\Omega \varphi_1.$$

In particular, $(\lambda_1(m) - \lambda)\int_\Omega \psi\varphi_1 > 0$ and $\lambda < \lambda_1(m)$. □

Now we characterize λ^* as the unique possible value of the parameter λ for which (E_λ) may admit a positive solution.

Proposition 2 *Assume that* $1 < q \le 2$, $0 \le \mu \in L^\infty(\Omega)$, *with* $\|\mu\|_{L^\infty(\Omega)} < 1$ *if* $q = 2$, *and* $0 < \eta \le m \in L^\infty(\Omega) \cap W^{1,\infty}_{loc}(\Omega)$. *If there exists a positive solution to* (E_λ), *then* $\lambda = \lambda^*$.

Proof Arguing by contradiction, if there exists a positive solution u to (E_λ) for some $\lambda > \lambda^*$ then, in particular, it is a supersolution and taking $(u + \varepsilon)^{2\gamma-1} - \varepsilon^{2\gamma-1}$ as test function and using Fatou lemma it is possible to prove that $u^\gamma \in H^1(\Omega)$ for every $\gamma > \frac{1}{2}$ if $q < 2$ or $\gamma > \frac{1+\|\mu\|_{L^\infty(\Omega)}}{2}$ if $q = 2$. This implies that $\lambda \in I_3$ and using Proposition 1 we have that

$$\lambda \le \sup I_3 = \lambda^* < \lambda.$$

On the other hand, if there exists a positive solution u to (E_λ) for some $\lambda < \lambda^*$, then tu is also a solution for every $t > 0$ and, using the characterization of λ^* given in Proposition 1, we have that (E_λ) admits a positive supersolution v. In the case $\mu \equiv 0$ this is a contradiction, since the comparison principle assures then that $tu \le v$ for every $t > 0$, which is not possible.

When $\mu \not\equiv 0$ we conclude the proof in a similar way once we generalize the comparison principle which in addition requires to prove stronger regularity of solutions (see Theorem 1 and Lemma 1 below, proved in [6]). □

In the next lemma, proved with the regularity theory developed by Ladyzhenskaya and Ural'tseva in [10], we summarize the main regularity properties of solutions to (P_λ) and, in particular, to (E_λ). Here we replace the $C^{1,1}$ regularity of $\partial\Omega$ by a less restrictive hypothesis.

Lemma 1 *Let* $1 < q \le 2$, $0 \le \mu \in L^\infty(\Omega)$, *with* $\|\mu\|_{L^\infty(\Omega)} < 1$ *if* $q = 2$, $0 < \eta \le m \in L^\infty(\Omega) \cap W^{1,\infty}_{loc}(\Omega)$, $0 \lneq f \in L^p(\Omega)$ *with* $p > \frac{N}{2}$, *and let* $u \in H_0^1(\Omega) \cap L^\infty(\Omega)$ *be a solution to* (P_λ) *for some* $\lambda \in \mathbb{R}$. *Assume also that there exist* $r_0, \theta_0 > 0$ *such that, if* $x \in \partial\Omega$ *and* $0 < r < r_0$, *then*

$$|\Omega_r| \le (1 - \theta_0)|B_r(x)|$$

for every connected component Ω_r *of* $\Omega \cap B_r(x)$, *where* $B_r(x)$ *denotes the ball centered at* x *with radius* r. *Then* $u \in C^{0,\alpha}(\overline{\Omega}) \cap W^{1,2p}_{loc}(\Omega)$ *for some* $\alpha \in (0, 1)$.

Now we state the main comparison principle that we have obtained in this context.

Theorem 1 *Let* $1 < q \le 2$, $\lambda \in \mathbb{R}$, $0 < \eta \le \mu \in L^\infty(\Omega)$, $0 \le h \in L^1_{loc}(\Omega)$, $0 < \eta \le m \in L^\infty_{loc}(\Omega)$, *and assume that* $u, v \in C(\Omega) \cap W^{1,N}_{loc}(\Omega)$ *are such that* $u, v > 0$ *in* Ω *and satisfy*

$$\limsup_{x \to x_0} \frac{u(x)}{v(x)} \le 1 \quad \forall x_0 \in \partial\Omega, \tag{8}$$

$$\int_\Omega m(x)\nabla u \cdot \nabla\phi \le \lambda \int_\Omega u\phi + \int_\Omega \mu(x)\frac{|\nabla u|^q}{u^{q-1}}\phi + \int_\Omega h(x)\phi, \tag{9}$$

and

$$\int_\Omega m(x)\nabla v \cdot \nabla\phi \ge \lambda \int_\Omega v\phi + \int_\Omega \mu(x)\frac{|\nabla v|^q}{v^{q-1}}\phi + \int_\Omega h(x)\phi, \tag{10}$$

for all $0 \le \phi \in H_0^1(\Omega) \cap L^\infty(\Omega)$ *with compact support. Then* $u \le v$ *in* Ω.

Proof The result is obtained arguing as in [1, Lemma 2.2] (see also the references therein) considering the equations satisfied by $u_1 = \log(u)$ and $v_1 = \log(v)$ and taking into account, using (8), that, for every $k > 0$, the function $(u_1 - v_1 - k)^+$ has compact support in Ω. □

3 Applications

3.1 Existence of Solution

In this subsection we will prove that problem (P_λ) admits at least a solution for every $\lambda < \lambda^*$. To this task, the characterization of λ^* as the only possible principal eigenvalue to (E_λ) will be the key point. In order to avoid the singularity we consider a sequence of approximating nonsingular problems.

$$\begin{cases} -\text{div}(m(x)\nabla u_n) = \lambda u_n + \mu(x) g_n(u_n)|\nabla u_n|^q + f_n(x) & \text{in } \Omega, \\ u_n = 0 & \text{on } \partial\Omega, \end{cases} \qquad (Q_n)$$

where $f_n(x) = \max\{-n, \min\{f(x), n\}\}$ and $g_n(s) = \dfrac{1}{|s + \frac{1}{n}|^{q-1}}$ when $0 \lneq f$, otherwise

$$g_n(s) \overset{\text{def}}{=} \begin{cases} \dfrac{1}{|s|^{q-1}} & |s| \geq \dfrac{1}{n}, \\ |s| n^q & |s| \leq \dfrac{1}{n}. \end{cases}$$

The role of λ^*, the only candidate to be the principal eigenvalue of (E_λ), is that it allows to prove an a priori estimate, for $\lambda < \lambda^*$, in the $L^\infty(\Omega)$ norm of this sequence of approximating solutions. Thanks to this estimate one can pass to the limit and prove the existence of solution to (P_λ).

The main result in [7] for the existence of solution, which includes the case where f may change sign, is the following (see also [6] for positive data).

Theorem 2 *Assume that* $1 < q \leq 2$, $f \in L^p(\Omega)$ *for some* $p > \frac{N}{2}$, $0 \leq \mu \in L^\infty(\Omega)$ *and* $0 < \eta \leq m \in L^\infty(\Omega) \cap W^{1,\infty}_{loc}(\Omega)$. *If* $q = 2$, *assume additionally that* $f \gneq 0$ *and* $\|\mu\|_{L^\infty(\Omega)} < 1$. *Then there exists at least a solution to problem* (P_λ) *for every* $\lambda < \lambda^*$.

Proof **Step 1**. First we deduce, by means of the subsolution and supersolution method in [5], the existence of $u_n \in H_0^1(\Omega) \cap L^\infty(\Omega)$ solution to (Q_n). Here, the supersolution to (Q_n) for any $\lambda < \lambda^*$ is given by a positive multiple of the supersolution to $(E_{\bar\lambda})$ for any $\bar\lambda \in (\lambda, \lambda^*) \subset I_1$. The subsolution is either a negative multiple of the same function or the zero function when $0 \lneq f$.

Step 2. Arguing as in [10, Theorem 1.1] at Sect. 4 (pp. 249–251) we deduce that $u_n \in C^{0,\alpha}(\overline{\Omega})$ for some $\alpha \in (0, 1)$ (see also [6, Appendix]).

Step 3. $\{u_n\}$ is uniformly bounded from below. This is deduced from the maximum principle. Indeed, $u_n \geq z$ with $z \in H_0^1(\Omega) \cap L^\infty(\Omega)$ and z satisfying $-\text{div}(m(x)\nabla z) = \lambda z - |f(x)|$ in Ω. Moreover, if $0 \lneq f$ we have that $u_n \geq w$ with $-\text{div}(m(x)\nabla w) = \lambda z + f_1(x)$ and the strong maximum principle assures that $\{u_n\}$ is uniformly bounded away from zero in compactly embedded subdomain of Ω.

Step 4. $\{u_n\}$ is bounded in $L^\infty(\Omega)$ and the proof finishes when $0 \lneq f$. Indeed, we argue by contradiction and take $\|u_n\|_{L^\infty(\Omega)} \to \infty$ (up to a subsequence). Then, we have that the function $z_n \equiv \frac{u_n}{\|u_n\|_{L^\infty(\Omega)}} \in H_0^1(\Omega) \cap L^\infty(\Omega)$ satisfies, for every n, that

$$\begin{cases} -\mathrm{div}(m(x)z_n) = \lambda z_n + \mu(x)\dfrac{|\nabla z_n|^q}{\left(z_n + \frac{1}{n\|u_n\|_{L^\infty(\Omega)}}\right)^{q-1}} + \dfrac{T_n(f(x))}{\|u_n\|_{L^\infty(\Omega)}} & \text{in } \Omega, \\ z_n = 0 & \text{on } \partial\Omega. \end{cases} \tag{11}$$

Taking z_n as test function and using that $\|z_n\|_{L^\infty(\Omega)} = 1$ we obtain that $\{z_n\}$ is bounded in $H_0^1(\Omega)$ and we deduce that there exists $0 \le z \in H_0^1(\Omega) \cap L^\infty(\Omega)$ such that, passing to a subsequence, $z_n \rightharpoonup z$ weakly in $H_0^1(\Omega)$ and $z_n \to z$ uniformly in $\overline{\Omega}$ (due to the compact embedding of $C^{0,\alpha}(\overline{\Omega})$ in $C^0(\overline{\Omega})$ and the uniform bound in $C^{0,\alpha}(\overline{\Omega})$ that regularity yields from the $L^\infty(\Omega)$ bound of z_n). In particular, $\|z\|_{L^\infty(\Omega)} = 1$ and as a consequence $z \gneq 0$ in Ω. Moreover, using weak limits $\int_\Omega m(x)\nabla z \nabla\phi - \lambda\int_\Omega z\phi \ge 0$, and the strong maximum principle ($\lambda < \lambda^* \le \lambda_1(m)$) leads to the facts that $z > 0$ in Ω and $\frac{|\nabla z|^q}{z^{q-1}} \in L^1_{\mathrm{loc}}(\Omega)$. Furthermore, the uniform convergence implies that z_n satisfies $z_n \ge c_\omega > 0$, $\forall \omega \subset\subset \Omega$, $\forall n \in \mathbb{N}$. This implies that $\{-\Delta z_n\}$ is bounded in $L^1_{\mathrm{loc}}(\Omega)$, that combined with the H^1 bound implies (see [4]) that

$$\nabla z_n \to \nabla z \quad \text{strongly in } L^r(\Omega)^N \text{ for any } r < 2.$$

The local lower bound and the convergence of the gradients will allow us to pass to the limit in (11). In this respect, the case $q = 2$ is special since in principle we do not have strong convergence of the gradients in $L^2(\Omega)^N$. Nevertheless, Fatou lemma can be applied to prove that z is both a subsolution and a supersolution to (E_λ), and here the assumptions $\|\mu\|_{L^\infty(\Omega)} < 1$ and $f \gneq 0$ are essential. In either case, we deduce that z is a solution to problem (E_λ), which is a contradiction with Proposition 2 since $\lambda < \lambda^*$.

The contradiction confirms that $\{u_n\}$ is bounded in $L^\infty(\Omega)$ and arguing as for the sequence $\{z_n\}$ we conclude the proof of the result by passing to the limit in (Q_n).

Step 5. $\{u_n\}$ is bounded in $L^\infty(\Omega)$ for sign-changing f.

Although we can not argue as in Step 4, we can use the proof for positive data (Step 4) in order to prove that u_n is uniformly bounded from above, which in addition to Step 3 implies that $\{u_n\}$ is bounded in $L^\infty(\Omega)$. Observe that, this uniform bound from above is trivial if the open set

$$\omega_n = \{x \in \Omega : u_n(x) > 0\}$$

is empty. Otherwise, since $u_n \in C^{0,\alpha}(\omega_n)$, then we deduce that $u_n \in W^{1,N}_{\mathrm{loc}}(\omega_n)$ and u_n is a subsolution to the problem

$$
\begin{cases}
-\mathrm{div}(m(x)\nabla\zeta) = \lambda\zeta + \mu(x)\dfrac{|\nabla\zeta|^q}{\zeta^{q-1}} + |f(x)| + 1 & \text{in } \omega_n, \\
\zeta > 0 & \text{in } \omega_n, \\
\zeta = 0 & \text{on } \partial\omega_n.
\end{cases} \tag{12}
$$

On the other hand, from Step 4, there exists a solution v to

$$
\begin{cases}
-\mathrm{div}(m(x)\nabla v) = \lambda v + \mu(x)\dfrac{|\nabla v|^q}{v^{q-1}} + |f(x)| + 1 & \text{in } \Omega, \\
v > 0 & \text{in } \Omega, \\
v = 0 & \text{on } \partial\Omega.
\end{cases}
$$

Moreover, $v \in C(\overline{\Omega}) \cap W^{1,N}_{\mathrm{loc}}(\Omega)$ reasoning as before. Then, v is a supersolution to (12) and applying Theorem 1 we deduce that $u_n \le v \le \|v\|_{L^\infty(\Omega)}$ in ω_n and as a consequence $u_n \le \|v\|_{L^\infty(\Omega)}$ in Ω.

Step 6. Passing to the limit for general data f and $1 < q < 2$. Arguing as in Step 4, we can deduce that there exists $u \in H_0^1(\Omega) \cap L^\infty(\Omega)$ such that $u_n \to u$ strongly in $H_0^1(\Omega)$ and in $L^r(\Omega)$ for every $r \in [1, \infty)$. Moreover, given $\phi \in H_0^1(\Omega) \cap L^\infty(\Omega)$,

$$
\begin{aligned}
\lim_{n\to\infty} \int_\Omega \mu(x) g_n(u_n)|\nabla u_n|^q \phi &= \lim_{n\to\infty} \left(\int_\Omega m(x)\nabla u_n \cdot \nabla\phi - \lambda \int_\Omega u_n\phi \right. \\
&\quad \left. - \int_\Omega f_n(x)\phi \right) = \int_\Omega m(x)\nabla u \cdot \nabla\phi - \lambda \int_\Omega u\phi - \int_\Omega f(x)\phi.
\end{aligned}
$$

The main difficulty is to prove that

$$
\lim_{n\to\infty} \int_\Omega \mu(x) g_n(u_n)|\nabla u_n|^q \phi = \int_{\{|u|>0\}} \mu(x)\frac{|\nabla u|^q}{|u|^{q-1}}\phi.
$$

In order to do that we choose a convenient decreasing sequence of positive real numbers $\delta_m \to 0$ and prove, using that $|u_n|^{\frac{1-\varepsilon}{q}}$ is bounded in $H_0^1(\Omega)$, that

$$
\lim_{m\to\infty} \left(\lim_{n\to\infty} \int_{\{|u_n|\le\delta_m\}} \mu(x) g_n(u_n)|\nabla u_n|^q \phi \right) = 0.
$$

On the other hand, using that $|u_n|^{\frac{1}{q}}$ is bounded in $H_0^1(\Omega)$ and Lebesgue Theorem

$$
\begin{aligned}
\lim_{m\to\infty} \left(\lim_{n\to\infty} \int_{\{|u_n|>\delta_m\}} \mu(x) g_n(u_n)|\nabla u_n|^q \phi \right) &= \lim_{m\to\infty} \left(\int_{\{|u|>\delta_m\}} \mu(x)\frac{|\nabla u|^q}{|u|^{q-1}}\phi \right) \\
&= \int_{\{|u|>0\}} \mu(x)\frac{|\nabla u|^q}{|u|^{q-1}}\phi.
\end{aligned}
$$

This concludes the proof that u is a solution to (P_λ). ☐

3.2 Nonexistence of Solution to (P_λ)

As in the proof of Proposition 2, when $0 \lneq f$, the main role of the principal eigenvalue λ^* for the nonexistence of positive solution is due to the characterization as $\lambda^* = \sup I_3$, since existence of positive solution for some λ implies $\lambda \in I_3$. As a consequence, no solution exists for $\lambda > \lambda^*$.

Proposition 3 *Assume that* $1 < q \leq 2$, $0 \lneq f \in L^p(\Omega)$ *for some* $p > \frac{N}{2}$, $0 \leq \mu \in L^\infty(\Omega)$ *and* $0 < \eta \leq m \in L^\infty(\Omega) \cap W^{1,\infty}_{loc}(\Omega)$. *If* $q = 2$, *assume additionally that* $\|\mu\|_{L^\infty(\Omega)} < 1$. *Then, there is no solution to* (P_λ) *for any* $\lambda > \lambda^*$.

Proof As commented above, the main difficulty lies on the proof of the existence of $\frac{1}{2} \leq \gamma(q) < 1$ such that $u^\gamma \in H^1(\Omega)$ for every $\gamma > \gamma(q)$ and for every u solution to (P_λ), which implies that $\lambda \in I_3$ and thus $\lambda \leq \lambda^*$. □

3.3 Uniqueness

The comparison principle given in Theorem 1 guarantees uniqueness of $C(\Omega) \cap W^{1,N}_{\mathrm{loc}}(\Omega)$ positive solution to (P_λ) when any pair of possible solutions satisfy (8). Since this last condition is hard to verify, we derive in [6] we derive another comparison result to improve the uniqueness result for (P_λ).

Theorem 3 *Let* $1 < q \leq 2$, $\lambda \in \mathbb{R}$, $0 \leq \mu \in L^\infty(\Omega)$, $0 \leq h \in L^1_{loc}(\Omega)$, $0 < \eta \leq m \in L^\infty_{loc}(\Omega)$. *Assume that* $u, v \in C(\Omega) \cap W^{1,N}_{loc}(\Omega)$, *with* $u, v > 0$ *in* Ω, *and satisfy* (9) *and* (10) *respectively. Assume also that, for all* $\varepsilon > 0$,

$$\limsup_{x \to x_0} \left(\frac{u(x)}{v(x) + \varepsilon} \right) \leq 1 \quad \forall x_0 \in \partial\Omega. \tag{13}$$

Furthermore, if $\lambda > 0$, *assume also that* h *is locally bounded away from zero and* $\lambda < \lambda^*$. *Then,* $u \leq v$ *in* Ω.

Next theorem summarizes the existence and uniqueness results we have introduced above for $f \gneq 0$.

Theorem 4 *Assume that* $1 < q \leq 2, 0 \lneq f \in L^p(\Omega)$ *with* $p > \frac{N}{2}, 0 \leq \mu \in L^\infty(\Omega)$, *with* $\|\mu\|_{L^\infty(\Omega)} < 1$ *if* $q = 2$, *and* $0 < \eta \leq m \in L^\infty(\Omega) \cap W^{1,\infty}_{loc}(\Omega)$. *Then* (P_λ) *has a unique solution if either* $\lambda \leq 0$, *or* f *is locally bounded away from zero and* $\lambda < \lambda^*$.

Proof We observe that if u, v are two solutions to (P_λ), then Lemma 1 implies that $u, v \in C(\overline{\Omega}) \cap W^{1,N}_{\mathrm{loc}}(\Omega)$. In particular, using the continuity up to the boundary of u, v and the fact that $u(x_0) = 0$ for any $x_0 \in \partial\Omega$, we have that u, v satisfy (13) for any $\varepsilon > 0$. Moreover, they obviously satisfy (9) and (10) respectively. Therefore, Theorem 3 implies that $u \leq v$ in Ω. The reverse inequality follows by interchanging the roles of u and v. Finally we observe that the existence of solution is deduced from Theorem 2.

□

3.4 Bifurcation

As in the semilinear case ($\mu \equiv 0$) we prove that λ^* is in fact a bifurcation point from infinity for (P_λ) when $f \gneq 0$. This, in addition, is useful to deduce that (E_{λ^*}) admits solution. This confirms that λ^* is the principal eigenvalue.

Theorem 5 *Assume that* $1 < q \leq 2, 0 \lneq f \in L^p(\Omega)$ *with* $p > \frac{N}{2}, 0 \leq \mu \in L^\infty(\Omega)$, *with* $\|\mu\|_{L^\infty(\Omega)} < 1$ *if* $q = 2$, *and* $0 < \eta \leq m \in L^\infty(\Omega) \cap W^{1,\infty}_{loc}(\Omega)$. *Then,* λ^* *is the unique possible bifurcation point from infinity of* (P_λ). *Moreover, if* f *is locally bounded away from zero, then the set*

$$\Sigma := \{(\lambda, u_\lambda) \in \mathbb{R} \times C(\overline{\Omega}) : u_\lambda \text{ is a solution to } (P_\lambda)\}$$

is a continuum. In this case, the continuum is unbounded and bifurcates from infinity at λ^* *to the left whenever* (P_λ) *has no solution for* $\lambda = \lambda^*$.

Remark 1 Observe that there are conditions on f such that there are no solutions to (P_{λ^*}). For instance, $f \geq c$ for some $c > 0$. See [6] for more details.

Proof For the first part we observe that, as before, if $\bar{\lambda} \in \mathbb{R}$ is a bifurcation point from infinity then the normalized sequence converges to a solution to $(E_{\bar{\lambda}})$ which implies that $\lambda = \lambda^*$.

For the existence of the continuum we observe that, when f is locally bounded away from zero, one has uniqueness of solution for all $\lambda < \lambda^*$. Then, we can define a map $\lambda \mapsto u_\lambda$, where u_λ is the unique solution to problem (P_λ) for all $\lambda < \lambda^*$. The proof that this map is continuous is deduced by deriving an L^∞ estimate and passing to the limit, as in Theorem 2.

Finally, for the global behavior of Σ we observe that, if for $\lambda_n \to \lambda^*$ the sequence $\{u_{\lambda_n}\}$ is bounded in $L^\infty(\Omega)$, then we can pass to the limit in (P_{λ_n}) to find a solution to (P_{λ^*}), which contradicts the assumption. □

Corollary 1 *Assume that* $1 < q \leq 2, 0 \leq \mu \in L^\infty(\Omega)$, *with* $\|\mu\|_{L^\infty(\Omega)} < 1$ *if* $q = 2$, *and* $0 < \eta \leq m \in L^\infty(\Omega) \cap W^{1,\infty}_{loc}(\Omega)$. *Then* (E_{λ^*}) *admits solution.*

Proof We may choose $\lambda_n \leq \lambda^*, \lambda_n \to \lambda^*$ such that $\|u_{\lambda_n}\|_{L^\infty(\Omega)} \to \infty$, where u_{λ_n} denotes, for any n, the unique solution to the problem

$$\begin{cases} -\Delta u = \lambda_n u + \mu(x)\dfrac{|\nabla u|^q}{u^{q-1}} + 1 & \text{in } \Omega, \\ u > 0 & \text{in } \Omega, \\ u = 0 & \text{on } \partial\Omega. \end{cases}$$

Then we prove that the normalized sequence converges to a solution to (E_{λ^*}). □

3.5 Homogenization

Here we state the homogenization result obtained in [7]. Following [8], we consider for every $\varepsilon > 0$ a finite number, $n(\varepsilon) \in \mathbb{N}$, of closed subsets $T_i^\varepsilon \subset \mathbb{R}^N$, $1 \le i \le n(\varepsilon)$, which are the holes. Let us denote $D^\varepsilon = \mathbb{R}^N \setminus \bigcup_{i=1}^{n(\varepsilon)} T_i^\varepsilon$. The domain Ω^ε is defined by removing the holes T_i^ε from Ω, that is

$$\Omega^\varepsilon = \Omega - \bigcup_{i=1}^{n(\varepsilon)} T_i^\varepsilon = \Omega \cap D^\varepsilon.$$

We assume that the sequence of domains Ω^ε is such that there exist a sequence of functions $\{w^\varepsilon\}$ and $\sigma \in H^{-1}(\Omega)$ such that

$$w^\varepsilon \in H^1(\Omega) \cap L^\infty(\Omega), \tag{14}$$

$$0 \le w^\varepsilon \le 1 \text{ a.e. } x \in \Omega, \tag{15}$$

$$w^\varepsilon \phi \in H_0^1(\Omega^\varepsilon) \cap L^\infty(\Omega^\varepsilon) \ \forall \phi \in H_0^1(\Omega) \cap L^\infty(\Omega), \tag{16}$$

$$w^\varepsilon \rightharpoonup 1 \text{ weakly in } H^1(\Omega), \tag{17}$$

and given $z^\varepsilon, \phi, z \in H^1(\Omega) \cap L^\infty(\Omega)$ such that $z^\varepsilon \phi \in H_0^1(\Omega^\varepsilon) \cap L^\infty(\Omega^\varepsilon)$ and z^ε weakly converges in $H^1(\Omega)$ to z, the following holds

$$\int_\Omega m(x) \nabla w^\varepsilon \cdot \nabla (z^\varepsilon \phi) \to \langle \sigma, z\phi \rangle_{H^{-1}(\Omega), H_0^1(\Omega)}. \tag{18}$$

For a function $u^\varepsilon \in H_0^1(\Omega^\varepsilon)$, we denote by $\widetilde{u^\varepsilon} \in H_0^1(\Omega)$ the extension of u by zero in $\Omega \setminus \Omega^\varepsilon$.

Theorem 6 *Assume that the sequence of perforated domains Ω^ε satisfies* (14), (15), (16), (17) *and* (18). *Suppose also that $1 < q < 2$, $f \in L^p(\Omega)$ for some $p > \frac{N}{2}$, $0 \le \mu \in L^\infty(\Omega)$, $0 < \eta \le m \in L^\infty(\Omega) \cap W^{1,\infty}_{loc}(\Omega)$, $\lambda < \lambda^*$ and that both Ω and D^ε satisfy the regularity condition of the domain in Lemma 1, where $D^\varepsilon = \mathbb{R}^N \setminus \bigcup_{i=1}^{n(\varepsilon)} T_i^\varepsilon$. Then, there exists a sequence $\{u^\varepsilon\}$ of solutions to problem*

$$\begin{cases} -\text{div}(m(x)\nabla u^\varepsilon) = \lambda u^\varepsilon + \mu(x)\dfrac{|\nabla u^\varepsilon|^q}{|u^\varepsilon|^{q-1}} + f(x) & \text{in } \Omega^\varepsilon, \\ u^\varepsilon = 0 & \text{on } \partial\Omega^\varepsilon, \end{cases}$$

such that $\{\widetilde{u^\varepsilon}\}$ *is bounded in* $L^\infty(\Omega)$ *and* $\widetilde{u^\varepsilon}$ *weakly converges in* $H_0^1(\Omega)$ *to a solution* u *to*

$$\begin{cases} -\operatorname{div}(m(x)\nabla u) + \sigma u = \lambda u + \mu(x)\dfrac{|\nabla u|^q}{|u|^{q-1}} + f(x) & \text{in } \Omega, \\ u = 0 & \text{on } \partial\Omega, \end{cases}$$

in the sense that $u \in H_0^1(\Omega) \cap L^\infty(\Omega)$, $\frac{|\nabla u|^q}{|u|^{q-1}} \in L^1(\{|u| > 0\})$ *and*

$$\int_\Omega m(x)\nabla u \cdot \nabla \phi + \langle \sigma, u\phi\rangle_{H^{-1}(\Omega), H_0^1(\Omega)} = \lambda \int_\Omega u\phi + \int_{\{|u|>0\}} \mu(x)\frac{|\nabla u|^q}{|u|^{q-1}}\phi + \int_\Omega f(x)\phi$$

for all $\phi \in H_0^1(\Omega) \cap L^\infty(\Omega)$.

References

1. Arcoya, D., de Coster, C., Jeanjean, L., Tanaka, L.: Remarks on the uniqueness for quasilinear elliptic equations with quadratic growth conditions. J. Math. Anal. Appl. **420**, 772–780 (2014)
2. Arcoya, D., Moreno-Mérida, L.: The effect of a singular term in a quadratic quasi-linear problem. J. Fixed Point Theory Appl. **19**, 815–831 (2017)
3. Berestycki, H., Nirenberg, L., Varadhan, S.R.S.: The principal eigenvalue and maximum principle for second-order elliptic operators in general domains. Comm. Pure Appl. Math. **47**, 47–92 (1994)
4. Boccardo, L., Murat, F.: Almost everywhere convergence of the gradients of solutions to elliptic and parabolic equations. Nonlinear Anal. **19**, 581–597 (1992)
5. Boccardo, L., Murat, F., Puel, J.-P.: Quelques propriétés des opérateurs elliptiques quasi linéaires. C. R. Acad. Sci. Paris Sér. I Math. **307**, 749–752 (1988)
6. Carmona, J., Leonori, T., López-Martínez, S., Martínez-Aparicio, P.J.: Quasilinear elliptic problems with singular and homogeneous lower order terms. Nonlinear Anal. **179**, 105–130 (2019)
7. Carmona, J., López-Martínez, S., Martínez-Aparicio, P.J.: Singular quasilinear elliptic problems with changing sign datum: existence and homogenization. Rev. Mat. Complut. https://doi.org/10.1007/s13163-019-00313-2
8. Cioranescu, D., Murat, F.: Un terme étrange venu d'ailleurs, I et II. In: Brezis, H., Lions, J.-L. (eds.) Nonlinear Partial Differential Equations and Their Applications, Collège de France Seminar, vols. II and III. Research Notes in Math. vols. 60 and 70, pp. 98–138 and 154–178. Pitman, London, (1982). English translation: Cioranescu, D., Murat, F.: A strange term coming from nowhere. In: Cherkaev, A., Kohn, R.V. (eds.) Topics in Mathematical Modeling of Composite Materials; Progress in Nonlinear Differential Equations and their Applications, vol. 31, pp. 44–93. Birkhäuger, Boston (1997)
9. Ferone, V., Posteraro, M.R., Rakotoson, J.M.: L^∞-estimates for nonlinear elliptic problems with p-growth in the gradient. J. Inequal. Appl. **3**(2), 109–125 (1999)
10. Ladyzhenskaya, O., Ural'tseva, N.: Linear and Quasilinear Elliptic Equations. Translated from the Russian by Scripta Technica, Academic Press, New York-London (1968), xviii+495 pp

Non-commutative Poisson Algebras Admitting a Multiplicative Basis

Antonio J. Calderón Martín, Boubacar Dieme and Francisco J. Navarro Izquierdo

Abstract A non-commutative Poisson algebra is a Lie algebra endowed with a, not necessarily commutative, associative product in such a way that the Lie and associative products are compatible via the Leibniz identity. If we consider a non-commutative Poisson algebra $\mathfrak{P}$ of arbitrary dimension, over an arbitrary base field $\mathbb{F}$, a basis $\mathfrak{B} = \{e_i\}_{i \in I}$ of $\mathfrak{P}$ is called multiplicative if for any $i, j \in I$ we have that $e_i e_j \in \mathbb{F}e_r$ and $[e_i, e_j] \in \mathbb{F}e_s$ for some $r, s \in I$. We show that if $\mathfrak{P}$ admits a multiplicative basis then it decomposes as the direct sum of well-described ideals admitting each one a multiplicative basis. Also the minimality of $\mathfrak{P}$ is characterized in terms of the multiplicative basis and it is shown that, under a mild condition, the previous decomposition is the direct sum of the family of its minimal ideals admitting a multiplicative basis.

Keywords Multiplicative basis · Non-commutative algebra · Poisson algebra · Infinite dimensional algebra · Structure theory

MSC2010 81R10 · 15A21 · 16xx · 17B63

The authors are supported by the PCI of the UCA 'Teoría de Lie y Teoría de Espacios de Banach', by the PAI with project numbers FQM298, FQM7156 and by the project of the Spanish Ministerio de Educación y Ciencia MTM2016-76327C31P.

A. J. Calderón Martín · F. J. Navarro Izquierdo (✉)
Department of Mathematics, University of Cádiz, Cádiz, Spain
e-mail: javi.navarroiz@uca.es

A. J. Calderón Martín
e-mail: ajesus.calderon@uca.es

B. Dieme
Department of Mathematics, University Cheikh Anta Diop of Dakar, Dakar, Senegal
e-mail: boubadieme@hotmail.com

M. Siles Molina et al. (eds.), *Associative and Non-Associative Algebras and Applications*, Springer Proceedings in Mathematics & Statistics 311,
https://doi.org/10.1007/978-3-030-35256-1_5

1 Introduction and Previous Definitions

The interest in the study of Poisson algebras have grown as consequence of their applications in geometry and mathematical physics (see for instance [1, 2, 11]). Respect to the study of structural properties of this class of algebras we have to mention the work of F. Kubo, who studied this problem in the finite-dimensional, and over an algebraically closed field of characteristic zero, case in [12, 14]; and began the study in the infinite dimensional case by centering his attention on the non-commutative Poisson algebraic structures over affine Kac-Moody algebras (also on an algebraically closed field of characteristic zero), in [13].

Throughout the present paper $\mathfrak{P}$ will denote a non-commutative Poisson algebra of arbitrary dimension and over an arbitrary base field $\mathbb{F}$. That is,

Definition 1 A *non-commutative Poisson algebra* $\mathfrak{P}$ is a Lie algebra $(\mathfrak{P}, [\cdot,\cdot])$ over an arbitrary base field $\mathbb{F}$, endowed with an associative product, denoted by juxtaposition, in such a way the following Leibniz identity

$$[xy, z] = [x, z]y + x[y, z]$$

holds for any $x, y, z \in \mathfrak{P}$.

Let us also recall some notions:

Definition 2 Let $\mathfrak{P}$ be a non-commutative Poisson algebra. A *subalgebra* of $\mathfrak{P}$ is a linear subspace closed by the Lie and the associative products. An *ideal* $\mathfrak{I}$ of $\mathfrak{P}$ is a subalgebra satisfying $[\mathfrak{I}, \mathfrak{P}] + \mathfrak{I}\mathfrak{P} + \mathfrak{P}\mathfrak{I} \subset \mathfrak{I}$. Finally, $\mathfrak{P}$ is called *simple* if $[\mathfrak{P}, \mathfrak{P}] \neq 0$, $\mathfrak{P}\mathfrak{P} \neq 0$ and its only ideals are $\{0\}$ and $\mathfrak{P}$.

In [7], multiplicative bases were considered for arbitrary algebras (with just one multiplication). In this reference, a basis $\mathfrak{B} = \{e_i\}_{i\in I}$ of an arbitrary algebra $\mathfrak{A}$ was called multiplicative if for any $i, j \in I$ we had either $e_ie_j = 0$ or $0 \neq e_ie_j \in \mathbb{F}e_k$ for some (unique) $k \in I$.

From here, we can introduce the concept of multiplicative basis for non-commutative Poisson algebras as follows:

Definition 3 A basis $\mathfrak{B} = \{e_i\}_{i\in I}$ of a non-commutative Poisson algebra $\mathfrak{P}$ is said to be *multiplicative* if for any $i, j \in I$ we have that $[e_i, e_j] \in \mathbb{F}e_r$ and that $e_ie_j \in \mathbb{F}e_s$ for some $r, s \in I$.

The present paper is devoted to the study of non-commutative Poisson algebras (of arbitrary dimension and over and arbitrary base field) admitting a multiplicative basis, by focussing on its structure.

In Sect. 2, and by inspiring in the connections in the support techniques developed for split non-commutative Poisson algebras in [4], we introduce connections techniques on the set of indexes I of the multiplicative basis so as to get a powerful tool for the study of this class of Poisson algebras. By making use of these techniques we show that any Poisson algebra $\mathfrak{P}$ admitting a multiplicative basis is of the form

$$\mathfrak{P} = \bigoplus_k \mathfrak{I}_k$$

with any $\mathfrak{I}_k$ a well described ideal of $\mathfrak{P}$ admitting also a multiplicative basis.

In Sect. 3 the minimality of $\mathfrak{P}$ is characterized in terms of the multiplicative basis and it is shown that, in case the basis is $\star$-multiplicative, the above decomposition of $\mathfrak{P}$ is actually by means of the family of its minimal ideals.

Finally, we would like to note that the present paper is the natural continuation of the reference by the first and third author [7], which also extends the paper by the first author [6]. In [6] it is studied (anti)commutative algebras having a multiplicative basis, later in [7] it is extended the obtained results to arbitrary algebras (not necessarily (anti)commutative) admitting also a multiplicative basis.

In the present paper we extend the results in the above two references in the sense that we study an algebraic system formed by two products (a Poisson algebra is an algebraic system formed by an associative product and by a Lie product which are compatible via a Leibniz identity), endowed with a multiplicative basis (which will be multiplicative for both products), while in [6, 7] are considered algebraic systems formed by just one product with a multiplicative basis. Hence, we will follow a development of the theory analogous to the one in [6, 7].

2 Connections in the Set of Indexes. Decompositions

In what follows $\mathfrak{B} = \{e_i\}_{i \in I}$ denotes the multiplicative basis of $\mathfrak{P}$, and $\mathcal{P}(I)$ the power set of I.

We begin this section by developing connection techniques among the elements in the set of indexes I as the main tool in our study.

First, we consider the mappings $\phi, \varphi : I \times I \to \mathcal{P}(I)$, which recover, in a sense, certain multiplicative relations among the elements of $\mathfrak{B}$. We define these two mappings as:

$$\phi(i, j) := \begin{cases} \emptyset & \text{if } e_i e_j = 0 \text{ and } e_j e_i = 0 \\ \{r\} & \text{if } 0 \neq e_i e_j \in \mathbb{F}e_r \text{ and } e_j e_i = 0 \\ \{s\} & \text{if } 0 \neq e_j e_i \in \mathbb{F}e_s \text{ and } e_i e_j = 0 \\ \{r, s\} & \text{if } 0 \neq e_i e_j \in \mathbb{F}e_r \text{ and } 0 \neq e_j e_i \in \mathbb{F}e_s. \end{cases}$$

and

$$\varphi(i, j) := \begin{cases} \emptyset & \text{if } [e_i, e_j] = 0 \\ \{k\} & \text{if } 0 \neq [e_i, e_j] \in \mathbb{F}e_k. \end{cases}$$

for any $i, j \in I$.

Remark 1 Let us observe that $\phi(i, j) = \phi(j, i)$ and $\varphi(i, j) = \varphi(j, i)$ for any $i, j \in I$.

Next, for each $i \in I$, a new variable $\bar{i} \notin I$ is introduced and we denote by

$$\bar{I} := \{\bar{i} : i \in I\}$$

the set consisting on all these new symbols. Given any $\bar{i} \in \bar{I}$ we will also denote $\overline{(\bar{i})} := i$.

Then, we introduce the following operation $\star : I \times (I \dot{\cup} \bar{I}) \to \mathcal{P}(I)$ given by:

$$i \star j := \phi(i, j) \cup \varphi(i, j)$$

and

$$i \star \bar{j} := \{k \in I : i \in \phi(k, j) \cup \varphi(k, j)\}$$

for any $i, j \in I$.

Lemma 1 *Let $i, j \in I$ and $k \in I \dot{\cup} \bar{I}$. Then $i \in j \star k$ if and only if $j \in i \star \bar{k}$.*

Proof Let $i, j \in I$ and $k \in I \dot{\cup} \bar{I}$ such that $i \in j \star k$. We have two cases to consider. In the first one $k \in I$. Then, $i \in \phi(j, k) \cup \varphi(j, k)$ and so $j \in i \star \bar{k}$. In the second case $k \in \bar{I}$ we have $j \in \phi(i, k) \cup \varphi(i, k)$, and then $j \in i \star \bar{k}$. The converse is proved similarly. □

Finally we consider the operation

$$\diamond : \mathcal{P}(I) \times (I \dot{\cup} \bar{I}) \to \mathcal{P}(I),$$

defined as $\emptyset \diamond (I \dot{\cup} \bar{I}) := \emptyset$ and

$$J \diamond i := \bigcup_{j \in J} (j \star i)$$

for any $\emptyset \neq J \in \mathcal{P}(I)$ and $i \in I \dot{\cup} \bar{I}$.

Lemma 2 *Let $J \in \mathcal{P}(I)$ and $i \in I \dot{\cup} \bar{I}$. Then $j \in J \diamond i$ if and only if $(\{j\} \diamond \bar{i}) \cap J \neq \emptyset$.*

Proof We have $j \in J \diamond i$ if and only if there exists $k \in J$ such that $j \in k \star i$. By Lemma 1, the last fact is equivalent to $k \in j \star \bar{i}$. From here the assertion follows. □

Definition 4 Let i and j be two elements in the set of indexes I. We say that i is *connected* to j if either $i = j$ or there exists a subset

$$\{i_1, i_2, \dots, i_{n-1}, i_n\} \subset I \dot{\cup} \bar{I}$$

such that the following conditions hold:

1. $\{i\} \diamond i_1 \neq \emptyset$,
   $(\{i\} \diamond i_1) \diamond i_2 \neq \emptyset$,
   $((\{i\} \diamond i_1) \diamond i_2) \diamond i_3 \neq \emptyset$,
   $\vdots$
   $((\cdots(\{i\} \diamond i_1)\cdots) \diamond i_{n-2}) \diamond i_{n-1} \neq \emptyset$.

2. $j \in ((\cdots(\{i\} \diamond i_1)\cdots) \diamond i_{n-1}) \diamond i_n$.

The subset $\{i_1, i_2, \ldots, i_{n-1}, i_n\}$ is called a *connection* from i to j.

Lemma 3 *Let $\{i_1, i_2, \ldots, i_{n-1}, i_n\}$ any connection from some i to some j where $i, j \in I$ with $i \neq j$. Then the set $\{\bar{i}_n, \bar{i}_{n-1}, \ldots, \bar{i}_2, \bar{i}_1\}$ is a connection from j to i.*

Proof Let us argue by induction on n.

If $n = 1$, then $j \in \{i\} \diamond i_1$. From here, we have $j \in i \star i_1$ and Lemma 1 gives us that $i \in j \star \bar{i}_1$. Hence the set $\{\bar{i}_1\}$ is a connection from j to i.

Suppose now the assertion holds for any connection with n elements and let us show this assertion also holds for any connection

$$\{i_1, i_2, \ldots, i_n, i_{n+1}\}$$

with $n + 1$ elements from some $i \in I$ to some $j \in I$ with $i \neq j$. From the fact

$$j \in ((\cdots(\{i\} \diamond i_1)\cdots) \diamond i_n) \diamond i_{n+1},$$

if we denote $J := (\cdots(\{i\} \diamond i_1)\cdots) \diamond i_n$, Lemma 2 gives us that $(\{j\} \diamond \bar{i}_{n+1}) \cap J \neq \emptyset$, which allow us to take some $k \in (\{j\} \diamond \bar{i}_{n+1}) \cap J$.

Then, we have that $k \in J$. So $\{i_1, i_2, \ldots, i_n\}$ is a connection from i to k and, by induction hypothesis we can take a connection

$$\{\bar{i}_n, \bar{i}_{n-1}, \ldots, \bar{i}_2, \bar{i}_1\}$$

from k to i.

From here, taking into account that $k \in \{j\} \diamond \bar{i}_{n+1}$, we can conclude that

$$\{\bar{i}_{n+1}, \bar{i}_n, \bar{i}_{n-1}, \ldots, \bar{i}_2, \bar{i}_1\}$$

is a connection from j to i, which completes the proof. □

Proposition 1 *The relation $\sim$ in I, defined by $i \sim j$ if and only if i is connected to j, is an equivalence relation.*

Proof By Definition 4 and Lemma 3 we have that $\sim$ is reflexive and symmetric.

To show the transitive character, consider $i, j, k \in I$ such that $i \sim j$ and $j \sim k$. If either $i = j$ or $j = k$ it is clear that $i \sim k$. So, let us suppose that $i \neq j$ and $j \neq k$. Then, we can find connections $\{i_1, \dots, i_m\}$ and $\{j_1, \dots, j_n\}$ from i to j and from j to k respectively. We can verify that

$$\{i_1, \dots, i_m, j_1, \dots, j_n\}$$

is a connection from i to k. So we get that $\sim$ is transitive and consequently an equivalence relation. □

By the above Proposition we can introduce the quotient set

$$I/\sim = \{[i] : i \in I\},$$

becoming $[i]$ the set of elements in I which are connected to i.

For any $[i] \in I/\sim$ we define the linear subspace

$$\mathfrak{P}_{[i]} := \bigoplus_{j\in[i]} \mathbb{F}e_j.$$

Lemma 4 *Let $[i], [j] \in I/\sim$ be. The following assertions hold:*

1. *If $[\mathfrak{P}_{[i]}, \mathfrak{P}_{[j]}] \neq \{0\}$ then $[\mathfrak{P}_{[i]}, \mathfrak{P}_{[j]}] \subset \mathfrak{P}_{[i]}$ and $[i] = [j]$.*
2. *If $\mathfrak{P}_{[i]}\mathfrak{P}_{[j]} \neq \{0\}$ then $\mathfrak{P}_{[i]}\mathfrak{P}_{[j]} \subset \mathfrak{P}_{[i]}$ and $[i] = [j]$.*

Proof (1) For any $r \in [i]$ and $s \in [j]$ such that $[e_r, e_s] \neq 0$ we have $[e_r, e_s] \in \mathbb{F}e_t$ for some $t \in I$. From here $t \in \{r\} \diamond s$ and then $\{s\}$ is a connection from r to t, which together with the transitivity of the connection relation, give us $[i] = [t]$. Consequently $[\mathfrak{P}_{[i]}, \mathfrak{P}_{[j]}] \subset \mathfrak{P}_{[i]}$.

Taking into account that $t \in \{r\} \diamond s = \{s\} \diamond r$, we have that the set $\{r\}$ a connection from s to t. Then, we conclude $[i] = [t] = [j]$.

(2) The proof is similar to the one for the Lie product. □

Definition 5 Let $\mathfrak{P}$ be a non-commutative Poisson algebra with a multiplicative basis $\mathfrak{B}$. It is said that a subalgebra $\mathfrak{P}'$ of $\mathfrak{P}$ admits a (multiplicative) basis $\mathfrak{B}'$ *inherited* from $\mathfrak{B}$, if $\mathfrak{B}'$ is a (multiplicative) basis of $\mathfrak{P}'$ satisfying $\mathfrak{B}' \subset \mathfrak{B}$.

Definition 6 A direct sum of linear subspaces of a non-commutative Poisson algebra

$$\mathfrak{P} = \bigoplus_{\alpha} \mathfrak{P}_\alpha,$$

is called *orthogonal* if $[\mathfrak{P}_\alpha, \mathfrak{P}_\beta] = \mathfrak{P}_\alpha\mathfrak{P}_\beta = \{0\}$ for any $\alpha \neq \beta$.

Theorem 1 *Let* $\mathfrak{P}$ *be a non-commutative Poisson algebra with a multiplicative basis* $\mathfrak{B}$. *Then* $\mathfrak{P}$ *is the orthogonal direct sum*

$$\mathfrak{P} = \bigoplus_{[i]\in I/\sim} \mathfrak{P}_{[i]},$$

being any $\mathfrak{P}_{[i]}$ *an ideal of* $\mathfrak{P}$ *admitting* $\mathfrak{B}_{[i]} = \{e_j : j \in [i]\}$ *as a multiplicative basis inherited from* $\mathfrak{B}$.

Proof As we can write $\mathfrak{P} = \bigoplus_{i\in I} \mathbb{F}e_i$ we have

$$\mathfrak{P} = \bigoplus_{[i]\in I/\sim} \mathfrak{P}_{[i]} \tag{1}$$

By Lemma 4, we have $[\mathfrak{P}_{[i]}, \mathfrak{P}_{[i]}] \subset \mathfrak{P}_{[i]}$ and $\mathfrak{P}_{[i]}\mathfrak{P}_{[i]} \subset \mathfrak{P}_{[i]}$ for any $[i] \in I/\sim$, but $[\mathfrak{P}_{[i]}, \mathfrak{P}_{[j]}] = \{0\}$ and $\mathfrak{P}_{[i]}\mathfrak{P}_{[j]} = \{0\}$ for any $[j] \in I/\sim$ such that $[i] \neq [j]$. From here,

$$[\mathfrak{P}_{[i]}, \mathfrak{P}] + \mathfrak{P}_{[i]}\mathfrak{P} + \mathfrak{P}\mathfrak{P}_{[i]} = [\mathfrak{P}_{[i]}, \mathfrak{P}_{[i]}] + \mathfrak{P}_{[i]}\mathfrak{P}_{[i]} + \mathfrak{P}_{[i]}\mathfrak{P}_{[i]} \subset \mathfrak{P}_{[i]}.$$

Then, we conclude that the decomposition given in Eq. (1) is, indeed, an orthogonal direct sum being any $\mathfrak{P}_{[i]}$ an ideal of $\mathfrak{P}$ which admits $\mathfrak{B}_{[i]} = \{e_j : j \in [i]\} \subset \mathfrak{B}$ as a multiplicative basis inherited from $\mathfrak{B}$. □

Corollary 1 *If* $\mathfrak{P}$ *is simple, then any pair of elements in* I *are connected.*

3 The Minimal Components

In this section our target is to characterize the minimality of the ideals which give rise to the decomposition of $\mathfrak{P}$ in Theorem 1, in terms of connectivity properties in the set of indexes I. Consequently we will state a context in which a second Wedderburn-type theorem (see for instance [10, pp. 137–139]) holds in the category of non-commutative Poisson algebras. We begin by recalling the next definition.

Definition 7 A non-commutative Poisson algebra $\mathfrak{P}$ admitting a multiplicative basis $\mathfrak{B}$ is called *minimal* if its only nonzero ideal admitting a multiplicative basis inherited from $\mathfrak{B}$ is $\mathfrak{P}$.

Let us introduce the notion of $\star$-multiplicativity in the framework of non-commutative Poisson algebras admitting a multiplicative basis, in a similar way to the ones of closed-multiplicativity for graded associative algebras, split Leibniz algebras, split Poisson algebras, or split Lie color algebras among other classes of algebras (see [3, 4, 8, 9] for these notions and examples).

Definition 8 We say that a multiplicative basis $\mathfrak{B} = \{e_i\}_{i \in I}$ of a non-commutative Poisson algebra $\mathfrak{P}$ if $\star$-multiplicative if $e_j \in [e_i, \mathfrak{P}] + e_i\mathfrak{P} + \mathfrak{P}e_i$ for any $i, j \in I$ such that $j \in i \star k$ for some $k \in I \dot{\cup} \bar{I}$.

We would like to note that the above concept appear in a natural way in the study of non-commutative Poisson algebras. For instance, the Poisson algebras considered in [4, Sect. 3] are examples of non-commutative Poisson algebras admitting $\star$-multiplicative bases. Indeed, any Σ-multiplicative extended graded Poisson algebra with symmetric support and of maximal length (see [5]) admits a $\star$-multiplicative basis. That is, any graded Poisson structure associate to the Cartan grading of a semisimple finite dimensional Lie algebra (in the finite-dimensional setting) or any graded Poisson structure defined either on the split grading of a semisimple separable L^*-algebra or on a semisimple locally finite split Lie algebra (in the infinite-dimensional setting) are examples of Poisson algebras admitting a $\star$-multiplicative basis.

Theorem 2 *Let $\mathfrak{P}$ be a non-commutative Poisson algebra admitting a $\star$-multiplicative basis $\mathfrak{B} = \{e_i\}_{i \in I}$. Then $\mathfrak{P}$ is minimal if and only if the set of indexes I has all of its elements connected.*

Proof Suppose that $\mathfrak{P}$ is minimal. By Theorem 1, $\mathfrak{P} = \mathfrak{P}_{[i]}$ for some $[i] \in I/\sim$. Hence $[i] = I$ and then, any couple of elements of I are connected.

To prove the converse, consider a nonzero ideal $\mathfrak{I}$ of $\mathfrak{P}$ admitting a basis inherited from $\mathfrak{B}$. Then, for a certain $\emptyset \neq I_{\mathfrak{I}} \subset I$, we can write

$$\mathfrak{I} = \bigoplus_{j \in I_{\mathfrak{I}}} \mathbb{F}e_j.$$

Fix some $i_0 \in I_{\mathfrak{I}}$ and, taking into account that $0 \neq e_{i_0} \in \mathfrak{I}$, let us show by induction on n that if $\{i_1, i_2, \ldots, i_n\}$ is a connection from i_0 to some $j \in I$, then $0 \neq e_j \in \mathfrak{I}$.

If $n = 1$, then $j \in \{i_0\} \diamond i_1$. From here, $j \in i_0 \star i_1$ and, by $\star$-multiplicativity of $\mathfrak{B}$,

$$e_j \in [e_{i_0}, \mathfrak{P}] + e_{i_0}\mathfrak{P} + \mathfrak{P}e_{i_0} \subset \mathfrak{I}.$$

Suppose now the assertion holds for any connection $\{i_1, i_2, \ldots, i_n\}$, with n elements, from i_0 to any element of I and consider an arbitrary connection

$$\{i_1, i_2, \ldots, i_n, i_{n+1}\},$$

with $n + 1$ elements, from i_0 to any $j \in I$. Denoting by $J := (\cdots(\{i_0\} \diamond i_1)\cdots) \diamond i_n$, we can take $k \in J$ such that $j \in \{k\} \diamond i_{n+1}$. Then, $j \in \{k\} \star i_{n+1}$ and

$$e_j \in [e_k, \mathfrak{P}] + e_k\mathfrak{P} + \mathfrak{P}e_k \tag{2}$$

by $\star$-multiplicativity of $\mathfrak{B}$.

Taking into account that $k \in J$, we have that $\{i_1, i_2, \ldots, i_n\}$ is a connection with n elements from i_0 to k and, by induction hypothesis, $e_k \in \mathfrak{I}$. From here and Eq. (2) we get $e_j \in \mathfrak{I}$ as desired.

Since given any $j \in I$ we know that i_0 is connected to j, we get by the above that $\mathbb{F}e_j \subset \mathfrak{I}$. Then, $\mathfrak{P} = \bigoplus_{j \in I} \mathbb{F}e_j \subset \mathfrak{I}$ and so $\mathfrak{I} = \mathfrak{P}$. □

Theorem 3 *Let $\mathfrak{P}$ be a non-commutative Poisson algebra admitting a $\star$-multiplicative basis $\mathfrak{B} = \{e_i\}_{i \in I}$. Then*

$$\mathfrak{P} = \bigoplus_k \mathfrak{I}_k$$

is the orthogonal direct sum of the family of its minimal ideals, each one admitting a $\star$-multiplicative basis inherited from $\mathfrak{B}$.

Proof By Theorem 1 we have that $\mathfrak{P}$ is the orthogonal direct sum of the ideals $\mathfrak{P}_{[i]}$.

We wish to apply Theorem 2 to any $\mathfrak{P}_{[i]}$. So, we are going to verify that for any $\mathfrak{P}_{[i]}$ the basis $\mathfrak{B}_{[i]} = \{e_j : j \in [i]\}$ is $\star$-multiplicative and that $[i]$ has all of its elements $[i]$-connected (connected through connections contained in $[i] \dot{\cup} \overline{[i]}$).

We clearly have that $\mathfrak{P}_{[i]}$ admits to $\mathfrak{B}_{[i]}$ as a $\star$-multiplicative basis as consequence of having a basis inherited from $\mathfrak{B}$ and the fact that $[\mathfrak{P}_{[i]}, \mathfrak{P}_{[j]}] = \{0\}$ and $\mathfrak{P}_{[i]}\mathfrak{P}_{[j]} = \{0\}$ whenever $[i] \neq [j]$.

Finally, consider any connection from i to j such that $i \neq j$,

$$\{i_1, i_2, \ldots, i_n\}. \tag{3}$$

Let us observe that any $k \in \{i\} \diamond i_1$ satisfies $k \in [i]$ through the connection $\{i_1\}$. Then we have two possibilities:

- If $i_1 \in I$ we have that $k \in \{i\} \diamond i_1 = \{i_1\} \diamond i$. So, the set $\{i\}$ is a connection from i_1 to k and then $i_1 \in [i]$.
- If $i_1 \in \bar{I}$, by Lemma 1 we have $i \in \{k\} \diamond \bar{i}_1$. By arguing as above we get $\bar{i}_1 \in [i]$ and so $i_1 \in \overline{[i]}$.

By iterating this process we obtain that all of the elements in the connection (3) are contained in $[i] \dot{\cup} \overline{[i]}$. That is, $[i]$ has all of its elements $[i]$-connected. From the above, we can apply Theorem 2 to any $\mathfrak{P}_{[i]}$ so as to conclude $\mathfrak{P}_{[i]}$ is minimal.

Then, we have that the decomposition

$$\mathfrak{P} = \bigoplus_{[i] \in I/\sim} \mathfrak{P}_{[i]}$$

satisfies the assertions of the theorem. □

Acknowledgements We would like to thank the referee for the detailed reading of this work and for the suggestions which have improved the final version of the same.

References

1. Aicardi, F.: Projective geometry from Poisson algebras. J. Geom. Phys. **61**(8), 1574–1586 (2011)
2. Avan, J., Doikou, A.: Boundary lax pairs from non-ultra-local Poisson algebras. J. Math. Phys. **50**(11), 113512–113519 (2009)
3. Calderón, A.J.: On graded associative algebras. Rep. Math. Phys. **69**(1), 75–86 (2012)
4. Calderón, A.J.: On the structure of split non-commutative Poisson algebras. Linear Multilinear Algebra **60**(7), 775–785 (2012)
5. Calderón, A.J.: On extended graded Poisson algerbras. Linear Algebra Appl. **439**, 879–892 (2013)
6. Calderón, A.J.: (Anti)commutative algebras with a multiplicative basis. Bull. Aust. Math. Soc. **91**(2), 211–218 (2015)
7. Calderón, A.J., Navarro, F.J.: Arbitrary algebras with a multiplicative basis. Linear Algebra Appl. **498**, 106–116 (2016)
8. Calderón, A.J., Sánchez, J.M.: Split Leibniz algebras. Linear Algebra Appl. **436**(6), 1648–1660 (2012)
9. Calderón, A.J., Sánchez, J.M.: On the structure of split Lie color algebras. Linear Algebra Appl. **436**(2), 307–315 (2012)
10. Cohn, P.M.: Basic Algebra: Groups, Rings, and Fields. Springer, London (2003)
11. Hone, A.N., Petrera, M.: Three-dimensional discrete systems of Hirota-Kimura type and deformed Lie-Poisson algebras. J. Geom. Mech. **1**(1), 55–85 (2009)
12. Kubo, F.: Finite-dimensional non-commutative Poisson algebras. J. Pure Appl. Algebra **113**(3), 307–314 (1996)
13. Kubo, F.: Non-commutative Poisson algebra structures on affine Kac-Moody algebras. J. Pure Appl. Algebra **126**(1–3), 267–286 (1998)
14. Kubo, F.: Finite-dimensional non-commutative Poisson algebras. II. Comm. Algebra **29**(10), 4655–4669 (2001)

Multiplication Algebras: Algebraic and Analytic Aspects

Miguel Cabrera García and Ángel Rodríguez Palacios

Abstract Applications of multiplication algebras to the algebraic and analytic strengthenings of primeness and semiprimeness of (possibly non-associative) algebras are fully surveyed, and complete normed complex algebras whose closed multiplication algebras satisfy the von Neumann inequality are studied in detail.

Keywords Non-associative algebras · Multiplication algebras · Semiprime algebras · Ultraprime normed algebras · von Neumann inequality

1 Introduction

In this chapter we survey several results appeared in the literature, which involve multiplication algebras of (possibly non-associative) algebras, and provide some new results in this regard. To be more precise, we deal with some strengthenings of (semi)primeness consisting of algebraic or analytic appropriate requirements, the latest in the normed setting, and with the complete normed complex algebras whose closed multiplication algebras satisfy the von Neumann inequality.

Section 2 is devoted to establish some classic algebraic notions (like those of annihilator, center, centroid, central closure, etc.). Moreover we review how the notion of annihilator of an ideal in an algebra A gives rise to an algebraic closure, called the π-closure, in the lattice of all ideals of A. This allows us to consider a first algebraic strengthening of the semipriminess, which gives birth to the so-called π-complemented algebras.

In Sect. 3 we deal with multiplicatively semiprime (respectively, prime) algebras (Definition 2). Without enjoying their name, these algebras were first

M. Cabrera García (✉) · Á. Rodríguez Palacios
Departamento de Análisis Matemático, Facultad de Ciencias, Universidad de Granada, 18071 Granada, Spain
e-mail: cabrera@ugr.es

Á. Rodríguez Palacios
e-mail: apalacio@ugr.es

M. Siles Molina et al. (eds.), *Associative and Non-Associative Algebras and Applications*, Springer Proceedings in Mathematics & Statistics 311,
https://doi.org/10.1007/978-3-030-35256-1_6

studied by Jacobson [30] and Albert [1] in a finite dimensional context. Nowadays, without any restriction on the dimension, these algebras are provided with a well-developed structure theory. Moreover they compose a very large class containing all semiprime (respectively, prime) associative algebras, many 'nearly associative' semiprime (respectively, prime) algebras, and all generalized annihilator (respectively, topologically simple) normed algebras. Recently, the interest in these algebras has been greatly increased thanks to the fact that they turned out to be the appropriate framework for translating to non-associative setting many results of the associative PI and GPI theories [15].

In Sect. 4 we provide a multiplicative non-associative characterization of (associative and commutative) integral domains, and obtain a normed variant of this characterization. Both results are new.

The starting point for our review of analytical strengthenings of primeness, included in Sect. 5, is Mathieu's work [32–34], where associative ultraprime normed algebras are introduced, and the fact that prime C^*-algebras are ultraprime is proved. Combining this fact with the Zel'manovian classification theorem of prime JB^*-algebras [27], it can be shown that prime non-commutative JB^*-algebras are ultraprime. It is also proved by Mathieu that ultraprime normed complex associative algebras are centrally closed. The extension of this result to the general non-associative setting needs to introduce the notion of a 'totally prime' normed algebra [20], which cannot be formulated without involving multiplication algebras. Indeed, as proved in [20], ultraprime normed algebras are totally prime, and totally prime normed complex algebras are centrally closed.

In Sect. 6 we consider the so-called 'totally multiplicatively prime' normed algebras, which are introduced as the natural normed strengthening of the multiplicative primeness. As proved in [18], topologically simple semi-H^*-algebras are totally multiplicatively prime (although they need not be ultraprime), and totally multiplicatively prime normed algebras are totally prime.

In Sect. 7 we discuss a classical topic in the theory of complete normed unital associative complex algebras, namely that of the von Neumann inequality. To be more precise, we deal with complete normed complex (possibly non-associative) algebras whose closed multiplication algebras satisfy the von Neumann inequality. In this regard we review in detail the last section of [22], where this matter is discussed for the first time, and we provide some new results. Among the results shown in [22], we emphasize the one asserting that a nonzero non-commutative JB^*-algebra A is actually a commutative C^*-algebra if and only if the closed multiplication algebra of A satisfies the von Neumann inequality. This is applied to prove a theorem according to which the commutative C^*-algebras are precisely those complete normed complex algebras having an approximate identity bounded by 1, and whose closed multiplication algebra is a C^*-algebra. This theorem is new.

For those notions on (possibly non-associative and occasionally normed) algebras, appearing in this chapter but not defined in it, the reader is referred to [23]. This book has been quoted often in the current chapter, although, as it happens with most books, not always the results taken from it are original. The reader interested in the precise paternity of each of these results may consult the corresponding subsections of 'Historical notes and comments' in [23].

2 (Semi)prime Algebras

Throughout this chapter, we consider (possibly non-associative and non-unital) algebras A over a field $\mathbb{K}$ of characteristic 0, and we understand that $\mathbb{K} = \mathbb{R}$ or $\mathbb{C}$ whenever A is a normed algebra.

Let A be an algebra. The *annihilator* of A is defined by

$$\mathrm{Ann}(A) := \{a \in A : aA = Aa = 0\}.$$

The algebra A is said to be *semiprime* if, for each ideal I of A, the condition $I^2 = 0$ implies $I = 0$. We recall also that A is said to be *prime* if $A \neq 0$ and the product of two arbitrary nonzero ideals of A is nonzero. A is said to be *simple* if A has nonzero product and 0 is the unique proper ideal of A. It is clear that

$$A \text{ simple} \;\Rightarrow\; A \text{ prime} \;\Rightarrow\; A \text{ semiprime} \;\Rightarrow\; \mathrm{Ann}(A) = 0.$$

Given a subspace X of A, the *core* of X in A is defined as the largest ideal of A contained in X. *Primitive ideals* of A are defined as the cores of maximal modular left ideals of A [23, Subsection 3.6.1], and A is said to be a *(left) primitive* algebra whenever $A \neq 0$ and zero is a primitive ideal of A [23, Definition 3.6.12]. As proved in [23, Proposition 3.6.16], we have

$$A \text{ primitive} \;\Rightarrow\; A \text{ prime}.$$

In the case that A is an associative algebra, the above definition is in agreement with the usual one in the literature: A is (left) primitive if and only if A has a faithful irreducible left A-module [23, Definition 3.6.35 and Theorem 3.6.38(i)]. It is also interesting to note that in the associative context the (semi)primeness can be described involving elements. Indeed, if A is an associative algebra, then, denoting by $M_{a,b}$ $(a, b \in A)$ the operator on A defined by $M_{a,b}(x) := axb$ (and for shortness $M_a := M_{a,a}$), we have

$$A \text{ is semiprime} \;\Leftrightarrow\; a \in A \text{ and } M_a = 0 \text{ imply } a = 0,$$

and

$$A \text{ is prime} \;\Leftrightarrow\; a, b \in A \text{ and } M_{a,b} = 0 \text{ imply either } a = 0 \text{ or } b = 0.$$

Let A be an algebra and let $L(A)$ stand for the algebra of all linear operators on A. Given a in A, we shall denote by L_a and R_a the operators of left and right multiplication by a on A, respectively. The *(unital) multiplication algebra* of A is defined as the subalgebra $\mathcal{M}(A)$ of $L(A)$ generated by the identity operator I_A and the set $\{L_a, R_b : a, b \in A\}$. It is clear that A is a left $\mathcal{M}(A)$-module for the module multiplication given by the evaluation, and that the submodules of A are nothing other than the ideals of A. Therefore, if A is a simple algebra, then A becomes a faithful irreducible left $\mathcal{M}(A)$-module, and hence

$$A \text{ simple} \quad \Rightarrow \quad \mathcal{M}(A) \text{ primitive.}$$

As we shall see immediately below, semiprimeness is closely related to the notion of the annihilator of ideals. Let I be an ideal of an algebra A. Then the set $\{a \in A : aI = 0 = Ia\}$ is a subspace of A whose core in A is called the *annihilator of* I relative to A, and is denoted by $\mathrm{Ann}(I)$.

§ 1 Let A be an algebra. The annihilator operation on the lattice of all ideals of A induces a 'closure operation' in that lattice. Indeed, for any ideal I of A, the *π-closure* $\overline{I}^{\pi}$ of I is defined by

$$\overline{I}^{\pi} := \mathrm{Ann}(\mathrm{Ann}(I)),$$

and moreover, for all ideals I, J of A, we have

(i) $I \subseteq J$ implies $\overline{I}^{\pi} \subseteq \overline{J}^{\pi}$.
(ii) $I \subseteq \overline{I}^{\pi}$.
(iii) $\mathrm{Ann}(I) = \mathrm{Ann}(\overline{I}^{\pi})$, and hence $\overline{I}^{\pi} = \overline{(\overline{I}^{\pi})}^{\pi}$.

The next fact is of straightforward verification.

Fact 1 *Let A be an algebra. Then the following conditions are equivalent:*

(i) A is semiprime.
(ii) For each π-closed proper ideal I of A, we have $\mathrm{Ann}(I) \neq 0$.
(iii) $A = \overline{I \oplus \mathrm{Ann}(I)}^{\pi}$ *for every ideal I of A.*

Given elements a, b, c in an algebra A, we set $[a, b] := ab - ba$ for the commutator and $[a, b, c] := (ab)c - a(bc)$ for the associator. Recall that the *centre* Z_A of A is defined by

$$Z_A := \{z \in A : [z, A] = [z, A, A] = [A, z, A] = [A, A, z] = 0\}.$$

Z_A is a commutative and associative subalgebra of A [23, Definition 2.5.31]. The *centroid* Γ_A of A is defined as the subalgebra of $L(A)$ consisting of all *centralizers* on A, that is the linear operators $T : A \to A$ satisfying

$$T(ab) = T(a)b = aT(b) \text{ for all } a, b \in A.$$

The algebra A is said to be *central* over $\mathbb{K}$ whenever $\Gamma_A = \mathbb{K}I_A$ [23, Definition 1.1.10]. The centre is closely related to the centroid. Indeed, by assigning to each

$z \in Z_A$ the operator of multiplication by z on A, we are provided with a natural algebra homomorphism $\Phi : Z_A \to \Gamma_A$ such that $\ker(\Phi) = \mathrm{Ann}(A)$. Moreover Φ is surjective if and only if A has a unit, if and only if Φ is bijective. Thus, when A is unital, A is central if and only if $Z_A = \mathbb{K}\mathbf{1}$ [23, Remark 2.5.34].

We need to introduce an enlargement of the centroid, namely the so-called extended centroid. To this end, we recall that an ideal D of an algebra A is said to be *essential* if $D \cap I \neq 0$ for every nonzero ideal I of A.

For semiprime associative algebras, the existence of algebras of quotients provide us with an elegant and easy way of introducing the concepts of extended centroid and central closure. Given a semiprime associative algebra A, the so called *symmetric Martindale algebra of quotients* $Q(A)$ of A is probably the most confortable algebra of quotients of A. This algebra has the following abstract characterization.

Proposition 1 ([23, Proposition 6.1.127]) *Let A be a semiprime associative algebra. Then $Q(A)$ is the largest associative algebra extension Q of A satisfying the following conditions:*

(1) for each $q \in Q$ there exists an essential ideal D of A such that

$$qD + Dq \subseteq A.$$

(2) if $q \in Q$ satisfies $qD = 0$ for some essential ideal D of A, then $q = 0$.

The *extended centroid* of A, denoted by C_A, is defined as the centre of $Q(A)$, and the *central closure* of A, denoted by Q_A, is defined as the C_A-subalgebra of $Q(A)$ generated by A.

In the general non-associative setting, the absence of algebras of quotients complicates the introduction of the above notions. Actually they were introduced (in a constructive way) and studied in [26] for prime algebras, and in [5] for semiprime algebras (see also [23]). To do the following abstract characterization understandable, recall that, by a *partially defined centralizer* on an algebra A we mean a linear mapping (say f) from a nonzero ideal of A (say $\mathrm{dom}(f)$) to A satisfying $f(xa) = f(x)a$ and $f(ax) = af(x)$ for all $x \in \mathrm{dom}(f)$ and $a \in A$.

Theorem 1 ([13, Theorem 2.4]) *Let A be a semiprime algebra. The central closure Q_A of A satisfies:*

(P1) Q_A is an algebra extension of A, and Q_A is generated by A as a Γ_{Q_A}-algebra. As a consequence, $\mathcal{M}(A)$ embeds into $\mathcal{M}(Q_A)$ in a natural way.

(P2) For each $q \in Q_A$, there exists an essential ideal D of A such that

$$D\mathcal{M}(A)(q) + \mathcal{M}(A)(q)D \subseteq A.$$

(P3) If $q \in Q_A$ satisfies $D\mathcal{M}(A)(q) = 0$ or $\mathcal{M}(A)(q)D = 0$, for some essential ideal D of A, then $q = 0$.

(P4) For each partially defined centralizer f from an ideal I of A into A, there exists an element $\lambda \in \Gamma_{Q_A}$ such that $f(x) = \lambda x$ for every $x \in I$.

Furthermore, Γ_{Q_A} *coincides with the extended centroid* C_A *of* A*, and properties* (P1)–(P4) *characterize* Q_A *up to isomorphism.*

Let A be a semiprime algebra. Then, according to the above theorem, we have natural embeddings

$$\Gamma_A \hookrightarrow C_A \text{ and } A \hookrightarrow Q_A.$$

A is called *centrally closed* if $C_A = \Gamma_A$, or equivalently if $Q_A = A$. Anyway, C_A becomes a von Neumann regular unital commutative and associative algebra, and Q_A is a centrally closed semiprime algebra with $C_{Q_A} = C_A$. Moreover, if A is a prime algebra, then C_A is a field extension of $\mathbb{K}$ [23, Proposition 2.5.44].

To conclude this section, let us introduce a first strengthening of semiprimeness.

Definition 1 An algebra A is said to be π*-complemented* if for every π-closed ideal I of A there exists a π-closed ideal J of A such that $A = I \oplus J$.

A structure theory of π-complemented algebras was developed in [9]. We include here the following characterization taken from [11, Theorem 1.8].

Theorem 2 *Let A be an algebra. Then the following conditions are equivalent:*

(i) *A is* π*-complemented.*
(ii) *A is semiprime, and every idempotent in* C_A *lies in* Γ_A.

3 Multiplicatively (Semi)prime Algebras

Definition 2 An algebra A is said to be *multiplicatively semiprime* (respectively *multiplicatively prime*) if both A and $\mathcal{M}(A)$ are semiprime (respectively prime).

Finite-dimensional algebras with a semiprime multiplication algebra were first considered by Jacobson [30], who proved the following theorem.

Theorem 3 (Jacobson's theorem, [23, Theorem 8.1.144]) *For a finite-dimensional algebra A, the following conditions are equivalent:*

(i) $\mathcal{M}(A)$ *is semiprime.*
(ii) $A = \oplus_{i=0}^{n} B_i$ *is a direct sum of ideals, where one of them is an algebra with zero product, and the others are simple algebras.*

Moreover, if the above conditions are fulfilled, then $\mathcal{M}(A) \cong \oplus_{i=0}^{n} \mathcal{M}(B_i)$.

As a consequence of Theorem 3, *a finite-dimensional algebra is a direct sum of simple ideals if and only if it is multiplicatively semiprime*. Motivated by this fact, Albert [1] (see also [31, pp. 1090–1091]) proposed a notion of radical of a finite-dimensional algebra, which is known today as the *Albert radical* (see [23, § 4.4.82]).

Theorem 4 (Albert's theorem) *If A is a finite-dimensional algebra, then there exists a smallest ideal R of A such that A/R is multiplicatively semiprime.*

§ 2 Moreover, Albert gave the first example of a prime (three-dimensional unital) algebra which is not multiplicatively semiprime [23, Example 8.1.24].

Without any restriction of dimension, multiplicatively semiprime algebras were considered in [13, 16, 17, 35], where the behavior of these algebras concerning extended centroid and central closure was studied, obtaining among others the following.

Theorem 5 *If A is a multiplicatively semiprime algebra, then* (in a natural way) *$C_A = C_{\mathcal{M}(A)}$, and $Q_{\mathcal{M}(A)} = \mathcal{M}_{C_A}(Q_A)$. As a consequence, Q_A is also a multiplicatively semiprime algebra.*

A structure theory for multiplicatively semiprime algebras was developed in [8]. We do not enter this theory, and limit ourselves to say that the cornerstone of this theory consists of the introduction of a new algebraic closure of the ideals in an algebra, called the ε-closure.

§ 3 Let A be an algebra. For subspaces S of A and $\mathcal{N}$ of $\mathcal{M}(A)$, we define

$$S^{\mathrm{ann}} := \{F \in \mathcal{M}(A) \,:\, F(S) = 0\} \text{ and } \mathcal{N}_{\mathrm{ann}} := \{a \in A \,:\, \mathcal{N}(a) = 0\}.$$

The *ε-closure* $\overline{S}^{\varepsilon}$ of a subspace S of A is defined by

$$\overline{S}^{\varepsilon} := (S^{\mathrm{ann}})_{\mathrm{ann}}.$$

For subspaces S, S_1, S_2 of A it is immediate to verify that

(1) $S_1 \subseteq S_2$ implies $\overline{S_1}^{\varepsilon} \subseteq \overline{S_2}^{\varepsilon}$.
(2) $S \subseteq \overline{S}^{\varepsilon}$.
(3) $S^{\mathrm{ann}} = (\overline{S}^{\varepsilon})^{\mathrm{ann}}$, and hence $\overline{S}^{\varepsilon} = \overline{(\overline{S}^{\varepsilon})}^{\varepsilon}$.

It is clear that, if I is an ideal of A, and if $\mathcal{P}$ is an ideal of $\mathcal{M}(A)$, then I^{ann} and $\mathcal{P}_{\mathrm{ann}}$ are ideals of $\mathcal{M}(A)$ and A, respectively. As a consequence, $\overline{I}^{\varepsilon}$ is an ideal of A.

Proposition 2 *The following assertions hold:*

(1) [23, Proposition 8.2.28(i)(a)] *For any ideal I of an algebra A, we have $\overline{I}^{\varepsilon} \subseteq \overline{I}^{\pi}$.*
(2) For any ideal I of a normed algebra A, we have $\overline{I} \subseteq \overline{I}^{\varepsilon}$.

Now the next theorem provides us with different characterizations of the multiplicative semiprimeness.

Theorem 6 ([23, Theorem 8.2.31]) *Let A be an algebra. Then the following conditions are equivalent:*

(i) *A is multiplicatively semiprime.*
(ii) *A is semiprime and* $\mathrm{Ann}(I^{\mathrm{ann}}) = \mathrm{Ann}(I)^{\mathrm{ann}}$ *for every ideal I of A.*
(iii) *A is semiprime and* $\overline{I}^{\varepsilon} = \overline{I}^{\pi}$ *for every ideal I of A.*
(iv) *A is semiprime and for each ε-closed proper ideal I of A, we have* $\mathrm{Ann}(I) \neq 0$.
(v) $A = \overline{I \oplus \mathrm{Ann}(I)}^{\varepsilon}$ *for every ideal I of A.*

As a matter of fact, *(semi)prime associative algebras are multiplicatively (semi) prime* [16, 17], and a similar result holds for most nearly associative algebras (sometimes under some strengthening of the (*semi*)primeness) [8, 10, 12, 14, 18, 24]. As samples, let us formulate Theorems 7 and 8 which follow.

Theorem 7 ([10]) *Let A be a semiprime associative algebra with an involution* $*$. *Then* K/Z_K *(the Lie algebra of all skew elements of A quotient by its annihilator) is a multiplicatively semiprime Lie algebra.*

One could expect the above theorem to remain true whenever 'semiprime associative' is replaced with 'prime associative' and 'multiplicatively semiprime Lie' is replaced with 'multiplicatively prime Lie'. However, this not correct. Indeed, the choice $A = M_4(\mathbb{C})$ with the transpose involution becomes a counter-example. 'Essentially', this counter-example is the unique possible.

Theorem 8 ([14]) *Let A be a non-commutative Jordan algebra, and assume that A is strongly prime in the sense of* [41]. *Then A is multiplicatively prime.*

Definition 3 A normed algebra A is said to be *generalized annihilator* if it is semiprime and if $\mathrm{Ann}(I) \neq 0$ for every closed proper ideal I of A.

Generalized annihilator normed algebras were introduced and studied in [25, 44] in the associative context, and in [28] (see also [8]) in the general non-associative context. As the following theorem shows, the consideration of generalized annihilator normed algebras becomes a first strengthening of semiprimeness through analytical requirements.

Theorem 9 ([23, Theorem 8.2.33]) *Let A be a normed algebra. Then, the following conditions are equivalent:*

(i) *A is a generalized annihilator normed algebra.*
(ii) *A is multiplicatively semiprime and every ε-dense ideal of A is dense.*

The 'atoms' in the structure theory of multiplicatively semiprime algebras are nothing other than the multiplicatively prime algebras. Examples of multiplicatively prime algebras are all simple algebras. In fact, the multiplication algebra of a simple algebra is primitive, hence prime. In the normed case, we have the following better result: *Topologically simple normed algebras are multiplicatively prime* [23, Corollary 8.1.26].

Let A be an algebra. For $F \in \mathcal{M}(A)$ and $a \in A$, we denote by $W_{F,a}$ the linear operator from $\mathcal{M}(A)$ to A defined by $W_{F,a}(T) := FT(a)$ for every $T \in \mathcal{M}(A)$. An useful ideal-free characterization of the multiplicative primeness is provided by the following.

Proposition 3 ([16], see also [23, Proposition 8.1.25]) *Let A be an algebra. Then the following conditions are equivalent:*

(i) A is multiplicatively prime.
(ii) A has nonzero product and, whenever $F \in \mathcal{M}(A)$ and $a \in A$ satisfy $W_{F,a} = 0$, we have either $F = 0$ or $a = 0$.
(iii) A has zero annihilator and $\mathcal{M}(A)$ is prime.
(iv) A is semiprime and $\mathcal{M}(A)$ is prime.

Recently, the interest in multiplicatively prime algebras has been greatly increased thanks to the fact that they turn out to be the appropriate framework to translate into the non-associative environment many results of the associative PI and GPI theories [15]. As a sample let us to conclude this section formulating Posner's prime PI-theorem and its non-associative extension.

Theorem 10 (Posner's theorem (1960) [Rowen (1974)]) *Let A be a prime associative algebra. Then the following conditions are equivalent:*

(i) A is a prime PI-algebra.
(ii) Q_A is a simple algebra of finite dimension over C_A.

In this case, C_A is the field of fractions of Γ_A, and Q_A is the algebra of fractions of A with respect to Γ_A.

As a byproduct of a theorem by Regev [36], it is proved in [15] that an associative algebra is PI if and only if so is its multiplication algebra. Therefore the next result becomes the non-associative extension of Posner's theorem.

Theorem 11 ([15]) *Let A be a prime algebra. Then the following conditions are equivalent:*

(i) $\mathcal{M}(A)$ is a prime PI-algebra.
(ii) Q_A is a simple algebra of finite dimension over C_A.

In this case, C_A is the field of fractions of Γ_A, and Q_A is the algebra of fractions of A with respect to Γ_A.

4 An Interlude: A Multiplicative Non-associative Characterization of Integral Domains, and a Normed Variant

One of the clearest strengthenings of the primeness of an algebra consists of requiring that the algebra lacks zero divisors. In this section we consider a natural multiplicative variant of this requirement.

Lemma 1 *Let A be an algebra such that every nonzero operator in $\mathcal{M}(A)$ is injective. Then A is associative and commutative.*

Proof Let a be in A. Since $(L_a - R_a)(a) = a^2 - a^2 = 0$, it follows that $L_a = R_a$, and hence A is commutative. Then, noticing that

$$(L_{a^2} - L_a^2)(a) = a^2a - aa^2 = 0,$$

we see that $L_{a^2} = L_a^2$, and hence A is alternative. Therefore, by [23, Fact 6.1.18], A is associative. □

Note that every vector space endowed with the zero product becomes an algebra A satisfying the requirement in the above lemma.

Proposition 4 *For an algebra A, the following conditions are equivalent:*

(i) $A \neq 0$, $\mathrm{Ann}(A) = 0$, *and every nonzero operator in* $\mathcal{M}(A)$ *is injective.*
(ii) *A is an (associative and commutative) integral domain.*

Proof The implication (i)⇒(ii) follows from Lemma 1.

Suppose that A is an integral domain. Then, clearly,

$$A \neq 0 \text{ and } \mathrm{Ann}(A) = 0.$$

If A is unital, then the proof that (i) holds is concluded by noticing that, by associativity and commutativity, we have $\mathcal{M}(A) = \mathcal{L} := \{L_a : a \in A\}$, and hence every nonzero operator in $\mathcal{M}(A)$ is injective. Otherwise we argue by contradiction. Assume that A is not unital and that (i) does not hold. Then $\mathcal{M}(A) = \mathbb{K}I_A \oplus \mathcal{L}$ and there is a nonzero $a \in A$ such that $I_A - L_a$ is not injective, hence there is $b \in A \setminus \{0\}$ such that $b - ab = 0$, and we have $(c - ca)b = 0$ for every $c \in A$. Since $b \neq 0$, and c is arbitrary in A, and A is an integral domain, we obtain that a is a unit for A, the desired contradiction. □

We recall that a linear operator T on a normed space X is said to be *bounded below* if there is $k > 0$ such that $k\|x\| \leq \|T(x)\|$ for every $x \in X$.

Proposition 5 *For a normed algebra A over* $\mathbb{K}$, *the following conditions are equivalent:*

(i) $A \neq 0$, $\mathrm{Ann}(A) = 0$, *and there is a positive number K such that*

$$K\|F\|\|a\| \leq \|F(a)\| \text{ for all } F \in \mathcal{M}(A) \text{ and } a \in A.$$

(ii) $A \neq 0$, $\mathrm{Ann}(A) = 0$, *and every nonzero operator in* $\mathcal{M}(A)$ *is bounded below.*
(iii) *A is isomorphic to* $\mathbb{C}$, *if* $\mathbb{K} = \mathbb{C}$, *and A is isomorphic to* $\mathbb{R}$ *or* $\mathbb{C}$, *if* $\mathbb{K} = \mathbb{R}$.

Proof The implications (i)⇒(ii) and (iii)⇒(i) are clear.

Suppose that condition (ii) is fulfilled. Then every nonzero operator in $\mathcal{M}(A)$ is injective, and hence, by Lemma 1, A is associative and commutative. Moreover, since A has zero annihilator, the operator L_a is bounded below whenever a is any nonzero element of A. But this means that A has no nonzero topological divisor of zero. Therefore condition (iii) holds thanks to [23, Proposition 2.5.49(i)] and [23, Theorem 2.5.50]. □

Note that condition (i) in the above proposition could be seen as the natural 'multiplicative' variant of the notion of a nearly absolute-valued algebra, which shall be introduced later.

5 Ultraprime Normed Algebras

To begin this section, let us summarize those aspects of the theory of ultraproducts which are needed for our purpose (cf. [23, § 2.8.58 and 2.8.61]).

§ 4 Let $\mathcal{U}$ be an ultrafilter on a non-empty set I. Given a family $\{X_i\}_{i\in I}$ of normed spaces, we may consider the normed space $\bigoplus_{i\in I}^{\ell_\infty} X_i$ ℓ_∞-sum of this family, and the closed subspace $N_{\mathcal{U}}$ of $\bigoplus_{i\in I}^{\ell_\infty} X_i$ given by

$$N_{\mathcal{U}} := \{\{x_i\} \in \bigoplus_{i\in I}^{\ell_\infty} X_i : \lim_{\mathcal{U}} \|x_i\| = 0\}.$$

The (*normed*) *ultraproduct* of the family $\{X_i\}_{i\in I}$ (relative to the ultrafilter $\mathcal{U}$) is defined as the quotient normed space $(\bigoplus_{i\in I}^{\ell_\infty} X_i)/N_{\mathcal{U}}$, and is denoted by $(X_i)_{\mathcal{U}}$. Denoting by (x_i) the element in $(X_i)_{\mathcal{U}}$ containing a given family $\{x_i\} \in \bigoplus_{i\in I}^{\ell_\infty} X_i$, it is easy to verify that $\|(x_i)\| = \lim_{\mathcal{U}} \|x_i\|$. If X_i is equal to a given normed space X for every $i \in I$, then the ultraproduct $(X_i)_{\mathcal{U}}$ is called the (*normed*) *ultrapower* of X (relative to $\mathcal{U}$) and is denoted by $X_{\mathcal{U}}$. In this case, the mapping $x \to x^\wedge$ from X into $X_{\mathcal{U}}$, where $x^\wedge = (x_i)$ with $x_i = x$ for all $i \in I$, is an isometric linear embedding.

§ 5 Let $\mathcal{U}$ be an ultrafilter on a non-empty set I. Given a family $\{A_i\}_{i\in I}$ of normed algebras, the normed space $(A_i)_{\mathcal{U}}$ shall be considered without notice as a new normed algebra under the (well-defined) product

$$(a_i)(b_i) := (a_i b_i).$$

Let us also note that, if A is any normed algebra, then the natural embedding $A \hookrightarrow A_{\mathcal{U}}$ becomes an algebra homomorphism.

An ultrafilter $\mathcal{U}$ on a set I is said to be *countably incomplete* if it is not closed under countable intersections of its elements. We say that a normed algebra A is *ultraprime* if some ultrapower $A_{\mathcal{U}}$, with respect to a countably incomplete ultrafilter $\mathcal{U}$, is a prime algebra.

Ultraprime normed algebras were introduced by Mathieu in [32] (see also [33]) in the associative setting. He provided the nice characterization of the ultraprimeness for normed associative algebras given by the equivalence (i)$\Leftrightarrow$(ii) in the following.

Proposition 6 ([23, Proposition 6.1.62]) *For a normed associative algebra A the following conditions are equivalent:*

(i) A is ultraprime.
(ii) There exists a positive number K such that $\|M_{a,b}\| \geq K\|a\|\|b\|$ for all a, b in A.
(iii) Every ultrapower of A is prime.

We recall that a nonzero normed (possibly non-associative) algebra A is said to be *nearly absolute-valued* if there exists $r > 0$ such that

$$r\|a\|\|b\| \leq \|ab\| \text{ for all } a, b \in A.$$

If $r = 1$, then A is said to be an *absolute-valued* algebra.

If $\mathcal{U}$ is an ultrafilter on a non-empty set I, and if A is a nearly absolute-valued algebra, then the equality $\|(a_i)\| = \lim_{\mathcal{U}} \|a_i\|$ shows that the ultrapower $A_{\mathcal{U}}$ of A is also a nearly absolute-valued algebra. Since nearly absolute-valued algebras are prime, it follows that they are in fact ultraprime.

From now on, we shall understand that a prime algebra A over $\mathbb{K}$ is centrally closed whenever $C_A = \mathbb{K}$. A relevant result obtained by Mathieu in [32] (see also [33]) is the following.

Fact 2 *Ultraprime normed associative complex algebras are centrally closed.*

In [20], with the aim of proving a general non-associative extension of Mathieu's Fact 2, we introduced the notion of a totally prime normed algebra as follows.

Given elements a, b in an algebra A, we denote by $N_{a,b}$, the bilinear mapping from $\mathcal{M}(A) \times \mathcal{M}(A)$ into A defined by $N_{a,b}(F, G) = F(a)G(b)$ for all F, G in $\mathcal{M}(A)$. Keeping in mind that the ideal generated by any element $a \in A$ agrees with $\mathcal{M}(A)(a)$, we realize that A is a prime algebra if and only if it is nonzero and, the conditions $a, b \in A$ and $N_{a,b} = 0$ imply $a = 0$ or $b = 0$. Therefore, for a normed algebra A, a reasonable strengthening of primeness consists of requiring that A be different from zero, and that there be a positive number L such that

$$\|N_{a,b}\| \geq L\|a\|\|b\| \text{ for all } a, b \text{ in } A.$$

(Here, for the computation of $\|N_{a,b}\|$ we consider $\mathcal{M}(A)$ as a normed space under the operator norm.) Normed algebras satisfying the above requirements shall be called *totally prime*.

The two theorems which follow become the main results in [20].

Theorem 12 ([23, Theorem 6.1.60]) *Every totally prime normed complex algebra is centrally closed.*

Theorem 13 ([23, Theorem 6.1.63]) *Every ultraprime normed algebra is totally prime.*

Now, as desired, as a straightforward consequence of Theorems 12 and 13 we obtain the following.

Corollary 1 *Ultraprime normed complex algebras are centrally closed.*

Since nearly absolute-valued algebras are ultraprime, we obtain the following.

Corollary 2 *Nearly absolute-valued complex algebras are centrally closed.*

In the particular case of absolute-valued algebras, an autonomous ultrapower-free proof of Corollary 2 can be seen in [38, pp. 34–36], where the reader can also find an example of an absolute-valued central real algebra which is not centrally closed.

One of the main results by Mathieu in [32] (see also [34]) is the following.

Theorem 14 ([23, Theorem 6.1.69]) *Let A be a prime C^*-algebra. Then A is ultraprime. More precisely, for all $a, b \in A$ we have that $\|M_{a,b}\| = \|a\|\|b\|$.*

Combining Corollary 1 with Theorem 14 we obtain that *prime C^*-algebras are centrally closed.* This result was rediscovered by Ara [2], who actually provided a characterization of the extended centroid of an arbitrary C^*-algebra as a certain ring of partially defined continuous functions on the primitive spectrum of the algebra (see also [3, pp. 32–33 and Theorem 2.2.8]).

§ 6 Given an element a in an algebra A, we denote by U_a the mapping

$$b \to a(ab + ba) - a^2 b$$

from A to A.

Definition 4 By a *non-commutative JB^*-algebra* we mean a complete normed non-commutative Jordan complex $*$-algebra (*say* A) such that the equality $\|U_a(a^*)\| = \|a\|^3$ holds for every $a \in A$. Non-commutative JB^*-algebras which are commutative are simply called *JB^*-algebras*.

It is easy to realize that C^*-algebras are non-commutative JB^*-algebras. More precisely, C^*-algebras are those non-commutative JB^*-algebras which are associative. Even more, according to [23, Theorems 2.3.32, 3.3.11, 5.9.9, and Corollary 5.9.12], non-commutative JB^*-algebras become the widest possible non-associative generalization of C^*-algebras.

§ 7 According to [23, §6.1.73], *the class of non-commutative JB^*-algebras (respectively, JB^*-algebras or C^*-algebras) is closed under ultrapowers.*

Using that the class of JB^*-algebras is closed under ultrapowers, Mathieu's Theorem 14, and the classification theorem of prime JB^*-algebras [27] (see also [23, Theorem 6.1.57]), the next result was obtained in [27].

Proposition 7 ([23, Proposition 6.1.74]) *Prime JB^*-algebras are ultraprime.*

We have unified Theorem 14 and Proposition 7 by means of the following.

Theorem 15 ([23, Theorem 6.1.78]) *Prime non-commutative JB^*-algebras are ultraprime.*

Combining Corollary 1 with Theorem 15 above, we obtain the following.

Corollary 3 *Prime non-commutative JB^*-algebras are centrally closed.*

In [39], Ara's work already commented was generalized to the setting of non-commutative JB^*-algebras, obtaining by the first time Corollary 3 above.

For any algebra A, let $(a, b) \to U_{a,b}$ stand for the unique symmetric bilinear mapping from $A \times A$ to $L(A)$ such that $U_{a,a} = U_a$ for every $a \in A$. It is easy to see that a sufficient condition for A to be prime is that,

$$\textit{the conditions } a, b \in A \textit{ and } U_{a,b} = 0 \textit{ imply } a = 0 \textit{ or } b = 0. \tag{1}$$

(By the way, according to [6], for a Jordan algebra A, condition (1) is equivalent to A be prime and non-degenerate.) Therefore, if A is a normed algebra, and if there is a positive constant K such that $\|U_{a,b}\| \geq K\|a\|\|b\|$ for all $a, b \in A$, then A is ultraprime. In this way, a refinement of Proposition 7 can be found in [21], where it is proved that *there is a universal positive constant K such that, for every prime JB^*-algebra A and all $a, b \in A$, we have $\|U_{a,b}\| \geq K\|a\|\|b\|$*. This implies that *every ultraproduct of prime JB^*-algebras is a prime JB^*-algebra.* Actually, Bunce et al. [7] prove that, for elements x, y in a prime JB^*-triple, the inequality $\|Q_{x,y}\| \geq \frac{1}{6}\|x\|\|y\|$ holds. This implies that $\frac{1}{6}$ is an admissible value of the universal constant K above. We emphasize this in the following

Corollary 4 *Let A be a prime non-commutative JB^*-algebra. Then*

$$\|U_{a,b}\| \geq \frac{1}{6}\|a\|\|b\| \textit{ for all } a, b \in A.$$

6 Totally Multiplicatively Prime Normed Algebras

Totally multiplicatively prime normed algebras were introduced in [35] (see also [18]) as the natural normed strengthening of multiplicative primeness. Indeed, in view of the equivalence (i)⇔(ii) in Proposition 3, for a normed algebra A, a reasonable strengthening of the multiplicative primeness consists of requiring that A have nonzero product and that there be a positive number K such that

$$K\|F\|\|a\| \leq \|W_{F,a}\| \text{ for all } F \in \mathcal{M}(A) \text{ and } a \in A.$$

Normed algebras satisfying the above requirements shall be called *totally multiplicatively prime*.

The next two results establish the relation between the class of totally multiplicatively prime normed algebras and the forerunner classes of normed algebras, namely that of ultraprime normed algebras and that of totally prime normed algebras, surveyed in the above section.

Proposition 8 ([23, Proposition 8.1.27]) *Let A be a totally multiplicatively prime normed algebra. Then A is totally prime.*

Proposition 9 ([23, Proposition 8.1.37]) *Let A be a normed algebra. We have:*

(1) If A is totally multiplicatively prime, then $\mathcal{M}(A)$ is ultraprime.
(2) If $\mathcal{M}(A)$ is ultraprime, and if there exists a positive constant K such that

$$K\|a\| \leq \max\{\|L_a\|, \|R_a\|\} \text{ for every } a \in A,$$

then A is totally multiplicatively prime.
(3) If both A and $\mathcal{M}(A)$ are ultraprime, then A is totally multiplicatively prime.

Remark 1 (a) It follows from the compactness of the unit sphere of finite-dimensional normed spaces that prime (respectively, multiplicatively prime) finite-dimensional algebras, endowed with any algebra norm [23, Proposition 1.1.7], are ultraprime (respectively, totally multiplicatively prime). Since there are prime finite-dimensional algebras which are not multiplicatively prime (cf. Sect. 3), we are provided with ultraprime normed algebras which are not multiplicatively prime (much less, totally multiplicatively prime).

(b) It is proved in [18] that the 'free normed non-associative algebra' generated by an arbitrary non-empty set [23, pp. 258–259 of Volume 1] is totally multiplicatively prime.

The theory of H^*-algebras is widely treated in [23, Sect. 8.1]. We recall that *semi-H^*-algebras* over $\mathbb{K}$ are introduced as those algebras A over $\mathbb{K}$ which are also Hilbert spaces and are endowed with a conjugate-linear vector space involution $*$ satisfying

$$(ab|c) = (b|a^*c) = (a|cb^*) \text{ for all } a, b, c \in A.$$

We also recall that *H^*-algebras* over $\mathbb{K}$ are those semi-H^*-algebras over $\mathbb{K}$ whose involution $*$ is an algebra involution. According to [23, Lemma 2.8.12(i)], up to the multiplication of the Hilbertian norm by a suitable positive number, every semi-H^*-algebra becomes a complete normed algebra.

All semi-H^*-algebras with zero annihilator are generalized annihilator [23, Fact 8.2.4]. The consequence that they are multiplicatively semiprime is, in their case, almost straightforward.

Given a normed space X, we denote by $BL(X)$ the norm-unital normed associative algebra of all bounded linear operators on X.

Given a nonzero complex Hilbert space H, the Schatten class $\mathcal{HS}(H)$ [40], of all Hilbert–Schmidt operators on H is a $*$-ideal of the C^*-algebra $BL(H)$, and has a natural structure of complex H^*-algebra [23, Example 8.1.3]. The H^*-algebras $\mathcal{HS}(H)$ above are essentially the unique topologically simple associative complex H^*-algebras [23, Theorem 8.1.151].

Proposition 10 ([23, Proposition 8.1.36]) *Let H be a nonzero complex Hilbert space, and let $|||\cdot|||$ denote the norm of H as well as the corresponding operator norm on $BL(H)$. Let A stand for the topologically simple complex H^*-algebra $\mathcal{HS}(H)$, and let $\|\cdot\|$ denote the norm of A as well as the corresponding operator norm on $BL(A)$. Then we have*

$$\|M_{a,b}\| = |||\, a \,|||\,|||\, b \,||| \text{ for all } a, b \in A.$$

As a consequence, A is ultraprime $(if\, and)$ only if H is finite-dimensional.

Now that we know that topologically simple semi-H^*-algebras need not be ultraprime, we could wonder whether at least they are totally prime. Even more, keeping in mind Proposition 8, we can wonder whether they are totally multiplicatively prime. The answer to this last question is affirmative. Indeed, we have the following.

Theorem 16 ([23, Theorem 8.1.107]) *Let A be a topologically simple semi-H^*-algebra. Then A is totally multiplicatively prime. More precisely, we have*

$$\|W_{F,a}\| = \|F\|\|a\| \text{ for all } F \in \mathcal{M}(A) \text{ and } a \in A.$$

Remark 2 As pointed out in Remark 1(a), *there exist ultraprime normed algebras which are not multiplicatively prime (much less, totally multiplicatively prime).* On the other hand, by Proposition 10 and Theorem 16 immediately above, *totally multiplicatively prime normed algebras need not be ultraprime.*

As a consequence of Proposition 8 and Theorem 16 we obtain.

Corollary 5 ([23, Corollary 8.1.108]) *Let A be a topologically simple semi-H^*-algebra. Then A is totally prime. More precisely, we have*

$$\|N_{a,b}\| = \rho\|a\|\|b\| \text{ for all } a, b \in A,$$

where ρ denotes the norm of the product of A.

The next corollary follows from Theorem 12 and Corollary 5 above, and had been shown earlier by the authors in [19, 20].

Corollary 6 *Every topologically simple complex semi-H^*-algebra is centrally closed.*

By means of the next proposition, Corollary 6 has outstanding applications to the automatic continuity theory in the setting of H^*-algebras. A detailed account of these applications can be found in [23, Theorems 8.1.41, 8.1.52, and 8.1.53, and Corollaries 8.1.82, 8.1.83, and 8.1.84].

Proposition 11 ([42], see also [23, Proposition 8.1.38]) *Let A be a centrally closed prime algebra over $\mathbb{K}$ such that* $\dim(T(A)) > 1$ *for every nonzero operator $T \in \mathcal{M}(A)$. Then there exist sequences b_n in A and T_n in $\mathcal{M}(A)$ such that*

$$(T_n \circ \cdots \circ T_1)(b_n) \neq 0 \, and \, (T_{n+1} \circ T_n \circ \cdots \circ T_1)(b_n) = 0 \, for \; every \; n \in \mathbb{N}.$$

To conclude our comments on analytic strengthenings of primeness, we describe in the following diagram the relationship between the different normed refinements of primeness. The diagram summarizes Proposition 8, and Theorems 13, 15, and 16 (cf. also Remark 2 and Proposition 10).

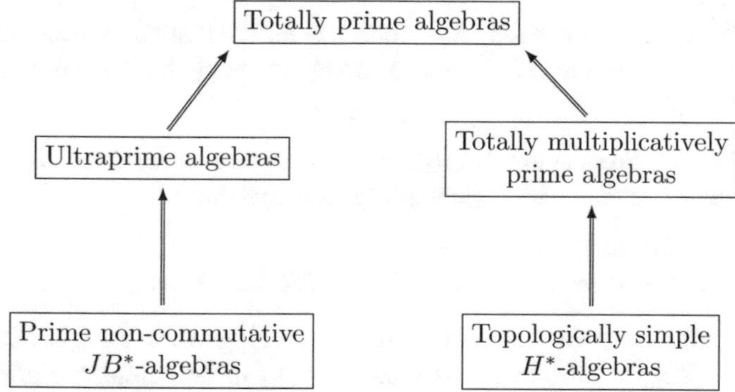

7 Complete Normed Algebras Whose Closed Multiplication Algebra Satisfies the von Neumann Inequality

We recall that a complete normed unital associative complex algebra A is said to satisfy the von Neumann inequality if, for every a in the open unit ball Δ_A of A, and for every complex polynomial $P \in \mathbb{C}[z]$, the inequality

$$\|P(a)\| \leq \sup\{|P(z)| : z \in \overline{\Delta}\}$$

holds, where Δ denotes the open unit disc in $\mathbb{C}$. The history and name of the von Neumann inequality goes back to von Neumann's paper [43] (see also [37, Sect. 153]), where it is proved that *the algebra of all bounded linear operators on any nonzero complex Hilbert space satisfies the von Neumann inequality.* It is worth mentioning that, concerning bounded linear operators, the study of the von Neumann inequality is completed by Foias' theorem [29] which proves the converse: *if the algebra of all* *bounded linear operators on a nonzero complex Banach space X satisfies the von Neumann inequality, then X is a Hilbert space.*

Let X be a nonzero complex Banach space, and let Δ_X stand for the open unit ball of X. We recall that a *holomorphic vector field* on Δ_X is nothing other than a holomorphic mapping from Δ_X to X. Let x be in Δ_X, and let Λ be a holomorphic vector field on Δ_X. An *integral curve* of Λ at x is a differentiable function $\psi_x : I \to \Delta_X$, where I is an open interval of $\mathbb{R}$ with $0 \in I$, satisfying

$$\psi_x(0) = x \text{ and } \frac{d}{dt}\psi_x(t) = \Lambda(\psi_x(t)) \text{ for every } t \in I.$$

Integral curves of Λ at x do exist, and two of them coincide on the intersection of their domains, and hence there exists a unique maximal integral curve, which is called the *flow* of Λ at x [23, Corollary 5.2.14 and Lemma 5.4.6]. The holomorphic vector field Λ on Δ_X is said to be *complete* if, for every $x \in \Delta_X$, the domain of the flow of Λ at x is all of $\mathbb{R}$.

In [4], Arazy studied in deep those complete normed unital associative complex algebras which satisfy the von Neumann inequality, being his crucial discovery the following.

Theorem 17 ([4, Proposition 2.2(ii)]) *For a complete normed unital associative complex algebra A, the following conditions are equivalent:*

(i) A satisfies the von Neumann inequality.
(ii) The holomorphic vector field $a \to \mathbf{1} - a^2$ is complete on Δ_A.

Since condition (ii) in Theorem 17 has a meaning when associativity of A is removed, in [22] we introduce *Arazy algebras* as those complete normed unital (possibly non-associative) complex algebras A satisfying (ii). It is worth mentioning that the non-associative discussion of Theorem 17 was already initiated by Arazy himself, when he proved that unital JB^*-algebras satisfy the von Neumann inequality [4, Proposition 4.2]. Actually, looking at the proof of [4, Proposition 4.2], we realized that Arazy was aware that Theorem 17 remained true when the associativity of A is relaxed to power-associativity. For this reason, we attributed the paternity of the power-associative generalization of Theorem 17 to him (see [23, Theorem 2.4.34]).

In [22] we developed a theory of (possibly non-associative) Arazy algebras, such as is done by Arazy in [4] in the associative particular case. Moreover, we devoted the last section of [22] to study those complete normed complex algebras whose closed multiplication algebras satisfy the von Neumann inequality. (Of course, we understand that the *closed multiplication algebra* $\overline{\mathcal{M}(A)}$ of a complete normed complex algebra A is nothing other than the closure of $\mathcal{M}(A)$ in $BL(A)$.) In the current section we review the content of the last section of [22], and prove some new results in this regard.

Actually, the discussion in [22] begins with the following.

Example 1 ([22, Example 8.1]) Let A be any nonzero complete normed complex algebra whose norm derives from an inner product. Then $\overline{\mathcal{M}(A)}$ is a closed subalgebra of $BL(A)$ containing the unit, and hence, since A is a Hilbert space, it follows from von Neumann's germinal theorem already reviewed (cf. also [22, Corollary 4.15]) that $\overline{\mathcal{M}(A)}$ satisfies the von Neumann inequality. This applies in particular to the semi-H^*-algebras (already introduced in Sect. 6) whenever they be nonzero and complex.

Then, involving Foias's theorem already reviewed, the next proposition is proved.

Proposition 12 ([22, Proposition 8.2]) *Let A be a normed finite-dimensional simple complex algebra such that $\mathcal{M}(A)$ satisfies the von Neumann inequality. Then the norm of A derives from an inner product. As a consequence, if A is norm-unital, then* $A = \mathbb{C}\mathbf{1}$.

Corollary 7 *For every $n \geq 2$, let A denote the algebra of all complex $n \times n$ matrices. Then there exists no norm on A in such a way that A is a norm-unital normed algebra, and that $\mathcal{M}(A)$ satisfies the von Neumann inequality.*

Proposition 13 ([22, Proposition 8.4]) *Let A be a complete normed unital associative and commutative complex algebra satisfying the von Neumann inequality. Then $\overline{\mathcal{M}(A)}$ satisfies the von Neumann inequality.*

The proof of the above proposition given in [22] shows that, if A is a norm-unital complete normed associative and commutative complex algebra, and if $\overline{\mathcal{M}(A)}$ satisfies the von Neumann inequality, then A satisfies the von Neumann inequality. We did not emphasize this converse of Proposition 13 because it is contained in the following.

Proposition 14 ([22, Proposition 8.5]) *Let A be a norm-unital complete normed complex algebra admitting power-associativity, and suppose that $\overline{\mathcal{M}(A)}$ satisfies the von Neumann inequality. Then A is an Arazy algebra.*

We recall that, according to [23, Definition 4.1.106], an algebra A is said to *admit power-associativity* if the algebra obtained by symmetrizing the product of A is power-associative. We note that power-associative algebras admit power-associativity. (Actually, an algebra A is power-associative if and only if it admits power-associativity and the subalgebra of A generated by each element of A is commutative [23, Corollary 2.4.18].) We also note that, since Jordan algebras are power-associative, Jordan-admissible algebras admit power-associativity.

Remark 3 Let A denote the C^*-algebra $\mathbb{C} \oplus_\infty \mathbb{C}$. Then, by von Neumann's theorem and Proposition 13, $\mathcal{M}(A)$ satisfies the von Neumann inequality. Moreover the norm of A does not derive from an inner product. Therefore the assumption of simplicity in Proposition 12 cannot be completely removed.

Let X be a nonzero complex Banach space. We denote by $\mathrm{aut}(\Delta_X)$ the set of all complete holomorphic vector fields on Δ_X, and recall that the *symmetric part* X_s of X, defined by the equality

$$X_s := \{\Lambda(0) : \Lambda \in \mathrm{aut}(\Delta_X)\},$$

is a closed complex subspace of X [23, Fact 5.6.28]. Now let u be a norm-one element of X. We denote by $H(X, u)$ the closed real subspace of X consisting of those elements $h \in X$ such that $f(h) \in \mathbb{R}$ for every norm-one bounded linear functional f on X satisfying $f(u) = 1$, and we denote by $J(X, u)$ the complex subspace of X defined by

$$J(X, u) := H(X, u) + iH(X, u).$$

According to [23, Corollary 2.1.13], if A is a norm-unital complete normed complex algebra, then we have $J(A, \mathbf{1}) = H(A, \mathbf{1}) \oplus iH(A, \mathbf{1})$.

The proof of the next theorem depends heavily on [22, Theorem 3.9], which among other things asserts that, if A is an Arazy algebra, then A is norm-unital, $A_s = J(A, \mathbf{1})$, and A_s is a subalgebra of A, which becomes naturally a non-commutative JB^*-algebra.

Theorem 18 ([22, Theorem 8.7]) *Let A be an Arazy algebra such that $\overline{\mathcal{M}(A)}$ satisfies the von Neumann inequality. Then we have:*

- *(i) $\overline{\mathcal{M}(A)}_s = J(\overline{\mathcal{M}(A)}, I_A)$, and $\overline{\mathcal{M}(A)}_s$ is a subalgebra of $\overline{\mathcal{M}(A)}$, which becomes a C^*-algebra under the restriction of the norm of $\overline{\mathcal{M}(A)}$ and the involution $(F + iG)^* := F - iG$ $(F, G \in H(\overline{\mathcal{M}(A)}, I_A))$.*
- *(ii) Every operator $T \in \overline{\mathcal{M}(A)}_s$ such that $T(\mathbf{1}) = 0$ is a Jordan derivation of A, which vanishes on A_s.*
- *(iii) A is norm-unital, $A_s = J(A, \mathbf{1})$, and A_s is a subalgebra of A, and is a commutative C^*-algebra under the restriction of the norm of A and the involution $(h + ik)^* := h - ik$ $(h, k \in H(A, \mathbf{1}))$.*

The next corollary follows from Proposition 14 and Theorem 18.

Corollary 8 *Let A be a norm-unital complete normed complex algebra admitting power-associativity, and suppose that $\overline{\mathcal{M}(A)}$ satisfies the von Neumann inequality. Then the conclusion in Theorem 18 holds.*

Given an algebra A, we denote by $A_{\mathbb{1}}$ the unital extension of A [23, § 1.1.104]. The next lemma is straightforward. For additional information see [13, Lemma 3.8].

Lemma 2 *Let A be an algebra. Then, as a subspace of $A_{\mathbb{1}}$, A is invariant under $\mathcal{M}(A_{\mathbb{1}})$, $T_{|A}$ (regarded as a mapping from A to A) lies in $\mathcal{M}(A)$ whenever T is in $\mathcal{M}(A_{\mathbb{1}})$, and the mapping $\Phi : T \to T_{|A}$ becomes a surjective algebra homomorphism from $\mathcal{M}(A_{\mathbb{1}})$ to $\mathcal{M}(A)$.*

Looking at [22, Proposition 8.10] and its proof, we are provided with the following.

Proposition 15 *Let A be a nonzero non-unital complete normed algebra having an approximate identity bounded by $M \geq 1$. Then there exists a complete algebra norm on $A_{\mathbb{1}}$ extending the norm of A, satisfying $\|\mathbb{1}\| \leq M$, and converting the surjective* *algebra homomorphism* $\overline{\Phi}: \overline{\mathcal{M}(A_{\mathbb{1}})} \to \overline{\mathcal{M}(A)}$ *given by Lemma 2 into an isometry. Therefore, when* $\overline{\mathcal{M}(A_{\mathbb{1}})}$ *is endowed with the corresponding operator norm, the algebras* $\overline{\mathcal{M}(A)}$ *and* $\overline{\mathcal{M}(A_{\mathbb{1}})}$ *are isometrically isomorphic.*

Remark 4 Let A be any non-unital finite-dimensional simple algebra (for example the Lie subalgebra of $M_2(\mathbb{C})$ consisting of all trace-zero matrices). Then, according to [16, Example 1], the surjective algebra homomorphism $\Phi : \mathcal{M}(A_{\mathbb{1}}) \to \mathcal{M}(A)$ given by Lemma 2 is not injective. Since every finite-dimensional algebra can be endowed with an algebra norm [23, Proposition 1.1.7], it follows that the assumption in Proposition 15 of the existence of a bounded approximate identity cannot be completely removed.

Now let A denote the non-unital complete normed associative and commutative algebra whose Banach space is ℓ_1 and whose product is defined coordinate-wise, and let $A_{\mathbb{1}}$ denote the normed unital extension of A [23, Proposition 1.1.107]. Then, according to the proof of Proposition 13, $\mathcal{M}(A_{\mathbb{1}}) = \overline{\mathcal{M}(A_{\mathbb{1}})}$ is isometrically isomorphic to $A_{\mathbb{1}}$. On the other hand, it is easy to see that $\overline{\mathcal{M}(A)}$ is isometrically isomorphic to the unital extension of c_0 endowed with the norm

$$\|\lambda\mathbb{1} + \{\mu_n\}_{n\in\mathbb{N}}\| := \sup\{|\lambda + \mu_n| : n \in \mathbb{N}\}.$$

(Thus, in the complex case, $\overline{\mathcal{M}(A)}$ is isometrically isomorphic to the C^*-algebra unital extension of the C^*-algebra c_0 [23, Proposition 1.2.44], and hence satisfies the von Neumann inequality.) It follows from the above identifications that the algebra homomorphism $\overline{\Phi} : T \to T_{|A}$ from $\overline{\mathcal{M}(A_{\mathbb{1}})}$ to $\overline{\mathcal{M}(A)}$ converts into the inclusion of $A_{\mathbb{1}}$ in the unital extension of c_0 induced by the inclusion $\ell_1 \subseteq c_o$. Therefore $\overline{\Phi}$ is injective but is not a topological imbedding. Since all complete algebra norms on $A_{\mathbb{1}}$ extending the norm of A are pairwise equivalent, it follows that there is no complete algebra norm on $A_{\mathbb{1}}$ extending the norm of A and converting the surjective algebra homomorphism $\Phi : \mathcal{M}(A_{\mathbb{1}}) \to \mathcal{M}(A)$ given by Lemma 2 into an isometry, nor even into a homeomorphism. But the sequence $a_m = \{\lambda_{mn}\}_{n\in\mathbb{N}}$ in A, defined by $\lambda_{mn} = 1$ if $n \leq m$ and $\lambda_{mn} = 0$ otherwise, is an approximate identity for A. Therefore the assumption of boundedness for the approximate identity, in Proposition 15, cannot be removed.

The fact that every non-commutative JB^*-algebra has an approximate identity bounded by 1 [23, Proposition 3.5.23] is involved in the next consequence of Proposition 15.

Corollary 9 [22, Corollary 8.12] *Let A be a nonzero non-unital non-commutative* JB^**-algebra. Then* $A_{\mathbb{1}}$ *becomes a non-commutative* JB^**-algebra for a unique norm and a unique involution which extend those of A* [23, Corollary 3.5.36]. *Moreover this norm on* $A_{\mathbb{1}}$ *converts the surjective algebra homomorphism* $\Phi : \mathcal{M}(A_{\mathbb{1}}) \to \mathcal{M}(A)$ *given by Lemma 2 into an isometry, and therefore the algebras* $\overline{\mathcal{M}(A)}$ *and* $\overline{\mathcal{M}(A_{\mathbb{1}})}$ *are isometrically isomorphic whenever* $\overline{\mathcal{M}(A_{\mathbb{1}})}$ *is endowed with the corresponding operator norm.*

The proof of the next theorem involves Proposition 13, Theorem 18, and Corollary 9.

Theorem 19 [22, Theorem 8.13] *A nonzero non-commutative JB^*-algebra A is actually a commutative C^*-algebra if and only if $\overline{\mathcal{M}(A)}$ satisfies the von Neumann inequality.*

As a straightforward consequence of the above theorem, we obtain the following.

Corollary 10 *A nonzero C^*-algebra A is commutative if and only if $\overline{\mathcal{M}(A)}$ satisfies the von Neumann inequality.*

Now we prove the following variant of Theorem 18.

Proposition 16 *Let A be a norm-unital complete normed complex algebra such that $\overline{\mathcal{M}(A)}$ satisfies the von Neumann inequality. Then we have:*

(i) $\overline{\mathcal{M}(A)}_s = J(\overline{\mathcal{M}(A)}, I_A)$, and $\overline{\mathcal{M}(A)}_s$ is a subalgebra of $\overline{\mathcal{M}(A)}$, and is a C^*-algebra under the restriction of the norm of $\overline{\mathcal{M}(A)}$ and the involution

$$(F + iG)^* := F - iG \quad (F, G \in H(\overline{\mathcal{M}(A)}, I_A)).$$

(ii) $J(A, \mathbf{1})$ is a subalgebra of A, and is a non-commutative JB^*-algebra under the restriction of the norm of A and the involution

$$(h + ik)^* := h - ik(h, k \in H(A, \mathbf{1})).$$

Proof Since $\overline{\mathcal{M}(A)}_s$ is an associative algebra, assertion (i) follows from [22, Proposition 3.7].

To prove (ii), note at first that, by [23, Proposition 2.1.11 and Lemma 3.3.14], $J(A, \mathbf{1})$ is a closed subspace of A. Let h, k be in $H(A, \mathbf{1})$, then, by [23, Lemma 2.1.10] and the consequence in [23, Corollary 2.1.2],

$$L_h, L_k \in H(BL(A), I_A) \cap \overline{\mathcal{M}(A)} = H(\overline{\mathcal{M}(A)}, I_A),$$

hence by assertion (i) already proved, $L_h \circ L_k \in J(\overline{\mathcal{M}(A)}, I_A)$, and therefore, by [23, Corollary 2.1.2(i)], $hk = (L_h \circ L_k)(\mathbf{1}) \in J(A, \mathbf{1})$. Since h, k are arbitrary in $H(A, \mathbf{1})$, and $J(A, \mathbf{1}) = H(A, \mathbf{1}) + iH(A, \mathbf{1})$, it follows that $J(A, \mathbf{1})$ is a subalgebra of A. Finally, since $H(J(A, \mathbf{1}), \mathbf{1}) = H(A, \mathbf{1})$ (by the consequence in [23, Corollary 2.1.2] again), assertion (ii) holds thanks to the non-associative Vidav–Palmer theorem [23, Theorem 3.3.11]. □

Now we prove the main new result in this section, the proof of which involves Proposition 15 and Theorem 19.

Theorem 20 *The commutative C^*-algebras are precisely those complete normed complex algebras having an approximate identity bounded by 1, and satisfying that $\overline{\mathcal{M}(A)}$ is a C^*-algebra.*

Proof Let A be a commutative C^*-algebra. Then A has an approximate identity bounded by 1 (see [23, Proposition 3.5.23] again). Moreover, by the associativity and commutativity of A, we have $\mathcal{M}(A) = \mathcal{L} := \{L_a : a \in A\}$ if A is unital, and $\mathcal{M}(A) = \mathbb{C}I_A \oplus \mathcal{L}$ otherwise. Thus, in the unital case, by [23, Lemma 2.1.10], $\mathcal{M}(A)$ is isometrically isomorphic to A, and hence is a C^*-algebra, whereas, in the non-unital case, $\mathcal{M}(A)$ is isometrically isomorphic to the C^*-algebra unital extension of A (see the proof of [23, Proposition 1.2.44]).

Now let A be a nonzero complete normed complex algebra having an approximate identity bounded by 1, and satisfying that $\overline{\mathcal{M}(A)}$ is a C^*-algebra. Since $\overline{\mathcal{M}(A)}$ is a C^*-algebra containing $I_A \neq 0$, it follows from von Neumann's theorem that $\overline{\mathcal{M}(A)}$ satisfies the von Neumann inequality, and moreover, by [23, Definition 2.2.7 and Lemma 2.2.8], $\overline{\mathcal{M}(A)}$ satisfies the Vidav–Palmer axiom, i.e. $\overline{\mathcal{M}(A)} = J(\overline{\mathcal{M}(A)}, I_A)$. Suppose that A is unital. Then, since A has an approximate identity bounded by 1, it is easily seen that A is norm-unital. Therefore, since for every $a \in A$ we have $L_a \in J(\overline{\mathcal{M}(A)}, I_A)$, it follows from [23, Corollary 2.1.2(i)] that $a = L_a(\mathbf{1}) \in J(A, \mathbf{1})$. Thus $A = J(A, \mathbf{1})$, and hence, by [23, Theorem 3.3.11] and Theorem 19, A is a commutative C^*-algebra. Finally, suppose that A is not unital. Then, since A has an approximate identity bounded by 1, it follows from Proposition 15 that there exists an algebra norm on $A_{\mathbb{1}}$ extending the norm of A, converting $A_{\mathbb{1}}$ into a norm-unital complete normed complex algebra, and satisfying that, when $\overline{\mathcal{M}(A_{\mathbb{1}})}$ is endowed with the corresponding operator norm, the algebras $\overline{\mathcal{M}(A)}$ and $\overline{\mathcal{M}(A_{\mathbb{1}})}$ are isometrically isomorphic. Now the sequence e_n in $A_{\mathbb{1}}$ defined by $e_n := \mathbb{1}$ for every n becomes an approximate identity in $A_{\mathbb{1}}$ bounded by 1, and, since $\overline{\mathcal{M}(A)}$ is a C^*-algebra, $\overline{\mathcal{M}(A_{\mathbb{1}})}$ is a C^*-algebra. Therefore, since $A_{\mathbb{1}}$ is unital, it follows from the unital case already discussed that $A_{\mathbb{1}}$ is a commutative C^*-algebra. Since A is the kernel of a character on $A_{\mathbb{1}}$, we conclude that A is a $*$-subalgebra of $A_{\mathbb{1}}$ [23, Corollary 1.2.22], and hence is a commutative C^*-algebra. □

We conclude this section with a byproduct of Proposition 15, which has nothing to do with Arazy algebras. Previously, we prove the following.

Lemma 3 *Let A be a normed unital algebra, and let B be a subalgebra of A. Suppose that there exists $b \in B$ such that $\|\mathbf{1} - b\| < 1$. Then $\mathbf{1}$ lies in the closure of B in A.*

Proof For every $n \in \mathbb{N}$ we have $\mathbf{1} - (I_A - L_b)^n(\mathbf{1}) \in B$ and

$$\|(I_A - L_b)^n(\mathbf{1})\| = \|(L_{\mathbf{1}-b})^n(\mathbf{1})\| \leq \|L_{\mathbf{1}-b}\|^n\|\mathbf{1}\| \leq \|\mathbf{1} - b\|^n\|\mathbf{1}\|.$$

Therefore, since $\|\mathbf{1} - b\| < 1$, the result follows by letting $n \to \infty$. □

Given an algebra A, we denote by $\mathcal{M}^\sharp(A)$ the subalgebra of $L(A)$ generated by all operators of left and right multiplication on A by elements of A.

Corollary 11 *Let A be a nonzero complete normed algebra having a bounded approximate identity* (for instance, a nonzero non-commutative JB^*-algebra), *and suppose that there exists $S \in \mathcal{M}^\sharp(A)$ such that $\|I_A - S\| < 1$. Then A is unital.*

Proof We argue by contradiction, and hence assume that A is not unital. Then, by Proposition 15, the mapping $\Phi : T \to T_{|A}$ becomes an isometric surjective algebra homomorphism from $\mathcal{M}(A_{\mathbb{1}})$ to $\mathcal{M}(A)$, whenever $A_{\mathbb{1}}$ has been suitably normed. Now we follow an argument in [16, Example 1]. Let R be in $\mathcal{M}^{\sharp}(A)$. Write R as a sum of products of multiplication operators on A by elements of A, and photocopy R into $\mathcal{M}(A_{\mathbb{1}})$ by simply seeing each of these multiplication operators acting on $A_{\mathbb{1}}$ in the natural way. Then the operator $\widehat{R} \in \mathcal{M}(A_{\mathbb{1}})$ obtained in this way has the property that $\widehat{R}(\mathbb{1})$ lies in A and $\Phi(\widehat{R}) = R$ (hence the correspondence $R \to \widehat{R}$ is nothing other than the restriction to $\mathcal{M}^{\sharp}(A)$ of the surjective linear isometry Φ^{-1}). But, by Lemma 3, I_A lies in $\overline{\mathcal{M}^{\sharp}(A)}$. Therefore, denoting by $\overline{\Phi} : \overline{\mathcal{M}(A_{\mathbb{1}})} \to \overline{\mathcal{M}(A)}$ the extension by continuity of Φ, it follows that $\left[\overline{\Phi}^{-1}(I_A)\right](\mathbb{1})$ lies in A. But this is a contradiction because $\left[\overline{\Phi}^{-1}(I_A)\right](\mathbb{1}) = I_{A_{\mathbb{1}}}(\mathbb{1}) = \mathbb{1}$, and $\mathbb{1}$ cannot lie in A by the definition itself of $A_{\mathbb{1}}$ as the unital extension of A. □

Acknowledgements The authors thank the organizers of the 3rd Moroccan Andalusian Meeting on Algebras and their Applications for inviting them to attend it and to deliver a lecture.
This chapter has been partially supported by the Junta de Andalucía and Spanish government grants FQM199 and MTMT2016-76327-C3-2-P.

References

1. Albert, A.A.: The radical of a non-associative algebra. Bull. Am. Math. Soc. **48**, 891–897 (1942)
2. Ara, P.: The extended centroid of C^*-algebras. Archiv. Math. **54**, 358–364 (1990)
3. Ara, P., Mathieu, M.: Local multipliers of C^*-algebras. Springer Monographs in Mathematics. Springer, London (2003)
4. Arazy, J.: Isometries of Banach algebras satisfying the von Neumann inequality. Math. Scand. **74**, 137–151 (1994)
5. Baxter, W.E., Martindale 3rd, W.S.: Central closure of semiprime nonassociative rings. Commun. Algebra **7**, 1103–1132 (1979)
6. Beidar, K.I., Mikhalev, A.V., Slinko, A.M.: A criterion for primeness of nondegenerate alternative and Jordan algebras. Trudy Moskov. Mat. Obshch. **50**, 130–137 (1988). (Translation in Trans. Moscow Math. Soc. **50**, 129–137 (1987))
7. Bunce, L.J., Chu, C.-H., Stachó, L.L., Zalar, B.: On prime JB^*-triples. Quart. J. Math. Oxford **49**, 279–290 (1998)
8. Cabello, J.C., Cabrera, M.: Structure theory for multiplicatively semiprime algebras. J. Algebra **282**, 386–421 (2004)
9. Cabello, J.C., Cabrera, M., Fernández, A.: π-complemented algebras through pseudocomplemented lattices. Order **29**, 463–479 (2012)
10. Cabello, J.C., Cabrera, M., López, G., Martindale 3rd, W.S.: Multiplicative semiprimeness of skew Lie algebras. Commun. Algebra **32**, 3487–3501 (2004)
11. Cabello, J.C., Cabrera, M., Rodríguez, Á., Roura, R.: A characterization of π-complemented algebras. Commun. Algebra **41**, 3067–3079 (2013)
12. Cabello, J.C., Cabrera, M., Roura, R.: A note on the multiplicative primeness of degenerate Jordan algebras. Sib. Math. J. **51**, 818–823 (2010)
13. Cabello, J.C., Cabrera, M., Roura, R.: π-complementation in the unitisation and multiplication algebras of a semiprime algebra. Commun. Algebra **40**, 3507–3531 (2012)

14. Cabrera, A.M., Cabrera, M.: Multiplicative primeness of strongly prime non-commutative Jordan algebras. J. Algebra **538**, 253–260 (2019)
15. Cabrera, M., Fernández, A., Golubkov, A.Yu., Moreno, A.: Algebras whose multiplication algebra is PI or GPI. J. Algebra **459**, 213–237 (2016)
16. Cabrera, M., Mohammed, A.A.: Extended centroid and central closure of the multiplication algebra. Commun. Algebra **27**, 5723–5736 (1999)
17. Cabrera, M., Mohammed, A.A.: Extended centroid and central closure of multiplicatively semiprime algebras. Commun. Algebra **29**, 1215–1233 (2001)
18. Cabrera, M., Mohammed, A.A.: Totally multiplicatively prime algebras. Proc. R. Soc. Edinb. Sect. A **132**, 1145–1162 (2002)
19. Cabrera, M., Rodríguez, Á.: Extended centroid and central closure of semiprime normed algebras: a first approach. Commun. Algebra **18**, 2293–2326 (1990)
20. Cabrera, M., Rodríguez, Á.: Nonassociative ultraprime normed algebras. Q. J. Math. Oxford **43**, 1–7 (1992)
21. Cabrera, M., Rodríguez, Á.: Non-degenerately ultraprime Jordan-Banach algebras: a Zel'manovian treatment. Proc. Lond. Math. Soc. **69**, 576–604 (1994)
22. Cabrera, M., Rodríguez, Á.: The von Neumann inequality in complete normed non-associative complex algebras. Math. Proc. R. Irish Acad. **118A**, 83–125 (2018)
23. Cabrera, M., Rodríguez, Á.: Non-associative normed algebras. Volume 1: The Vidav-Palmer and Gelfand-Naimark Theorems, and Volume 2: Representation Theory and the Zel'manov Approach, Encyclopedia of Mathematics and Its Applications 154 and 167. Cambridge University Press, Cambridge, 2014 and 2018
24. Cabrera, M., Villena, A.R.: Multiplicative-semiprimeness of nondegenerate Jordan algebras. Commun. Algebra **32**, 3995–4003 (2004)
25. Civin, P., Yood, B.: Lie and Jordan structures in Banach algebras. Pac. J. Math. **15**, 775–797 (1965)
26. Erickson, T.S., Martindale III, W.S., Osborn, J.M.: Prime nonassociative algebras. Pac. J. Math. **60**, 49–63 (1975)
27. Fernández, A., García, E., Rodríguez, Á.: A Zel'manov prime theorem for JB^*-algebras. J. Lond. Math. Soc. **46**, 319–335 (1992)
28. Fernández, A., Rodríguez, Á.: A Wedderburn theorem for nonassociative complete normed algebras. J. Lond. Math. Soc. **33**, 328–338 (1986)
29. Foias, C.: Sur certains théorèmes de J. von Neumann concernant les ensembles spectraux. Acta Sci. Math. Szeged. **18**, 15–20 (1957)
30. Jacobson, N.: A note on non-associative algebras. Duke Math. J. **3**, 544–548 (1937)
31. Jacobson, N.: Abraham Adrian Albert 1905–1972. Bull. Am. Math. Soc. **80**, 1075–1100 (1974)
32. Mathieu, M.: Applications of ultraprime Banach algebras in the theory of elementary operators. Ph.D. thesis, Universität Tübingen, Tübingen (1986)
33. Mathieu, M.: Rings of quotients of ultraprime Banach algebras, with applications to elementary operators. In: Loy, R.J. (ed.) Conference on Automatic Continuity and Banach Algebras (Canberra, 1989). Proceedings of the Centre for Mathematical Analysis, Australian National University. vol. 21, pp. 297–317. Australian National University, Canberra (1989)
34. Mathieu, M.: Elementary operators on prime C^*-algebras, I. Math. Ann. **284**, 223–244 (1989)
35. Mohammed, A.A.: Álgebras multiplicativamente primas: visión algebraica y analítica. Ph.D. thesis, Granada (2000)
36. Regev, A.: Existence of identities in $A \otimes_F B$. Israel J. Math. **11**, 131–152 (1972)
37. Riesz, F., Sz-Nagy, B.: Leçons d'analyse fonctionelle. Cinquième édition. Gauthier-Villars, Paris; Akadémiai Kiadó, Budapest (1968)
38. Rodríguez, Á.: Números hipercomplejos en dimensión infinita. Discurso de ingreso en la Academia de Ciencias Matemáticas, Físico-Químicas y Naturales de Granada, Granada (1993)
39. Rodríguez, Á., Villena, A.R.: Centroid and extended centroid of JB^*-algebras. In: González, S., Myung, HCh. (eds.) Nonassociative algebraic models. Proceedings of the workshop held at the Universidad de Zaragoza, Zaragoza, April 1989, pp. 223–232. Nova Science Publishers Inc, Commack, NY (1992)

40. Schatten, R.: Norm ideals of completely continuous operators. Second printing. In: Ergebnisse der Mathematik und ihrer Grenzgebiete, vol. 27. Springer, Berlin (1970)
41. Skosyrskii, V.G.: Strongly prime noncommutative Jordan algebras. Trudy Inst. Mat. (Novosibirsk) **16**, 131–164 (1989)
42. Villena, A.R.: Continuity of derivations on H^*-algebras. Proc. Am. Math. Soc. **122**, 821–826 (1994)
43. von Neumann, J.: Eine Spektraltheorie für allgemeine Operatoren eines unitären Raumes. Math. Nach. **4**, 258–281 (1951)
44. Yood, B.: Closed prime ideals in topological rings. Proc. Lond. Math. Soc. **24**, 307–323 (1972)

Generalized Drazin Inverse and Commuting Riesz Perturbations

Mourad Oudghiri and Khalid Souilah

Abstract In this note, we provide necessary and sufficient conditions for the stability of generalized Drazin invertible operators under commuting Riesz perturbation. We also focus on the commuting perturbation class of meromorphic operators.

Keywords Drazin inverse · Riesz perturbations

1 Introduction

Throughout this paper, X denotes an infinite-dimensional complex Banach space, and $\mathcal{B}(X)$ denotes the algebra of all bounded linear operators on X.

In general, perturbation theory is concerned with the following question: Given an operator $T \in \mathcal{B}(X)$ what can we say about the properties of an operator $T + S$? The statements vary, depending on which operator class S is taken from. Perturbation theory for linear operators and their spectra is one of the main objectives in operator theory and functional analysis, with numerous applications in mathematics, physics and engineering sciences.

In many approaches, quasi-nilpotent, compact or Riesz perturbations are investigated, e.g. when stability properties of semi-Fredholm and Browder operators are analyzed [12]. In recent years, some results on commuting Riesz perturbations of generalized Drazin invertible operators have appeared [9, 13, 14].

Recall that $T \in \mathcal{B}(X)$ is called a *Riesz* operator if $\pi(T)$ is quasi-nilpotent in the Calkin algebra $C(X) = \mathcal{B}(X)/\mathcal{K}(X)$ where $\pi : \mathcal{B}(X) \to C(X)$ is the natural homomorphism and $\mathcal{K}(X)$ is the ideal of all compact operators on X. Clearly, a sum of a quasi-nilpotent operator and a compact one is a Riesz operator. Furthermore,

M. Oudghiri (✉) · K. Souilah
Département de Mathématiques, Labo LAGA, Faculté des Sciences, Université Mohammed Premier, 60000 Oujda, Morocco
e-mail: m.oudghiri@ump.ac.ma

K. Souilah
e-mail: k.souilah@ump.ac.ma

M. Siles Molina et al. (eds.), *Associative and Non-Associative Algebras and Applications*, Springer Proceedings in Mathematics & Statistics 311,
https://doi.org/10.1007/978-3-030-35256-1_7

in the context of Hilbert spaces, every Riesz operator is a sum of a quasi-nilpotent operator and a compact one; the so-called West decomposition [15]. For general Banach spaces, the West decomposition is an open problem.

Following [9], recall that an operator $T \in \mathcal{B}(X)$ is said to be *generalized Drazin invertible* if there exists $S \in \mathcal{B}(X)$ such that

$$TS = ST, \ \ STS = S \ \text{ and } \ T^2S - T \text{ is quasi-nilpotent.}$$

Such element S is unique, it is called the *generalized Drazin inverse* of T and denoted by T^{d}. Note that if T is generalized Drazin invertible and $T^2T^{\mathrm{d}} - T$ is nilpotent, then T^{d} is the conventional Drazin inverse of T (see [2, 5]). Clearly, every quasi-nilpotent operator $Q \in \mathcal{B}(X)$ is generalized Drazin invertible and $Q^{\mathrm{d}} = 0$, and every invertible operator $S \in \mathcal{B}(X)$ is generalized Drazin invertible and $S^{\mathrm{d}} = S^{-1}$. For more details and characterizations of generalized Drazin inverse, we refer the reader to [4, 9, 14] and references therein.

The generalized Drazin inverse for operators arises naturally in the context of isolated spectral points and becomes an important tool in the spectral theory. It was applied to systems of singular linear differential equations, perturbed differential equations and semi-groups of linear operators (see [7, 10]).

For $T \in \mathcal{B}(X)$, we denote by $\sigma_{\mathrm{gD}}(T)$ its corresponding generalized Drazin spectrum; that is the set of all complex numbers λ for which $T - \lambda$ is not generalized Drazin invertible. In [9, Theorem 5.6], the author has studied the stability of generalized Drazin invertibility under commuting quasi-nilpotent perturbations, showing that if $S \in \mathcal{B}(X)$ is quasi-nilpotent and $T \in \mathcal{B}(X)$ is generalized Drazin invertible commuting with S then $T + S$ is also generalized Drazin invertible. Hence, for such operator S, one can easily see that

$$\sigma_{\mathrm{gD}}(T) = \sigma_{\mathrm{gD}}(T + S) \ \text{ for every } T \in \mathcal{B}(X) \text{ commuting with } S. \tag{1}$$

Later, in [13, Corollary 2.6] the authors considered commuting power finite-rank perturbations of generalized Drazin invertibility. They establish that if S^n is a finite-rank operator for some integer $n \geq 1$, then (1) holds. Note that, in this case, the operator S can be written as a some of a nilpotent operator and a finite-rank operator, and hence it is a Riesz operator.

From these results, it is a natural question whether or not the class of quasi-nilpotent operators or operators having some iterate with finite-dimensional range is characterized by the perturbation property (1). Motivated by these ideas, we propose in this paper to characterize all Riesz operators $S \in \mathcal{B}(X)$ satisfying (1).

Recall that an operator $T \in \mathcal{B}(X)$ is *meromorphic* if its non-zero points spectrum consist of poles if its resolvent. It is a classical fact that every compact, or more generally, Riesz operator is meromorphic. Furthermore, there is a crucial link between meromorphic operators and generalized Drazin invertibility. Indeed, it is easily seen that an operator

$$T \in \mathcal{B}(X) \text{ is meromorphic } \quad \Leftrightarrow \quad \sigma_{\mathrm{gD}}(T) \subseteq \{0\}.$$

Given a subset $\Lambda \subset \mathcal{B}(X)$, the *commuting perturbation class* $\mathcal{P}_c(\Lambda)$ of Λ is defined by

$$\mathcal{P}_c(\Lambda) = \{S \in \mathcal{B}(X) : T + S \in \Lambda \text{ for every } T \in \Lambda \text{ commuting with } S\}.$$

The concept of perturbation classes has been considered in other situations. For example, it is well-known that the perturbation class of all invertible elements in a Banach algebra is the radical of that algebra; and its commuting perturbation class is the set of all quasinilpotent elements (see [11, 16]). For some expositions on perturbation classes problem, the reader is referred to [1, 6, 8, 11, 16] and the references therein.

Let $\mathcal{M}(X)$ denote the set of all meromorphic operators in $\mathcal{B}(X)$. In this paper, we consider also the commuting perturbation class problem of meromorphic operators. More precisely, we establish that $\mathcal{P}_c(\mathcal{M}(X))$ consists of power finite-rank operators.

2 Riesz Perturbations of Generalized Drazin Invertibility

Let $\mathcal{Q}(X)$ denote the set of all quasi-nilpotent operators in $\mathcal{B}(X)$, and let $\mathcal{F}(X)$ denote the set of all finite rank operators in $\mathcal{B}(X)$.

The main result of this section is the following theorem that gives necessary and sufficient conditions for the stability of generalized Drazin spectrum under the commuting Riesz perturbation class.

Theorem 1 *Let $R \in \mathcal{B}(X)$ be a Riesz operator. Then the following assertions are equivalent:*

1. $R = Q + F$ *where* $Q \in \mathcal{Q}(X)$, $F \in \mathcal{F}(X)$ *and* $QF = FQ$;
2. $\sigma_{gD}(T) = \sigma_{gD}(T + R)$ *for every* $T \in \mathcal{B}(X)$ *commuting with* R.

Before proving this theorem, we need to establish the following lemma.

Lemma 1 *Let $R \in \mathcal{B}(X)$ be a Riesz operator such that $R = Q_o + F_o$ where $Q_o \in \mathcal{Q}(X)$, $F_o \in \mathcal{F}(X)$ and $Q_oF_o = F_oQ_o$. Then, there exist $Q \in \mathcal{Q}(X)$ and $F \in \mathcal{F}(X)$ such that $R = Q + F$ and $QF = FQ = 0$.*

Proof Since F_o is a finite-rank operator, we can write $X = X_1 \oplus X_2$ where $X_1 = \ker(F_o^n)$ and $X_2 = \text{ran}(F_o^n)$ for some integer $n \geq 1$. Since $Q_oF_o = F_oQ_o$, one can easily check that X_1 and X_2 are Q_o-invariant. Set $Q_i = Q_{o|X_i}$ and $F_i = F_{o|X_i}$ for $i \in \{1, 2\}$. It follows that $Q_iF_i = F_iQ_i$, and so $Q_1 + F_1$ is quasi-nilpotent and $Q_2 + F_2$ is finite-rank and invertible. If we take $Q = (Q_1 + F_1) \oplus 0$ and $F = 0 \oplus (Q_2 + F_2)$ with respect to the decomposition of X, we get that $Q \in \mathcal{Q}(X)$, $F \in \mathcal{F}(X)$, $R = Q + F$ and $QF = FQ = 0$ as desired. □

Remark 1 Let $A, B \in \mathcal{B}(X)$ be non-zero generalized Drazin invertible. We have the following well-known properties (see [9]):

1. A^{d} commutes with every element in $\mathcal{B}(X)$ commuting with A;
2. If $AB = BA = 0$, then $A + B$ is generalized Drazin invertible and $(A + B)^{\mathrm{d}} = A^{\mathrm{d}} + B^{\mathrm{d}}$;

For an operator $T \in \mathcal{B}(X)$, write $\ker(T)$ for its kernel, $\operatorname{ran}(T)$ for its range, and $\sigma(T)$ for its spectrum.

Recall that an operator $T \in \mathcal{B}(X)$ is said to be *Fredholm* if $\dim\ker(T)$ and $\operatorname{codim}\operatorname{ran}(T)$ are finite. The *essential spectrum* $\sigma_{\mathrm{e}}(T)$ of T is defined as the set of all complex number λ such that $T - \lambda$ is not a Fredholm operator. It is well-known that $T \in \mathcal{B}(X)$ is a Riesz operator if and only if $\sigma_{\mathrm{e}}(T) = \{0\}$.

The main result, Theorem 1, in [4] asserts that for two commuting generalized Drazin invertible operators $T, R \in \mathcal{B}(X)$ the sum $T + R$ is generalized Drazin invertible if and only if $I + R^{\mathrm{d}}T$ is generalized Drazin invertible.

Proof (Proof of Theorem 1*)* (1) $\Rightarrow$ (2). According to the previous lemma, we may assume that $R = Q + F$ where $Q \in \mathcal{Q}(X)$, $F \in \mathcal{F}(X)$ and $QF = FQ = 0$. Let $T \in \mathcal{B}(X)$ be a generalized Drazin invertible operator such that $TR = RT$, and let us show that $T + R$ is generalized Drazin invertible. Since $QF = FQ = 0$, then R is generalized Drazin invertible and $R^{\mathrm{d}} = Q^{\mathrm{d}} + F^{\mathrm{d}} = F^{\mathrm{d}}$. So that $I + R^{\mathrm{d}}T = I + F^{\mathrm{d}}T$ is generalized Drazin invertible because F and F^{d} are finite-rank operators. Now, it follows from [4, Theorem 1] that $T + R$ is generalized Drazin invertible.

(2) $\Rightarrow$ (1). In particular, for $T = 0$, we have $\sigma_{\mathrm{gD}}(R) = \sigma_{\mathrm{gD}}(0) = \emptyset$. So that the spectrum $\sigma(R)$ consists of isolated points, and hence it is finite. If we set $\sigma(R) = \{\alpha_1, \ldots, \alpha_n\}$, then there exist closed R-invariant subspaces X_i, $1 \le i \le n$, such that $X = X_1 \oplus \cdots \oplus X_n$ and $\sigma(R_{|X_i}) = \{\alpha_i\}$ for $1 \le i \le n$. Since R is a Riesz operator, then $\sigma_{\mathrm{e}}(R) = \cup_i \sigma_{\mathrm{e}}(R_{|X_i}) = \{0\}$, and once $\sigma_{\mathrm{e}}(R_{|X_i})$ is empty then X_i is finite-dimensional. Thus, we conclude that X_i is a finite-dimensional subspace whenever $\alpha_i \neq 0$. Without loss of generality, we may assume that $\alpha_1 = 0$ and $\alpha_i \neq 0$ for $2 \le i \le n$. With respect to the decomposition of X, consider the operators $Q, F \in \mathcal{B}(X)$ given by

$$Q = R_{|X_1} \oplus 0 \oplus \cdots \oplus 0 \quad \text{and} \quad F = 0 \oplus R_{|X_2} \oplus \cdots \oplus R_{|X_n}.$$

Clearly, Q is a quasi-nilpotent operator, F is a finite-rank operator, $R = Q + F$ and $QF = FQ$. This completes the proof. □

3 Commuting Perturbation Class of Meromorphic Operators

In this section, we completely characterize the commuting perturbations leaving invariant the set of meromorphic operators as stated in the following theorem:

Theorem 2 *We have*

$$\mathcal{P}_c(\mathcal{M}(X)) = \{F \in \mathcal{B}(X) : F^n \text{ has finite rank for some } n \geq 1\}.$$

Before presenting the proof of the above theorem, we shall first establish the following lemma.

Lemma 2 *The commutant of every bounded nilpotent operator on an infinite-dimensional complex Banach space contains a compact operator of infinite spectrum.*

Proof Let $S \in \mathcal{B}(X)$ be a nilpotent operator. We shall discuss two cases.

Case 1. $\operatorname{ran}(S)$ is infinite-dimensional. Let n_0 be the nilpotence index of S, and let $x_0 \in X$ be such that $S^{n_0-1}x_0 \neq 0$. Let $X = \operatorname{Span}\{S^k x_0 : 0 \leq k \leq n_0 - 1\} \oplus X_0$, where X_0 is a closed S-invariant subspace, and let $P_0 \in \mathcal{B}(X)$ the finite-rank projection onto $\operatorname{Span}\{S^k x_0 : 0 \leq k \leq n_0 - 1\}$. Clearly $SP_0 = P_0 S$. Since $\operatorname{ran}(S)$ is infinite-dimensional, then $S_{|X_1}$ is nilpotent of index $n_1 \geq 2$. Take $x_1 \in X_0$ with $S^{n_1-1}x_1 \neq 0$, and denote by X_1 an S-invariant subspace such that $X_0 = \operatorname{Span}\{S^k x_1 : 0 \leq k \leq n_1 - 1\} \oplus X_1$. Consider the projection $P_1 \in \mathcal{B}(X)$ onto $\operatorname{Span}\{S^k x_1 : 0 \leq k \leq n_1 - 1\}$. Then P_1 is finite-rank and commutes with S.

Repeating the same argument, we get linearly independent sets

$$\{S^k x_m : 0 \leq k \leq n_m - 1 \text{ and } m \geq 0\} \subset X \quad \text{and} \quad \{P_m : m \geq 0\} \subset \mathcal{B}(X)$$

such that P_m is a finite-rank projection onto $\operatorname{Span}\{S^k x_m : 0 \leq k \leq n_m - 1\}$ and $SP_m = P_m S$ for $m \geq 0$.

Now, consider the compact operator $K \in \mathcal{B}(X)$ given by $K = \sum_{i\geq 0} \alpha_i P_i$, where α_i are distinct complex numbers for which $\sum_{i\geq 0} |\alpha_i| \|P_i\|$ is finite. For every integer $n \geq 1$, we have $Kx_i = \alpha_i P_i x_i = \alpha_i x_i$. Thus, $\{0, \alpha_i : i \geq 0\} \subseteq \sigma(K)$.

Case 2. $\operatorname{ran}(S)$ is finite-dimensional. Since

$$\operatorname{codim}\,(\operatorname{ran}(S) + \ker(S)) = \dim \operatorname{ran}(S)/\operatorname{ran}(S^2) < \infty,$$

then there exist a finite-dimensional subspace Y containing $\operatorname{ran}(S)$ and a closed subspace $Z \subseteq \ker(S)$ such that $X = Y \oplus Z$. With respect to this decomposition, consider the operator $K \in \mathcal{B}(X)$ given by $K_{|Y} = 0$ and $K_{|Z}$ is an arbitrary compact operator of infinite spectrum. Then K is compact of infinite spectrum and commutes with S. This finishes the proof. □

The following question arises in a natural way from the previous result.

Question: Does the commutant of a bounded quasi-nilpotent operator on an infinite-dimensional complex Banach space contains an operator of infinite spectrum?

Recall that the *descent* $\mathrm{d}(T)$ of $T \in \mathcal{B}(X)$ is defined by

$$\mathrm{d}(T) = \inf\{n \geq 0 \colon \operatorname{ran}(T^n) = \operatorname{ran}(T^{n+1})\},$$

where the infimum over the empty set is taken to be infinite (see [12]). The descent spectrum, $\sigma_{\text{des}}(T)$, is defined as those complex numbers λ for which $\mathrm{d}(T-\lambda)$ is not finite.

Note that meromorphic operators are characterized by the descent spectrum as follows (see [3, Corollary 1.9]): An operator

$$T \in \mathcal{B}(X) \text{ is meromorphic if and only if } \sigma_{\text{des}}(T) \subseteq \{0\}.$$

Proof (Proof of Theorem 2*)* Let $F \in \mathcal{B}(X)$ be such that F^n is of finite rank for some $n \geq 1$. Hence, we get from [3, Theorem 3.1] that

$$\sigma_{\text{des}}(T) = \sigma_{\text{des}}(T+F)$$

for every $T \in \mathcal{B}(X)$ commuting with F. Hence, we infer that $T+F$ is a meromorphic operator for every meromorphic operator $T \in \mathcal{B}(X)$ commuting with F. This implies that $F \in \mathcal{P}_{\text{c}}(\mathcal{M}(X))$.

Conversely, let $F \in \mathcal{P}_{\text{c}}(\mathcal{M}(X))$ be non-zero. Then, for every $\lambda \in \mathbb{C}$, the operator $F-\lambda$ is meromorphic, and so $\sigma_{\text{des}}(F-\lambda) \subseteq \{0\}$. Thus, we infer that $\sigma_{\text{des}}(F) = \emptyset$, and therefore F is algebraic by [3, Theorem 1.5]. Write $X = X_1 \oplus \cdots \oplus X_n$ where $X_k = \ker(F-\lambda_k)^{m_k}$ for $1 \leq k \leq n$, and the scalars λ_k are distinct. Clearly, the restriction F_k of $F-\lambda_k$ to X_k is nilpotent for every $1 \leq k \leq n$.

It suffices to show that $\dim X_i < \infty$ for every $\lambda_i \neq 0$. Suppose that there exists $\lambda_i \neq 0$ such that $\dim X_i = \infty$. It follows from the previous lemma that the commutant of F_i contains a compact operator K_i of infinite spectrum. Consider the operator $K \in \mathcal{B}(X)$ given by $K_{|X_j} = 0$ for $j \neq i$ and $K_{|X_i} = K_i$. Clearly, K is compact of infinite spectrum and commutes with F. Thus, K is a non-algebraic meromorphic operator, and so $\sigma_{\text{des}}(K) = \{0\}$. Since

$$\sigma_{\text{des}}(K+F-\lambda_i) = \sigma_{\text{des}}(K_i+F_i-\lambda_i) = \sigma_{\text{des}}(K_i) = \sigma_{\text{des}}(K) = \{0\},$$

then $\sigma_{\text{des}}(K+F) = \{\lambda_i\}$, that is $K+F$ is non-meromorphic. This contradiction finishes the proof. □

References

1. Aiena, P., González, M.: Intrinsic characterizations of perturbation classes on some Banach spaces. Arch. Math. **94**, 373–381 (2010)
2. Ben-Israel, A., Greville, T.N.E.: Generalised Inverses: Theory and Applications. Springer, New York (2003)
3. Burgos, M., Kaidi, A., Mbekhta, M., Oudghiri, M.: The descent spectrum and perturbations. J. Oper. Theory **56**, 259–271 (2006)
4. Deng, C., Wei, Y.: New additive results for the generalized Drazin inverse. J. Math. Anal. Appl. **370**, 313–321 (2010)

5. Drazin, M.P.: Pseudo-inverses in associative rings and semigroups. Am. Math. Mon. **65**, 506–514 (1958)
6. González, M.: The perturbation classes problem in Fredholm theory. J. Funct. Anal. **200**, 65–70 (2003)
7. González, N.C., Koliha, J.J.: Perturbation of the Drazin inverse for closed linear operators. Integral Equ. Oper. Theory **36**, 92–106 (2000)
8. González, M., Martínez-Abejón, A., Pello, J.: A survey on the perturbation classes problem for semi-Fredholm and Fredholm operators. Funct. Anal. Approx. Comput. **7**, 75–87 (2015)
9. Koliha, J.J.: A generalized Drazin inverse. Glasg. Math. J. **38**, 367–381 (1996)
10. Koliha, J.J., Tran, T.D.: Semistable operators and singularly perturbed differential equations. J. Math. Anal. Appl. **231**, 446–458 (1999)
11. Lebow, A., Schechter, M.: Semigroups of operators and measures of noncompactness. J. Funct. Anal. **7**, 1–26 (1971)
12. Müller, V.: Spectral Theory of Linear Operators and Spectral Systems in Banach Algebras. Operator Theory: Advances and Applications, vol. 139, 2nd edn. Birkhäuser Verlag, Basel (2007)
13. Ounadjela, D., Hocine, K.M., Messirdi, B.: The perturbation classes problem for generalized Drazin invertible operators I. Rend. Circ. Mat. Palermo **67**(2), 159–172 (2018)
14. Rakočević, V.: Koliha-Drazin invertible operators and commuting Riesz perturbations. Acta Sci. Math. (Szeged) **68**, 291–301 (2002)
15. West, T.T.: The decomposition of Riesz operators. Proc. Lond. Math. Soc. **16**, 737–752 (1966)
16. Živković-Zlatanović, S.Č., Djordjević, D.S., Harte, R.: Ruston, Riesz and perturbation classes. J. Math. Anal. Appl. **389**, 871–886 (2012)

Generalized Rigid Modules and Their Polynomial Extensions

Mohamed Louzari and Armando Reyes

Abstract Let R be a ring with unity, σ an endomorphism of R and M_R a right R-module. In this paper, we study some connections between rigid, σ-rigid, semicommutative, σ-semicommutative, abelian and σ-reduced modules. Also, we show that the class of σ-rigid modules is not closed under homomorphic images and some module extensions. Moreover, we examine the transfer of σ-reducibly, σ-semicommutative and σ-rigidness from a module M_R to its extensions of the form $M[x]/M[x](x^n)$ where $n \geq 2$ is an integer, and vice versa.

Keywords Generalized rigid modules · Rigid modules · Reduced modules · Semicommutative modules

1 Introduction

Let us fix some useful notations and recall the main notions that we shall use in the sequel. In this paper, R denotes an associative ring with unity and modules are unitary. We write M_R to mean that M is a right module. The tensor product between right and left R-modules, is denoted by $M \otimes_R N$. Let $\mathbf{N}$ be the set of all natural integers. Throughout, σ is an endomorphism of R with $\sigma(1) = 1$, and $R[x; \sigma]$ the associated skew polynomial extension, that is, an extension of R, whose rule of multiplication is given by $rx = \sigma(r)x$, for every $r \in R$. We also consider its localization $R[x, x^{-1}; \sigma]$ at the multiplicative set generated by x, that is, the attached skew Laurent polynomial ring of R. The set of all endomorphisms (respectively, automorphisms) of R is denoted by $End(R)$ (respectively, $Aut(R)$). For a subset X of a module M_R, $r_R(X) = \{a \in R | Xa = 0\}$ and $\ell_R(X) = \{a \in R | aX = 0\}$ will

M. Louzari (✉)
Department of Mathematics, Abdelmalek Essaadi University, Tetouan, Morocco
e-mail: mlouzari@yahoo.com

A. Reyes
Departamento de Matemáticas, Universidad Nacional de Colombia, Bogotá, Colombia
e-mail: mareyesv@unal.edu.co

M. Siles Molina et al. (eds.), *Associative and Non-Associative Algebras and Applications*, Springer Proceedings in Mathematics & Statistics 311,
https://doi.org/10.1007/978-3-030-35256-1_8

stand for the right and the left annihilator of X in R respectively. A ring R is called *semicommutative* if for every $a \in R$, $r_R(a)$ is an ideal of R (equivalently, for any $a, b \in R, ab = 0$ implies $aRb = 0$). In [6], a module M_R is called semicommutative, if for any $m \in M$ and $a \in R$, $ma = 0$ implies $mRa = 0$. A module M_R is called σ-semicommutative [7] if, for any $m \in M$ and $a \in R$, $ma = 0$ implies $mR\sigma(a) = 0$.

Following Annin [4], a module M_R is *σ-compatible*, if for any $m \in M$ and $a \in R$, $ma = 0$ if and only if $m\sigma(a) = 0$. An σ-compatible module M is called σ-reduced if for any $m \in M$ and $a \in R$, $ma = 0$ implies $mR \cap Ma = 0$. The module M_R is called reduced if it is id_R-reduced, where id_R is the identity endomorphism of R (equivalently, for any $m \in M$ and $a \in R$, $ma^2 = 0$ implies $mRa = 0$) [14].

According to Krempa [13], an endomorphism σ of a ring R is called *rigid*, if $a\sigma(a) = 0$ implies $a = 0$ for any $a \in R$, and R is called a *σ-rigid* ring if the endomorphism σ is rigid. Motivated by the properties of σ-rigid rings that have been studied in [10–13], Guner et al. [9], introduced the notion of σ-rigid module as a generalization of σ-rigid ring. A module M_R is said to be *σ-rigid*, if $ma\sigma(a) = 0$ implies $ma = 0$ for any $m \in M$ and $a \in R$. Clearly, σ-reduced modules are σ-rigid, but the converse need not be true [9, Example 2.18]. Recall that, a module M_R is σ-reduced if and only if it is σ-semicommutative and σ-rigid [9, Theorem 2.16]. Thus, the concept of σ-reduced modules is connected to the σ-semicommutative and σ-rigid modules. A module M_R is called *rigid*, if it is id_R-rigid. Recall that, a module M is called *abelian* [1], if for any $m \in M$ and any $e^2 = e, r \in R$, we have $mer = mre$.

Given a right R-module M, Lee–Zhou [14] introduced the following notations:

$$M[x;\sigma] := \left\{ \sum_{i=0}^{s} m_i x^i \,:\, s \geq 0, m_i \in M \right\},$$

$$M[x, x^{-1};\sigma] := \left\{ \sum_{i=-s}^{t} m_i x^i \,:\, t \geq 0, s \geq 0, m_i \in M \right\}.$$

Each of these is an abelian group under an obvious addition operation, and thus $M[x;\sigma]$ becomes a module over $R[x;\sigma]$ under the following scalar product operation. For $m(x) = \sum_{i=0}^{n} m_i x^i \in M[x;\sigma]$ and $f(x) = \sum_{j=0}^{m} a_j x^j \in R[x;\sigma]$,

$$m(x)f(x) = \sum_{k=0}^{n+m} \left(\sum_{k=i+j} m_i \sigma^i(a_j) \right) x^k \tag{1}$$

The module $M[x;\sigma]$ is called the *skew polynomial extension* of M. If $\sigma \in Aut(R)$, then with a scalar product similar to that given in Equation 1, $M[x, x^{-1};\sigma]$ becomes a module over $R[x, x^{-1};\sigma]$, this is called the *skew Laurent polynomial extension* of M. In particular, if $\sigma = id_R$ then we get the classical extensions $M[x]$ and $M[x, x^{-1}]$. Both modules are in fact particular case of the tensor product. Namely considering

$R[x;\sigma]$ as an $(R, R[x;\sigma])$-bimodule (here R acts on the left while $R[x;\sigma]$ acts on the right), then for any right R-module M, we have the right $R[x;\sigma]$-module $M \otimes_R R[x;\sigma]$, which is naturally isomorphic to $M[x;\sigma]$. A similar natural isomorphism occurs between $M \otimes_R R[x, x^{-1};\sigma]$ and $M[x, x^{-1};\sigma]$.

According to Zhang and Chen [7], a module M_R is called *σ-skew Armendariz*, if $m(x)f(x) = 0$ where $m(x) = \sum_{i=0}^{n} m_i x^i \in M[x;\sigma]$ and $f(x) = \sum_{j=0}^{m} a_j x^j \in R[x;\sigma]$ implies $m_i\sigma^i(a_j) = 0$ for all i and j. Also, M_R is an Armendariz module if and only if it is id_R-skew Armendariz. A ring R is skew-Armendariz if and only if R_R is a skew-Armendariz module.

In this work, we continue studding σ-rigid modules that were introduced by Guner et al. [9]. Indeed, We give some results on σ-rigid modules and related concepts as semicommutative, σ-semicommutative, abelian and σ-reduced modules. Also, we show that the class of σ-rigid modules is not closed under homomorphic images and some module extensions. Also, we study the transfer of σ-rigidness from a module M_R to its extensions, as triangular matrix modules and polynomial modules. In fact, if $M[x]/M[x](x^n)$ is $\overline{\sigma}$-reduced (respectively, $\overline{\sigma}$-rigid, $\overline{\sigma}$-semicommutative) as a right $R[x]/(x^n)$-module, then M_R is σ-reduced (respectively, σ-rigid, σ-semicommutative). However, the converse is not true by Examples 3, 4 and 5. Furthermore, for a module M_R, we study the behavior of σ-rigidness regarding to its skew polynomial extensions.

2 Generalized Rigid Modules and Related Modules

In this section, we give some connections between rigid, σ-rigid, semicommutative, σ-semicommutative. We begin with the subsequent definition.

Definition 1 Let M_R be a module, $m \in M$ and $a \in R$. We say that M_R satisfies the condition:

($\mathcal{C}_1$) if $ma = 0$ implies $m\sigma(a) = 0$.
($\mathcal{C}_2$) if $m\sigma(a) = 0$ implies $ma = 0$.

Lemma 1 *Given a module M_R. Then*

1. *Every rigid (respectively, σ-semicommutative) module satisfying ($\mathcal{C}_2$) is σ-rigid (respectively, semicommutative).*
2. *Every σ-rigid (respectively, semicommutative) module satisfying ($\mathcal{C}_1$) is rigid (respectively, σ-semicommutative).*
3. *For σ-compatible modules, the notions of rigidness (respectively, semicommutativity) and σ-rigidness (respectively, σ-semicommutativity) coincide.*

Proof Straightforward. □

Lemma 2 *Let M_R be a module and σ an endomorphism of R. Then*

1. *If M_R is rigid, then it is abelian,*
2. *If M_R is σ-rigid which satisfies the condition $(\mathcal{C}_1)$, then it is abelian,*
3. *If M_R is σ-semicommutative, then it is abelian.*

Proof (1) Let $e^2 = e, r \in R$ and $m \in M$, we have $[er(1-e)]^2 = 0$ then $m[er(1-e)]^2 = 0$, since M_R is rigid then $m[er(1-e)] = 0$ which gives $mer = mere$. On the other side, we have $[(1-e)re]^2 = 0$, with the same manner as above, it gives $mre = mere$. Thus $mer = mre$ for all $r \in R$.
(2) If M_R is σ-rigid with the condition $(\mathcal{C}_1)$, then it is rigid by Lemma 1(2).
(3) Suppose that M_R is σ-semicommutative, then M_R satisfies the condition $(\mathcal{C}_1)$. Let $m \in M$ and $e^2 = e \in R$, we have $me(1-e) = 0$ then $meR\sigma((1-e)) = 0$, so $mer(1-e) = 0$ for all $r \in R$, by [14, Lemma 2.9]. Also, we have $m(1-e)e = 0$ which gives $m(1-e)re = 0$ for all $r \in R$. Therefore $mer = mre$ for all $r \in R$. □

As the following example shows, the class of σ-rigid modules need not be closed under homomorphic images.

Example 1 Consider $R = \mathbf{Z}$, the ring of all integer numbers. Take $M = \mathbf{Z}_{\mathbf{Z}}$ and $N = 12\mathbf{Z}$ a submodule of M. It is clear that M is rigid, however M/N is not rigid. Consider $m = 3 + N \in M/N$ and $a = 2 \in R$. We have $ma^2 = 0$ and $ma \neq 0$.

Recall that a module M_R is said to be *torsion-free* if for every non zero-divisor $a \in R$ and $0 \neq m \in M$, we have $ma \neq 0$. A submodule N of a module M is called a *prime submodule* of M if whenever $ma \in N$ for $m \in M$ and $a \in R$, then $m \in N$ or $Ma \subseteq N$.

Proposition 1 *Let M_R be a module over a ring R that has no zero divisors, and N be a prime submodule of M such that M/N is torsion-free. Then M is σ-rigid if and only if N and M/N are σ-rigid.*

Proof ($\Rightarrow$). Clearly, N is σ-rigid. Now, let $\overline{m} \in M/N$ such that $\overline{m}a\sigma(a) = 0$, hence $ma\sigma(a) \in N$. Since N is prime then $m \in N$ or $Ma\sigma(a) \subseteq N$. If $m \in N$ then $ma \in N$, so $\overline{m}a = 0$. Now, if $Ma\sigma(a) \subseteq N$, then $(M/N)a\sigma(a) = 0$ in M/N. Since M/N is torsion-free and R without zero-divisors, we get $a\sigma(a) = 0$. Then $ma\sigma(a) = 0$, which implies $ma = 0$ because M is σ-rigid. Thus $\overline{m}a = 0$. Hence M/N is σ-rigid.
($\Leftarrow$). Let $m \in M$ and $a \in R$ such that $ma\sigma(a) = 0$. If $m \in N$, we have $ma = 0$ because N is σ-rigid. Now, suppose that $m \notin N$. Since N is prime, then from $ma\sigma(a) \in N$, we get $Ma\sigma(a) \subseteq N$. Hence $(M/N)a\sigma(a) = 0$ in M/N. But M/N is σ-rigid, then $(M/N)a = 0$ which implies $a = 0$, since M/N is torsion-free and R has no zero divisors. Therefore $ma = 0$. □

The next example shows that the condition "R has no zero divisors" is not superfluous in Proposition 1.

Example 2 Let F be a field and $R = \begin{pmatrix} F & F \\ 0 & F \end{pmatrix}$ its associated upper triangular matrix ring. Let σ be the endomorphism of R defined by:

$$\sigma\left(\begin{pmatrix} a & b \\ 0 & c \end{pmatrix}\right) = \begin{pmatrix} a & -b \\ 0 & c \end{pmatrix}.$$

Consider the right R-module $M = \begin{pmatrix} 0 & F \\ F & F \end{pmatrix}$ and the submodule $K = \begin{pmatrix} 0 & F \\ 0 & F \end{pmatrix}$. Then we can make the following observations:

(**1**) By [9, Example 2.6], we have that K and M/K are σ-rigid but M is not σ-rigid.

(**2**) K is a prime submodule of M. Namely, let $m = \begin{pmatrix} 0 & \beta \\ \alpha & \gamma \end{pmatrix} \in M$ and $a = \begin{pmatrix} x & y \\ 0 & z \end{pmatrix} \in R$. We have $ma = \begin{pmatrix} 0 & \beta z \\ \alpha x & \alpha y + \gamma z \end{pmatrix}$. Then $ma \in K \Leftrightarrow \alpha = 0$ or $x = 0$. If $\alpha = 0$, we have $m \in K$. Now, suppose that $\alpha \neq 0$, so $x = 0$. We can easily see that $Ma \subseteq K$. Thus K is prime.

(**3**) The ring R has zero divisors. In fact, for $a = \begin{pmatrix} 0 & 1 \\ 0 & -1 \end{pmatrix} \in R$, we can take any element of R of the form $b = \begin{pmatrix} \alpha & \alpha \\ 0 & 0 \end{pmatrix}$, with $\alpha \neq 0$. We have $ab = ba = 0$. Hence, a is a divisor of zero.

3 Triangular Matrix Modules and Polynomial Modules

In this section, we observe the σ-rigidness of some module extensions, as triangular matrix modules, ordinary polynomial modules and skew polynomial modules. As we will see in Examples 4 and 5, the class of σ-rigid modules need not be closed under module extensions.

For a nonnegative integer $n \geq 2$. Consider

$$V_n(M) := \left\{ \begin{pmatrix} m_0 & m_1 & m_2 & m_3 & \dots & m_{n-1} \\ 0 & m_0 & m_1 & m_2 & \dots & m_{n-2} \\ 0 & 0 & m_0 & m_1 & \dots & m_{n-3} \\ \vdots & \vdots & \vdots & \vdots & \ddots & \vdots \\ 0 & 0 & 0 & 0 & \dots & m_1 \\ 0 & 0 & 0 & 0 & \dots & m_0 \end{pmatrix} \;\middle|\; m_0, m_1, m_2, \cdots, m_{n-1} \in M \right\}$$

and

$$V_n(R) := \left\{ \begin{pmatrix} a_0 & a_1 & a_2 & a_3 & \dots & a_{n-1} \\ 0 & a_0 & a_1 & a_2 & \dots & a_{n-2} \\ 0 & 0 & a_0 & a_1 & \dots & a_{n-3} \\ \vdots & \vdots & \vdots & \vdots & \ddots & \vdots \\ 0 & 0 & 0 & 0 & \dots & a_1 \\ 0 & 0 & 0 & 0 & \dots & a_0 \end{pmatrix} \;\middle|\; a_0, a_1, a_2, \cdots, a_{n-1} \in R \right\}$$

We denote elements of $V_n(R)$ by $(a_0, a_1, \cdots, a_{n-1})$, and elements of $V_n(M)$ by $(m_0, m_1, \cdots, m_{n-1})$. Clearly, $V_n(M)$ is a right $V_n(R)$-module under the usual matrix addition operation and the following scalar product operation.

For $U = (m_0, m_1, \cdots, m_{n-1}) \in V_n(M)$ and $A = (a_0, a_1, \cdots, a_{n-1}) \in V_n(R)$, we have

$$UA = (m_0a_0, m_0a_1 + m_1a_0, \cdots, m_0a_{n-1} + m_1a_{n-2} + \cdots + m_{n-1}a_0) \in V_n(M).$$

Furthermore, $V_n(M) \cong M[x]/M[x](x^n)$ where $M[x](x^n)$ is a submodule of $M[x]$ generated by x^n and $V_n(R) \cong R[x]/(x^n)$ where (x^n) is an ideal of $R[x]$ generated by x^n. The endomorphism σ of R can be extended to $V_n(R)$ and $R[x]$, and we will denote it in both cases by $\overline{\sigma}$.

From now on, we will freely use the isomorphisms

$$V_n(M) \cong M[x]/M[x](x^n) \text{ and } V_n(R) \cong R[x]/(x^n).$$

Proposition 2 *Let M_R be a module and an integer $n \geq 2$. Then*

1. *If $M[x]/M[x](x^n)$ is $\overline{\sigma}$-semicommutative (as a right $R[x]/(x^n)$-module), then M_R is σ-semicommutative,*
2. *If $M[x]/M[x](x^n)$ is $\overline{\sigma}$-rigid (as a right $R[x]/(x^n)$-module), then M_R is σ-rigid.*

Proof Let $U = (m, 0, 0, \cdots, 0) \in V_n(M)$ and $A = (a, 0, 0, \cdots, 0) \in V_n(R)$ with $m \in M_R$ and $a \in R$.
(1) Suppose that $ma = 0$. Then $MA = (ma, 0, 0, \cdots, 0) = 0$, since $V_n(M)$ is $\overline{\sigma}$-semicommutative, we have $UB\overline{\sigma}(A) = 0$ for any $B = (r, 0, \cdots, 0) \in V_n(R)$ with r be an arbitrary element of R. But $UB\overline{\sigma}(A) = (mr\sigma(a), 0, \cdots, 0) = 0$, hence $mr\sigma(a) = 0$ for any $r \in R$. Therefore M_R is σ-semicommutative.
(2) Suppose that $ma\sigma(a) = 0$. Then $UA\overline{\sigma}(A) = (ma\sigma(a), 0, 0, \cdots, 0) = 0$. Since $V_n(M)_{V_n(R)}$ is $\overline{\sigma}$-rigid, we get $UA = 0 = (ma, 0, 0, \cdots, 0)$, so $ma = 0$. Therefore M_R is σ-rigid. □

The following are immediate corollaries of Proposition 2.

Corollary 1 *Let M_R be a module and $n \geq 2$ an integer. If $M[x]/M[x](x^n)$ is $\overline{\sigma}$-reduced (as a right $R[x]/(x^n)$-module), then M_R is σ-reduced.*

Corollary 2 ([3, Proposition 2.14]) *If the ring $R[x]/(x^n)$ is semicommutative for any integer $n \geq 2$. Then R is a semicommutative ring.*

All reduced rings are Armendariz. There are non-reduced Armendariz rings. For a positive integer n, let $\mathbf{Z}/n\mathbf{Z}$ denotes ring of integers modulo n. For each positive integer n, the ring $\mathbf{Z}/n\mathbf{Z}$ is Armendariz but not reduced whenever n is not square free [8, Proposition 2.1].

Corollary 3 *If $M[x]/M[x](x^n)$ is $\overline{\sigma}$-reduced as a right $R[x]/(x^n)$-module, then M_R is σ-skew Armendariz.*

Proof If M_R is σ-reduced then it is σ-skew Armendariz (see [2, Theorem 2.20]). □

Note that, the converse of Corollary 3 need not be true, since $M = \mathbf{Z}/n\mathbf{Z}$ is Armendariz but not reduced (n is not square free) and so $V_n(M)$ is not reduced, by Corollary 1. Furthermore, the converse of Proposition 2 and Corollary 1 may not be true, as the subsequent examples show. For the first one, if M_R is σ-semicommutative then the right $R[x]/(x^n)$-module $M[x]/M[x](x^n)$ need not be $\overline{\sigma}$-semicommutative.

Example 3 Consider the ring $R = \left\{\begin{pmatrix} a & b \\ 0 & a \end{pmatrix} \mid a, b \in \mathbf{Z}\right\}$, with an endomorphism σ defined by $\sigma\left(\begin{pmatrix} a & b \\ 0 & a \end{pmatrix}\right) = \begin{pmatrix} a & -b \\ 0 & a \end{pmatrix}$. The module R_R is σ-semicommutative by [5, Example 2.5(1)]. Take

$$A = \begin{pmatrix} \begin{pmatrix} 0 & 1 \\ 0 & 0 \end{pmatrix} & \begin{pmatrix} -1 & 1 \\ 0 & -1 \end{pmatrix} \\ \begin{pmatrix} 0 & 0 \\ 0 & 0 \end{pmatrix} & \begin{pmatrix} 0 & 1 \\ 0 & 0 \end{pmatrix} \end{pmatrix}, \quad B = \begin{pmatrix} \begin{pmatrix} 0 & 1 \\ 0 & 0 \end{pmatrix} & \begin{pmatrix} 1 & 1 \\ 0 & 1 \end{pmatrix} \\ \begin{pmatrix} 0 & 0 \\ 0 & 0 \end{pmatrix} & \begin{pmatrix} 0 & 1 \\ 0 & 0 \end{pmatrix} \end{pmatrix} \in V_2(R).$$

We have $AB = 0$, but for $C = \begin{pmatrix} \begin{pmatrix} 1 & 0 \\ 0 & 1 \end{pmatrix} & \begin{pmatrix} 0 & 0 \\ 0 & 0 \end{pmatrix} \\ \begin{pmatrix} 0 & 0 \\ 0 & 0 \end{pmatrix} & \begin{pmatrix} 1 & 0 \\ 0 & 1 \end{pmatrix} \end{pmatrix} \in V_2(R)$, we get

$$AC\overline{\sigma}(B) = \begin{pmatrix} \begin{pmatrix} 0 & 0 \\ 0 & 0 \end{pmatrix} & \begin{pmatrix} 0 & 2 \\ 0 & 0 \end{pmatrix} \\ \begin{pmatrix} 0 & 0 \\ 0 & 0 \end{pmatrix} & \begin{pmatrix} 0 & 0 \\ 0 & 0 \end{pmatrix} \end{pmatrix} \neq 0.$$

Thus, $V_2(R)$ is not $\overline{\sigma}$-semicommutative as a right $V_2(R)$-module.

Now, if M_R is σ-reduced (respectively, σ-rigid) then the right $R[x]/(x^n)$-module $M[x]/M[x](x^n)$ is not $\overline{\sigma}$-reduced (respectively, $\overline{\sigma}$-rigid).

Example 4 Let $\mathbf{K}$ be a field, then $\mathbf{K}$ is reduced as a right $\mathbf{K}$-module. Consider the polynomial $p = \overline{x} \in \mathbf{K}[x]/x^2\mathbf{K}[x]$. We have $p \neq \overline{0}$, because $x \notin x^2\mathbf{K}[x]$ but $p^2 = \overline{0}$. Therefore, $\mathbf{K}[x]/x^2\mathbf{K}[x] \simeq V_2(\mathbf{K})$ is not reduced as a right $V_2(\mathbf{K})$-module

Example 5 Let $R = \mathbf{K}\langle x, y\rangle$ be the ring of polynomials in two non-commuting indeterminates over a field $\mathbf{K}$. Consider the right R-module $M = R/xR$. The module M is rigid but not semicommutative by [9, Example 2.18].

Now, for $U = \begin{pmatrix} 1 + xR & xR \\ 0 & 1 + xR \end{pmatrix} \in V_2(M)$ and $A = \begin{pmatrix} 0 & 1 \\ 0 & 0 \end{pmatrix} \in V_2(R)$. We have $UA^2 = 0$, however $UA = \begin{pmatrix} 0 & 1 + xR \\ 0 & 0 \end{pmatrix} \neq 0$. Thus $V_2(M)$ is not rigid as a right $V_2(R)$-module.

Let $n \geq 2$ be an integer, we have:

$$\begin{array}{c} V_n(M)\ is\ \overline{\sigma}-rigid \\ \not\Uparrow\ \Downarrow \\ M_R\ is\ \sigma-rigid \\ \Uparrow \\ M[x]\ is\ \overline{\sigma}-rigid \end{array}$$

Proposition 3 *Let $n \geq 2$ be an integer and M_R a σ-rigid module. If M_R is σ-semicommutative, then $M[x]/M[x](x^n)$ is $\overline{\sigma}$-semicommutative as a right $R[x]/(x^n)$-module.*

Proof Let $U = (m_0, m_1, \cdots, m_{n-1}) \in V_n(M)$ and $A = (a_0, a_1, \cdots, a_{n-1})$, $B = (b_0, b_1, \cdots, b_{n-1}) \in V_n(R)$ such that $UA = 0$. We will show $UB\overline{\sigma}(A) = 0$. We have $UA = (m_0a_0, m_0a_1 + m_1a_0, \cdots, m_0a_{n-1} + \cdots + m_{n-1}a_0)$, $UB = (m_0b_0, m_0b_1 + m_1b_0, \cdots, m_0b_{n-1} + \cdots + m_{n-1}b_0) = (\beta_0, \beta_1, \cdots, \beta_{n-1})$ and $UB\overline{\sigma}(A) = (\beta_0\sigma(a_0), \beta_0\sigma(a_1) + \beta_1\sigma(a_0), \cdots, \beta_0\sigma(a_{n-1}) + \beta_1\sigma(a_{n-2}) + \cdots + \beta_{n-1}\sigma(a_0))$. From $UA = 0$, we get the following system of equations.

$$0 = m_0a_0 \tag{0}$$

$$= m_0a_1 + m_1a_0 \tag{1}$$

$$\vdots$$

$$= m_0a_{n-1} + m_1a_{n-2} + \cdots + m_{n-1}a_0 \tag{$n-1$}$$

Equation (0) implies $m_0a_0 = 0$, then $m_0R\sigma(a_0) = 0$. Multiplying Equation (1) on the right side by $\sigma(a_0)$, we get $m_0a_1\sigma(a_0) + m_1a_0\sigma(a_0) = 0$ $(1')$. Since $m_0a_1\sigma(a_0) = 0$ because $m_0R\sigma(a_0) = 0$, we have $m_1a_0\sigma(a_0) = 0$. But M_R is σ-rigid then $m_1a_0 = 0$. From Equation (1) we have also $m_0a_1 = 0$. For both cases, we have $m_1R\sigma(a_0) = m_0R\sigma(a_1) = 0$. Summarizing at this point: $m_0a_0 = m_0a_1 = m_1a_0 = 0$ and $m_0R\sigma(a_0) = m_0R\sigma(a_1) = m_1R\sigma(a_0) = 0$. Suppose that the result is true until i, multiplying Equation (i+1) on the right side by $\sigma(a_0)$, we get

$$0 = m_0a_{i+1}\sigma(a_0) + m_1a_i\sigma(a_0) + \cdots + m_ia_1\sigma(a_0) + m_{i+1}a_0\sigma(a_0) \tag{$(i+1)'$}$$

By the inductive hypothesis, we have

$$m_0a_{i+1}\sigma(a_0) = m_1a_i\sigma(a_0) = \cdots = m_ia_1\sigma(a_0) = 0.$$

Then $m_{i+1}a_0\sigma(a_0) = 0$, which gives $m_{i+1}a_0 = 0$ and so $m_{i+1}R\sigma(a_0) = 0$. Continuing with the same manner by multiplying Equation $(i+1)'$ on the right side, respectively by $\sigma(a_1), \sigma(a_2), \cdots, \sigma(a_i)$ we get $m_ka_{i+1-k} = 0$ for all $k = 0, 1, \cdots, i+1$. Hence, we get the result for $i+1$. That is $m_ia_j = 0$, and so $m_iR\sigma(a_j) = 0$ for all integers i and j with $i + j \leq n - 1$. Therefore, all components of $UB\overline{\sigma}(A)$ are zero. Thus $V_n(M)$ is $\overline{\sigma}$-semicommutative. □

Corollary 4 *Let M be a right R-module.*

1. *If M_R is σ-reduced then $M[x]/M[x](x^n)$ is $\overline{\sigma}$-semicommutative as a right $R[x]/(x^n)$-module.*
2. *If M_R is reduced then $M[x]/M[x](x^n)$ is semicommutative as a right $R[x]/(x^n)$-module.*
3. *If R is reduced then $R[x]/(x^n)$ is semicommutative.*

Corollary 5 ([7, Theorem 3.9]) *Let M be a right R-module.*

1. *If M_R is a σ-rigid module. Then M_R is σ-semicommutative if and only if $M[x]/M[x](x^n)$ is $\overline{\sigma}$-semicommutative as a right $R[x]/(x^n)$-module.*
2. *If M_R is a rigid module. Then M_R is semicommutative if and only if $M[x]/M[x](x^n)$ is semicommutative as a right $R[x]/(x^n)$-module.*

Proof It follows from Propositions 2(1) and 3. □

Corollary 6 *Let $n \geq 2$ be an integer. If $M[x]/M[x](x^n)$ is Armendariz as a right $R[x]/(x^n)$-module, then $M[x]/M[x](x^n)$ is semicommutative as a right $R[x]/(x^n)$-module.*

Proof It follows from [14, Theorem 1.9] and Corollary 4(2). □

For $U = (m_0, m_1, \cdots, m_{n-1}) \in V_n(M)$, $A = (a_0, a_1, \cdots, a_{n-1}) \in V_n(R)$. Consider $\alpha_i \in V_n(M)$, such that $\alpha_i = m_0a_i + m_1a_{i-1} + \cdots + m_ia_0$ for all $i \in \{0, 1, \cdots, n-1\}$. We have $UA = (\alpha_0, \alpha_1, \cdots, \alpha_{n-1})$ and $UA\overline{\sigma}(A)=(\alpha_0\sigma(a_0), \alpha_0\sigma(a_1) + \alpha_1\sigma(a_0), \cdots, \alpha_0\sigma(a_{n-1}) + \alpha_1\sigma(a_{n-2}) + \cdots + \alpha_{n-1}\sigma(a_0))$.

Proposition 4 *Let $n \geq 2$ be an integer and M_R a σ-reduced module. If $UA\overline{\sigma}(A) = 0$ then $\alpha_i\sigma(a_j) = 0$ for all nonnegative integers i, j with $i + j = 0, 1, \cdots, n-1$.*

Proof From $UA\overline{\sigma}(A) = 0$, we get the following system of equations.

$$0 = m_0\sigma(a_0) \quad (0)$$

$$= m_0\sigma(a_1) + m_1\sigma(a_0) \quad (1)$$

$$\vdots$$

$$= m_0\sigma(a_{n-1}) + m_1\sigma(a_{n-2}) + \cdots + m_{n-1}\sigma(a_0) \quad (n-1)$$

For $i + j = 0$, Equation (0) implies $\alpha_0\sigma(a_0) = 0$. Assume that $k \geq 0$, and suppose that $\alpha_i\sigma(a_j)$ for all i, j with $i + j \leq k$. Now, multiplying Equation $(k + 1)$ on the right side by $\sigma^2(a_0)$ we get:

$$\alpha_0\sigma(a_{k+1})\sigma^2(a_0) + \alpha_1\sigma(a_k)\sigma^2(a_0) + \cdots + \alpha_{k+1}\sigma(a_0)\sigma^2(a_0) = 0 \qquad (k+1)'$$

By the inductive hypothesis, we have $\alpha_i\sigma(a_0) = 0$ for all $0 \leq i \leq k$, then $\alpha_i R\sigma^2(a_0) = 0$ for all $0 \leq i \leq k$ because M_R is σ-semicommutative. Thus, Equation $(k+1)'$ gives $\alpha_{k+1}\sigma(a_0)\sigma^2(a_0) = 0$ which implies $\alpha_{k+1}\sigma(a_0) = 0$, by the σ-rigidness of M_R. So, Equation $(k+1)$ becomes:

$$\alpha_0\sigma(a_{k+1}) + \alpha_1\sigma(a_k) + \cdots + \alpha_k\sigma(a_1) = 0 \qquad (k+1)''$$

Multiplying Equation $(k+1)''$ on the right side by $\sigma^2(a_1)$ and use the fact that $\alpha_i R\sigma^2(a_1) = 0$ for all $0 \leq i \leq k-1$, we get $\alpha_k\sigma(a_1)\sigma^2(a_1) = 0$, which implies $\alpha_k\sigma(a_1) = 0$. Continuing this procedure yields $\alpha_i(a_j) = 0$ for all i, j such that $i + j = k+1$. Therefore, $\alpha_i\sigma(a_j) = 0$ for all nonnegative integers i, j with $i + j = 0, 1, \cdots, n-1$. □

Let R be a commutative domain. The set $T(M) = \{m \in M \mid r_R(m) \neq 0\}$ is called the *torsion submodule* of M_R. If $T(M) = M$ (respectively, $T(M) = 0$) then M_R is *torsion* (respectively, *torsion-free*).

Corollary 7 *Let R be a commutative domain, M_R a torsion-free module. Then for any integer $n \geq 0$, we have:*

i. M_R is σ-reduced if and only if $M[x]/M[x](x^n)$ is $\overline{\sigma}$-reduced as $R[x]/(x^n)$-module.
ii. M_R is reduced if and only if $M[x]/M[x](x^n)$ is reduced as $R[x]/(x^n)$-module.

Proof (i) ($\Leftarrow$) Obvious from Corollary 1. ($\Rightarrow$) From Proposition 3, $V_n(M)$ is $\overline{\sigma}$-semicommutative. On the other hand, if we take U and A as in Proposition 4 such that $UA\overline{\sigma}(A) = 0$, then we get $\alpha_i\sigma(a_j) = 0$ for all i, j, then $\alpha_i = 0$ for all i, because M_R is torsion-free and so $UA = 0$. Therefore $V_n(M)$ is $\overline{\sigma}$-rigid. Hence $V_n(M)$ is $\overline{\sigma}$-reduced. (ii) is obvious from (i). □

A regular element of a ring R means a nonzero element which is not zero divisor. Let S be a multiplicatively closed subset of R consisting of regular central elements. We may localize R and M at S. Let σ be an endomorphism of R, consider the map $S^{-1}\sigma : S^{-1}R \to S^{-1}R$ defined by $S^{-1}\sigma(a/s) = \sigma(a)/s$ with $\sigma(s) = s$ for any $s \in S$. Then $S^{-1}\sigma$ is an endomorphism of the ring $S^{-1}R$. Clearly $S^{-1}\sigma$ extends σ, we will denote it by σ. In the next, we will discuss when the localization $S^{-1}M$ is σ-rigid as a right $S^{-1}R$-module.

Lemma 3 *For a multiplicatively closed subset S of a ring R consisting of all central regular elements. A right R-module M is σ-rigid if and only if $S^{-1}M$ is σ-rigid (as a right $S^{-1}R$-module).*

Proof Clearly, if $S^{-1}M$ is σ-rigid then M is σ-rigid, because the class of σ-rigid modules is closed under submodules. Conversely, suppose that M_R is σ-rigid, let $(m/s) \in S^{-1}M$ and $(a/t) \in S^{-1}R$ such that $(m/s)(a/t)\sigma((a/t)) = 0$ in $S^{-1}M$. Then $ma\sigma(a) = 0$, by hypothesis $ma = 0$. Hence $(m/s)(a/t) = 0$, so that $S^{-1}M$ is σ-rigid. □

Proposition 5 *Let M be a right R-module. Then $M[x]_{R[x]}$ is σ-rigid if and only if $M[x, x^{-1}]_{R[x,x^{-1}]}$ is σ-rigid.*

Proof Consider $S = \{1, x, x^2, \cdots\} \subseteq R[x]$. It is clear that S is a multiplicatively closed subset of the ring $R[x]$ consisting of all central regular elements. Also, we have $S^{-1}M[x] = M[x, x^{-1}]$ and $S^{-1}R[x] = R[x, x^{-1}]$. By Lemma 3, we get the result. □

Corollary 8 *Let R be a ring and M a right R-module. Then*
(1) $M[x]_{R[x]}$ is rigid if and only if $M[x, x^{-1}]_{R[x,x^{-1}]}$ is rigid,
(2) $R[x]$ is rigid if and only if $R[x, x^{-1}]$ is rigid.

Proof Clearly from Proposition 5. □

Proposition 6 *If $M[x; \sigma]_{R[x;\sigma]}$ is rigid then M_R is σ-rigid.*

Proof Assume that $M[x; \sigma]$ is rigid. Consider $m(x) = m \in M[x; \sigma]$ and $f(x) = ax \in R[x; \sigma]$ such that $ma\sigma(a) = 0$. We have $m(x)[f(x)]^2 = ma\sigma(a)x^2 = 0$, then $m(x)f(x) = 0$ because $M[x; \sigma]$ is rigid. But $m(x)f(x) = 0$ implies $ma = 0$. Which completes the proof. □

According to Krempa [13], if a ring R is σ-rigid then $R[x; \sigma]$ is reduced, hence $R[x; \sigma]$ is rigid. Also, if a module M_R is σ-rigid and σ-semicommutative, then $M[x; \sigma]_{R[x;\sigma]}$ is reduced [14, Theorem 1.6]. However, we do not know if we can drop the σ-semicommutativity from this implication?

Acknowledgements The authors are deeply indebted to the referee for many helpful comments and suggestions for the improvement of this paper.

References

1. Agayev, N., Gungoroglu, G., Harmanci, A., Halicioğlu, S.: Abelian modules. Acta Math. Univ. Comen. **78**(2), 235–244 (2009)
2. Agayev, N., Harmanci, A., Halicioğlu, S.: On reduced modules. Commun. Fac. Sci. Univ. Ank. Series A1 **58**(1), 9–16 (2009)
3. Agayev, N., Harmanci, A.: On semicommutative modules and rings. Kyungpook Math. J. **47**, 21–30 (2007)
4. Annin, S.: Associated primes over skew polynomials rings. Comm. Algebra **30**, 2511–2528 (2002)
5. Başer, M., Harmanci, A., Kwak, T.K.: Generalized semicommutative rings and their extensions. Bull. Korean Math. Soc. **45**(2), 285–297 (2008)
6. Buhphang, A.M., Rege, M.B.: Semicommutative modules and Armendariz modules. Arab J. Math. Sci. **8**, 53–65 (2002)
7. Chen, J.L., Zhang, C.P.: σ-skew Armendariz modules and σ-semicommutative modules. Taiwanese J. Math. **12**(2), 473–486 (2008)
8. Chhawchharia, S., Rege, M.B.: Armendariz rings. Proc. Japan. Acad. Ser. A Math. Sci. **73**, 14–17 (1997)

9. Guner, E., Halicioglu, S.: Generalized rigid modules. Revista Colomb. Mat. **48**(1), 111–123 (2014)
10. Hirano, Y.: On the uniqueness of rings of coefficients in skew polynomial rings. Publ. Math. Debrecen **54**, 489–495 (1999)
11. Hong, C.Y., Kim, N.K., Kwak, T.K.: Ore extensions of Baer and p.p.-rings. J. Pure Appl. Algebra **151**(3), 215–226 (2000)
12. Hong, C.Y., Kim, N.K., Kwak, T.K.: On skew armendariz rings. Comm. Algebra **31**(1), 103–122 (2003)
13. Krempa, J.: Some examples of reduced rings. Algebra Colloq. **3**(4), 289–300 (1996)
14. Lee, T.K., Zhou, Y.: Reduced Modules. In: Rings, modules, algebras and abelian groups, pp. 365–377. Lecture Notes in Pure and Applied Mathematics, vol. 236. Dekker, New York (2004)

n-Ary k-Actions Between Sets and Their Applications

Antonio J. Calderón Martín, Babacar Gaye and Francisco J. Navarro Izquierdo

Abstract We consider families F of n-ary k-actions

$$f : \mathfrak{A} \times \overset{k)}{\cdots} \times \mathfrak{A} \times \mathfrak{B} \times \overset{n-k)}{\cdots} \times \mathfrak{B} \to \mathfrak{A}$$

between arbitrary non-empty sets $\mathfrak{A}$ and $\mathfrak{B}$ and show that if every $f \in F$ fixes some element in $\mathfrak{A}$, then this family induces an adequate decomposition of $\mathfrak{A}$ as the (orthogonal) disjoint-pointed union of well-described F-invariant subsets (F-submodules). If $\mathfrak{A}$ is furthermore a division F-module, it is shown that the above decomposition is by means of the family of its pointed simple F-submodules. The obtained results are applied to the structure theory of arbitrarily graded n-linear k-modules by stating a second Wedderburn type theorem for the class of n-linear k-modules with an arbitrary division grading.

Keywords Set · Application · k-module over a linear space · Grading · Graded k-module · Structure theory

MSC2010 03E75 · 03E20 · 16W50 · 16D80

The authors are supported by the PCI of the UCA 'Teoría de Lie y Teoría de Espacios de Banach', by the PAI with project numbers FQM298, FQM7156 and by the project of the Spanish Ministerio de Educación y Ciencia MTM2016-76327C31P.

A. J. Calderón Martín (✉) · F. J. Navarro Izquierdo
Department of Mathematics, University of Cádiz, Cádiz, Spain
e-mail: ajesus.calderon@uca.es

F. J. Navarro Izquierdo
e-mail: javi.navarroiz@uca.es

B. Gaye
Department of Mathematics, University Cheikh Anta Diop of Dakar, Dakar, Senegal
e-mail: bacargay@gmail.com

M. Siles Molina et al. (eds.), *Associative and Non-Associative Algebras and Applications*, Springer Proceedings in Mathematics & Statistics 311,
https://doi.org/10.1007/978-3-030-35256-1_9

1 Introduction and Previous Definitions

We begin this paper by fixing two natural numbers $n, k \in \mathbb{N}$ such that $0 < k < n$. Given a couple of arbitrary non-empty sets $\mathfrak{A}$ and $\mathfrak{B}$, any n-ary map

$$f : \mathfrak{A} \times \overset{k)}{\cdots} \times \mathfrak{A} \times \mathfrak{B} \times \overset{n-k)}{\cdots} \times \mathfrak{B} \to \mathfrak{A}. \tag{1}$$

is said to be an *n-ary k-action between* $\mathfrak{A}$ *and* $\mathfrak{B}$.

Since we are going to work in the category of pointed sets (sets with a distinguish element and morphisms preserving the distinguish elements), we will just consider n-ary k-actions (1) between sets satisfying that, for some fixed $\epsilon \in \mathfrak{A}$, we have that

$$f(\mathfrak{A}, \ldots, \mathfrak{A}, \epsilon^{(i)}, \mathfrak{A}, \ldots, \mathfrak{A}, \mathfrak{B}, \ldots, \mathfrak{B}) = \{\epsilon\}$$

for any $1 \leq i \leq k$, where $\epsilon^{(i)}$ denotes that the element ϵ occupies the i-th position.

Definition 1 Let $(\mathfrak{A}, \epsilon)$ be a pointed set, $\mathfrak{B}$ a non-empty set and $F = \{f_\alpha : \alpha \in \Upsilon\}$ a non-empty family of n-ary k-actions between $\mathfrak{A}$ and $\mathfrak{B}$ satisfying

$$f_\alpha(\mathfrak{A}, \ldots, \mathfrak{A}, \epsilon^{(i)}, \mathfrak{A}, \ldots, \mathfrak{A}, \mathfrak{B}, \ldots, \mathfrak{B}) = \{\epsilon\}$$

for any $f_\alpha \in F$ and $1 \leq i \leq k$. Then, the triplet $(\mathfrak{A}, \epsilon, F)$ is called an *n-ary pointed set-module* over $\mathfrak{B}$ by F or just a *pointed F-module* over $\mathfrak{B}$.

In a natural way we can introduce the following concepts.

Definition 2 Let (X, ϵ) and (Y, ϵ) two pointed subsets of a pointed set $(\mathfrak{A}, \epsilon)$, that is $X, Y \subset \mathfrak{A}$ with $\epsilon \in X \cap Y$. We will say that X and Y are *disjoint-pointed* if $X \cap Y = \{\epsilon\}$.

Definition 3 Let $(\mathfrak{A}, \epsilon, F)$ be a pointed F-module over $\mathfrak{B}$ with $k \geq 2$. A pair of subsets X and Y of $\mathfrak{A}$ are said to be *orthogonal* if

$$f_\alpha(\mathfrak{A}, \ldots, \mathfrak{A}, X^{(i)}, \mathfrak{A}, \ldots, \mathfrak{A}, Y^{(j)}, \mathfrak{A}, \ldots, \mathfrak{A}, \mathfrak{B}, \ldots, \mathfrak{B}) = \{\epsilon\}$$

for any $f_\alpha \in F$ and any $1 \leq i, j \leq k$ such that $i \neq j$.

Definition 4 Let $(\mathfrak{A}, \epsilon, F)$ be a pointed F-module over $\mathfrak{B}$. A union

$$\bigcup_{i \in I} X_i$$

of pointed subsets of $(\mathfrak{A}, \epsilon)$ is called an *orthogonal disjoint-pointed union* if X_i and X_j are disjoint-pointed and orthogonal for any $i, j \in I$ such that $i \neq j$.

Remark 1 Let us observe that the concept of orthogonality only has sense when $k \geq 2$. From here, for the case $k = 1$ we will understand that an orthogonal disjoint-pointed union just means a disjoint-pointed union.

Definition 5 Let $(\mathfrak{A}, \epsilon, F)$ be a pointed F-module over $\mathfrak{B}$. A subset X of $\mathfrak{A}$ is called a *pointed F-submodule* if $\epsilon \in X$ and

$$f_\alpha(\mathfrak{A}, \ldots, \mathfrak{A}, X^{(i)}, \mathfrak{A}, \ldots, \mathfrak{A}, \mathfrak{B}, \ldots, \mathfrak{B}) \subset X$$

for any $f_\alpha \in F$ and any $1 \leq i \leq k$.

Definition 6 Let $(\mathfrak{A}, \epsilon, F)$ be a pointed F-module over $\mathfrak{B}$. We say that $\mathfrak{A}$ is *pointed simple* if its only pointed F-submodules are $\{\epsilon\}$ and $\mathfrak{A}$.

This paper is devoted to the study of the structure of pointed F-modules so as to obtain decomposition results for graded k-modules. These results will follow the spirit of the the second Wedderburn theorem for associative algebras (which asserts that any finite-dimensional associative semisimple algebra, over a base field $\mathbb{F}$, is isomorphic to a direct sum

$$\bigoplus_{i=1}^{k} \mathcal{M}_{n_i}(D_i)$$

where the n_i are natural numbers, the D_i are finite dimensional division algebras over $\mathbb{F}$ and $\mathcal{M}_{n_i}(D_i)$ is the associative algebra of $n_i \times n_i$ matrices over D_i, see for instance [7, pp. 137–139]

We will organize our study as follows. In Sect. 2 we develop connections techniques among the elements of an F-module $(\mathfrak{A}, \epsilon, F)$, to show that $\mathfrak{A}$ is the orthogonal disjoint-pointed union of a family of pointed F-submodules $\{\mathfrak{I}_i : i \in I\}$. Then, in Chap. 3 we will prove that if $\mathfrak{A}$ is a division F-module, then the above decomposition of $\mathfrak{A}$ is by means of the family of its pointed simple F-submodules.

Chapter 4 is devoted to apply the results obtained in Sects. 2 and 3 to the structure theory of graded k-modules over linear spaces induced by n-linear maps, considered in their widest sense. That is, there will not be any restriction on the dimensions of the linear spaces, or on the base field or on the grading sets, and any identity will not be supposed. To do that, we will consider the grading sets as adequate set modules and then translate the results obtained for set modules in the previous sections into structural theorems of the initial graded linear k-module. We note that this is a very general structure which contains as particular cases algebras, superalgebras, triple systems, pairs and n-algebras.

Finally, we note that the present paper is the natural extension of the Refs. [3] and [4] by the same authors, and so we will follow a similar development of the paper than the one in this reference.

In [4] it is studied a map of the form

$$f : \mathfrak{A} \to \mathfrak{B} \tag{2}$$

between nonempty sets $\mathfrak{A}$ and $\mathfrak{B}$, which is called an *action between sets*. However, in the preset paper the results in [4] are extended in two ways:

- First, we study maps involving the sets $\mathfrak{A}$ and $\mathfrak{B}$ but allowing that both sets appear not just once (see Eq. (2)), but a finite number of times. That is, it is considered maps of the form

$$f : \mathfrak{A} \times \overset{k)}{\cdots} \times \mathfrak{A} \times \mathfrak{B} \times \overset{n-k)}{\cdots} \times \mathfrak{B} \to \mathfrak{A},$$

  called *n-ary k-actions between sets* (action between sets are the particular case in which $n = 2$ and $k = 1$.)
- Second, we consider arbitrary families F of n-ary k-actions between sets. That is,

$$F = \{f_\alpha : \alpha \in \Upsilon\}$$

  a non-empty family of n-ary k-actions between $\mathfrak{A}$ and $\mathfrak{B}$, while in [4] it is studied just one action between sets. That is, the particular case in which the cardinal of F is one, $n = 2$ and $k = 1$.

2 Connections in $\mathfrak{A}$ Techniques

From now on and throughout this paper, $(\mathfrak{A}, \epsilon, F)$ will denote a (pointed) F-module over $\mathfrak{B}$, and we will refer to this F-module just as $\mathfrak{A}$ if there is not possible confusion. We will also denote by S_k and S_{n-k} the permutation groups of k and $n - k$ elements respectively.

Let us introduce some of notation. We define

$$f_\alpha^{(\sigma,\nu)} : \mathfrak{A} \times \overset{k)}{\cdots} \times \mathfrak{A} \times \mathfrak{B} \times \overset{n-k)}{\cdots} \times \mathfrak{B} \to \mathfrak{A}$$

as

$$f_\alpha^{(\sigma,\nu)}(a_1, \dots, a_k, b_1, \dots, b_{n-k}) = f_\alpha(a_{\sigma(1)}, \dots, a_{\sigma(k)}, b_{\nu(1)}, \dots, b_{\nu(n-k)})$$

for any $f_\alpha \in F$ and $(\sigma, \nu) \in S_k \times S_{n-k}$.

Remark 2 Let us observe that given two nonempty subsets X, Y of $\mathfrak{A}$, we have that X is an F-submodule of $\mathfrak{A}$ orthogonal to Y (when $k \geq 2$), if and only if

$$f_\alpha^{(\sigma,\nu)}(X, \mathfrak{A}, \dots, \mathfrak{A}, \mathfrak{B}, \dots, \mathfrak{B}) \subset X$$

and

$$f_\alpha^{(\sigma,\nu)}(X, Y, \mathfrak{A}, \dots, \mathfrak{A}, \mathfrak{B}, \dots, \mathfrak{B}) = \{\epsilon\}$$

respectively, for any $f_\alpha \in F$ and $(\sigma, \nu) \in S_k \times S_{n-k}$.

Next, we will denote by $\mathfrak{A}^*$ the set $\mathfrak{A} \setminus \{\epsilon\}$ and by $\mathcal{P}(\mathfrak{A}^*)$ the power set of $\mathfrak{A}^*$. We will also introduce the set

$$\Omega := \mathfrak{A}^* \times \overset{k-1)}{\cdots} \times \mathfrak{A}^* \times \mathfrak{B} \times \overset{n-k)}{\cdots} \times \mathfrak{B}$$

being any $\omega \in \Omega$ a tuple $\omega = (a_2, \ldots, a_k, b_1, \ldots, b_{n-k})$.

Then, we can consider the operation $\star : \mathfrak{A}^* \times \Omega \to \mathcal{P}(\mathfrak{A}^*)$ defined by

$$x \star \omega = \bigcup_{(\sigma,\nu)\in S_k\times S_{n-k}} \{y \in \mathfrak{A}^* : f_\alpha^{(\sigma,\nu)}(x, a_2 \ldots, a_k, b_1, \ldots, b_{n-k}) = y,\ f_\alpha \in F\}$$

for any $x \in \mathfrak{A}^*$ and $\omega \in \Omega$.

Remark 3 Let us observe that, for any $a_1, \ldots, a_k \in \mathfrak{A}$ and $b_1, \ldots, b_{n-k} \in \mathfrak{B}$,

$$a_1 \star (a_2 \ldots, a_k, b_1, \ldots, b_{n-k}) = a_{\sigma(1)} \star (a_{\sigma(2)} \ldots, a_{\sigma(k)}, b_{\nu(1)}, \ldots, b_{\nu(n-k)})$$

for any $\sigma \in S_k$ and $\nu \in S_{n-k}$.

Now, for each $\omega \in \Omega$ a new variable $\bar{\omega} \notin \Omega$ is introduced. We denote by

$$\bar{\Omega} := \{\bar{\omega} : \omega \in \Omega\}$$

the set of all these new symbols, and write $\overline{(\bar{\omega})} := \omega \in \Omega$.

Then, we introduce the following operation

$$\diamond : \mathcal{P}(\mathfrak{A}^*) \times (\Omega \mathbin{\dot\cup} \bar{\Omega}) \to \mathcal{P}(\mathfrak{A}^*)$$

defined as $\emptyset \diamond (\Omega \mathbin{\dot\cup} \bar{\Omega}) = \emptyset$, and by

$$X \diamond \omega := \bigcup_{x\in X} \{y \in \mathfrak{A}^* : y \in x \star \omega\}$$

and

$$X \diamond \bar{\omega} := \bigcup_{x\in X} \{y \in \mathfrak{A}^* : x \in y \star \omega\}$$

for any $\emptyset \neq X \subset \mathfrak{A}^*$ and $\omega \in \Omega$.

Lemma 1 *Let $x, y \in \mathfrak{A}^*$ and $\omega \in \Omega \mathbin{\dot\cup} \bar{\Omega}$. Then, $x \in \{y\} \diamond \omega$ if and only if $y \in \{x\} \diamond \bar{\omega}$.*

Proof Suppose $x \in \{y\} \diamond \omega$ and let us distinguish two cases. First, if $\omega \in \Omega$, then $x \in y \star \omega$ and so $y \in \{x\} \diamond \bar{\omega}$. Second, if $\omega \in \bar{\Omega}$ then $y \in x \star \bar{\omega}$, so $y \in \{x\} \diamond \bar{\omega}$.

Finally, note that the converse can be proved similarly. □

Lemma 2 *Let* $\omega \in \Omega \dot{\cup} \bar{\Omega}$ *and* $X \in \mathcal{P}(\mathfrak{A}^*)$. *Then,* $y \in X \diamond \omega$ *if and only if* $(\{y\} \diamond \bar{\omega}) \cap X \neq \emptyset$.

Proof Let us suppose $y \in X \diamond \omega$. Then there exists $x \in X$ such that $y \in \{x\} \diamond \omega$. By Lemma 1 we have $x \in \{y\} \diamond \bar{\omega}$. So $x \in (\{y\} \diamond \bar{\omega}) \cap X \neq \emptyset$.

The converse can be proved in a similar way. □

Definition 7 Let $x, y \in \mathfrak{A}^*$ be. We say that x is *connected* to y if either $x = y$ or there exists a subset $\{\omega_1, \omega_2, \ldots, \omega_m\} \subset \Omega \dot{\cup} \bar{\Omega}$, such that

$$y \in ((\ldots((\{x\} \diamond \omega_1) \diamond \omega_2) \ldots) \diamond \omega_{m-1}) \diamond \omega_m.$$

In this case we say that $\{\omega_1, \omega_2, \ldots, \omega_m\}$ is a *connection* from x to y.

We will also say that x is *strongly connected* to y if

$$\{\omega_1, \omega_2, \ldots, \omega_m\} \subset \Omega.$$

Then we will call $\{\omega_1, \omega_2, \ldots, \omega_m\}$ a *strong connection* from x to y.

Lemma 3 *Let* $\{\omega_1, \omega_2, \ldots, \omega_{m-1}, \omega_m\}$ *be any connection from some* x *to some* y, *where* $x, y \in \mathfrak{A}^*$ *with* $x \neq y$. *Then the set* $\{\bar{\omega}_m, \bar{\omega}_{m-1}, \ldots, \bar{\omega}_2, \bar{\omega}_1\}$ *is a connection from* y *to* x.

Proof Let us prove it by induction on m. For $m = 1$ we have that $y \in \{x\} \diamond \omega_1$. By Lemma 1, $x \in \{y\} \diamond \bar{\omega}_1$ and then, $\{\bar{\omega}_1\}$ is a connection from y to x.

Let us suppose that the assertion holds for any connection with $m \geq 1$ elements and let us show that this assertion also holds for any connection

$$\{\omega_1, \omega_2, \ldots, \omega_m, \omega_{m+1}\}.$$

By denoting the set $X := ((\ldots((\{x\} \diamond \omega_1) \diamond \omega_2) \ldots) \diamond \omega_{m-1}) \diamond \omega_m$ and taking into the account the Definition 7 we have that $y \in X \diamond \omega_{m+1}$. Then, by Lemma 2, we can take $z \in (\{y\} \diamond \bar{\omega}_{m+1}) \cap X$.

Since $z \in X$ we have that $\{\omega_1, \omega_2, \ldots, \omega_{m-1}, \omega_m\}$ is a connection from x to z. Hence $\{\bar{\omega}_m, \bar{\omega}_{m-1}, \ldots, \bar{\omega}_2, \bar{\omega}_1\}$ connects z with x. From here, taking into account that $z \in \{y\} \diamond \bar{\omega}_{m+1}$, we obtain

$$x \in (\ldots((\{y\} \diamond \bar{\omega}_{m+1}) \diamond \bar{\omega}_m) \ldots) \diamond \bar{\omega}_2) \diamond \bar{\omega}_1.$$

So $\{\bar{\omega}_{m+1}, \bar{\omega}_m, \ldots, \bar{\omega}_2, \bar{\omega}_1\}$ connects y with x. □

Proposition 1 *The relation* $\sim$ *in* $\mathfrak{A}^*$, *defined by* $x \sim y$ *if and only if* x *is connected to* y, *is an equivalence relation.*

Proof The reflexive and the symmetric character of $\sim$ are given by Definition 7, and Lemma 3, respectively.

Hence, let us verify the transitivity of $\sim$. Consider $x, y, z \in \mathfrak{A}^*$ such that $x \sim y$ and $y \sim z$. If either $x = y$ or $y = z$ it is clear that $x \sim z$. So, let us suppose $x \neq y$ and $y \neq x$. Then we can find connections $\{\omega_1, \ldots, \omega_r\}$ and $\{\omega'_1, \ldots, \omega'_s\}$ from x to y and from y to z respectively, being clear that $\{\omega_1, \ldots, \omega_r, \omega'_1, \ldots, \omega'_s\}$ is a connection from x to z. So $\sim$ is transitive and consequently an equivalence relation. □

By the above Proposition we can introduce the quotient set

$$\mathfrak{A}^*/ \sim := \{[a] : a \in \mathfrak{A}^*\},$$

where $[a]$ denotes the set of elements in $\mathfrak{A}^*$ which are connected to a.

Proposition 2 *Let* $[x], [y] \in \mathfrak{A}^*/ \sim$ *such that* $[x] \neq [y]$ *and* $k \geq 2$. *Then, the sets* $[x]$ *and* $[y]$ *are orthogonal.*

Proof Let us suppose $[x] \neq [y]$ and take some $x_0 \in [x]$, $y_0 \in [y]$, $a_3, \ldots, a_k \in \mathfrak{A}$ and $b_1, \ldots, b_{n-k} \in \mathfrak{B}$ such that

$$\epsilon \neq f_\alpha^{(\sigma,\nu)}(x_0, y_0, a_3, \ldots, a_k, b_1, \ldots, b_{n-k}) = z$$

for some $f_\alpha \in F$ and $(\sigma, \nu) \in S_k \times S_{n-k}$, with $z \in \mathfrak{A}^*$. Then, by the Remark 3

$$z \in \{x_0\} \diamond \omega_y = \{y_0\} \diamond \omega_x$$

with $\omega_y := (y_0, a_3, \ldots, a_k, b_1, \ldots, b_{n-k})$ and $\omega_x := (x_0, a_3, \ldots, a_k, b_1, \ldots, b_{n-k})$.

Then, $x_0 \sim z$ and $y_0 \sim z$. Hence, by simetry and transitivity we get $[x] = [y]$, a contradiction. □

Proposition 3 *For any* $[a] \in \mathfrak{A}^*/ \sim$, *we have that the subset* $[a] \cup \{\epsilon\}$ *of* $\mathfrak{A}$ *is an* F*-submodule of* $\mathfrak{A}$.

Proof Since for any $f_\alpha \in F$ and $(\sigma, \nu) \in S_k \times S_{n-k}$ we have that

$$f_\alpha^{(\sigma,\nu)}(\epsilon, \mathfrak{A}, \ldots, \mathfrak{A}, \mathfrak{B}, \ldots, \mathfrak{B}) = \{\epsilon\},$$

we just need to check that $f_\alpha^{(\sigma,\nu)}([a], \mathfrak{A}, \ldots, \mathfrak{A}, \mathfrak{B}, \ldots, \mathfrak{B}) \subset [a] \cup \{\epsilon\}$.

So, let $a_1 \in [a]$, $a_2, \ldots, a_k \in \mathfrak{A}$ and $b_1, \ldots, b_{n-k} \in \mathfrak{B}$ be such that

$$f_\alpha^{(\sigma,\nu)}(a_1, a_2 \ldots, a_k, b_1, \ldots, b_{n-k}) = x$$

for some $f_\alpha \in F$ and $(\sigma, \nu) \in S_k \times S_{n-k}$, with $x \in \mathfrak{A}^*$. Hence $x \in \{a_1\} \diamond \omega$ with $\omega = (a_2 \ldots, a_k, b_1, \ldots, b_{n-k})$. So $\{\omega\}$ is a connection from a_1 to x. By transitivity $[a] = [x]$ and then,

$$f_\alpha^{(\sigma,\nu)}(a_1, a_2 \ldots, a_k, b_1, \ldots, b_{n-k}) \in [a]$$

as we wanted to prove. □

Taking into account that

$$\mathfrak{A}^* = \dot{\bigcup_{a \in \mathfrak{A}^*}} a = \dot{\bigcup_{[a] \in \mathfrak{A}^*/\sim}} [a],$$

Propositions 2 and 3 allow us to assert:

Theorem 1 *Let* $(\mathfrak{A}, \epsilon, F)$ *be a pointed* F*-module over* $\mathfrak{B}$*. Then*

$$\mathfrak{A} = \bigcup_{[a] \in \mathfrak{A}^*/\sim} ([a] \cup \{\epsilon\})$$

is the orthogonal disjoint-pointed union of the family $\{[a] \cup \{\epsilon\} : [a] \in \mathfrak{A}^*/\sim\}$ *of the pointed* F*-submodules of* $\mathfrak{A}$*.*

Corollary 1 *If* $\mathfrak{A}$ *is pointed simple, then any couple of elements in* $\mathfrak{A}^*$ *are connected.*

Proof The pointed-simplicity of $\mathfrak{A}$ applies to get that $[a] \cup \{\epsilon\} = \mathfrak{A}$ for some $[a] \in \mathfrak{A}^*/\sim$. So $[a] = \mathfrak{A}^*$, and any couple of elements in $\mathfrak{A}^*$ are connected. □

3 Division Set Modules

In this section we show that if an F-module $\mathfrak{A}$ is furthermore a division F-module, then we can characterize the (pointed) simplicity of $\mathfrak{A}$ in terms of a connectivity property, and that the decomposition of $\mathfrak{A}$ given in Theorem 1 is actually through the family of its pointed simple F-submodules, stating so a second Wedderburn-type theorem for this class of F-modules.

Definition 8 We say that an F-module $\mathfrak{A}$ over $\mathfrak{B}$ is a *division* F*-module* if for any $x, y \in \mathfrak{A}$ such that $y \in f_\alpha(\mathfrak{A}, \ldots, \mathfrak{A}, x^{(i)}, \mathfrak{A}, \ldots, \mathfrak{A}, \mathfrak{B}, \ldots, \mathfrak{B})$ for some $f_\alpha \in F$ and $1 \leq i \leq k$ then

$$x \in f_\beta(\mathfrak{A}, \ldots, \mathfrak{A}, y^{(j)}, \mathfrak{A}, \ldots, \mathfrak{A}, \mathfrak{B}, \ldots, \mathfrak{B})$$

for some $f_\beta \in F$ and $1 \leq j \leq k$.

Remark 4 Let us observe that $\mathfrak{A}$ is a division F-module over $\mathfrak{B}$ if and only if given $x, y, a_2, \ldots, a_k \in \mathfrak{A}$ and $b_1, \ldots, b_{n-k} \in \mathfrak{B}$ such that

$$y = f_\alpha^{(\sigma,\nu)}(x, a_2, \ldots, a_k, b_1, \ldots, b_{n-k})$$

for some $f_\alpha \in F$ and $(\sigma, \nu) \in S_k \times S_{n-k}$, then there exist $a'_2, \ldots, a'_k \in \mathfrak{A}$ and $b'_1, \ldots, b'_{n-k} \in \mathfrak{B}$ such that

$$x = f_\beta^{(\tau,\pi)}(y, a'_2, \ldots, a'_k, b'_1, \ldots, b'_{n-k})$$

for some $f_\beta \in F$ and $(\tau, \pi) \in S_k \times S_{n-k}$.

Now, observe that a subset $\{\omega_1, \omega_2, \ldots, \omega_m\} \subset \Omega \mathbin{\dot\cup} \bar{\Omega}$ is a connection from x to y in $\mathfrak{A}^*$ if and only if there exits $\{z_0, z_1, \ldots, z_m\} \subset \mathfrak{A}^*$ such that $z_0 = x$, $z_m = y$ and $z_i \in \{z_{i-1}\} \diamond \omega_i$ for any $1 \leq i \leq m$. A representation of this fact as

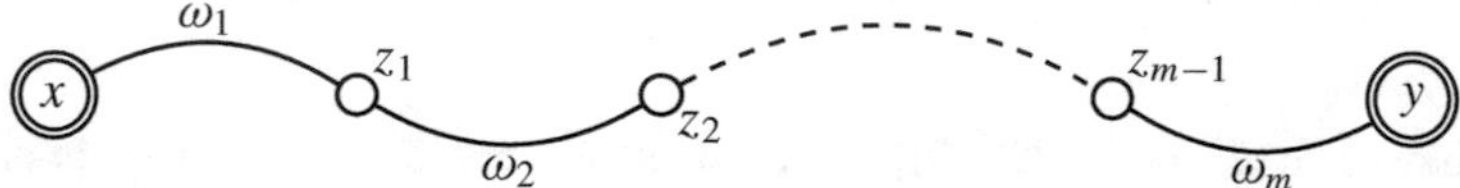

leads us to introduce the next concept.

Definition 9 Let $\{\omega_1, \omega_2, \ldots, \omega_m\} \subset \Omega \mathbin{\dot\cup} \bar{\Omega}$ be a connection from x to y in $\mathfrak{A}^*$. Then any $\{z_0, z_1, \ldots, z_m\} \subset \mathfrak{A}^*$ satisfying $z_0 = x$, $z_m = y$ and $z_i \in \{z_{i-1}\} \diamond \omega_i$ for any $1 \leq i \leq m$, is called a *set of nodes* of the connection $\{\omega_1, \omega_2, \ldots, \omega_m\}$.

Lemma 4 *A pair of distinct elements $x, y \in \mathfrak{A}^*$ of a division F-module $\mathfrak{A}$ are connected if and only if they are strongly connected (see Definition 7).*

Proof Let $\{\omega_1, \omega_2, \ldots, \omega_m\} \subset \Omega \mathbin{\dot\cup} \bar{\Omega}$ be a connection from x to y with a set of nodes $\{z_0, z_1, \ldots, z_m\} \subset \mathfrak{A}^*$. If some $\omega_i \in \bar{\Omega}$, by Lemma 1 we get $z_{i-1} \in \{z_i\} \diamond \bar{\omega}_i$ and, taking into account Remark 4, there exists $\omega_i' \in \Omega$ such that $z_i \in \{z_{i-1}\} \diamond \omega_i'$. Then we can replace $\omega_i \in \bar{\Omega}$ by $\omega_i' \in \Omega$ in the connection $\{\omega_1, \omega_2, \ldots, \omega_m\}$, by getting a new connection $\{\omega_1, \ldots, \omega_{i-1}, \omega_i', \omega_{i+1}, \ldots, \omega_m\}$ from x to y. Then, we conclude there exists a a strong connection from x to y. □

Theorem 2 *Let $(\mathfrak{A}, \epsilon, F)$ be a division F-module over $\mathfrak{B}$. Then $\mathfrak{A}$ is pointed simple if and only if $\mathfrak{A}^*$ has all of its elements (strongly) connected.*

Proof The first implication is similar to Corollary 1. To prove the converse, consider a pointed F-submodule $\mathfrak{I} \neq \{\epsilon\}$ of $\mathfrak{A}$ and fix some $x \in \mathfrak{I} \setminus \{\epsilon\}$.

Since for any $y \in \mathfrak{A}^*$ we have that x is connected to y, then Lemma 4 allows us to assert that there exists a strong connection $\{\omega_1, \omega_2, \ldots, \omega_m\} \subset \Omega$ from x to y satisfying $y \in (\ldots((\{x\} \diamond \omega_1) \diamond \omega_2) \ldots) \diamond \omega_m$.

From here, taking into account $\{\omega_1, \omega_2, \ldots, \omega_m\} \subset \Omega$ and $x \in \mathfrak{I}$ we have that

$$y \in (\ldots((x \star \omega_1) \star \omega_2) \ldots) \star \omega_m \subset \mathfrak{I}$$

as wished. □

The following example illustrates that the division condition is necessary in Theorem 2.

Example 1 Example Consider the F-module $\mathfrak{A} := \mathbb{R}^2$ over $\mathfrak{B} := \mathbb{Z}^2 \setminus \{(0, 0)\}$ with $\epsilon = (0, 0)$ and the n-ary k-actions

$$f_\alpha : \mathfrak{A} \times \overset{k)}{\cdots} \times \mathfrak{A} \times \mathfrak{B} \times \overset{n-k)}{\cdots} \times \mathfrak{B} \to \mathfrak{A}$$

defined by

$$f_\alpha\big((x_1, y_1), \ldots, (x_k, y_k), (u_{k+1}, v_{k+1}), \ldots, (u_n, v_n)\big) =$$

$$= \Big(\frac{x_1 \cdots x_k u_{k+1} \cdots u_n}{\alpha}, \frac{y_1 \ldots y_k v_{k+1} \cdots v_n}{\alpha}\Big)$$

for any $\alpha \in \mathbb{N}$.

We have that $\mathfrak{A}^* = \mathbb{R} \setminus \{(0, 0)\}$ has all of its elements connected. This is consequence of the fact that for any $0 \neq x \in \mathbb{R}$ and $y \in \mathbb{R}$, the element (x, y) is connected to $(1, 0)$ and to $(0, 1)$ (in particular $(1, 0) \sim (0, 1)$) through the connections

$$\Big\{\underbrace{\big((x^{-1}, 0), (1, 0), \overset{n-2)}{\ldots}, (1, 0)\big)}_{\omega_1}\Big\}$$

and

$$\Big\{\underbrace{\overline{\big((x, y), (1, 1), \overset{k-2)}{\ldots}, (1, 1), (1, 0), \overset{n-k)}{\ldots}, (1, 0)\big)}}_{\omega_1}, \underbrace{\big((0, 1), \overset{n-1)}{\ldots}, (0, 1)\big)}_{\omega_2}\Big\}$$

respectively, and the fact that

$$\Big\{\underbrace{\big((0, x^{-1}), (0, 1), \overset{n-2)}{\ldots}, (0, 1)\big)}_{\omega_1}\Big\}$$

is a connection from (y, x) to $(0, 1)$.

However, $\mathfrak{A}$ is not a pointed simple F-module since, for instance

$$\mathfrak{I} := \mathbb{R} \oplus \{0\}$$

is an F-submodule of $\mathfrak{A}$. This happens because $\mathfrak{A}$ is not a division F-module. Indeed, for any $0 \neq x_1, \ldots, x_k, y_1, \ldots, y_k \in \mathbb{R}$ and $0 \neq u_{k+1}, \ldots, u_n, v_{k+1}, \ldots, v_{n-1} \in \mathbb{Z}$ we have that

$$f_\alpha\big((x_1, y_1), \ldots, (x_k, y_k), (u_{k+1}, v_{k+1}), \ldots, (u_{n-1}, v_{n-1}), (u_n, 0)\big) =$$

$$= \Big(\frac{x_1 \cdots x_k u_{k+1} \cdots u_n}{\alpha}, 0\Big)$$

for every $f_\alpha \in F$, but for any $1 \leq i \leq k$ we have that

$$(x_i, y_i) \notin f_\beta\left(\mathfrak{A}, \overset{j-1)}{\ldots}, \mathfrak{A}, \Big(\frac{x_1 \cdots x_k u_{k+1} \cdots u_n}{\alpha}, 0\Big), \mathfrak{A}, \overset{k-j)}{\ldots}, \mathfrak{A}, \mathfrak{B}, \overset{n-k)}{\ldots}, \mathfrak{B}\right)$$

for any $f_\beta \in F$ and $1 \leq j \leq k$.

By Proposition 3 we know that for any $[a] \in \mathfrak{A}^*/\sim$, the subset $[a] \cup \{\epsilon\}$ is an F-submodule of $\mathfrak{A}$. Then, we can consider the family $F_{[a]}$ of n-ary k-actions

$$f_{\alpha,[a]} : ([a] \cup \{\epsilon\}) \times \overset{k)}{\cdots} \times ([a] \cup \{\epsilon\}) \times \mathfrak{B} \times \overset{n-k)}{\cdots} \times \mathfrak{B} \to [a] \cup \{\epsilon\}$$

given by $f_{\alpha,[a]}(a_1, \ldots, a_k, b_1, \ldots, b_{n-k}) = f_\alpha(a_1, \ldots, a_k, b_1, \ldots, b_{n-k})$. By considering any $([a] \cup \{\epsilon\}, \epsilon, F_{[a]})$ as a pointed $F_{[a]}$-module by itself we can state that.

Lemma 5 *If $(\mathfrak{A}, \epsilon, F)$ is a division F-module over $\mathfrak{B}$ then $([a] \cup \{\epsilon\}, \epsilon, F_{[a]})$ is a division $F_{[a]}$-module over $\mathfrak{B}$.*

Proof Suppose $x, y, a_2, \ldots, a_n \in [a]$ and $b_1, \ldots, b_{n-k} \in \mathfrak{B}$ in such a way that

$$y = f_\alpha^{(\sigma,\nu)}(x, a_2, \ldots, a_k, b_1, \ldots, b_{n-k})$$

for some $f_\alpha \in F$ and $(\sigma, \nu) \in S_k \times S_{n-k}$. By the division property of $\mathfrak{A}$, there exist $a'_2, \ldots, a'_k \in \mathfrak{A}$ and $b'_1, \ldots, b'_{n-k} \in \mathfrak{B}$ such that

$$x = f_\beta^{(\tau,\pi)}(y, a'_2, \ldots, a'_k, b'_1, \ldots, b'_{n-k})$$

for some $f_\beta \in F$ and $(\tau, \pi) \in S_k \times S_{n-k}$. But, taking into account Remark 3, for any $2 \le i \le k$ we have that

$$x \in \{a'_i\} \diamond \omega_i$$

with $\omega_i = (a'_2, \ldots, a'_{i-1}, y, a'_{i+1}, \ldots, a'_k, b'_1, \ldots, b'_{n-k})$. So $a'_2, \ldots, a'_k \in [a]$ and then $[a] \cup \{\epsilon\}$ is a division $F_{[a]}$-module over $\mathfrak{B}$. □

Now, let us denote the subsets

$$\Omega_{[a]} := [a] \times \overset{k-1)}{\cdots} \times [a] \times \mathfrak{B} \times \overset{n-k)}{\cdots} \times \mathfrak{B} \subset \Omega$$

and $\bar{\Omega}_{[a]} := \{\bar{\omega} \in \bar{\Omega} : \omega \in \Omega_{[a]}\}$, for any $[a] \in \mathfrak{A}^*/\sim$. Then, we can state the following result.

Lemma 6 *If $\{\omega_1, \omega_2, \ldots, \omega_m\} \subset \Omega \,\dot{\cup}\, \bar{\Omega}$ is a connection from x to y with $x, y \in \mathfrak{A}^*$ such that $x \neq y$. Then $\{\omega_1, \omega_2, \ldots, \omega_m\} \subset \Omega_{[x]} \,\dot{\cup}\, \bar{\Omega}_{[x]}$.*

Proof Consider any connection $\{\omega_1, \omega_2, \ldots, \omega_m\} \subset \Omega \,\dot{\cup}\, \bar{\Omega}$ from x to y with $x, y \in \mathfrak{A}^*$ such that $x \neq y$. We begin by observing that any $z \in \{x\} \diamond \omega_1$ satisfies $z \in [x]$. Now we can distinguish two cases:

- If $\omega_1 = (a_2, \ldots, a_k, b_1, \ldots, b_{n-k}) \in \Omega$, then we have by Remark 3 that for any $2 \le i \le k$ we get $z \in \{a_i\} \diamond \omega'_i$ with

$$\omega'_i = (a_2, \ldots, a_{i-1}, x, a_{i+1}, \ldots, a_k, b_1, \ldots, b_{n-k}).$$

From here, the set $\{\omega'_i\}$ is a connection from a_i to z. Hence $a_i \in [x]$, and then $\omega_1 \in \Omega_{[x]}$.

- If $\omega_1 \in \bar{\Omega}$, then Lemma 2 gives us $x \in \{z\} \diamond \bar{\omega}_1$. We obtain as above that $\bar{\omega}_1 \in \Omega_{[z]} = \Omega_{[x]}$ so $\omega_1 \in \bar{\Omega}_{[x]}$.

By iterating this process we obtain that all of the elements in the connection $\{\omega_1, \omega_2, \ldots, \omega_m\}$ are contained in $\Omega_{[x]} \dot{\cup} \bar{\Omega}_{[x]}$. □

Theorem 3 (Second Wedderburn Theorem) *Let* $(\mathfrak{A}, \epsilon, F)$ *be a division* F*-module over* $\mathfrak{B}$. *Then*

$$\mathfrak{A} = \bigcup_{[a]\in\mathfrak{A}^*/\sim} ([a] \cup \{\epsilon\})$$

is the orthogonal disjoint-pointed union of the family of all its pointed simple F*-submodules.*

Proof By Theorem 1 we have that

$$\mathfrak{A} = \bigcup_{[a]\in\mathfrak{A}^*/\sim} ([a] \cup \{\epsilon\})$$

is the (orthogonal disjoint-pointed) union of the family of F-submodules $\{[a] \cup \{\epsilon\} : [a] \in \mathfrak{A}/\sim\}$. From Lemma 5 any $([a] \cup \{\epsilon\}, \epsilon, F_{[a]})$ is a division $F_{[a]}$-module over $\mathfrak{B}$ and by Lemma 6 the set $[a]$ has all its elements connected through connections contained in $\Omega_{[a]} \dot{\cup} \bar{\Omega}_{[a]}$. Hence, we can apply Theorem 2 to conclude that any $[a] \cup \{\epsilon\}$ is pointed simple. So, the above decomposition satisfies the assertions of the theorem.

Finally, consider any pointed simple F-submodule $\mathfrak{I} \neq \{\epsilon\}$ of $\mathfrak{A}$. Then we can fix some $x \in \mathfrak{I} \setminus \{\epsilon\}$ and so $x \in \mathfrak{I} \cap [x]$. Since $\{\epsilon\} \neq (\mathfrak{I} \cap [x]) \cup \{\epsilon\}$ is a pointed F-submodule of $\mathfrak{I}$ and of $[x] \cup \{\epsilon\}$ we get

$$\mathfrak{I} = \mathfrak{I} \cap ([x] \cup \{\epsilon\}) = [x] \cup \{\epsilon\}$$

and so any pointed simple F-submodule $\mathfrak{I}$ of $\mathfrak{A}$ appears in the decomposition given in Theorem 1. □

4 Arbitrary Graded k-Modules over Linear Spaces Induced by n-Linear Maps

In this section we will apply the results given in Sects. 2 and 3 to the study of the structure theory of arbitrarily graded k-modules over linear spaces.

We begin by noting that throughout this section we will consider two linear spaces $\mathcal{V}$ and $\mathcal{W}$ of arbitrary dimensions and over a same arbitrary base field $\mathbb{F}$. We also note that our k-modules will be considered in their widest sense. That is, there is not any restriction on dimensions or on the base field or on the grading sets, and any identity is not supposed.

Definition 10 Let $\mathcal{V}$ be a vector space over an arbitrary base field $\mathbb{F}$. It is said that $\mathcal{V}$ is k-moduled by a linear space $\mathcal{W}$ (over the same base field $\mathbb{F}$), or just that $\mathcal{V}$ is a *k-module* over $\mathcal{W}$ if it is endowed with a nonempty family of n-linear maps

$$\langle \cdot, \ldots, \cdot \rangle_\alpha : \mathcal{V} \times \overset{k)}{\cdots} \times \mathcal{V} \times \mathcal{W} \times \overset{n-k)}{\cdots} \times \mathcal{W} \to \mathcal{V},$$

$\alpha \in \Upsilon$. A linear subspace $\mathcal{U} \subset \mathcal{V}$ is said to be a *k-submodule* of $\mathcal{V}$ if

$$\langle \mathcal{U}, \overset{k)}{\ldots}, \mathcal{U}, \mathcal{W}, \overset{n-k)}{\ldots}, \mathcal{W} \rangle_\alpha \subset \mathcal{U}$$

for any $\alpha \in \Upsilon$.

Remark 5 The definition of k-modules given in Definition 10 is more general than the given in [1] in the sense that there is not a number of operations determined by n and k, nor any identity imposed. In fact, in that reference, given an arbitrary n-linear map

$$[\cdot, \ldots, \cdot] : \mathcal{V} \times \overset{k)}{\cdots} \times \mathcal{V} \times \mathcal{W} \times \overset{n-k)}{\cdots} \times \mathcal{W} \to \mathcal{V},$$

the linear space $\mathcal{V}$ is called a k-module over $\mathcal{W}$ if it is endowed by a number C_n^k of n-linear maps such that for every n-linear map there is a suitable element $\sigma \in S_n$, such that this n-linear map may be realized as $[x_{\sigma(1)}, \ldots, x_{\sigma(n)}] \in \mathcal{V}$.

This is a wide concept which allows us to treat in an unifying way many different algebraic structures. For instance (by denoting the cardinal of Υ by $|\Upsilon|$), if we suppose $|\Upsilon| = 1$, we have an usual module over a algebra and also algebras and superalgebras (see [6, Sect. 1]) when $n = 2$, arbitrary triple systems when $n = 3$, and arbitrary n-algebras when $n \in \mathbb{N}$.

As well as an arbitrary algebra is just a k-module over itself with $|\Upsilon| = 1$ and $n = 2$, a Poisson algebra is a k-module over itself but with $|\Upsilon| = 2$ and $n = 2$.

The following example shows that algebraic pairs can be considered as an adequate k-module with $|\Upsilon| = 2$ and $n = 3$.

Example 2 Example We recall that an algebraic pair (P^+, P^+) is a couple of $\mathbb{F}$-linear spaces endowed with two trilinear maps

$$\langle \cdot, \cdot, \cdot \rangle_+ : P^+ \times P^- \times P^+ \to P^+ \text{ and } \langle \cdot, \cdot, \cdot \rangle_- : P^- \times P^+ \times P^- \to P^-$$

called the triple products of (P^+, P^+). In the literature have been considered different classes of algebraic pairs, depending on the identities satisfied by the triple products: associative, Jordan, alternative, etc. (see [2, 9]). It is interesting to mention the high impact of the theory of Jordan pairs initiated by O. Loos and K. McCrimmon (see for instance [8, 10]). Any kind of algebraic pair (P^+, P^+) can be considered as the k-module $P = P^+ \cup P^-$ over itself under the actions

$$\langle \cdot, \cdot, \cdot \rangle_1 : P \times P \times P \to P \text{ and } \langle \cdot, \cdot, \cdot \rangle_2 : P \times P \times P \to P$$

determined by

$$\langle x, y, z \rangle_1 := \begin{cases} \langle x, y, z \rangle_+ & \text{if } x, z \in P^+ \text{ and } y \in P^- \\ 0 & \text{otherwise} \end{cases}$$

and

$$\langle x, y, z \rangle_2 := \begin{cases} \langle x, y, z \rangle_- & \text{if } x, z \in P^- \text{ and } y \in P^+ \\ 0 & \text{otherwise} \end{cases}$$

respectively.

We recall that a linear space $\mathcal{V}$ is said to be *graded* by means of an arbitrary non-empty set I, or just I*-graded*, if it decomposes as the direct sum

$$\mathcal{V} = \bigoplus_{i \in I} \mathcal{V}_i$$

of linear subspaces $\mathcal{V}_i$. We also recall that the set $\Sigma := \{i \in I : \mathcal{V}_i \neq \{0\}\}$ is said to be the *support* of the grading.

Definition 11 Let $\mathcal{V}$ be a k-module over a J-graded linear space $\mathcal{W}$ and I a nonempty set. We say that $\mathcal{V}$ is an I*-graded* k*-module* over $\mathcal{W}$ if it decomposes as a direct sum

$$\mathcal{V} = \bigoplus_{i \in I} \mathcal{V}_i$$

of linear subspaces satisfying that, for any $i_1, \ldots, i_k \in I$ and $j_1, \ldots, j_{n-k} \in J$ we have that for any $\alpha \in \Upsilon$ such that $\langle \mathcal{V}_{i_1}, \ldots, \mathcal{V}_{i_k}, \mathcal{W}_{j_1}, \ldots, \mathcal{W}_{j_{n-k}} \rangle_\alpha \neq \{0\}$ then

$$\langle \mathcal{V}_{i_1}, \ldots, \mathcal{V}_{i_k}, \mathcal{W}_{j_1}, \ldots, \mathcal{W}_{j_{n-k}} \rangle_\alpha \subset \mathcal{V}_{i'}$$

for some (unique) $i' \in I$.

Let us recall that, given an I-graded k-module $\mathcal{V}$ over a J-graded linear space $\mathcal{W}$, a k-submodule $\mathcal{U}$ of $\mathcal{V}$ is said to be a *homogeneous* k*-submodule* if it is of the form

$$\mathcal{U} = \bigoplus_{i' \in I'} \mathcal{V}_{i'}$$

with $I' \subset I$. We also recall that $\mathcal{V}$ is called *homogeneous-simple* if its only homogeneous k-submodules are $\{0\}$ and $\mathcal{V}$.

Let us fix an I-graded module $\mathcal{V}$ over a J-graded linear space $\mathcal{W}$ and consider the supports Σ_I and Σ_J of the gradings of $\mathcal{V}$ and $\mathcal{W}$ respectively. Let us fix an external element $\epsilon \notin \Sigma_I$ and introduce the pointed F-module

$$(\Sigma_I \cup \{\epsilon\}, \epsilon, F)$$

over Σ_J with $F = \{f_\alpha : \alpha \in \Upsilon\}$, where any $f_\alpha \in F$ is the n-ary map

$$f_\alpha : (\Sigma_I \cup \{\epsilon\}) \times \overset{k)}{\cdots} \times (\Sigma_I \cup \{\epsilon\}) \times \Sigma_J \times \overset{n-k)}{\cdots} \times \Sigma_J \to (\Sigma_I \cup \{\epsilon\})$$

defined by

$$f_\alpha(\Sigma_I \cup \{\epsilon\}, \ldots, \Sigma_I \cup \{\epsilon\}, \epsilon^{(r)}, \Sigma_I \cup \{\epsilon\}, \ldots, \Sigma_I \cup \{\epsilon\}, \Sigma_J \ldots, \Sigma_J) = \{\epsilon\}$$

for any $1 \leq r \leq k$, and

$$f_\alpha(i_1, \ldots, i_k, j_1, \ldots, j_{n-k}) = \begin{cases} \epsilon \text{ if } \langle \mathcal{V}_{i_1}, \ldots, \mathcal{V}_{i_k}, \mathcal{W}_{j_1}, \ldots, \mathcal{W}_{j_{n-k}} \rangle_\alpha = \{0\} \\ i' \text{ if } \{0\} \neq \langle \mathcal{V}_{i_1}, \ldots, \mathcal{V}_{i_k}, \mathcal{W}_{j_1}, \ldots, \mathcal{W}_{j_{n-k}} \rangle_\alpha \subset \mathcal{V}_{i'} \end{cases}$$

for any $i_1, \ldots, i_k \in \Sigma_I$ and $j_1, \ldots, j_{n-k} \in \Sigma_J$.

By Theorem 1, we can write $\Sigma_I \cup \{\epsilon\} = \bigcup\limits_{[x] \in \Sigma_I/\sim} ([x] \cup \{\epsilon\})$ in such a way that

$$f_\alpha\big(\Sigma_I, \ldots, \Sigma_I, [x]^{(r)}, \Sigma_I, \ldots, \Sigma_I, \Sigma_J, \ldots, \Sigma_J\big) \subset [x] \cup \{\epsilon\} \tag{3}$$

and

$$f_\alpha\big(\Sigma_I, \ldots, \Sigma_I, [x]^{(r)}, \Sigma_I, \ldots, \Sigma_I, [y]^{(s)}, \Sigma_I, \ldots, \Sigma_I, \Sigma_J, \ldots, \Sigma_J\big) = \{\epsilon\} \tag{4}$$

when $[x] \neq [y]$ for any $f_\alpha \in F$ and $1 \leq r, s \leq k$ such that $r \neq s$. We also have that $[x] \cap [y] = \emptyset$ when $[x] \neq [y]$. If we introduce the linear subspace

$$\mathcal{V}_{[i]} := \bigoplus_{i' \in [i]} \mathcal{V}_{i'} \tag{5}$$

of $\mathcal{V}$ for any $[i] \in \Sigma_I/\sim$, Eq. (3) gives us that $\mathcal{V}_{[i]}$ is a homogeneous k-submodule of $\mathcal{V}$ while Eq. (4) gives us

$$\langle \mathcal{V}, \ldots, \mathcal{V}, \mathcal{V}_{[x]}^{(r)}, \mathcal{V}, \ldots, \mathcal{V}, \mathcal{V}_{[y]}^{(s)}, \mathcal{V}, \ldots, \mathcal{V}, \mathcal{W}, \ldots, \mathcal{W} \rangle_\alpha = \{0\}$$

for any $\alpha \in \Upsilon$ and $1 \leq r, s \leq k$ such that $r \neq s$, when $[x] \neq [y]$. Since

$$\mathcal{V} = \bigoplus_{i \in I} \mathcal{V}_i = \bigoplus_{i \in \Sigma_I} \mathcal{V}_i = \bigoplus_{[i] \in \Sigma_I/\sim} \mathcal{V}_{[i]}$$

we can assert.

Theorem 4 *Let $\mathcal{V}$ be an I-graded k-module over a J-graded linear space $\mathcal{W}$. Then $\mathcal{V}$ decomposes as the orthogonal direct sum*

$$\mathcal{V} = \bigoplus_{[i]\in\Sigma_I/\sim} \mathcal{V}_{[i]},$$

being any $\mathcal{V}_{[i]}$ a nonzero homogeneous k-submodule of $\mathcal{V}$.

We can introduce the concept of division grading for modules in a similar way to the concept of division basis for certain classes of algebras (see [5]).

Definition 12 We will say that a grading

$$\mathcal{V} = \bigoplus_{i\in I} \mathcal{V}_i$$

on a k-module $\mathcal{V}$ over a J-graded linear space $\mathcal{W}$ is a *division* grading if for any $i, i' \in I$ such that

$$\{0\} \neq \langle \mathcal{V}, \ldots, \mathcal{V}, \mathcal{V}_i^{(r)}\mathcal{V}, \ldots, \mathcal{V}, \mathcal{W}, \ldots, \mathcal{W}\rangle_\alpha \subset \mathcal{V}_{i'}$$

for some $\alpha \in \Upsilon$ and $1 \leq r \leq k$, then

$$\{0\} \neq \langle \mathcal{V}, \ldots, \mathcal{V}, \mathcal{V}_{i'}^{(s)}, \mathcal{V}, \ldots, \mathcal{V}, \mathcal{W}, \ldots, \mathcal{W}\rangle_\beta \subset \mathcal{V}_i$$

for some $\beta \in \Upsilon$ and $1 \leq s \leq k$.

Proposition 4 *Let $\mathcal{V}$ be a k-module with a division grading over $\mathcal{W}$. Then the F-module $(\Sigma_I \cup \{\epsilon\}, \epsilon, F)$ over Σ_J is a division F-module.*

Proof Let us take $i, i' \in \Sigma_I$ such that $i' \in f_\alpha(\Sigma_I, \ldots, \Sigma_I, i^{(r)}, \Sigma_I, \ldots, \Sigma_I, \Sigma_J, \ldots, \Sigma_J)$ for some $f_\alpha \in F$ and $1 \leq r \leq k$. Then

$$\{0\} \neq \langle \mathcal{V}, \ldots, \mathcal{V}, \mathcal{V}_i^{(r)}\mathcal{V}, \ldots, \mathcal{V}, \mathcal{W}, \ldots, \mathcal{W}\rangle_\alpha \subset \mathcal{V}_{i'}.$$

Taking now into account that we have a division grading, we can assert that there exist $\beta \in \Upsilon$ and $1 \leq s \leq k$ such that

$$\{0\} \neq \langle \mathcal{V}, \ldots, \mathcal{V}, \mathcal{V}_{i'}^{(s)}, \mathcal{V}, \ldots, \mathcal{V}, \mathcal{W}, \ldots, \mathcal{W}\rangle_\beta \subset \mathcal{V}_i$$

From here $i \in f_\beta(\Sigma_I, \ldots, \Sigma_I, i'^{(s)}, \Sigma_I, \ldots, \Sigma_I, \Sigma_J, \ldots, \Sigma_J)$. We have shown that the set $\Sigma_I \cup \{\epsilon\}$ is a division F-module. □

Proposition 5 *A pointed F-submodule $[i] \cup \{\epsilon\}$ of $(\Sigma_I \cup \{\epsilon\}, \epsilon, F)$ is pointed simple if and only if $\mathcal{V}_{[i]}$ is a homogeneous-simple graded k-submodule.*

Proof Let us suppose that the F-submodule $[i] \cup \{\epsilon\}$ is pointed simple in the set-modules sense. If we consider any nonzero homogeneous k-submodule

$$\mathcal{U} = \bigoplus_{i' \in I'} \mathcal{V}_{i'}$$

of $\mathcal{V}_{[i]}$, where $\emptyset \neq I' \subset [i]$. Given $1 \leq r \leq k$, for any $i_1, \ldots, i_{r-1}, i_{r+1}, \ldots, i_k \in [i]$ and $j_1, \ldots, j_{n-k} \in J$ we have that

$$\langle \mathcal{V}_{i_1}, \ldots, \mathcal{V}_{i_{r-1}}, \mathcal{V}_{i_r}, \mathcal{V}_{i_{r+1}}, \ldots, \mathcal{V}_{i_k}, \mathcal{W}_{j_1}, \ldots, \mathcal{W}_{j_{n-k}} \rangle_\alpha \subset \mathcal{U}$$

for any $i_r \in I'$ and $\alpha \in \Upsilon$. So either

$$f_\alpha(i_1, \ldots, i_{r-1}, i_r, i_{r+1}, \ldots, i_k, j_1, \ldots, j_{n-k}) = \epsilon$$

if $\langle \mathcal{V}_{i_1}, \ldots, \mathcal{V}_{i_{r-1}}, \mathcal{V}_{i_r}, \mathcal{V}_{i_{r+1}}, \ldots, \mathcal{V}_{i_k}, \mathcal{W}_{j_1}, \ldots, \mathcal{W}_{j_{n-k}} \rangle_\alpha = \{0\}$ or

$$f_\alpha(i_1, \ldots, i_{r-1}, i_r, i_{r+1}, \ldots, i_k, j_1, \ldots, j_{n-k}) \in I'$$

if $\langle \mathcal{V}_{i_1}, \ldots, \mathcal{V}_{i_{r-1}}, \mathcal{V}_{i_r}, \mathcal{V}_{i_{r+1}}, \ldots, \mathcal{V}_{i_k}, \mathcal{W}_{j_1}, \ldots, \mathcal{W}_{j_{n-k}} \rangle_\alpha \neq \{0\}$.

From here, $f_\alpha([i], \ldots, [i], I'^{(r)}, [i], \ldots, [i], \Sigma_J \ldots, \Sigma_J) \subset I' \cup \{\epsilon\}$, and so $\{\epsilon\} \neq I' \cup \{\epsilon\}$ is an F-submodule of $[i] \cup \{\epsilon\}$. Consequently $I' = [i]$ and then

$$\mathcal{U} = \bigoplus_{i' \in I'} \mathcal{V}_j = \mathcal{V}_{[i]}.$$

So $\mathcal{V}_{[i]}$ is an homogeneous-simple graded k-submodule.

Conversely, if $\mathcal{V}_{[i]}$ is homogeneous-simple and we take some F-submodule $\{\epsilon\} \neq I' \cup \{\epsilon\}$ of $[i] \cup \{\epsilon\}$, being $\emptyset \neq I' \subset [i]$, we get as above that

$$\mathcal{U} := \bigoplus_{i' \in I'} \mathcal{V}_{i'}$$

is a nonzero homogeneous k-submodule of $\mathcal{V}_{[i]}$ and so $\mathcal{U} = \mathcal{V}_{[i]}$. Hence $I' = [i]$ and $[i] \cup \{\epsilon\}$ is a pointed simple F-submodule of $\Sigma_I \cup \{\epsilon\}$. □

Remark 6 We note, respect to Proposition 5, that we actually can state a bijective correspondence from the pointed F-submodules of $([i] \cup \{\epsilon\}, \epsilon, F_{[i]})$ to the homogeneous k-submodules of $\mathcal{V}_{[i]}$. That is, if

$$\mathcal{V} = \bigoplus_{i \in I} \mathcal{V}_i$$

is an I-graded k-module over a J-graded linear subspace $\mathcal{W}$ and consider some $[i] \in \Sigma_I / \sim$ then, by denoting $\mathcal{F}$ the family of all of the pointed F-submodules of $[i] \cup \{\epsilon\}$ (different to $\{\epsilon\}$), and by $\mathcal{G}$ the family of all of the homogeneous k-submodules of $\mathcal{V}_{[i]}$, we can define the following map $\phi : \mathcal{F} \to \mathcal{G}$ as

$$\phi(I') := \bigoplus_{i' \in (I' \setminus \{\epsilon\})} \mathcal{V}_{i'}$$

for any pointed F-submodule $\{\epsilon\} \neq I'$ of $[i] \cup \{\epsilon\}$. Observe that the fact

$$f_\alpha(\Sigma_I \cup \{\epsilon\}, \ldots, \Sigma_I \cup \{\epsilon\}, I'^{(r)}, \Sigma_I \cup \{\epsilon\}, \ldots, \Sigma_I \cup \{\epsilon\}, \Sigma_J, \ldots, \Sigma_J) \subset I'$$

for any $f_\alpha \in F$ and $1 \leq r \leq k$, ensures $\phi(I') \in \mathcal{G}$ for any $I' \in \mathcal{F}$. We also have that ϕ is bijective as consequence of $[i] \subset \Sigma_I$.

Theorem 5 (Second Wedderburn-type theorem) *Let $\mathcal{V}$ be a k-module with a division grading over $\mathcal{W}$. Then*

$$\mathcal{V} = \bigoplus_{[i] \in \Sigma_I/\sim} \mathcal{V}_{[i]}$$

is the orthogonal direct sum of the family of all its nonzero homogeneous-simple k-submodules.

Proof By Theorem 4 we can write

$$\mathcal{V} = \bigoplus_{[i] \in \Sigma_I/\sim} \mathcal{V}_{[i]}, \tag{6}$$

being any $\mathcal{V}_{[i]}$ (see Eq. (5)) a nonzero homogeneous k-submodule of $\mathcal{V}$. By Proposition 4, the F-module $(\Sigma_I \cup \{\epsilon\}, \epsilon, F)$ is a division F-module over Σ_J. Hence, Theorem 3 gives us that any $[i] \cup \{\epsilon\}$ for $[i] \in \Sigma_I/\sim$ is a pointed simple F-submodule of $\Sigma_I \cup \{\epsilon\}$. From here, we get by Proposition 5 that the homogeneous k-submodule $\mathcal{V}_{[i]}$ is homogeneous-simple.

Furthermore, if we take some nonzero homogeneous-simple k-submodule $\mathcal{U}$ of $\mathcal{V}$, we can write

$$\mathcal{U} = \bigoplus_{i' \in I'} \mathcal{V}_{i'}$$

with $\emptyset \neq I' \subset \Sigma_I$. Then we can fix some $i' \in I'$, being so $\mathcal{V}_{i'} \subset \mathcal{U}$. Since $i' \in [i']$ we have $\{0\} \neq \mathcal{V}_{i'} \subset \mathcal{U} \cap \mathcal{V}_{[i']}$. Taking now into account that $\mathcal{U} \cap \mathcal{V}_{[i']}$ is a nonzero homogeneous k-submodule of $\mathcal{V}$, we get by homogeneous-simplicity that

$$\mathcal{U} = \mathcal{U} \cap \mathcal{V}_{[i']} = \mathcal{V}_{[i']},$$

and so any nonzero homogeneous-simple k-submodule $\mathcal{U}$ appears in the decomposition given by Eq. (6). □

Let us the illustrate this result with the next example.

Example 3 Example Consider the linear spaces $\mathcal{V} = \mathbb{R}^4$ and $\mathcal{W} = \mathbb{Z}^3$ in which we define a number $N \in \mathbb{N}$ of n-linear maps,

$$\langle \cdot, \ldots, \cdot \rangle_\alpha : \mathcal{V} \times \overset{k)}{\cdots} \times \mathcal{V} \times \mathcal{W} \times \overset{n-k)}{\cdots} \times \mathcal{W} \to \mathcal{V}$$

defined by

$$\big\langle (x_1, y_1, z_1, t_1), \ldots, (x_k, y_k, z_k, t_k), (x_{k+1}, y_{k+1}, z_{k+1}), \ldots, (x_n, y_n, z_n) \big\rangle_\alpha =$$

$$= \left(\frac{t_1 \cdots t_k z_{k+1} \cdots z_n}{\alpha}, \frac{y_1 \cdots y_n + z_1 \cdots z_n}{\alpha}, \frac{y_1 \cdots y_n + z_1 \cdots z_n}{\alpha}, \frac{x_1 \cdots x_n}{\alpha} \right)$$

for any $1 \leq \alpha \leq N$.

Then $\mathcal{V}$ becomes a k-module over $\mathcal{W}$ with a division grading

$$\mathcal{V} = \mathcal{V}_a \oplus \mathcal{V}_b \oplus \mathcal{V}_c \text{ and } \mathcal{W} = \mathcal{W}_d \oplus \mathcal{W}_e,$$

where $\mathcal{V}_a = \{(x, 0, 0, 0) : x \in \mathbb{R}\}, \mathcal{V}_b = \{(0, x, y, 0) : x, y \in \mathbb{R}\}, \mathcal{V}_c = \{(0, 0, 0, x) : x \in \mathbb{R}\}$, $\mathcal{W}_d = \{(x, 0, 0) : x \in \mathbb{Z}\}$ and $\mathcal{W}_e = \{(0, x, y) : x, y \in \mathbb{Z}\}$.

In order to verify this is a division grading, observe that the only nonzero products among the homogeneous components are $\{0\} \neq \langle \mathcal{V}_a, \ldots, \mathcal{V}_a, \mathcal{W}_d, \ldots, \mathcal{W}_d \rangle_\alpha \subset \mathcal{V}_c$, $\{0\} \neq \langle \mathcal{V}_b, \ldots, \mathcal{V}_b, \mathcal{W}_e, \ldots, \mathcal{W}_e \rangle_\alpha \subset \mathcal{V}_b$ and $\{0\} \neq \langle \mathcal{V}_c, \ldots, \mathcal{V}_c, \mathcal{W}_e, \ldots, \mathcal{W}_e \rangle_\alpha \subset \mathcal{V}_a$. From here, it is clear that we have a division grading.

It is straightforward to verify that $[a] = \{a, c\}$ and $[b] = \{b\}$ and so, by Theorem 5, $\mathcal{V}$ is the direct sum of its family of homogeneous-simple k-submodules

$$\mathcal{V} = \mathcal{V}_{[a]} \oplus \mathcal{V}_{[b]}$$

with $\mathcal{V}_{[a]} = \{(x, 0, 0, y) : x, y \in \mathbb{R}\}$ and $\mathcal{V}_{[b]} = \{(0, x, y, 0) : x, y \in \mathbb{R}\}$.

Acknowledgements The authors thank the referee for an exhaustive review of the paper as well as for suggestions that helped to improve the work.

References

1. Barreiro, E., Kaygorodov, I., Sánchez, J.M.: k-Modules over linear spaces by n-linear maps admitting a multiplicative basis. Algebr. Represent. Theory **22** (2019). https://doi.org/10.1007/s10468-018-9790-8
2. Bingjun, L.: Lifting property of the Jacobson radical in associative pairs. Bull. Malays. Math. Sci. Soc. **2**(34), 521–528 (2011)
3. Calderón A.J.: Extended magmas and their applications. J. Algebra Appl. **16**(8), 1750150, 11 (2017)
4. Calderón, A.J., Gaye, B., Navarro, F.J.: Actions between sets. Arbitrarily graded linear modules. Comm. Algebra. **47**(9), 3849–3858 (2019)

5. Calderón, A.J., Hegazi, A.S., Hani, A.: A characterization of the semisimplicity of Lie-type algebras through the existence of certain linear bases. Linear Multilinear Algebr. **65**(9), 1781–1792 (2017)
6. Calderón, A.J., Navarro, F.J., Sánchez, J.M.: Modules over linear spaces admitting a multiplicative basis. Linear Multilinear Algebr. **65**(1), 156–165 (2017)
7. Cohn, P.M.: Basic Algebra: Groups, Rings, and Fields. Springer, London (2003)
8. McCrimmon, K.: Involutions on rectangular Jordan pairs. J. Algebr. **225**(2), 885–903 (2000)
9. Montaner, F., Paniello, I.: On polynomial identities in associative and Jordan pairs. Algebr. Represent. Theory **13**, 189–205 (2010)
10. Loos, O.: Jordan pairs. Lecture Notes in Mathematics, vol. 460 (1975)

Primary Group Rings

Mohammed El Badry, Mostafa Alaoui Abdallaoui and Abdelfattah Haily

Abstract A ring R is said to be primary if the Jacobson radical $J(R)$ is nilpotent and the factor ring $R/J(R)$ is simple artinian. The main result of this note is the characterization of the primary group rings of not necessary abelian groups. This generalizes the work of Chin and Qua (Rendiconti del Seminario Matematico della Università di Padova 137:223–228 2017, [1]) in which the author characterizes the primary group rings of abelian groups.

Keywords Group · Ring · Primary group ring

Throughout this note all rings are associative with identity element and all the modules are left modules. For terminology and results in ring theory, we refer to [2].

Definition 1 Let R be a ring.
(1) R is said to be an artinian ring (on the left side), if it satisfies the descending chain on the left ideals.
(2) R is said to be a semiprimary ring, if the Jacobson radical $J(R)$ of R is nilpotent and the ring $R/J(R)$ is artinian.
(3) R is said to be a primary ring, if the Jacobson radical $J(R)$ of R is nilpotent and the factor ring $R/J(R)$ is simple artinian.

Definition 2 Let R be a ring with Jacobson radical $J(R)$.
(i) $J(R)$ is left T-nilpotent, if for every sequences $a_1, a_2, \cdots$ of elements of $J(R)$, there exists an integer n such that $a_1 a_2 \cdots a_n = 0$.
(ii) The ring R is said to be left perfect, if $R/J(R)$ is artinian and $J(R)$ is left T-nilpotent. artinian.

M. El Badry (✉) · M. Alaoui Abdallaoui · A. Haily
Department of Mathematics, Chouaib Doukkali University, BP 20 El jadida, Morocco
e-mail: elbadrymohammed2@gmail.com

M. Alaoui Abdallaoui
e-mail: abdallaouialaoui@hotmail.com

A. Haily
e-mail: afhaily@yahoo.fr

M. Siles Molina et al. (eds.), *Associative and Non-Associative Algebras and Applications*, Springer Proceedings in Mathematics & Statistics 311,
https://doi.org/10.1007/978-3-030-35256-1_10

Let R be a ring and G a multiplicative group (not necessary finite).

The group ring $R[G]$ is the free R-module with basis the elements of G. Every element x in $R[G]$ has the form $x = \sum_{g\in G} x_g g$, where x_g in R are all zero except a finite number.

Let $x = \sum_{g\in G} x_g g$ and $y = \sum_{g\in G} y_g g$. The multiplication in $R[G]$ is defined by:

$$xy = \sum_{g\in G} c_g g, \quad \text{where} \quad c_g = \sum_{hk=g} x_h y_k$$

It is easy to see that $R[G]$ is an associative ring and R is a subring of $R[G]$.

1 Primary Group Rings

In this section, we give necessary and sufficient conditions on a group G and a ring R for the group ring $R[G]$ to be a primary ring. This generalizes [1]. For this purpose we need the following results.

Theorem 1 ([3]) *Let R be a ring and G be a group. Then $R[G]$ is semiprimary ring, if and only if R is semiprimary and G is finite.*

Lemma 1 *Let R be a primary ring, if I is a two sided ideal contained in $J(R)$. Then R/I is primary.*

Proof Since R is primary, $J(R)$ is nilpotent and $R/J(R)$ is simple artinian. Then $J(R/I) = J(R)/I$ is nilpotent and $(R/I)/(J(R/I)) \simeq R/J(R)$ is simple artinian. □

Now, we can state our main result.

Theorem 2 *Let R be a ring and G a nontrivial group. Then $R[G]$ is primary if and only if R is primary and G is a finite p-group, where p is the characteristic of $R/J(R)$.*

Proof **The sufficient condition**: Since R is primary and G is finite, by Theorem 1, $R[G]$ is semiprimary and $J(R[G])$ is nilpotent. $J(R)[G]$ is contained in $J(R[G])$ (because $J(R)$ is nilpotent). Hence

$$\begin{aligned} R[G]/J(R[G]) &\simeq (R[G]/J(R)[G]) \,/\, (J(R[G])/J(R)[G]) \\ &\simeq (R[G]/J(R)[G]) \,/ J\, (R[G]/J(R)[G]) \\ &\simeq ((R/J(R))[G]) \,/ J\, ((R/J(R))[G]) \end{aligned}$$

Since $R/J(R)$ is simple artinian, then $R/J(R) \simeq M_n(D)$ for some positive integer n and division ring D of characteristic p. Therefore $(R/J(R))[G] \simeq M_n(D[G])$ this implies that $R[G]/J(R[G]) \simeq M_n(D[G]/J(D[G]))$.

We shall show that $D[G]$ is a local ring. For that, let $k = Z(D)$ the center of D. Clearly k is a field. Since $D[G] \simeq D \otimes_k k[G]$. Then $J(D[G]) \simeq D \otimes_k J(k[G])$. By [4, Chap. 3, Lemma 1.6], $k[G]$ is a local ring. It follows that $D[G]/J(D[G]) \simeq D \otimes_k k[G]/J(k[G]) \simeq D \otimes_k k \simeq D$ is a division ring and $D[G]$ is local.

The necessary condition: Since $R[G]$ is primary, by Theorem 1, R is semiprimary and G is finite. $R/J(R)$ is simple artinian, because it is a homomorphic image of the simple artinian ring $R[G]/J(R[G])$. This shows that R is primary ($J(R)$ is nilpotent because R is semiprimary). Clearly $J(R)[G]$ is contained in $J(R[G])$. Using the last lemma, $R[G]/J(R)[G] \simeq (R/J(R))[G]$ is primary and $(R/J(R))[G] \simeq \mathcal{M}_n(D[G])$ is primary, where n is integer and D is a division ring of characteristic p. This implies that $D[G]$ is primary. Let I be the augmentation ideal of $D[G]$. Obviously I is a maximal ideal of $D[G]$. Since $J(D[G])$ is contained in I and $D[G]/J(D[G])$ is simple, we have $I = J(D[G])$ (because $I/J(D[G])$ is a proper ideal of $D[G]/J(D[G])$) and $D[G]$ is local, then G is a finite p-group. □

References

1. Chin, A.Y.M., Qua, K.T.: Primary group rings. Rendiconti del Seminario Matematico della Università di Padova **137**, 223–228 (2017)
2. Lam, T.Y.: A First Course in Non Commutative Rings. Springer, New York (1991)
3. Tan, K.T.: A note on semiprimary group rings. Acta Math. Hungarica **33**(3–4), 261 (1979)
4. Passman, D.S.: The Algebraic Structure of Group Rings. Wiley-Interscience, New York (1977)

Semi-ring Based Gröbner–Shirshov Bases over a Noetherian Valuation Ring

Yatma Diop, Laila Mesmoudi and Djiby Sow

Abstract Commutative and non commutative Gröbner–Shirshov bases were first studied over fields and after extended to some particular rings. In theses works, the monomials are in a monoid. Recently, Bokut and al. gave a new extension of Gröbner–Shirshov bases over a field by choosing the monomials in a semi-ring rather in a monoid. In this paper, we study Gröbner–Shirshov bases where the monomials are in a semi-ring and the coefficients are in a noetherian valuation ring and we establish the relation between weak and strong Gröbner bases.

Keywords Gröbner–Shirshov bases · Monomial order · Reduction · Composition · A-polynomial · Least common multiple · Greatest common divisor · Standard expression

1 Introduction

Gröbner bases were independently invented by B. Buchberger in commutative algebra [4], H. Hironaka in algebraic geometry [5] and A. I. Shirshov in Lie algebras [12] between the end of years 50s and the middle of the years 60s. During the following years, many other authors extended them to many other algebras: Bokut [3], Bergman [1], Mikhalev [8], ...

Commutative and non commutative Gröbner bases were widely developed over fields. One fundamental difference between commutative and non commutative Gröbner bases is that the Buchberger's algorithm does not always terminate in the non commutative case. The notion of Gröbner bases was generalized to some

Y. Diop (✉) · L. Mesmoudi · D. Sow
Cheikh Anta Diop University of Dakar, Dakar, Senegal
e-mail: yatma.diop@ucad.edu.sn

L. Mesmoudi
e-mail: lmesmoudi@gmail.com

D. Sow
e-mail: sowdjibab@yahoo.fr

M. Siles Molina et al. (eds.), *Associative and Non-Associative Algebras and Applications*, Springer Proceedings in Mathematics & Statistics 311,
https://doi.org/10.1007/978-3-030-35256-1_11

particular rings such as valuation rings [15], Dedekind rings, [16, 17], D-A rings [7], gaussian rings, reduction rings [13], euclidean domain [6], principal ideal domain [11].

More generally, Gröbner bases over rings were developped and studied by Zacharias in her thesis entitled "Generalized gröbner bases in commutative polynomial rings". Later, Möller showed how to construct Gröbner bases by using szygies [9]. In all theses works, the set of monomials is a monoid.

In [2], L. A. Bokut, Yuqun Chen and Qiuhui Mo introduced a new extension of Gröbner–Shirshov bases over a field where the set of monomials is a semi-ring rather a monoid. To highlight the difference between these two approaches, we call the first one by monoid-based Gröbner–Shirshov bases and the second one by semiring-based Gröbner–Shirshov bases.

In this paper, we extend semiring-Gröbner–Shirshov bases over a noetherian valuation ring and we establish the relation between weak and strong Gröbner bases.

Let $X = \{X_1, X_2, ..., X_n\}$ be an alphabet, $X^* = \{X_{i_1}X_{i_2}...X_{i_s}, X_i \in X\}$, $Rig\langle X\rangle = \{u_1 \circ u_2 \circ ... \circ u_t, u_i \in X^*\}$. $(Rig\langle X\rangle, \circ)$ is a commutative monoid with neutral element θ. We extend the concatenation over X^* to an operation "." over $Rig\langle X\rangle$ which is distributive from left and right relative to $\circ$. So $(Rig\langle X\rangle, ., \circ)$ is a semi-ring. We call it the free semi-ring generated bt X^* and denote it by $(X^*, ., \circ)$.

In this paper:

1. the set of coefficients is a valuation ring denoted by V;
2. the set of monomials is the semi-ring $Rig\langle X\rangle = (X^*, ., \circ)$ instead of the monoid $(X^*, .)$. Thus, a monomial is a composition of words (elements of X^*) by the product $\circ$.

This paper is organized as follows:

- in Sect. 2, we give notions and notations that we use in this paper;
- in Sect. 3, we define Gröbner–Shirshov bases over a valuation ring and we characterize them by giving and proving the corresponding Composition-Diamond-Lemma.

2 Preliminaries

In this section, we give basic notions necessary for the definition and the characterization of semiring Gröbner–Shirshov bases. This section also contains the notations we will use in the following parts.

For more information, refer to [2].

Definition 1 • A ring V is called a valuation ring if for any $a, b \in V \setminus \{0\}$, a divides b or b divides a. For example, for any prime number p and any natural number α, $\frac{\mathbb{Z}}{p^\alpha\mathbb{Z}}$ is a valuation ring.

- Let V be a valuation ring and $V Rig\langle X\rangle$ the groupoid over V; it means the V-semi-module with V-basis $Rig\langle X\rangle$ and multi-linear products $\omega \in \{., \circ\}$ that are

extended by linearity from $Rig\langle X\rangle$ to $VRig\langle X\rangle$. We will respectively call monomial and polynomial any element u of $Rig\langle X\rangle$ and any element f of $VRig\langle X\rangle$. Any ideal of $VRig\langle X\rangle$ is called an Ω-ideal where $\Omega = \{., \circ\}$.
A monomial u is expressed $u = u_1 \circ u_2 \circ ... \circ u_t$ with $u_i = u_{i_1}u_{i_2}...u_{i_s} \in X^*$. For any monomial $u = u_1 \circ u_2 \circ ... \circ u_t \in Rig\langle X\rangle$, we will denote $|u|_\circ = t$ which is the length of u with respect to $\circ$. If $u_i = u_j$ for any $i, j \in [1, t]$ then $u = u_1 \circ u_2 \circ ... \circ u_t$ will also be denoted $u = u'^{|t|_\circ}$ where $u' = u_1 = u_2 = ... = u_t$. By convention $u = \theta$ if and only if $t = 0$.
A polynomial f can be expressed:

$$f = \sum_{i=1}^{n} \alpha_i U_i,\ \alpha_i \in V,\ U_i \in Rig\langle X\rangle$$

Example 1 Let's take $X = \{x, y, z\}$, $Rig\langle X\rangle = (X^*, \cdot, \circ)$, $V = \frac{\mathbb{Z}}{8\mathbb{Z}}$.
Then $x \circ xz$, $zxy \circ xz \circ yx \circ z \circ y$, $y^{|3|_\circ}$, $z^{|4|_\circ}yz^{|2|_\circ}$, z are monomials and $f = 2x \circ xz + zxy \circ xz \circ yx \circ z \circ y + 3y^{|3|_\circ} + z^{|4|_\circ}yz^{|2|_\circ} + 3z$ is a polynomial.

Definition 2 A total order $\preceq$ on X^* is admissible if it satisfies the following condition: for any $u, v, w, t \in X^*$, $u \preceq v \Rightarrow wut \preceq wvt$.

Example 2 Let $X = \{x_1, x_2, ..., x_n\}$ a total ordered alphabet such that $x_i \succ x_{i+1}$ for any $i \in \{1, 2, ..., n-1\}$.

Let $u = u_1u_2...u_s \in X^* \setminus \{1\}$. The natural number s is called the degree of u with respect to the law ".", we denote $deg(u) = s$. By convention, $deg(u) = 0$ if $u = 1$.

The degree-lexicographic order defined by

$$u \prec v \Leftrightarrow \begin{cases} deg(u) < deg(v) \\ \quad or \\ deg(u) = deg(v)\ and\ u \prec_{lex} v \end{cases}$$

is an admissible ordering on X^*.

Remark 1 If we fix an admissible order on X^* then any element $u \in Rig\langle X\rangle$ has a unique form $u = u_1 \circ u_2 \circ ... \circ u_l$, with $u_i \in X^*$ and $u_i \preceq u_{i+1}$ for any $i = 1, ..., l-1$.

Definition 3 Let $u = w_1 \circ w_2 \circ ... \circ w_m \circ u_{m+1} \circ ... \circ u_n$, $v = w_1 \circ w_2 \circ ... \circ w_m \circ v_{m+1} \circ ... \circ v_t \in Rig\langle X\rangle$ satisfying $u_i \neq v_j$ for any $(i, j) \in [m+1, n] \times [m+1, t]$.

Then $d = w_1 \circ w_2 \circ ... \circ w_m \circ u_{m+1} \circ ... \circ u_n \circ v_{m+1} \circ ... \circ v_t$ and $w = w_1 \circ w_2 \circ ... \circ w_m$ will be respectively called the least common multiple and the greatest common divisor of u and v with respect to $\circ$. We will denote them by $d = lcm_\circ(u, v)$ and $w = gcd_\circ(u, v)$.

Example 3 If $u = x^{|4|_\circ} \circ yz \circ zxy \circ y$ and $v = x \circ yz \circ z^{|2|_\circ} \circ yx \circ zx \circ y^{|3|_\circ}$ then $lcm_\circ(u, v) = x^{|4|_\circ} \circ yz \circ zxy \circ z^{|2|_\circ} \circ yx \circ zx \circ y^{|3|_\circ} = u \circ z^{|2|_\circ} \circ yx \circ zx \circ y^{|2|_\circ} = v \circ x^{|3|_\circ} \circ zxy$ and $gcd \circ (u, v) = x \circ y \circ yz$.

Definition 4 An ordering $\prec$ on $Rig\langle X\rangle$ is called a monomial ordering if:

1. it is a total ordering: it means u, v are always comparable under $\prec$ for any $u, v \in Rig\langle X\rangle$;
2. it is a well ordering: there's no infinite decreasing sequence in $Rig\langle X\rangle$ with respect to $\preceq$;
3. it is compatible with the semi-ring structure: for any $u, v, w \in Rig\langle X\rangle$, t', $w' \in X^*$, we have

$$u \prec v \Rightarrow \begin{cases} u \circ w \prec v \circ w \\ t'uw' \prec t'vw' \end{cases}$$

4. $\forall u \in Rig\langle X\rangle \setminus \{\theta\}, \theta \prec u$.

Example 4 Let $u = u_1 \circ u_2 \circ ... \circ u_n \in Rig\langle X\rangle$ with $u_i \preceq u_{i+1}$, we define $vect(u)$ by $vect(u) = (u_1, u_2, ..., u_n)$. In order to compare two monomials u and v lexicographically, we compare $vect(u)$ and $vect(v)$ lexicographically by using an admissible order on X^*.

The length-lexicographic order defined by

$$u \prec_{ll} v \Leftrightarrow \begin{cases} |u|_\circ < |v|_\circ \\ or \\ |u|_\circ = |v|_\circ \; and \; u \prec_{lex} v \end{cases}$$

is a monomial ordering on $Rig\langle X\rangle$.

Proposition 1 *Let $\prec$ be a monomial order on $Rig\langle X\rangle$. Then for any $u, v, w \in Rig\langle X\rangle$, we have:*

1. $u \circ w = u \circ v \Rightarrow v = w$. *We say that $Rig\langle X\rangle$ is cancellative.*
2. $u \circ v = \theta \Rightarrow u = v = \theta$.

The proof is obvious.

Let us fix a monomial ordering on $Rig\langle X\rangle$. Then:

1. The greatest monomial occurring in $f \in VRig\langle X\rangle$ will be called the leading monomial of f and denoted by $LM(f)$.
2. The coefficient of $LM(f)$ is called the leading coefficient of f and denoted by $LC(f)$.
3. $LC(f)LM(f)$ is the leading term of f. It is denoted by $LT(f)$.
4. $LM(A) = \{LM(f),\ f \in A\}$
5. $LT(A) = \{LT(f),\ f \in A\}$
6. $NonLM(A) = Rig\langle X\rangle \setminus LM(A)$
7. Any polynomial f has a unique form: $f = \sum_{i=1}^{n} \alpha_i u_i,\ \alpha_i \in V \setminus \{0\},\ u_i \in Rig\langle X\rangle$ satisfying $u_i \prec u_{i+1}$ for any $i \in [1, n-1]$.

Example 5 Let $V = \frac{\mathbb{Z}}{8\mathbb{Z}}$, $f = 4yy \circ zyzx \circ zyyx + 2z \circ xy + y \circ x + 1 \in VRig\langle X\rangle$. Let's consider the length lexicographic order on $Rig\langle X\rangle$.

Then: $LM(f) = yy \circ zyzx \circ zyyx$,
$LC(f) = 4,\ LT(f) = 4yy \circ zyzx \circ zyyx$.

Definition 5 Let V be a ring and $f,\ g,\ h \in VRig\langle X\rangle$, $G \subset VRig\langle X\rangle$ and $\prec$ a monomial order on $Rig\langle X\rangle$.

1. We will say that f strongly reduces to h modulo g if
   - $LT(f) = \alpha a LT(g) b \circ u$ for some $a, b \in X^*,\ u \in Rig\langle X\rangle, \alpha \in V$
   - $h = f - \alpha a g b \circ u$.

   We will denote $f \xrightarrow{g}_s h$ or $h = SRed(f, g)$.
   Otherwise, f is said to be strongly irreducible modulo h.
2. We will say that f weakly reduces to h modulo G if there exists $\{g_1, g_2, ..., g_r\} \in G$ such that
   - $LT(f) = \sum_{i=1}^{r} \alpha_i a_i LT(g_i) b_i \circ u_i,\ (\alpha_i, a_i, b_i) \in V \times X^* \times X^*$
   - $h = f - \sum_{i=1}^{r} \alpha_i a_i g_i b_i \circ u_i$

   We will denote $f \xrightarrow{G}_w h$ or $h = WRed(f, G)$.
   Otherwise, f is said to be weakly irreducible modulo G.

Proposition 2 *If f is strongly reducible modulo $g \in G$ then f is weakly reducible modulo G. The assertion"f is weakly reducible modulo G" does not imply that f is strongly reducible modulo an element g of G.*

Proof The proof of the first claim is obvious. For the second one, it is sufficient to consider the example $G = \{2x, 3y\} \subset \mathbb{Z}[x, y]$ taken from [7]. The polynomial $f = 3xy - 2xy = xy \in \langle G\rangle$ is weakly reducible modulo G even if it is neither strongly reducible modulo $2x$ nor modulo $3y$. □

Further, we will see that if the considered set of coefficients is a valuation ring then the weak reduction implies the strong reduction.

The following algorithm which takes as input a polynomial f, a set G of polynomials and a monomial order on $Rig\langle X\rangle$ returns a polynomial r which strongly irreducible modulo any element of G.

Algorithm 1: TotRed

Input: $(f, G, \prec)$
Output : r a reduced polynomial
$r \leftarrow 0$;
While $f \neq 0$ **:**
  If *f is strongly reducible modulo g for some $g \in G$* **Then:**
    $f \longleftarrow SRed(f, g)$
  Else:
    $r \leftarrow r + LT(f)$;
    $f \leftarrow f - LT(f)$;
Return r

Proposition 3 *The previous algorithm terminates.*

Proof At the first step of the execution of the algorithm, we obtain a polynomial f_1 satisfying $LM(f_1) \prec LM(f)$.

At the second step, we obtain a polynomial f_2 satisfying $LM(f_2) \prec LM(f_1) \prec LM(f)$.

At the ith step, the polynomial f_i which is the reduced of f_{i-1} satisfies $LM(f_i) \prec LM(f_{i-1}) \prec ... \prec LM(f_2) \prec LM(f_1) \prec LM(f)$.

Since $\prec$ is a monomial ordering, the decreasing sequence of monomials f_{i-1} satisfies $LM(f_i) \prec LM(f_{i-1}) \prec ... \prec LM(f_2) \prec LM(f_1) \prec LM(f)$ is finite. So the algorithm terminates. □

Example 6 Let $V = \frac{\mathbb{Z}}{8\mathbb{Z}}$, $f = 4yy \circ zyzx \circ zyyx + 2z \circ xy + y \circ x + 1 \in VRig \langle X \rangle$, $G = \{g_1 = 2zx \circ yx + 3x \circ y, g_2 = 5zy \circ y + y \circ 1, g_3 = xy + 2\} \subset VRig \langle X \rangle$. We want to reduce f by G by using the length lexicographic order on $Rig\langle X \rangle$.

First step, we initialize: $r := 0$

$LT(f) = 4yy \circ zyzx \circ zyyx = 2zyLT(g_1) \circ yy$. So f is reducible by g_1 and we have:

$f \xrightarrow{g_1}_s f_1 = f - 2zyg_1 \circ yy = 2zyx \circ zyy \circ yy + y \circ x + 2z \circ xy + 1$

$LT(f_1) = 2zyx \circ zyy \circ yy = 2LT(g_2)y \circ zyx$. So f_1 is reducible by g_2 and we have:

$f_1 \xrightarrow{g_2}_s f_2 = f_1 - 2g_2y \circ zyx = 6y \circ yy \circ zyx + y \circ x + 2z \circ xy + 1$

$LT(f_2) = 6y \circ yy \circ zyx$ which is not divisible by a $LT(g_i)$ for an $i \in \{1, 2, 3\}$.

So $r := 6y \circ yy \circ zyx$ and $f_3 = f_2 - LT(f_2) = y \circ x + 2z \circ xy + 1$

$LT(f_3) = y \circ x$ which is not divisible by a $LT(g_i)$ for an $i \in \{1, 2, 3\}$.

So $r := 6y \circ yy \circ zyx + y \circ x$ and $f_4 = f_3 - LT(f_3) = 2z \circ xy + 1$

$LT(f_4) = 2z \circ xy = 2z \circ LT(g_3)$. So f_4 is reducible by g_3 and we have:

$f_4 \xrightarrow{g_3}_s f_5 = f_4 - 2z \circ g_3 = 4z \circ 1 + 1$

$LT(f_5) = 4z \circ 1$ which is not divisible by an $LT(g_i)$ for an $i \in \{1, 2, 3\}$.

So $r := 6yy \circ y \circ zyx + 4z \circ 1$ and $f_6 = f_5 - LT(f_5) = 1$

$LT(f_6) = 1$ which is not divisible by an $LT(g_i)$ for an $i \in \{1, 2, 3\}$.

So $r := 6yy \circ y \circ zyx + 4z \circ 1 + 1$ and $f_7 = f_6 - LT(f_6) = 0$. So the algorithm stops.

$r := 6yy \circ y \circ zyx + 4z \circ 1 + 1$ is a reduced of f modulo G.

The notion of composition is very important for the theory of Gröbner–Shirshov bases. It usually allows to say wheither a set is a Gröbner–Shirshov basis or not. In the following, we give it in the commutative and non commutative cases according to a semiring (as the set of monomials) and a valuation ring (as the set of coefficients).

Definition 6

1. Commutative case: we assume that $VRig\langle X \rangle$ is a commutative semi-ring. Let $f, g \in VRig\langle X \rangle$, $a, c \in X^*$. Let $(u, v) \in Rig\langle X \rangle \times Rig\langle X \rangle$ the unique pair of monomials such that: $lcm_\circ(aLM(f), cLM(g)) = aLM(f) \circ u = cLM(g) \circ v$. The polynomial

$$O(f, g, a, c, u, v) = \begin{cases} af \circ u - \frac{LC(f)}{LC(g)} cg \circ v \; if \; LC(g) \; divides \; LC(f) \\ \frac{LC(g)}{LC(f)} af \circ u - cg \circ v \; else \end{cases}$$

is called a composition of f and g with respect to $w = lcm_{\circ}(aLM(f), cLM(g))$.

2. Non commutative case: we assume that $VRig\langle X\rangle$ is a non commutative semi-ring.
   Let $f, g \in VRig\langle X\rangle$, $a, b, c, d \in X^*$. Let $(u, v) \in Rig\langle X\rangle \times Rig\langle X\rangle$ the unique pair of monomials such that:
   $lcm_{\circ}(aLM(f)b, cLM(g)d) = aLM(f)b \circ u = cLM(g)d \circ v$. The polynomial

$$O(f, g, a, b, c, d, u, v) = \begin{cases} afb \circ u - \frac{LC(f)}{LC(g)} cgd \circ v \; if \; LC(g) \; divides \; LC(f) \\ \frac{LC(g)}{LC(f)} afb \circ u - cgd \circ v \; else \end{cases}$$

   is called a composition of f and g with respect to $w = lcm_{\circ}(aLM(f)b, cLM(g)d)$.
3. w is called an ambiguity between f and g. We denote $O(f, g)$ the set of all compositions of f and g and $O(G)$ the set of all compositions defined in G

Remark 2

1. For any $f, g \in VRig\langle X\rangle, O(f, g) \neq \emptyset$.
2. Let $f, g \in VRig\langle X\rangle$ such that $LM(f) = LM(g)$. Then $O(f, g, 1, 1, 1, 1, \theta, \theta) \in O(f, g)$. We will call it a *simple composition* of f and g and denote it by $\widetilde{O}(f, g)$.
3. In the non commutative case, our definition generalizes the one given in [2, p. 5, def 3.1]. It is enough to take respectively $(a, d) = (1, 1)$ and $(a, b) = (1, 1)$ to have composition of intersection and composition of inclusion similarly in [2]. But in [2], the authors only consider compositions such that $|lcm_{\circ}(aLM(f)b, cLM(g)d)|_{\circ} < |aLM(f)b|_{\circ} + |cLM(g)d|_{\circ}$.
   This restriction is due to the fact that they take coefficients in a field. Thus, if $|lcm_{\circ}(aLM(f)b, cLM(g)d)|_{\circ} = |aLM(f)b|_{\circ} + |cLM(g)d|_{\circ}$ or equivalently $gcd_{\circ}(aLM(f)b, cLM(g)d) = \theta$ then $O(f, g, a, b, c, d, u, v)$ is always reduced to zero by $\{f, g\}$. In general, this is not the case in a valuation ring.

Example 7 Let us consider $f = 5yz \circ zx \circ x \circ 1 + 2xy \circ z + z \circ 1$, $g = 2xy \circ y \circ z \circ 1 + 3x \circ z \circ 1 \in \frac{\mathbf{Z}}{8\mathbf{Z}} Rig\langle X\rangle$ and the length lexicographic order. We assume that $VRig\langle X\rangle$ is not commutative.
Then $LM(f) = yz \circ zx \circ x \circ 1$, $LM(g) = xy \circ y \circ z \circ 1$.

1. Let's take $(a, b, c, d) = (xy, z, 1, xz)$.
   $xyLm(f)z = xyyzz \circ xyzxz \circ xyxz \circ xyz$ and $1LM(g)xz = xyxz \circ yxz \circ zxz \circ xz$.

$$\begin{aligned} lcm_{\circ}(xyLM(f)z, LM(g)xz) &= xyyzz \circ xyzxz \circ xyz \circ xyxz \circ yxz \circ zxz \circ xz \\ &= xyLM(f)z \circ yxz \circ zxz \circ xz \\ &= LM(g)xz \circ xyyzz \circ xyzxz \circ xyz \end{aligned}$$

$$\begin{aligned} O(f, g, xy, z, 1, xz, u, v) &= 2xyfz \circ yxz \circ zxz \circ xz - gxz \circ xyyzz \circ xyzxz \circ xyz \\ &= 2xyyzz \circ xyzxz \circ xyz \circ xyxz \circ yxz \circ zxz \circ xz + \\ &\quad 4xyxyz \circ xyzz \circ yxz \circ zxz \circ xz + 2xyzz \circ xyz \circ yxz \circ \\ &\quad zxz \circ xz - 2xyyzz \circ xyzxz \circ xyz \circ xyxz \circ yxz \circ zxz \circ xz \end{aligned}$$

$$
\begin{aligned}
&- 3xxz \circ zxz \circ xz \circ xyyzz \circ xyzxz \circ xyz \\
&= 4xyxyz \circ xyzz \circ yxz \circ zxz \circ xz + 2xyzz \circ xyz \circ yxz \circ \\
&\quad zxz \circ xz + 5xxz \circ zxz \circ xz \circ xyyzz \circ xyzxz \circ xyz
\end{aligned}
$$

2. Let's take $(a, b, c, d) = (zx, xy, z, xy)$.
   $zxLM(g)xy = zxxyxy \circ zxyxy \circ zxzxy \circ zxxy$ and
   $zLM(g)xy = zxyxy \circ zyxy \circ zzxy \circ zxy$

$$
\begin{aligned}
lcm_\circ(zxLM(g)xy, zLM(g)xy) &= zxxyxy \circ zxyxy \circ zxzxy \circ zxxy \circ zyxy \circ zzxy \circ zxy \\
&= zxLM(g)xy \circ zyxy \circ zzxy \circ zxy \\
&= zLM(g)xy \circ zxxyxy \circ zxzxy \circ zxxy \\
O(g; g, zx, xy, z, xy, u, v) &= zxgxy \circ zyxy \circ zzxy \circ zxy - zgxy \circ zxxyxy \circ zxzxy \circ zxxy \\
&= 2zxxyxy \circ zxyxy \circ zxzxy \circ zxxy \circ zyxy \circ zzxy \circ zxy + \\
&\quad 3zxxxy \circ zxzxy \circ zxxy \circ zyxy \circ zzxy \circ zxy \\
&\quad - 2zxxyxy \circ zxyxy \circ zxzxy \circ zxxy \circ zyxy \circ zzxy \circ zxy \\
&\quad - 3zxxy \circ zzxy \circ zxy \circ zxxyxy \circ zxzxy \circ zxxy \\
&= 3zxxxy \circ zxzxy \circ zxxy \circ zyxy \circ zzxy \circ zxy \\
&\quad + 5zxxy \circ zzxy \circ zxy \circ zxxyxy \circ zxzxy \circ zxxy
\end{aligned}
$$

3 Semi-ring Based Gröbner–Shirshov Bases over a Noetherian Valuation Ring

In this section, we assume that $VRig\langle X\rangle$ is a noncommutative polynomial ring.

Definition 7 Let V be a ring, $G \subseteq VRig\langle X\rangle$, $\mathcal{I}$ the two-sided ideal of $VRig\langle X\rangle$ generated by G and $\preceq$ a monomial ordering on $Rig\langle X\rangle$.

1. G is a weak Semi-ring based Gröbner–Shirshov basis (Weak-GS) of $\mathcal{I}$ with respect to $\preceq$ if $\langle LT(G)\rangle = \langle LT(\mathcal{I})\rangle$.
2. G is a strong semi-ring based Gröbner–Shirshov basis (Strong-GS) of $\mathcal{I}$ with respect to $\preceq$ if for any $f \in I$ there exit $g \in G,\ a, b\ X^*,\ u \in Rig\langle X\rangle,\ \alpha \in V$ such that $LT(f) = \alpha a LT(g) b \circ u$.

The proof of the following proposition is obvious.

Proposition 4 *Any strong-GS basis is a weak-GS basis.*

Proposition 5 *If V is a valuation ring then any weak-GS basis is strong-GS basis.*

Proof

1. Let G be a strong-GS basis and $f \in \mathcal{I}$. Then $LT(f) \in \langle(\mathcal{I})\rangle$. Since G is a strong Gröbner basis, there exists $g \in G$ such that $LT(f) = \alpha a Lt(g) b \circ u$ for some $a, b \in X^*,\ u \in Rig\langle X\rangle$. Thus $LT(f) \in \langle LT(G)\rangle$.
   So $\langle LT(\mathcal{I})\rangle \subseteq \langle LT(G)\rangle$. That means $\langle LT(\mathcal{I})\rangle = \langle LT(G)\rangle$.
2. Let V be a valuation ring, G a weak-GS basis and $f \in \mathcal{I}$.
   Since $f \in \mathcal{I}$ we have $LT(f) \in LT(\mathcal{I}) \subset \langle LT(\mathcal{I})\rangle = \langle LT(G)\rangle$.
   Thus: $LT(f) = \sum_{i,j} \alpha_{i,j} a_{i,j} LT(g_i) b_{i,j} \circ u_{i,j},\quad \alpha_{i,j} \in V, a_{i,j}, b_{i,j} \in X^*, u_{i,j} \in Rig\langle X\rangle, g_i \in G$. It is possible to choose the expression of $LT(f)$ such that

$a_{i,j}LM(g_i)b_{i,j} \circ u_{i,j} = LM(f)\forall(i, j)$.
Let (i_0, j_0) such that

$$LM(f) = a_{i_0,j_0}LM(g_{i_0})b_{i_0,j_0} \circ u_{i_0,j_0} = a_{i,j}LM(g_i)b_{i,j} \circ u_{i,j}$$

and $LC(g_{i_0})$ divides $LC(g_i)$ $\forall(i, j)$. That is for any i there exists $\beta_i \in V$ such that $LC(g_i) = \beta_i LC(g_{i_0})$.

$$\begin{aligned} LT(f) &= \textstyle\sum_{i,j} \alpha_{i,j}a_{i,j}LT(g_i)b_{i,j} \circ u_{i,j} \\ &= \textstyle\sum_{i,j} \alpha_{i,j}LC(g_i)a_{i,j}LM(g_i)b_{i,j} \circ u_{i,j} \\ &= \textstyle\sum_{i,j} \alpha_{i,j}\beta_i LC(g_{i_0})a_{i,j}LM(g_i)b_{i,j} \circ u_{i,j} \\ &= \textstyle\sum_{i,j} \alpha_{i,j}\beta_i LC(g_{i_0})a_{i_0,j_0}LM(g_{i_0})b_{i_0,j_0} \circ u_{i_0,j_0} \\ &= (\textstyle\sum_{i,j} \alpha_{i,j}\beta_i)a_{i_0,j_0}LT(g_{i_0})b_{i_0,j_0} \circ u_{i_0,j_0} \\ &= \alpha a_{i_0,j_0}LT(g_{i_0})b_{i_0,j_0} \circ u_{i_0,j_0} \ with\ \alpha = \textstyle\sum_{i,j} \alpha_{i,j}\beta_i \end{aligned}$$

So G is a strong-GS basis. □

Right now, V is always a valuation ring. It follows that weak and strong Gröbner bases are the same. Equivalently, the weak reduction implies the strong reduction. Thus, we note $f \xrightarrow{g} h$ or $h = Red(f, g)$ instead of $f \xrightarrow{g}_s h$ or $h = SRed(f, g)$ and $f \xrightarrow{G} h$ or $h = Red(f, G)$ instead of $f \xrightarrow{G}_w h$ or $h = WRed(f, G)$.

Example 8

1. If G is a set of terms then G is a strong-GS basis of its ideal relative to a monomial ordering.
2. $G = \{g_1 = 2z \circ x \circ xy \circ y + 3x \circ z \circ 1,\quad g_2 = 3x \circ z \circ 1\} \subseteq \frac{\mathbb{Z}}{4\mathbb{Z}}Rig\langle X\rangle$ is a strong-GS basis of $\langle G\rangle$ with respect to the length lexicographic order.
   Let $G' = \{g_1' = g_1 - g_2,\ g_2\}$. One can easily check that $\langle G\rangle = \langle G'\rangle$ and $LT(G) = LT(G')$. Since G' is a strong-GS basis, for any $f \in \langle G\rangle$ there exist $g' \in G',\ a, b\ X^*,\ u \in Rig\langle X\rangle,\ \alpha \in V$ such that $LT(f) = \alpha a LT(g')b \circ u$. Equivalently, there exist $g \in G,\ a, b\ X^*,\ u \in Rig\langle X\rangle,\ \alpha \in V$ such that $LT(f) = \alpha a LT(g)b \circ u$.

The *Buchberger* test/completion is well known in the theory of *Gröbner* bases over a field. But when we deal with a ring with zero divisors, the fact that any composition is reduced to zero is not enough to characterize *Gröbner* bases.

In the case of a noetherian valuation ring, by adding the notion of *a-polynomial*, we get a complete characterization of strong-GS bases.

Definition 8 Let V be a noetherian valuation ring and $0 \neq f \in VRig\langle X\rangle$. We define:

1. $a\text{-}pol^1(f) = a_1 f$ where $\langle a_1\rangle = Ann(LC(f))$
   $a\text{-}pol^i(f) = a\text{-}pol(a\text{-}pol^{i-1}(f)) = a_i(a\text{-}pol^{i-1}(f))$ where $\langle a_i\rangle = Ann(LC(a\text{-}pol^{i-1}(f))$.

2. $A\text{-}pol(f) = \{a\text{-}pol^i(f)\}$
3. $A\text{-}pol(G) = \bigcup_{f \in G} A\text{-}pol(f)$

Example 9 Let $f = 18xy \circ z \circ 1 \circ 1 + 3zz \circ x \circ y + 5x \circ x \circ 1 + 1 \circ 1 + 6y \in \frac{\mathbb{Z}}{27\mathbb{Z}} Rig\langle X\rangle$ and $\prec_{ll}$ the length lexicographic order. Then:
$a\text{-}pol^1(f) = 3f = 9zz \circ x \circ y + 15x \circ x \circ 1 + 3(1 \circ 1) + 18y$
$a\text{-}pol^2(f) = 3a\text{-}pol^1(f) = 18x \circ x \circ 1 + 9(1 \circ 1)$
$a\text{-}pol^3(f) = 3a\text{-}pol^2(f) = 0$
$A\text{-}pol(f) = \{9zz \circ x \circ y + 15x \circ x \circ 1 + 3(1 \circ 1) + 18y, 18x \circ x \circ 1 + 9(1 \circ 1), 0\}$.

Remark 3 Let us remark that if $a\text{-}pol^i(f) = 0$ then $a\text{-}pol^j(f) = 0$ for any $j > i$.

Lemma 1 *Let $g \in G$ such that any element of $A\text{-}pol(g)$ is reduced to zero modulo G. Then for any $\alpha \in V$ such that $\alpha g \neq 0$, we have:*
$\alpha g = \sum_i \beta_i a_i g_i b_i \circ u_i,\ \beta_i \in V, a_i, b_i \in X^*, u_i \in Rig\langle X\rangle, g_i \in G,\ \beta_i LC(g_i) \neq 0 \forall i$ *and* $LM(\alpha g) = \max\{LM(\beta_i a_i g_i b_i \circ u_i)\}$.

Proof Let $g \in G$ such that any element of $A\text{-}pol(g)$ is reduced to zero modulo G and $\alpha \in V$.

If $\alpha LC(g) \neq 0$ then the expression αg is ok.

Else $\alpha g \in A\text{-}pol(g)$. So, there exists $h_1 \in G$ such that $LT(\alpha g) = \beta_1 a_1 LT(h_1) b_1 \circ u_1$ for some $\beta_1 \in V, a_1, b_1 \in X^*, u_1 \in Rig\langle X\rangle$. So $\beta_1 LC(h_1) \neq 0$

We reduce αg by h_1. We obtain: $g_1 = \alpha g - \beta_1 a_1 h_1 b_1 \circ u_1$.

$\alpha g = \beta_1 a_1 h_1 b_1 \circ u_1 + g_1$ $(\star)$

If $g_1 = 0$ then $\alpha g = \beta_1 a_1 h_1 b_1 \circ u_1$

If $g_1 \neq 0$ then $LM(g_1) \prec LM(\alpha g) = LM(\beta_1 a_1 h_1 b_1 \circ u_1) = a_1 LM(h_1) b_1 \circ u_1$

Also, there exists $h_2 \in G$ such that $LT(g_1) = \beta_2 a_2 LT(h_2) b_2 \circ u_2$. By reducing g_1 by h_2, we get: $g_2 = g_1 - \beta_2 a_2 h_2 b_2 \circ u_2$ which satisfies $LM(g_2) \prec LM(g_1)$.

The relation $(\star)$ becomes: $\alpha g = \beta_1 a_1 h_1 b_1 \circ u_1 + \beta_2 a_2 LT(h_2) b_2 \circ u_2 + g_2$.

When the reduction terminates, we get: $\alpha g = \sum_i \beta_i a_i g_i b_i \circ u_i,\ \beta_i \in V, a_i, b_i \in X^*, u_i \in Rig\langle X\rangle, g_i \in G$. □

Lemma 2 *Let G be a set of polynomials such that any a-polynomial defined in G is reduced to zero modulo G. Then*

$$f = \sum_{i,j} \alpha_{i,j} a_{i,j} g_i b_{i,j} \circ u_{i,j},\ \alpha_{i,j} \in V,\ a_{i,j}, b_{i,j} \in X^*,\ g_i \in G\ , u_{i,j} \in Rig\langle X\rangle$$

can be rewritten:

$$f = \sum_{k,l} \lambda_{k,l} c_{k,l} g_k d_{k,l} \circ v_{k,l},\ \lambda_{k,l} \in V,\ c_{k,l}, d_{k,l} \in X^*,\ g_k \in G\ , v_{k,l} \in Rig\langle X\rangle$$

which satisfies

$$\lambda_{k,l} LC(g_k) \neq 0 \text{ and } \max\{LM(\alpha_{i,j} a_{i,j} g_i b_{i,j} \circ u_{i,j})\} = \max\{LM(\lambda_{k,l} c_{k,l} g_k d_{k,l} \circ v_{k,l})\}$$

Proof Let $f = \sum_{i,j} \alpha_{i,j} a_{i,j} g_i b_{i,j} \circ u_{i,j},\ \alpha_{i,j} \in V,\ a_{i,j}, b_{i,j} \in X^*,\ g_i \in G,\ u_{i,j} \in Rig\langle X\rangle$,
$\gamma = \max\{LM(\alpha_{i,j} a_{i,j} g_i b_{i,j} \circ u_{i,j})\}$, $\Gamma = \{(i,j) : LM(\alpha_{i,j} a_{i,j} g_i b_{i,j} \circ u_{i,j}) = \gamma\}$

Let $(i,j) \in \Gamma$ such that $\alpha_{i,j} LC(g_i) = 0$.

$\alpha_{i,j} LC(g_i) = 0 \Rightarrow \alpha_{i,j} g_i \in A\text{-}pol(g_i)$. Since any a-polynomial is reduced to zero, by the previous lemma, we have: $\alpha_{i,j} g_i = \sum_{k,l} \beta_{k,l} c_{k,l} g_k b_{k,l} \circ v_{k,l}$ with $\beta_{k,l} LC(g_k) \neq 0$ and $LM(\beta_{k,l} c_{k,l} g_k b_{k,l} \circ v_{k,l}) = LM(c_{k,l} g_k b_{k,l} \circ v_{k,l}) \preceq LM(\alpha_{i,j} g_i)$.

So $LM(\beta_{k,l} a_{i,j} c_{k,l} g_k b_{k,l} b_{i,j} \circ a_{i,j} v_{k,l} b_{i,j} \circ u_{i,j}) = LM(a_{i,j} c_{k,l} g_k b_{k,l} b_{i,j} \circ a_{i,j} v_{k,l} b_{i,j} \circ u_{i,j}) \preceq LM(\alpha_{i,j} a_{i,j} g_i b_{i,j} \circ u_{i,j}) = \gamma$.

Thus, $LM(\alpha_{i,j} a_{i,j} g_i b_{i,j} \circ u_{i,j}) = \gamma = \max\{LM(\beta_{k,l} a_{i,j} c_{k,l} g_k b_{k,l} b_{i,j} \circ a_{i,j} v_{k,l} b_{i,j} \circ u_{i,j})\}$.

Finally, the expression:

$\alpha_{i,j} a_{i,j} g_i b_{i,j} \circ u_{i,j} = \sum_{k,l} \beta_{k,l} a_{i,j} c_{k,l} g_k b_{k,l} b_{i,j} \circ a_{i,j} v_{k,l} b_{i,j} \circ u_{i,j}$ verifies

$\beta_{k,l} LC(g_k) \neq 0$ and $LM(\alpha_{i,j} a_{i,j} g_i b_{i,j} \circ u_{i,j}) = \gamma = \max\{LM(\beta_{k,l} a_{i,j} c_{k,l} g_k b_{k,l} b_{i,j} \circ a_{i,j} v_{k,l} b_{i,j} \circ u_{i,j})\}$.

By applying this procedure to any $(i,j) \in \Gamma$ such that $\alpha_{i,j} LC(g_i) = 0$ we obtain an expression of f as desired. □

Definition 9 An expression of f:

$$f = \sum_{i,j} \alpha_{i,j} a_{i,j} g_i b_{i,j} \circ u_{i,j},\ \alpha_{i,j} \in V,\ a_{i,j}, b_{i,j} \in X^*,\ g_i \in G,\ u_{i,j} \in Rig\langle X\rangle$$

which satisfies $LM(\alpha_{i,j} a_{i,j} g_i b_{i,j} \circ u_{i,j}) \preceq LM(f)$ is called a standard expression of f in G.

Theorem 1 *Let $G \subseteq V Rig\langle X\rangle$ such that any a-polynomial defined in G is reduced to zero. The following statements are equivalent:*

1. *G is a strong-GS basis.*
2. *Any f in $\langle G\rangle$ has a standard expression.*

Proof (1)⇒ (2):

Suppose G is a strong (Semi-ring based) Gröbner–Shirshov basis. Then there exits $g_1 \in G$ such that $LT(f) = \alpha_1 a_1 LT(g_1) b_1 \circ u_1$ for some $a_1, b_1 \in X^*,\ u_1 \in Rig\langle X\rangle,\ \alpha_1 \in V$. We reduce f by such g_1. We get $f_1 = f - \alpha_1 a_1 g_1 b_1 \circ u_1$.

$f_1 \in \langle G\rangle$ and $LM(f_1) \prec LM(f)$. If $f_1 \neq 0$ then we repeat this by replacing f by f_1. By reducing f_1 we get $f_2 = f_1 - \alpha_2 a_2 g_2 b_2 \circ u_2$.

We have $f = f_1 + \alpha a_1 g_1 b_1 \circ u_1 = \alpha a_1 g_1 b_1 \circ u_1 + \alpha_2 a_2 g_2 b_2 \circ u_2 + f_2$.

Also $LM(f_2) \prec LM(f_1) = LM(\alpha_2 a_2 g_2 b_2 \circ u_2) \prec LM(f) = LM(\alpha agb \circ u)$.

Since f is reduced to zero, at the last step we obtain:

$f = \alpha_1 a_1 g_1 b_1 \circ u_1 + \alpha_2 a_2 g_2 b_2 \circ u_2 + ... + \alpha_n a_n g_n b_n \circ u_n$. We obtain f as desired.

(2)⇒(1):

Suppose (2) is satisfied. By Lemma 1, we choose the expression $f = \sum_{i,j} \alpha_{i,j} a_{i,j} g_i b_{i,j} \circ u_{i,j}$ such that $\alpha_{i,j} LC(g_i) \neq 0$ for any $(i, j) \in \Gamma$ which is defined as above.

So $LM(f) = \gamma = \max\{LM(\alpha_{i,j} a_{i,j} g_i b_{i,j} \circ u_{i,j}),\ (i, j) \in \Gamma\}$

Then $LT(f) = \sum_{(i,j)\in\Gamma} \alpha_{i,j} LC(g_i) a_{i,j} LM(g_i) b_{i,j} \circ u_{i,j}$ (1).

Since V is a valuation ring there exits $(i_0, j_0) \in \Gamma$ such that $\alpha_{i_0,j_0} LC(g_{i_0})$ divides $\alpha_{i,j} LC(g_i)$ for any $(i, j) \in \Gamma$. So equation (1) becomes: $LT(f) = \alpha a_{i_0} LT(g_{i_0}) b_{i_0} \circ u_{i_0} \in \langle LT(G)\rangle$. □

The two following lemmas are very important for the proof of Proposition 6.

Lemma 3 *Let $f_1, f_2, ..., f_n \in VRig\langle X\rangle$ such that $LM(f_i) = LM(f_j) = \gamma\ \forall i, j$. If there exist $\alpha_1, \alpha_2, ..., \alpha_n \in V$ such that $LM(\sum_{i=1}^{n} \alpha_i f_i) \prec \gamma$ then $\sum_{i=1}^{n} \alpha_i f_i$ is a linear combination of simple compositions of f_i added to an element of A-pol(f_{i_0}) for some $i_0 \in \{1, 2, ..., n\}$*

Proof This proof is similar to that in [14, p. 121]. But for shake of completeness, we give it in the following.

Since V is a valuation ring, we can suppose that $LC(f_n)|LC(f_{n-1})|...|LC(f_1)$. Since $LM(f_i) = LM(f_j)$ for $i < j$ then $\widetilde{O}(f_i, f_j) = f_i - \frac{LC(f_i)}{LC(f_j)} f_j$.

$$\sum_{i=1}^{n} \alpha_i f_i = \alpha_1 (f_1 - \frac{LC(f_1)}{LC(f_2)} f_2) + (\alpha_2 + \frac{LC(f_1)}{LC(f_2)} \alpha_1)(f_2 - \frac{LC(f_2)}{LC(f_3)} f_3)$$

$$+ ... + (\alpha_{n-1} + \frac{LC(f_{n-2})}{LC(f_{n-1})} \alpha_{n-2} + ... + \frac{LC(f_1}{LC(f_{n-1}} \alpha_1)(f_{n-1} - \frac{LC(f_{n-1})}{LC(f_n)} f_n)$$

$$+ (\alpha_n + \frac{LC(f_{n-1})}{LC(f_n)} \alpha_{n-1} + ... + \frac{LC(f_1)}{LC(f_n)} \alpha_1) f_n$$

Let $\beta_i = \sum_{j=1}^{i} \frac{LC(f_j)}{LC(f_i)} \alpha_j,\ i = 1, 2, ..., n$

Since $LM(\sum_{i=1}^{n} \alpha_i f_i =) \prec \gamma$ we get $\beta_n LC(f_n) = 0$. Therefore $\beta_n f_n \in A\text{-}pol(f_n)$

$$\sum_{i=1}^{n} \alpha_i f_i = \sum_{1=1}^{n-1} \beta_i \widetilde{O}(f_i, f_{i+1}) + \beta_n f_n$$ □

Lemma 4 *Let $f, g \in VRig\langle X\rangle$, $a, b, c, d \in X^*$, $u, v \in Rig\langle X\rangle$ such that $aLM(f)b \circ u = cLM(g)d \circ v$ and $LC(g)$ divides $LC(f)$. Then there exits $t \in Rig\langle X\rangle$ such that the simple composition hold:*
$\widetilde{O}(afb \circ u, cgd \circ v) = O(f, g, a, b, c, d, u_1, v_1) \circ t$ *for some* $u_1, v_1 \in Rig\langle X\rangle$.

Proof Suppose that $aLM(f)b \circ u = cLM(g)d \circ v$ and $LC(g)$ divides $LC(f)$.
Let $t = gcd_\circ(u, v)$. So $u = u_1 \circ t$, $v = v_1 \circ t$ for some $u_1, v_1 \in Rig\langle X\rangle$.
Then:

$$\begin{aligned}\widetilde{O}(afb \circ u, cgd \circ v) &= afb \circ u - \frac{LC(f)}{LC(g)}cgd \circ v \\ &= afb \circ u_1 \circ t - \frac{LC(f)}{LC(g)}cgd \circ v_1 \circ t \\ &= (afb \circ u_1 - \frac{LC(f)}{LC(g)}cgd \circ v_1) \circ t\end{aligned}$$

Now, we must prove that

$$lcm_\circ(aLM(f)b, cLM(g)d) = aLM(f) \circ u_1 = cLM(g) \circ v_1$$

Let $u', v' \in Rig\langle X\rangle$ such that

$$lcm_\circ(aLM(f)b, cLM(g)d) = aLM(f)b \circ u' = cLM(g)d \circ v'$$

We will show that $u = u'$ and $v = v'$.
Let us remark that $gcd_\circ(u_1, v_1) = gcd_\circ(u', v') = \theta$.

$$\begin{aligned}aLM(f)b \circ u_1 = cLM(g)d \circ v_1 &\Rightarrow aLM(f)b \circ u_1 \circ u' = cLM(g)d \circ v_1 \circ u' \\ &\Rightarrow (aLM(f)b \circ u') \circ u_1 = cLM(g)d \circ v_1 \circ u' \\ &\Rightarrow cLM(g)d \circ v' \circ u_1 = cLM(g)d \circ v_1 \circ u' \\ &\Rightarrow v' \circ u_1 = u' \circ v_1\end{aligned}$$

$gcd_\circ(u', v') = \theta \Rightarrow \exists w_1, w_2 \in Rig\langle X\rangle$ such that $u_1 = u' \circ w_1$ and $v_1 = v' \circ w_2$ (i).
$gcd_\circ(u_1, v_1) = \theta \Rightarrow \exists w_1', w_2' \in Rig\langle X\rangle$ such that $u' = u_1 \circ w_1'$ and $v' = v_1 \circ w_2'$ (ii).
(i) and (ii) imply: $u_1 = u' \circ w_1 = u_1 \circ w_1' \circ w_1$ $(3i)$ and $v_1 = v' \circ w_2 = v_1 \circ w_2' \circ w_2$ $(4i)$
$(3i) \Rightarrow w_1' \circ w_1 = \theta$. By Proposition 2.1, $w_1' = w_1 = \theta$
$(4i) \Rightarrow w_2' \circ w_2 = \theta$. Thus $w_2' = w_2 = \theta$ It follows that $u_1 = u_1'$ and $v_1 = v_1'$.
We conclude that $lcm_\circ(aLM(f)b, cLM(g)d) = aLM(f)b \circ u_1 = cLM(g)d \circ v_1$
$\widetilde{O}(afb \circ u, cgd \circ v) = (afb \circ u_1 - \frac{LC(f)}{LC(g)}cgd \circ v_1) \circ t$
$= O(f, g, a, b, c, d, u_1, v_1) \circ t$ □

The following theorem usually called Buchberger test/completion or Composition-Diamond-Lemma is fundamental for the theory of Gröbner–Shirshov bases. It gives a complete characterization of Gröbner–Shirshov bases; we use it:

- to check if a set of polynomials is a Gröbner–Shirshov basis relatively to a monomial ordering;
- to complete a set into Gröbner–Shirshov basis.

Proposition 6 *Let $G \subseteq VRig\langle X\rangle$. The following statements are equivalent:*

1. *G is a strong-GS basis.*
2. *$O(G)$ and $A\text{-}pol(G)$ are reduced to zero modulo G.*

Proof (1)⇒ (2):

We know that all compositions and all $a\text{-}plynomials$ are in $\langle G\rangle$. Then they are all reduced to zero modulo G.

(2) ⇒ (1) :

We will show that f has a standard expression in G.

Let $f = \sum_{i,j} \alpha_{i,j} a_{i,j} g_i b_{i,j} \circ u_{i,j},\ \alpha_{i,j} \in V,\ a_{i,j}, b_{i,j} \in X^*,\ g_i \in G\ , u_{i,j} \in Rig\langle X\rangle$

Among all the expressions of f in this form, there exits at least one which minimizes $\gamma = \max\{LM(\alpha_{i,j} a_{i,j} g_i b_{i,j} \circ u_{i,j})\}$ and $\alpha_{i,j} LC(g_i) \neq 0$

Let $f = \sum_{i,j} \alpha_{i,j} a_{i,j} g_i b_{i,j} \circ u_{i,j},\ \alpha_{i,j} \in V,\ a_{i,j}, b_{i,j} \in X^*,\ g_i \in G\ , u_{i,j} \in Rig\langle X\rangle$ such that

$\max\{LM(\alpha_{i,j} a_{i,j} g_i b_{i,j} \circ u_{i,j})\} = \gamma, \alpha_{i,j} LC(g_i) \neq 0.$

Let $h_{i,j} = a_{i,j} g_i b_{i,j} \circ u_{i,j}\ \forall (i,j), \Gamma = \{(i,j) :\ LM(\alpha_{i,j} h_{i,j}) = \gamma\}$

If $\gamma \succ LM(f)$ then $LM(\sum_{(i,j)\in\Gamma} \alpha_{i,j} h_{i,j}) \prec \gamma$. Then by Lemma 3, we can write:

$$\sum_{(i,j)\in\Gamma} \alpha_{i,j} h_{i,j} = \sum_{(k,l),(m,n)\in\Gamma} \beta_{k,l,m,n} \widetilde{O}(h_{k,l}, h_{m,n}) + h$$

with $h \in A\text{-}pol(h_{i_0,j_0})$ for some $(i_0, j_0) \in \Gamma$

For any $(k,l),\ (m,n) \in \Gamma$, we have:

$$\gamma_{k,l,m,n} = LM(\widetilde{O}(h_{k,l}, h_{m,n})) \prec \gamma.$$

By Lemma 4, we have: for any $(k,l),\ (m,n) \in \Gamma$ there exist $u'_{k,l},\ v'_{k,l},\ t_{k,l,m,n}$ such that $\widetilde{O}(h_{k,l}, h_{m,n}) = \widetilde{O}(a_{k,l} g_k b_{k,l} \circ u_{k,l}, a_{m,n} g_m b_{m,n} \circ u_{m,n})$

$$= O(g_k, g_m, a_{k,l}, b_{k,l}, a_{m,n}, b_{m,n}, u'_{k,l}, v'_{m,n}) \circ t_{k,l,m,n}$$

Since $O(g_k, g_m, a_{k,l}, b_{k,l}, a_{m,n}, b_{m,n}, u'_{k,l}, v'_{m,n})$ and h are reduced to zero modulo G, we have:

$O(g_k, g_m, a_{k,l}, b_{k,l}, a_{m,n}, b_{m,n}, u'_{k,l}, v'_{m,n}) = \sum_{s,t} \lambda_{s,t} c_{s,t} g_s d_{s,t} \circ v_{s,t}$ and

$h = \sum_{p,q} \gamma_{p,q} c'_{p,q} g_p d'_{p,q} \circ v'_{p,q}$

with $LM(\lambda_{s,t} c_{s,t} g_s d_{s,t} \circ v_{s,t}) \preceq LM(O(g_k, g_l, a_{k,l}, b_{k,l}, a_{m,n}, b_{m,n}, u'_{k,l}, v'_{m,n}))$

So

$$
\begin{aligned}
\sum_{(i,j)\in\Gamma} \alpha_{i,j} h_{i,j} &= \sum_{(k,l),(m,n)} \beta_{k,l,m,n} \Big(\sum_{s,t} \lambda_{s,t} c_{s,t} g_s d_{s,t} \circ v_{s,t}\Big) \circ t_{k,l,m,n} \\
&\quad + \sum_{p,q} \gamma_{p,q} c'_{p,q} g_p d'_{p,q} \circ v'_{p,q} \\
&= \sum_{(k,l),(m,n)} \sum_{s,t} \lambda_{k,l,m,n} \beta_{s,t} c_{s,t} g_s d_{s,t} \circ v_{s,t} \circ t_{k,l,m,n} \\
&\quad + \sum_{p,q} \gamma_{p,q} c'_{p,q} g_p d'_{p,q} \circ v'_{p,q}
\end{aligned}
$$

Thus $LM(\lambda_{s,t} c_{s,t} g_s d_{s,t} \circ v_{s,t}) \circ t_{k,l,m,n} \preceq LM(O(g_k, g_l, a_{k,l}, b_{k,l}, a_{m,n}, b_{m,n}, u'_{k,l}, v'_{m,n})) \circ t_{k,l,m,n} = \gamma_{k,l,m,n} \prec \gamma$

Also: $LM(\gamma_{p,q} c'_{p,q} g_p d'_{p,q} \circ v'_{p,q}) \prec LM(h_{i_0,j_0}) = \gamma$

Therefore

$$
\begin{aligned}
\sum_{(i,j)\in\Gamma} \alpha_{i,j} h_{i,j} &= \sum_{(k,l),(m,n)} \sum_{s,t} \lambda_{k,l,m,n} \beta_{s,t} c_{s,t} g_s d_{s,t} \circ v_{s,t} \circ t_{k,l,m,n} \\
&\quad + \sum_{p,q} \gamma_{p,q} c'_{p,q} g_p d'_{p,q} \circ v'_{p,q} \\
&= \sum_{(x,y)} \alpha'_{x,y} a'_{x,y} g_{x,y} b'_{x,y} \circ w_{x,y}
\end{aligned}
$$

This expression satisfies $LM(\alpha'_{x,y} a'_{x,y} g_{x,y} b'_{x,y} \circ w_{x,y}) \prec \gamma = LM(\alpha_{i,j} h_{i,j})$, $(i,j) \in \Gamma$.

This contradicts the minimality of γ. We conclude that $\gamma = \max\{LM(\alpha_{i,j} h_{i,j})\} \preceq LM(f)$. □

Acknowledgements This work was partially supported by CEA-MITIC (Centre d'Excellence Africain en Mathématiques, Informatique et TIC).

References

1. Bergman, G.M.: The diamond lemma for ring theory. Adv. Math. **29**, 178–218 (1978)
2. Bokut, L.A., Chen, Y., Mo, Q.: Gröbner-Shirshov bases for semirings. J. Algebra **385**, 47–63 (2013)
3. Bokut, L.A.: Imbedding into simple associative algebras. Algebra Logic **15**, 117–142 (1976)
4. Bucherger, B.: An algorithm for finding the basis elements of the residue class of a zero dimensional ideal. J. Symb. Comput. **41**, 475–511 (2006)
5. Hironaka, H.: Resolution of singularities of an algebraic variety over a field of characteristic zero: I. Ann. Math. **79**(1), 109–203 (1964)
6. Kandri-Rody, A., Kapur, D.: Computing a Gröbner basis of a polynomial ideal over a Euclidean domain. J. Symb. Comput. **7**, 37–57 (1988)
7. Kapur, D., Yongyang, C.: An algorithm for computing a Gröbner basis of a polynomial ideal over a ring with zero divisors. Math. Comput. Sci., 601-634 (2009)

8. Mikhalev, A.A.: A composition lemma and the word problem for color Lie superalgebras. Moscow Univ. Math. Bull. **44**(5), 87–90 (1989)
9. Möller, H.M.: The construction of Gröbner bases using syzygies. J. Symb. Comput. **6**, 345–359 (1988)
10. Mora, T.: An introduction to commutative and non-commutative Gröbner Bases. J. Theor. Comput. Sci. **13**, 131–173 (1994)
11. Pan, L.: On the D-bases of polynomial ideals over principal ideal domain. J. Symb. Comput. **7**, 55–69 (1988)
12. Shirshov, A.I.: On free Lie rings. Mat. Sb. (N.S.) **45**(87):2, 113–122 (1958)
13. Stifter, S.: Gröbner bases of modules over reduction rings. J. Algebra **159**, 54–63 (1993)
14. Yengui, I.: Constructive Commutative Algebra. Lecture Notes in Mathematics, vol. 2138. Springer series 304 (2015)
15. Yengui, I.: Dynamical Gröbner bases. J. Algebra **301**, 447–458 (2006)
16. Yengui, I.: Dynamical Gröbner bases over Dedekind rings. J. Algebra **324**, 12–24 (2010)
17. Yengui, I.: Corrigendum to "Dynamical Gröbner bases". J. Algebra **301**, 447–458 (2006) and to "Dynamical Grobener bases over Dedekind rings". J. Algebra **324**, 12–24 (2010), J. of Algebra **339**, 370–375 (2011)

The 4-Rank of the Class Group of Some Real Pure Quartic Number Fields

Mbarek Haynou and Mohammed Taous

Abstract Let $K = \mathbb{Q}(\sqrt[4]{4pq^2})$ be a real pure quartic number field and $k = \mathbb{Q}(\sqrt{p})$ its real quadratic subfield, where $p \equiv 5 \pmod 8$ and $q \equiv 1 \pmod 4$ are two different odd prime numbers such that $(\frac{p}{q}) = 1$. In this work, we are interested in studying the 2-rank and the 4-rank of the class group of K.

Keywords Class groups · Pure quartic number field · Ambiguous class number formula

1 Introduction and Notations

Let K be a number field and let $Cl_2(K)$ denote its 2-class group that is the 2-Sylow subgroup of its class group $Cl(K)$. We define the 2-rank and the 4-rank of $Cl(K)$ respectively as follows: $r_2(K) = \dim_{\mathbb{F}_2}(Cl_2(K)/Cl_2(K)^2)$ and $r_4(K) = \dim_{\mathbb{F}_2}(Cl_2(K)^2/Cl_2(K)^4)$ where $\mathbb{F}_2$ is the finite field with 2 elements.

Several authors calculated, for small degree and by using genus theory, $r_2(K)$ and $r_4(K)$ whenever K is a quadratic extension of a number field having an odd class number. For example, C. Parry has determined (see, [10]) the exact power of 2 dividing the class number of the pure quartic number field $K = \mathbb{Q}(\sqrt[4]{m})$ in the fowling cases:

- $m = p$ or $2p^2$ where p is a prime number $\equiv 7 \pmod 8$.
- $m = p$ or $2p^2$ where p is a prime number $\equiv 1 \pmod 8$.
- $m = pq^2$ or $4pq^2$ where $\left(\frac{p}{q}\right) = -1$ show that p, q are a primes numbers such that $p \equiv 5 \pmod 8$ and $q \equiv 3 \pmod 4$.

M. Haynou (✉) · M. Taous
Faculty of Sciences and Technology, Moulay Ismail University, P.O. Box 509-Boutalamine, 52 000 Errachidia, Morocco
e-mail: haynou_mbarek@hotmail.com

M. Taous
e-mail: taousm@hotmail.com

M. Siles Molina et al. (eds.), *Associative and Non-Associative Algebras and Applications*, Springer Proceedings in Mathematics & Statistics 311,
https://doi.org/10.1007/978-3-030-35256-1_12

- $m = 2n^2$ where $n = p_1 p_2 \cdots p_t$ and $\left(\frac{2}{p_i}\right) = \left(\frac{-2}{p_i}\right) = -1$ for $i = 1, \ldots, t$, show that $p_1, \cdots, p_t$ are primes numbers $\equiv 1 \pmod 4$.
- $m = pn^2$ or $4pn^2$ where $n = q_1 \cdots q_t$ and $\left(\frac{p}{q_i}\right) = \left(\frac{-p}{q_i}\right) = -1$ for $i = 1, \ldots, t$, show that p is a prime number $\equiv 3 \pmod 8$ and $q_1, \cdots, q_t$ are primes numbers $\equiv 3 \pmod 4$.
- $m = pn^2$ or $4pn^2$ where $n = q_1 \cdots q_t$ and $\left(\frac{p}{q_i}\right) = \left(\frac{-p}{q_i}\right) = -1$ for $i = 1, \ldots, t$. show that p is a prime number $\equiv 7 \pmod 8$ and $q_1, \cdots, q_t$ are primes numbers $\equiv 3 \pmod 4$.
- $m = pn^2$ or $4pn^2$ where $n = q_1 \cdots q_t$ and $\left(\frac{p}{q_i}\right) = \left(\frac{-p}{q_i}\right) = -1$ for $i = 1, \ldots, t$ show that p is a prime number $\equiv 5 \pmod 8$ and $q_1, \cdots, q_t$ are primes numbers $\equiv 3 \pmod 4$.

Y. Qin [11] provided a method to calculate the 4-rank of a quadratic extension $k(\sqrt{\delta})$ of number fields k with odd class, which gives a relation between $r_4(k(\sqrt{\delta}))$ and the rank of the following matrix:

$$R_{k(\sqrt{\delta})/k} = \begin{pmatrix} \left(\frac{a_1;\,\delta}{\mathcal{P}_1}\right) \cdots \left(\frac{a_n;\,\delta}{\mathcal{P}_1}\right) \cdots \left(\frac{a_{n+r};\,\delta}{\mathcal{P}_1}\right) \\ \cdots \qquad \cdots \\ \left(\frac{a_1;\,\delta}{\mathcal{P}_t}\right) \cdots \left(\frac{a_n;\,\delta}{\mathcal{P}_t}\right) \cdots \left(\frac{a_{n+r};\,\delta}{\mathcal{P}_t}\right) \end{pmatrix}.$$

It is a matrix of type $t \times (n + r)$ with coefficients in $\mathbb{F}_2$, called the generalized Rédei-matrix, where $(\mathcal{P}_i)_{1 \le i \le t}$ are primes (finite and infinite) of k which ramify in $k(\sqrt{\delta})$, $(a_j)_{1 \le j \le n+r}$ is a family of elements of k determined by Y. Qin in [11, §2, Lemma 2.4, p.27] and $\left(\frac{-;\,\delta}{\mathcal{P}_i}\right)$ is the Hilbert symbol on k. Note that this matrix is given by replacing the 1's by 0's and the -1's by 1's. For more details concerning the generalized Rédei-matrix, we refer to [11], where one can find some applications for calculating the rank of some biquadratic fields.

In this paper, our first goal is to count the 2-rank of the class group of $K = \mathbb{Q}(\sqrt[4]{4pq^2}) = k(\sqrt{2q\sqrt{p}})$, using the ambiguous class number formula in K/k [5], with $k = \mathbb{Q}(\sqrt{p})$, such that p and q satisfying the conditions alluded below. The second goal is to calculate the 4-rank of the class group of fields K satisfying $r_2(K) = 2$, by using technics inspired from works of Y. Qin.

Let $p \equiv 5 \pmod 8$ and $q \equiv 1 \pmod 4$ be two different odd prime numbers such that $\left(\frac{p}{q}\right) = 1$. During this paper, we adopt the following notations:

- $\mathcal{O}_F$: the ring of integers of a number field F.
- $\mathcal{O}_k^*$: the unit group of $\mathcal{O}_k$.
- $Cl(K)$: the class group of K.
- $Cl_2(K)$: the 2-class group of K.
- $\delta = 2q\sqrt{p}$.
- $\mathcal{P}_i$: a prime ideal of k which ramify in K.

- $N(a)$: denotes the absolute norm of a number a.
- $N_{K/k}(-)$: the relative norm of a number K/k.
- $R_{K/k}$: the generalized Rédei-matrix of K.
- $\left(\frac{m}{n}\right)$: the Legendre symbol.
- $\left[\frac{-}{\mathcal{P}_i}\right]$: the quadratic symbol over k.
- $(-;\delta)_{\mathcal{P}_i}$: the local Hilbert symbol over k.
- $\left(\frac{-,-}{\mathcal{P}_i}\right)$: the Hilbert symbol over k.
- $\left(\frac{m}{n}\right)_4$: the rational biquadratic symbol.

Let $\varepsilon_p = \frac{1}{2}(u + v\sqrt{p})$ be the fundamental unit of k with u, v positive rational integers. It is well-known that k has odd class number with $N(\varepsilon_p) = -1$.

The aim of this paper is to prove the following theorems.

Theorem 1 *The 2-rank of the class group of K is $r_2(K) = 2$.*

Theorem 2 *Keep the previous notations. Then we have the following cases:*

(i) If $q \equiv 1 \pmod 8$, then

$$r_4(K) = \begin{cases} 0 & \text{if } \left(\frac{p}{q}\right)_4 = -1, \\ 1 & \text{if } \left(\frac{p}{q}\right)_4 = -\left(\frac{q}{p}\right)_4 = 1, \\ 2 & \text{if } \left(\frac{p}{q}\right)_4 = \left(\frac{q}{p}\right)_4 = 1. \end{cases}$$

(ii) If $q \equiv 5 \pmod 8$, then

$$r_4(K) = \begin{cases} 0 & \text{if } \left(\frac{p}{q}\right)_4 = 1, \\ 1 & \text{if } \left(\frac{p}{q}\right)_4 = -1. \end{cases}$$

2 The 2-Rank of $Cl(K)$

In this section, we will determine the 2-rank of the class groups of K, for this reason, we have to know the primes of k which ramify in K. The relative discriminant of K/k is $\Delta_{K/k} = (8q\sqrt{p})$ (see [7]). As $\left(\frac{p}{q}\right) = 1$, then the finite primes ramified in K/k are $\sqrt{p}$, 2_I, π_1 and $\tilde{\pi}_1$, where $qO_k = \pi_1\tilde{\pi}_1$, $pO_k = (\sqrt{p})^2$ and $2_I = 2O_k$.

In regards to infinite primes ramifying in K/k, k has two infinite primes p_∞ and $\tilde{p}_\infty$ which correspond respectively to both $\mathbb{Q}$-embeddings $i_{p_\infty} : \sqrt{p} \longmapsto -\sqrt{p}$ and $i_{\tilde{p}_\infty} : \sqrt{p} \longmapsto \sqrt{p}$, so i_{p_∞} has a complex extension to K, then p_∞ is the unique infinite prime of k ramifying in K.

Lemma 1 *Keep the previous hypotheses and notations, then*

$$\left(\frac{-1;\delta}{p_\infty}\right)=\left(\frac{\varepsilon_p;\delta}{\sqrt{p}}\right)=\left(\frac{-\varepsilon_p;\delta}{\sqrt{p}}\right)=-1.$$

Proof

- By definition of the Hilbert symbol given in [6, Chap. II, § 7, Definition 7.1, p. 195], we have

$$\left(\frac{-1;\delta}{p_\infty}\right)=i_{p_\infty}^{-1}((i_{p_\infty}(-1),i_{p_\infty}(\delta))_{p_\infty})=i_{p_\infty}^{-1}((-1,-\delta)_{p_\infty})=i_{p_\infty}^{-1}(-1)=-1.$$

- Since the $\sqrt{p}$-valuation $v_{\sqrt{p}}(\delta)=1$ by the property [9, lemma V.p105],

$$\left(\frac{\varepsilon_p;\delta}{\sqrt{p}}\right)=\left[\frac{\varepsilon_p}{\sqrt{p}}\right]^{v_{\sqrt{p}}(\delta)}=\left[\frac{\varepsilon_p}{\sqrt{p}}\right]=-1.$$

- We have,

$$\left(\frac{-1;\delta}{\sqrt{p}}\right)=\left[\frac{-1}{\sqrt{p}}\right]=\left(\frac{-1}{p}\right)=p^{\frac{p-1}{2}}=1.$$

  Then by applying the multiplicative property of Hilbert symbol, we get

$$\left(\frac{-\varepsilon_p,\delta}{\sqrt{p}}\right)=\left(\frac{\varepsilon_p,\delta}{\sqrt{p}}\right)\left(\frac{-1,\delta}{\sqrt{p}}\right)=-1.$$

Proof (**Proof of Theorem** 1) As the class number of k is odd, then by the ambiguous class number formula (see [5]), we have: $r_2(K)=t-e-1$, where t is the number of primes of k that ramify in K and e is determined by $2^e=[O_k^*:O_k^*\cap N_{K/k}((K^\times)]$. According to Lemma 1 and Hasse's theorem [6, Chap. II § 4, Theorem 4.2, p. 112] the units $-1,\varepsilon_p,-\varepsilon_p$ of k are not norms of elements in K, so $e=2$. On the other hand, the primes ramified in K/k are $\sqrt{p}$, p_∞, 2_I, π_1 and $\tilde{\pi}_1$, then $t=5$. Finally, $r_2(K)=5-1-2=2$. □

3 The 4-Rank of $Cl(K)$

Let $p\equiv q\equiv 1 \pmod 4$ be two different odd prime numbers such that $\left(\frac{p}{q}\right)=1$ and $\pi_1,\bar{\pi}_1$ are prime ideals of O_k, lying above q, the ideals $\pi_1^h,\bar{\pi}_1^h$ are principal ideals of O_k, with $\pi_1^h=(x+y\sqrt{p})/2)$ and $\bar{\pi}_1^h=(x-y\sqrt{p})/2)$. To prove the second main result, we need the following lemmas.

Lemma 2 *Keep the previous hypotheses and notations, then*

(*a*) $\left(\frac{\varepsilon_p;\,\delta}{p_\infty}\right) = \left(\frac{\sqrt{p};\,\delta}{p_\infty}\right) = -1.$

(*b*) $\left(\frac{2;\,\delta}{p_\infty}\right) = \left(\frac{2(x+y\sqrt{p});\,\delta}{p_\infty}\right) = \left(\frac{2(x-y\sqrt{p});\,\delta}{p_\infty}\right) = 1.$

Proof By definition of the Hilbert symbol given in [6, Chap. II § 7, Definition 7.1, p.195], we have

(*a*) $\left(\frac{\varepsilon_p;\,\delta}{p_\infty}\right) = i^{-1}_{p_\infty}((i_{p_\infty}(\varepsilon_p), i_{p_\infty}(\delta))_{p_\infty}) = i^{-1}_{p_\infty}((\varepsilon'_p, -\delta)_{p_\infty}) = i^{-1}_{p_\infty}(-1) = -1.$

where $\varepsilon_p = \frac{u+v\sqrt{p}}{2} > 0$ is the fundamental unit of k, $\varepsilon'_p = \frac{u-v\sqrt{p}}{2}$ and $N(\varepsilon_p) = \varepsilon_p\varepsilon'_p = -1$, then $\varepsilon'_p < 0$. We have also $\left(\frac{\sqrt{p};\,\delta}{p_\infty}\right) = i^{-1}_{p_\infty}((i_{p_\infty}(\sqrt{p}), i_{p_\infty}(\delta))_{p_\infty})= i^{-1}_{p_\infty}((-\sqrt{p}, -\delta)_{p_\infty}) = i^{-1}_{p_\infty}(-1) = -1.$

(*b*) We can prove the results of (*b*) in the same way as in the proof of the case (*a*). □

Lemma 3 *Keep the previous hypotheses and notations, then*

$$\left(\frac{x}{p}\right) = \left(\frac{2}{p}\right)\left(\frac{q}{p}\right)_4 \text{ and } \left(\frac{x}{q}\right) = \left(\frac{p}{q}\right)_4.$$

Proof Recall that $\pi_1^h = ((x + y\sqrt{p})/2))$, then $4q^h = x^2 - py^2$ such that $x, y \in \mathbb{Z}$. Hence $4q^h \equiv x^2 \pmod{p}$. As $p \equiv 1 \pmod 4$, then

$$\begin{aligned}
\left(\frac{x}{p}\right) &\equiv (x^2)^{\frac{p-1}{4}} \pmod{p}\\
&\equiv \left(\frac{x^2}{p}\right)_4 \pmod{p}\\
&\equiv \left(\frac{4q^h}{p}\right)_4 \pmod{p}\\
&\equiv \left(\frac{4}{p}\right)_4\left(\frac{q}{p}\right)_4 \pmod{p}\\
&\equiv \left(\frac{2}{p}\right)\left(\frac{q}{p}\right)_4 \pmod{p}.
\end{aligned}$$

So

$$\left(\frac{x}{p}\right) = \left(\frac{x^2}{p}\right)_4 = \left(\frac{2}{p}\right)\left(\frac{q}{p}\right)_4.$$

If we put $y = 2^e y'$ such that $2 \nmid y'$, then

$$\begin{aligned}\left(\frac{x}{q}\right) &= \left(\frac{x^2}{q}\right)_4\\ &= \left(\frac{4q^h+y^2p}{q}\right)_4\\ &= \left(\frac{y^2p}{q}\right)_4\\ &= \left(\frac{y^2}{q}\right)_4\left(\frac{p}{q}\right)_4\\ &= \left(\frac{y}{q}\right)\left(\frac{p}{q}\right)_4\\ &= \left(\frac{2}{q}\right)^e\left(\frac{y'}{q}\right)\left(\frac{p}{q}\right)_4\\ &= \left(\frac{2}{q}\right)^e\left(\frac{q}{y'}\right)\left(\frac{p}{q}\right)_4\\ &= \left(\frac{2}{q}\right)^e\left(\frac{4q^h}{y'}\right)\left(\frac{p}{q}\right)_4\\ &= \left(\frac{2}{q}\right)^e\left(\frac{x^2-y^2p}{y'}\right)\left(\frac{p}{q}\right)_4\\ &= \left(\frac{2}{q}\right)^e\left(\frac{x^2}{y'}\right)\left(\frac{p}{q}\right)_4\\ &= \left(\frac{2}{q}\right)^e\left(\frac{p}{q}\right)_4.\end{aligned}$$

Moreover, the fact that $\left(\frac{2}{q}\right) = 1$ if $q \equiv 1 \pmod 8$ and $e = 1$ or $e = 2$ if $q \equiv 5 \pmod 8$. Hence,

$$\left(\frac{x}{q}\right) = \left(\frac{p}{q}\right)_4$$

Lemma 4 *Keep the previous hypotheses and notations, then*

(*a*) $\left(\frac{2;\,\delta}{\sqrt{p}}\right) = \left(\frac{-\sqrt{p};\,\delta}{\sqrt{p}}\right) = -1.$

(*b*) $\left(\frac{2(x+y\sqrt{p});\,\delta}{\sqrt{p}}\right) = \left(\frac{2(x-y\sqrt{p});\,\delta}{\sqrt{p}}\right) = \left(\frac{q}{p}\right)_4.$

Proof (*a*)

$$\left(\frac{2;\,\delta}{\sqrt{p}}\right) = \left[\frac{2}{\sqrt{p}}\right] = \left(\frac{2}{p}\right) = -1.$$

(*b*)

$$\left(\frac{-\sqrt{p};\,\delta}{\sqrt{p}}\right) = \left[\frac{q}{\sqrt{p}}\right] = \left(\frac{q}{p}\right) = 1.$$

(*c*)

$$\left(\frac{2(x+y\sqrt{p});\,\delta}{\sqrt{p}}\right) = \left[\frac{2(x+y\sqrt{p})}{\sqrt{p}}\right] = \left(\frac{2x}{p}\right) = \left(\frac{2}{p}\right)\left(\frac{x}{p}\right) = \left(\frac{q}{p}\right)_4.$$

However, a similar argument shows that

$$\left(\frac{2(x-y\sqrt{p});\,\delta}{\sqrt{p}}\right) = \left(\frac{q}{p}\right)_4.$$

Lemma 5 *Keep the previous hypotheses and notations, then*

(*a*) $\left(\frac{-1;\,\delta}{\pi_1}\right) = \left(\frac{-1;\,\delta}{\tilde{\pi_1}}\right) = 1.$

(*b*) $\left(\frac{\varepsilon_p;\,\delta}{\pi_1}\right) = \left(\frac{\varepsilon_p;\,\delta}{\tilde{\pi_1}}\right) = \left(\frac{p}{q}\right)_4\left(\frac{q}{p}\right)_4.$

(*c*) $\left(\frac{2;\,\delta}{\pi_1}\right) = \left(\frac{2;\,\delta}{\tilde{\pi_1}}\right) = \left(\frac{2}{q}\right).$

(*d*) $\left(\frac{\sqrt{p};\,\delta}{\pi_1}\right) = \left(\frac{\sqrt{p};\,\delta}{\tilde{\pi_1}}\right) = \left(\frac{p}{q}\right)_4.$

(*e*) $\left(\frac{2(x+y\sqrt{p});\,\delta}{\pi_1}\right) = \left(\frac{2(x-y\sqrt{p});\,\delta}{\tilde{\pi_1}}\right) = \left(\frac{2}{q}\right).$

(*f*) $\left(\frac{2(x-y\sqrt{p});\,\delta}{\pi_1}\right) = \left(\frac{2(x+y\sqrt{p});\,\delta}{\tilde{\pi_1}}\right) = \left(\frac{p}{q}\right)_4.$

Proof The main tools of the proof of those equalities are the properties of the Hilbert symbol (see [6, Chap. II § 7, proposition 7.4.3, p. 205]), we get

(*a*)

$$\left(\frac{-1;\,\delta}{\tilde{\pi_1}}\right) = \left(\frac{-1;\,\delta}{\pi_1}\right) = \left[\frac{-1}{\pi_1}\right] = (-1)^{\frac{N(\pi_1)-1}{2}} = (-1)^{\frac{(q-1)}{2}} = \left(\frac{-1}{q}\right) = 1.$$

(*b*) Since $p \equiv q \equiv 1 \pmod 4$ and as $\left(\frac{p}{q}\right) = 1$, so by Scholz's reciprocity law (see [8, Chap. 5 § 5.2 Proposition 5.8, p. 160]), we have

$$\left(\frac{\varepsilon_p;\,\delta}{\pi_1}\right) = \left[\frac{\varepsilon_p}{\pi_1}\right] = \left(\frac{p}{q}\right)_4\left(\frac{q}{p}\right)_4.$$

Moreover, $\left[\frac{\varepsilon_p}{\pi_1}\right]\left[\frac{\varepsilon_p}{\tilde{\pi_1}}\right] = \left(\frac{-1}{q}\right) = 1$, then

$$\left(\frac{\varepsilon_p;\,\delta}{\tilde{\pi_1}}\right) = \left(\frac{p}{q}\right)_4\left(\frac{q}{p}\right)_4.$$

(*c*)

$$\left(\frac{2;\,\delta}{\pi_1}\right) = \left[\frac{2}{\pi_1}\right] = \left(\frac{2}{q}\right).$$

$$\left(\frac{2;\,\delta}{\tilde{\pi_1}}\right) = \left[\frac{2}{\tilde{\pi_1}}\right] = \left(\frac{2}{q}\right).$$

(*d*) We have

$$\left(\frac{\sqrt{p};\delta}{\pi_1}\right) = \left[\frac{\sqrt{p}}{\pi_1^h}\right]$$
$$= \left[\frac{2y}{\pi_1^h}\right]\left[\frac{2y\sqrt{p}}{\pi_1^h}\right]$$
$$= \left[\frac{2y}{\pi_1}\right]\left[\frac{-2x+2(x+y\sqrt{p})}{\tilde{\pi_1}^h}\right]$$
$$= \left[\frac{2y}{\pi_1}\right]\left[\frac{-2x}{\tilde{\pi_1}}\right]$$
$$= \left(\frac{-1}{q}\right)\left(\frac{y}{q}\right)\left(\frac{x}{q}\right)$$
$$= \left(\frac{p}{q}\right)_4.$$

and

$$\left(\frac{\sqrt{p};\delta}{\tilde{\pi_1}}\right) = \left[\frac{\sqrt{p}}{\tilde{\pi_1}^h}\right]$$
$$= \left[\frac{2y}{\tilde{\pi_1}^h}\right]\left[\frac{2y\sqrt{p}}{\tilde{\pi_1}^h}\right]$$
$$= \left[\frac{2y}{\tilde{\pi_1}}\right]\left[\frac{2x-2(x-y\sqrt{p})}{\tilde{\pi_1}^h}\right]$$
$$= \left[\frac{2y}{\tilde{\pi_1}}\right]\left[\frac{2x}{\tilde{\pi_1}}\right]$$
$$= \left(\frac{y}{q}\right)\left(\frac{x}{q}\right)$$
$$= \left(\frac{p}{q}\right)_4.$$

(e) We have

$$\left(\frac{2(x+y\sqrt{p});\delta}{\pi_1}\right) = \left(\frac{2(x+y\sqrt{p});\delta}{\pi_1}\right)\left(\frac{-\sqrt{p}(x-y\sqrt{p});\delta}{\pi_1}\right)\left(\frac{-\sqrt{p}(x-y\sqrt{p});\delta}{\pi_1}\right)$$
$$= \left(\frac{-8q^h\sqrt{p};\delta}{\pi_1}\right)\left(\frac{-\sqrt{p}(x-y\sqrt{p});\delta}{\pi_1}\right)$$
$$= \left(\frac{-\delta;\delta}{\pi_1}\right)\left(\frac{-\sqrt{p}(x-y\sqrt{p});\delta}{\pi_1}\right)$$
$$= \left[\frac{-\sqrt{p}(x-y\sqrt{p})}{\pi_1}\right]$$
$$= \left[\frac{-\sqrt{p}(x-y\sqrt{p})}{\pi_1^h}\right]$$
$$= \left[\frac{-2\sqrt{p}}{\pi_1^h}\right]\left[\frac{2(x-y\sqrt{p})}{\pi_1^h}\right]$$
$$= \left[\frac{-\sqrt{p}}{\pi_1^h}\right]\left[\frac{4x-2(x+y\sqrt{p})}{\pi_1^h}\right]$$
$$= \left[\frac{-2\sqrt{p}}{\pi_1^h}\right]\left[\frac{4x}{\pi_1^h}\right]$$
$$= \left[\frac{-2\sqrt{p}}{\pi_1^h}\right]\left[\frac{x}{\pi_1}\right]$$
$$= \left(\frac{2}{q}\right)\left(\frac{p}{q}\right)_4\left(\frac{x}{q}\right)$$
$$= \left(\frac{p}{q}\right)_4\left(\frac{p}{q}\right)_4$$
$$= \left(\frac{2}{q}\right).$$

and

$$\begin{aligned}
\left(\frac{2(x-y\sqrt{p});\delta}{\tilde{\pi}_1}\right) &= \left(\frac{2(x-y\sqrt{p});\delta}{\tilde{\pi}_1}\right)\left(\frac{-\sqrt{p}(x+y\sqrt{p});\delta}{\tilde{\pi}_1}\right)\left(\frac{-\sqrt{p}(x+y\sqrt{p});\delta}{\tilde{\pi}_1}\right)\\
&= \left(\frac{-\delta;\delta}{\tilde{\pi}_1}\right)\left(\frac{-\sqrt{p}(x+y\sqrt{p});\delta}{\tilde{\pi}_1}\right)\\
&= \left[\frac{-\sqrt{p}(x+y\sqrt{p})}{\tilde{\pi}_1}\right]\\
&= \left[\frac{-\sqrt{p}}{\tilde{\pi}_1}\right]\left[\frac{(x+y\sqrt{p})}{\tilde{\pi}_1}\right]\\
&= \left[\frac{-2\sqrt{p}}{\tilde{\pi}_1^h}\right]\left[\frac{2(x+y\sqrt{p})}{\tilde{\pi}_1{}^h}\right]\\
&= \left[\frac{-2\sqrt{p}}{\tilde{\pi}_1{}^h}\right]\left[\frac{4x-2(x-y\sqrt{p})}{\tilde{\pi}_1{}^h}\right]\\
&= \left[\frac{-\sqrt{p}}{\tilde{\pi}_1^h}\right]\left[\frac{4x}{\tilde{\pi}_1{}^h}\right]\\
&= \left[\frac{-2\sqrt{p}}{\tilde{\pi}_1^h}\right]\left[\frac{x}{\tilde{\pi}_1}\right]\\
&= \left(\frac{2}{q}\right)\left(\frac{p}{q}\right)_4\left(\frac{x}{q}\right)\\
&= \left(\frac{p}{q}\right)_4\left(\frac{2}{q}\right)\left(\frac{p}{q}\right)_4\\
&= \left(\frac{2}{q}\right).
\end{aligned}$$

(f)

$$\begin{aligned}
\left(\frac{2(x+y\sqrt{p});\delta}{\tilde{\pi}_1}\right) &= \left[\frac{2(x+y\sqrt{p})}{\tilde{\pi}_1}\right]\\
&= \left[\frac{4x}{\tilde{\pi}_1}\right]\\
&= \left(\frac{x}{q}\right)\\
&= \left(\frac{p}{q}\right)_4.
\end{aligned}$$

and

$$\begin{aligned}
\left(\frac{2(x-y\sqrt{p});\delta}{\pi_1}\right) &= \left[\frac{2(x-y\sqrt{p})}{\pi_1}\right]\\
&= \left[\frac{4x}{\pi_1}\right]\\
&= \left(\frac{x}{q}\right)\\
&= \left(\frac{p}{q}\right)_4.
\end{aligned}$$

Lemma 6 *Keep the previous hypotheses and notations, then*

(a) $\left(\frac{\varepsilon_p;\delta}{2_I}\right) = \left(\frac{\sqrt{p};\delta}{2_I}\right) = 1.$

(b) $\left(\frac{2;\delta}{2_I}\right) = \left(\frac{-1;\delta}{2_I}\right) = -1.$

(c) $\left(\frac{x+y\sqrt{p};\delta}{2_I}\right) = \left(\frac{x-y\sqrt{p};\delta}{2_I}\right) = \left(\frac{2}{q}\right)\left(\frac{p}{q}\right)_4\left(\frac{q}{p}\right)_4.$

Proof The product formulas for the Hilbert symbol and previous lemmas imply the results. □

Remark 1

- If $q \equiv 1 \pmod 4$, then $\left(\frac{-1;\,\delta}{\pi_1}\right) = \left(\frac{-1;\,\delta}{\tilde{\pi_1}}\right)$. We summarize the previous results in the following character table:

Element	Prime				
	π_1	$\tilde{\pi_1}$	2_I	$\sqrt{p}$	p_∞
$2(x+y\sqrt{p})$	$\left(\frac{2}{q}\right)$	$\left(\frac{p}{q}\right)_4$	$\left(\frac{2}{q}\right)\left(\frac{p}{q}\right)_4\left(\frac{q}{p}\right)_4$	$\left(\frac{q}{p}\right)_4$	1
$2(x-y\sqrt{p})$	$\left(\frac{p}{q}\right)_4$	$\left(\frac{2}{q}\right)$	$\left(\frac{2}{q}\right)\left(\frac{p}{q}\right)_4\left(\frac{q}{p}\right)_4$	$\left(\frac{q}{p}\right)_4$	1
2	$\left(\frac{2}{q}\right)$	$\left(\frac{2}{q}\right)$	-1	-1	1
$\sqrt{p}$	$\left(\frac{p}{q}\right)_4$	$\left(\frac{p}{q}\right)_4$	1	-1	-1
-1	1	1	-1	1	-1
ε_p	$\left(\frac{p}{q}\right)_4\left(\frac{q}{p}\right)_4$	$\left(\frac{p}{q}\right)_4\left(\frac{q}{p}\right)_4$	1	-1	-1

Proof (**Proof of Theorem** 2) By assumption, $q \equiv 1 \pmod 4$ is a prime number and $\left(\frac{p}{q}\right) = 1$. Then all finite primes of k ramifying in K are: π_1,$\tilde{\pi}_1$,2_I and $\sqrt{p}$ where $\pi_1\tilde{\pi}_1 = qO_k$, $2_I = 2O_k$ and $(\sqrt{p})^2 = pO_k$ and also the ideals π_1^h and $\tilde{\pi}_1^h$ are principal then $\pi_1^h = ((x+y\sqrt{p})/2))$, $\tilde{\pi}_1^h = ((x-y\sqrt{p})/2))$.
So there exist primes ideals, $\mathcal{H}_1, \mathcal{H}_2, \mathcal{H}_3, \mathcal{H}_4$ of of O_K, such that

$$\mathcal{H}_1^2 = \pi_1 O_K, \mathcal{H}_2^2 = \tilde{\pi}_1 O_K, \mathcal{H}_3^2 = 2_I O_K, \mathcal{H}_4^2 = \sqrt{p}O_K.$$

Now, we have

- $\mathcal{H}_1^h\sigma(\mathcal{H}_1^h) = \mathcal{H}_1^h\mathcal{H}_1^h O_K = \pi_1^h O_K = ((x+y\sqrt{p})/2)O_K.$
- $\mathcal{H}_2^h\sigma(\mathcal{H}_2^h) = \mathcal{H}_2^h\mathcal{H}_2^h O_K = \tilde{\pi}_1^h O_K = ((x-y\sqrt{p})/2)O_K.$
- $\mathcal{H}_3^h\sigma(\mathcal{H}_3^h) = \mathcal{H}_3^h\mathcal{H}_3^h O_K = 2_I^h O_K = 2O_K.$
- $\mathcal{H}_4^h\sigma(\mathcal{H}_4^h) = \mathcal{H}_4^h\mathcal{H}_4^h O_K = \sqrt{p}^h O_K = \sqrt{p}O_K.$

Moreover, we know from Lemma 1 that -1; ε_p; $-\varepsilon_p \notin O_k^* \cap N(K)$.

Then the $5\times(4+2)$ generalized Rédei's matrix of local Hilbert symbols $R_{K/k}$ given in [11, § 2 Lemma 2.4. p. 27] can be written as

$$R_{K/k} = \begin{pmatrix} \left(\frac{2(x+y\sqrt{p});\,\delta}{\pi_1}\right) & \left(\frac{2(x-y\sqrt{p});\,\delta}{\pi_1}\right) & \left(\frac{2;\,\delta}{\pi_1}\right) & \left(\frac{\sqrt{p};\,\delta}{\pi_1}\right) & \left(\frac{-1;\,\delta}{\pi_1}\right) & \left(\frac{\varepsilon_p;\,\delta}{\pi_1}\right) \\ \left(\frac{2(x+y\sqrt{p});\,\delta}{\tilde{\pi}_1}\right) & \left(\frac{2(x-y\sqrt{p});\,\delta}{\tilde{\pi}_1}\right) & \left(\frac{2;\,\delta}{\tilde{\pi}_1}\right) & \left(\frac{\sqrt{p};\,\delta}{\tilde{\pi}_1}\right) & \left(\frac{-1;\,\delta}{\tilde{\pi}_1}\right) & \left(\frac{\varepsilon_p;\,\delta}{\tilde{\pi}_1}\right) \\ \left(\frac{2(x+y\sqrt{p});\,\delta}{2_I}\right) & \left(\frac{2(x-y\sqrt{p});\,\delta}{2_I}\right) & \left(\frac{2;\,\delta}{2_I}\right) & \left(\frac{\sqrt{p};\,\delta}{2_I}\right) & \left(\frac{-1;\,\delta}{2_I}\right) & \left(\frac{\varepsilon_p;\,\delta}{2_I}\right) \\ \left(\frac{2(x+y\sqrt{p});\,\delta}{\sqrt{p}}\right) & \left(\frac{2(x-y\sqrt{p});\,\delta}{\sqrt{p}}\right) & \left(\frac{2;\,\delta}{\sqrt{p}}\right) & \left(\frac{\sqrt{p};\,\delta}{\sqrt{p}}\right) & \left(\frac{-1;\,\delta}{\sqrt{p}}\right) & \left(\frac{\varepsilon_p;\,\delta}{\sqrt{p}}\right) \\ \left(\frac{2(x+y\sqrt{p});\,\delta}{p_\infty}\right) & \left(\frac{2(x-y\sqrt{p});\,\delta}{p_\infty}\right) & \left(\frac{2;\,\delta}{p_\infty}\right) & \left(\frac{\sqrt{p};\,\delta}{p_\infty}\right) & \left(\frac{-1;\,\delta}{p_\infty}\right) & \left(\frac{\varepsilon_p;\,\delta}{p_\infty}\right) \end{pmatrix}.$$

We consider the above matrix with coefficients in $\mathbb{F}_2$ by replacing the 1 by 0 and the -1 by 1. Consequently, we can fill the Rédei matrix following the values of prime number q:

Suppose that $q \equiv 1 \pmod 8$, then $\left(\frac{2}{q}\right) = 1$.

If $\left(\frac{p}{q}\right)_4 = \left(\frac{q}{p}\right)_4 = 1$ then the generalized Rédei matrix $R_{K/k}$ is

$$\begin{pmatrix} 0\,0\,0\,0\,0\,0 \\ 0\,0\,0\,0\,0\,0 \\ 0\,0\,1\,0\,1\,0 \\ 0\,0\,1\,1\,0\,1 \\ 0\,0\,0\,1\,1\,1 \end{pmatrix}.$$

whose rank is 2. Hence, $r_4(K) = 5 - 1 - rank R_{K/k} = 5 - 1 - 2 = 2$.

By a similar calculation in other cases we get:

If $\left(\frac{p}{q}\right)_4 = \left(\frac{q}{p}\right)_4 = -1$ Then $r_4(K) = 0$.

If $\left(\frac{p}{q}\right)_4 = -\left(\frac{q}{p}\right)_4 = 1$ Then $r_4(K) = 1$.

If $\left(\frac{p}{q}\right)_4 = -\left(\frac{q}{p}\right)_4 = -1$. Then $r_4(K) = 0$.

Suppose that $q \equiv 5 \pmod 8$, then $\left(\frac{2}{q}\right) = -1$.

If $\left(\frac{p}{q}\right)_4 = \left(\frac{q}{p}\right)_4 = 1$ Then $r_4(K) = 0$.

If $\left(\frac{p}{q}\right)_4 = \left(\frac{q}{p}\right)_4 = -1$ Then $r_4(K) = 1$.

If $\left(\frac{p}{q}\right)_4 = -\left(\frac{q}{p}\right)_4 = 1$ Then $r_4(K) = 0$.

If $\left(\frac{p}{q}\right)_4 = -\left(\frac{q}{p}\right)_4 = -1$ Then $r_4(K) = 1$.

4 Some Numerical Examples

Example 1 Keep the previous hypotheses and notations, we have

$$k = \mathbb{Q}(\sqrt{p}) \text{ and } K = \mathbb{Q}(\sqrt[4]{4pq^2})$$

(i) For $p = 13$ and $q = 113$, $13 \equiv 5 \pmod 8$, $\left(\frac{13}{113}\right) = 1$, by Theorem 1, then $r_2(K) = 2$.
also $113 \equiv 1 \pmod 8$, $\left(\frac{13}{113}\right)_4 = -1$, so the condition of Theorem (2-(i)) are satisfied. Hence, $r_4(K) = 0$.
In fact, $Cl(K) \cong \mathbb{Z}/2\mathbb{Z} \times \mathbb{Z}/2\mathbb{Z}$ by using the Pari-GP calculator (version 2.9.3) [8].

(ii) For $p = 13$ and $q = 17$, $13 \equiv 5 \pmod 8$, $\left(\frac{13}{17}\right) = 1$, by Theorem 1, then $r_2(K) = 2$.

also $17 \equiv 1 \pmod 8$, $\left(\frac{13}{17}\right)_4 = -\left(\frac{17}{13}\right)_4 = 1$, so the condition of Theorem (2-(i)) are satisfied. Hence, $r_4(K) = 1$.
In fact, $Cl(K) \cong \mathbb{Z}/8\mathbb{Z} \times \mathbb{Z}/2\mathbb{Z}$ by using the Pari-GP calculator (version 2.9.3) [8].

(iii) For $p = 29$ and $q = 401$, $29 \equiv 5 \pmod 8$,$\left(\frac{29}{401}\right) = 1$, by Theorem 1, then $r_2(K) = 2$.
also $401 \equiv 1 \pmod 8$, $\left(\frac{29}{401}\right)_4 = \left(\frac{401}{29}\right)_4 = 1$, so the condition of Theorem (2-(i)) are satisfied. Hence, $r_4(K) = 2$.
In fact, $Cl(K) \cong \mathbb{Z}/16\mathbb{Z} \times \mathbb{Z}/4\mathbb{Z}$ by using the Pari-GP calculator (version 2.9.3) [8]. □

Example 2 Keep the previous hypotheses and notations, we have

$$k = \mathbb{Q}(\sqrt{p}) \text{ and } K = \mathbb{Q}(\sqrt[4]{4pq^2})$$

(i) For $p = 13$ and $q = 53$, $13 \equiv 5 \pmod 8$, $\left(\frac{13}{53}\right) = 1$, by Theorem 1, then $r_2(K) = 2$.
also $53 \equiv 5 \pmod 8$, $\left(\frac{13}{53}\right)_4 = 1$, so the condition of Theorem (2-(ii)) are satisfied. Hence, $r_4(K) = 0$.
In fact, $Cl(K) \cong \mathbb{Z}/2\mathbb{Z} \times \mathbb{Z}/2\mathbb{Z}$ by using the Pari-GP calculator (version 2.9.3) [8].

(ii) For $p = 13$ and $q = 29$, $13 \equiv 5 \pmod 8$, $\left(\frac{13}{29}\right) = 1$, by Theorem 1, then $r_2(K) = 2$.
also $29 \equiv 5 \pmod 8$, $\left(\frac{13}{29}\right)_4 = -1$, so the condition of Theorem (2-(ii)) are satisfied. Hence, $r_4(K) = 1$.
In fact, $Cl(K) \cong \mathbb{Z}/4\mathbb{Z} \times \mathbb{Z}/2\mathbb{Z}$ by using the Pari-GP calculator (version 2.9.3) [8]. □

References

1. Azizi, A.: Sur la capitulation des 2-classes d'idéaux de $\mathbb{Q}(\sqrt{2pq}, i)$ où $p \equiv -q \equiv 1 \bmod 4$. Acta Arithmetica **94**, 383–399 (2000)
2. Azizi, A., Taous, M.: Détermination des corps $k = \mathbb{Q}(\sqrt{d}, i)$ dont les 2-groupes de classes sont de type (2, 4) ou (2, 2, 2) Rend. Istit. Mat. Univ. Trieste. **40**, 93–116 (2008)
3. Azizi, A., Taous, M., Zekhnini, A.: On the rank of the 2-class group of $\mathbb{Q}(\sqrt{p}, \sqrt{q}, \sqrt{-1})$. Period. Math. Hungar. **69**(2), 231–238 (2014)
4. Azizi, A., Taous, M., Zekhnini, A.: On the quartic residue symbols of certain fundamental units. In: An International Journal Edited by The Hassan II Academy of Science and Technology, vol.**6**, no. 1 (2016)
5. Chevalley, C.: Sur la théorie du corps de classes dans les corps finis et les corps locaux. J. Fac. Sc. Tokyo Sect. 1 t.**2**, 365–476 (1933)
6. Gras, G.: Class Field Theory, from Theory to Practice. Springer, Berlin (2003)
7. Hymo, J.A., Parry, C.J.: On relative integral bases for pure quartic fields. Indian J. Pure Appl. Math. **23**, 359–376 (1992)

8. Lemmermeyer, F.: Reciprocity Laws. From Euler to Eisenstein. Springer Monographs in Math (2000)
9. Parry, C.J.: Pure quartic number fields whose class numbers are even. J. Reine Angew. Math. **264**, 102–112 (1975)
10. Parry, C.J.: A genus theory for quartic fields. J. Reine Angew. Math. **314**, 40–71 (1980)
11. Qin, Y.: The generalized Rédei-matrix. Math. Z. **261**, 23–37 (2009)
12. Taous, M.: Capitulation des 2-classes d'idéaux de certains corps $\mathbb{Q}(\sqrt{d}, i)$ de type $(2, 4)$, thèse. Université, Mohammed Premier Faculté des Science, Oujda (2008)
13. The PARI Group, PARI/GP version 2.10.0. University Bordeaux, 2018. http://pari.math.u-bordeaux.fr/
14. Taous, M.: On the 2-class group of $\mathbb{Q}(\sqrt{5pF_p})$ where F_p is a prime Fibonacci number. Fibonacci. Quart. **55**(5), 192–200 (2017)

Homological and Categorical Methods in Algebra

This part contains five contributions. The most common factors between them are perhaps the categorical and/or homological methods that were employed in the different contexts that gather such papers. The first contribution gives an exhaustive review on the topic of Frobenius structures in the framework of categories with multiplications (monoidal categories). The second one makes some homological observations on the localization functor (after its extension to the category of unbonded complexes, or the category of functors from the poset of the integer numbers to the category of modules) and its relation with the derived functors of the tensor and the Hom functors, respectively. In the same lines, the last contribution can be seen as well known facts about relative homological algebra, in the sense of Eilenberg-Moore. The third contribution reports on the folkloric relation, at the level of bicategories, between corings and Bocses. Lastly, the fourth contribution announces new results on the representation theory of quadratic color Hom-Lie algebra and its cohomology.

Frobenius Monoidal Algebras and Related Topics

Daniel Bulacu and Blas Torrecillas

Abstract We survey results on Frobenius algebras and illustrate their importance to the structure of some generalizations of the notion of Hopf algebra, as well as their connections to topics like monoidal categories, 2-categories, functors, topological quantum field theories, etc.

Keywords Frobenius algebra · Hopf algebra · Frobenius functor · TQFT

1 Introduction

We survey results that we obtained so far in the study of (monoidal) Frobenius algebras, with emphasis to Frobenius algebra extensions in monoidal categories, Frobenius (monoidal) functors, Frobenius algebras in categories of (co)representations of a Hopf algebra, and Frobenius wreaths and cowreaths. Most of the results will be presented without proofs; instead we will always indicate a place where they can be found. Proofs will be presented only for results that are either new or can be considered alternatives to the proofs that already exist.

As it is explained by Lam in [46, Sect. 16G], the comparison of the two modules A_A and A^*_A originated with the work of Frobenius on the so called "hypercomplex systems", an older term for algebras A. More precisely, let A be a finite dimensional algebra over a field k and fix $\{a_1, \cdots, a_n\} \subset A$ a k-basis. Then $l, r : A \to M_n(k)$ given, for all $a \in A$, by

Work supported by the project MTM2017-86987-P "Anillos, modulos y algebra de Hopf". DB thanks the University of Almeria (Spain) for its warm hospitality.

D. Bulacu
Faculty of Mathematics and Computer Science, University of Bucharest, Str. Academiei 14, RO-010014 Bucharest 1, Romania
e-mail: daniel.bulacu@fmi.unibuc.ro

B. Torrecillas (✉)
Department of Mathematics, Universidad de Almería, E-04071 Almería, Spain
e-mail: btorreci@ual.es

M. Siles Molina et al. (eds.), *Associative and Non-Associative Algebras and Applications*, Springer Proceedings in Mathematics & Statistics 311,
https://doi.org/10.1007/978-3-030-35256-1_13

$$
\begin{aligned}
l(a) &= (\alpha^a_{ij})_{1\le i,j\le n} \text{ with } aa_i = \sum_{j=1}^{n} \alpha^a_{ji} a_j, \ \forall\, 1 \le i \le n, \\
r(a) &= (\beta^a_{ij})_{1\le i,j\le n} \text{ with } a_i a = \sum_{j=1}^{n} \beta^a_{ij} a_j, \ \forall\, 1 \le i \le n,
\end{aligned} \tag{1}
$$

are representations of A. The problem of Frobenius was to find necessary and sufficient conditions for l, r to be equivalent representations, and this led to the notion of Frobenius algebra. More exactly, if we write $a_i a_j = \sum_{l=1}^{n} \gamma^l_{ij} a_l$, for all $1 \le i,\ j \le n$, and for $\lambda = (\lambda_1, \cdots, \lambda_n) \in k^n$ define $P_\lambda = (\sum_{l=1}^{n} \lambda_l \gamma^l_{ij})_{1\le i,j\le n}$ then A is a Frobenius algebra (i.e. l, r are equivalent representations of A) if and only if $P_\lambda \in GL_n(k)$ for some $\lambda \in k^n$; we refer to [46, (16.82)] for a proof of this fact.

Frobenius algebras started to be studied in the thirties of the last century by Brauer and Nesbitt, who named these algebras after Frobenius; they also introduced the concept of symmetric algebra in [7]. Nakayama discovered the beginnings of the duality property of a Frobenius algebra in [54, 55], and Dieudonné used this to characterize Frobenius algebras in [28]. Nakayama also studied symmetric algebras in [53]; the automorphism that carries his name was defined in [55]. Besides representation theory, Frobenius algebras play an important role in number theory, algebraic geometry, combinatorics, coding theory, geometry, Hopf algebra and quantum group theory, in finding solutions for the quantum Yang–Baxter equation, the Jones polynomials, etc. More details about the connections between Frobenius algebras and the classical, respectively modern, directions mentioned above can be found in the books [37, 46].

Kasch [40] extended the notion of Frobenius algebra to an arbitrary algebra extension. Frobenius extensions have a well-developed theory of induced representations, investigated by Kasch [39–41], Pareigis [59, 60], Nakayama and Tzuzuku [56–58], Morita [50, 51], to name a few. Frobenius extensions, and so Frobenius algebras as well, can be characterized in terms of Frobenius functors, first introduced by Morita [50]. The terminology is based on the fact that an algebra extension $i : R \to S$ is Frobenius if and only if the restriction of scalars functor is Frobenius. When the restriction of scalars functor and the induction functor are Frobenius in arbitrary monoidal categories C was investigated in [21]: if C has coequalizers and any object in C is coflat and robust then the two mentioned functors are Frobenius if and only if the algebra extension is Frobenius, provided that the unit object $\underline{1}$ of C is a left $\otimes$-generator. We should point out that the result still holds in Vect^G although in this case we have a left $\otimes$-generator different from the unit object of the category (see [24, Sect. 7]).

Recently, the interest for Frobenius algebras has been renewed due to connections to monoidal categories and topological quantum field theory (TQFT for short), see [66]. Roughly speaking, if **nCob** is the category of n cobordisms then a TQFT is a (symmetric) monoidal functor from **nCob** to ${}_k\mathcal{M}$, the category of k-vector spaces. For $n = 2$ there exists a complete classification of surfaces, and so the cobordism category **2Cob** is described completely. Furthermore, the relations that hold in **2Cob** correspond exactly to the axioms of a commutative Frobenius algebra and this leads to the fact that **2TQFT** is equivalent to the category of commutative Frobenius algebras.

For more details on this topic we invite the reader to consult [1, 43]. We also note that the Frobenius equation expresses the compatibility between the algebra and coalgebra structure on a Frobenius algebra. It makes sense in any monoidal category, and therefore the notion of Frobenius algebra can be defined in any such category. This has already been done, see for instance [21, 31, 43, 66, 68]. Furthermore, in the monoidal categorical framework Frobenius algebras appear in different contexts. Apart from the TQFT case mentioned above, we have a correspondence between Frobenius algebras in monoidal categories and weak monoidal Morita equivalence of monoidal categories [52], Frobenius functors and Frobenius algebras in the category of endofunctors [65], and Frobenius monads in 2-categories and Frobenius algebras in some suitable monoidal categories [21].

This survey is an attempt to summarize results obtained so far by the authors in the theory of Frobenius algebra extensions, and respectively of Frobenius wreaths and cowreaths, with emphasis to Hopf and quasi-Hopf algebra theory. In Sect. 2 we present preliminary results on monoidal categories, categories of bimodules, 2-categories and (quasi-)Hopf algebras. In Sect. 3 we present the classical theory of Frobenius algebras (characterizations, the uniqueness of the Frobenius system, the Nakayama automorphism) and to illustrate its importance we give examples coming from various domains of mathematics like module theory, algebraic geometry, de Rham cohomology, (quasi-)Hopf algebra theory, etc. The monoidal case is treated as well, and we insist here in highlighting what can and can not be generalized from the category of vector spaces to an arbitrary monoidal category (for instance, to define for a Frobenius algebra A its Nakayama morphism as an algebra automorphism we need A to be a so called sovereign object). Also, plenty of examples are given; and we should mention here the ones in categories of graded vector spaces (like Cayley–Dickson and Clifford algebras, see [9, 10]) or, more generally, in categories of corepresentations of a Hopf algebra (like Hopf smash product algebras for instance).

In Sect. 4 we discuss the connection provided by Morita between Frobenius algebra extensions and Frobenius functors. A theory parallel to the one performed by Morita can be done in arbitrary monoidal categories, provided (again) that $\underline{1}$ is a left $\otimes$-generator. Apart from this, we give also concrete examples of Frobenius algebra extensions (such like Hopf Galois extensions or, more generally, cleft extensions), the connection with the Frobenius algebra theory and with the theory of monoidal Frobenius functors [23, 64]. More examples of Frobenius algebra extensions arise from the canonical algebra extension associated to a wreath. We explain this in detail in Sect. 5, which reproduces the results obtained in [17, 18, 21]. By studying the categories of Doi-Hopf modules, two-sided Hopf modules and Yetter–Drinfeld modules over a quasi-Hopf algebra we were able to obtain new examples of Frobenius cowreaths (and therefore of Frobenius algebra extensions) which are not ordinary entwining structures.

2 Preliminaries

2.1 Rigid Monoidal Categories

For the definition of a monoidal category C and related topics we refer to [15, 42, 48]. Usually, for a monoidal category C, we denote by $\otimes$ the tensor product, by $\underline{1}$ the unit object, and by a, l, r the associativity constraint and the left and right unit constraints, respectively. The monoidal category $(C, \otimes, \underline{1}, a, l, r)$ is called strict if all the natural isomorphisms a, l and r are defined by identity morphisms.

It is well known that every monoidal category is monoidally equivalent to a strict monoidal category, and this enables us to assume without loss of generality that C is strict. Sometimes we will delete the tensor symbol $\otimes$, and write $X \otimes Y = XY$, $f \otimes g = fg$ or X^n for the tensor product of n copies of X. The identity morphism of an object $X \in C$ will be denoted by Id_X or simply X.

If $(C, \otimes, \underline{1}, a, l, r)$ is a monoidal category, by $\overline{C} := (C, \overline{\otimes} := \otimes \circ \tau, \overline{a} = (\overline{a}_{X,Y,Z} := a^{-1}_{Z,Y,X})_{X,Y,Z \in C}, \underline{1}, \overline{l} := r, \overline{r} := l)$ we denote the reverse monoidal category associated to C, where $\tau : C \times C \to C \times C$ stands for the twist functor, that is $\tau(X, Y) = (Y, X)$, for any $X, Y \in C$, and $\tau(f, g) = (g, f)$, for any morphisms $X \xrightarrow{f} X'$ and $Y \xrightarrow{g} Y'$ in C.

An object $X \in C$ has a left dual if there exist $X^* \in C$ and morphisms $\mathrm{ev}_X : X^* \otimes X \to \underline{1}$ and $\mathrm{coev}_X : \underline{1} \to X \otimes X^*$ in C such that

$$r_X(X\mathrm{ev}_X)a_{X,X^*,X}(\mathrm{coev}_X X) = Xl_X, \quad l_{X^*}(\mathrm{ev}_X X^*)a^{-1}_{X^*,X,X^*}(X^*\mathrm{coev}_X) = X^* r_{X^*}.$$

In this situation we also say that we have an adjunction $(\mathrm{coev}_X, \mathrm{ev}_X) : X^* \dashv X$.

If $X, Y \in C$ admit left duals then $Y^* \otimes X^*$ is a left dual for $X \otimes Y$. Indeed, for simplicity assume that C is strict monoidal and then define $\eta := (X\mathrm{coev}_Y X^*)\mathrm{coev}_X : \underline{1} \to XYY^*X^*$ and $\varepsilon := \mathrm{ev}_Y(Y^*\mathrm{ev}_X Y) : Y^*X^*XY \to \underline{1}$. From the definition of a left dual it follows that $(\eta, \varepsilon) : Y^* \otimes X^* \dashv X \otimes Y$ is an adjunction in C, hence $Y^* \otimes X^*$ is a left dual for $X \otimes Y$.

If any object of C has a left dual then we call C left rigid; if this is the case, we have a well defined functor $(-)^* : C \to C^{\mathrm{op}}$ that maps X to X^* and $f : X \to Y$ to $f^* := (\mathrm{ev}_Y X^*)a^{-1}_{Y^*,Y,X^*}(Y^*(fX^*))(Y^*\mathrm{coev}_X)r^{-1}_{Y^*} : Y^* \to X^*$, the left transpose of f. Here C^{op} is the opposite category associated to C. Furthermore, $(-)^*$ is strong monoidal if it is regarded as a functor from C to $\overline{C^{\mathrm{op}}} := (C^{\mathrm{op}}, \otimes^{\mathrm{op}} := \otimes \circ \tau, \underline{1}, (a_{Z,Y,X})_{X,Y,Z \in C}, r^{-1}, l^{-1})$. The functor $(-)^*$ is called the left dual functor.

By taking $\overline{C}$ instead of C we get the notion of right duality, in which case we have a strong monoidal functor ${}^*(-) : C \to \overline{C^{\mathrm{op}}}$ called the right dual functor.

2.2 Bimodule Categories

Let C be a monoidal category with coequalizers. Take an algebra A in C, $\mathfrak{M} \in C_A$ and $X \in {}_AC$, with structure morphisms $\mu_X^A : A \otimes X \to X$ and $\nu_{\mathfrak{M}}^A : \mathfrak{M} \otimes A \to \mathfrak{M}$, respectively. From now on, $(\mathfrak{M} \otimes_A X, q_{\mathfrak{M},X}^A)$ stands for the coequalizer of the parallel morphisms $\nu_{\mathfrak{M}}^A \otimes \mathrm{Id}_X$ and $\mathrm{Id}_{\mathfrak{M}} \otimes \mu_X^A$ in C.

For $f : X \to Y$ in ${}_AC$, let $\mathrm{Id}_{\mathfrak{M}} \otimes_A f : \mathfrak{M} \otimes_A X \to \mathfrak{M} \otimes_A Y$ be the unique morphism in C satisfying $(\mathrm{Id}_{\mathfrak{M}} \otimes_A f)q_{\mathfrak{M},X}^A = q_{\mathfrak{M},Y}^A(\mathrm{Id}_{\mathfrak{M}} \otimes f)$. For $X \xrightarrow{f} Y \xrightarrow{g} Z$ in ${}_AC$ we have that $\mathrm{Id}_{\mathfrak{M}} \otimes_A gf = (\mathrm{Id}_{\mathfrak{M}} \otimes_A g)(\mathrm{Id}_{\mathfrak{M}} \otimes_A f)$.

Now let $g : \mathfrak{M} \to \mathfrak{N}$ in C_A and $Y \in {}_AC$. Then $g \otimes_A \mathrm{Id}_Y : \mathfrak{M} \otimes_A Y \to \mathfrak{N} \otimes_A Y$ denotes the unique morphism in C obeying $(g \otimes_A \mathrm{Id}_Y)q_{\mathfrak{M},Y}^A = q_{\mathfrak{N},Y}^A(g \otimes \mathrm{Id}_Y)$. For $\mathfrak{M} \xrightarrow{f} \mathfrak{N} \xrightarrow{g} \mathfrak{P}$ in C_A, we have that $(gf \otimes_A \mathrm{Id}_Y) = (g \otimes_A \mathrm{Id}_Y)(f \otimes_A \mathrm{Id}_Y)$.

For $\mathfrak{M} \in C_A$, $X \in C$ and $Y \in {}_AC$, we have isomorphisms $\Upsilon_{\mathfrak{M}}$ and Υ'_Y:

- $\Upsilon_{\mathfrak{M}} : \mathfrak{M} \otimes_A A \xrightarrow{\cong} \mathfrak{M}$, uniquely determined by the property $\Upsilon_{\mathfrak{M}} q_{\mathfrak{M},A}^A = \nu_{\mathfrak{M}}^A$;
- $\Upsilon'_Y : A \otimes_A Y \xrightarrow{\cong} Y$, uniquely determined by the equality $\Upsilon'_Y q_{A,Y}^A = \mu_Y^A$.

$X \in C$ is called left (right) coflat if the functor $- \otimes X$ $(X \otimes -)$ preserves coequalizers. An object of C is called coflat if it is both left and right coflat.

If A is a left coflat algebra in C then for any $X \in C_A$ and $Y \in {}_AC_A$ the morphism $\nu_{X\otimes_A Y}^A : X \otimes_A Y \otimes A \to X \otimes_A Y$ determined by $\nu_{X\otimes_A Y}^A(q_{X,Y}^A \otimes \mathrm{Id}_A) = q_{X,Y}^A(\mathrm{Id}_X \otimes \nu_Y^A)$ makes $X \otimes_A Y$ object of C_A, where, as before, by $\nu_Y^A : Y \otimes A \to Y$ we denote the morphism structure of a right A-module Y in C.

Also, if A is an algebra in C we say that $Y \in {}_AC$ is left robust if for any $\mathfrak{M} \in C$, $X \in C_A$ the morphism $\theta'_{\mathfrak{M},X,Y} : (\mathfrak{M} \otimes X) \otimes_A Y \to \mathfrak{M} \otimes (X \otimes_A Y)$ defined by $\theta'_{\mathfrak{M},X,Y} q_{\mathfrak{M}\otimes X,Y}^A = \mathrm{Id}_{\mathfrak{M}} \otimes q_{X,Y}^A$ is an isomorphism. For instance, if A is a left coflat algebra and X a left coflat object then $A \otimes X$ is left coflat and left robust.

Notice that the left robustness of an object $Y \in {}_AC$ is needed in order to define a left A-module structure on $X \otimes_A Y$ in C, for any $X \in {}_AC_A$. Namely, if we define $\overline{\mu}_{X\otimes_A Y} : (A \otimes X) \otimes_A Y \to X \otimes_A Y$ as being the unique morphism in C obeying $\overline{\mu}_{X\otimes_A Y} q_{A\otimes X,Y}^A = q_{X,Y}^A(\mu_X^A \otimes \mathrm{Id}_Y)$ then the morphism $\mu_{X\otimes_A Y}^A := \overline{\mu}_{X\otimes_A Y}\theta'^{-1}_{A,X,Y}$ defines on $X \otimes_A Y$ a left A-module structure in C. In this way $X \otimes_A Y$ becomes an A-bimodule, provided that $Y \in {}^!_AC_A$ and A is left coflat, where ${}^!_AC_A$ is the category of A-bimodules in C that are left coflat and left robust objects in C.

The robustness condition is needed also to define a morphism $\Gamma'_{\mathfrak{M},X,Y} : \mathfrak{M} \otimes_A (X \otimes_A Y) \to (\mathfrak{M} \otimes_A X) \otimes_A Y$, for all $\mathfrak{M}, X, Y \in {}_AC_A$. It is uniquely determined by $\Gamma'_{\mathfrak{M},X,Y} q_{\mathfrak{M},X\otimes_A Y}^A = (q_{\mathfrak{M},X}^A \otimes_A \mathrm{Id}_Y)\theta'^{-1}_{\mathfrak{M},X,Y}$. It is actually an isomorphism with inverse $\Sigma'_{\mathfrak{M},X,Y}$ uniquely determined by the property that $\Sigma'_{\mathfrak{M},X,Y} q_{\mathfrak{M}\otimes_A X,Y}^A(q_{\mathfrak{M},X}^A \otimes \mathrm{Id}_Y) = q_{\mathfrak{M},X\otimes_A Y}^A(\mathrm{Id}_{\mathfrak{M}} \otimes q_{X,Y}^A)$.

For A left coflat, ${}^!_AC_A$ is monoidal with tensor product $\otimes_A$, associativity constraint $\Sigma'_{-,-,-}$, unit object A, and left and right unit constraints Υ'_- and Υ_-. Furthermore, the functor $\otimes_A$ provides a right ${}^!_AC_A$-category structure on C_A; full details can be found in [13, 61, 63]. For simplicity, in most of the cases we assume that C is monoidal and such that any object of it is coflat and left robust; in other words, we identify ${}^!_AC_A$ and ${}_AC_A$ as monoidal categories.

We call a (co)algebra $\mathfrak{C}$ in ${}^{!}_{A}C_A$ an A-(co)ring. The category of right (co)representations over $\mathfrak{C}$ in C, denoted by ($C^{\mathfrak{C}}$) $C_{\mathfrak{C}}$, is just the category of right (co)modules and right (co)linear maps in C_A over the (co)algebra $\mathfrak{C}$ in ${}^{!}_{A}C_A$.

2.3 Monads in 2-Categories

Let $\mathcal{K}$ be a 2-category; its objects (or 0-cells) will be denoted by capital letters. 1-cell between two 0-cells U and V will be denoted as $U \xrightarrow{f} V$, the identity morphism of a 1-cell f by 1_f and a 2-cell by $f \overset{\rho}{\Longrightarrow} f'$. We also denote by $\circ$ (or juxtaposition) the vertical composition of 2-cells $f \overset{\rho}{\Longrightarrow} f' \overset{\tau}{\Longrightarrow} f''$ in $\mathcal{K}(U, V)$ and by $\odot$ the horizontal composition of 2-cells. In what follows, ($U \xrightarrow{1_U} U$, $1_U \overset{i_U}{\Longrightarrow} 1_U$) is the pair defined by the image of the unit functor from $\mathbf{1}$ to $\mathcal{K}(U, U)$, where $\mathbf{1}$ is the terminal object of the category of small categories. For more details we refer to [6, Chap. 7] or [47, Chap. XII].

Let U be 0-cell in a 2-category $\mathcal{K}$. Then $\mathcal{K}(U) := \mathcal{K}(U, U)$ is a monoidal category. The objects are 1-cells $U \to U$, morphisms are 2-cells, and the tensor product is given by horizontal composition of 2-cells. The unit is 1_U.

A monad in a 2-category $\mathcal{K}$ is an algebra in a monoidal category $\mathcal{K}(A)$, for a certain 0-cell A of $\mathcal{K}$. So a monad in $\mathcal{K}$ is (A, t, μ, η) consisting of an object A of $\mathcal{K}$, a 1-cell $A \xrightarrow{t} A$ in $\mathcal{K}$ and 2-cells $t \circ t \overset{\mu}{\Longrightarrow} t$ and $1_A \overset{\eta}{\Longrightarrow} t$ in $\mathcal{K}$ such that $\mu \circ (\mu \odot 1_t) = \mu \circ (1_t \odot \mu)$ and $\mu \circ (1_t \odot \eta) = 1_t = \mu \circ (\eta \odot 1_t)$.

If $\mathbb{A} = (A, t, \mu_t, \eta_t)$ and $\mathbb{B} = (B, s, \mu_s, \eta_s)$ are monads in a 2-category $\mathcal{K}$ then a monad morphism between $\mathbb{A}$ and $\mathbb{B}$ is a pair (f, ψ) with $A \xrightarrow{f} B$ a 1-cell in $\mathcal{K}$ and $s \circ f \overset{\psi}{\Longrightarrow} f \circ t$ a 2-cell in $\mathcal{K}$ such that the following equalities hold:

$$(1_f \odot \mu_t) \circ (\psi \odot 1_t) \circ (1_s \odot \psi) = \psi \circ (\mu_s \odot 1_f)\,,\ \psi \circ (\eta_s \odot 1_f) = 1_f \odot \eta_t.$$

To a 2-category $\mathcal{K}$ we can associate a new 2-category $EM(\mathcal{K})$, the Eilenberg–Moore category associated to $\mathcal{K}$; see [45]. The 0-cells in $EM(\mathcal{K})$ are monads in $\mathcal{K}$, 1-cells are the monad morphisms and 2-cells $(f, \psi) \overset{\rho}{\Longrightarrow} (g, \phi)$ are 2-cells $f \overset{\rho}{\Longrightarrow} gt$ in $\mathcal{K}$ obeying $(1_g \odot \mu_t)(\rho \odot 1_t)\psi = (1_g \odot \mu_t)(\phi \odot 1_t)(1_s \odot \rho)$.

The vertical composition of two 2-cells $(f, \psi) \overset{\rho}{\Longrightarrow} (g, \phi) \overset{\rho'}{\Longrightarrow} (h, \gamma)$ is given by $(f, \psi) \overset{\rho' \circ \rho}{\Longrightarrow} (h, \gamma)$, $\rho' \circ \rho := (1_h \odot \mu_t)(\rho' \odot 1_t)\rho$, while their horizontal composition is defined by $(g, \phi)(f, \psi){=}(gf, (1_g \odot \psi) \circ (\phi \odot 1_f))$, etc. and $gf \overset{\rho' \oslash \rho}{\Longrightarrow} g'f't$ given by $\rho' \oslash \rho{:=}(1_{g'} \odot 1_{f'} \odot \mu_t)(1_{g'} \odot \rho \odot 1_t)(1_{g'} \odot \psi)(\rho' \odot 1_f)$.

The identity morphism of the 1-cell (f, ψ) is $1_f \odot \eta_t$. For any monad $\mathbb{A} = (A, t, \mu_t, \eta_t)$ in $\mathcal{K}$ we have $(1_{\mathbb{A}}, i_{\mathbb{A}}) = ((1_A, 1_t), \eta_t)$.

2.4 *Quasi-Hopf Algebras*

We present, briefly, the definition an the basic properties of a quasi-Hopf algebra. For more information we refer to [15] or [30, 42, 48]. We work over a field k. All algebras, linear spaces, etc. will be over k; unadorned $\otimes$ means $\otimes_k$.

A quasi-bialgebra is a 4-tuple $(H, \Delta, \varepsilon, \Phi)$, where H is an associative algebra with unit 1_H, Φ is an invertible element in $H \otimes H \otimes H$, and $\Delta : H \to H \otimes H$ and $\varepsilon : H \to k$ are algebra homomorphisms such that Δ is coassociative up to conjugation by Φ and ε is counit for Δ; furthermore, Φ is a normalized 3-cocycle.

In what follows we denote $\Delta(h) = h_1 \otimes h_2$, for all $h \in H$, the tensor components of Φ by capital letters and the ones of Φ^{-1} by lower case letters.

H is called a quasi-Hopf algebra if, moreover, there exists an anti-morphism S of the algebra H and elements $\alpha, \beta \in H$ such that, for all $h \in H$, we have:

$$S(h_1)\alpha h_2 = \varepsilon(h)\alpha,\ h_1\beta S(h_2) = \varepsilon(h)\beta,\ X^1\beta S(X^2)\alpha X^3 = 1_H = S(x^1)\alpha x^2\beta S(x^3).$$

A Hopf algebra is a quasi-Hopf algebra with $\Phi = 1_H \otimes 1_H \otimes 1_H$ and $\alpha = \beta = 1_H$.

For a quasi-Hopf algebra H we introduce the following elements in $H \otimes H$:

$$p_R = p^1 \otimes p^2 := x^1 \otimes x^2\beta S(x^3) \text{ and } q_R = q^1 \otimes q^2 := X^1 \otimes S^{-1}(\alpha X^3)X^2. \quad (2)$$

Our definition of a quasi-Hopf algebra is different from the one of Drinfeld [30], as we do not ask S to be bijective. Anyway, the bijectivity of S will be implicitly understood in the case when S^{-1}, the inverse of S, appears is formulas or computations. By [12], S is always bijective, provided that H is finite dimensional.

The antipode S of a Hopf algebra is an anti-morphism of coalgebras. For a quasi-Hopf algebra H there is an invertible element $f = f^1 \otimes f^2 \in H \otimes H$, called the Drinfeld twist, such that $\varepsilon(f^1)f^2 = \varepsilon(f^2)f^1 = 1_H$ and $f\Delta(S(h))f^{-1} = (S \otimes S)(\Delta^{\mathrm{cop}}(h))$, for all $h \in H$, where $\Delta^{\mathrm{cop}}(h) = h_2 \otimes h_1$.

For H a quasi-bialgebra, the category of left H-modules ${}_H\mathcal{M}$ is monoidal. The tensor product is defined by $\otimes$ endowed with the H-module structure given by Δ and unit object k, considered as an H-module via ε; the associativity constraint is determined by Φ and the left and right unit constraints are given by the canonical isomorphisms in the category of k-vector spaces.

3 Frobenius (Monoidal) Algebras

3.1 The Classical Case

Throughout this subsection A is a finite dimensional algebra with unit 1_A over a field k and A^* is the k-linear dual of A; A^* is an A-bimodule via the A-actions $\langle a \rightharpoonup a^*, a'\rangle = \langle a^*, a'a\rangle$, $\langle a^* \leftharpoonup a, a'\rangle = \langle a^*, aa'\rangle$, for all $a^* \in A^*$ and $a, a' \in A$. Also, the multiplication of A induces on A an A-bimodule structure as well.

Recall that A is Frobenius if $l, r : A \to M_n(k)$ from (1) are equivalent representations. When A is Frobenius is recorded in the next proposition, whose proof can be found in [15, 46].

Proposition 1 *For a k-algebra A the following assertions are equivalent:*

(i) A is a Frobenius algebra;

(ii) There exists a pair (ϕ, e) consisting of a k-linear map $\phi : A \to k$ and an element $e = e^1 \otimes e^2 \in A \otimes A$ (formal notation, summation implicitly understood) such that $ae^1 \otimes e^2 = e^1 \otimes e^2 a$, for all $a \in A$, and $\phi(e^1)e^2 = \phi(e^2)e^1 = 1_A$;

(iii) A is isomorphic to A^* as a right A-module;

(iv) A is isomorphic to A^* as a left A-module;

(v) A has a k-coalgebra structure (Δ, ε) such that $\Delta : A \to A \otimes A$ is an A-bimodule morphism, where $A \otimes A \in {}_A\mathcal{M}_A$ via the multiplication of A;

(vi) There exists a bilinear map $B_r : A \times A \to k$ which is right non-degenerate and associative, that is $B_r(x, a) = 0$, for all $a \in A$, implies $x = 0$, and $B_r(ab, c) = B_r(a, bc)$, for all $a, b, c \in A$, respectively;

(vii) There exists a bilinear map $B_l : A \times A \to k$ which is left non-degenerate (i.e. $B_l(a, x) = 0$ for all $a \in A$ implies $x = 0$) and associative;

(viii) There exists a hyperplane (linear subspace of codimension 1) in A that does not contain either left or right non-zero ideals.

If (ϕ, e) is a Frobenius system for A (i.e. a pair as in Proposition 1 (ii)) we refer to ϕ as being a Frobenius morphism for A, and call e a Casimir element for A. A Frobenius system (ϕ, e) for A is unique up to an invertible element, in the sense that any other Frobenius system for A is of the form $(\phi \leftharpoonup d, e^1 \otimes d^{-1}e^2)$, for some invertible element $d \in A$ with inverse d^{-1}, or equivalently of the form $(d' \rightharpoonup \phi, e^1 d'^{-1} \otimes e^2)$, for some invertible element $d' \in A$ with inverse d'^{-1}; see [38]. The elements d, d' are related through the Nakayama automorphism of A.

If A is Frobenius via (ϕ, e), $\chi : A \to A$ defined by $a \rightharpoonup \phi = \phi \leftharpoonup \chi(a)$, for all $a \in A$, is an algebra morphism called the Nakayama automorphism. By [38],

$$\chi(a) = \phi(e^1 a)e^2, \ \ \forall\ a \in A, \tag{3}$$

and is invertible with inverse χ^{-1} given by $\chi^{-1}(a) = \phi(ae^2)e^1$, for all $a \in A$.

Frobenius algebras appear in different domains of mathematics. We next supply a list of examples that support this statement; see also [43, Sect. 2.2].

Examples 1 (1) Let R be a finite dimensional k-algebra and R^* the linear dual space of R. The k-vector space $A := R \oplus R^*$ endowed with the multiplication given by $(r, \varphi)(r', \varphi') = (rr', r \rightharpoonup \varphi' + \varphi \leftharpoonup r')$, for all $r, r' \in R$ and $\varphi, \varphi' \in R^*$, is a k-algebra with unit $(1_R, 0)$, where, as before, $\rightharpoonup$, $\leftharpoonup$ are the regular actions of R on R^* that turns R^* into an R-bimodule. According to [46, (16.62)], A is a Frobenius k-algebra via the bilinear form $B : A \times A \to k$ given by $B((r, \varphi), (r', \varphi')) = \varphi'(r) + \varphi(r')$, for all $(r, \varphi), (r', \varphi') \in A$; see Proposition 1 (vi).

Thus Frobenius k-algebras can be build out of any finite dimensional k-algebra. Actually, any finite dimensional k-vector space V becomes a Frobenius algebra if it is considered with multiplication $v_i v_j = \delta_{i,j} v_i$, for all $1 \leq i, j \leq n$, extended by linearity ($\{v_1, \cdots, v_n\}$ is a basis of V and $\delta_{i,j}$ stands for Kronecker's delta). Indeed, V is associative and unital with unit $\sum_{i=1}^{n} v_i$, and $B : V \times V \ni (v_i, v_j) \mapsto \delta_{i,j} \in k$ is associative and non-degenerate.

(2) For G a finite group of order $\mid G \mid$ denote by $R(G)$ the class functions on G, i.e. functions from G to the field $\mathbb{C}$ of complex numbers that are constant on each conjugacy class of G. Then $R(G)$ is a $\mathbb{C}$-algebra with multiplication given by the convolution product and, moreover, a Frobenius algebra via the bilinear form $B : R(G) \times R(G) \to \mathbb{C}$ given, for $\chi, \chi' \in R(G)$, by $B(\chi, \chi') = \frac{1}{|G|} \sum_{g \in G} \chi(g)\chi'(g^{-1})$.

(3) The matrix algebra $M_n(k)$ is Frobenius. Consequently, for A a k-algebra, the matrix algebra $M_n(A)$ is Frobenius if and only if A is Frobenius; this implies that semisimple algebras are Frobenius. We recall that a Frobenius system for $M_n(k)$ is (ϕ=Tr, $e = \sum_{i,j=1}^{n} E_{ij} \otimes E_{ji}$), where $\mathrm{Tr} : M_n(k) \to k$ is the trace morphism and $\{E_{ij} \mid 1 \leq i, j \leq n\}$ is the canonical basis of $M_n(k)$.

If A is Frobenius then so is $M_n(A)$ since $M_n(A) \simeq M_n(k) \otimes A$ as algebras and the tensor product algebra behaves well with respect to the Frobenius property. The converse follows from [24, Proposition 7.2].

Any finite dimensional division k-algebra D is Frobenius, see [46, (16.59)]. Thus by the results above and the Artin–Wedderburn theorem we get that any semisimple algebra is Frobenius.

(4) Any finite dimensional (quasi-)Hopf algebra H is Frobenius. By [12, 36], a Frobenius system for H is $(\lambda \circ S^{-1}, q^1 t_1 p^1 \otimes S(q^2 t_2 p^2))$, with λ a non-zero left cointegral on H and t a non-zero left integral in H satisfying $\lambda(S^{-1}(t)) = 1$. We refer to [15, Proposition 7.59] for definitions and details.

Note that, if H is a finite dimensional weak Hopf algebra then H is not necessarily Frobenius, unless the semisimple base algebra of H has all its matrix blocks of the same dimension; see [35].

(5) Suppose that the zero set $Z(f) \subset \mathbb{C}^n$ of a polynomial $f \in \mathbb{C}[X_1, \cdots, X_n]$ has an isolated singularity at $0 \in \mathbb{C}^n$, that is the quotient $A = \frac{\mathbb{C}[X_1, \cdots, X_n]}{(f_1, \cdots, f_n)}$ is finite dimensional, where $f_i = \frac{\partial f}{\partial X_i}$, for all $1 \leq i \leq n$. By integrating around the singularity along the real n-ball $B := \{z \mid f_i(z) = \rho\}$, for some small $\rho > 0$, we get a functional (called the residue of f),

$$\mathrm{res}_f : A \ni g \mapsto \left(\frac{1}{2\pi i}\right)^{2n} \int_B \frac{g(z) \cdot dX_1 \wedge \cdots \wedge dX_n}{f_1(z) \cdots f_n(z)} \in \mathbb{C}.$$

To res_f one can associate a bilinear map $B_f : A \times A \to \mathbb{C}$, given by $B_f(g, h) = \mathrm{res}_f(gh)$, for all $g, h \in A$. By the local duality theorem [32, p. 659] we have that B is an associative and non-degenerate bilinear form, and therefore A is a (commutative) Frobenius algebra.

(6) Let $H^*(X) = \bigoplus_{i=0}^n H^i(X)$ be the de Rham cohomology of a real compact oriented n-dimensional manifold X, where $H^i(X)$ consists of closed differentiable i-forms modulo the exact ones. Then the wedge product $\wedge$ induces on $H^*(X)$ an $\mathbb{R}$-algebra structure, and the integration $\int_X$ provides a nontrivial linear map from $H^*(X)$ to $\mathbb{R}$ which can be used to construct, for all $1 \le i \le n$,

$$H^i(X) \times H^{n-i}(X) \ni (\alpha, \beta) \mapsto \int_X \alpha \wedge \beta \in \mathbb{R},$$

a bilinear morphism. It induces $B : H^*(X) \times H^*(X) \to \mathbb{R}$, a bilinear morphism which is associative since $\wedge$ is so. Furthermore, B is non-degenerate since the Poincaré duality guarantees that $H^i(X)$ is isomorphic to the dual space of $H^{n-i}(X)$, for all $1 \le i \le n$. Thus $H^*(X)$ is a real Frobenius algebra.

For instance, $A_1 := \mathbb{R}\langle u, v \mid u^2 = v^2 = 0,\ uv = -vu\rangle$ and $A_2 := \mathbb{R}\langle h \mid h^{n+1} = 0\rangle$ are real Frobenius algebras since A_1 identifies to de Rham cohomology of a torus, while A_2 identifies to de Rham cohomology of $\mathbb{CP}^n := \frac{\mathbb{C}^{n+1}\backslash\{0\}}{\mathbb{C}\backslash\{0\}}$, where the group $\mathbb{C}\backslash\{0\}$ acts canonically on $\mathbb{C}^{n+1}\backslash\{0\}$.

3.2 The Monoidal Case

Many of the characterizations in Proposition 1 extend to monoidal categories C, not necessarily rigid. In fact, the rigidity of the algebra follows from its property of being Frobenius, a result parallel to one stating that a Frobenius k-algebra is necessarily finite dimensional. Nevertheless, "locally" we need that if A is an algebra in a monoidal category C such that A has a left dual object A^* (resp. a right dual object ${}^*A$) then A^* (resp. ${}^*A$) is a right (resp. left) A-module via

$$\begin{aligned} &\nu^A_{A^*} = (\mathrm{ev}_A A^*)(A^* \underline{m}_A A^*)((A^*)A\mathrm{coev}_A) \\ &(\text{resp. } \mu^A_{{}^*A} = ({}^*A\mathrm{ev}'_A)({}^*A\underline{m}_A {}^*A)(\mathrm{coev}'_A A({}^*A))), \end{aligned} \tag{4}$$

where for simplicity we assumed that C is strict monoidal.

The next result is [21, Theorem 5.1], and an algebra A in C satisfying one of the equivalences below is called Frobenius algebra (in C). Recall that for X an arbitrary object of C, a morphism $\mathbb{B} : A \otimes A \to X$ in C is called associative if, up to the associativity constraints of C, we have $\mathbb{B}(\underline{m}_A \otimes \mathrm{Id}_A) = \mathbb{B}(\mathrm{Id}_A \otimes \underline{m}_A)$.

Theorem 1 *For an algebra A in C the following assertions are equivalent:*

(i) There exist morphisms $\vartheta : A \to \underline{1}$ and $e : \underline{1} \to A \otimes A$ such that, up to the constraints of C, the following equalities hold: $(\underline{m}_A \otimes \mathrm{Id}_A)(\mathrm{Id}_A \otimes e) = (\mathrm{Id}_A \otimes \underline{m}_A)(e \otimes \mathrm{Id}_A)$, $(\vartheta \otimes \mathrm{Id}_A)e = \underline{\eta}_A = (\mathrm{Id}_A \otimes \vartheta)e$.

(ii) A admits a left dual A^* and A is isomorphic to A^* as a right A-module;

(iii) A admits a right dual ${}^*A$ and A is isomorphic to ${}^*A$ as a left A-module;

(iv) A admits a coalgebra structure $(A, \underline{\Delta}_A, \underline{\varepsilon}_A)$ in C such that $\underline{\Delta}_A$ is an A-bimodule map, where $A, A \otimes A \in {}_AC_A$ via the multiplication $\underline{m}_A$ of A;

(v) A has a right dual ${}^*A$ and there exists $B : A \otimes A \to \underline{1}$ that is associative and such that $\Phi^r_B := (\mathrm{Id}_{{}^*A} \otimes B)(\mathrm{coev}'_A \otimes \mathrm{Id}_A) : A \to {}^*A$ is an isomorphism;

(vi) A has a left dual A^* and there exists $B : A \otimes A \to \underline{1}$ in C that is associative and such that $\Phi^l_B := (B \otimes \mathrm{Id}_{A^*})(\mathrm{Id}_A \otimes \mathrm{coev}_A) : A \to A^*$ is an isomorphism;

(vii) We have an adjunction $(\rho, \lambda) : A \dashv A$ with $\lambda : A \otimes A \to \underline{1}$ associative;

(viii) We have an adjunction $(\rho, \vartheta\underline{m}_A) : A \dashv A$, for some $\vartheta : A \to \underline{1}$.

By analogy to the classical case, a pair (ϑ, e) as in Theorem 1 (i) is called a Frobenius system for A; e, ϑ will be called the Casimir, respectively Frobenius, morphism of A. As we will next see (fact pointed out also in [31]), a Frobenius system is unique up to a convolution invertible morphism from $\underline{1}$ to A in C.

Denote $A_0 := \mathrm{Hom}_C(\underline{1}, A)$. The algebra structure on A induces an algebra structure (monoid) on A_0: $\underline{m}_{A_0}(\widetilde{a}, \widetilde{b}) := \underline{m}_A(\widetilde{a} \otimes \widetilde{b}) : \underline{1} \to A$, for all $\widetilde{a}, \widetilde{b} \in A_0$; the unit is $\underline{\eta}_A$. The elements of A_0 invertible with respect to $\underline{m}_{A_0}$ form a group, which will be denoted by $A_0^\times$. For $\widetilde{a} \in A_0^\times$ we denote its inverse by $\widetilde{a}^{-1}$.

To $\widetilde{a} \in A_0$ we associate $l_{\widetilde{a}}, r_{\widetilde{a}} : A \to A$ given by $l_{\widetilde{a}} = \underline{m}_A(\widetilde{a} \otimes \mathrm{Id}_A)$ and $r_{\widetilde{a}} = \underline{m}_A(\mathrm{Id}_A \otimes \widetilde{a})$. Then the assigments $l, r : A_0 \to \mathrm{End}_C(A) := \mathrm{Hom}_C(A, A)$ mapping $\widetilde{a} \in A_0$ to $l_{\widetilde{a}}$ and respectively $r_{\widetilde{a}}$ are the analogous of the left and right regular representations l, r of a k-algebra A. If $\widetilde{a} \in A_0^\times$, $l_{\widetilde{a}}$ and $r_{\widetilde{a}}$ are isomorphisms in C; furthermore, $\mathrm{ad}_{\widetilde{a}} := l_{\widetilde{a}^{-1}} r_{\widetilde{a}} = r_{\widetilde{a}} l_{\widetilde{a}^{-1}}$ is an algebra automorphism of A.

Proposition 2 *Let (ϕ, e), (ψ, f) be two Frobenius systems for a Frobenius algebra A in C. Then $\widetilde{a} := (\psi \otimes \mathrm{Id}_A)e : \underline{1} \to A$ is invertible in A_0 with inverse given by $\widetilde{a}^{-1} := (\phi \otimes \mathrm{Id}_A)f : \underline{1} \to A$ and $f = (\mathrm{Id}_A \otimes l_{\widetilde{a}^{-1}})e$, $\psi = \phi l_{\widetilde{a}}$.*

To define the Nakayama automorphism $\mathcal{N}$ for a monoidal Frobenius algebra A we need extra conditions on A. These are needed since A^* (or ${}^*A$) is not, in general, an A-bimodule in C. This problem can be overcome if we assume ${}^*A = A^*$; then we are looking for conditions on A for which $A^* = {}^*A$ becomes an A-bimodule in C via the left and right A-actions considered in (4). The assumption ${}^*A = A^*$ means that we have adjunctions $(\mathrm{coev}_A, \mathrm{ev}_A) : A^* \dashv A$ and $(\mathrm{coev}'_A, \mathrm{ev}'_A) : A \dashv A^*$ in C, and this allows to define, for (ϑ, e) a Frobenius system of A, $\mathcal{N} = (\mathrm{ev}_A A)(A^* e)(A^*(\vartheta\underline{m}_A))(\mathrm{coev}'_A A) : A \to A$ which has the property that $\vartheta\underline{m}_A(\mathcal{N}A) = \mathrm{ev}_A(A^*(\vartheta\underline{m}_A)A)(\mathrm{coev}'_A A^2)$. When $C = {}_k\mathcal{M}$ the two equalities above read as (3) and the defining property of $\mathcal{N}$, respectively.

Coming back to the general case, $\mathcal{N}$ is an isomorphism in C with inverse given by $\mathcal{N}^{-1} = (A\mathrm{ev}'_A)(eA^*)((\vartheta\underline{m}_A)A^*)(A\mathrm{coev}_A)$, fact that follows from the defining properties of the Frobenius system (ϑ, e); so no extra condition on A is required.

We should impose conditions on A only to make $\mathcal{N}$ an algebra morphism in C, and therefore an algebra automorphism of A. Namely,

$$\begin{aligned}(a)\ & (A^{*2}\mathrm{ev}_A)(A^{*2}\underline{m}_A A^*)(A^*\mathrm{coev}'_A AA^*)(\mathrm{coev}'_A A^*) \\ & \qquad = (\mathrm{ev}_A A^{*2})(A^*\underline{m}_A A^{*2})((A^*)A\mathrm{coev}_A A^*)A^*\mathrm{coev}_A, \qquad (5) \\ (b)\ & \mathrm{ev}_A(A^*\mathrm{ev}_A A)(A^{*2}e) = \mathrm{ev}'_A(A\mathrm{ev}'_A A^*)(eA^{*2}).\end{aligned}$$

The conditions above appear from the following consistency problem: if $X \in C$ has X^* as both left and right dual object then for any $f : X \to A$ (resp. $g : A \to X$) one has ${}^*f, f^* : A^* \to X^*$ (resp. ${}^*g, g^* : X^* \to A^*$) morphisms in C, and so we should ask for ${}^*f = f^*$ (resp. ${}^*g = g^*$). So, from now on, an object with these properties will be called a sovereign object. We chose this terminology because if C is a sovereign category (i.e. C is left and right rigid such that the corresponding left and right dual functors $(-)^*, {}^*(-) : C \to \overline{C^{op}}$ are equal as strong monoidal functors) then any object of C is sovereign in our sense. Now, if A is a sovereign object of C then (5.a) follows from $\underline{m}_A^* = {}^*\underline{m}_A$, while (5.b) follows from $e^* = {}^*e$ (clearly $\underline{1}$ is a self left and right dual object of C).

The proof of the next result is based on the equations in (5); further details can be found in the proof of [31, Proposition 18].

Proposition 3 *Let A be a Frobenius algebra in C with Frobenius system (ϑ, e) and $\mathcal{N} : A \to A$ the associated Nakayama isomorphism. If A is a sovereign object of C then $\mathcal{N}$ is an algebra automorphism of A in C. Furthermore, in this hypothesis $A^* = {}^*A \in {}_AC_A$ via the actions defined in* (4).

With notation as in Proposition 2, let $\mathcal{N}_{\widetilde{a}}$ be the Nakayama morphism of A associated to (ψ, f). If A is a sovereign, by applying this to $l_{\widetilde{a}} : A \to A$ we get $\mathcal{N}_{\widetilde{a}} = \mathrm{ad}_{\widetilde{a}}\mathcal{N}$; more details can be found in [31].

Remark 1 For an algebra A in C which is a sovereign object, one can check when $A \simeq A^*$ in ${}_AC_A$. This led in [31] to the notion of symmetric algebra in C: A is symmetric iff A is Frobenius with $\mathcal{N}$ inner, i.e. $\mathcal{N} = \mathrm{ad}_{\widetilde{a}}$ for some $\widetilde{a} \in A_0^\times$.

We end this section by presenting examples of monoidal Frobenius algebras.

Examples 2 (1) The category Sets of sets and functions is monoidal with the tensor product the Cartesian product and the unit object $\{*\}$, a one element set. An algebra in Sets is a monoid, and the only monoid carrying out a Frobenius structure is the trivial one.

(2) Let k be a commutative ring and ${}_k\mathcal{M}$ the category of left k-modules and left k-linear maps. ${}_k\mathcal{M}$ is monoidal with $\otimes = \otimes_k$, unit object $\underline{1} = k$ and the obvious a, l, r. An algebra in ${}_k\mathcal{M}$ is an algebra R over k. R is Frobenius in the monoidal sense if and only if it is Frobenius in the classical sense.

(3) Let k be a commutative ring and R a k-algebra. The category ${}_R\mathcal{M}_R$ of R-bimodules and R-bilinear morphisms is monoidal with $\otimes = \otimes_R$, $\underline{1} = R$ and the natural morphisms a, l, r. An algebra in ${}_R\mathcal{M}_R$ is a so called R-ring.

There is a one to one correspondence between R-rings and k-algebra morphisms $i : R \to S$. Furthermore, the R-ring S is Frobenius if and only if the corresponding extension $i : R \to S$ is so (see the next section).

(4) Denote by **FdHilb** the category of complex finite dimensional Hilbert spaces and continuous linear maps. Since the objects of **FdHilb** are finite dimensional, the morphisms in **FdHilb** are actually linear maps over the field of complex numbers $\mathbb{C}$. Also, the completeness condition with respect to the norm defined by the inner product is insignificant for finite dimensional Hilbert spaces.

FdHilb is monoidal with the monoidal structure inherited from the monoidal structure of ${}_{\mathbb{C}}\mathcal{M}$: it is strict monoidal with $\otimes = \otimes_{\mathbb{C}}$ and $\underline{1} = \mathbb{C}$. Particular classes of Frobenius algebras in **FdHilb** were studied in [22]. Note that any basis of a finite dimensional complex vector space gives rise to a commutative special Frobenius algebra, and that the orthogonal bases of a finite dimensional complex Hilbert space $\mathcal{H}$ are in a one to one correspondence with the so called commutative †-Frobenius algebra structures on $\mathcal{H}$; for details see [22].

(5) To a category $\mathcal{C}$ we can associate a monoidal category $[\mathcal{C}, \mathcal{C}]$, called the category of endo-functors. Frobenius algebras in $[\mathcal{C}, \mathcal{C}]$ are called Frobenius monads, and they can be characterized as adjunctions with some additional structures.

The category $[\mathcal{C}, \mathcal{C}]$ has as objects functors $F : \mathcal{C} \to \mathcal{C}$ and morphisms natural transformations; the composition $\nu \circ \mu$ of $F \overset{\mu}{\to} G \overset{\nu}{\to} H$ is the "vertical" composition, i.e. $(\nu \circ \mu)_X := \nu_X \circ \mu_X$, for any $X \in \mathcal{C}$, and the identity of F is the identity natural transformation 1_F. The composition of functors defines a tensor product $\otimes$ on $[\mathcal{C}, \mathcal{C}]$, and for $\mu : F \to G$ and $\mu' : F' \to G'$ we have $\mu \otimes \mu' : F \otimes F' = F \circ F' \to G \otimes G' = G \circ G'$ defined by $(\mu \otimes \mu')_X = \mu_{G'(X)} F(\mu'_X) = G(\mu'_X)\mu_{F'(X)} : FF'(X) \to GG'(X)$, for all $X \in \mathcal{C}$; $\mu \otimes \mu'$ is called the "horizontal" composition of μ and μ'. The identity for the composition $\otimes$ is $1_{\mathcal{C}} : \mathrm{Id}_{\mathcal{C}} \to \mathrm{Id}_{\mathcal{C}}$, the identity natural transformation of the identity functor $\mathrm{Id}_{\mathcal{C}}$ to itself. All these structures turn $[\mathcal{C}, \mathcal{C}]$ into a strict monoidal category.

An algebra in $[\mathcal{C}, \mathcal{C}]$, called a monad in $\mathcal{C}$, is a triple $(T, \underline{m}, \underline{\eta})$ with $T : \mathcal{C} \to \mathcal{C}$ a functor and $\underline{m} : T \circ T \to T$ and $\underline{\eta} : \mathrm{Id}_{\mathcal{C}} \to T$ natural transformations such that $\underline{m}_X \underline{m}_{T(X)} = \underline{m}_X T(\underline{m}_X)$ and $\underline{m}_X \underline{\eta}_{T(X)} = \underline{m}_X T(\underline{\eta}_X) = \mathrm{Id}_{T(X)}$, for all $X \in \mathcal{C}$.

Any monad T is completely determined by a pair of adjoint functors. Indeed, if $(\eta, \varepsilon) : F \dashv G$ is an adjunction between $\mathcal{C} \underset{G}{\overset{F}{\rightleftarrows}} \mathcal{D}$ then $T := GF : \mathcal{C} \to \mathcal{C}$ is a monad with unit η and multiplication $\underline{m} = 1_G \otimes \varepsilon \otimes 1_F$, that is $\underline{m}_X = G(\varepsilon_{F(X)})$, for all $X \in \mathcal{C}$. Conversely, let $\mathcal{C}^T$ be the category of T-algebras in $\mathcal{C}$: objects are T-algebras, i.e. pairs (M, μ_M) consisting of $M \in \mathcal{C}$ and a morphism $\mu_M : T(M) \to M$ such that $\mu_M T(\mu_M) = \mu_M \underline{m}_M$ and $\mu_M \underline{\eta}_M = \mathrm{Id}_M$, while the morphisms $f : (M, \mu_M) \to (N, \mu_N)$ are $f : M \to N$ in $\mathcal{C}$ satisfying $\mu_N T(f) = f \mu_M$.

By [47, Chap. VI Theorem 1], the forgetful functor $G^T : \mathcal{C}^T \to \mathcal{C}$ admits a left adjoint $F^T : \mathcal{C} \to \mathcal{C}^T$ defined as follows. For $X \in \mathcal{C}$, $F^T(X) = (T(X), \underline{m}_X)$; F^T sends the morphism $f : X \to Y$ in $\mathcal{C}$ to $T(f)$, a morphism now from $F^T(X)$ to $F^T(Y)$. The unit of the adjunction $F^T \dashv G^T$ is the unit $\underline{\eta}$ of the monad $\mathcal{C}$ and the counit $\varepsilon^T : F^T G^T \to C^T$ is defined by $\varepsilon^T_{(M,\mu_M)} = \mu_M$, for all $(M, \mu_M) \in C^T$. Then the monad associated to the adjunction $F^T \dashv G^T$ is just T.

A monad T is Frobenius if and only if there exists natural transformations $e : 1_C \to T^2$ (the Casimir morphism) and $\vartheta : T \to 1_C$ (the Frobenius morphism) obeying $\underline{m}_{T(X)}T(e_X) = T(\underline{m}_X)e_{T(X)}$, $\vartheta_{T(X)}e_X = \underline{\eta}_X = T(\vartheta_X)e_X$, for $X \in C$.

(6) A monad (A, t, μ, η) in a 2-category $\mathcal{K}$ is Frobenius if (t, μ, η) is a Frobenius algebra in the monoidal category $\mathcal{K}(A)$. Thus (A, t, μ, η) in $\mathcal{K}$ is Frobenius if and only if there exist 2-cells $\vartheta : t \Rightarrow 1_A$ and $e : 1_A \Rightarrow tt$ such that $(\mu \odot 1_t)(1_t \odot e) = (1_t \odot \mu)(e \odot 1_t)$, $(\vartheta \odot 1_t)e = \eta = (1_t \odot \vartheta)e$.

The existence of a left (right) dual for an object $u : A \to A$ in the monoidal category $\mathcal{K}(A)$ reduces to the existence of a left (right) adjunction for u in the 2-categorical sense. Namely, u has a left dual if there is a 1-cell $v : A \to A$ and 2-cells $\iota : 1_A \Rightarrow uv$ and $j : vu \Rightarrow 1_A$ such that $(1_u \odot j)(\iota \odot 1_u) = 1_u$ and $(j \odot 1_v)(1_v \odot \iota) = 1_v$. In this case we say also that u is a right adjoint to v and denote this adjunction as before, $(\iota, j) : v \dashv u$.

By Theorem 1 we have that the assertions below are characterizations for a monad $\mathbb{A} = (A, t, \mu, \eta)$ in $\mathcal{K}$ to be Frobenius:

(i) (t, μ, η) is a Frobenius algebra in the monoidal category $\mathcal{K}(A)$;

(ii) t admits a coalgebra structure in the monoidal category $\mathcal{K}(A)$, say $(t, \delta : t \Rightarrow tt, \varepsilon : t \Rightarrow 1_A)$, such that $(1_t \odot \mu)(\delta \odot 1_t) = \delta\mu = (\mu \odot 1_t)(1_t \odot \delta)$;

(iii) There exists an adjunction $(\rho, \lambda) : t \dashv t$ such that $\lambda(\mu \odot 1_t) = \lambda(1_t \odot \mu)$;

(iv) There exists an adjunction $(\rho, \vartheta\mu) : t \dashv t$, with $\vartheta : t \Rightarrow 1_A$ a 2-cell in $\mathcal{K}$. □

The next goal is to study Frobenius algebras in categories of (co)representations, so the next subsections can be seen as a continuation of Examples 2.

3.3 Frobenius Algebras in Vect^G

Let G be a group and let k be a field. If we denote by Vect^G the category of G-graded k-vector spaces and k-linear morphisms that preserve the gradings then Vect^G is a monoidal category. For $V = \oplus_{g\in G} V_g$ and $W = \oplus_{g\in G} W_g$ two G-graded k-vector spaces their tensor product is the same as in ${}_k\mathcal{M}$, and it is considered as a G-graded vector space via the grading given by $(V \otimes W)_g = \oplus_{\sigma\tau=g} V_\sigma W_\tau$, for all $g \in G$. On morphisms the tensor product in Vect^G is defined exactly as in ${}_k\mathcal{M}$. The unit object is k viewed as a G-graded vector space in the trivial way: $k_e = k$ and $k_g = 0$, for all $G \ni g \neq e$, where e is the neutral element of G. The remarkable fact is that for Vect^G the associativity constraints a are in a one to one correspondence with the normalized 3-cocycles ϕ on G, while the left and right unit constraints are always equal with those of ${}_k\mathcal{M}$. This monoidal structure on Vect^G will be denoted by Vect^G_ϕ.

An algebra $A = \oplus_{g\in G} A_g$ in Vect^G_ϕ is Frobenius if there exist a G-graded morphism $\vartheta : A \to k$ (so $\vartheta(A_g) = 0$, for all $G \ni g \neq e$) and an element $e = \sum_{g\in G} a_g \otimes b_g \in \oplus_{g\in G} A_g \otimes A_{g^{-1}}$ such that $\vartheta(a_e)b_e = \vartheta(b_e)a_e = 1$ and for all homogeneous elements $x \in A$ we have that $\sum_{g\in G} \phi^{-1}(|\, x\,|, g, g^{-1})xa_g \otimes b_g = \sum_{g\in G} \phi(g, g^{-1},$

$| x |)a_g \otimes b_g x$. Particular examples of such algebras are Cayley–Dickson and Clifford algebras; see [9, 10]. In fact, they are Frobenius algebras in $\mathrm{Vect}^G_{\Delta_2(F^{-1})}$ of type $k_F[G]$, where $k_F[G]$ is the k-vector space $k[G]$ endowed with the multiplication $x \bullet y = F(x, y)xy$, for all $x, y \in G$, extended by linearity, and $\Delta_2(F^{-1})$ is the (coboundary) 3-cocycle on G given, for all $x, y, z \in G$, by $\Delta_2(F^{-1})(x, y, z) = F(y, z)^{-1}F(xy, z)F(x, yz)^{-1}F(x, y)$. Also, $F : G \times G \to k\backslash\{0\}$ is a pointwise invertible map satisfying $F(e, x) = F(x, e) = 1$, for all $x \in G$, and G is a finite group having the order $| G |$ non-zero in k.

If $\{P_g\}_{g\in G}$ is the basis in $k_F[G]^*$ dual to the basis $\{g \mid g \in G\}$ of $k_F[G]$ then (ϑ, e) consisting of $\vartheta =| G | P_e$ and $e = \frac{1}{|G|}\sum_{g\in G} F(g, g^{-1})^{-1}g \otimes g^{-1}$ is a Frobenius system for $k_F[G]$ in $\mathrm{Vect}^G_{\Delta_2(F^{-1})}$.

We should mention that Frobenius algebras in Vect^G (the category Vect^G_ϕ with ϕ equals the trivial 3-cocycle on G) have their own theory, deeply developed in [24]. In this case Vect^G is just the category of G-graded vector spaces and an algebra in Vect^G is a G-graded k-algebra. The expectation is to have an algebra A in Vect^G Frobenius if and only if A_e is Frobenius as a k-algebra; this happens only if A is strongly graded, i.e. $A_g A_h = A_{gh}$, for all $g, h \in G$ (see [24, Corollary 4.2] or the alternative proof that will be given in the next section). In general, by [24, Corollary 4.4], an algebra A in Vect^G is Frobenius if and only if A_e is a Frobenius k-algebra and A is left (or right) e-faithful: for any non-zero $a_g \in A_g$, $g \in G$, we have $A_{g^{-1}}a_g \neq 0$ (resp. $a_g A_{g^{-1}} \neq 0$). Thus G-graded semisimple k-algebras (i.e. semisimple algebras in Vect^G) are G-graded Frobenius (Frobenius algebras in Vect^G); see [24, Corollary 4.5]. Actually, by [26, Theorem A] we have that G-graded semisimple k-algebras are, moreover, G-graded symmetric k-algebras (this means symmetric algebras in the sovereign category Vect^G).

3.4 *Frobenius Algebras in Categories of Representations*

Let H be a bialgebra over a field k and ${}_H\mathcal{M}$ the category of its left representations. We have seen that ${}_H\mathcal{M}$ is a monoidal category.

An algebra in ${}_H\mathcal{M}$ is a k-algebra A with the multiplication and the unit morphism left H-linear morphisms. Hence A is Frobenius in ${}_H\mathcal{M}$ if there exist $\vartheta : A \to k$ in ${}_H\mathcal{M}$ and $e = e^1 \otimes e^2 \in A \otimes A$ such that $\vartheta(e^1)e^2 = \vartheta(e^2)e^1 = 1$ and

$$h_1 \cdot e^1 \otimes h_2 \cdot e^2 = \varepsilon(h)e,\ \forall\, h \in H\ ;\ ae^1 \otimes e^2 = e^1 \otimes e^2 a,\ \forall\, a \in A. \tag{6}$$

Clearly a Frobenius algebra in ${}_H\mathcal{M}$ is a Frobenius k-algebra. The converse is not true since for $H = k[G]^*$, the dual of a finite group algebra, the categories ${}_{k[G]^*}\mathcal{M}$ and Vect^G identifies as monoidal categories, and so Frobenius algebras in ${}_{k[G]^*}\mathcal{M}$ are precisely G-graded Frobenius algebras. By [24, Example 6.1], the algebra A from Examples 1 (1) is a Frobenius k-algebra, $\mathbb{Z}_2$-graded algebra but not $\mathbb{Z}_2$-graded

Frobenius algebra since A is not left (or right) $\widehat{0}$-faithful. Nevertheless, we can prove the following.

Proposition 4 *If H is a Hopf algebra with the underlying algebra semisimple then an algebra A in ${}_H\mathcal{M}$ is Frobenius if and only if A is k-Frobenius via a Frobenius system (ϕ, e) with ϕ an H-linear morphism.*

Proof We must prove only the converse. For this, let $(\phi, e^1 \otimes e^2)$ be a Frobenius system for A in ${}_k\mathcal{M}$ such that $\phi : A \to k$ obeys $\phi(h \cdot a) = \varepsilon(h)\phi(a)$, for all $h \in H$, $a \in A$. As H is semisimple, one can find a left integral $t \in H$ (which is at the same time right integral in H) such that $\varepsilon(t) = 1$; see [27]. So $ht = th = \varepsilon(h)t$,

$$t_1 \otimes \mathfrak{g}S(t_2) = t_2 \otimes S^{-1}(t_1) \text{ and } ht_1 \otimes t_2 = t_1 \otimes S^{-1}(h)t_2, \tag{7}$$

for all $h \in H$, where S^{-1} is the inverse antipode S (which always exists, see [27, Lemma 5.3.1]) and $\mathfrak{g} := \lambda(S^{-1}(t_2))S^{-1}(t_1)$ is the modular element of H, and where λ is a left integral on H obeying $\lambda(S^{-1}(t)) = 1$. Recall that λ is determined by $\lambda(h_2)h_1 = \lambda(h)1_H$, for all $h \in H$, the space of left (right) integrals in/on H is one dimensional, and $\lambda(h_1)h_2 = \lambda(h)\mathfrak{g}$, for all $h \in H$. Also, $\mathfrak{g}$ is a grouplike element, i.e. $\Delta(\mathfrak{g}) = \mathfrak{g} \otimes \mathfrak{g}$, $\varepsilon(\mathfrak{g}) = 1$ and $\mathfrak{g}$ is invertible with $\mathfrak{g}^{-1} = S(\mathfrak{g})$.

We can prove now that $(\phi, t_1 \cdot e^1 \otimes t_2\mathfrak{g}^{-1} \cdot e^2)$ is a Frobenius system for A in ${}_H\mathcal{M}$, and so A is indeed a Frobenius algebra in ${}_H\mathcal{M}$. In fact, for $a \in A$, we have

$$\begin{aligned}
a(t_1 \cdot e^1) \otimes t_2\mathfrak{g}^{-1} \cdot e^2 &= (t_2 S^{-1}(t_1) \cdot a)(t_3 \cdot e^1) \otimes t_4\mathfrak{g}^{-1} \cdot e^2 \\
&= t_2 \cdot [(S^{-1}(t_1) \cdot a)e^1] \otimes t_3\mathfrak{g}^{-1} \cdot e^2 \\
&= t_2 \cdot e^1 \otimes t_3\mathfrak{g}^{-1} \cdot [e^2(S^{-1}(t_1) \cdot a)] \\
&= t_2 \cdot e^1 \otimes (t_3\mathfrak{g}^{-1} \cdot e^2)(t_4\mathfrak{g}^{-1}S^{-1}(t_1) \cdot a) \\
&\overset{(5)}{=} (t_1)_1 \cdot e^1 \otimes ((t_1)_2\mathfrak{g}^{-1} \cdot e^2)((t_1)_3 S(t_2) \cdot a) \\
&= t_1 \cdot e^1 \otimes (t_2\mathfrak{g}^{-1} \cdot e^2)(t_3 S(t_4) \cdot a) \\
&= t_1 \cdot e^1 \otimes (t_2\mathfrak{g}^{-1} \cdot e^2)a.
\end{aligned}$$

Hence $t_1 \cdot e^1 \otimes t_2\mathfrak{g}^{-1} \cdot e^2$ is a Casimir element (because this time it is actually an element) for A in ${}_H\mathcal{M}$; the fact that it satisfies the first relation in (6) follows from the definition of a left integral in H. So whether or not H is semisimple to any Casimir element of A in ${}_k\mathcal{M}$ one can associate a Casimir element for A in ${}_H\mathcal{M}$. We need H semisimple (i.e. $\varepsilon(t) = 1$) to have $\phi(t_1 \cdot e^1)t_2\mathfrak{g}^{-1} \cdot e^2 = \phi(e^1)t\mathfrak{g}^{-1} \cdot e^2 = \varepsilon(\mathfrak{g}^{-1})t \cdot 1_A = \varepsilon(t)1_A = 1_A$, and similarly $\phi(t_2\mathfrak{g}^{-1} \cdot e^2)t_1 \cdot e^1 = \varepsilon(t)1_A = 1_A$. $\square$

We end with this class of examples by mentioning that a finite dimensional Hopf algebra H is not only a k-Frobenius algebra, it is also a Frobenius algebra in the category of left representations of its linear dual space. For this, recall that H^* is as well a Hopf algebra and that H is an algebra in ${}_{H^*}\mathcal{M}$ via the underlying algebra structure of H and the left H^*-action $\rightharpoonup$ defined by $h^* \rightharpoonup h = h^*(h_2)h_1$.

Proposition 5 *If H is finite dimensional, H is a Frobenius algebra in ${}_{H^*}\mathcal{M}$.*

Proof The second relation in (7) guarantees that $t_1 \otimes S(t_2)$ is a Casimir element for H in ${}_k\mathcal{M}$, provided that t is a non-zero left integral in H; it is also a Casimir element for H in ${}_{H^*}\mathcal{M}$ since

$$\begin{aligned} h_1^* \rightharpoonup t_1 \otimes h_2^* \rightharpoonup S(t_2) &= h^*(t_2S(t_3))t_1 \otimes S(t_4) \\ &= h^*(1_H)t_1 \otimes S(t_2) = \varepsilon_{H^*}(h^*)t_1 \otimes S(t_2), \end{aligned}$$

for all $h^* \in H^*$. It follows that $(\lambda \circ S^{-1}, t_1 \otimes S(t_2))$ is a Frobenius system for H in ${}_{H^*}\mathcal{M}$, whenever t (resp. λ) is a left integral in (resp. on) H such that $\lambda(S^{-1}(t)) = 1$. Note that $\lambda \circ S^{-1} = \lambda \leftharpoonup \mathrm{g}$ implies $\lambda(t) = 1$, too. □

3.5 Frobenius Algebras in Categories of Corepresentations

For a Hopf algebra H we consider the category of right H-corepresentations (or right H-comodules), $\mathcal{M}^H$. $\mathcal{M}^H$ is monoidal with the same tensor product, unit object and constraints as those of ${}_k\mathcal{M}$; the multiplication of H endows the tensor product of two right H-comodules with a right H-comodule structure, and the unit object k is considered right H-comodule via the unit of H. For $M \in \mathcal{M}^H$ denote by $\rho_M : M \ni m \mapsto m_{(0)} \otimes m_{(1)} \in M \otimes H$ the right coaction of H on M.

An algebra in $\mathcal{M}^H$ is called a right H-comodule algebra. It is a k-algebra A which is also a right H-comodule such that $\rho_A : A \to A \otimes H$ is an algebra map. It follows from Theorem 1 that a right H-comodule algebra A is Frobenius in $\mathcal{M}^H$ if and only if there exists a pair (ϑ, e) consisting of an element $e = e^1 \otimes e^2 \in A \otimes A$ and a k-linear map $\vartheta : A \to k$ obeying the following rules:

$$\begin{aligned} &e^1_{(0)} \otimes e^2_{(1)} \otimes e^1_{(1)}e^2_{(1)} = e^1 \otimes e^2 \otimes 1_H; \quad \vartheta(a_{(0)})a_{(1)} = \vartheta(a)1_H, \ \forall\, a \in A, \\ &ae^1 \otimes e^2 = e^1 \otimes e^2a, \ \forall\, a \in A; \quad \vartheta(e^1)e^2 = 1_A = \vartheta(e^2)e^1. \end{aligned}$$

If $A \in \mathcal{M}^H$ is Frobenius then so is A in ${}_k\mathcal{M}$; the converse is not true. Recall that H is cosemisimple if there is λ a left (and so right, too) integral with $\lambda(1_H) = 1$.

Proposition 6 *If H is a finite dimensional cosemisimple Hopf algebra then an algebra A in $\mathcal{M}^H$ is Frobenius if and only if A is Frobenius as a k-algebra via a Frobenius system (ϑ, e) with ϑ an H-colinear morphism.*

Proof If $(\vartheta, e^1 \otimes e^2)$ is a Frobenius system for A in ${}_k\mathcal{M}$ with ϑ in $\mathcal{M}^H$ then $(\vartheta, \mu^{-1}(e^2_{(1)})\lambda(e^1_{(1)}e^2_{(2)})e^1_{(0)} \otimes e^2_{(0)})$ is a Frobenius system for A in $\mathcal{M}^H$, where λ is a left integral on H and μ is the modular element of H^* defined by $th = \mu(h)t$, for any left integral t in H and $h \in H$; μ^{-1} is the convolution inverse of μ. As $\mu : H \to k$ is an algebra map, $\mu^{-1} = \mu \circ S = \mu \circ S^{-1}$. □

For the rest of this subsection H is a finite dimensional Hopf algebra, $t \in H$ is a left integral in H, $\lambda \in H^*$ is a left integral on H such that $\lambda(S^{-1}(t)) = 1$ (and so $\lambda(t) = 1$, too), $\mathfrak{g}$ is the modular element of H and A is a finite dimensional algebra in ${}_H\mathcal{M}$. Recall that the smash product of A with H, denoted by $A\#H$, is the k-vector space $A \otimes H$ equipped with the multiplication $(a\#h)(a'\#h') = a(h_1 \cdot a')\#h_2h'$, where $a, a' \in A$ and $h, h' \in H$. $A\#H$ is a k-algebra with unit $1_A\#1_H$ that contains A, H as subalgebras. Moreover, $A\#H$ is an algebra in $\mathcal{M}^H$ if it is considered a right H-comodule via $A\#H \ni a\#h \mapsto a\#h_1 \otimes h_2 \in A\#H \otimes H$.

By the dual version of [25, Theorem 3.1], if $A \in {}_H\mathcal{M}$ is Frobenius then so is $A\#H$ in $\mathcal{M}^H$. We show that this stills true in a weaker hypothesis.

Theorem 2 *Let H be a finite dimensional Hopf algebra and A a finite dimensional algebra in ${}_H\mathcal{M}$. Then the smash product algebra $A\#H$ is Frobenius as an algebra in $\mathcal{M}^H$ if and only if A is a Frobenius k-algebra.*

Proof If $A\#H$ is Frobenius in $\mathcal{M}^H$ then it is a Frobenius k-algebra, and it follows from [3, Theorem] or [24, Theorem 7.5] that A is a Frobenius k-algebra. An alternative proof for the latest result will be presented in the next section.

Conversely, assume A is a Frobenius k-algebra and let $(\vartheta, e^1 \otimes e^2)$ be a Frobenius system for A. With notation as above, if we define $\widetilde{\vartheta} : A\#H \to k$ by $\widetilde{\vartheta}(a\#h) = \lambda(\mathfrak{g}h)\vartheta(\mathfrak{g} \cdot a)$, for all $a \in A$, $h \in H$, then $(\widetilde{\vartheta}, \widetilde{e} := t_1 \cdot e^1\#t_2 \otimes \mathfrak{g}^{-1} \cdot e^2\#S(t_3))$ is a Frobenius system for $A\#H$ in $\mathcal{M}^H$. Indeed,

$$\begin{aligned}\widetilde{e}^1_{(0)} \otimes \widetilde{e}^2_{(0)} \otimes \widetilde{e}^1_{(1)}\widetilde{e}^2_{(1)} &= t_1 \cdot e^1\#t_2 \otimes \mathfrak{g}^{-1} \cdot e^2\#S(t_5) \otimes t_3S(t_4)\\ &= t_1 \cdot e^1\#t_2 \otimes \mathfrak{g}^{-1} \cdot e^2\#S(t_3) \otimes 1_H = \widetilde{e}^1 \otimes \widetilde{e}^2 \otimes 1_H,\end{aligned}$$

and the fact that $\widetilde{\vartheta}$ is in $\mathcal{M}^H$ follows from, $a \in A$ and $h \in H$,

$$\begin{aligned}\widetilde{\vartheta}(a\#h_1)h_2 &= \lambda(\mathfrak{g}h_1)\vartheta(\mathfrak{g} \cdot a)h_2 = \lambda(S^{-1}(h_1))\vartheta(\mathfrak{g} \cdot a)h_2 = \lambda(S^{-1}(h))\vartheta(\mathfrak{g} \cdot a)1_H\\ &= \lambda(\mathfrak{g}h)\vartheta(\mathfrak{g} \cdot a)1_H = \widetilde{\vartheta}(a\#h)1_H,\end{aligned}$$

where we used $\lambda \leftharpoonup \mathfrak{g} = \lambda \circ S^{-1}$ and the fact that λ is a left integral. Likewise,

$$\begin{aligned}&(a\#h)(t_1 \cdot e^1\#t_2) \otimes \mathfrak{g}^{-1} \cdot e^2\#S(t_3)\\ &\qquad = (a\#h)(1_A\#t_1)(e^1\#1_H) \otimes \mathfrak{g}^{-1} \cdot e^2\#S(t_2)\\ &\qquad \overset{(5)}{=} (a\#t_1)(e^1\#1_H) \otimes \mathfrak{g}^{-1} \cdot e^2\#S(t_2)h\\ &\qquad = (t_2S^{-1}(t_1) \cdot a)(t_3 \cdot e^1)\#t_4 \otimes \mathfrak{g}^{-1} \cdot e^2\#S(t_5)h\\ &\qquad = t_2 \cdot e^1\#t_3 \otimes \mathfrak{g}^{-1} \cdot [e^2(S^{-1}(t_1) \cdot a)]\#S(t_4)h\\ &\qquad = t_2 \cdot e^1\#t_3 \otimes (\mathfrak{g}^{-1} \cdot e^2)(\mathfrak{g}^{-1}S^{-1}(t_1) \cdot a)\#S(t_4)h\\ &\qquad \overset{(5)}{=} t_1 \cdot e^1\#t_2 \otimes (\mathfrak{g}^{-1} \cdot e^2)(S(t_4) \cdot a)\#S(t_3)h\\ &\qquad = t_1 \cdot e^1\#t_2 \otimes (\mathfrak{g}^{-1} \cdot e^2\#S(t_3))(a\#h),\end{aligned}$$

valid for all $a \in A$, $h \in H$, shows that $\widetilde{e}$ is a Casimir element for $A\#H$ in $\mathcal{M}^H$. Finally, we can see that

$$\begin{aligned}
\widetilde{\vartheta}(t_1 \cdot e^1\#t_2)\mathfrak{g}^{-1} \cdot e_2\#S(t_3) &= \lambda(\mathfrak{g}t_2)\vartheta(\mathfrak{g}t_1 \cdot e^1)\mathfrak{g}^{-1} \cdot e^2\#S(t_3) \\
&= \lambda(S^{-1}(t_2))\vartheta(\mathfrak{g}t_1 \cdot e^1)\mathfrak{g}^{-1} \cdot e^2\#S(t_3) \\
&= \lambda(S^{-1}(t_2))\vartheta(\mathfrak{g}t_1 \cdot e^1)\mathfrak{g}^{-1} \cdot e^2\#1_H \\
&= \lambda(\mathfrak{g}t_2)\vartheta(\mathfrak{g}t_1 \cdot e^1)\mathfrak{g}^{-1} \cdot e^2\#1_H \\
&= \lambda(\mathfrak{g}t)\vartheta(1_H \cdot e^1)\mathfrak{g}^{-1} \cdot e^2\#1_H \\
&= \mathfrak{g}^{-1} \cdot \vartheta(e^1)e^2\#1_H = \mathfrak{g}^{-1} \cdot 1_A\#1_H = 1_A\#1_H,
\end{aligned}$$

and similarly $\widetilde{\vartheta}(\mathfrak{g}^{-1} \cdot e^2\#S(t_3))t_1 \cdot e^1\#t_2 = \lambda(t_3)t_1 \cdot \vartheta(e^2)e^1\#t_2 = 1_A\#1_H$. □

The next result uncovers a situation when being Frobenius in $\mathcal{M}^H$ is equivalent to being Frobenius in ${}_k\mathcal{M}$ (recall that this is not true in general). Its proof follows easily from Theorem 2 and [3, Theorem] or [24, Theorem 7.5].

Corollary 1 *Let A be an algebra in ${}_H\mathcal{M}$. If A, H are finite dimensional then $A\#H$ is a Frobenius algebra in $\mathcal{M}^H$ if and only if it is a Frobenius k-algebra.*

Corollary 2 ([25, Corollary 3.2]) *Any finite dimensional Hopf algebra H is a Frobenius algebra in $\mathcal{M}^H$.*

Proof Take $A = k$ in Theorem 2 to get that $(\lambda \circ S^{-1}, t_1 \otimes S(t_2))$ is a Frobenius system for H in the monoidal category $\mathcal{M}^H$. □

If $(\vartheta, e^1 \otimes e^2)$ is a Frobenius system for A in ${}_k\mathcal{M}$, $FS{:=}(\phi, t_2 \cdot e^1\#t_3 \otimes e^2\#S^{-1}(t_1))$ with $\phi(a\#h) = \lambda(S^{-1}(h))\vartheta(a)$, for all $a \in A$, $h \in H$, is a Frobenius system for $A\#H$ in ${}_k\mathcal{M}$. This follows by using that $FS_\mathfrak{g} := (\widetilde{\vartheta}, \widetilde{e} := t_1 \cdot e^1\#t_2 \otimes \mathfrak{g}^{-1} \cdot e^2\#S(t_3))$ from the proof of Theorem 2 is as well a Frobenius system for $A\#H$ in ${}_k\mathcal{M}$, and the fact that FS is obtained from $FS_\mathfrak{g}$ through the invertible element $1_A\#\mathfrak{g}^{-1}$ of $A\#H$ (see Sect. 3.1). This can be viewed as an alternative proof for one implication in [24, Theorem 7.5].

4 Frobenius Algebra Extensions and Frobenius (Monoidal) Functors

4.1 Frobenius Algebra Extensions

As we mentioned, Kasch [40] extended the notion of Frobenius algebra to an arbitrary algebra extension, i.e. to a k-algebra morphism $i : R \to S$.

In what follows, $\mathrm{Hom}_R(S, R)$ and ${}_R\mathrm{Hom}(S, R)$ are the set of right, respectively left, R-linear morphisms from S to R. $\mathrm{Hom}_R(S, R)$ is an (R, S)-bimodule via $(r \cdot f \cdot s)(x) = rf(sx)$ while ${}_R\mathrm{Hom}(S, R)$ is a (S, R)-bimodule via $sfr(x) = f(xs)r$, for all $r \in R$, $f \in \mathrm{Hom}_R(S, R)$ (resp. $f \in {}_R\mathrm{Hom}(S, R)$) and $s, x \in S$.

Definition 1 An algebra extension $i : R \to S$ is called Frobenius if S is finitely generated and projective as a right R-module and $\mathrm{Hom}_R(S, R)$ and S are isomorphic as (R, S)-bimodules or, equivalently, if S is finitely generated and projective as a left R-module and ${}_R\mathrm{Hom}(S, R)$ and S are isomorphic as (S, R)-bimodules.

It turns out that the two equivalent conditions in Definition 1 are both equivalent to the existence of a pair $(\vartheta, e = e^1 \otimes_R e^2)$ consisting of an element $e \in S \otimes_R S$ and an R-bimodule morphism $\vartheta : S \to R$ obeying $se^1 \otimes_R e^2 = e^1 \otimes_R e^2 s$, for all $s \in S$, and $\vartheta(e^1)e^2 = 1_S = e^1\vartheta(e^2)$; see [37, Theorem 1.2]. Furthermore, by [2, Theorem 1.1], they are equivalent also to giving a k-bilinear map $B : S \times S \to R$ such that B is R-bilinear and associative, i.e. $B(rs, s') = rB(s, s')$, $B(s, s'r) = B(s, s')r$, $B(ss', s'') = B(s, s's'')$, for all $r \in R$ and $s, s', s'' \in S$, and relative to which a dual projective pair $\{x_1, \cdots, x_n\}$, $\{y_1, \cdots, y_n\}$ for S exists. The latter means that $s = \sum_{i=1}^n x_i B(y_i, s) = \sum_{i=1}^n B(s, x_i)y_i$, for all $s \in S$.

$i : R \to S$ is a Frobenius algebra extension if and only if S is a Frobenius algebra within $({}_R\mathcal{M}_R, \otimes_R, R)$. Thus from Theorem 1 one can obtain some other characterizations for $i : R \to S$ to be Frobenius; we postpone this for the general case when the algebra extensions are considered in arbitrary monoidal categories.

We call the pair (ϑ, e) the Frobenius system of the Frobenius extension $i : R \to S$; we call e the Casimir element and ϑ the Frobenius morphism of i.

Let $C_R(S) := \{s \in S \mid rs = sr,\ \forall\ r \in R\}$, a k-subalgebra of S containing the center of R. The set of invertible elements of $C_R(S)$ will be denoted by $C_R(S)^\times$. Parallel to Frobenius k-algebras, a Frobenius system (ϑ, e) for $i : R \to S$ is unique up to an element of $C_R(S)^\times$: any other Frobenius system for $i : R \to S$ is of the form $(\vartheta \leftharpoonup d, e^1 \otimes_A d^{-1}e^2)$, for some invertible element $d \in C_R(S)$ with inverse d^{-1}, or equivalently of the form $(d' \rightharpoonup \vartheta, e^1 d'^{-1} \otimes_A e^2)$, for some invertible element $d' \in C_R(S)$ with inverse d'^{-1}; see [37].

When (ϑ, e) is a Frobenius system for R, $\Phi_r : {}_R S_S \to \mathrm{Hom}_R(S, R)$, $s \mapsto \vartheta \leftharpoonup s = \vartheta(s\cdot)$, is an (R, S)-bimodule isomorphism with inverse given by $\Psi_r^{-1}(f) = f(e^1)e^2$, for all $f \in \mathrm{Hom}_R(S, R)$. Similarly, $\Phi_l : {}_S S_R \ni s \mapsto s \rightharpoonup \vartheta := \vartheta(\cdot s) \in {}_R\mathrm{Hom}(S, R)$ is a (S, R)-bimodule isomorphism whose inverse is given by $\Psi_l^{-1}(f) = e^1 f(e^2)$, for all $f \in {}_R\mathrm{Hom}(S, R)$.

If $d \in C_R(S)$ then $d \rightharpoonup \vartheta \in {}_R\mathrm{Hom}_R(S, R)$ (the set of R-bimodule morphisms from S to R) and, moreover, lies in the image of Φ_r. To see this, if $\mathcal{N}(d) := \vartheta(e^1 d)e^2$, we have $\mathcal{N}(d) \in C_R(S)$ and $d \rightharpoonup \vartheta = \vartheta \leftharpoonup \mathcal{N}(d)$. Thus for any $d \in C_R(S)$ there is a unique $\mathcal{N}(d) \in C_R(S)$ satisfying $d \rightharpoonup \vartheta = \vartheta \leftharpoonup \mathcal{N}(d)$; in this way we have a well defined k-algebra isomorphism $\mathcal{N} : C_R(S) \to C_R(S)$, called the Nakayama automorphism of the Frobenius algebra extension i. The inverse of $\mathcal{N}$ is $\mathcal{N}^{-1}$ given by $\mathcal{N}^{-1}(d) = e^1\vartheta(de^2)$, for all $d \in C_R(S)$.

A Frobenius algebra extension $i : R \to S$ is called symmetric if its Nakayama automorphism is an inner automorphism of $C_R(S)$.

Examples 3 (1) Let $i : R \to S$, $j : S \to T$ be Frobenius algebra extensions with Frobenius systems $(\vartheta, e^1 \otimes_R e^2)$ and $(\phi, f^1 \otimes_S f^2)$, respectively. Then $ji : R \to T$ is Frobenius with Frobenius system $(\vartheta\phi, f^1 e^1 \otimes_R e^2 f^2)$. Hence, if $i : R \to S$ is Frobenius and R is a Frobenius k-algebra then S is also a Frobenius k-algebra.

(2) Let $H \leq G$ be a subgrup of finite index and take $g_1 = e, g_2, \cdots, g_n$ left coset representatives. Then the group algebra $k[H]$ is a subalgebra of the group algebra $k[G]$, and the inclusion i produces a k-algebra extension. i is Frobenius with Frobenius system defined by the Casimir element $\sum_{i=1}^{n} g_i \otimes g_i^{-1}$ and Frobenius morphism $\vartheta : k[G] \to k$ given by $\vartheta(\sum_{g\in G} k_g g) = \sum_{h\in H} k_h h$, for all $\sum_{g\in G} k_g g \in k[G]$. A proof for this can be found in [37, Example 1.7]; for short, it follows since $g_1^{-1} = e, g_2^{-1}, \cdots, g_n^{-1}$ are right coset representatives for $H \leq G$.

(3) For A a k-algebra and n a non-zero natural number, $i : A \to M_n(A)$ is a Frobenius algebra extension; the Casimir element is $e = \sum_{i,j=1}^{n} E_{ij} \otimes_A E_{ji}$ and the Frobenius morphism is the usual trace morphism $\mathrm{tr} : M_n(A) \to A$. By $\{E_{ij}\}_{i,j}$ we denoted the matrix having 1_A in the position (i, j) and zero elsewhere. □

Other examples of Frobenius algebra extensions come from Hopf Galois theory.

4.2 Hopf Galois Extensions

For H a Hopf algebra and A an algebra in $\mathcal{M}^H$ we define $B := \{a \in A \mid \rho_A(a) = a \otimes 1_H\}$, the subalgebra of coinvariants of A, and $\mathrm{can} : A \otimes_B A \ni a \otimes_B a' \mapsto aa'_{(0)} \otimes a'_{(1)}$, the canonical map. The inclusion map $i : B \to A$ is called H-extension; an H-extension $i : B \to A$ is called Galois if can is bijective.

Kreimer and Takeuchi [44] proved that Hopf Galois extensions are Frobenius.

Theorem 3 *Let H be a finite dimensional Hopf algebra and $i : B \to A$ a Hopf Galois H-extension. Then i is a Frobenius algebra extension.*

Proof The Frobenius system is $(\mathrm{can}^{-1}(1_A \otimes S^{-1}(t)), \mathrm{tr})$, where t is a non-zero left integral in H and $\mathrm{tr} : A \ni a \mapsto \lambda(a_{(1)})a_{(0)} \in B$ is the trace map (defined by a left integral λ on H such that $\lambda(S^{-1}(t)) = 1$ or, equivalently, $\lambda(t) = 1$). □

The result of Kreimer and Takeuchi allows to see, in a particular case, how for a Galois extension the Frobenius property transfers from one algebra to another.

Proposition 7 *Let H be a finite dimensional cosemisimple Hopf algebra and $i : B \to A$ a Hopf Galois H-extension. Then B is a Frobenius k-algebra if and only if A is Frobenius in $\mathcal{M}^H$.*

Proof By Proposition 6, it is suffices to prove that B is Frobenius in ${}_k\mathcal{M}$ if and only if so is A in ${}_k\mathcal{M}$ via a Frobenius system (ϑ, e) with ϑ in $\mathcal{M}^H$.

Assume that B is a Frobenius k-algebra with Frobenius system $(\phi, f^1 \otimes f^2)$. Theorem 3 and Examples 3 (1) tell us that A is a Frobenius k-algebra with Frobenius system $(\phi\mathrm{tr}, e^1 f^1 \otimes f^2 e^2)$, where $e^1 \otimes_B e^2 = \mathrm{can}^{-1}(1_A \otimes S^{-1}(t))$ and $\mathrm{tr} : A \ni a \mapsto \lambda(a_{(1)})a_{(0)} \in B$. We see that, for all $a \in A$,

$$\phi\mathrm{tr}(a_{(0)})a_{(1)} = \phi(a_{(0)})\lambda(a_{(1)_1})a_{(1)_2} = \lambda(a_{(1)})\phi(a_{(0)})1_H = \phi\mathrm{tr}(a)1_H,$$

and so $\phi\mathrm{tr} : A \to k$ is in $\mathcal{M}^H$, as needed. Note that H being cosemisimple implies that the space of left and right integrals on H are equal (equivalently, $\mathfrak{g} = 1$).

Conversely, let (ϑ, e) be a Frobenius system for A with ϑ in $\mathcal{M}^H$. Then $(\vartheta|_B, \mathrm{tr}(e^1) \otimes \mathrm{tr}(e^2))$ is a Frobenius system for the k-algebra B since tr is B-bilinear, $\vartheta(\mathrm{tr}(a)) = \vartheta(a_{(0)})\lambda(a_{(1)}) = \vartheta(a)\lambda(1_H) = \vartheta(a)$, for all $a \in A$, and $\mathrm{tr}(1_A) = 1_B$ (H is cosemisimple, and so one can take $\lambda(1_H) = 1$). □

Proposition 7 generalizes [24, Corollary 4.2]. Indeed, for G a finite group the group Hopf algebra $k[G]$ is finite dimensional and cosemisimple, as $P_e : k[G] \to k$ determined by $P_e(g) = \delta_{g,e}$, for all $g \in G$, is an integral on $k[G]$ with $P_e(e) = 1$. $\mathcal{M}^{k[G]} = \mathrm{Vect}^G$ as monoidal categories, and $i : A_e \to A$ is a Hopf Galois $k[G]$-extension if and only if A is strongly graded. Hence a strongly G-graded finite dimensional algebra A is G-graded Frobenius iff A_e is a Frobenius k-algebra.

One can remove the cosemisimple property of H in Proposition 7 but instead we have to ask for $i : B \to A$ to be, moreover, a cleft H-extension.

An H-extension $i : B \to A$ is cleft if there exists a map $\gamma : H \to A$ in $\mathcal{M}^H$ which is convolution invertible. Any cleft H-extension is of the form $i : A \to A\#_\sigma H$ for some 2-cocycle $\sigma : H \otimes H \to A$, where $A\#_\sigma H$ is the crossed product algebra of A with H via σ from [4, 29, 49]. When σ is trivial (equivalently, γ is an algebra morphism) we have that A is an algebra in ${}_H\mathcal{M}$ and $A\#_\sigma H$ equals $A\#H$; so the embedding $i : A \to A\#H$ is a cleft H-extension, too.

Corollary 3 *If H is a finite dimensional Hopf algebra then a crossed product algebra $A\#_\sigma H$ is Frobenius in ${}_k\mathcal{M}$ if and only if A is a Frobenius k-algebra.*

Proof Any cleft H-extension is Galois, so by the result of Kreimer and Takeuchi [44] the algebra extension $i : A \to A\#_\sigma H$ is Frobenius with Frobenius system

$$(\vartheta, \sigma^{-1}(t_3 \otimes S^{-1}(t_2))\#_\sigma t_4 \otimes_A 1_A\#_\sigma S^{-1}(t_1)), \tag{8}$$

where σ^{-1} is the convolution inverse of σ and $\vartheta(a\#_\sigma h) = \lambda(h)a\#_\sigma 1_H$, for all $a \in A$, $h \in H$. As before, λ is a left integral on H and t is a left integral in H such that $\lambda(S^{-1}(t)) = 1$. When A is Frobenius, from Examples 3 (1) we deduce that $A\#_\sigma H$ is a Frobenius k-algebra with Frobenius system given by

$$(\widetilde{\vartheta}, (\sigma^{-1}(t_3 \otimes S^{-1}(t_2))\#_\sigma t_4)(e^1\#_\sigma 1_H) \otimes (e^2\#_\sigma S^{-1}(t_1))) \tag{9}$$

with $\widetilde{\vartheta}(a\#_\sigma h) = \lambda(h)\phi(a)$, for all $a \in A$ and $h \in H$.

Conversely, if a crossed product algebra $A\#_\sigma H$ is Frobenius as a k-algebra then so is A. Indeed, $A\#_\sigma H \ni a\#_\sigma h \mapsto a\#_\sigma h_1 \otimes h_1 \in A\#_\sigma H \otimes H$ makes $A\#_\sigma H$ an algebra in $\mathcal{M}^H$ or, equivalently, an algebra in ${}_{H^*}\mathcal{M}$. By the previous paragraph it follows that $(A\#_\sigma H)\#H^*$ is a Frobenius k-algebra, too. Theorem 2.2 in [5] tells us that $(A\#_\sigma H)\#H^* \simeq M_n(A)$ as algebras, where $n = \dim_k(H)$. We use now [24, Proposition 7.2] to conclude that A is a Frobenius k-algebra. □

Remark 2 For σ trivial we recover the Frobenius system for the k-algebra $A\#H$ presented at the end of Sect. 3.

Next, we generalize Theorem 2 from smash to crossed product algebras.

Theorem 4 *Let H be a finite dimensional Hopf algebra and $A\#_\sigma H$ a crossed product algebra of H with a finite dimensional k-algebra A, via a certain 2-cocycle σ. Then $A\#_\sigma H$ is a Frobenius algebra in $\mathcal{M}^H$ if and only if A is a Frobenius k-algebra, if and only if $A\#_\sigma H$ is a Frobenius k-algebra.*

Consequently, if $i : B \to A$ is a cleft H-extension then A is Frobenius in $\mathcal{M}^H$ if and only if B is Frobenius in ${}_k\mathcal{M}$, if and only if A is Frobenius in ${}_k\mathcal{M}$.

Proof Note that $1_A\#_\sigma g$ is invertible in $A\#_\sigma H$, whenever g is a grouplike element of H. Indeed, if $\gamma : A \ni a \mapsto a\#_\sigma h \in A\#_\sigma H$ then γ is convolution invertible with inverse γ^{-1} given by $\gamma^{-1}(h) = \sigma^{-1}(S(h_2) \otimes h_3)\#_\sigma S(h_1)$, for all $h \in H$; see [5, (1.11)]. Therefore, if g is grouplike in H then $\gamma(g) = 1_A\#_\sigma g$ is invertible in $A\#_\sigma H$ with inverse $\gamma^{-1}(g) = \sigma^{-1}(g^{-1} \otimes g)\#_\sigma g^{-1}$.

By taking $g = \mathfrak{g}^{-1}$, $\mathfrak{g}$ the modular element of H, we get $d^{-1} := 1_A\#_\sigma \mathfrak{g}^{-1}$ invertible in $A\#_\sigma H$ with inverse d given by $d = \gamma^{-1}(\mathfrak{g}^{-1})$. Thus, from (9) and the comments made after Proposition 1, we derive from the Frobenius system $(\vartheta, e^1 \otimes e^2)$ of $A \in {}_k\mathcal{M}$ the following Frobenius system for the k-algebra $A\#_\sigma H$:

$$\mathbf{FS}_{\mathfrak{g}} := (\phi, \sigma^{-1}(t_3 \otimes S^{-1}(t_2))(t_4 \cdot e^1)\#_\sigma t_5 \otimes (1_A\#_\sigma \mathfrak{g}^{-1})(e^2\#_\sigma S^{-1}(t_1))).$$

We consider, for all $a \in A$ and $h \in H$,

$$\begin{aligned}\phi(a\#_\sigma h) &= \widetilde{\vartheta}((\sigma^{-1}(\mathfrak{g} \otimes \mathfrak{g}^{-1})\#_\sigma \mathfrak{g})(a\#_\sigma h)) = \lambda(\mathfrak{g}h_2)\\ &\times \vartheta(\sigma^{-1}(\mathfrak{g} \otimes \mathfrak{g}^{-1})(\mathfrak{g} \cdot a)\sigma(\mathfrak{g} \otimes h_1)) = \lambda(\mathfrak{g}h)\vartheta(\sigma^{-1}(\mathfrak{g} \otimes \mathfrak{g}^{-1})(\mathfrak{g} \cdot a)\sigma(\mathfrak{g} \otimes \mathfrak{g}^{-1})).\end{aligned}$$

So the converse ends if we prove that $\mathbf{FS}_{\mathfrak{g}}$ is a Frobenius system for $A\#_\sigma H$ in $\mathcal{M}^H$. To this end, observe that the Casimir element is right H-colinear since the right H-comodule structure morphism of $A\#_\sigma H$ is an algebra map, and so

$$\begin{aligned}&\sigma^{-1}(t_4 \otimes S^{-1}(t_3))(t_5 \cdot e^1)\#_\sigma t_6 \otimes (1_A\#_\sigma \mathfrak{g}^{-1})(e^2\#_\sigma S^{-1}(t_2)) \otimes t_7\mathfrak{g}^{-1}S^{-1}(t_1)\\ &\overset{(5)}{=}\sigma^{-1}(t_3 \otimes S^{-1}(t_2))(t_4 \cdot e^1)\#_\sigma t_5 \otimes (1_A\#_\sigma \mathfrak{g}^{-1})(e^2\#_\sigma S^{-1}(t_1)) \otimes t_6S(t_7)\\ &= \sigma^{-1}(t_3 \otimes S^{-1}(t_2))(t_4 \cdot e^1)\#_\sigma t_5 \otimes (1_A\#_\sigma \mathfrak{g}^{-1})(e^2\#_\sigma S^{-1}(t_1)) \otimes 1_H,\end{aligned}$$

as desired. Finally, the fact that ϕ is right H-colinear follows from the equality $\lambda(\mathfrak{g}h_1)h_2 = \lambda(\mathfrak{g}h)\mathfrak{g}^{-1}\mathfrak{g} = \lambda(\mathfrak{g}h)1_H$, valid for all $h \in H$. As being Frobenius in $\mathcal{M}^H$ implies Frobenius in ${}_k\mathcal{M}$, the direct implication follows from Corollary 3. □

4.3 Frobenius Functors and Frobenius Monads

Morita [50] called a functor F Frobenius if F has a left adjoint G which is as well a right adjoint for it; we say also that (F, G) is a Frobenius pair of functors.

Let $i : R \to S$ be an extension of k-algebras and $F : \mathcal{M}_S \to \mathcal{M}_R$ the functor restriction of scalars induced by i. It is well known that the induction functor $G = -\otimes_R S : \mathcal{M}_R \to \mathcal{M}_S$ is a left adjoint for F. In addition, G is as well a right adjoint for F if and only if $i : R \to S$ is a Frobenius algebra extension; see [37, Theorem 1.2]. So F is Frobenius if and only if i is a Frobenius algebra extension, and this justifies the terminology used by Morita.

On the other hand, as we explained in Examples 2 (5), giving an adjunction is equivalent to giving a monad. Thus, a natural question arises: is there a connection between Frobenius functors and Frobenius monads? The answer was given by Street in [65, Theorem 1.6]; for short, it says that a monad is Frobenius if and only if it is determined by a Frobenius pair of functors.

Theorem 5 *A monad $(T, \underline{m}, \underline{\eta})$ in $\mathcal{C}$ is Frobenius if and only if (F^T, G^T) is a Frobenius pair of functors.*

Proof With notation as in Examples 2 (5), F^T is always a left adjoint to G^T. So we must show that F^T is a right adjoint to G^T if and only if T is a Frobenius monad. This is just the equivalence between (a) and (f) in [65, Theorem 1.6]. □

4.4 Frobenius Algebra Extensions in Monoidal Categories

At this moment it is not known whether the result of Morita is true in arbitrary monoidal categories $\mathcal{C}$. We see that if $\underline{1}$ is a left $\otimes$-generator then this is true.

With notation as in Sect. 2.2 we have the following (see [21]):

Definition 2 An algebra morphism $i : R \to S$ in a monoidal category $\mathcal{C}$ is called Frobenius algebra extension if there exist $\vartheta : S \to R$ in ${}_R\mathcal{C}_R$ and $e : \underline{1} \to S \otimes_R S$ in $\mathcal{C}$ such that $\mu^S_{S\otimes_R S}(Se) = \nu^S_{S\otimes_R S}(eS)$, $(S \otimes_R \vartheta)e = \Upsilon_S^{-1}\underline{\eta}_S$, $(\vartheta \otimes_R S)e = \Upsilon_S'^{-1}\underline{\eta}_S$.

For $i : R \to S$ an algebra extension in $\mathcal{C}$, the functor restriction of scalars F is a right adjoint to the induction functor $G = -\otimes_R S : \mathcal{C}_R \to \mathcal{C}_S$.

Remark 3 (due to the referee) $T := FG = -\otimes_R S : \mathcal{C}_R \to \mathcal{C}_R$ is a monad whose category $(\mathcal{C}_R)^T$ of T-algebras has as objects pairs (M, μ_M) consisting of $M \in \mathcal{C}_R$ and $\mu_M : M \otimes_R S \to M$ in $\mathcal{C}_R$ which is associative and unital; morphisms are $f : M \to N$ in $\mathcal{C}_R$ obeying $f\mu_M = \mu_N(f \otimes_R S)$. μ_M induces a right S-module structure on M, by considering $\mu^S_M = \mu_M q^R_{M,S}$, and so we have a well defined functor $H : (\mathcal{C}_R)^T \to \mathcal{C}_S$. It can be seen easily that H is an isomorphism of categories, its inverse being the comparison functor $K : \mathcal{C}_S \to (\mathcal{C}_R)^T$ given by $K(\mathfrak{M}) = (F(\mathfrak{M}), \mu_{F(\mathfrak{M})})$ with $\mu_{F(\mathfrak{M})} : F(\mathfrak{M}) \otimes_R S \to F(\mathfrak{M})$ in $\mathcal{C}_R$ uniquely determined by $\mu_{F(\mathfrak{M})} q^R_{F(\mathfrak{M}),S} = \mu^S_{\mathfrak{M}}$, for all $\mathfrak{M} \in \mathcal{C}_S$, and $K(g) = F(g)$ for any morphism g in $\mathcal{C}_S$. Thus F is monadic, and so (G, F) is a Frobenius pair iff so is (F^T, G^T), iff the monad T is Frobenius (see Theorem 5); we refer to [47, VI.3] for more details about the comparison functor of an adjunction.

To see when F is also a left adjoint to G we need $\underline{1}$ to be a left $\otimes$-generator: $P \in C$ is a left $\otimes$-generator of C if wherever we consider $f, g : Y \otimes Z \to W$ in C such that $f(\epsilon \otimes \mathrm{Id}_Z) = g(\epsilon \otimes \mathrm{Id}_Z)$, for all $\epsilon : P \to Y$ in C, we have $f = g$. If this is the case then we have isomorphisms $\mathrm{Nat}(F \circ (- \otimes_R S), -) \simeq {}_R\mathrm{Hom}_R(S, R)$ and $\mathrm{Nat}(-, (- \otimes_R S) \circ F) \simeq \{e : \underline{1} \to S \otimes_R S \mid \mu^S_{S\otimes_R S}(\mathrm{Id}_S \otimes e) = \nu^S_{S\otimes_R S}(e \otimes \mathrm{Id}_S)\}$. Consequently, assuming $\underline{1}$ a left $\otimes$-generator, we have that $i : R \to S$ in C is Frobenius iff $F : C_S \to C_R$, the restriction of scalars functor induced by i, is a Frobenius functor in the sense of Morita [21, Theorem 3.5]; this applies to $\underline{\mathrm{Sets}}$, ${}_k\mathcal{M}$, ${}_R\mathcal{M}_R$ with R an Azumaya algebra or **FdHilb**, see [21, Examples 3.2].

It turns out that $i : R \to S$ in C is Frobenius iff S is a Frobenius algebra in ${}_RC_R$ [21, Proposition 4.8]. This fact allows to give a bunch of characterizations for an algebra extension to be Frobenius. The next result is [21, Corollary 5.2].

Theorem 6 *Let $i : R \to S$ be an algebra extension in C. Then the following assertions are equivalent:*

(i) The extension $i : R \to S$ is Frobenius;

(ii) S admits a left dual object $S^{\surd}$ in ${}_RC_R$ and S and $S^{\surd}$ are isomorphic as right S-modules in ${}_RC_R$;

(iii) S admits a right dual object ${}^{\surd}S$ in ${}_RC_R$ and S and ${}^{\surd}S$ are isomorphic as left S-modules in ${}_RC_R$;

(iv) S admits a coalgebra structure in ${}_RC_R$, that is an R-coring structure, such that the comultiplication morphism is S-bilinear in ${}_RC_R$.

(v) S admits a right dual ${}^{\surd}S$ in ${}_RC_R$ and there exists $B : S \otimes_R S \to R$ in ${}_RC_R$ which is associative and such that $\Phi^r_B := \Upsilon_{{}^{\surd}S}(\mathrm{Id}_S \otimes_R B)\Sigma'_{{}^{\surd}S,S,S}(\mathrm{coev}'_S \otimes_R \mathrm{Id}_S)\Upsilon'^{-1}_S : S \to {}^{\surd}S$ is an isomorphism in ${}_RC_R$;

(vi) S admits a left dual $S^{\surd}$ in ${}_RC_R$ and there exists $B : S \otimes_R S \to R$ in ${}_RC_R$ which is associative and such that $\Phi^l := \Upsilon'_{S^{\surd}}(B \otimes_R \mathrm{Id}_S)\Gamma'_{S,S,S^{\surd}}(\mathrm{Id}_S \otimes_R \mathrm{coev}_S)\Upsilon^{-1}_S : S \to S^{\surd}$ is an isomorphism in ${}_RC_R$;

(vii) There exists an adjunction $(\rho, \lambda) : S \dashv S$ in ${}_RC_R$ for which $\lambda : S \otimes_R S \to R$ is associative;

(viii) There exists an adjunction $(\rho, \lambda) : S \dashv S$ in ${}_RC_R$ such that $\lambda = \vartheta \underline{m}^R_S$ for some $\vartheta : S \to R$ morphism in ${}_RC_R$.

If R is a left $\otimes_R$-generator for ${}_RC_R$ then (i)–(viii) are also equivalent to

(ix) $- \otimes_R S : {}_RC_R \to {}_RC_S$ is a right adjoint of the forgetful functor $F : {}_RC_S \to {}_RC_R$ or, otherwise stated, F is a Frobenius functor.

Let H be a Hopf algebra with bijective antipode and ${}_H\mathcal{M}^H_H = ({}_H\mathcal{M}_H)^H$ the category of right H-comodules in ${}_H\mathcal{M}_H$. Owing to [62], ${}_H\mathcal{M}^H_H$ and ${}_H\mathcal{M}$ are monoidally equivalent, so to an algebra A in ${}_H\mathcal{M}$ corresponds an algebra $\mathfrak{A}$ in ${}_H\mathcal{M}^H_H$; by [8], $\mathfrak{A} = A\#H$. Since a monoidal equivalence preserves the Frobenius property, we get $A \in {}_H\mathcal{M}$ Frobenius iff so is $A\#H \in {}_H\mathcal{M}^H_H$. As the monoidal structure on ${}_H\mathcal{M}^H_H$ is designed in such a way that the forgetful functor from ${}_H\mathcal{M}^H_H$ to $({}_H\mathcal{M}_H, \otimes_H, H)$ is strict monoidal, by Examples 2 (3) we get that $A\#H$ is Frobenius in ${}_H\mathcal{M}^H_H$ if and only if the extension $j : H \ni h \mapsto 1_A\#h \in A\#H$ is Frobenius in $\mathcal{M}^H$. Directly, any Frobenius system (ψ, f) for $A\#H$ in ${}_H\mathcal{M}^H_H$ is of the form $f = e^1\#1_H \otimes_H e^2\#1_H$ and

$\psi(a\#h) = \vartheta(a)h$, for all $a \in A, h \in H$, for some Frobenius system $(\vartheta, e = e^1 \otimes e^2)$ of A in ${}_H\mathcal{M}$.

We summarize all these in the following.

Example 1 Let H be a Hopf algebra with bijective antipode and $A \in {}_H\mathcal{M}$ an algebra. Then $A \in {}_H\mathcal{M}$ is Frobenius iff $j : H \to A\#H$ is Frobenius in $\mathcal{M}^H$.

4.5 Monoidal Frobenius Functors

If an algebra extension $i : R \to S$ in C is Frobenius then S has left and right dual objects in ${}_RC_R$. Next, we explain when the existence of the dual object of S in ${}_RC_R$ implies the existence of the dual object in C, and vice-versa. The final aim is to characterize the Frobenius property of i in terms given by R and S. We will see that this is possible in the case when R is Frobenius and separable.

Recall from [21, Proposition 4.10] that a Frobenius algebra R in C is separable (i.e. $\underline{m}_R$ splits in ${}_RC_R$) iff there exists $\alpha : \underline{1} \to R$ in C such that

$$\underline{m}_R(\underline{m}_R \otimes \mathrm{Id}_R)(\mathrm{Id}_R \otimes \alpha \otimes \mathrm{Id}_R)e = \underline{\eta}_R. \tag{10}$$

Here $e : \underline{1} \to R \otimes R$ is the Casimir morphism of the Frobenius algebra R in C.

A Frobenius algebra in ${}_RC_R$ remains Frobenius in C although the forgetful functor $\mathfrak{U} : {}_RC_R \to C$ is not strong monoidal. And this is because there are weaker conditions on a functor to preserve the Frobenius property for algebras (or the duals). For instance, by [23, Theorem 2] monoidal Frobenius functors have such properties and [64, Definition 6.1] or [23, Definition 1] ensures that Frobenius monoidal functor is a concept weaker than the strong monoidal functor; we refer to [15, Definition 1.22] for the concept of (op)monoidal functor.

Definition 3 $\mathfrak{F} : (C, \otimes, \underline{1}) \to (\mathcal{D}, \square, \underline{I})$ is functor between monoidal categories.

(i) We call $\mathfrak{F}$ Frobenius monoidal if $\mathfrak{F}$ is equipped with a monoidal $(\phi_2 = (\phi_{X,Y} : \mathfrak{F}(X)\square\mathfrak{F}(Y) \to \mathfrak{F}(X \otimes Y))_{X,Y\in C}, \phi_0 : \underline{I} \to \mathfrak{F}(\underline{1}))$ and comonoidal $(\psi_2 = (\psi_{X,Y} : \mathfrak{F}(X \otimes Y) \to \mathfrak{F}(X)\square\mathfrak{F}(Y))_{X,Y\in C}, \psi_0 : \mathfrak{F}(\underline{1}) \to \underline{I})$ structure such that

$$(\mathrm{Id}_{\mathfrak{F}(X)}\square\phi_{Y,Z})(\psi_{X,Y}\square\mathrm{Id}_{\mathfrak{F}(Z)}) = \psi_{X,Y\otimes Z}\phi_{X\otimes Y,Z},$$
$$(\phi_{X,Y}\square\mathrm{Id}_{\mathfrak{F}(Z)})(\mathrm{Id}_{\mathfrak{F}(X)}\square\psi_{Y,Z}) = \psi_{X\otimes Y,Z}\phi_{X,Y\otimes Z},\ \forall\ X,\ Y,\ Z \in C.$$

(ii) $\mathfrak{F}$ is called a strong monoidal functor if it is monoidal (or opmonoidal) and, moreover, φ_0 and φ_2 (resp. ψ_0 and ψ_2) are defined by isomorphisms in $\mathcal{D}$.

One can check that $\phi_2 = (\phi_{X,Y} = q^R_{X,Y} : X \otimes Y \to X \otimes_R Y)_{X,Y\in {}_RC_R}$ and $\phi_0 = \underline{\eta}_R : \underline{1} \to R$ define on the forgetful functor $\mathfrak{U} : {}_RC_R \to C$ a monoidal structure. We refer to it as being the trivial monoidal structure of $\mathfrak{U}$.

The next result gives the connection between Frobenius monoidal functors and Frobenius monoidal algebras. Its first part generalizes [64, Lemmas 6.3 and 6.4]. Its second part is [21, Theorem 6.2] and says that the opmonoidal structures of $\mathfrak{U}$ are uniquely determined by the Frobenius structures of R.

Theorem 7 *For R an algebra in $\mathcal{C}$, the forgetful functor $\mathfrak{U} : {}_R\mathcal{C}_R \to \mathcal{C}$ endowed with the trivial monoidal structure $(q^R_{-,-}, \underline{\eta}_R)$ is Frobenius monoidal if and only if R is a Frobenius algebra. Furthermore, if this is the case then the opmonoidal (ψ_2, ψ_0) structure of $\mathfrak{U}$ is uniquely determined by a Frobenius structure of R, in the sense that there exists (ϑ, e) a Frobenius pair for R such that $\psi_0 = \vartheta$ and*

$$\psi_2 q^R_{-,-} := (\psi_{X,Y} q^R_{X,Y})_{X,Y\in {}_R\mathcal{C}_R} = \left((\nu^R_X \otimes \mu^R_Y)(\mathrm{Id}_X \otimes e \otimes \mathrm{Id}_Y)\right)_{X,Y\in {}_R\mathcal{C}_R}. \quad (11)$$

A Frobenius monoidal functor carries Frobenius algebras to Frobenius algebras and behaves well with respect to the duals. So, for R a Frobenius algebra,
• if $i : R \to S$ is a Frobenius algebra extension then S is a Frobenius algebra;
• if $X \in {}_R\mathcal{C}_R$ admits a (left) right dual object in ${}_R\mathcal{C}_R$ then it admits a (left) right dual object in $\mathcal{C}$, too.

The second assertion has a converse, provided that R is, moreover, a separable algebra in $\mathcal{C}$. By [21, Proposition 7.1] we have the following:

Proposition 8 *Let R be a Frobenius separable algebra in $\mathcal{C}$, (ϑ, e) a Frobenius pair for R, $\alpha : \underline{1} \to R$ as in* (10) *and $(\rho, \lambda) : Y \dashv X$ an adjunction in $\mathcal{C}$.*

(i) If $Y \in {}_R\mathcal{C}_R$ then $X \in {}_R\mathcal{C}_R$, too, and via these structures the morphisms $\rho_0 = q^R_{X,Y}(X(\mu^R_Y(R\nu^R_Y)))(X\alpha Y R)(\rho R) : R \to X \otimes_R Y$ and $\lambda_0 : Y \otimes_R X \to R$ uniquely determined by $\lambda_0 q^R_{Y,X} = (R(\lambda(\mu^R_Y X)))(eYX)$ define an adjunction $(\rho_0, \lambda_0) : Y \dashv X$ in ${}_R\mathcal{C}_R$.

(ii) Similarly, if $X \in {}_R\mathcal{C}_R$ then $Y \in {}_R\mathcal{C}_R$, too, and via these structures the morphisms $\rho^0 = q^R_{X,Y}((\mu^R_X(R\nu^R_X)Y)(RX\alpha Y)(R\rho) : R \to X \otimes_R Y$ and $\lambda^0 : Y \otimes_R X \to R$ uniquely determined by $\lambda^0 q^R_{X,Y} = ((\lambda(Y\nu^R_X))R)(YXe)$ define an adjunction $(\rho^0, \lambda^0) : Y \dashv X$ in ${}_R\mathcal{C}_R$.

For R a Frobenius separable algebra and $X \in {}_R\mathcal{C}_R$, X admits a (left) right dual in ${}_R\mathcal{C}_R$ if and only if X admits a (left) right dual in $\mathcal{C}$. This allows to characterize a Frobenius extensions i in terms of R and S, see [21, Theorem 7.3].

Theorem 8 *Let R be a Frobenius separable algebra and $i : R \to S$ an algebra extension in $\mathcal{C}$. Then $i : R \to S$ is a Frobenius algebra extension if and only if S is a Frobenius algebra in $\mathcal{C}$ and the following equality holds*

$$\widetilde{\vartheta}\underline{m}_S\vartheta(Si) = \vartheta\underline{m}_R(R(\widetilde{\vartheta}\underline{m}_S)R)(RiSR)(eSR), \quad (12)$$

where (ϑ, e) is a Frobenius pair for R and $(\widetilde{\vartheta}, \widetilde{e})$ is a Frobenius pair for S.

When $\mathcal{C} = {}_k\mathcal{M}$ the condition (12) reads: the restriction at R of the Nakayama automorphism of S is equal to the Nakayama automorphism of R. According to [21, Theorem 7.4], (12) has the same meaning in any other sovereign category.

Theorem 9 *Let C be a sovereign monoidal category, R a Frobenius separable algebra and $i : R \to S$ an algebra extension in C. Then $i : R \to S$ is Frobenius if and only if S is Frobenius in C and $\widetilde{N} \circ i = i \circ N$, where $\widetilde{N}$ and N are the Nakayama automorphisms of S and R, respectively.*

5 Frobenius Wreaths and Cowreaths

5.1 Frobenius Wreath Extensions

Motivated by the theory of entwined modules in C-categories [17, 18], we were interested by those "algebra" extensions produced by a monad in $EM(\mathcal{K})$ that are Frobenius. Note that a (co)monad in $EM(\mathcal{K})$ is called a (co)wreath, so we next see when an algebra extension defined by a wreath is Frobenius.

According to [45], a wreath is a monad $\mathbb{A} = (A, t, \mu, \eta)$ in $\mathcal{K}$ together with a 1-cell $A \xrightarrow{s} A$ and 2-cells $ts \stackrel{\psi}{\Longrightarrow} st$, $1_A \stackrel{\sigma}{\Longrightarrow} st$ and $ss \stackrel{\zeta}{\Longrightarrow} st$ satisfying the following conditions:

$$(1_s \odot \mu)(\psi \odot 1_t)(1_t \odot \psi) = \psi(\mu \odot 1_s) \, , \; \psi(\eta \odot 1_s) = 1_s \odot \eta \, ; \tag{13}$$

$$(1_s \odot \mu)(\psi \odot 1_t)(1_t \odot \sigma) = (1_s \odot \mu)(\sigma \odot 1_t) \, ; \tag{14}$$

$$(1_s \odot \mu)(\psi \odot 1_t)(1_t \odot \zeta) = (1_s \odot \mu)(\zeta \odot 1_t)(1_s \odot \psi)(\psi \odot 1_s) \, ; \tag{15}$$

$$(1_s \odot \mu)(\zeta \odot 1_t)(1_s \odot \zeta) = (1_s \odot \mu)(\zeta \odot 1_t)(1_s \odot \psi)(\zeta \odot 1_s) \, ; \tag{16}$$

$$(1_s \odot \mu)(\zeta \odot 1_t)(1_s \odot \sigma) = 1_s \odot \eta \, ; \tag{17}$$

$$(1_s \odot \mu)(\zeta \odot 1_t)(1_s \odot \psi)(\sigma \odot 1_s) = 1_s \odot \eta \, . \tag{18}$$

As we already have mentioned several times, since $EM(\mathcal{K})$ is a 2-category it follows that for any 0-cell $\mathbb{A} = (A, t, \mu, \eta)$ of $EM(\mathcal{K})$ we have a monoidal structure on the category $EM(\mathcal{K})(\mathbb{A})$. Furthermore, since a wreath is a monad in the 2-category $EM(\mathcal{K})$ it follows that a wreath in $\mathcal{K}$ is nothing but an algebra in a monoidal category of the form $EM(\mathcal{K})(\mathbb{A})$, where $\mathbb{A}$ is a 0-cell in $EM(\mathcal{K})$, that is, a monad in $\mathcal{K}$. Similarly, a cowreath in $\mathcal{K}$, that is a comonad in the 2-category $EM(\mathcal{K})$, is nothing but a coalgebra in a monoidal category of the form $EM(\mathcal{K})(\mathbb{A})$, where $\mathbb{A}$ is a suitable monad in $\mathcal{K}$.

From now on we denote a wreath in $EM(\mathcal{K})$ by $(A, t, \mu, \eta, s, \psi, \zeta, \sigma)$ or, shortly, by $(\mathbb{A}, s, \psi, \zeta, \sigma)$ in the case when the structure of the monad $\mathbb{A} = (A, t, \mu, \eta)$ is fixed from the beginning. By [45] to such a wreath we can associate the so called wreath product. That is the monad $(A, st, \mu_{st}, \sigma)$ in $\mathcal{K}$, with $\mu_{st} : stst \stackrel{1_s \odot \psi \odot 1_t}{\Longrightarrow} sstt \stackrel{1_s \odot 1_s \odot \mu}{\Longrightarrow} sst \stackrel{\zeta \odot 1_t}{\Longrightarrow} stt \stackrel{1_s \odot \mu}{\Longrightarrow} st$. The wreath product is an algebra in $\mathcal{K}(A)$. The same is (t, μ, η) and we have $\iota := (1_s \odot \mu)(\sigma \odot 1_t) : t \Rightarrow st$ an algebra morphism in $\mathcal{K}(A)$.

Definition 4 The canonical monad extension associated to $(\mathbb{A}, s, \psi, \zeta, \sigma)$, a wreath in $\mathcal{K}$, is the monad morphism $(1_A, \iota) : \mathbb{A} \to (A, st, \mu_{st}, \sigma)$ in $\mathcal{K}$. We call it Frobenius if $\mathcal{K}(A)$ admits coequalizers and any object of it is coflat and, moreover, the algebra extension $\iota : t \Rightarrow st$ is Frobenius in $\mathcal{K}(A)$.

Due to the monoidal flavor of the above definition we have the following characterizations for $(1_A, \iota)$ to be Frobenius (see [21, Theorem 8.6]).

Theorem 10 *For $(\mathbb{A}, s, \psi, \zeta, \sigma)$ a wreath in $\mathcal{K}$ the following are equivalent:*

(i) $(\mathbb{A}, s, \psi, \zeta, \sigma)$ is a Frobenius monad in $EM(\mathcal{K})$, i.e. a Frobenius wreath;

(ii) (s, ψ) is a Frobenius algebra in the monoidal category $EM(\mathcal{K})(\mathbb{A})$;

(iii) (s, ψ) admits a coalgebra structure in $EM(\mathcal{K})(\mathbb{A})$ with the comultiplication structure morphism (s, ψ)-bilinear, that is there exists a cowreath structure in $\mathcal{K}$ of the form $(\mathbb{A}, s, \psi, s \overset{\delta}{\Longrightarrow} sst\ ,\ s \overset{\varepsilon}{\Longrightarrow} t\ )$ such that

$$
\begin{aligned}
(1_s \odot 1_s \odot \mu)(\delta \odot 1_t)\zeta &= (1s \odot 1_s \odot \mu)(1_s \odot \zeta \odot 1_t)(1_s \odot 1_s \odot \psi)(\delta \odot 1_s) \\
&= (1_s \odot 1_s \odot \mu)(1_s \odot \psi \odot 1_t)(\zeta \odot 1_s \odot 1_t)(1_s \odot \delta);
\end{aligned}
$$

(iv) There exists an adjunction $(\rho, \lambda) : (s, \psi) \dashv (s, \psi)$ in $EM(\mathcal{K})(\mathbb{A})$ such that λ is associative, this means, $\mu(\lambda \odot 1_t)(1_s \odot \psi)(\zeta \odot 1_s) = \mu(\lambda \odot 1_t)(1_s \odot \zeta)$.

(v) There is an adjunction $(\rho, \lambda) : (s, \psi) \dashv (s, \psi)$ in $EM(\mathcal{K})(\mathbb{A})$ with λ having the form $\lambda = \mu(\vartheta \odot 1_t)\zeta : ss \Rightarrow t$, for some 2-cell $\vartheta : s \Rightarrow t$ in $\mathcal{K}$.

If $\mathcal{K}(A)$ admits coequalizers and any object of it is coflat then (i)–(v) above are also equivalent to

(vi) The canonical monad extension of $(A, t, \mu, \eta, s, \psi, \zeta, \sigma)$ is Frobenius.

We specialize Theorems 10 to the case when $\mathcal{K} = C$, a monoidal category regarded as an one object 2-category. It this situation a wreath in C is a pair (A, X) with A an algebra in C and (X, ψ, ζ, σ) an algebra $\mathcal{T}_A^\#$, see [14]. Here $\mathcal{T}_A^\#$ denotes the monoidal category $EM(C)(A)$, and the notation is justified by the fact that $\mathcal{T}_A^\#$ is a generalization of the category of transfer morphisms through the algebra A in C, $\mathcal{T}_A$, previously introduced by Tambara in [67]. Note that $\psi : X \otimes A \to A \otimes X$, $\zeta : X \otimes X \to A \otimes X$ and $\sigma : \underline{1} \to A \otimes X$ are morphisms in C satisfying seven compatibility relations that come from (13)–(18).

For a wreath (A, X) in C we denote by $A\#_{\psi,\zeta,\sigma}X$ the associated wreath product. $A\#_{\psi,\zeta,\sigma}X$ is an algebra in C and σ induces an algebra morphism $\iota = \underline{m}_A(\mathrm{Id}_A \otimes \sigma) : A \to A\#_{\psi,\zeta,\sigma}X$ in C. Furthermore, the wreath algebra $A\#_{\psi,\zeta,\sigma}X$ is also an A-ring in C, that is an algebra in the monoidal category ${}^{!}_{A}C_A$, provided that A, X are left coflat objects (we refer to Sect. 2.2 for more details). In fact, we have an A-ring structure on $A \otimes X$ with the left A-module structure given by $\underline{m}_A$ if and only if (A, X) is a wreath in C, cf. [14]. When $A\#_{\psi,\zeta,\sigma}X$ is considered as an A-ring we will denote it by $A \otimes X$.

We can now characterize Frobenius wreaths in C (the order switches, as $\circ$ is given by the tensor product; so $(X, Y) \mapsto YX = Y \circ X = Y \otimes X$, if $X, Y \in C$).

Corollary 4 *Let C be a monoidal category and $(A, X, \psi, \zeta, \sigma)$ a wreath in C. Then the following assertions are equivalent:*

(i) $(A, X, \psi, \zeta, \sigma)$ is a Frobenius wreath in C;

(ii) $(X, \psi) \in \mathcal{T}_A^\#$ is a Frobenius monoidal algebra;

(iii) (X, ψ) admits a coalgebra structure (X, ψ, δ, f) in $\mathcal{T}_A^\#$ such that δ is X-bilinear, that is

$$(\underline{m}_A X^2)(A\zeta X)(\psi X^2)(X\delta) = (\underline{m}_A X^2)(A\delta)\zeta = (\underline{m}_A X^2)(A\psi X)(AX\zeta)(\delta X).$$

(iv) There is an adjunction $(\rho, \lambda) : (X, \psi) \dashv (X, \psi)$ in $\mathcal{T}_A^\#$ with λ associative or, otherwise stated, there exist morphisms $\rho : \underline{1} \to A \otimes X \otimes X$ and $\lambda : X \otimes X \to A$ in C such that the following relations are satisfied:

$$(\underline{m}_A X^2)(A\psi X)(AX\psi)(\rho A) = (\underline{m}_A X^2)(A\rho),$$
$$\underline{m}_A(\lambda A) = \underline{m}_A(A\lambda)(\psi X)(X\psi)\ ,\ \underline{m}_A(A\lambda)(\zeta A) = \underline{m}_A(A\lambda)(\psi X)(X\zeta),$$
$$(\underline{m}_A X)(A\lambda X)(\psi X^2)(X\rho) = \underline{\eta}_A X = (\underline{m}_A X)(A\psi)(AX\lambda)(\rho X).$$

(v) There is an adjunction $(\rho, \lambda) : (X, \psi) \dashv (X, \psi)$ in $\mathcal{T}_A^\#$ with λ of the form $\lambda = \underline{m}_A(\mathrm{Id}_A \otimes \varsigma)\zeta$, for some morphism $\varsigma : X \to A$ in C, or, in other words, there exist morphisms $\rho : \underline{1} \to A \otimes X \otimes X$ and $\varsigma : X \to A$ in C such that the conditions in (iv) above are fulfilled if we keep ρ and replace λ with $\underline{m}_A(\mathrm{Id}_A \otimes \varsigma)\zeta$.

The assertions (i)–(v) above are also equivalent to

(vi) $\iota : A \to A\#_{\psi,\zeta,\sigma}X$ is an algebra Frobenius extension in C;

(vii) $A \otimes X$ is a Frobenius algebra in ${}^{!}_{A}C_A$, i.e., a Frobenius A-ring.

If $\underline{1}$ is a left $\otimes$-generator for C then (i)–(vii) above are also equivalent to

(viii) The functor restriction of scalars $F : C_{A\#_{\psi,\zeta,\sigma}X} \to C_A$ is Frobenius.

In [18] we applied the results in the last corollary to the wreath extensions produced by generalized entwining structures, previously introduced in [17]. The aim of the next subsections is to recall part of these results.

5.2 Frobenius Functors for the Category of Generalized Entwined Modules

It has been proved in [14] that pairs (A, X) with A an algebra in C and (X, ψ) a coalgebra in $\mathcal{T}_A^\#$ are in a one to one correspondence with the so called cowreaths in C, that is comonads in the 2-category $EM(C)$, cf. [45]. Thus a right generalized entwining structure in C is nothing but a cowreath in C. Explicitly, if A is an algebra in C then a coalgebra in $\mathcal{T}_A^\#$ is a 4-tuple (X, ψ, δ, f) with $X \in C$ and $\psi : XA \to AX$, $\delta : X \to AX^2$ and $f : X \to A$ such that $(X, \psi) \in \mathcal{T}_A^\#$ and

$$(\underline{m}_A X^2)(A\psi X)(AX\psi)(\delta A) = (\underline{m}_A X^2)(A\delta)\psi,$$
$$(\underline{m}_A X^3)(A\delta X)\delta = (\underline{m}_A X^3)(A\psi X^2)(AX\delta)\delta,$$
$$\underline{m}_A(Af)\psi = \underline{m}_A(fA),\ \ ((\underline{m}_A(Af))X)\delta = \underline{\eta}_A f = (\underline{m}_A X)(A\psi)(AXf)\delta.$$

To distinguish this generalized entwining structure, from now on we will denote it by (A, X, ψ, δ, f) or, when no confusion is possible, by (A, X, ψ). We call it entwining structure if $\delta = \underline{\eta}_A \otimes \Delta$ and $f = \underline{\eta}_A\varepsilon$ for some morphisms $\Delta : X \to X \otimes X$ and $\varepsilon : X \to \underline{1}$. Note that examples of generalized entwining structures which are not entwining structures can be produced by using actions and coactions of quasi-bialgebras; see [14, Proposition 5.3 and Corollary 5.4].

The connections between wreaths and cowreaths in C was established in [17, Theorem 3.1]. The category ${}^{\#}_{A}\mathcal{T}$ of left transfer morphisms through A has as objects pairs $X = (X, \varphi)$, with $\varphi : A \otimes X \to X \otimes A$ such that $(X, \varphi) \in \mathrm{EM}(C^{\mathrm{op}})(A, A)$; and similarly for morphisms.

Theorem 11 *Let A be an algebra in a (strict) monoidal category C. Take $(X, \psi) \in \mathcal{T}^{\#}_{A}$, and assume that $X \dashv {}^*X$ in C. Then $({}^*X, \overline{\psi}) \in {}^{\#}_{A}\mathcal{T}$, where*

$$\overline{\psi} = ({}^*X(A\mathrm{ev}'_X)(\psi{}^*X))(\mathrm{coev}'_X A{}^*X) \ :\ A{}^*X \to {}^*XA. \tag{19}$$

If C has right duality, we have a strong monoidal functor ${}^*(-) : \mathcal{T}^{\#}_{A} \to \overline{{}^{\#}_{A}\mathcal{T}^{\mathrm{op}}}$.

For (A, X, ψ, δ, f) a right generalized entwining structure in C, a right generalized entwined module in C is a right module $\mathfrak{M}$ in C over A (denote by $\nu_{\mathfrak{M}} : \mathfrak{M}A \to \mathfrak{M}$ the right A-module morphism structure of $\mathfrak{M}$ in C) for which there exists $\rho_{\mathfrak{M}} : \mathfrak{M} \to \mathfrak{M}X$ in C_A satisfying $\nu_{\mathfrak{M}}(\mathfrak{M}f)\rho_{\mathfrak{M}} = \mathrm{Id}_{\mathfrak{M}}$ and $(\rho_{\mathfrak{M}}X)\rho_{\mathfrak{M}} = (\nu_{\mathfrak{M}}X^2)(\mathfrak{M}\delta)\rho_{\mathfrak{M}}$. A morphism between two right generalized entwined modules $\mathfrak{M}$, $\mathfrak{N}$ is a morphism $f : \mathfrak{M} \to \mathfrak{N}$ in C_A such that $\rho_{\mathfrak{N}}f = (fX)\rho_{\mathfrak{M}}$. By $C(\psi)^X_A$ we denote the category of right generalized entwined modules and right colinear morphisms in C_A over the coalgebra (X, ψ, δ, f) in $\mathcal{T}^{\#}_{A}$.

Under some coflatness conditions on A, X it was proved in [14] that $\mathfrak{C} := AX$ has an A-coring structure in C, and that the category $C(\psi)^X_A$ is isomorphic to the category of right corepresentations in C over the A-coring $\mathfrak{C}$.

If X has a right dual in C then $C(\psi)^X_A$ can be identified also with a category of representations. More exactly, according to [17, Proposition 3.3], if (A, X, ψ) is a cowreath and $X \dashv {}^*X$ in C, then $(A, {}^*X, \overline{\psi})$, with $\overline{\psi}$ given by (19), is a right wreath, with multiplication $\underline{m}_{{}^*X}$ and unit $\underline{\eta}_{{}^*X}$ given by the formulas $\underline{m}_{{}^*X} = ({}^*XA\mathrm{ev}'_X(X\mathrm{ev}'_X{}^*X))(({}^*X\delta)\mathrm{coev}'_X{}^*X^2)$ and $\underline{\eta}_{{}^*X} = ({}^*Xf)\mathrm{coev}'_X$, respectively. The wreath product, denoted by ${}^*X\#A$, is an algebra in C with unit $\underline{\eta}_{\#} = ({}^*Xf)\mathrm{coev}'_X$ and multiplication $\underline{m}_{\#}$ given by

$$\underline{m}_{\#} = \big({}^*X\underline{m}_A(\underline{m}_A\mathrm{ev}'_X A)(A\psi{}^*XA)(AX\mathrm{ev}'_X A{}^*XA)(\delta({}^*XA)^2)\big)(\mathrm{coev}'_X({}^*XA)^2).$$

Now, [17, Theorem 3.4] ensures that, for any cowreath (A, X, ψ), if $X \dashv {}^*X$ in C then the categories $C(\psi)^X_A$ and $C_{{}^*X\#A}$ are isomorphic.

The above descriptions for $C(\psi)_A^X$ help us to find necessary and sufficient conditions for the forgetful functor $\mathfrak{F} : C(\psi)_A^X \to C_A$ to be Frobenius. Recall that a coalgebra C in a monoidal category C is Frobenius if it is Frobenius as an algebra in the opposite monoidal category associated to C. This is equivalent to the existence of a Frobenius system (t, B), consisting of morphisms $t : \underline{1} \to C$ (the Frobenius morphism or element when it can be identified with an element in the target) and $B : C \otimes C \to \underline{1}$ (the Casimir morphism) in C satisfying appropriate conditions; see [17, Definition 4.2].

The next result is [17, Theorems 4.8 and 5.2]. For the equivalence between (ii) and (iii) below we did not need $\underline{1}$ to be a left $\otimes$-generator; instead, we ask C to be with coequalizers and such that any object of it is left coflat and robust.

Theorem 12 *Let C be a monoidal category for which its unit object is a left $\otimes$-generator. If (A, X, ψ) is a cowreath in C then the following assertions are equivalent:*

(i) the forgetful functor $\mathfrak{F} : C(\psi)_A^X \to C_A$ is Frobenius;

(ii) (X, ψ) is a Frobenius coalgebra in $\mathcal{T}_A^\#$;

(iii) $A \otimes X$ is a Frobenius coalgebra in ${}_AC_A$, this means, a Frobenius A-coring.

More can be said if X has a right dual object, see [17, Theorem 5.6].

Theorem 13 *Let (A, X, ψ) be a cowreath, and assume that $X \dashv {}^*X$ in C. Then the following assertions are equivalent:*

(i) (X, ψ) is a Frobenius coalgebra in $\mathcal{T}_A^\#$;

(ii) ${}^*X \otimes A$ is a Frobenius A-ring;

(iii) $({}^*X, \overline{\psi})$ is a Frobenius algebra in ${}_A^\#\mathcal{T}$;

(iv) the wreath algebra extension $A \to {}^*X\#A$ is Frobenius;

(v) $A \otimes X$ and ${}^*X \otimes A$ are isomorphic as left A, right ${}^*X\#A$-modules in C.

(vi) $A \otimes X$ and ${}^*X \otimes A$ are isomorphic in $C(\psi)_A^X$ and as left A-modules;

(vii) there exists $t : \underline{1} \to X$ in $\mathcal{T}_A^\#$ (a Frobenius morphism for (X, ψ)) such that $\Phi := (\underline{m}_A(A\underline{m}_A)X)(A^2\psi(X\mathrm{ev}'_X A))((A\delta)t{}^*XA) : {}^*XA \to AX$ is an isomorphism in C;

(viii) there exists $\mathbf{B} : X \otimes X \to \underline{1}$ in $\mathcal{T}_A^\#$ (a Casimir morphism for (X, ψ)) such that $\Psi := ({}^*X\underline{m}_A)({}^*XA\mathbf{B})({}^*X\psi X)(\mathrm{coev}'_X AX) : AX \to {}^*XA$ is an isomorphism in C.

If $\underline{1}$ is a left $\otimes$-generator of C, then the statements (i)–(viii) are equivalent to

(ix) the functor $F : C(\psi)_A^X \to C_A$ is a Frobenius functor.

5.3 Hopf Modules over Quasi-Hopf Algebras

Let H be a quasi-Hopf algebra. We apply the results obtained in the previous subsection to give necessary and sufficient conditions for the forgetful functor from Doi-Hopf modules, two-sided Hopf modules or Yetter–Drinfeld modules over H to representations of the underlying algebra to be Frobenius.

5.3.1 Doi-Hopf Modules

Let H be a quasi-bialgebra, $(\mathfrak{A}, \rho, \Phi_\rho)$ a right H-comodule algebra as in [33] and C a coalgebra in the monoidal category of right H-modules $\mathcal{M}_H$. Then $(\mathfrak{A}, C)$ is a cowreath in ${}_k\mathcal{M}$ via the following structure:

- $\psi : C \otimes \mathfrak{A} \to \mathfrak{A} \otimes C$, $\psi(c \otimes \mathfrak{a}) = \mathfrak{a}_{(0)} \otimes c \cdot \mathfrak{a}_{(1)}$;
- $\delta : C \to \mathfrak{A} \otimes C \otimes C$, $\delta(c) = \tilde{X}^1_\rho \otimes c_{\underline{1}} \cdot \tilde{X}^2_\rho \otimes c_{\underline{2}} \cdot \tilde{X}^3_\rho$, where $\underline{\Delta}(c) = c_{\underline{1}} \otimes c_{\underline{2}}$ is our Sweedler notation for the comultiplication on C and $\Phi_\rho = \tilde{X}^1_\rho \otimes \tilde{X}^2_\rho \otimes \tilde{X}^3_\rho \in \mathfrak{A} \otimes H \otimes H$ is the reassociator of $\mathfrak{A}$;
- $f : C \to A$, $f(c) = \underline{\varepsilon}(c)1_{\mathfrak{A}}$, where $\underline{\varepsilon}$ is the counit of C.

The category of right corepresentations of the induced $\mathfrak{A}$-coring $\mathfrak{A} \otimes C$ in ${}_k\mathcal{M}$ is isomorphic to the category of right Doi-Hopf modules $\mathcal{M}(H)^C_{\mathfrak{A}}$ from [11]. If $\mathfrak{A} = H$, $\mathcal{M}(H)^C_{\mathfrak{A}} = \mathcal{M}^C_H$ is the category of relative Hopf modules from [19].

By (vii) of Theorem 13, the forgetful functor $\mathfrak{F} : \mathcal{M}(H)^C_{\mathfrak{A}} \to \mathcal{M}_{\mathfrak{A}}$ is Frobenius if and only if there exists $t = \mathfrak{a}_i \otimes c_i \in \mathfrak{A} \otimes C$ such that $\mathfrak{a}\mathfrak{a}_i \otimes c_i = \mathfrak{a}_i \mathfrak{a}_{(0)} \otimes c_i \cdot \mathfrak{a}_{(1)}$, for all $\mathfrak{a} \in \mathfrak{A}$, and the map

$$ {}^*C \otimes \mathfrak{A} \ni {}^*c \otimes \mathfrak{a} \mapsto {}^*c((c_i)_{\underline{2}} \cdot \tilde{X}^3_\rho)\mathfrak{a}_i \tilde{X}^1_\rho \mathfrak{a}_{(0)} \otimes (c_i)_{\underline{1}} \cdot \tilde{X}^2_\rho \mathfrak{a}_{(1)} \in \mathfrak{A} \otimes C $$

is a k-linear isomorphism, where ${}^*C$ is the linear dual of C.

If C is Frobenius in $\mathcal{M}_H$ with Casimir element t then the element $1_{\mathfrak{A}} \otimes t \in \mathfrak{A} \otimes C$ has the above two properties, so $\mathfrak{F} : \mathcal{M}(H)^C_{\mathfrak{A}} \to \mathcal{M}_{\mathfrak{A}}$ is Frobenius; see [18, Proposition 2.3]. Note that the converse is also true, provided that $\mathfrak{A}$ is an augmented algebra; the next result is [18, Theorem 2.4].

Theorem 14 *Let H be a quasi-Hopf algebra, $\mathfrak{A}$ a right H-comodule algebra and C a right H-module coalgebra. If the forgetful functor $\mathfrak{F} : \mathcal{M}(H)^C_{\mathfrak{A}} \to \mathcal{M}_{\mathfrak{A}}$ is Frobenius then for any algebra map $\mathfrak{E} : \mathfrak{A} \to k$ the morphism*

$$ {}^*C \ni {}^*c \mapsto \mathfrak{E}(\tilde{x}^1_\rho){}^*c(\mathfrak{t}_{\underline{2}} \cdot (\tilde{x}^2_\rho)_2 p^2 S(\tilde{x}^3_\rho))\mathfrak{t}_{\underline{1}} \cdot (\tilde{x}^2_\rho)_1 p^1 \in C $$

is a k-linear isomorphism, where $\mathfrak{t} = \mathfrak{E}(\mathfrak{a}_i)c_i \in C$ with $t = \mathfrak{a}_i \otimes c_i \in \mathfrak{A} \otimes C$ as above; $p_R = p^1 \otimes p^2$ is the element defined in (2) while $\tilde{x}^1_\rho \otimes \tilde{x}^2_\rho \otimes \tilde{x}^3_\rho$ is our notation for Φ^{-1}_ρ, the inverse of Φ_ρ. Consequently, the forgetful functor $\mathfrak{F} : \mathcal{M}^C_H \to \mathcal{M}_H$ is Frobenius if and only if C is a Frobenius coalgebra in $\mathcal{M}_H$.

Assume that C is a Frobenius coalgebra in $\mathcal{M}_H$; by [18, Proposition 2.2], C is finite dimensional and therefore $\mathcal{M}^C_H$ is isomorphic to $\mathcal{M}_{{}^*C\#H}$, the category of right representations of the wreath algebra ${}^*C\#H$ (which in this case is the right smash product of ${}^*C$ with H). It follows then that $\mathfrak{F}$ is Frobenius if and only if the extension $H \to {}^*C\#H$ is a Frobenius algebra extension in ${}_k\mathcal{M}$. Now, if H is a Hopf algebra, an alternative proof for the last assertion in Theorem 14 can be obtained with the help of Example 1.

5.3.2 Two-Sided Hopf Modules

Let H be a quasi-bialgebra, $\mathfrak{A}$ a right H-comodule algebra and C an H-bimodule coalgebra, that is a coalgebra within ${}_H\mathcal{M}_H$. Then C is right $H \otimes H^{\rm op}$-module coalgebra, $\mathfrak{A} \otimes H^{\rm op}$ is a right $H \otimes H^{\rm op}$-comodule algebra and the category of Doi-Hopf modules $\mathcal{M}(H \otimes H^{\rm op})^C_{\mathfrak{A}\otimes H^{\rm op}}$ identifies with the category of two-sided $(H, \mathfrak{A})$-bimodules over C, ${}_H\mathcal{M}^C_{\mathfrak{A}}$, defined in [11].

If H is a quasi-Hopf algebra, by using part (vii) of Theorem 13 we deduce that the forgetful functor $\mathfrak{F} : {}_H\mathcal{M}^C_{\mathfrak{A}} \to {}_H\mathcal{M}_{\mathfrak{A}}$ is Frobenius if and only if there exists an element $t = \mathfrak{a}_i \otimes h_i \otimes c_i \in \mathfrak{A} \otimes H \otimes C$ such that $\mathfrak{a}\mathfrak{a}_i \otimes h_i h \otimes c_i = \mathfrak{a}_i \mathfrak{a}_{(0)} \otimes h_i h_1 \otimes h_2 \cdot c_i \cdot \mathfrak{a}_{(1)}$, for all $\mathfrak{a} \in \mathfrak{A}$, $h \in H$, and the map ${}^*c \otimes \mathfrak{a} \otimes h \mapsto {}^*c(x^3 \cdot (c_i)_{\underline{2}} \cdot \tilde{X}^3_\rho)\mathfrak{a}_i \tilde{X}^1_\rho \mathfrak{a}_{(0)} \otimes h_1 x^1 h_i \otimes h_2 x^2 \cdot (c_i)_{\underline{1}} \cdot \tilde{X}^2_\rho \mathfrak{a}_{(1)}$ is an isomorphism.

Theorem 14, specialized for the cowreath $(\mathfrak{A} \otimes H^{\rm op}, C)$, gives the following.

Theorem 15 ([18, Theorem 3.4]) *Let H be a quasi-Hopf algebra with bijective antipode, $\mathfrak{A}$ a right H-comodule algebra and C an H-bimodule coalgebra. Then:*

(i) If C is a Frobenius coalgebra in ${}_H\mathcal{M}_H$ then so is $\mathfrak{F} : {}_H\mathcal{M}^C_{\mathfrak{A}} \to {}_H\mathcal{M}_{\mathfrak{A}}$.

(ii) If the forgetful functor $\mathfrak{F} : {}_H\mathcal{M}^C_{\mathfrak{A}} \to {}_H\mathcal{M}_{\mathfrak{A}}$ is Frobenius then for any algebra morphism $\mathfrak{E} : \mathfrak{A} \to k$ there exists and element $\mathfrak{t} \in C$ such that $h \cdot \mathfrak{t} = \varepsilon(h)\mathfrak{t}$, for all $h \in H$, and $\mathfrak{E}(\mathfrak{a}_{\langle 0\rangle})\mathfrak{t} \cdot \mathfrak{a}_{\langle 1\rangle} = \mathfrak{E}(\mathfrak{a})\mathfrak{t}$, for all $\mathfrak{a} \in \mathfrak{A}$, and, moreover, the map ${}^*C \ni {}^*c \mapsto \mathfrak{E}(\tilde{x}^1_\rho){}^*c(q^2 \cdot \mathfrak{t}_{\underline{2}} \cdot (\tilde{x}^2_\rho)_2 p^2 S(\tilde{x}^3_\rho))q^1 \cdot \mathfrak{t}_{\underline{1}} \cdot (\tilde{x}^2_\rho)_1 p^1 \in C$ is a k-linear isomorphism. Consequently,

(iii) The forgetful functor $\mathfrak{F} : {}_H\mathcal{M}^C_H \to {}_H\mathcal{M}_H$ is Frobenius if and only if C is a Frobenius coalgebra within the monoidal category ${}_H\mathcal{M}_H$.

The underlying quasi-coalgebra structure of H yields a coalgebra structure on H in ${}_H\mathcal{M}_H$, where H is viewed an H-bimodule via its multiplication. Owing to the proof of [18, Theorem 3.4], H is a Frobenius coalgebra in ${}_H\mathcal{M}_H$ if and only if H is finite dimensional and unimodular. Hence, $\mathfrak{F} : {}_H\mathcal{M}^H_H \to {}_H\mathcal{M}_H$ is Frobenius if and only if H is finite dimensional and unimodular.

5.3.3 Yetter–Drinfeld Modules

Let H be a quasi-Hopf algebra, C an H-bimodule coalgebra and $\mathbb{A}$ an H-bicomodule algebra as in [33]. For the definition of the category of right Yetter–Drinfeld modules $\mathcal{YD}(H)^C_{\mathbb{A}}$ over the Yetter–Drinfeld datum $(H, \mathbb{A}, C)$ we refer to [20]. Due to [16, 18], the category $\mathcal{YD}(H)^C_{\mathbb{A}}$ is isomorphic to the category of Doi-Hopf modules $\mathcal{M}(H^{\rm op} \otimes H)^C_{\underline{\mathbb{A}}^2}$, where $\underline{\mathbb{A}}^2 = \mathbb{A}$ as a k-algebra and it is a right $H^{\rm op} \otimes H$-comodule algebra with coaction determined by $\underline{\rho}^2 : \mathbb{A} \to \mathbb{A} \otimes (H^{\rm op} \otimes H)$, $\underline{\rho}^2(u) = u_{[0]_{(0)}} \otimes (S(u_{[-1]}) \otimes u_{[0]_{(1)}})$, for all $u \in \mathbb{A}$, and $\Phi_{\underline{\rho}^2} = (\tilde{x}^3_\lambda)_{(0)} \tilde{X}^1_\rho \Theta^2_{(0)} \otimes (S(\tilde{x}^2_\lambda \Theta))^1 f^1 \otimes (\tilde{x}^3_\lambda)_{(1)_1} \tilde{X}^2_\rho \Theta^2_{(1)}) \otimes (S(\tilde{x}^1_\lambda) f^2 \otimes (\tilde{x}^3_\lambda)_{(1)_2} \tilde{X}^3_\rho \Theta^3)$.

C is a right $H^{\rm op} \otimes H$-module coalgebra, with right $H^{\rm op} \otimes H$-action given by $c \cdot (h \otimes h') = h \cdot c \cdot h'$, for all $h, h' \in H$ and $c \in C$. We denoted by $(\mathbb{A}, \lambda_{\mathbb{A}} : \mathbb{A} \ni$

$u \mapsto u_{[-1]} \otimes u_{[0]} \in H \otimes \mathbb{A}$, $\Phi_\lambda = \tilde{X}^1_\lambda \otimes \tilde{X}^2_\lambda \otimes \tilde{X}^3_\lambda \in H \otimes H \otimes \mathbb{A}$) the left H-comodule algebra structure of $\mathbb{A}$, by $\tilde{x}^1_\lambda \otimes \tilde{x}^2_\lambda \otimes \tilde{x}^3_\lambda$ the inverse of Φ_λ and by $\theta^1 \otimes \theta^2 \otimes \theta^3$ the inverse of the element $\Phi_{\lambda,\rho} \in H \otimes \mathbb{A} \otimes H$ that makes the connection between the left and right H-comodule structures on $\mathbb{A}$; for more details we refer to [15, 33].

Proposition 9 ([18, Proposition 4.2]) *Let H be a quasi-Hopf algebra, $C \in {}_H\mathcal{M}_H$ a coalgebra and $\mathbb{A}$ an H-bicomodule algebra. The forgetful functor $\mathfrak{F} : \mathcal{YD}(H)^C_{\mathbb{A}} \to \mathcal{M}_{\mathbb{A}}$ is Frobenius if and only if C is finite dimensional and there exists $t = u_i \otimes c_i \in \mathbb{A} \otimes C$ with $uu_i \otimes c_i = u_i u_{[0]_{(0)}} \otimes S(u_{[-1]}) c_i u_{[0]_{(1)}}$, for all $u \in \mathbb{A}$, and such that*

$$\begin{aligned}\kappa :\ \mathbb{A} \otimes {}^*C \ni u \otimes {}^*c \mapsto \langle {}^*c, \tilde{X}^1_\lambda S(\theta^1_1 (\tilde{X}^2_\lambda)_1 \mathfrak{p}^1) f^2 \cdot (c_i)_{\underline{2}} \cdot \theta^2_{(1)_2} (\tilde{x}^2_\rho)_2 p^2 S(\theta^3 \tilde{x}^3_\rho) \rangle \\ uu_i \theta^2_{(0)} \tilde{x}^1_\rho (\tilde{X}^3_\lambda)_{(0)} \otimes S(\theta^1_2 (\tilde{X}^2_\lambda)_2 \mathfrak{p}^2) f^1 \cdot (c_i)_{\underline{1}} \cdot \theta^2_{(1)_1} (\tilde{x}^2_\rho)_1 p^1 (\tilde{X}^3_\lambda)_{(1)} \in \mathbb{A} \otimes C,\end{aligned}$$

is an isomorphism. So $\mathfrak{F}$ is Frobenius if C is a Frobenius coalgebra in ${}_H\mathcal{M}_H$.

The natural question arises whether the converse of the final statement in Proposition 9 holds: is C a Frobenius coalgebra in ${}_H\mathcal{M}_H$ if $\mathfrak{F}$ is Frobenius? Theorem 14 provides an answer to questions of this type, but, unfortunately, it cannot be applied in our situation. However, if $C = H$ an affirmative answer to the question has been done in [18, Theorem 4.5]; recall that $\mathcal{YD}^H_H$ is $\mathcal{YD}(H)^C_{\mathbb{A}}$ with $\mathbb{A} = C = H$.

Theorem 16 *For a quasi-Hopf algebra H, the following are equivalent:*
(i) The forgetful functor $\mathfrak{F} : \mathcal{YD}^H_H \to \mathcal{M}_H$ is Frobenius;
(ii) H is finite dimensional and unimodular;
(iii) H is finite dimensional and Frobenius as a coalgebra in ${}_H\mathcal{M}_H$.

Replacing H by $H^{\rm op}$, we obtain necessary and sufficient conditions for the forgetful functor $\mathfrak{U} : {}_H\mathcal{YD}^H \to {}_H\mathcal{M}$ to be Frobenius. Unimodularity of H and $H^{\rm op}$ are equivalent, and since H is finite dimensional we have ${}_H\mathcal{YD}^H \simeq {}_{D(H)}\mathcal{M}$, where $D(H)$ is the quantum double of H; see [15, 33, 34]. So our results imply that the algebra extension $H \hookrightarrow D(H)$ is Frobenius if and only if H is unimodular.

Acknowledgements The authors are very grateful to the referee for careful reading of the paper and valuable suggestions and comments.

References

1. Abrams, L.: Two-dimensional topological quantum field theories and Frobenius algebras. J. Knot Theory Ramif. **5**, 569–587 (1996)
2. Bell, A.D., Farnsteiner, R.: On the theory of Frobenius extensions and its application to Lie superalgebras. Trans. Am. Math. Soc. **335**, 407–424 (1993)
3. Bergen, J.: A note on smash products over Frobenius algebras. Comm. Algebra **21**, 4021–4024 (1993)

4. Blattner, R.J., Cohen, M., Montgomery, S.: Cross products and inner actions of Hopf algebras. Trans. Am. Math. Soc. **298**, 671–711 (1986)
5. Blattner, R.J., Montgomery, S.: Crossed products and Galois extensions of Hopf algebras. Pacific J. Math. **137**, 37–54 (1989)
6. Borceux, F.: Handbook of Categorical Algebra I: Basic Category Theory. Encyclopedia Mathematics and Its Applications, vol. 50. Cambridge University Press, Cambridge (1994)
7. Brauer, R., Nesbitt, C.: On the regular representations of algebras. Proc. Nat. Acad. Sci. USA **23**, 236–240 (1937)
8. Bulacu, D.: A structure theorem for quasi-Hopf bimodule coalgebras. Theory Appl. Categ. (TAC) **32**, 1–30 (2017)
9. Bulacu, D.: The weak braided Hopf algebra structures of some Cayley-Dickson algebras. J. Algebra **322**, 2404–2427 (2009)
10. Bulacu, D.: A Clifford algebra is a weak Hopf algebra in a suitable symmetric monoidal category. J. Algebra **332**, 244–284 (2011)
11. Bulacu, D., Caenepeel, S.: Two-sided two-cosided Hopf modules and Doi-Hopf modules for quasi-Hopf algebras. J. Algebra **270**, 55–95 (2003)
12. Bulacu, D., Caenepeel, S.: Integrals for (dual) quasi-Hopf algebras. Appl. J. Algebra **266**, 552–583 (2003)
13. Bulacu, D., Caenepeel, S.: Corings in monoidal categories. In: New Techniques in Hopf Algebras and Graded Ring Theory, pp. 53–78. K. Vlaam. Acad. België Wet. Kunsten (KVAB), Brussels (2007)
14. Bulacu, D., Caenepeel, S.: Monoidal ring and coring structures obtained from wreaths and cowreaths. Algebr. Represent. Theory **17**, 1035–1082 (2014)
15. Bulacu, D., Caenepeel, S., Panaite, F., Van Oystaeyen, F.: Quasi-Hopf Algebras: A Monoidal Approach. Encyclopedia Mathematics and its Application, vol. 171. Cambridge University Press, Cambridge (2019)
16. Bulacu, D., Caenepeel, S., Torrecillas, B.: Doi-Hopf modules and Yetter-Drinfeld modules for quasi-Hopf algebras. Comm. Algebra **34**, 3413–3449 (2006)
17. Bulacu, D., Caenepeel, S., Torrecillas, B.: Frobenius and separable functors for the category of generalized entwined modules, I: general theory. Algebr. Represent. Theory. arXiv:1612.09540v3
18. Bulacu, D., Caenepeel, S., Torrecillas, B.: Frobenius and separable functors for the category of generalized entwined modules, II: applications. J. Algebra **515**, 236–277 (2018)
19. Bulacu, D., Nauwelaerts, E.: Relative Hopf modules for (dual) quasi-Hopf algebras. J. Algebra **229**, 632–659 (2000)
20. Bulacu, D., Panaite, F., Van Oystaeyen, F.: Generalized diagonal crossed products and smash products for (quasi) Hopf algebras. Comm. Math. Phys. **266**, 355–399 (2006)
21. Bulacu, D., Torrecillas, B.: On Frobenius and separable algebra extensions in monoidal categories: applications to wreaths. J. Noncommutative Geom. **9**, 707–774 (2015)
22. Coecke, B., Pavlovic, D., Vicary, J.: A new description of orthogonal bases. Math. Struct. Comput. Sci. **23**, 555–567 (2013)
23. Day, B., Pastro, C.: Note on Frobenius monoidal functors. New York J. Math. **14**, 733–742 (2008)
24. Dăscălescu, S., Năstăsescu, C., Năstăsescu, L.: Frobenius algebras of corepresentations and group-graded vector spaces. J. Algebra **406**, 226–250 (2014)
25. Dăscălescu, S., Năstăsescu, C., Năstăsescu, L.: Symmetric algebras in categories of corepresentations and smash products. J. Algebra **465**, 62–80 (2016)
26. Dăscălescu, S., Năstăsescu, C., Năstăsescu, L.: Graded semisimple algebras are symmetric. J. Algebra **491**, 207–218 (2017)
27. Dăscălescu, S., Năstăsescu, C., Raianu, Ş.: Hopf Algebras. An Introduction. Monographs Textbooks in Pure and Applied Mathematics, vol. 235. Marcel Dekker, New York (2001)
28. Dieudonné, J.: Remarks on quasi-Frobenius rings. Illinois J. Math. **2**, 346–354 (1958)
29. Doi, Y., Takeuchi, M.: Cleft comodule algebras for a bialgebra. Comm. Alg. **14**, 801–817 (1986)

30. Drinfeld, V.G.: Quasi-Hopf algebras. Leningrad Math. J. **1**, 1419–1457 (1990)
31. Fuchs, J., Stigner, C.: On Frobenius algebras in rigid monoidal categories. Arab. J. Sci. Eng. Sect. C Theme Issues **33**(2), 175–191 (2008)
32. Griffits, P., Harris, J.: Principles of Algebraic Geometry. Wiley, New York (1978)
33. Hausser, F., Nill, F.: Diagonal crossed products by duals of quasi-quantum groups. Rev. Math. Phys. **11**, 553–629 (1999)
34. Hausser, F., Nill, F.: Doubles of quasi-quantum groups. Comm. Math. Phys. **199**, 547–589 (1999)
35. Iovanov, C.M., Kadison, L.: When weak Hopf algebras are Frobenius. Proc. Amer. Math. Soc. **138**, 837–845 (2010)
36. Kadison, L.: An approach to quasi-Hopf algebras via Frobenius coordinates. J. Algebra **295**, 27–43 (2006)
37. Kadison, L.: New Examples of Frobenius Extensions. University Lecture Series, vol. 14. American Mathematical Society, Providence (1999)
38. Kadison, L., Stolin, A.A.: An approach to Hopf algebras via Frobenius coordinates. Beiträge Algebra Geom. **42**, 359–384 (2001)
39. Kasch, F.: Grundlagen einer Theorie der Frobeniuserweiterungen. Math. Annalen **127**, 453–474 (1954)
40. Kasch, F.: Projective Frobenius-Erweiterungen; Sitzungsber. Heidelberger Akad. Wiss. (Math.-Naturw. Kl.) **61**, 89–109
41. Kasch, F.: Dualitätseigenschaften von Frobenius Erweiterungen. Math. Zeit. **77**, 219–227 (1961)
42. Kassel, C.: Quantum groups. Graduate Texts in Mathematics, vol. 155. Springer, Berlin (1995)
43. Kock, J.: Frobenius Algebras and 2D Topological Quantum Field Theories. London Mathematical Society Student Texts, vol. 59. Cambridge University Press, Cambridge (2004)
44. Kreimer, H.F., Takeuchi, M.: Hopf algebras and Galois extensions of an algebra. Indiana Math J. **30**, 675–692 (1981)
45. Lack, S., Street, R.: The formal theory of monads II. J. Pure Appl. Algebra **175**, 243–265 (2002)
46. Lam, T.Y.: Lectures on Modules and Rings. Graduate Texts in Mathematics, vol. 189. Springer, New York (1999)
47. Mac Lane, S.: Categories for the Working Mathematician. Graduate Texts in Mathematics, vol. 5. Springer, New York (1974)
48. Majid, S.: Foundations of Quantum Group Theory. Cambridge University Press, Cambridge (1995)
49. Montgomery, S.: Hopf Algebras and Their Actions on Rings. CBMS Regional Conference Series in Mathematics, vol. 82. American Mathematical Society, Providence, (1993)
50. Morita, K.: Adjoint pairs of functors and Frobenius extensions. Sc. Rep. T.K.D. Sect. A **9**, 40–71 (1965)
51. Morita, K.: The endomorphism ring theorem for Frobenius extensions. Math. Zeit. **102**, 385–404 (1967)
52. Müger, M.: From sufactors to categories and topology I. Frobenius algebras in and Morita equivalence of tensor categories. J. Pure Appl. Alg. **180**, 81–157 (2003)
53. Nakayama, T.: Note on symmetric algebras. Ann. Math. **39**, 659–668 (1938)
54. Nakayama, T.: On Frobeniusean algebras. Ann. Math. Second Ser. **40**, 611–633 (1939)
55. Nakayama, T.: On Frobeniusean algebras II. Ann. Math. Second Ser. **42**, 1–21 (1941)
56. Nakayama, T., Tsuzuku, T.: A Remark on Frobenius extensions and endomorphism rings I. Nagoya Math. J. **17**, 89–110 (1960)
57. Nakayama, T., Tsuzuku, T.: A Remark on Frobenius extensions and endomorphism rings II. Nagoya Math. J. **19**, 127–148 (1961)
58. Nakayama, T., Tsuzuku, T.: Correction to our paper "On Frobenius extensions II". Nagoya Math. J. **20**, 205 (1962)
59. Pareigis, B.: Einige Bemerkung über Frobeniuserweiterungen. Math. Ann. **153**, 1–13 (1964)

60. Pareigis, B.: Vergessende Funktoren and Ringhomomorphismen. Math. Zeit. **93**, 265–275 (1966)
61. Pareigis, B.: Non-additive ring and module theory V. Projective and coflat objects. Algebra Ber. **40** (1980)
62. Schauenburg, P.: Hopf modules and Yetter-Drinfel'd modules. J. Algebra **169**, 874–890 (1994)
63. Schauenburg, P.: Actions on monoidal categories and generalized Hopf smash products. J. Algebra **270**, 521–563 (2003)
64. Szlachányi, K.: Adjointable Monoidal Functors and Quantum Grupoids. Lecture Notes in Pure and Apllied Mathematics, vol. 239, pp. 291–308 (2004)
65. Street, R.: Frobenius monads and pseudomonoids. J. Math. Phys. **45**, 3930–3948 (2004)
66. Street, R.: Frobenius Algebras and Monoidal Categories. Annual Meeting Australian Mathematical Society (2004)
67. Tambara, D.: The coendomorphism bialgebra of an algebra. J. Fac. Sci. Univ. Tokyo Sect. IA Math. **37**, 425–456 (1990)
68. Yamagami, S.: Frobenius algebras in tensor categories and bimodule extensions. In: George Janelidze, et al. (Eds.), Galois Theory, Hopf Algebras, and Semiabelian Categories, Papers from the Workshop on Categorical Structures for Descent and Galois Theory, Hopf Algebras, and Semiabelian Categories, Toronto, ON, Canada, September 23–28, 2002. In: Fields Institute Communications, vol. 43, pp. 551–570 (2004)

The Functor $S_C^{-1}()$ and Its Relationship with Homological Functors Tor_n and $\overline{EXT}^n$

Bassirou Dembele, Mohamed Ben Faraj ben Maaouia and Mamadou Sanghare

Abstract In this paper we construct a functor that we call localization functor defined in the category of complexes of left A-modules where A is a not necessarily commutative ring and we study some of its properties. Besides, we study its relationship with the homological functors Tor_n and $\overline{EXT}^n$.

Keywords Ring, Module · Saturated multiplicative part · Category of complexes · Localization functors $s^{-1}()$ and $S_C^{-1}()$ · Tensor product functor · Homological functors H^n, Tor_n and $\overline{EXT}^n$

1 Introduction

Let A be an unitary ring and assume that all (left or right) A-modules are unitary ones. In this paper we study the localization functor in the category of complexes of left A-modules $S_C^{-1}() : Comp(A - Mod) \longrightarrow Comp(S^{-1}A - Mod)$, where A is a ring, which is not necessarily commutative and S a saturated multiplicative part of A satisfying the left Ore conditions. Some of the properties of this functor and its relationship with the functors Tor_n and $\overline{EXT}^n$, are also considered.

It is noteworthy to mention that, similar studies have been carried out by several authors in the the context of the category of A-modules $A - Mod$, where A is a commutative ring (see, for instance, [13, Chap. 4], [2, Chap. 3], among others).

B. Dembele (✉) · M. B. Faraj ben Maaouia
Laboratory of Algebra, Codes and Cryptography Applications (LACCA) of UFR of Applied Sciences and Technologies of Gaston Berger University, Saint-Louis (UGB), Senegal
e-mail: bassirou.dembele@aims-senegal.org

M. B. Faraj ben Maaouia
e-mail: maaouiaalg@hotmail.com

M. Sanghare
Cheikh Anta Diop University of Dakar (UCAD), Doctoral School of Mathematics-Computer Sciences, UCAD, Senegal
e-mail: mamadou.sanghare@ucad.edu.sn

M. Siles Molina et al. (eds.), *Associative and Non-Associative Algebras and Applications*, Springer Proceedings in Mathematics & Statistics 311,
https://doi.org/10.1007/978-3-030-35256-1_14

The case where the ring A is not necessarily commutative and endowed with a saturated multiplicative part of A satisfying the left Ore conditions, was treated in [12, Theorem 2.8], where the same functor was constructed and some of its properties were established (see also [8, pages 190, 191], where the relationship of that functor with homological functors Tor_n^A and Ext_A^n, was settled).

The main aim of this paper is to explore the previous results in the framework of the category of complexes over the ring A. Explicitly, our aim is to see if the quoted results on the localization functor can be reached when we change the category of left A-modules $A - Mod$ to the category of complexes of left A-modules $Comp(A - Mod)$.

Note that, in order to recover, in the category of complexes $Comp(A - Mod)$, the results that have been established in the domain of modules $A - Mod$, we have to define a new tensor product in the category $Comp(A - Mod)$. That tensor product allows us also to define a flat object in $Comp(A - Mod)$ and the Tor_n complex. We have to define also a new functor $HOm^{\bullet}$ different from the usual one $Hom_{Comp(A-Mod)}$, with the aim of giving the adjoint functor of our new localization functor, since with the usual one we can't have a complex structure. We will denote by $\overline{EXT}^n$, which in turn is the nth right derived functor of the $HOm^{\bullet}$ functor.

The paper is organized as follows. In the first section we give some useful definitions and results. In the second section we show the following results:

- $S_C^{-1}()$ preserves the projectivity and the flatness of complexes of left A-modules,
- $S_C^{-1}()$ and $S_C^{-1}(\bar{A}) \otimes - : Comp(A - Mod) \longrightarrow Comp(S^{-1}A - Mod)$ are isomorphic, where $\bar{A}$ is the complex of A-modules with A in the 0-th place and 0 in the other places.
- The functors $S_C^{-1}(\bar{A}) \otimes -$ and $HOm^{\cdot\bullet}(S_C^{-1}(\bar{A}), -) : Comp(S^{-1}A - Mod) \longrightarrow Comp(A - Mod)$ are adjoint functors,
- $S_C^{-1}()$ and $HOm^{\cdot\bullet}(S_C^{-1}(\bar{A}), -)$ are adjoint functors.

And finally in the last section we have showed the following results:

- $S_C^{-1}\mathcal{H}_n(C_*) \cong \mathcal{H}_n(S_C^{-1}(C_*))$ for all $n \geq 0$. Where C_* is a projective resolution of a complex of left A-modules C and $\mathcal{H}_n$ is the nth homological functor.
- $S_C^{-1}Tor_n^{Comp(A-Mod)}(C, D) \cong Tor_n^{Comp(S^{-1}A-Mod)}(S_C^{-1}(C), S_C^{-1}(D))$ for all $n \geq 0$ and for all complexes of left A-modules C, D.
- $S_C^{-1}HOm^{\bullet}(P, C) \cong HOm^{\bullet}(S_C^{-1}(P), S_C^{-1}(C))$ for all complexes of left A-modules C and P such that P is of finite type.
- $S_C^{-1}\overline{EXT}^n_{Comp(A-Mod)}(P, C) \cong \overline{EXT}^n_{Comp(S^{-1}A-Mod)}(S_C^{-1}(P), S_C^{-1}(C))$ for all $n \geq 0$ and for all complexes of left A-modules C, P such that P is of type FP_{∞}.

2 Definitions and Preliminary Results

Definition 1 Let C and $\mathcal{D}$ be two categories, F and G two functors with same variance from C to $\mathcal{D}$. A natural transformation or functorial isomorphism from F to G is a map $\Phi : F \longrightarrow G$ so that:

- If F and G are covariant, then

$$\begin{aligned} \Phi : Ob(C) &\longrightarrow Mor(\mathcal{D}) \\ M &\longmapsto \Phi_M \end{aligned}$$

 is a map such that $\Phi_M : F(M) \longrightarrow G(M)$ and for any $f \in Mor(C)$ so that $f : M \longrightarrow N$, then the following diagram is commutative:

$$\begin{array}{ccc} F(M) & \xrightarrow{F(f)} & F(N) \\ {\scriptstyle \Phi_M}\downarrow & & \downarrow{\scriptstyle \Phi_N} \\ G(M) & \xrightarrow{G(f)} & G(N) \end{array}$$

- If F and G are contravariant then the following diagram is commutative:

$$\begin{array}{ccc} F(N) & \xrightarrow{F(f)} & F(M) \\ {\scriptstyle \Phi_N}\downarrow & & \downarrow{\scriptstyle \Phi_M} \\ G(N) & \xrightarrow{G(f)} & G(M) \end{array}$$

Definition 2 Let C and $\mathcal{D}$ two categories, $F : C \longrightarrow \mathcal{D}$ and $G : \mathcal{D} \longrightarrow C$ two functors. It is said that the couple (F, G) is adjoint if for any $A \in Ob(C)$ and for any $B \in Ob(\mathcal{D})$, there is an isomorphism:

$$r_{A,B} : Hom_C(A, G(B)) \longrightarrow Hom_{\mathcal{D}}(F(A), B)$$

so that:

(a) For any $f \in Hom_C(A, A')$, the following diagram is commutative:

$$\begin{array}{ccc} Hom_C(A, G(B)) & \xrightarrow{Hom(f,G(B))} & Hom_C(A', G(B)) \\ {\scriptstyle r_{A,B}}\downarrow & & \downarrow{\scriptstyle r_{A',B}} \\ Hom_{\mathcal{D}}(F(A), B) & \xrightarrow{Hom(F(f),B)} & Hom_{\mathcal{D}}(F(A'), B) \end{array}$$

(b) For any $g \in Hom_{\mathcal{D}}(B, B')$, the following diagram is commutative:

$$\begin{array}{ccc} Hom_{\mathcal{C}}(A, G(B)) & \xrightarrow{Hom(A,G(g))} & Hom_{\mathcal{C}}(A, G(B')) \\ \downarrow{r_{A,B}} & & \downarrow{r_{A,B'}} \\ Hom_{\mathcal{D}}(F(A), B) & \xrightarrow{Hom(F(A),g)} & Hom_{\mathcal{D}}(F(A), B') \end{array}$$

Definition and proposition 1 We define the category of complexes of left A-modules to be the category denoted by $Comp(A - Mod)$ such that:

1. objects are complexes of left A-modules. A complex of left A-modules C is a sequence of homomorphisms of left A-modules $(C^n \xrightarrow{d_C^n} C^{n+1})_{n\in\mathbb{Z}}$ such that $d^{n+1} \circ d^n = 0$, for all $n \in \mathbb{Z}$.
2. Morphisms are maps of complexes of left A-modules. Let C and D be two complexes, a map of complexes of left A-modules $f : C \longrightarrow D$ is a sequence of homomorphisms of left A-modules $(f^n : C^n \longrightarrow D^n)_{n\in\mathbb{Z}}$ such that $f^{n+1} \circ d_C^n = d_D^n \circ f^n$ for $n \in \mathbb{Z}$.

We define in the same way $Comp(Comp(A - Mod))$ the category of double complexes of left A-modules to be the category such as:

- Objets are double complexes of left A-modules.
- Morphisms are maps of double complexes of left A-modules.

Definition 3 1. A complex of left A-modules C is bounded if $C^n = 0$ for $\mid n \mid$ large.
2. C is of finite type if and only if C is bounded and C^n is of finite type for all $n \in \mathbb{Z}$.
3. C is of finite presentation if and only if C is bounded and C^n is of finite presentation for all $n \in \mathbb{Z}$.
4. A complex of left A-modules C is of type FP_∞ if C has a projective resolution:

$$\ldots \longrightarrow P_n \xrightarrow{d_n} \cdots \xrightarrow{d_2} P_1 \xrightarrow{d_1} P_0 \xrightarrow{\epsilon} C \longrightarrow 0$$

where P_n are finite type complexes of left A-modules for all $n \geq 0$.

Proposition 1 *Let A be a ring and S a saturated multiplicative part of A verifying the left Ore conditions. Then the correspondence:*
$S_C^{-1}() : Comp(A - Mod) \longrightarrow Comp(S^{-1}A - Mod)$ *such that*

1. if $C := \ldots \longrightarrow C^n \xrightarrow{d_C^n} C^{n+1} \longrightarrow \ldots$ is an object of $Comp(A - Mod)$ then:

$$S_C^{-1}(C) := \ldots \longrightarrow S^{-1}C^n \xrightarrow{S^{-1}d_C^n} S^{-1}C^{n+1} \longrightarrow \ldots$$

is an object of $Comp(S^{-1}A - Mod)$
2. if $f : C \longrightarrow D$ is a morphism of $Comp(A - Mod)$ then $S_C^{-1}(f) = (S^{-1}(f^n))_{n\in\mathbb{Z}} : S_C^{-1}(C) \longrightarrow S_C^{-1}(D)$ is a morphism of $Comp(S^{-1}A - Mod)$

Then $S_C^{-1}()$ is an exact covariant functor.

Proof • Let $C := \ldots \longrightarrow C^n \xrightarrow{d_C^n} C^{n+1} \longrightarrow \ldots$ be a complex of left A-modules, then

$S_C^{-1}(C) := \ldots \longrightarrow S^{-1}C^n \xrightarrow{S^{-1}(d_C^n)} S^{-1}C^{n+1} \longrightarrow \ldots$ is a complex of left S^{-1} A-modules, since according to ([12, Theorem 2.8]), $S^{-1}C^n$ is a $S^{-1}A$-module for all n and for $\frac{m}{s} \in S^{-1}(C^n)$ we have:

$S^{-1}(d_C^{n+1}) \circ S^{-1}(d_C^n)(\frac{m}{s}) = \frac{d_C^{n+1} \circ d_C^n(m)}{s} = \frac{0}{s} = 0.$

• Let $f : C \longrightarrow D$, be a map of complexes, then we have $\forall n \in \mathbb{Z},\ f^{n+1} \circ d_C^n = d_D^n \circ f^n$.

$S_C^{-1}(f) : S_C^{-1}(C) \longrightarrow S_C^{-1}(D)$ is a morphism of $Comp(S^{-1}A - Mod)$.

Indeed we have:

$$S^{-1}(f^{n+1} \circ d_C^n) = S^{-1}(d_D^n \circ f^n)$$

$$\Longrightarrow S^{-1}(f^{n+1}) \circ S^{-1}(d_C^n) = S^{-1}(d_D^n) \circ S^{-1}(f^n)$$

since $S^{-1}()$ is a covariant functor.

• it is obvious that $S_C^{-1}(1_C) = 1_{S_C^{-1}(C)}$.

• $S_C^{-1}()$ is covariant.

Let $f : C \longrightarrow D$ and $g : D \longrightarrow E$ be two complexes of left A-modules.

We have:

$$\begin{aligned} S_C^{-1}(g \circ f) &= (S^{-1}(g^n \circ f^n))_{n\in\mathbb{Z}} = (S^{-1}(g^n) \circ S^{-1}(f^n))_{n\in\mathbb{Z}} \\ &= (S^{-1}(g^n))_{n\in\mathbb{Z}} \circ (S^{-1}(f^n))_{n\in\mathbb{Z}}, \\ &\Longrightarrow S_C^{-1}(g \circ f) = S_C^{-1}(g) \circ S_C^{-1}(f) \end{aligned}$$

• Let $0 \longrightarrow C \longrightarrow D \longrightarrow E \longrightarrow 0$ be a short exact sequence of complexes of left A-modules. Then $\forall\, n \in \mathbb{Z}$, $0 \longrightarrow C^n \longrightarrow D^n \longrightarrow E^n \longrightarrow 0$ is exact and since the functor $S^{-1}()$ is exact (see [12, Theorem 4.3]) then

$$0 \longrightarrow S^{-1}(C^n) \longrightarrow S^{-1}(D^n) \longrightarrow S^{-1}(E^n) \longrightarrow 0$$

is exact. Thus

$$0 \longrightarrow S_C^{-1}(C) \longrightarrow S_C^{-1}(D) \longrightarrow S_C^{-1}(E) \longrightarrow 0$$

is exact.

□

Definition and Proposition 2 Let X be a complex of $A - B$- bimodules and let be the following correspondence:

$$X \bigotimes - : Comp(B - Mod) \longrightarrow Comp(A - Mod)$$

such that:

- If $Y \in Ob(Comp(B - Mod))$ then $X \bigotimes Y$ is a complex of left A-modules such that:

$$(X \bigotimes Y)^n = \bigoplus_{t \in \mathbb{Z}} X^t \otimes Y^{n-t}$$

$$\delta^n_{(X \bigotimes Y)}(x \otimes y) = d^t_X(x) \otimes y + (-1)^t x \otimes d^{n-t}_Y(y)$$

- If $f : Y_1 \longrightarrow Y_2$ is a map of complexes of $Comp(B - Mod)$ then $(X \bigotimes -)(f) : X \bigotimes Y_1 \longrightarrow X \bigotimes Y_2$ such that:

$$\begin{aligned}(X \bigotimes -)(f)^n : (X \bigotimes Y_1)^n &\longrightarrow (X \bigotimes Y_2)^n \\ x \otimes y &\longmapsto x \otimes f^{n-t}(y)\end{aligned}$$

is a map of complexes of $Comp(A - Mod)$.

Then $X \bigotimes -$ is a covariant functor that is right exact.

Proof $X \bigotimes Y$ is a complex according to [3, Definition 5.7]. It is a complex of left A-modules according to [13, Proposition 2.51]. $(X \bigotimes -)(f)$ is a map of complexes according to [3, Definition 5.7] by setting $X' = X$ and $f = id_X$. Finally $X \bigotimes -$ is right exact by right exactness of tensor product functor in $A - Mod$. □

Corollary 1 *Let A be a ring and S a saturated multiplicative part of A verifying the left Ore conditions. Let $\bar{A} : \ldots \longrightarrow 0 \longrightarrow A \longrightarrow 0 \longrightarrow \ldots$ be the complex of left A-modules such that all its components are 0 except the 0th component that is the A-module A. Then*

$$S_C^{-1}(\bar{A}) \bigotimes - : Comp(A - Mod) \longrightarrow Comp(S^{-1}A - Mod)$$

is an exact covariant functor.

Proof Remark that $S^{-1}A$ is a $S^{-1}A - A$-bimodule. □

Definition and Proposition 3 Let X be a complex of $A - B$-bimodules. Let be the following correspondence:

$$HOm^{\bullet}(X, -) : Comp(A - Mod) \longrightarrow Comp(B - Mod)$$

such that

- If Y is a complex of left A-modules then $HOm^{\bullet}(X, -)(Y) = HOm^{\bullet}(X, Y)$ is a complex of left B-modules such that:

$$HOm^{\bullet}(X, Y)^n = \prod_{t \in \mathbb{Z}} Hom_A(X^t, Y^{n+t})$$

and its $\delta_{HOm^\bullet(X,Y)}$ is defined as follows:

$$\left(\delta^n_{HOm^\bullet(X,Y)}\right)_t : Hom_A(X^t, Y^{n+t}) \longrightarrow Hom_A(X^t, Y^{n+t+1})$$
$$g^t \longmapsto d_Y^{n+t} g^t + (-1)^{n+1} g^{t+1} d_X^t$$

- If $f : Y_1 \longrightarrow Y_2$ is a morphism of $Comp(A)$ then:

$$HOm^\bullet(X,-)(f)^n : HOm^\bullet(X,Y_1)^n \longrightarrow HOm^\bullet(X,Y_2)^n$$
$$(g^t)_t \longmapsto (f^{n+t} \circ g^t)_t$$

  is morphism of $Comp(B - Mod)$.

Then $HOm^\bullet(X,-)$ is a covariant functor that is left exact.

Proof $HOm^\bullet(X,Y)$ is a complex of left B-modules, see [3, Definition 5.2] and [13, Proposition 2.54]. Also by setting $X = X'$, $Y_1 = Y$ and $Y_2 = Y'$ in [3, Definition 5.2], we see that $HOm^\bullet(X,-)(f)$ is a morphism of $Comp(B - Mod)$. So, $HOm^\bullet(X,-)$ is a covariant functor. $HOm^\bullet(X,-)$ is left exact by left exactness of $Hom(X^t,-)$ for all $t \in \mathbb{Z}$. □

Corollary 2 *The functor*

$$HOm^\bullet(S_C^{-1}(\bar{A}),-) : Comp(S^{-1}A - Mod) \longrightarrow Comp(A - Mod)$$

is an exact covariant functor.

Proof Observe that $S^{-1}A$ is an $S^{-1}A - A$ bimodule. □

3 Functor $S_C^{-1}()$, Isomorphism and Adjoint Isomorphism

Definition 4 Let P be a complex of left A-modules. Then P is projective if for all epimorphism of complexes of left A-modules $\alpha : C \longrightarrow D$ and for all map of complexes $\beta : P \longrightarrow D$, there exists $\phi : P \longrightarrow C$ such that $\beta = \alpha \circ \phi$ otherwise the following diagram is commutative:

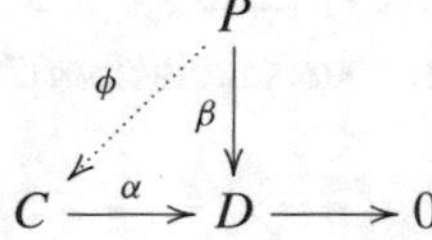

Dually we have the concept of injective complex of left A-modules.

Proposition 2 *Let A be a ring and S a saturated multiplicative part of A verifying the left Ore conditions. Then $S_C^{-1}()$ preserve the projectivity of complexes of left A-modules.*

Proof Let P be an projective object of $Comp(A - Mod)$ and the following diagram:

$$\begin{array}{ccccc} & & S_C^{-1}(P) & & \\ & & \big\downarrow{\scriptstyle g} & & \\ C & \xrightarrow{f} & D & \longrightarrow & 0 \end{array}$$

where f is an epimorphism of $Comp(S^{-1}A - Mod)$ and g is a map of complexes of $Comp(S^{-1}A - Mod)$.

Then f and g are also maps of complexes of $Comp(A - Mod)$ and since P is projective and we have:

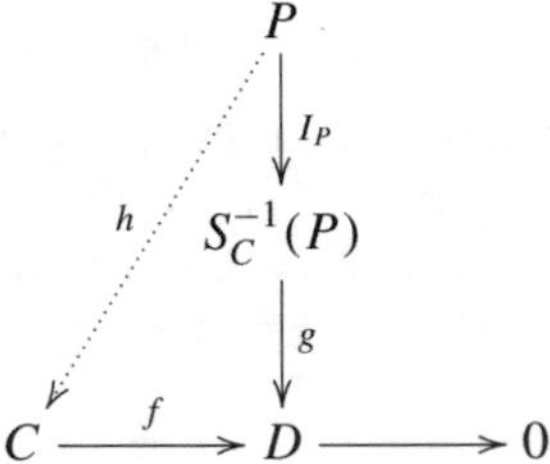

where $I_P : P \longrightarrow S_C^{-1}(P)$ is the canonical map of complexes.

Then we have:

$$f \circ h = g \circ I_P$$

$$\Longrightarrow S_C^{-1}(f \circ h) = S_C^{-1}(g \circ I_P)$$

$$\Longrightarrow S_C^{-1}(f) \circ S_C^{-1}(h) = S_C^{-1}(g) \circ S_C^{-1}(I_P) = S_C^{-1}(g) = g$$

$$\Longrightarrow f \circ S_C^{-1}(h) = g$$

Thus $S_C^{-1}(P)$ is projective. □

Theorem 1 *Let A be a ring and S a saturated multiplicative part of A verifying the left Ore conditions. Then the functors $S_C^{-1}()$ and $S_C^{-1}(\bar{A}) \bigotimes -$ are isomorphic.*

Proof Let $C : \ldots \longrightarrow C^{n-1} \xrightarrow{d_C^{n-1}} D^n \xrightarrow{d_C^n} C^{n+1} \longrightarrow \ldots$ be a complex of left A-modules; let us show that there exists an isomorphism:

$$\Phi_C : S_C^{-1}(\bar{A}) \bigotimes C \longrightarrow S_C^{-1}(C)$$

According to the proof of [12, Theorem 4.7], for all $n \in \mathbb{Z}$ there exists an isomorphism:

$$\Phi_{C^n} : S^{-1}(A) \otimes C^n \longrightarrow S^{-1}C^n$$
$$(\sum \frac{a_i}{s_i} \otimes m_i) \longmapsto \sum \frac{(a_i m_i)}{s_i}$$

Let us set $\Phi_C = (\Phi_{C^n})_{n \in \mathbb{Z}}$ and show that it is a map of complexes. By definition, Φ_C is a map of complexes if and only if the following diagram is of square commutative:

$$\begin{array}{ccccccc}
\ldots \longrightarrow & S^{-1}A \otimes C^{n-1} & \xrightarrow{1_{S^{-1}A} \otimes d_C^{n-1}} & S^{-1}A \otimes C^n & \longrightarrow \ldots \\
& \downarrow \Phi_{C^{n-1}} & & \downarrow \Phi_{C^n} & \\
\ldots \longrightarrow & S^{-1}C^{n-1} & \xrightarrow{S^{-1}(d_C^{n-1})} & S^{-1}C^n & \longrightarrow \ldots
\end{array}$$

Indeed, let $\sum \frac{a_i}{s_i} \otimes m_i \in S^{-1}A \otimes C^{n-1}$ then:

(a)

$$\begin{aligned}
\Phi_{C^n} \circ \left(1_{S^{-1}A} \otimes d_C^{n-1}\right) (\sum \frac{a_i}{s_i} \otimes m_i) &= \Phi_{C^n} \left(\sum \frac{a_i}{s_i} \otimes d_C^{n-1}(m_i)\right) \\
&= \sum \frac{a_i d^{n-1}(m_i)}{s_i}
\end{aligned}$$

(b)

$$\begin{aligned}
S^{-1}(d_C^{n-1}) \circ \Phi_{C^n} (\sum \frac{a_i}{s_i} \otimes m_i) &= S^{-1}(d_C^{n-1}) (\sum \frac{a_i m_i}{s_i}) \\
&= \sum \frac{a_i d_C^{n-1}(m_i)}{s_i}
\end{aligned}$$

(a) and (b) imply that the diagram is of square commutative. Then Φ_C is a map of complexes and also for all $n \in \mathbb{Z}$, Φ_{C^n} is an isomorphism. Therefore, for all complex of left A-modules C, Φ_C is an isomorphism.

On the other hand, let $C : \ldots \longrightarrow C^{n-1} \xrightarrow{d_C^{n-1}} C^n \xrightarrow{d_C^n} C^{n+1} \longrightarrow \ldots$ and $D : \ldots \longrightarrow D^{n-1} \xrightarrow{d_D^{n-1}} D^n \xrightarrow{d_C^n} D^{n+1} \longrightarrow \ldots$ be two complexes of left A-modules and $f : C \longrightarrow D$ be a map of complexes. Then the following diagram is commutative:

$$\begin{array}{ccc}
S_C^{-1}(\bar{A}) \otimes C & \xrightarrow{\Phi_C} & S_C^{-1}(C) \\
\downarrow 1_{S_C^{-1}(\bar{A})} \otimes f & & \downarrow S_C^{-1}(f) \\
S_C^{-1}(\bar{A}) \otimes D & \xrightarrow{\Phi_D} & S_C^{-1}(D)
\end{array}$$

because according to the proof of [12, Theorem 4.7], for all $n \in \mathbb{Z}$, we have that the following diagram is commutative:

$$\begin{array}{ccc} S^{-1}A \otimes C^n & \xrightarrow{\Phi_{C^n}} & S^{-1}C^n \\ \downarrow {\scriptstyle 1_{S^{-1}A}\otimes f^n} & & \downarrow {\scriptstyle S^{-1}(f^n)} \\ S^{-1}A \otimes D^n & \xrightarrow{\Phi_{D^n}} & S^{-1}D^n \end{array}$$

Hence, the functors $S_C^{-1}()$ and $S_C^{-1}(\bar{A}) \bigotimes -$ are isomorphic. □

Definition 5 Let F be a complex of right A-modules. We say that F is flat if and only if the functor $F \bigotimes -$ is exact.

Corollary 3 *Let A be a ring and S a saturated multiplicative part of A verifying the left Ore conditions. Then $S_C^{-1}(\bar{A})$ is a flat complex of A-modules.*

Proof Let $0 \longrightarrow C \xrightarrow{f} D \xrightarrow{g} E \longrightarrow 0$ be an exact sequence of complexes of left A-modules. According to the proof of [12, Theorem 4.7], the following diagram is commutative and with exact rows.

$$\begin{array}{ccccccccc} 0 & \longrightarrow & S_C^{-1}(C) & \xrightarrow{S_C^{-1}(f)} & S_C^{-1}(D) & \xrightarrow{S_C^{-1}(g)} & S_C^{-1}(E) & \longrightarrow & 0 \\ & & \downarrow {\scriptstyle \Phi_C^{-1}} & & \downarrow {\scriptstyle \Phi_D^{-1}} & & \downarrow {\scriptstyle \Phi_E^{-1}} & & \\ 0 & \longrightarrow & S_C^{-1}(\bar{A}) \otimes C & \xrightarrow{1_{S_C^{-1}(C)} \otimes f} & S_C^{-1}(\bar{A}) \otimes D & \xrightarrow{1_{S_C^{-1}(D)} \otimes g} & S_C^{-1}(\bar{A}) \otimes E & \longrightarrow & 0 \end{array}$$

Then the functor $S_C^{-1}(\bar{A}) \otimes -$ is exact; therefore the complex of A-modules $S_C^{-1}(\bar{A})$ is flat. □

Corollary 4 *Let A be a ring and S a saturated multiplicative part of A verifying the left Ore conditions and C a flat complex of right A-modules. Then, $S_C^{-1}(C)$ is a flat complex of right A-modules.*

Proof Let $0 \longrightarrow C_1 \xrightarrow{f} C_2$ be an exact sequence of complexes of right A-modules. Since C is flat then the sequence

$$0 \longrightarrow C \bigotimes C_1 \xrightarrow{1_{S_C^{-1}(C)} \otimes f} C \bigotimes C_2$$

is actually an exact sequence of complexes of $\mathbb{Z}$-modules. Now, since the following diagram is commutative:

$$\begin{array}{ccccc} 0 & \longrightarrow & S_C^{-1}(C) \bigotimes C_1 & \xrightarrow{1_{S_C^{-1}(C)} \otimes f} & S_C^{-1}(C) \bigotimes C_2 \\ & & \downarrow {\scriptstyle \Phi_C^{-1} \otimes 1_{C_1}} & & \downarrow {\scriptstyle \Phi_C^{-1} \otimes 1_{C_2}} \\ 0 & \longrightarrow & S_C^{-1}(\bar{A}) \otimes C \bigotimes C_1 & \xrightarrow{1_{S_C^{-1}(\bar{A})} \otimes 1_C \otimes f} & S_C^{-1}(\bar{A}) \otimes C \bigotimes C_1, \end{array}$$

and seeing that the row below is exact, then the first row is also exact. This means that

$$0 \longrightarrow S_C^{-1}(C) \bigotimes C_1 \overset{1_{S_C^{-1}(C)} \otimes f}{\longrightarrow} S_C^{-1}(C) \bigotimes C_2$$

is an exact sequence. Therefore, $S_C^{-1}(C)$ is a flat complex of left A-modules. □

Theorem 2 *Let A be a ring, S a saturated multiplicative part of A verifying the left Ore conditions and C a flat complex of left A-modules. Then $S_C^{-1}(C)$ is a flat complex of left $S^{-1}A$-modules.*

Proof Let $0 \longrightarrow C_1 \longrightarrow C_2$ be an exact sequence of complexes of right $S^{-1}A$-modules. Then for all n and t, $0 \longrightarrow C_1^{n-t} \longrightarrow C_2^{n-t}$ is an exact sequence of right $S^{-1}A$-modules. Furthermore, by [12, Corollaire 4.11], we know that $S^{-1}C^t$ is flat, therefore $0 \longrightarrow S^{-1}C^t \otimes C_1^{n-t} \longrightarrow S^{-1}C^t \otimes C_2^{n-t}$ is exact. Thus $0 \longrightarrow S_C^{-1}(C) \otimes C_1 \longrightarrow S_C^{-1}(C) \otimes C_2$ is exact as well. Hence $S_C^{-1}(C)$ is a flat complex of left $S^{-1}A$-modules. □

Corollary 5 *Let A be a ring and S a saturated multiplicative part of A verifying the left Ore conditions. Then $S_C^{-1}()$ preserve the flatness.*

Lemma 1 *Let X be a complex of $A - B$-bimodules. Then the functors*

$$HOm^\bullet(X, -) : Comp(A - Mod) \longrightarrow Comp(B - Mod)$$

and

$$X \bigotimes - : Comp(B - Mod) \longrightarrow Comp(A - Mod)$$

are adjoint functors.

Proof See [3, p. 180]. □

Theorem 3 *Let A be a ring and S a saturated multiplicative part of A verifying the left Ore conditions. Then the functors $S_C^{-1}(\bar{A}) \otimes -$ and $Hom^\bullet(S_C^{-1}(\bar{A}), -)$ are adjoint functors.*

Proof It follows from Lemma 1 and the fact that $S^{-1}A$ is a $S^{-1}A - A$ bimodule. □

Theorem 4 *Let A be a ring and S a saturated multiplicative part of A verifying the left Ore conditions. Then the functors $S_C^{-1}()$ and $Hom^\bullet(S_C^{-1}(\bar{A}), -)$ are adjoint functors.*

Proof Theorem 3 and the fact that $S_C^{-1}()$ and $S_C^{-1}(\bar{A}) \otimes -$ are isomorphic, give the result. □

4 The Action of $S_C^{-1}()$ on the Homological Functors Tor_n and $\overline{EXT}^n$

Definition 6 Let $n \in \mathbb{Z}$ and $C_* \in Ob(Comp(Comp(A-Mod)))$. Then we define the nth homology of C_* by:

$$\mathcal{H}_n(C_*) = \left(kerd_n^m / imd_{n+1}^m\right)_{m\in\mathbb{Z}} = \left(H_n(C_*^m)\right)_{m\in\mathbb{Z}},$$

where C_*^m is the mth row of the double complex C_*.

Theorem 5 *Let A be a ring and S a saturated multiplicative part of A verifying the left Ore conditions and C a complex of left A-modules. Then for all projective resolution of C, $C_\bullet : \ldots \longrightarrow P_n \xrightarrow{d_n} \ldots \longrightarrow P_1 \xrightarrow{d_1} P_0 \longrightarrow C \longrightarrow O$ we have:*

$$\mathcal{H}_n(S_C^{-1}(C_\bullet)) \cong S_C^{-1}\mathcal{H}_n(C_\bullet)$$

Proof Let $C_\bullet : \ldots \longrightarrow P_n \xrightarrow{d_n} \ldots \longrightarrow P_1 \xrightarrow{d_1} P_0 \longrightarrow C \longrightarrow O$ be a projective resolution of the complex of left A-modules C. At the first part, we have:

$$\begin{aligned} S_C^{-1}\mathcal{H}_n(C_\bullet) &= S_C^{-1}(Kerd_n/Imd_{n+1}) \\ &=(S^{-1}(Kerd_n^m/Imd_{n+1}^m))_{m\in\mathbb{Z}} \\ &= (S^{-1}H_n(C_*^m))_{m\in\mathbb{Z}} \end{aligned}$$

And secondly we have:

$$\begin{aligned} \mathcal{H}_n(S_C^{-1}(C_\bullet)) &= Ker(S_C^{-1}d_n)/Im(S_C^{-1}d_{n+1}) \\ &= (KerS^{-1}d_n^m/ImS^{-1}d_{n+1}^m)_{m\in\mathbb{Z}} \\ &= (H_n(S^{-1}C_*^m))_{m\in\mathbb{Z}}. \end{aligned}$$

Now since according to [8, Lemma 4.2] by taking $T = S^{-1}()$, we have that $H_n(S^{-1}C_*^m) \cong S^{-1}H_n(C_*^m)$. Then

$$\mathcal{H}_n(S_C^{-1}(C_\bullet)) \cong (S^{-1}H_n(C^m))_{m\in\mathbb{Z}} = S_C^{-1}\mathcal{H}_n(C_\bullet)$$

Hence,

$$\mathcal{H}_n(S_C^{-1}(C)) \cong S_C^{-1}\mathcal{H}_n(C)$$

and this finishes the proof. □

Theorem 6 *Let A be a ring and S a saturated multiplicative part of A verifying the left Ore conditions and C and D two complexes of left A-modules. Then for all $n \geq 0$ we have the following isomorphism:*

$$S_C^{-1} Tor_n^{Comp(A-Mod)}(C, D) \cong Tor_n^{Comp(S^{-1}A-Mod)}(S_C^{-1}(C), S_C^{-1}(D)),$$

where Tor_n is the nth left derived of the tensor product functor.

Proof Let C and D two complexes of left A-modules. We will consider the following two cases: $n = 0$ and $n > 0$.

- For $n = 0$, let us show that the result is true, that is,

$$S_C^{-1}(C \bigotimes D) \simeq S_C^{-1} C \bigotimes S_C^{-1} D.$$

By [13, Theorem 6.29] this is equivalent to show that

$$S_C^{-1} Tor_0^{Comp(A-Mod)}(C, D) \cong Tor_0^{Comp(S^{-1}A-Mod)}(S_C^{-1}(C), S_C^{-1}(D)).$$

So take $m \in \mathbb{Z}$, and let us check that:

$$S_C^{-1}(C \bigotimes D)^m = S^{-1}(\bigoplus C^t \otimes D^{m-t}) \cong (S_C^{-1} C \bigotimes S_C^{-1} D)^m = \bigoplus S^{-1} C^t \otimes S^{-1} D^{m-t}$$

The morphism

$$\begin{aligned} \Phi_1 : S^{-1}(\bigoplus C^t \otimes D^{m-t}) &\longrightarrow \bigoplus S^{-1}(C^t \otimes D^{m-t}) \\ \frac{\sum c_t \otimes p_{m-t}}{s} &\longmapsto \sum \frac{c_t \otimes p_{m-t}}{s} \end{aligned}$$

is an isomorphism and since:

$$\begin{aligned} (\Phi_2)_t : S^{-1}(C^t \otimes D^{m-t}) &\longrightarrow S^{-1} C^t \otimes S^{-1} D^{m-t} \\ \frac{c_t \otimes p_{m-t}}{s} &\longmapsto \frac{c_t}{s} \otimes \frac{p_{m-t}}{s} \end{aligned}$$

is an isomorphism according to the proof of [8, Theorem 4.4], we get that:

$$\begin{aligned} \Phi_2 : \bigoplus S^{-1}(C^t \otimes D^{m-t}) &\longrightarrow \bigoplus S^{-1} C^t \otimes S^{-1} D^{m-t} \\ \sum \frac{c_t \otimes p_{m-t}}{s} &\longmapsto \sum \frac{c_t}{s} \otimes \frac{p_{m-t}}{s} \end{aligned}$$

is an isomorphism. This implies that $\Phi_D^m = \Phi_2 \circ \Phi_1$ is an isomorphism too, and we have:

$$\begin{aligned} \Phi_D^m : S^{-1}(\bigoplus C^t \otimes D^{m-t}) &\longrightarrow \bigoplus S^{-1} C^t \otimes S^{-1} D^{m-t} \\ \frac{\sum c_t \otimes p_{m-t}}{s} &\longmapsto \sum \frac{c_t}{s} \otimes \frac{p_{m-t}}{s} \end{aligned}$$

Let us set $\Phi_D = (\Phi_D^m)$ and show that Φ_D is a map of complexes, that is, to show that the following diagram is commutative:

$$\begin{array}{ccc}
\dots \longrightarrow S^{-1}(\bigoplus C^t \otimes D^{m-t}) & \xrightarrow{\delta^m_{S_C^{-1}(C\otimes D)}} & S^{-1}(\bigoplus C^t \otimes D^{m+1-t}) \longrightarrow \\
\downarrow \Phi_D^m & & \downarrow \Phi_D^{m+1} \\
\dots \longrightarrow \bigoplus(S^{-1}C^t \otimes S^{-1}D^{m-t}) & \xrightarrow{\delta^m_{(S_C^{-1}(C)\otimes S_C^{-1}(D))}} & \bigoplus(S^{-1}C^t \otimes S^{-1}D^{m+1-t}) \longrightarrow
\end{array}$$

On one hand we have:

$$\begin{aligned}
\Phi_D^{m+1} \circ \delta^m_{S_C^{-1}(C\bigotimes D)}\left(\frac{\sum c_t \otimes p_{m-t}}{s}\right) &= \Phi_D^{m+1}\left(\frac{\sum \delta_C^t(c_t)\otimes p_{m-t} + (-1)^t c_t \otimes \delta_D^{m-t}(p_{m-t})}{s}\right) \\
&= \sum \frac{\delta_C^t(c_t)}{s} \otimes \frac{p_{m-t}}{s} + (-1)^t \frac{c_t}{s} \otimes \frac{\delta_D^{m-t}(p_{m-t})}{s}
\end{aligned}$$

On the other hand, we have:

$$\begin{aligned}
\delta^m_{(S_C^{-1}(C)\bigotimes S_C^{-1}(D))} \circ \Phi_D^m\left(\frac{\sum c_t \otimes p_{m-t}}{s}\right) &= \delta^m_{(S_C^{-1}(C)\bigotimes S_C^{-1}(D))}\left(\sum \frac{c_t}{s} \otimes \frac{p_{m-t}}{s}\right) \\
&= \sum \frac{\delta_C^t(c_t)}{s} \otimes \frac{p_{m-t}}{s} + (-1)^t \frac{c_t}{s} \otimes \frac{\delta_D^{m-t}(p_{m-t})}{s}
\end{aligned}$$

Therefore, Φ_D is a map of complexes. Thus the result is true for $n = 0$.

- For $n > 0$, let $D_\bullet \dots \longrightarrow P_n \longrightarrow \dots \longrightarrow P_1 \longrightarrow P_0 \longrightarrow D \longrightarrow 0$ be a projective resolution of the complex of left A-modules D. Since for all $n \geq 0$, $S_C^{-1}(C \bigotimes P_n) \cong S_C^{-1}(C) \bigotimes S_C^{-1}(P_n)$, then the following diagram is square commutative:

$$\begin{array}{ccccccc}
\dots \longrightarrow S_C^{-1}(C\bigotimes P_1) & \longrightarrow & S_C^{-1}(C\bigotimes P_0) & \longrightarrow & S_C^{-1}(C\bigotimes D) & \longrightarrow & 0 \\
\downarrow \Phi_{P_1} & & \downarrow \Phi_{P_0} & & \downarrow \Phi_D & & \\
\dots \longrightarrow S_C^{-1}C\bigotimes S_C^{-1}P_1 & \longrightarrow & S_C^{-1}C\bigotimes S^{-1}P_0 & \longrightarrow & S_C^{-1}C\bigotimes S_C^{-1}D & \longrightarrow & 0
\end{array}$$

Because the following diagram is square commutative:

$$\begin{array}{ccccc}
\ldots \longrightarrow S^{-1}(\bigoplus C^t \otimes P_0^{m-t}) & \longrightarrow & S^{-1}(\bigoplus C^t \otimes D^{m-t}) & \longrightarrow & 0 \\
\downarrow \Phi^m_{P0} & & \downarrow \Phi^m_D & & \\
\ldots \longrightarrow \bigoplus(S^{-1}C^t \otimes S^{-1}P_0^{m-t}) & \longrightarrow & \bigoplus(S^{-1}C^t \otimes S^{-1}D^{m-t}) & \longrightarrow & 0
\end{array}$$

For verification, we do the same thing that we have done with the first diagram of this proof. So, we obtain the isomorphism:

$$S_C^{-1}(C \bigotimes D_\bullet) \cong S_C^{-1}(C) \bigotimes S_C^{-1}(D_\bullet)$$

Now, since, by Proposition 2, $S_C^{-1}()$ preserve the projectivity of complexes of left A-modules, we have that

$$\begin{aligned}
\mathcal{H}_n(S_C^{-1}(C \otimes D_\bullet)) &\cong \mathcal{H}_n(S_C^{-1}(C) \otimes S_C^{-1}(D_\bullet)) \\
&= Tor_n^{Comp(S^{-1}A-Mod)}(S_C^{-1}(C), S_C^{-1}(D))
\end{aligned}$$

According to Theorem 5, we have:

$$\mathcal{H}_n(S_C^{-1}(C \bigotimes D_\bullet)) \cong S_C^{-1}\mathcal{H}_n(C \bigotimes D_\bullet) = S_C^{-1}Tor_n^{Comp(A-Mod)}(C, D)$$

So,

$$S_C^{-1}Tor_n^{Comp(A-Mod)}(C, D) \cong Tor_n^{Comp(S^{-1}A-Mod)}(S_C^{-1}(C), S_C^{-1}(D))$$

and this finishes the proof. □

Theorem 7 *Let A be a ring and S a saturated multiplicative part of A verifying the left Ore conditions, P and C two complexes of left A-modules such that P is of finite type. Then we have:*

$$S_C^{-1}HOm^\bullet(P, C) \cong HOm^\bullet(S_C^{-1}(P), S_C^{-1}(C))$$

Proof Let $n \in \mathbb{Z}$, we need to show that:

$$(S_C^{-1}HOm^\bullet(P, C))^n \overset{\Phi^n_{(P,C)}}{\cong} (HOm^\bullet(S_C^{-1}(P), S_C^{-1}(C)))^n.$$

That is:

$$S^{-1}\prod_{t\in\mathbb{Z}} Hom_A(P^t, C^{n+t}) \cong \prod_{t\in\mathbb{Z}} Hom_{S^{-1}A}(S^{-1}P^t, S^{-1}C^{n+t}).$$

Since $S^{-1}\prod_{t\in\mathbb{Z}} Hom_A(P^t, C^{n+t}) \cong \prod_{t\in\mathbb{Z}} S^{-1}Hom_A(P^t, C^{n+t})$ and according to [9, p. 82], we have that

$$S^{-1}Hom_A(P^t, C^{n+t}) \overset{\Phi_{(P^t,C^{n+t})}}{\cong} Hom_{S^{-1}A}(S^{-1}P^t, S^{-1}C^{n+t}).$$

In this way, we get:

$$S^{-1}\prod_{t\in\mathbb{Z}} Hom_A(P^t, C^{n+t}) \cong \prod_{t\in\mathbb{Z}} Hom_{S^{-1}A}(S^{-1}P^t, S^{-1}C^{n+t}).$$

Setting $\Phi^n_{(P,C)} = (\Phi_{(P^t,C^{n+t})})$, let us show that $\Phi_{(P,C)} = (\Phi^n_{(P,C)})$ is a map of complexes, that is the following diagram is square commutative:

$$\begin{array}{ccccccc}
\cdots \longrightarrow & S_C^{-1}(HOm^\bullet(P,C))^n & \xrightarrow{\delta^n_{S_C^{-1}HOm^\bullet(P,C)}} & S_C^{-1}(HOm^\bullet(P,C))^{n+1} & \longrightarrow \cdots \\
& \downarrow \Phi^n_{(P,C)} & & \downarrow \Phi^{n+1}_{(P,C)} & \\
\cdots \longrightarrow & HOm^\bullet(S_C^{-1}(P), S_C^{-1}(C))^n & \xrightarrow{\delta^n_{HOm^\bullet(S_C^{-1}(P),S_C^{-1}(C))}} & HOm^\bullet(S_C^{-1}(P), S_C^{-1}(C))^{n+1} & \longrightarrow \cdots
\end{array}$$

otherwise, for all $t \in \mathbb{Z}$ the following diagram is square commutative:

$$\begin{array}{ccccccc}
\cdots \longrightarrow & S^{-1}(HOm_A(P^t, C^{n+t})) & \xrightarrow{\alpha_{n,t}} & S^{-1}(HOm_A(P^t, C^{n+t+1})) & \longrightarrow \cdots \\
& \downarrow \Phi_{(P^t,C^{n+t})} & & \downarrow \Phi_{(P^t,C^{n+t+1})} & \\
\cdots \longrightarrow & HOm_{S^{-1}A}(S^{-1}P^t, S^{-1}C^{n+t}) & \xrightarrow{(\beta_{n,t}} & HOm^\bullet(S_C^{-1}(P), S_C^{-1}(C))_n & \longrightarrow \cdots
\end{array}$$

where $(\delta^n_{S_C^{-1}HOm^\bullet(P,C)})_t = \alpha_{n,t}$ and $(\delta^n_{HOm^\bullet(S_C^{-1}(P),S_C^{-1}(C))})_t = \beta_{n,t}$

On the one hand, since:

$$\Phi_{(C^t,P^{n+t})}(\frac{g_t}{\sigma})(\frac{p}{s}) = \frac{1}{s}.\frac{g_t(p)}{\sigma}$$

and

$$\alpha_{n,t}(\frac{g_t}{\sigma}) = \frac{\delta_C^{n+t} \circ g_t + (-1)^{n+1} g_{n+1} \circ \delta_P^t}{\sigma},$$

then we have:

$$\Phi_{(P^t,C^{n+t+1})} \circ \alpha_{n,t}(\frac{g_t}{\sigma}) = \Phi_{(P^t,C^{n+t+1})}(\frac{\delta_C^{n+t} \circ g_t + (-1)^{n+1} g_{t+1} \circ \delta_P^t}{\sigma})$$

$$\Longrightarrow \Phi_{(P^t,C^{n+t+1})} \circ \alpha_{n,t}(\frac{g_t}{\sigma})(\frac{p}{s}) = \Phi_{(P^t,C^{n+t+1})}(\frac{\delta_C^{n+t} \circ g_t + (-1)^{n+1} g_{t+1} \circ \delta_P^t}{\sigma})(\frac{p}{s})$$
$$= (\Phi_{(P^t,C^{n+t+1})}(\frac{\delta_C^{n+t} g_t}{\sigma}))(\frac{p}{s})$$
$$+(-1)^{n+1}(\Phi_{(P^t,C^{n+t+1})}(\frac{g_{t+1}\delta_P^t}{\sigma}))(\frac{p}{s})$$
$$= ((\widehat{\delta_C^{n+t} g_t})_\sigma)(\frac{p}{s}) + (-1)^{n+1}((\widehat{g_{t+1}\delta_P^t})_\sigma)(\frac{p}{s})$$
$$= \frac{1}{s}\frac{(\delta_C^{n+t} g_t)(p)}{\sigma} + (-1)^{n+1}\frac{1}{s}.\frac{(g_{t+1}\delta_P^t)(p)}{\sigma}$$

On the other hand:

$$(\beta_{n,t} \circ \Phi_{(P^t,C^{n+t})})(\frac{g_t}{\sigma})(\frac{p}{s}) = \beta_{n,t}(\widehat{(g_t)_\sigma})(\frac{p}{s})$$
$$= (\delta^{n+t}_{S_C^{-1}(C)}(\widehat{(g_t)_\sigma}) + (-1)^{n+1}\widehat{(g_{t+1})_\sigma}\delta^t_{S_C^{-1}(P)})(\frac{p}{s})$$
$$= (\delta^{n+t}_{S_C^{-1}(C)}(\widehat{(g_t)_\sigma}))(\frac{p}{s}) + (-1)^{n+1}(\widehat{(g_{t+1})_\sigma}\delta^t_{S_C^{-1}(P)})(\frac{p}{s})$$
$$= \delta^{n+t}_{S_C^{-1}(C)}(\frac{\frac{g_t(p)}{\sigma}}{s}) + (-1)^{n+1}\widehat{(g_{t+1})_\sigma}(\frac{\delta^t_{S_C^{-1}(P)}}{s})$$
$$= \frac{1}{s}\frac{(\delta_C^{n+t} g_t)(p)}{\sigma} + (-1)^{n+1}\frac{1}{s}\frac{(g_{t+1}\delta_P^t)(p)}{\sigma}$$

So $\Phi_{(P,C)}$ is a chain map. Thus we obtain the claimed result. □

Theorem 8 *Let A be a ring and S a saturated multiplicative part of A verifying the left Ore conditions, P and C two complexes of left A-modules such that P is of type FP_∞. Then we have:*

$$S_C^{-1}\overline{EXT}^n_{Comp(A-Mod)}(P, C) \cong \overline{EXT}^n_{Comp(S^{-1}A-Mod)}(S_C^{-1}(P), S_C^{-1}(C))$$

where $\overline{EXT}^n$ is the nth right derived functor of $HOm^\bullet$.

Proof According to Theorem 7 we have:

$$S_C^{-1}HOm^\bullet(P, C) \cong HOm^\bullet(S_C^{-1}(P), S_C^{-1}(C)).$$

Since P is of type FP_∞, there exists a projective resolution of P, $P_\bullet$ of finite type projective components:

$$P_\bullet : \ldots \longrightarrow P_n \longrightarrow P_{n-1} \longrightarrow \ldots \longrightarrow P_0 \longrightarrow P \longrightarrow 0.$$

So, the following diagram is square commutative:

$$\begin{array}{ccc} 0 \longrightarrow S_C^{-1}(HOm^\bullet(P,C)) & \longrightarrow & S_C^{-1}(HOm^\bullet(P_0,C)) \longrightarrow \dots \\ \Phi(P,C) \downarrow & & \downarrow \Phi(P_0,C) \\ 0 \longrightarrow HOm^\bullet(S_C^{-1}(P), S_C^{-1}(C)) & \longrightarrow & HOm^\bullet(S_C^{-1}(P_0), S_C^{-1}(C)) \longrightarrow \dots \end{array}$$

since according to ([9], proof of Theorem 3.12) for all integer m and t the following one is square commutative:

$$\begin{array}{ccc} 0 \longrightarrow S^{-1}(Hom_A(P^t, C^{m+t})) & \longrightarrow & S^{-1}(Hom_A(P_0^t, C^{m+t})) \longrightarrow \dots \\ \Phi(P^t,C^{m+t}) \downarrow & & \downarrow \Phi(P_0^t,C^{m+t}) \\ 0 \longrightarrow Hom_{S^{-1}A}(S^{-1}(P^t), S^{-1}(C^{m+t})) & \longrightarrow & Hom_{S^{-1}A}(S^{-1}(P_0^t), S^{-1}(C^{m+t})) \longrightarrow \dots \end{array}$$

Thus

$$S_C^{-1}HOm^\bullet(P_\bullet, C) \cong HOm^\bullet(S_C^{-1}(P_\bullet), S_C^{-1}(C)),$$

and since, by Proposition 2, $S_C^{-1}()$ preserve the projectivity of complexes of left A-modules, we have that

$$\begin{aligned} &\mathcal{H}^n(S_C^{-1}HOm^\bullet(P_\bullet, C)) \cong \mathcal{H}^n(HOm^\bullet(S_C^{-1}(P_\bullet), S_C^{-1}(C))) \\ &\quad \Longrightarrow \mathcal{H}^n(S_C^{-1}HOm^\bullet(P_\bullet, C)) \cong \overline{EXT}^n_{Comp(S^{-1}A-Mod)}(S_C^{-1}(P), S_C^{-1}(C)). \end{aligned}$$

By Theorem 5, we get

$$\begin{aligned} &S_C^{-1}\mathcal{H}^n(HOm^\bullet(P_\bullet, C)) \cong \overline{EXT}^n_{Comp(S^{-1}A-Mod)}(S_C^{-1}(P), S_C^{-1}(C)) \\ &\quad \Longrightarrow S_C^{-1}\overline{EXT}^n_{Comp(A-Mod)}(P, C) \cong \overline{EXT}^n_{Comp(S^{-1}A-Mod)}(S_C^{-1}(P), S_C^{-1}(C)), \end{aligned}$$

and this finishes the proof. □

References

1. Anderson, F.W., Fuller, K.R.: Rings and categories of modules. Springer, New York (1973)
2. Atiyah, M.F., Macdonald, I.G.: Introduction to commutative algebra, vol. 361. Addison-Wesly Pub. Co., Boston (1969)
3. Beck, V: Algèbre des invariants relatifs pour les groupes de réflexion - Catégorie stable. Thèse de doctorat, Université Denis Diderot-Paris 7 (2008)
4. Diallo, E.O., Maaouia, M.B., Sanghare, M.: Cohopfian objects in the category of complexes of left A-modules. Int. Math. Forum **8**(39), 1903–1920 (2013)
5. Diallo, E.O.: Hopficité et Co-HopficitÃl dans la catégorie COMP des complexes. Thèse de doctorat, Université Cheikh Anta Diop, Dakar (2013)

6. Enochs, E.E., Garcia Rozas, J.R.: Flat covers of complexes. J. Algebra **210**, 86–102 (1998)
7. Enochs, E.E., Garcia Rozas, J.R.: Tensor products of complexes. Math. J. Okayama Univ. **39**, 17–39 (1997)
8. Faye, D., Maaouia, M.B., Sanghare, M.: Localization in duo-ring and polynomials Algebra. In: Gueye, C.T., Siles Molina, M. (eds.) Non Associative, Non-commutative Algebra and Operator Theory. Springer Proceedings in Mathematics and Statistics, vol. 160, pp. 183-191. Springer, Berlin (2016)
9. Faye, D.: Localisation et algébre des polynmes dans un duo-anneau-propriétés homologiques du foncteur $S^{-1}(-)$. Thèse de Doctorat, Université Cheikh Anta Diop, Dakar (2016)
10. Garcia Rozas, J.R.: Covers and envelopes in the category of complexes of modules. Research Notes in Mathematics, vol. 407. Chapman & Hall/CRC, Boca Raton, FL (1999)
11. Maaouia, M.B.: Thèse d'État anneaux et modules de fractions-envoloppes et couvertures plates dans les Duo-Anneaux. Université Cheikh Anta Diop, Dakar (2011)
12. Maaouia, M.B., Sangharé, M.: Modules de fractions, sous-modules S-saturés et foncteur $S^{-1}()$. Int. J. Algebra **6**(16), 775–798 (2012)
13. Rotman, J.J.: An introduction to homological algebra. Academic Press, New York (1972)
14. Rotman, J.J.: Advanced Modern Algebra. Prentice Hall, Upper Saddle River (2003)
15. Weibel, C.A.: An Introduction to Homological Algebra. Cambridge Studies in Advanced Mathematics, vol 38. Cambridge University Press, Cambridge (1995)

BOCSES over Small Linear Categories and Corings

Laiachi El Kaoutit

Abstract This note does not claim anything new, since the material exposed here is somehow folkloric. We provide the main steps in showing the equivalence of categories between the category of BOCSES over a small linear category and the category of corings over the associated ring with enough orthogonal idempotents.

Keywords BOCSES · Small linear categories · Monoidal categories · Corings · Unital modules

1 Introduction

The term BOCS, was used as a terminology for a certain object that have been introduced by Roiter (see for instance [20]) with the aim of systematizing the study of a wide class of *matrix problems* that often appears in representation theory. Inconspicuously, the notion of BOCS is not far from that of cotriple (dual notion of that introduced in [7]). Namely, as it was corroborated by Bautista et al. [4], BOCSES can be realized as a special kind of cotriples (or comonads) over certain functor categories. In this way, the category of BOCSES representations over a full subcategory of all vector spaces turns out to be equivalent to some full subcategory of the Kleisli category attached to the corresponding comonad (see [15] for the precise definition of this category). We refer to [4] for more details on this equivalence of categories.

Research supported by grant MTM2016-77033-P from the Spanish Ministerio de Economía y Competitividad and the European Union FEDER

L. El Kaoutit (✉)
Departamento de Álgebra and IEMath-Granada, Universidad de Granada, Granada, Spain
e-mail: kaoutit@ugr.es
URL: http://www.ugr.es/~kaoutit/

Facultad de Educación, Econonía y Tecnología de Ceuta, Cortadura del Valle s/n, E-51001 Ceuta, Spain

M. Siles Molina et al. (eds.), *Associative and Non-Associative Algebras and Applications*, Springer Proceedings in Mathematics & Statistics 311,
https://doi.org/10.1007/978-3-030-35256-1_15

A more general equivalence between comonads and corings was provided by the author in [8]. Specifically, in [8] we established a bi-equivalence of bicategories between the bicategory of corings over rings with local units and the bicategory of comonads in (right) unital modules with continuous underlying functors (i.e., functors which are right exact and preserve direct sums). Roughly speaking, corings are comonoids in a suitable category of bimodules (either over ring with unit or without). Apparently they appeared for the first time in the literature, under this name, in Sweedler's work [21] about Jacobson–Bourbaki–Hochschild's Theorem (a kind of Galois correspondence first Theorem for division rings. Specifically, a bijection between the set of all coideals of a Sweedler's canonical coring of a division rings extension and the set of all intermediate division rings). Apart from the interest that corings generated in the late sixties and eighties [12–14, 17], in the last decades these objects proved to be crucial in the study of relative modules (representation theories over entwined structures, or distributive laws, see [5, 6]).

In this note we give the main steps to show that for a given small linear category (linear over a commutative base ring), the category of BOCSES over this category is equivalent to the category of corings over the associated ring with enough orthogonal idempotents. As a final conclusion, we can claim that both notions of BOCS (over small linear categories) and comonad (with continuous functor over unital modules), are equivalent to the notion of coring (over ring with enough orthogonal idempotents).

Notations and Conventions: We work over a commutative base ring with 1 (or $1_{\Bbbk}$) denoted by $\Bbbk$. The category of (central) $\Bbbk$-modules is denoted by $\mathrm{Mod}_{\Bbbk}$. A morphism in this category is referred to as a $\Bbbk$-*linear map*. A hom-sets category $\mathcal{C}$ is said to be *small* if its class of objects is actually a set. The notation $c \in \mathcal{C}$ stands for: c is an object of $\mathcal{C}$. The set of all morphisms from an object $c \in \mathcal{C}$ to another one $c' \in \mathcal{C}$, is denoted by $\mathrm{Hom}_{\mathcal{C}}\big(c,\ c'\big)$. The category $\mathcal{C}$ is said to be $\Bbbk$-*linear* if it is additive [16, pp. 192], each of its hom-sets is a $\Bbbk$-module and the composition law consists of $\Bbbk$-bilinear maps. For instance, the category of modules $\mathrm{Mod}_{\Bbbk}$ is a $\Bbbk$-linear category (not necessarily small). A covariant functor $F : \mathcal{C} \to \mathcal{D}$ between two $\Bbbk$-linear categories is said to be $\Bbbk$-*linear* provided that the maps $\mathrm{Hom}_{\mathcal{C}}\big(c,\ c'\big) \to \mathrm{Hom}_{\mathcal{D}}\big(F(c),\ F(c')\big)$ are $\Bbbk$-linear for every pair of objects $(c, c') \in \mathcal{C} \times \mathcal{C}$. Given two $\Bbbk$-linear functors $F, F' : \mathcal{C} \to \mathcal{D}$, where $\mathcal{C}$ and $\mathcal{D}$ are small $\Bbbk$-linear categories, we denote by $\mathrm{Nat}(F, F')$ the set of all natural transformations from F to F'. An element α of this set is a family of morphisms $\{\alpha_c\}_{c \in \mathcal{C}}$, where $\alpha_c \in \mathrm{Hom}_{\mathcal{D}}\big(F(c),\ F'(c)\big)$ are such that the following diagrams commute

$$\begin{array}{ccc} F(c) & \xrightarrow{\quad \alpha_c \quad} & F'(c) \\ {\scriptstyle F(f)}\downarrow & & \downarrow{\scriptstyle F'(f)} \\ F(c') & \xrightarrow{\quad \alpha_{c'} \quad} & F'(c'), \end{array}$$

for every morphism $f \in \mathrm{Hom}_{\mathcal{C}}\big(c,\ c'\big)$. For more basic notions on categories, functors and adjunctions, we refer to the first chapters of Mac Lane's book [16].

In this paper, we shall consider rings without identity element. Nevertheless, we will consider a class of rings (or $\Bbbk$-algebras) which have enough orthogonal idempotents, in the sense of [10, 11], and that are mainly constructed from small categories. Specifically, given any small, hom-sets and $\Bbbk$-linear category $\mathcal{D}$, we can consider the *path algebra*, or the *Gabriel's ring, of* $\mathcal{D}$: Its underlying $\Bbbk$-module is the direct sum $R = \bigoplus_{x, x' \in \mathcal{D}} \mathrm{Hom}_{\mathcal{D}}(x, x')$ of $\Bbbk$-modules. The multiplication of this ring is given by the composition of $\mathcal{D}$. This means that, for any two homogeneous generic elements $r \in \mathrm{Hom}_{\mathcal{D}}(x, x'), r' \in \mathrm{Hom}_{\mathcal{D}}(y, y')$, the multiplication $(1_{\Bbbk}r) . (1_{\Bbbk}r')$ is defined by the rule:

$$(1_{\Bbbk}r) . (1_{\Bbbk}r') = 1_{\Bbbk}(rr'),$$

the image of the composition of r and r', when they are composable, otherwise we set $(1_{\Bbbk}r) . (1_{\Bbbk}r') = 0$, see [11, pp. 346]. For any $x \in \mathcal{D}$, we denote by 1_x the image of the identity arrow of x in the $\Bbbk$-module R.

In general, the ring R has no unity, unless the set of objects of $\mathcal{D}$ is finite. Instead of that, it has a set of local units (for the precise definition see for instance [1–3] or [8]). Namely, the local units are given by the set of idempotents elements:

$$\Big\{1_{x_1} + \cdots + 1_{x_n} \in R \mid \; x_i \in \mathcal{D}, \; i = 1, \cdots, n, \text{ and } n \in \mathbb{N} \setminus \{0\}\Big\}.$$

For example, if we have a discrete category $\mathcal{D}$ with X as a set of objects, that is, the only morphisms are the identities ones, then $R = \Bbbk^{(X)}$ is the ring defined as the direct sum of X-copies of the base ring $\Bbbk$.

An *unital right R-module* is a right R-module M such $MR = M$; *left unital modules* are similarly defined (see [11, pp. 347]). For instance, the previous ring R attached to a given small hom-set $\Bbbk$-linear category $\mathcal{D}$ decomposes as a direct sum of left, and also of right, unital R-modules:

$$R = \bigoplus_{x \in \mathcal{D}} R\, 1_x = \bigoplus_{x \in \mathcal{D}} 1_x\, R.$$

Following [10], a ring which satisfies these two equalities is referred to as a *ring with enough orthogonal idempotents*, whose complete set of idempotents is the set $\{1_x\}_{x \in \mathcal{D}}$. An *unital bimodule* is a bimodule which is unital on both sides.

2 The Monoidal Structure of the Category of Functors on Small Categories

Let $\mathcal{A}$ be a $\Bbbk$-linear small category; we shall use the letters $\mathfrak{p}, \mathfrak{q}, \mathfrak{r}, \mathfrak{s}, \mathfrak{p}_1, \mathfrak{q}_2, \ldots$ to denote the objects of $\mathcal{A}$. The identity morphism of $\mathfrak{p} \in \mathcal{A}$ will be denoted by $1_{\mathfrak{p}}$. The opposite category of $\mathcal{A}$ is denoted as usual by $\mathcal{A}^{op}$, and its objects by $\mathfrak{p}^o, \mathfrak{q}^o, \mathfrak{r}^o, \mathfrak{s}^o, \mathfrak{p}_1^o, \mathfrak{q}_2^o, \ldots$. The object set of $\mathcal{A}^{op}$ is that of $\mathcal{A}$, while morphisms of $\mathcal{A}^{op}$ are the reversed ones in $\mathcal{A}$, that is, a morphism $f^o : \mathfrak{p}^o \to \mathfrak{q}^o$ in $\mathcal{A}^{op}$ stands for a morphism $f : \mathfrak{q} \to \mathfrak{p}$ in $\mathcal{A}$.

We will work with the categories of $\Bbbk$-module valued functors. These are the categories of all $\Bbbk$-linear covariant functors $\mathsf{Funct}\big(\mathcal{A}, \mathrm{Mod}_{\Bbbk}\big)$ and $\mathsf{Funct}\big(\mathcal{A}^{op}, \mathrm{Mod}_{\Bbbk}\big)$. Functors from the product category $\mathcal{A}^{op} \times \mathcal{A}$ to $\mathrm{Mod}_{\Bbbk}$ are also invoked, for instance, the functor: $(\mathfrak{p}^o, \mathfrak{q}) \to \mathrm{Hom}_{\mathcal{A}}\big(\mathfrak{p}, \mathfrak{q}\big)$.

It is well known from Freyd Theorem's [18, pp. 99 and 109] that the set $\big\{\mathrm{Hom}_{\mathcal{A}}\big(\mathfrak{p}, -\big)\big\}_{\mathfrak{p} \in \mathcal{A}}$ forms a generating set of small objects in the category $\mathsf{Funct}\big(\mathcal{A}, \mathrm{Mod}_{\Bbbk}\big)$. Analogously, the set $\{\mathrm{Hom}_{\mathcal{A}^{op}}\big(-, \mathfrak{q}^o\big)\}_{\mathfrak{q}^o \in \mathcal{A}^{op}}$ is a generating set of small objects in the category $\mathsf{Funct}\big(\mathcal{A}^{op}, \mathrm{Mod}_{\Bbbk}\big)$.

For simplicity, we shall denote by $H(-, \mathfrak{p}) := \mathrm{Hom}_{\mathcal{A}}\big(-, \mathfrak{p}\big)$ and by $H(\mathfrak{q}, -) := \mathrm{Hom}_{\mathcal{A}}\big(\mathfrak{q}, -\big)$, for every pair of objects $\mathfrak{p}, \mathfrak{q} \in \mathcal{A}$. The forthcoming lemma, known in the literature as Yoneda lemma, is a well known fact in functor categories, whose proof is based up on the following observation: For every $\mathfrak{p}, \mathfrak{q} \in \mathcal{A}$ and $\alpha \in \mathrm{Nat}\Big(H(-, \mathfrak{p}),\ H(-, \mathfrak{q})\Big)$, we have that

$$\alpha_{\mathfrak{p}'}(f) = \alpha_{\mathfrak{p}}(1_{\mathfrak{p}}) \circ f, \quad \text{for every } f \in \mathrm{Hom}_{\mathcal{A}}\big(\mathfrak{p}', \mathfrak{p}\big). \tag{1}$$

Lemma 1 *Let $\mathfrak{p}$ and $\mathfrak{q}$ be two objects in the category $\mathcal{A}$. Then, there is a bijective map*

$$\begin{array}{ccc} \mathrm{Nat}\Big(H(-, \mathfrak{p}),\ H(-, \mathfrak{q})\Big) & \xrightarrow{\quad \zeta_{\mathfrak{p},\mathfrak{q}} \quad} & \mathrm{Hom}_{\mathcal{A}}\big(\mathfrak{p}, \mathfrak{q}\big) \\ \alpha_- & \longmapsto & \alpha_{\mathfrak{p}}(1_{\mathfrak{p}}), \end{array}$$

which is a natural isomorphism on the category $\mathcal{A}^{op} \times \mathcal{A}$.

Proof Straightforward. □

Let $F : \mathcal{A} \to \mathrm{Mod}_{\Bbbk}$ be an object in the category $\mathsf{Funct}\big(\mathcal{A}, \mathrm{Mod}_{\Bbbk}\big)$. Define the following functor

$$\begin{array}{rcl} \overline{F} : \mathrm{Mod}_{\Bbbk} & \longrightarrow & \mathsf{Funct}\big(\mathcal{A}^{op}, \mathrm{Mod}_{\Bbbk}\big) \\ M & \longmapsto & \left[\begin{array}{c} \mathcal{A}^{op} \longrightarrow \mathrm{Mod}_{\Bbbk} \\ \mathfrak{p}^o \mapsto \mathrm{Hom}_{\mathrm{Mod}_{\Bbbk}}\big(F(\mathfrak{p}),\ M\big) \end{array} \right] \end{array}$$

Next we will check that $\overline{F}$ has a left adjoint functor, which we denote by:

$$F^* : \mathsf{Funct}\big(\mathcal{A}^{op}, \mathrm{Mod}_{\Bbbk}\big) \longrightarrow \mathrm{Mod}_{\Bbbk}.$$

First we define as follows the action of F^* over the full subcategory whose objects set is given by $\Big\{H(-, \mathfrak{q})\Big\}_{\mathfrak{q} \in \mathcal{A}}$, that is, over the set of small generators. This is given by the following assignments:

- $F^*\Big(H(-, \mathfrak{q})\Big) = F(\mathfrak{q})$, for every object $\mathfrak{q} \in \mathcal{A}$;
- $F^*(\alpha_-) = F(\alpha_{\mathfrak{p}}(1_{\mathfrak{p}}))$, for every morphism $\alpha_- : H(-, \mathfrak{p}) \to H(-, \mathfrak{q})$ (i.e., natural transformation).

Using Eq. (1), one can easily show that F^* is a well defined functor over the aforementioned subcategory. Applying now Mitchell's Theorem [18, Theorem 4.5.2], one shows that F^* extends uniquely (up to natural isomorphisms) to the whole category. We denote this extension functor by F^* too, thus, we have a functor $F^* : \mathsf{Funct}\Big(\mathcal{A}^{op}, \mathrm{Mod}_{\Bbbk}\Big) \longrightarrow \mathrm{Mod}_{\Bbbk}$.

Let M be any $\Bbbk$-module and $\mathfrak{q}$ any object of $\mathcal{A}$, we consider the mutually inverse maps

$$\begin{array}{ccc}
\mathrm{Hom}_{\Bbbk}\big(F(\mathfrak{q}),\, M\big) & \xrightarrow{\Phi_{H(-,\mathfrak{q}),\, M}} & \mathrm{Hom}_{\mathsf{Funct}(\mathcal{A}^{op},\, \mathrm{Mod}_{\Bbbk})}\Big(\mathrm{Hom}_{\mathcal{A}}(-,\, \mathfrak{q}),\, \mathrm{Hom}_{\Bbbk}\big(F(-),\, M\big)\Big) \\
f & \xrightarrow{\widehat{\ }} & \left[\begin{array}{c} \widehat{f_{\mathfrak{p}}} : \mathrm{Hom}_{\mathcal{A}}(\mathfrak{p},\, \mathfrak{q}) \longrightarrow \mathrm{Hom}_{\Bbbk}\big(F(\mathfrak{p}),\, M\big) \\ \gamma \longmapsto f \circ F(\gamma) \end{array}\right] \\
\sigma_{\mathfrak{q}}(1_{\mathfrak{q}}) & \xleftarrow{\ \widetilde{\ }\ } & \Big[\sigma_- : \mathrm{Hom}_{\mathcal{A}}(-,\, \mathfrak{q}) \longrightarrow \mathrm{Hom}_{\Bbbk}\big(F(-),\, M\big)\Big]
\end{array}$$

By Lemma 1, $\Phi_{-,-}$ is natural over the class of objects of the form $(H(-, \mathfrak{q}), M) \in \mathsf{Funct}\Big(\mathcal{A}^{op}, \mathrm{Mod}_{\Bbbk}\Big) \times \mathrm{Mod}_{\Bbbk}$. Again by Mitchell's Theorem [18, Theorem 4.5.2], the natural transformation $\Phi_{-,-}$ extends to a natural isomorphism on the whole category $\mathsf{Funct}\Big(\mathcal{A}^{op}, \mathrm{Mod}_{\Bbbk}\Big) \times \mathrm{Mod}_{\Bbbk}$:

$$\mathrm{Hom}_{\Bbbk}\big(F^*(-),\, -\big) \xrightarrow{\Phi_{-,-}} \mathrm{Hom}_{\mathsf{Funct}(\mathcal{A}^{op},\, \mathrm{Mod}_{\Bbbk})}\Big(-,\, \overline{F}(-)\Big), \tag{2}$$

which proves the claimed adjunction.

Now, we define the following two variable correspondence (see also [9, Chap. 5, Exercise I] and [19, pp. 26]):

$$\begin{array}{rcl}
-\underset{\mathcal{A}}{\otimes}- : \mathsf{Funct}\Big(\mathcal{A}^{op},\, \mathrm{Mod}_{\Bbbk}\Big) \times \mathsf{Funct}\Big(\mathcal{A},\, \mathrm{Mod}_{\Bbbk}\Big) & \longrightarrow & \mathrm{Mod}_{\Bbbk} \\
(T,\, F) & \longrightarrow & T \underset{\mathcal{A}}{\otimes} F \;=\; F^*(T) \\
\Big([\alpha : H(-, \mathfrak{q}_1) \to H(-, \mathfrak{q}_2)],\, [\beta : F_1 \to F_2]\Big) & \longrightarrow & \alpha \underset{\mathcal{A}}{\otimes} \beta \;=\; F_2(\alpha_{\mathfrak{q}_1}(1_{\mathfrak{q}_1})) \circ \beta_{\mathfrak{q}_1}.
\end{array}$$

This in fact establishes a functor. Let us check, for instance, its compatibility with the componentwise composition. So take four natural transformation

$$T_1 \xrightarrow{\alpha_1} T_2 \xrightarrow{\alpha_2} T_3, \qquad\qquad F_1 \xrightarrow{\beta_1} F_2 \xrightarrow{\beta_2} F_3,$$

and assume that $T_i = H(-, \mathfrak{q}_i)$, $i = 1, 2, 3$. By definition we obtain the following diagram

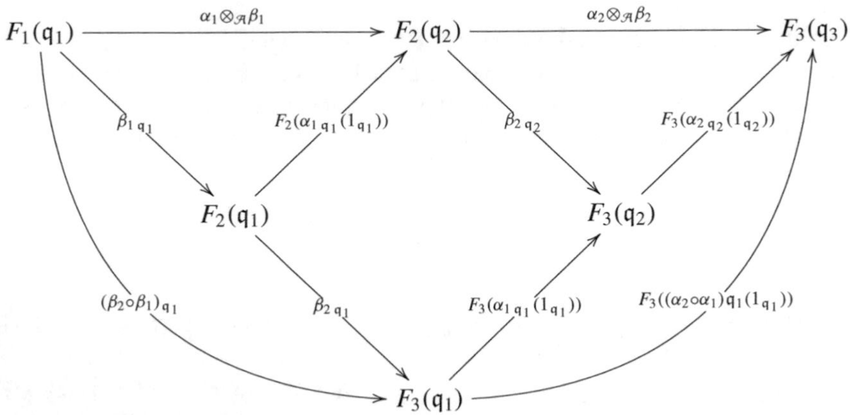

which commutes by the naturality of β_2 (the middle square) and the equality

$$(\alpha_2 \circ \alpha_1)_{\mathfrak{q}_1}(1_{\mathfrak{q}_1}) = \alpha_{2\,\mathfrak{q}_2}(1_{\mathfrak{q}_2}) \circ \alpha_{1\,\mathfrak{q}_1}(1_{\mathfrak{q}_1}),$$

that can be deduced from Eq. (1). The remaining verifications are left to the reader. The bi-functor $- \underset{\mathcal{A}}{\otimes} -$ is referred to as *the tensor product functor* and enjoys the following properties:

(a) For every object $F \in \mathsf{Funct}\Big(\mathcal{A},\, \mathrm{Mod}_{\Bbbk}\Big)$, the functor $- \underset{\mathcal{A}}{\otimes} F : \mathsf{Funct}\Big(\mathcal{A}^{op},\, \mathrm{Mod}_{\Bbbk}\Big) \to \mathrm{Mod}_{\Bbbk}$ is a right exact and direct sums preserving functor.

(b) For every object $T \in \mathsf{Funct}\Big(\mathcal{A}^{op},\, \mathrm{Mod}_{\Bbbk}\Big)$, the functor $T \underset{\mathcal{A}}{\otimes} - : \mathsf{Funct}\Big(\mathcal{A},\, \mathrm{Mod}_{\Bbbk}\Big) \to \mathrm{Mod}_{\Bbbk}$ is a right exact and direct sums preserving functor.

(c) For every object $\mathfrak{q} \in \mathcal{A}$, $F \in \mathsf{Funct}\Big(\mathcal{A},\, \mathrm{Mod}_{\Bbbk}\Big)$ and $T \in \mathsf{Funct}\Big(\mathcal{A}^{op},\, \mathrm{Mod}_{\Bbbk}\Big)$, we have

$$T \underset{\mathcal{A}}{\otimes} \mathrm{Hom}_{\mathcal{A}}\big(\mathfrak{q},\, -\big) = T(\mathfrak{q}^o), \;\; \mathrm{Hom}_{\mathcal{A}}\big(-,\, \mathfrak{q}\big) \underset{\mathcal{A}}{\otimes} F = F(\mathfrak{q}).$$

(d) The following property is the one expressed in [4, Theorem 2.2]. Let $T \in \mathsf{Funct}\Big(\mathcal{A}^{op},\, \mathrm{Mod}_{\Bbbk}\Big)$, $F \in \mathsf{Funct}\Big(\mathcal{A},\, \mathrm{Mod}_{\Bbbk}\Big)$ and $M \in \mathrm{Mod}_{\Bbbk}$. Then an *$\mathcal{A}$-balanced transformation* $\nu_- : T(-) \otimes_{\Bbbk} F(-) \to M$ is a family of $\Bbbk$-linear maps

$$\Big\{ \nu_{\mathfrak{q}} : T(\mathfrak{q}^o) \otimes_{\Bbbk} F(\mathfrak{q}) \longrightarrow M \Big\}_{\mathfrak{q} \in \mathcal{A}}$$

such that, for every morphism $f : \mathfrak{p} \to \mathfrak{q}$ in $\mathcal{A}$, the following diagram

$$\begin{array}{ccc} T(\mathfrak{q}^o) \otimes_{\Bbbk} F(\mathfrak{p}) & \xrightarrow{T(\mathfrak{q}^o)\otimes_{\Bbbk} F(f)} & T(\mathfrak{q}^o) \otimes_{\Bbbk} F(\mathfrak{q}) \\ {\scriptstyle T(f^o)\otimes_{\Bbbk} F(\mathfrak{p})}\downarrow & & \downarrow{\scriptstyle \nu_{\mathfrak{q}}} \\ T(\mathfrak{p}^o) \otimes_{\Bbbk} F(\mathfrak{p}) & \xrightarrow[\nu_{\mathfrak{p}}]{} & M \end{array}$$

commutes. There is a universal $\mathcal{A}$-balanced transformation $\varphi : T(-) \otimes_{\Bbbk} F(-) \to T \otimes_{\mathcal{A}} F$ in the sense that for any other $\mathcal{A}$-balanced transformation $\nu_{-} : T(-) \otimes_{\Bbbk} F(-) \to M$, there exists a $\Bbbk$-linear map $\overline{\nu} : T \otimes_{\mathcal{A}} F \to M$ such that $\overline{\nu} \circ \varphi_{\mathfrak{q}} = \nu_{\mathfrak{q}}$, for every object $\mathfrak{q} \in \mathcal{A}$. Furthermore, if $\xi : T_1 \to T_2$ and $\chi : F_1 \to F_2$ are two natural transformations, then $(\xi \otimes_{\mathcal{A}} \chi) \circ \varphi^1_{\mathfrak{q}} = \varphi^2_{\mathfrak{q}} \circ (\xi_{\mathfrak{q}} \otimes_{\Bbbk} \chi_{\mathfrak{q}})$, for every object $\mathfrak{q} \in \mathcal{A}$, where φ^i_{-} is the universal $\mathcal{A}$-balanced transformation associated with the pair (T_i, F_i).

Remark 1 Properties (a)–(d) listed above are indeed consequences of the universal property of *coend* (see [16, pp. 222] for the pertinent definition of this object). Namely, the tensor product $T \underset{\mathcal{A}}{\otimes} F$ of two functors $T \in \mathsf{Funct}\Big(\mathcal{A}^{op}, \mathrm{Mod}_{\Bbbk}\Big)$ and $F \in \mathsf{Funct}\Big(\mathcal{A}^{op}, \mathrm{Mod}_{\Bbbk}\Big)$, can be realized as the following coend:

$$T \underset{\mathcal{A}}{\otimes} F \;\cong\; \frac{\bigoplus_{(\mathfrak{p}^o,\, \mathfrak{p}) \,\in\, \mathcal{A}^{op} \times \mathcal{A}} T(\mathfrak{q}^o) \otimes_{\Bbbk} F(\mathfrak{p})}{\mathcal{J}_{T,F}},$$

where $\mathcal{J}_{T,F}$ is the sub $\Bbbk$-module of the numerator generated by the set

$$\Big\{T(f^o)(u) \otimes_{\Bbbk} v - u \otimes_{\Bbbk} F(f)(v) \,|\; u \in T(\mathfrak{q}^o),\ v \in F(\mathfrak{p}),\ f \in \mathrm{Hom}_{\mathcal{A}}\big(\mathfrak{p},\, \mathfrak{q}\big)\Big\}_{(\mathfrak{p}^o,\, \mathfrak{p}) \,\in\, \mathcal{A}^{op} \times \mathcal{A}}.$$

The interested reader can consult [19, pp. 26, 27], for more details.

From now on, we denote by A the ring with enough orthogonal idempotents attached to the small $\Bbbk$-linear category $\mathcal{A}$. As we have mentioned above, its underlying $\Bbbk$-module is the direct sum

$$A = \bigoplus_{(\mathfrak{p},\, \mathfrak{q}) \,\in\, \mathcal{A} \times \mathcal{A}} \mathrm{Hom}_{\mathcal{A}}\big(\mathfrak{p},\, \mathfrak{q}\big)$$

with an associative multiplication induced by the composition law. The categories of right and left unital A-modules are, respectively, denoted by Mod_A and ${}_A\mathrm{Mod}$. The tensor product over this ring will be denoted by $- \otimes_A -$. Notice that any (left) unital A-module M decomposes as a direct sum of $\Bbbk$-modules of the form:

$$M = \oplus_{\mathfrak{p} \in \mathcal{A}} 1_{\mathfrak{p}} M, \quad \text{where } 1_{\mathfrak{p}} M := \Big\{ y \in M \,|\, y = 1_{\mathfrak{p}} z, \ \text{ for some } z \in M \Big\}, \quad \mathfrak{p} \in \mathcal{A}. \tag{3}$$

Given an homogeneous element $a \in A$, which belongs to $\mathrm{Hom}_{\mathcal{A}}(\mathfrak{p}, \mathfrak{q})$, the left multiplication by this a leads to a $\Bbbk$-linear map $\ell_a : 1_{\mathfrak{p}}M \to 1_{\mathfrak{q}}M,\ 1_{\mathfrak{p}}y \mapsto 1_{\mathfrak{q}}(ay)$.

It is well know from Gabriel's classical result [11, Proposition 2, pp. 347] that the following functors establish an equivalence of categories

$$\begin{array}{ccccccc} \mathsf{Funct}\big(\mathcal{A}, \mathrm{Mod}_{\Bbbk}\big) & \xrightarrow{\mathscr{L}} & {}_A\mathrm{Mod}, & \mathsf{Funct}\big(\mathcal{A}^{op}, \mathrm{Mod}_{\Bbbk}\big) & \xrightarrow{\mathscr{R}} & \mathrm{Mod}_A \\ F & \longrightarrow & \bigoplus_{\mathfrak{p} \in \mathcal{A}} F(\mathfrak{p}) & T & \longrightarrow & \bigoplus_{\mathfrak{p} \in \mathcal{A}} T(\mathfrak{p}^o) \\ \theta & \longrightarrow & \bigoplus_{\mathfrak{p} \in \mathcal{A}} \theta_{\mathfrak{p}}, & \vartheta & \longrightarrow & \bigoplus_{\mathfrak{p} \in \mathcal{A}} \vartheta_{\mathfrak{p}^o} \end{array} \tag{4}$$

These equivalences of categories can also be checked directly by hand. For instance, using the previous notations, one can easily check that the following functor

$$\begin{array}{ccc} {}_A\mathrm{Mod} & \xrightarrow{\mathscr{P}} & \mathsf{Funct}\big(\mathcal{A}, \mathrm{Mod}_{\Bbbk}\big) \\ M & \longrightarrow & \Big[\mathfrak{p} \longrightarrow 1_{\mathfrak{p}}M, \quad a \longrightarrow \ell_a\Big], \end{array}$$

whose action on morphism is the obvious one, is an inverse of $\mathscr{L}$. Using the functors displayed in Eq. (4), we establish the subsequent useful lemma:

Lemma 2 *Let $\mathcal{A}$ be a small $\Bbbk$-linear category and A it associated ring with enough orthogonal idempotents. Then the following diagram is commutative up to natural isomorphism*

$$\begin{array}{ccc} \mathsf{Funct}\big(\mathcal{A}^{op}, \mathrm{Mod}_{\Bbbk}\big) \times \mathsf{Funct}\big(\mathcal{A}, \mathrm{Mod}_{\Bbbk}\big) & \xrightarrow{-\otimes_{\mathcal{A}}-} & \mathrm{Mod}_{\Bbbk} \\ \downarrow{\scriptstyle \mathscr{R} \times \mathscr{L}} & \nearrow{\scriptstyle -\otimes_A -} & \\ \mathrm{Mod}_A \times {}_A\mathrm{Mod} & & \end{array}$$

Proof Let $T \in \mathsf{Funct}\big(\mathcal{A}^{op}, \mathrm{Mod}_{\Bbbk}\big)$ and $F \in \mathsf{Funct}\big(\mathcal{A}, \mathrm{Mod}_{\Bbbk}\big)$. By Mitchell's Theorem [18, Theorem 4.5.2], there is no loss of generality in assuming that T is of the form $\mathrm{Hom}_{\mathcal{A}}(-, \mathfrak{q})$, for some object $\mathfrak{q} \in \mathcal{A}$. In this case, we know that $T \otimes_{\mathcal{A}} F = F^*(T) = F(\mathfrak{q})$ and $\mathscr{R}(T) = \oplus_{\mathfrak{p} \in \mathcal{A}} \mathrm{Hom}_{\mathcal{A}}(\mathfrak{p}, \mathfrak{q}) = 1_{\mathfrak{q}}A$. Therefore,

$$\begin{aligned} 1_{\mathfrak{q}} A \otimes_A \mathscr{L}(F) &= 1_{\mathfrak{q}} A \otimes_A \Big(\oplus_{\mathfrak{p} \in \mathcal{A}} F(\mathfrak{p}) \Big) \\ &\cong 1_{\mathfrak{q}} \Big(\oplus_{\mathfrak{p} \in \mathcal{A}} F(\mathfrak{p}) \Big) \\ &\cong F(\mathfrak{q}) \;=\; T \otimes_{\mathcal{A}} F, \end{aligned}$$

and this finishes the proof. □

Let $\mathbb{T}$ and $\mathbb{F}$ be two objects of the category of functors $\mathsf{Funct}\Big(\mathcal{A}^{op} \times \mathcal{A},\, \mathrm{Mod}_{\Bbbk}\Big)$. For every object $\mathfrak{q} \in \mathcal{A}$, we then get two functors

$$\begin{aligned} \mathbb{T}(-, \mathfrak{q}) :\ & \mathcal{A}^{op} \longrightarrow \mathrm{Mod}_{\Bbbk} \\ \mathbb{F}(\mathfrak{q}^o, -) :\ & \mathcal{A} \longrightarrow \mathrm{Mod}_{\Bbbk} \end{aligned}$$

In this way, using the the bifunctor $- \underset{\mathcal{A}}{\otimes} -$ defined above, we can define a new object in that category, which is given as follows:

$$\begin{aligned} \mathbb{T} \otimes_{\mathcal{A}} \mathbb{F} : \mathcal{A}^{op} \times \mathcal{A} &\longrightarrow \mathrm{Mod}_{\Bbbk} \\ (\mathfrak{p}^o, \mathfrak{q}) &\longrightarrow \mathbb{T}(-, \mathfrak{q}) \underset{\mathcal{A}}{\otimes} \mathbb{F}(\mathfrak{p}^o, -) \\ (f^o, g) &\longrightarrow \mathbb{T}(-, g) \underset{\mathcal{A}}{\otimes} \mathbb{F}(f^o, -). \end{aligned}$$

The previous discussion serves to announce the following result:

Proposition 1 *Let $\mathcal{A}$ be a small $\Bbbk$-linear category. Then the category $\mathfrak{M} :=$ $\mathsf{Funct}\Big(\mathcal{A}^{op} \times \mathcal{A},\, \mathrm{Mod}_{\Bbbk}\Big)$ is a strict monoidal category with tensor product given by the bifunctor $- \otimes_{\mathcal{A}} -$ and with identity object $\mathbb{I}_{\mathfrak{M}} := \mathrm{Hom}_{\mathcal{A}}\big(-,\, -\big)$ the functor which acts on objects by sending $(\mathfrak{p}^o, \mathfrak{q}) \to \mathrm{Hom}_{\mathcal{A}}\big(\mathfrak{p},\, \mathfrak{q}\big)$ and on morphisms by sending $(f^o, g) \to \mathrm{Hom}_{\mathcal{A}}\big(f^o,\, g\big)$.*

Proof We only sketch the main steps of the proof. The left unit constraint is given as follows:

$$\big(\mathbb{I}_{\mathfrak{M}} \otimes_{\mathcal{A}} \mathbb{F}\big)(\mathfrak{p}^o, \mathfrak{q}) \;=\; \mathbb{I}_{\mathfrak{M}}(-, \mathfrak{q}) \underset{\mathcal{A}}{\otimes} \mathbb{F}(\mathfrak{p}^o, -) \;=\; H(-, \mathfrak{q}) \underset{\mathcal{A}}{\otimes} \mathbb{F}(\mathfrak{p}^o, -) \;=\; \mathbb{F}(\mathfrak{p}^o, \mathfrak{q}),$$

for every pair of objects $(\mathfrak{p}^o, \mathfrak{q}) \in \mathcal{A}^{op} \times \mathcal{A}$. As for the right unit constraint, given an object $\mathbb{T} \in \mathfrak{M}$, we know that

$$\big(\mathbb{T} \otimes_{\mathcal{A}} \mathbb{I}_{\mathfrak{M}}\big)(\mathfrak{p}^o, \mathfrak{q}) \;=\; \mathbb{T}(-, \mathfrak{q}) \underset{\mathcal{A}}{\otimes} \mathbb{I}_{\mathfrak{M}}(\mathfrak{p}^o, -),$$

for every pair of objects $(\mathfrak{p}^o, \mathfrak{q})$. Now, without loss of generality we can assume that $\mathbb{T}(-, \mathfrak{q}) = H(-, \mathfrak{p}')$ for some $\mathfrak{p}' \in \mathcal{A}$, then we get that

$$\begin{aligned} \mathbb{T}(-, \mathfrak{q}) \underset{\mathcal{A}}{\otimes} \mathbb{I}_{\mathfrak{M}}(\mathfrak{p}^o, -) &= \mathbb{I}_{\mathfrak{M}}(\mathfrak{p}^o, \mathfrak{p}') \\ &= \mathrm{Hom}_{\mathcal{A}}\big(\mathfrak{p},\, \mathfrak{p}'\big) \\ &= \mathbb{T}(\mathfrak{p}^o, \mathfrak{q}). \end{aligned}$$

This gives the desired unit constraints.

Now let us sketch the associativity constraint. Given $\mathbb{G}$ another object of $\mathfrak{M}$ and fixing a pair of objects $(\mathfrak{p}^o, \mathfrak{q}) \in \mathcal{A}^{op} \times \mathcal{A}$, we have the following equalities:

$$\Big(\big(\mathbb{T} \otimes_{\mathcal{A}} \mathbb{F}\big) \otimes_{\mathcal{A}} \mathbb{G}\Big)(\mathfrak{p}^o, \mathfrak{q}) = \big(\mathbb{T} \otimes_{\mathcal{A}} \mathbb{F}\big)(-, \mathfrak{q}) \underset{\mathcal{A}}{\otimes} \mathbb{G}(\mathfrak{p}^o, -)$$

$$\Big(\mathbb{T} \otimes_{\mathcal{A}} \big(\mathbb{F} \otimes_{\mathcal{A}} \mathbb{G}\big)\Big)(\mathfrak{p}^o, \mathfrak{q}) = \mathbb{T}(-, \mathfrak{q}) \underset{\mathcal{A}}{\otimes} \Big((\mathbb{F} \otimes_{\mathcal{A}} \mathbb{G})(\mathfrak{p}^o, -)\Big).$$

Without loss of generality, we may assume that $\mathbb{T}(-, \mathfrak{q}) = \mathrm{Hom}_{\mathcal{A}}\big(-,\, \mathfrak{p}'\big)$, for some $\mathfrak{p}' \in \mathcal{A}$. This implies, on the one hand, that

$$\mathbb{T}(-, \mathfrak{q}) \underset{\mathcal{A}}{\otimes} \Big((\mathbb{F} \otimes_{\mathcal{A}} \mathbb{G})(\mathfrak{p}^o, -)\Big) \;=\; (\mathbb{F} \otimes_{\mathcal{A}} \mathbb{G})(\mathfrak{p}^o, \mathfrak{p}') \;=\; \mathbb{F}(-, \mathfrak{p}') \underset{\mathcal{A}}{\otimes} \mathbb{G}(\mathfrak{p}^o, -). \tag{5}$$

On the other hand, we also have

$$(\mathbb{T} \otimes_{\mathcal{A}} \mathbb{F})(\mathfrak{r}^o, \mathfrak{q}) \;=\; \mathbb{T}(-, \mathfrak{q}) \underset{\mathcal{A}}{\otimes} \mathbb{F}(\mathfrak{r}^o, -) \;=\; \mathbb{F}(\mathfrak{r}^o, \mathfrak{p}'),$$

for every object $\mathfrak{r} \in \mathcal{A}$, which implies that

$$\big(\mathbb{T} \otimes_{\mathcal{A}} \mathbb{F}\big)(-, \mathfrak{q}) \underset{\mathcal{A}}{\otimes} \mathbb{G}(\mathfrak{p}^o, -) \;=\; \mathbb{F}(-, \mathfrak{p}') \underset{\mathcal{A}}{\otimes} \mathbb{G}(\mathfrak{p}^o, -). \tag{6}$$

Comparing Eqs. (5) and (6), we arrive to the equality

$$\Big(\big(\mathbb{T} \otimes_{\mathcal{A}} \mathbb{F}\big) \otimes_{\mathcal{A}} \mathbb{G}\Big)(\mathfrak{p}^o, \mathfrak{q}) \;=\; \Big(\mathbb{T} \otimes_{\mathcal{A}} \big(\mathbb{F} \otimes_{\mathcal{A}} \mathbb{G}\big)\Big)(\mathfrak{p}^o, \mathfrak{q}),$$

for every pair of objects $(\mathfrak{p}^o, \mathfrak{q}) \in \mathcal{A}^{op} \times \mathcal{A}$. Therefore, we have that $\Big(\big(\mathbb{T} \otimes_{\mathcal{A}} \mathbb{F}\big) \otimes_{\mathcal{A}} \mathbb{G}\Big) = \Big(\mathbb{T} \otimes_{\mathcal{A}} \big(\mathbb{F} \otimes_{\mathcal{A}} \mathbb{G}\big)\Big)$, and this shows the associativity constraint. $\square$

The objects of the monoidal category $\mathfrak{M}$ are referred to, in the literature, as $\mathcal{A}$-*bimodules*. Here we also adopt this terminology.

3 BOCSES are Equivalent to Corings

The notation is that of the previous section. Thus $\mathcal{A}$ still denotes a small $\Bbbk$-linear category, A is its associated ring with enough orthogonal idempotents, and $\mathfrak{M}$ denotes the category of bi-functors $\mathfrak{M} := \mathsf{Funct}\Big(\mathcal{A}^{op} \times \mathcal{A},\, \mathrm{Mod}_{\Bbbk}\Big)$. We denote by ${}_A\mathrm{Mod}_A$ the category of all unital A-bimodules (i.e., bimodule which are left and right unital).

Definition 1 (*A. V. Roiter*) A *BOCS* over the category $\mathcal{A}$ is a comonoid in the monoidal category of $\mathcal{A}$-bimodules $\mathfrak{M} = \mathsf{Funct}\Big(\mathcal{A}^{op} \times \mathcal{A},\, \mathrm{Mod}_{\Bbbk}\Big)$ of Proposition 1.

Thus a BOCS is three-tuple $(\mathbb{C}, \delta, \xi)$ consisting of an $\mathcal{A}$-bimodule $\mathbb{C}$ (i.e., $\mathbb{C} \in \mathfrak{M}$) and two natural transformations

$$\underbrace{\delta : \mathbb{C} \longrightarrow \mathbb{C} \otimes_{\mathcal{A}} \mathbb{C}}_{\text{comultiplication}}, \qquad \underbrace{\xi : \mathbb{C} \longrightarrow \mathbb{I}_{\mathfrak{M}}}_{\text{counit}},$$

satisfying the usual counitary and coassociativity properties.

A *morphism of BOCSES* is a natural transformation $\phi : \mathbb{C} \to \mathbb{C}'$ such that

$$\delta' \circ \phi \;=\; (\phi \otimes_{\mathcal{A}} \phi) \circ \delta, \qquad \xi' \circ \phi \;=\; \xi.$$

BOCSES over $\mathcal{A}$ and their morphisms form a category which we denote by $\mathcal{A}$-Bocses.

Our next goal is to show that the category $\mathcal{A}$-Bocses is equivalent to the category of corings over the ring with enough orthogonal idempotents A, which we denote by A-$\mathsf{corings}$ (i.e., the category of comonoids in the monoidal category of bimodules ${}_A\mathrm{Mod}_A$), see [8] for additional facts on corings over rings with local units. We start with the following lemma, which can be seen as a variation of the equivalences of categories given explicitly in Eq. (4).

Lemma 3 *Let $\mathcal{A}$ be a small $\Bbbk$-linear category and A its associated ring with enough orthogonal idempotents. Then, the following functor establishes an equivalence of categories*

$$\begin{array}{ccc}
\mathfrak{M} := \mathsf{Funct}\Big(\mathcal{A}^{op} \times \mathcal{A},\, \mathrm{Mod}_{\Bbbk}\Big) & \xrightarrow{\qquad \mathscr{F} \qquad} & {}_A\mathrm{Mod}_A \\
\mathbb{F} & \longrightarrow & \bigoplus_{(\mathfrak{p}^o,\, \mathfrak{q}) \in \mathcal{A}^{op} \times \mathcal{A}} \mathbb{F}(\mathfrak{p}^o, \mathfrak{q}) \\
\theta & \longrightarrow & \bigoplus_{(\mathfrak{p}^o,\, \mathfrak{q}) \in \mathcal{A}^{op} \times \mathcal{A}} \theta_{(\mathfrak{p}^o, \mathfrak{q})}
\end{array}$$

Proof Using the functors $\mathscr{R}$ and $\mathscr{L}$ of Eq. (4), one get the A-bimodule structure on $\mathscr{F}(\mathbb{F})$, for a given $\mathbb{F} \in \mathfrak{M}$. A direct verification shows that $\mathscr{F}$ is a well defined functor. The inverse functor is defined as follows: To every unital A-bimodule X, one corresponds the bi-functor $\mathscr{G}(X) : \mathcal{A}^{op} \times \mathcal{A} \to \mathrm{Mod}_{\Bbbk}$ which acts over objects by $(\mathfrak{p}^o, \mathfrak{q}) \to 1_{\mathfrak{q}} X 1_{\mathfrak{p}}$ (the $\Bbbk$-submodule of X generated by the set of elements $\{1_{\mathfrak{q}} x 1_{\mathfrak{p}} \,|\, x \in X\}$), and acts over morphisms as follows:

$$\Big[(f^o, g) : (\mathfrak{p}_1^o, \mathfrak{q}_1) \to (\mathfrak{p}_2^o, \mathfrak{q}_2)\Big] \longrightarrow \Big[\mathscr{G}(f^o, g) : 1_{\mathfrak{q}_1} X 1_{\mathfrak{p}_1} \to 1_{\mathfrak{q}_2} X 1_{\mathfrak{p}_2}, \\ \Big(1_{\mathfrak{q}_1} x 1_{\mathfrak{p}_1} \mapsto 1_{\mathfrak{q}_2} g x f 1_{\mathfrak{p}_2}\Big)\Big],$$

where the notation gxf stands for the bi-action of A on X, after identifying the elements g, f with their images in A. The rest of the proof is left to the reader. □

Proposition 2 *Let $\mathcal{A}$ be a small $\Bbbk$-linear category and A its associated ring with enough orthogonal idempotents. The functor $\mathscr{F}$ defined in Lemma 3 establishes a monoidal equivalence between the monoidal categories* $\mathsf{Funct}\Big(\mathcal{A}^{op} \times \mathcal{A}, \mathrm{Mod}_{\Bbbk}\Big)$ *and ${}_A\mathrm{Mod}_A$.*

Proof Fix $\mathbb{T}$ and $\mathbb{F}$ two objects of the category $\mathfrak{M}$. For every pair of objects $(\mathfrak{p}^o, \mathfrak{q}) \in \mathcal{A}^{op} \times \mathcal{A}$, we have by Lemma 2 an isomorphism of $\Bbbk$-modules

$$\mathbb{T}(-, \mathfrak{q}) \otimes_{\mathcal{A}} \mathbb{F}(\mathfrak{p}^o, -) \cong \mathscr{R}\Big(\mathbb{T}(-, \mathfrak{q})\Big) \otimes_A \mathscr{L}\Big(\mathbb{F}(\mathfrak{p}^o, -)\Big) \\ \cong \Big(\bigoplus_{\mathfrak{r} \in \mathcal{A}} \mathbb{T}(\mathfrak{r}^o, \mathfrak{q})\Big) \otimes_A \Big(\bigoplus_{\mathfrak{t} \in \mathcal{A}} \mathbb{F}(\mathfrak{p}^o, \mathfrak{t})\Big).$$

Applying the functor $\bigoplus_{(\mathfrak{p}^o, \mathfrak{q}) \in \mathcal{A}^{op} \times \mathcal{A}}(-)$ (i.e., the direct sum functor) to the latter isomorphism, we obtain an isomorphism of an A–bimodules as follows:

$$\bigoplus_{(\mathfrak{p}^o, \mathfrak{q}) \in \mathcal{A}^{op} \times \mathcal{A}} \Big(\mathbb{T}(-, \mathfrak{q}) \otimes_{\mathcal{A}} \mathbb{F}(\mathfrak{p}^o, -)\Big) \cong \bigoplus_{(\mathfrak{p}^o, \mathfrak{q}) \in \mathcal{A}^{op} \times \mathcal{A}} \Big(\Big(\bigoplus_{\mathfrak{r} \in \mathcal{A}} \mathbb{T}(\mathfrak{r}, \mathfrak{q})\Big) \otimes_A \Big(\bigoplus_{\mathfrak{t} \in \mathcal{A}} \mathbb{F}(\mathfrak{p}^o, \mathfrak{t})\Big)\Big) \\ \cong \Big(\bigoplus_{(\mathfrak{r}^o, \mathfrak{q}) \in \mathcal{A}^{op} \times \mathcal{A}} \mathbb{T}(\mathfrak{r}^o, \mathfrak{q})\Big) \otimes_A \Big(\bigoplus_{(\mathfrak{p}^o, \mathfrak{t}) \in \mathcal{A}^{op} \times \mathcal{A}} \mathbb{F}(\mathfrak{p}^o, \mathfrak{t})\Big).$$

Therefore, we have an isomorphism $\mathscr{F}(\mathbb{T} \otimes_{\mathcal{A}} \mathbb{F}) \cong \mathscr{F}(\mathbb{T}) \otimes_A \mathscr{F}(\mathbb{F})$ of an A-bimodules, for every pair of objects $\mathbb{T}, \mathbb{F} \in \mathfrak{M}$. It is not difficult to check that this isomorphism is in fact a natural isomorphism. Lastly, the image of the identity object $\mathbb{I}_{\mathfrak{M}}$, is

$$\mathscr{F}(\mathbb{I}_{\mathfrak{M}}) = \bigoplus_{(\mathfrak{p}^o, \mathfrak{q}) \in \mathcal{A}^{op} \times \mathcal{A}} \mathbb{I}_{\mathfrak{M}}(\mathfrak{p}^o, \mathfrak{q}) = \bigoplus_{(\mathfrak{p}^o, \mathfrak{q}) \in \mathcal{A}^{op} \times \mathcal{A}} \mathrm{Hom}_{\mathcal{A}}\big(\mathfrak{p}, \mathfrak{q}\big) = A$$

is the identity object of the monoidal category ${}_A\mathrm{Mod}_A$. The rest of constraints that $\mathscr{F}$ should satisfy are immediate and this finishes the proof. □

Now, we state the promised equivalence of categories:

Corollary 1 *Let $\mathcal{A}$ be a small $\Bbbk$-linear category and A its associated ring with enough orthogonal idempotents. The functor $\mathscr{F}$ of Lemma 3 induces an equivalence of categories between A-corings and $\mathcal{A}$-Bocses such that the following diagram*

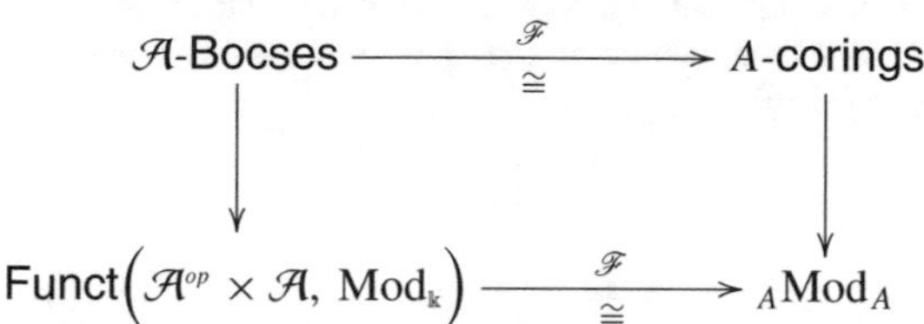

commutes, where the vertical arrows are the canonical forgetful functors.

Proof It is a consequence of Lemma 3 and Proposition 2. □

Acknowledgements The author would like to thank the referee for her/his careful reading of the first version of this note.

References

1. Abrams, G.D.: Morita equivalence for rings with local units. Comm. Algebra **11**(8), 801–837 (1983)
2. Ánh, P.N., Márki, L.: Rees matrix rings. J. Algebra **81**, 340–369 (1983)
3. Ánh, P.N., Márki, L.: Morita equivalence for rings without identity. Tsukuba J. Math. **11**(1), 1–16 (1987)
4. Bautista, R., Colavita, L., Salmerón, L.: On adjoint functors in representation theory. Representation Theory. Lecture Notes in Mathematics, vol. 903, pp. 9–25. Springer, Berlin (1981)
5. Beck, J.: Distributive laws. Seminar on Triples and Categorical Homology Theory. Lecture Notes in Mathematics, vol. 80, pp. 119–140. Springer, Berlin (1969)
6. Brzeziński, T.: The structure of corings. Induction functors, Maschke-type theorem, and Frobenius and Galois-type properties. Algebra Represent. Theory **5**, 389–410 (2002)
7. Eilenberg, S., Moore, J.C.: Adjoint functors and triples. Illinois J. Math. **9**, 381–398 (1965)
8. El Kaoutit, L.: Corings over rings with local units. Math. Nachr. **282**(5), 726–747 (2009)
9. Freyd, P.: Abelian Categories. Introduction An to the Theory of Functors. Harper's Series in Modern Mathematics. Harper & Row Publishers, New York (1964) ; Reprints in Theory and Applications of Categories. No. **3**, 23–164 (2003)
10. Fuller, K.R.: On rings whose left modules are direct sums of finitely generated modules. Proc. Amer. Math. Soc. **54**, 39–44 (1976)
11. Gabriel, P.: Des catégories abéliennes. Bulletin de la S. M. F., tome **90**, 323–448 (1962)
12. Guzman, F.: Cointegration, relative cohomology for comdules, and coseparable coring. J. Algebra **126**, 211–224 (1989)
13. Jonah, D.W.: Cohomology of coalgebras. Mem. Amer. Math. Soc. **82** (1968)
14. Kleiner, M.: The dual ring to a coring with a grouplike. Proc. Amer. Math. Soc. **91**, 540–542 (1984)
15. Kleisli, H.: Every standard construction is induced by a pair of adjoint functors. Proc. Amer. Math. Soc. **16**, 544–546 (1965)
16. Mac Lane, S.: Categories for the working mathematician. Graduate Texts in Mathematics, vol. 5. Springer, New York (1998)

17. Masuoka, A.: Corings and invertible bimodules. Tsukuba. J. Math. **13**(2), 353–362 (1989)
18. Mitchell, B.: Theory of Categories. New York and London ed., Academic Press (1965)
19. Mitchell, B.: Rings with several objects. Adv. Math. **8**, 1–161 (1972)
20. Roiter, A.V.: Matrix problems and representations of BOCS's. Representation Theory I. Lecture Notes in Mathematics, vol. 831, pp. 288–324. Springer, Berlin (1980)
21. Sweedler, M.: The predual theorem to the Jacobson-Bourbaki theorem. Trans. Amer. Math. Soc. **213**, 391–406 (1975)

Quadratic Color Hom-Lie Algebras

Faouzi Ammar, Imen Ayadi, Sami Mabrouk and Abdenacer Makhlouf

Abstract The purpose of this paper is to study quadratic color Hom-Lie algebras. We present some constructions of quadratic color Hom-Lie algebras which we use to provide several examples. We describe T^*-extensions and central extensions of color Hom-Lie algebras and establish some cohomological characterizations.

Keywords Hom-Lie algebra · Color Hom-Lie algebra · Color Hom-associative algebra · Quadratic structure · Nondegenerate bilinear form · T^*-extension · Representation · Cohomology · Color Hom-Leibniz algebra

1 Introduction

The aim of this paper is to introduce and study quadratic color Hom-Lie algebras which are graded Hom-Lie algebras with ε-symmetric, invariant and nondegenerate bilinear forms. Color Lie algebras, originally introduced in [27, 28], can be seen as a direct generalization of Lie algebras bordering Lie superalgebras. The grading is determined by an abelian group Γ and the definition involves a bicharacter function. Hom-Lie algebras are a generalization of Lie algebras, where the classical Jacobi identity is twisted by a linear map. Quadratic Hom-Lie algebras were studied in [11].

F. Ammar
Sfax University, Sfax, Tunisia
e-mail: Faouzi.Ammar@fss.rnu.tn

I. Ayadi · A. Makhlouf (✉)
IRIMAS - Département de Mathématiques, 6, rue des frères Lumière, 68093 Mulhouse, France
e-mail: abdenacer.makhlouf@uha.fr

I. Ayadi
e-mail: Imen.ayadi@uha.fr

S. Mabrouk
Faculty of Sciences, University of Gafsa, 2112 Gafsa, Tunisia
e-mail: mabrouksami00@yahoo.fr

M. Siles Molina et al. (eds.), *Associative and Non-Associative Algebras and Applications*, Springer Proceedings in Mathematics & Statistics 311,
https://doi.org/10.1007/978-3-030-35256-1_16

Γ-graded Lie algebras with quadratic-algebraic structures, that is Γ-graded Lie algebras provided with homogeneous, symmetric, invariant and nondegenerate bilinear forms, have been extensively studied specially in the case where $\Gamma = \mathbb{Z}_2$ (see for example [2, 6–9, 25, 29]). These algebras are called homogeneous (even or odd) quadratic Lie superalgebras. One of the fundamental results connected to homogeneous quadratic Lie superalgebras is to give its inductive description. The main tool used to obtain these inductive descriptions is to develop some concept of double extensions. This concept was introduced by Medina and Revoy (see [25]) to give a classification of quadratic Lie algebras. The concept of T^*-extension was introduced by Bordemann [12]. Recently a generalization to the case of quadratic (even quadratic) color Lie algebras was obtained in [26, 34]. They mainly generalized the double extension notion and its inductive descriptions.

Hom-algebraic structures appeared first as a generalization of Lie algebras in [1, 13, 14] were the authors studied q-deformations of Witt and Virasoro algebras. A general study and construction of Hom-Lie algebras were considered in [19, 21]. Since then, other interesting Hom-type algebraic structures of many classical structures were studied as Hom-associative algebras, Hom-Lie admissible algebras and more general G-Hom-associative algebras [22], n-ary Hom-Nambu-Lie algebras [4], Hom-Lie admissible Hom-coalgebras and Hom-Hopf algebras [23], Hom-alternative algebras, Hom-Malcev algebras and Hom-Jordan algebras [16, 23, 37]. Hom-algebraic structures were extended to the case of Γ-graded Lie algebras by studying Hom-Lie superalgebras and Hom-Lie admissible superalgebras in [5]. Recently, the study of Hom-Lie algebras provided with quadratic-algebraic structures was initiated by S. Benayadi and A. Makhlouf in [11] and our purpose in this paper is to generalize this study to the case of color Hom-Lie algebras.

The paper is organized as follows: in Sect. 2 we give a definition of quadratic color Hom-Lie algebras and generalize to the case of quadratic color Hom-Lie algebras the result of [38]. More precisely, we prove that the category of color Hom-Lie algebras (resp. the subcategory of quadratic color Hom-Lie algebras) is closed under self weak morphisms (resp. self symmetric automorphisms). In Sect. 3, we describe in one hand some constructions of quadratic color Hom-Lie algebras and in the other hand we use them to provide examples. In Sect. 4, we give some elements of representation theory of color Hom-Lie algebras and describe a semi-direct product. Section 5 includes the main results dealing with extensions and their relationships to cohomology. We introduce some elements of a cohomology of color Hom-Lie algebras controlling central extensions. Moreover, we establish the T^*-extension for color Hom-Lie algebras. Finally, in Sect. 6 we construct color Hom-Leibniz algebras arising from Faulkner construction introduced in [18].

2 Quadratic Color Hom-Lie Algebras

Throughout this paper $\mathbb{K}$ denotes a commutative field of characteristic zero and Γ stands for an abelian group. A vector space V is said to be a Γ-graded if we are given a family $(V_\gamma)_{\gamma\in\Gamma}$ of vector subspace of V such that

$$V = \bigoplus_{\gamma \in \Gamma} V_\gamma .$$

An element $x \in V_\gamma$ is said to be homogeneous of degree γ. For simplicity the degree of an element x is denoted by x. If the base field is considered as a graded vector space, it is understood that the graduation of $\mathbb{K}$ is given by

$$\mathbb{K}_0 = \mathbb{K} \quad \text{and} \quad \mathbb{K}_\gamma = \{0\}, \quad \text{if } \gamma \in \Gamma \setminus \{0\}.$$

Let V and W be two Γ-graded vector spaces. A linear map $f : V \longrightarrow W$ is said to be homogeneous of degree $\xi \in \Gamma$ if $f(x)$ is homogeneous of degree $\gamma + \xi$ whenever the element $x \in V_\gamma$. The set of all such maps is denoted by $(Hom(V, W))_\xi$. It is a subspace of $Hom(V, W)$, the vector space of all linear maps from V into W.

We mean by algebra (resp. Γ-graded algebra) $(A, .)$ a vector space (resp. Γ-graded vector space) with a multiplication which we denote by the juxtaposition and such that $A_\gamma A_{\gamma'} \subseteq A_{\gamma+\gamma'}$, for all $\gamma, \gamma' \in \Gamma$. In the graded case, a map $f : A \longrightarrow B$, where A and B are Γ-graded algebras, is called a homomorphism of Γ-graded algebras if it is a homomorphism of algebras which is homogeneous of degree zero.

We mean by a Hom-algebra (resp. Γ-graded Hom-algebra) a triple $(A, ., \alpha)$ consisting of an algebra (resp. Γ-graded algebra) and a map $\alpha : A \to A$ (an even linear map), called a twist map.

For more detail about graded algebraic structures, we refer to [31, 32]. In the following, we study a particular case of Γ-graded algebras which are quadratic color Hom-Lie algebras.

Definition 1 Let Γ be an abelian group. A map $\varepsilon : \Gamma \times \Gamma \to \mathbb{K} \setminus \{0\}$ is called a *bicharacter* on Γ if the following identities are satisfied

$$\varepsilon(a, b)\varepsilon(b, a) = 1, \tag{1}$$

$$\varepsilon(a, b + c) = \varepsilon(a, b)\varepsilon(a, c), \tag{2}$$

$$\varepsilon(a + b, c) = \varepsilon(a, c)\varepsilon(b, c), \quad \forall a, b, c \in \Gamma. \tag{3}$$

The definition above implies, in particular, the following relations

$$\varepsilon(a, 0) = \varepsilon(0, a) = 1, \ \varepsilon(a, a) = \pm 1, \ \text{for all } a \in \Gamma.$$

If x and x' are two homogeneous elements of degree γ and γ' respectively and ε is a bicharacter, then we shorten the notation by writing $\varepsilon(x, x')$ instead of $\varepsilon(\gamma, \gamma')$.

Unless stated, in the sequel all the graded spaces are over the same abelian group Γ and the bicharacter will be the same for all the structures.

Definition 2 A *color Hom-Lie* algebra is a tuple $(\mathfrak{g}, [\cdot, \cdot], \alpha, \varepsilon)$ consisting of a Γ-graded vector space $\mathfrak{g}$, a bicharacter ε, an even bilinear map $[\cdot, \cdot] : \mathfrak{g} \times \mathfrak{g} \to \mathfrak{g}$ (i.e. $[\mathfrak{g}_a, \mathfrak{g}_b] \subset \mathfrak{g}_{a+b}$) and an even homomorphism $\alpha : \mathfrak{g} \to \mathfrak{g}$ such that for homogeneous elements $x, y, z \in \mathfrak{g}$ we have

$$[x, y] = -\varepsilon(x, y)[y, x] \quad (\varepsilon - \text{skew symmetry}), \tag{4}$$

$$\circlearrowleft_{x,y,z} \varepsilon(z, x)[\alpha(x), [y, z]] = 0 \quad (\varepsilon - \text{Hom-Jacobi identity}), \tag{5}$$

where $\circlearrowleft_{x,y,z}$ denotes summation over the cyclic permutation on x, y, z. In particular, if α is a morphism of Lie algebras (i.e. $\alpha \circ [\cdot, \cdot] = [\cdot, \cdot] \circ \alpha^{\otimes 2}$), then we call $(\mathfrak{g}, [\cdot, \cdot], \varepsilon, \alpha)$ multiplicative color Hom-Lie algebra.

A Γ-graded subspace I of $\mathfrak{g}$ is said to be an ideal (resp. subalgebra) if $[I, \mathfrak{g}] \subset I$ (resp. $[I, I] \subset I$).

We prove easily that the ε – Hom-Jacobi identity is equivalent to the following condition

$$[\alpha(x), [y, z]] - \varepsilon(x, y)[\alpha(y), [x, z]] = [[x, y], \alpha(z)]. \tag{6}$$

We recover color Lie algebra when $\alpha = id_{\mathfrak{g}}$. Color Lie algebra is a generalization of Lie algebra and Lie superalgebra (if $\Gamma = \{0\}$, we have $\mathfrak{g} = \mathfrak{g}_0$ is a Lie algebra and if $\Gamma = \mathbb{Z}_2 = \{\bar{0}, \bar{1}\}$ and $\varepsilon(\bar{1}, \bar{1}) = -1$, then $\mathfrak{g}$ is a Lie superalgebra).

Example 1 Let Γ be the group $\mathbb{Z}_2^3$, and $\varepsilon : \Gamma \times \Gamma \longrightarrow \mathbb{C}$ defined by the matrix

$$[\varepsilon(i, j)] = \begin{pmatrix} 1 & -1 & -1 \\ -1 & 1 & -1 \\ -1 & -1 & 1 \end{pmatrix}.$$

Elements of this matrix specify in the natural way values of ε on the set $\{(1, 1, 0), (1, 0, 1), (0, 1, 1)\} \times \{(1, 1, 0), (1, 0, 1), (0, 1, 1)\} \subset \mathbb{Z}_2^3 \times \mathbb{Z}_2^3$ (The elements $(1, 1, 0)$, $(1, 0, 1)$, $(0, 1, 1)$ are ordered and numbers $1, 2, 3$ are respectively associated to them). The values of ε on other elements from $\mathbb{Z}_2^3 \times \mathbb{Z}_2^3$ do not affect the multiplication $[., .]$.

The graded analogue of $\mathfrak{sl}(2, \mathbb{C})$ is defined as complex algebra with three generators e_1, e_2 and e_3 satisfying the commutation relations $e_1e_2 + e_2e_1 = e_3,\ e_1e_3 + e_3e_1 = e_2,\ e_2e_3 + e_3e_2 = e_1$. Let $\mathfrak{sl}(2, \mathbb{C})$ be a $\mathbb{Z}_2^3$-graded vector space $\mathfrak{sl}(2, \mathbb{C}) = \mathfrak{sl}(2, \mathbb{C})_{(1,1,0)} \oplus \mathfrak{sl}(2, \mathbb{C})_{(1,0,1)} \oplus \mathfrak{sl}(2, \mathbb{C})_{(0,1,1)}$ with basis $e_1 \in \mathfrak{sl}(2, \mathbb{C})_{(1,1,0)}$, $e_2\ \mathfrak{sl}(2, \mathbb{C})_{(1,0,1)}$, $e_3 \in \mathfrak{sl}(2, \mathbb{C})_{(0,1,1)}$. The homogeneous subspaces of $\mathfrak{sl}(2, \mathbb{C})$ graded by the elements of $\mathbb{Z}_2^3$ different from $(1, 1, 0), (1, 0, 1)$ and $(1, 0, 1)$ are zero and so are omitted. The bilinear multiplication $[., .] : \mathfrak{sl}(2, \mathbb{C}) \times \mathfrak{sl}(2, \mathbb{C}) \longrightarrow \mathfrak{sl}(2, \mathbb{C})$ defined, with respect to the basis $\{e_1, e_2, e_3\}$, by the formulas

$$\begin{aligned} [e_1, e_1] &= e_1e_1 - e_1e_1 = 0, \quad [e_1, e_2] = e_1e_2 + e_2e_1 = e_3, \\ [e_2, e_2] &= e_2e_2 - e_2e_2 = 0, \quad [e_1, e_3] = e_1e_3 + e_3e_1 = e_2, \\ [e_3, e_3] &= e_3e_3 - e_3e_3 = 0, \quad [e_2, e_3] = e_2e_3 + e_3e_2 = e_1, \end{aligned}$$

makes $\mathfrak{sl}(2, \mathbb{C})$ into a three-dimensional Lie color algebra.

On the same space, we have a color Hom-Lie algebra by considering the nontrivial brackets

$$[e_1, e_2]_\alpha = -e_2,\ [e_1, e_3]_\alpha = -e_3,\ [e_2, e_3]_\alpha = -e_1,$$

together with a linear map defined by

$$\alpha(e_1) = -e_1,\ \alpha(e_2) = -e_3,\ \alpha(e_3) = -e_2.$$

Definition 3 Let $(\mathfrak{g}, [\cdot,\cdot], \alpha)$ and $(\mathfrak{g}', [\cdot,\cdot]', \alpha')$ be two color Hom-Lie algebras. An even linear map $f : \mathfrak{g} \to \mathfrak{g}'$ is called:

(i) a *weak morphism* of color Hom-Lie algebras if it satisfies $f([x, y]) = [f(x), f(y)]', \forall x, y \in \mathfrak{g}$.
(ii) a *morphism* of color Hom-Lie algebras if f is a weak morphism and $f \circ \alpha = \alpha' \circ f$.
(iii) an *automorphism* of color Hom-Lie algebras if f is a bijective morphism of color Hom-Lie algebras.

In [38], the author applied the twisting principle introduced by Yau and proved that starting from a color Lie algebra $(\mathfrak{g}, [\cdot,\cdot], \varepsilon)$ and a weak morphism $\beta : \mathfrak{g} \to \mathfrak{g}$, one obtains a color Hom-Lie algebra $(\mathfrak{g}, [\cdot,\cdot]_\beta, \beta, \varepsilon)$, where $[\cdot,\cdot]_\beta = \beta \circ [\cdot,\cdot]$. In the following, we generalize the result obtained in [38] in the following sense. We consider the category of color Hom-Lie algebras consisting of color Hom-Lie algebras and morphisms of color Hom-Lie algebras. We give a color Hom-Lie version of [38, Theorem 2.5].

Proposition 1 *The category of color Hom-Lie algebras is closed under self weak morphisms.*

Proof Let $(\mathfrak{g}, [\cdot,\cdot], \alpha, \varepsilon)$ be a color Hom-Lie algebra and $\beta : \mathfrak{g} \to \mathfrak{g}$ be a color Hom-Lie algebras weak morphism. Consider a bracket $[\cdot,\cdot]_\beta$ on $\mathfrak{g}$ defined by $[\cdot,\cdot]_\beta = \beta \circ [\cdot,\cdot]$. Then, for any homogeneous elements $x, y, z \in \mathfrak{g}$ and using the ε-Hom-Jacobi identity (5), we have

$$\begin{aligned}
\circlearrowleft_{x,y,z} \varepsilon(z,x)[\beta\circ\alpha(x), [y,z]_\beta]_\beta &= \circlearrowleft_{x,y,z} \varepsilon(z,x)\beta\circ[\beta\circ\alpha(x), \beta\circ[y,z]] \\
&= \circlearrowleft_{x,y,z} \varepsilon(z,x)\beta^2\circ[\alpha(x), [y,z]] \\
&= \beta^2(\circlearrowleft_{x,y,z} \varepsilon(z,x)[\alpha(x), [y,z]]) = 0.
\end{aligned}$$

In addition, the fact that $[\cdot,\cdot]_\beta$ is ε-skew-symmetric implies that $(\mathfrak{g}, [\cdot,\cdot]_\beta, \beta\circ\alpha, \varepsilon)$ is a color Hom-Lie algebra. □

As a particular case we obtain the following examples:

Example 2 1. Let $(\mathfrak{g}, [\cdot,\cdot], \varepsilon)$ be a color Lie algebra and α be a color Lie algebra morphism, then $(\mathfrak{g}, [\cdot,\cdot]_\alpha, \alpha, \varepsilon)$ is a multiplicative color Hom-Lie algebra.
2. Let $(\mathfrak{g}, [\cdot,\cdot], \alpha, \varepsilon)$ be a multiplicative color Hom-Lie algebra then $(\mathfrak{g}, \alpha^n \circ [\cdot,\cdot], \alpha^{n+1}, \varepsilon)$ is a multiplicative color Hom-Lie algebra for any $n \geq 0$.

In the following, we extend to color Hom-Lie algebras the study of quadratic Hom-Lie algebras introduced in [11].

Definition 4 Let $(\mathfrak{g}, [\cdot,\cdot], \alpha, \varepsilon)$ be a color Hom-Lie algebra and $B : \mathfrak{g} \times \mathfrak{g} \to \mathbb{K}$ be a bilinear form on $\mathfrak{g}$.

(i) B is said ε-symmetric if $B(x, y) = \varepsilon(x, y)B(y, x)$, (called also color-symmetry).
(ii) B is said invariant if $B([x, y], z) = B(x, [y, z]), \ \forall x, y, z \in \mathfrak{g}$.

Definition 5 A color Hom-Lie algebra $(\mathfrak{g}, [\cdot,\cdot], \alpha, \varepsilon)$ is called *quadratic* color Hom-Lie algebra if there exists a nondegenerate, ε-symmetric and invariant bilinear form B on $\mathfrak{g}$ such that α is B-symmetric (i.e. $B(\alpha(x), y) = B(x, \alpha(y))$). It is denoted by $(\mathfrak{g}, [\cdot,\cdot], \alpha, \varepsilon, B)$ and B is called invariant scalar product.

Remark 1 We recover quadratic color Lie algebras when $\alpha = id_{\mathfrak{g}}$ and quadratic Hom-Lie superalgebras when $\Gamma = \mathbb{Z}_2$ and $\varepsilon(\bar{1}, \bar{1}) = -1$.

Definition 6 A color Hom-Lie algebra $(\mathfrak{g}, [\cdot,\cdot], \alpha, \varepsilon)$ is called *Hom-quadratic* if there exists (B, β), where B is a nondegenerate and ε-symmetric bilinear form on $\mathfrak{g}$ and $\beta : \mathfrak{g} \longrightarrow \mathfrak{g}$ is an even homomorphism such that

$$B(\alpha(x), y) = B(x, \alpha(y)) \quad \text{and} \quad B(\beta(x), [y, z]) = B([x, y], \beta(z)), \ \ \forall\, x, y, z \in \mathfrak{g}.$$

We call the second condition the β-invariance of B.

Remark 2 We recover quadratic color Hom-Lie algebras when $\beta = id_{\mathfrak{g}}$.

In the following, we give a similar result to Theorem 1, in the case of quadratic color Hom-Lie algebras. To this end, we consider $(\mathfrak{g}, [\cdot,\cdot], \alpha, \varepsilon, B)$ a quadratic color Hom-Lie algebra and define:

• $Aut_s(\mathfrak{g}, B)$ as the set consisting of all automorphisms φ of $\mathfrak{g}$ which satisfy $B(\varphi(x), y) = B(x, \varphi(y)), \forall x, y \in \mathfrak{g}$. We call $Aut_s(\mathfrak{g}, B)$ the set of symmetric automorphisms of $(\mathfrak{g}, [\cdot,\cdot], \alpha, \varepsilon, B)$.

• The subcategory of quadratic color Hom-Lie algebras, in which the objets are quadratic color Hom-Lie algebras and the morphisms are isometry morphisms of quadratic color Hom-Lie algebras. A morphism f between two quadratic color Hom-Lie algebras $(\mathfrak{g}_1, B_1)$ and $(\mathfrak{g}_2, B_2)$ is said to be an isometry if $f : \mathfrak{g}_1 \to \mathfrak{g}_2$ satisfies $B_1(x, y) = B_2(f(x), f(y)), \ \forall x, y \in \mathfrak{g}_1$.

Proposition 2 *The subcategory of quadratic color Hom-Lie algebras is closed under symmetric automorphisms.*

Proof Let $(\mathfrak{g}, [\cdot,\cdot], \alpha, \varepsilon, B)$ be a quadratic color Hom-Lie algebra and β be an element of $Aut_s(\mathfrak{g}, B)$. Consider the bilinear form B_β defined on $\mathfrak{g}$ by $B_\beta(x, y) = B(\beta(x), y), \forall x, y \in \mathfrak{g}$. Using the fact that β is B-symmetric and $\beta \circ \alpha = \alpha \circ \beta$, we obtain

$$\begin{aligned} B_\beta(\beta \circ \alpha(x), y) &= B(\beta \circ \beta \circ \alpha(x), y) \\ &= B(\alpha \circ \beta(x), \beta(y)) = B(\beta(x), \beta \circ \alpha(y)) = B_\beta(x, \beta \circ \alpha(y)), \end{aligned}$$

which means that $\beta \circ \alpha$ is B_β-symmetric. In addition

$$B_\beta([x, y]_\beta, z) = B(\beta \circ \beta[x, y], z)$$
$$= B([\beta(x), \beta(y)], \beta(z)) = B(\beta(x), [\beta(y), \beta(z)]) = B_\beta(x, [y, z]_\beta),$$

which means that B_β is invariant with respect to $[\cdot, \cdot]_\beta$. The color symmetry and the non-degeneracy of B_β follow from the symmetry of B and the fact that β is bijective. Consequently, $(\mathfrak{g}, [\cdot, \cdot]_\beta, \beta \circ \alpha, \varepsilon, B_\beta)$ is a quadratic color Hom-Lie algebra. □

Corollary 1 *Let* $(\mathfrak{g}, [\cdot, \cdot], \alpha, \varepsilon)$ *be a multiplicative color Hom-Lie algebra. If* B *is an invariant scalar product on* $\mathfrak{g}$*, then* $(\mathfrak{g}, \alpha^n \circ [\cdot, ..., \cdot], \alpha^{n+1}, B_{\alpha^n})$ *is a quadratic multiplicative color Hom-Lie algebra and* $(\mathfrak{g}, \alpha^n \circ [\cdot, ..., \cdot], \alpha^{n+1}, B_\alpha)$*, where* $B_\alpha(x, y) = B(\alpha^n(x), y) = B(x, \alpha^n(y))$*, is a Hom-quadratic multiplicative color Hom-Lie algebra.*

3 Constructions and Examples of Quadratic Color Hom-Lie Algebras

In this section, we describe in one hand some constructions of quadratic color Hom-Lie algebras and in the other hand provide examples using these constructions.

In the following we construct color Hom-Lie algebras involving elements of the centroid of color Lie algebras. We first generalize the definition of centroid structures introduced in [11].

Let $(\mathfrak{g}, [\cdot, \cdot], \varepsilon)$ be a color Lie algebra and $End(\mathfrak{g})$ be the set of all homogeneous linear endomorphisms on the vector space $\mathfrak{g}$. Clearly $End(\mathfrak{g})$ is Γ-graded and considered with the color-commutator introduced in [38], defined by

$$[f, g]_\varepsilon = f \circ g - \varepsilon(f, g)\, g \circ f, \quad \forall f \in (End(\mathfrak{g}))_f,\ g \in (End(\mathfrak{g}))_g,$$

one obtains a color Lie algebra.

Definition 7 The vector subspace $Cent(\mathfrak{g})$ of $End(\mathfrak{g})$ defined by

$$Cent(\mathfrak{g}) = \{\theta \in End(\mathfrak{g}) :\ \theta([x, y]) = [\theta(x), y] = \varepsilon(\theta, x)[x, \theta(y)],\ \forall x, y \in \mathfrak{g}\}.$$

is a subalgebra of $End(\mathfrak{g})$ which we call the centroid of $\mathfrak{g}$.

In the following corollary, we construct color Hom-Lie algebras starting from a color Hom-Lie algebra and an even element in its centroid. This result is a graded version of a result obtained in [11].

Proposition 3 *Let* $(\mathfrak{g}, [\cdot, \cdot], \alpha, \varepsilon)$ *be a color Hom-Lie algebra and* θ *be an even element in the centroid of* $\mathfrak{g}$*. Then we have the following color Hom-Lie algebras*

(i) $(\mathfrak{g}, [\cdot,\cdot], \theta\circ\alpha, \varepsilon)$, $(\mathfrak{g}, [\cdot,\cdot]_1^\theta, \theta\circ\alpha, \varepsilon)$, $(\mathfrak{g}, [\cdot,\cdot]_2^\theta, \theta\circ\alpha, \varepsilon)$,
(ii) $(\mathfrak{g}, [\cdot,\cdot], \alpha\circ\theta, \varepsilon)$, $(\mathfrak{g}, [\cdot,\cdot]_1^\theta, \alpha\circ\theta, \varepsilon)$, $(\mathfrak{g}, [\cdot,\cdot]_2^\theta, \alpha\circ\theta, \varepsilon)$,

where the two even brackets $[\cdot,\cdot]_1^\theta$ *and* $[\cdot,\cdot]_2^\theta$ *on* $\mathfrak{g}$ *are defined by*

$$[x,y]_1^\theta = [\theta(x), y] \quad \textit{and} \quad [x,y]_2^\theta = [\theta(x), \theta(y)] \quad \forall x,y\in\mathfrak{g}.$$

Proof (i) Let x, y and z be homogeneous elements in $\mathfrak{g}$. The fact that θ is an even element of the centroid of $\mathfrak{g}$ implies

$$\varepsilon(z,x)\big[\theta\circ\alpha(x), [y,z]\big] = \theta\big(\varepsilon(z,x)\big[\alpha(x), [y,z]\big]\big).$$

Consequently,

$$\circlearrowleft_{x,y,z} \varepsilon(z,x)\big[\theta\circ\alpha(x), [y,z]\big] = \theta\big(\circlearrowleft_{x,y,z} \varepsilon(z,x)\big[\alpha(x), [y,z]\big]\big) = 0.$$

So, we deduce that $(\mathfrak{g}, [\cdot,\cdot], \theta\circ\alpha, \varepsilon)$ is a color Hom-Lie algebra. To prove that $(\mathfrak{g}, [\cdot,\cdot]_i^\theta, \theta\circ\alpha, \varepsilon)$, $i=1,2$, are color Hom-Lie algebras, we need only to check that $(\mathfrak{g}, [\cdot,\cdot]_i^\theta, \alpha, \varepsilon)$, $i=1,2$, are color Hom-Lie algebras. Using again the fact that θ is an even element in the centroid of $\mathfrak{g}$, it comes

$$\begin{aligned}\circlearrowleft_{x,y,z} \varepsilon(z,x)[\alpha(x), [y,z]_i^\theta]_i^\theta &= \circlearrowleft_{x,y,z} [\theta\circ\alpha(x), [\theta(y), z]] \\ &= \theta^2\big(\circlearrowleft_{x,y,z} \varepsilon(z,x)[\alpha(x),[y,z]]\big) = 0, \quad i=1,2.\end{aligned}$$

(ii) Applying the ε-Hom-Jacobi identity to $\theta(x)$, y and z in $\mathfrak{g}$, it follows

$$\varepsilon(z,x)\big[\alpha\circ\theta(x), [y,z]\big] = -\theta\big(\varepsilon(x,y)\big[\alpha(y), [z,x]\big] + \varepsilon(y,z)\big[\alpha(z), [x,y]\big]\big).$$

Consequently

$$\begin{aligned}\circlearrowleft_{x,y,z} \varepsilon(z,x)\big[\alpha\circ\theta(x), [y,z]\big] = &- \theta\big(\varepsilon(x,y)\big[\alpha(y), [z,x]\big] + \varepsilon(y,z)\big[\alpha(z), [x,y]\big]\big) \\ &- \theta\big(\varepsilon(y,z)\big[\alpha(z), [x,y]\big] + \varepsilon(z,x)\big[\alpha(x), [y,z]\big]\big) \\ &- \theta\big(\varepsilon(z,x)\big[\alpha(x), [y,z]\big] + \varepsilon(x,y)\big[\alpha(y), [z,x]\big]\big).\end{aligned}$$

Hence, we deduce that $(\mathfrak{g}, [\cdot,\cdot], \alpha\circ\theta, \varepsilon)$ is a color Hom-Lie algebra. Since $(\mathfrak{g}, [\cdot,\cdot]_i^\theta, \alpha, \varepsilon)$, $i=1,2$, are color Hom-Lie algebra, reasoning similarly as above proves the result. □

As an application of Proposition 3, we construct a color Hom-Lie algebra such that the twisted map is an element of its centroid. We assume that $(\mathfrak{g}, [\cdot,\cdot], \alpha, \varepsilon)$ is a color Hom-Lie algebra and define the following vector subspace $\mathfrak{U}$ of $End(\mathfrak{g})$ consisting of even linear maps σ on $\mathfrak{U}$ as follows:

$$\mathfrak{U} = \{u\in End(\mathfrak{g}), \quad u\circ\alpha = \alpha\circ u\}$$

and

$$\sigma : \mathfrak{U} \longrightarrow \mathfrak{U} \ ; \ \sigma(u) = \alpha \circ u.$$

Then, $\mathfrak{U}$ is a color subalgebra of $End(\mathfrak{g})$. Moreover, $(\mathfrak{U}, \{\cdot,\cdot\}_1^{\sigma}, \sigma)$ and $(\mathfrak{U}, \{\cdot,\cdot\}_2^{\sigma}, \sigma)$ are two color Hom-Lie algebras. Indeed, by a simple computation, it comes that $\mathfrak{U}$ is a color subalgebra of $End(\mathfrak{g})$. In addition, for two homogeneous elements u and v in $\mathfrak{U}$, we have

$$\sigma([u, v]) = \alpha(u \circ v - \varepsilon(u, v)v \circ u) = (\alpha \circ u) \circ v - \varepsilon(u, v)v \circ (\alpha \circ u) = [\sigma(u), v].$$

Now applying Proposition 3, we conclude that $(\mathfrak{U}, [\cdot,\cdot]_1^{\sigma}, \sigma)$ and $(\mathfrak{U}, [\cdot,\cdot]_2^{\sigma}, \sigma)$ are two color Hom-Lie algebras.

We consider now constructions using elements of the centroid in the quadratic case.

Proposition 4 *Let* $(\mathfrak{g}, [\cdot,\cdot], \varepsilon, B)$ *be a quadratic color Lie algebra and* θ *be an invertible and* B*-symmetric element in the centroid of* $\mathfrak{g}$*. Then, the two color Hom-Lie algebras* $(\mathfrak{g}, [\cdot,\cdot]_1^{\theta}, \theta, \varepsilon)$ *and* $(\mathfrak{g}, [\cdot,\cdot]_2^{\theta}, \theta, \varepsilon)$ *defined in Proposition 3 provided with the even bilinear form* B_θ *such that* $B_\theta(x, y) = B(\theta(x), y), \forall x, y \in \mathfrak{g}$ *are two quadratic color Hom-Lie algebras.*

Proof Since B is nondegenerate color symmetric bilinear form and θ is invertible then β_θ is a nondegenerate color symmetric bilinear form on $\mathfrak{g}$. Moreover for homogeneous elements x, y, z in $\mathfrak{g}$ we have

$$\begin{aligned} B_\theta([x, y]_1^{\theta}, z) &= B(\theta[\theta(x), y], z) = B([\theta(x), y], \theta(z)) \\ &= B(\theta(x), [y, \theta(z)]) = B_\theta(x, [\theta(y), z]) = B_\theta(x, [y, z]_1^{\theta}), \end{aligned}$$

and

$$\begin{aligned} B_\theta([x, y]_2^{\theta}, z) &= B(\theta[\theta(x), \theta(y)], z) = B([\theta(x), \theta(y)], \theta(z)) \\ &= B(\theta(x), [\theta(y), \theta(z)]) = B_\theta(x, [y, z]_2^{\theta}). \end{aligned}$$

□

Definition 8 A *color Hom-associative* algebra is a tuple $(A, \mu, \alpha, \varepsilon)$ consisting of a Γ-graded vector space A, an even bilinear map $\mu : A \times A \to A$ (i.e. $\mu(A_a, A_b) \subset A_{a+b}$) and an even homomorphism $\alpha : A \to A$ such that for homogeneous elements $x, y, z \in A$ we have

$$\mu(\alpha(x), \mu(y, z)) = \mu(\mu(x, y), \alpha(z)).$$

In the case where $\mu(x, y) = \varepsilon(x, y)\mu(y, x)$, the color Hom-associative algebra (A, μ, α) is called commutative.

Definition 9 A color Hom-associative algebra $(A, \mu, \alpha, \varepsilon)$ is called *quadratic* if there exists a color-symmetric, invariant and nondegenerate bilinear form B on $\mathfrak{g}$ such that α is B-symmetric.

Theorem 1 *Given a quadratic color Hom-associative algebra* $(A, \mu, \alpha, \varepsilon, B)$*, one can define color commutator on homogeneous elements* $x, y \in A$ *by*

$$[x, y] = \mu(x, y) - \varepsilon(x, y)\mu(y, x),$$

and then extended by linearity to all elements. The tuple $(A, [\cdot, \cdot], \alpha, \varepsilon, B)$ *determines a quadratic color Hom-Lie algebra.*

Proof The fact that $(A, [\cdot, \cdot], \alpha, \varepsilon)$ is a color Hom-Lie algebra was proved in [38, Proposition 3.13]. By using the fact that ε is bicharacter, we have for homogeneous elements x, y, z in A

$$\begin{aligned}
B([x, y], z) &= B(\mu(x, y) - \varepsilon(x, y)\mu(y, x), z)\\
&= B(\mu(x, y), z) - \varepsilon(x, y)B(\mu(y, x), z)\\
&= B(x, \mu(y, z)) - \varepsilon(x, y)\varepsilon(x + y, z)B(z, \mu(y, x))\\
&= B(x, \mu(y, z)) - \varepsilon(x, y)\varepsilon(x + y, z)B(\mu(z, y), x)\\
&= B(x, \mu(y, z)) - \varepsilon(x, y)\varepsilon(x + y, z)\varepsilon(z + y, x)B(x, \mu(z, y))\\
&= B(x, \mu(y, z)) - \varepsilon(y, z)B(x, \mu(z, y))\\
&= B(x, \mu(y, z)) - B(x, \varepsilon(y, z)\mu(z, y)) = B(x, [y, z]).
\end{aligned}$$

Thus B is invariant, hence $(A, [\cdot, \cdot], \alpha, \varepsilon, B)$ is a quadratic color Hom-Lie algebra. □

Recall that if V and V' are two Γ-graded vector spaces, then the tensor product $V \otimes V'$ is still a Γ-graded vector space such that for $\delta \in \Gamma$ we have

$$(V \otimes V')_\delta = \sum_{\delta=\gamma+\gamma'} V_\gamma \otimes V_{\gamma'},$$

where $\gamma, \gamma' \in \Gamma$.

Theorem 2 *Let* (A, μ, α_A) *be a commutative color Hom-associative algebra and* $(\mathfrak{g}, [\cdot, \cdot]_\mathfrak{g}, \alpha_\mathfrak{g}, \varepsilon)$ *be a color Hom-Lie algebra.*

(i) The tensor product $(\mathfrak{g} \otimes A, [\cdot, \cdot], \alpha, \varepsilon)$ *is a color Hom-Lie algebra such that*

$$\begin{aligned}
\alpha(x \otimes a) &= \alpha_\mathfrak{g}(x) \otimes \alpha_A(a),\\
[x \otimes a, y \otimes b] &= \varepsilon(a, y)\, [x, y]_\mathfrak{g} \otimes \mu(a, b), \quad \forall x,\ y \in \mathfrak{g},\ a, b \in A.
\end{aligned}$$

(ii) If B_A *and* $B_\mathfrak{g}$ *are respectively associative scalar product on* A *and invariant scalar product on* $\mathfrak{g}$*, then* $(\mathfrak{g} \otimes A, [\cdot, \cdot], \alpha, \varepsilon)$ *is a quadratic color Hom-Lie algebra such that the invariant scalar product* B *on* $\mathfrak{g} \otimes A$ *is given by*

$$B(x \otimes a, y \otimes b) := \varepsilon(a, y)\ B_\mathfrak{g}(x, y)B_A(a, b) \quad \forall x,\ y \in \mathfrak{g},\ a, b \in A.$$

Proof By using the definition of bracket and the commutativity of the product on the color Hom-associative algebra A, we have

$$\begin{aligned}&\varepsilon(z+c,x+a)\big[\alpha(x\otimes a),[y\otimes b,z\otimes c]\big]\\&=\varepsilon(z+c,x+a)\varepsilon(b,z)\varepsilon(a,y+z)\big[\alpha_{\mathfrak{g}}(x),[y,z]\big]\otimes\mu(\alpha_A(a),\mu(b,c)).\end{aligned}$$

$$\begin{aligned}&\varepsilon(x+a,y+b)\big[\alpha(y\otimes b),[z\otimes c,x\otimes a]\big]\\&=\varepsilon(x+a,y+b)\varepsilon(c,x)\varepsilon(b,z+x)\varepsilon(b+c,a)\big[\alpha_{\mathfrak{g}}(y),[z,x]\big]\otimes\mu(\alpha_A(a),\mu(b,c)).\end{aligned}$$

$$\begin{aligned}&\varepsilon(y+b,z+c)\big[\alpha(z\otimes c),[x\otimes a,y\otimes b]\big]\\&=\varepsilon(y+b,z+c)\varepsilon(a,y)\varepsilon(c,x+y)\varepsilon(c,a+b)\big[\alpha_{\mathfrak{g}}(z),[x,y]\big]\otimes\mu(\alpha_A(a),\mu(b,c)).\end{aligned}$$

Hence, using the identities (1), (2) and (3), we prove the ε-Hom-Jacobi identity. Moreover, we have

$$\begin{aligned}B(\big[x\otimes a,y\otimes b\big],z\otimes c)&=\varepsilon(a,y)\varepsilon(a+b,z)B_{\mathfrak{g}}([x,y],z)B_A(\mu(a,b),c)\\&=\varepsilon(a,y+z)\varepsilon(b,z)B_{\mathfrak{g}}(x,[y,z])B_A(a,\mu(b,c))\\&=\varepsilon(b,z)B(x\otimes a,[y,z]\otimes\mu(b,c))=B(x\otimes a,[y\otimes b,z\otimes c]).\end{aligned}$$

In addition,

$$\begin{aligned}B(\alpha(x\otimes a),y\otimes b)&=\varepsilon(a,y)B_{\mathfrak{g}}(\alpha_{\mathfrak{g}}(x),y)B_A(\alpha_A(a),b)\\&=\varepsilon(a,y)B_{\mathfrak{g}}(x,\alpha_{\mathfrak{g}}(y))B_A(a,\alpha_A(b))=B(x\otimes a,\alpha(y\otimes b)).\end{aligned}$$

Consequently, we obtain $(\mathfrak{g}\otimes A,\alpha,[\cdot,\cdot],\varepsilon,B)$ is a quadratic color Hom-Lie algebra. □

Color Hom-Lie algebras have been studied in [38] where many examples are provided. In the following, we give some examples of quadratic color Hom-Lie algebras by using methods of construction mentioned above.

Example 3 The twisting maps which make $\mathfrak{sl}_2$ a Hom-Lie algebra were given in [24]. Then $\mathfrak{g}$ defined, with respect to a basis $\{x_1,x_2,x_3\}$, by

$$[x_1,x_2]=2x_2,\quad [x_1,x_3]=-2x_3,\quad [x_2,x_3]=x_1,$$

and linear maps α defined, with respect to the previous basis, by:

$$\begin{pmatrix}a&d&c\\2c&b&f\\2d&e&b\end{pmatrix}$$

are Hom-Lie algebras.

In the case where the matrix is the identity matrix, one gets the classical Lie algebra $\mathfrak{sl}_2$. It is well known that the Lie algebra $\mathfrak{sl}_2$ provided with its Killing form

$\mathfrak{K}$ is a quadratic Lie algebra. Moreover, a straightforward computation shows that α are $\mathfrak{K}$-symmetric (i.e. $\mathfrak{K}(\alpha(x), y) = \mathfrak{K}(x, \alpha(y))$). Consequently, $(\mathfrak{g}, [\cdot, \cdot], \alpha, \mathfrak{K})$ are quadratic Hom-Lie algebras.

Example 4 Nilpotent Lie superalgebras up to dimension 5 are classified in [20]. Among these Lie superalgebras, there is a class of quadratic 2-nilpotent Lie superalgebras $\mathfrak{L} = \mathfrak{L}_{\bar{0}} \oplus \mathfrak{L}_{\bar{1}}$ where $\mathfrak{L}_{\bar{0}}$ is generated by $< l_0, k_0 >$ and $\mathfrak{L}_{\bar{1}}$ is generated by $< l_1, k_1 >$. The bracket being defined as $[l_0, l_1] = k_1$, $[l_1, l_1] = k_0$. The quadratic structures are given by bilinear forms $B_{p,q}$ which are defined with respect to the basis $\{l_0, k_0, l_1, k_1\}$ by the matrices

$$\begin{pmatrix} p & -q & 0 & 0 \\ -q & 0 & 0 & 0 \\ 0 & 0 & 0 & q \\ 0 & 0 & -q & 0 \end{pmatrix}$$

where $p \in \mathbb{K}$ and $q \in \mathbb{K} \setminus \{0\}$.

Now, let us consider the even linear map $\alpha : \mathfrak{L} \longrightarrow \mathfrak{L}$ defined, with respect to the basis $\{l_0, k_0, l_1, k_1\}$, by matrices:

$$\begin{pmatrix} a & c & 0 & 0 \\ b & a + \frac{p}{q}c & 0 & 0 \\ 0 & 0 & d & 0 \\ 0 & 0 & 0 & d \end{pmatrix}.$$

Direct calculations show that super-Hom-Jacobi identity is satisfied and

$$\begin{aligned} B(\alpha(l_0), k_0) &= B(al_0 + bk_0, k_0) = -aq = cp - q(a + \frac{p}{q}c) \\ &= B(l_0, cl_0 + (a + \frac{p}{q}c)k_0) = B(l_0, \alpha(k_0)), \end{aligned}$$

$$\begin{aligned} B(\alpha(l_1), l_1) &= B(l_1, \alpha(l_1)) = d\ B(l_1, l_1) = 0 \\ &= d\ B(k_1, k_1) = B(\alpha(k_1), k_1) = B(k_1, \alpha(k_1)), \end{aligned}$$

$$B(\alpha(l_1), k_1) = B(dl_1, k_1) = dq = B(l_1, dk_1) = B(l_1, \alpha(k_1)).$$

So, we deduce that $(\mathfrak{L}, [\cdot, \cdot], \alpha, B)$ are quadratic Hom-Lie superalgebras.

Example 5 The two-dimensional superalgebra $A = A_{\bar{1}}$ is a symmetric associative super-commutative superalgebra. Applying Theorem 2, where $\Gamma = \mathbb{Z}_2$, to previous examples

- Let $(\mathfrak{g}, [\cdot, \cdot], \alpha, \mathfrak{K})$ be a quadratic Lie superalgebra defined in Example 3. Then $\mathfrak{g} \otimes A$ is a quadratic Hom-Lie superalgebra where the twist map is given by $\alpha \otimes id$.

• Let $(\mathfrak{L}, [\cdot,\cdot], \alpha, B)$ be a quadratic Lie superalgebra defined in Example 4. Then $\mathfrak{L} \otimes A$ is a quadratic Hom-Lie superalgebra where the twist map is given by $\alpha \otimes id$.

Example 6 Let us consider the class of symmetric associative super-commutative superalgebras defined by $A = A_{\bar{0}} \oplus A_{\bar{1}}$, where $dimA_{\bar{0}} = dimA_{\bar{1}} = 2$, such that $A_{\bar{0}}.A_{\bar{1}} = \{0\}$, $A_{\bar{1}}.A_{\bar{1}} = \{0\}$ and if we assume that $A_{\bar{0}} = \mathbb{K}e_0 \oplus \mathbb{K}f_0$, then we have $e_0 \in Ann(A_{\bar{0}})$ and $f_0 \cdot f_0 = ae_0$, where $a \neq 0$. In addition, the symmetric structures are expressed, with respect to the basis $\{e_0, f_0, e_1, f_1\}$, by the following matrices

$$\begin{pmatrix} 0 & \alpha & 0 & 0 \\ \alpha & \beta & 0 & 0 \\ 0 & 0 & 0 & \gamma \\ 0 & 0 & -\gamma & 0 \end{pmatrix}$$

where $\alpha, \gamma \in \mathbb{K} \setminus \{0\}$ and $\beta \in \mathbb{K}$.

Theorem 2, where $\Gamma = \mathbb{Z}_2$, leads to $\mathfrak{g} \otimes A$ (resp. $\mathfrak{L} \otimes A$), where $\mathfrak{g}$ is defined in Example 3 (resp. $\mathfrak{L}$ defined in Example 4) is a quadratic Hom-Lie superalgebra where the twist map is given by $\alpha \otimes id$.

Example 7 Let A be an n-dimensional vector space and $\wedge A$ be a Grassmann algebra of A. We know that $\wedge A$ is a super-commutative associative superalgebra with

$$(\wedge A)_{\bar{0}} = \bigoplus_{i\in\mathbb{Z}} \wedge^{2i} A \quad \text{and} \quad (\wedge A)_{\bar{1}} = \bigoplus_{i\in\mathbb{Z}} \wedge^{2i+1} A.$$

Moreover, it has been proved in [10] that $\wedge A$ is a symmetric superalgebra if and only if the dimension n of A is even. Here we consider $\wedge A$ where A is a vector space with even dimension. We denote by θ an associative symmetric structure on $\wedge A$.

Similarly, Theorem 2 with $\Gamma = \mathbb{Z}_2$, leads to $\mathfrak{g} \otimes A$ (resp. $\mathfrak{L} \otimes A$), where $\mathfrak{g}$ is defined in Example 3 (resp. $\mathfrak{L}$ defined in Example 4) is a quadratic Hom-Lie superalgebra where the twist map is given by $\alpha \otimes id$.

4 Representations of Color Hom-Lie Algebras

In this section we extend representation theory of Hom-Lie algebras introduced in [11, 33] to color Hom-Lie case. Moreover, we discuss semi-direct product.

Definition 10 Let $(\mathfrak{g}, [\cdot,\cdot]_{\mathfrak{g}}, \alpha, \varepsilon)$ be a color Hom-Lie algebra and (M, β) be a pair consisting of Γ-graded vector space M and an even homomorphism $\beta : M \to M$. The pair (M, β) is said to be a *Hom-module* over $\mathfrak{g}$ (or $\mathfrak{g}$-Hom-module) if there exists an even color skew-symmetric bilinear map $[\cdot,\cdot] : \mathfrak{g} \times M \longrightarrow M$, that is $[x, m] = -\varepsilon(x, m)\,[m, x]$, $\forall x \in \mathfrak{g}$ and $\forall m \in M$, such that

$$\varepsilon(m, x)\big[\alpha(x), [y, m]\big] + \varepsilon(m, x)\big[\alpha(y), [m, x]\big] + \varepsilon(y, m)\,\big[\beta(m), [x, y]_{\mathfrak{g}}\big] = 0, \quad \forall\, x, y \in \mathfrak{g}, \forall m \in M.$$

Definition 11 Let $(\mathfrak{g}, [\cdot,\cdot], \alpha, \varepsilon)$ be a color Hom-Lie algebra and (M, β) be a pair consisting of a Γ-graded vector space M and an even homomorphism $\beta : M \to M$. A *representation* of $\mathfrak{g}$ on (M, β) is an even linear map $\rho : \mathfrak{g} \to End(M)$ satisfying

$$\rho([x, y]) \circ \beta = \rho(\alpha(x)) \circ \rho(y) - \varepsilon(x, y)\rho(\alpha(y)) \circ \rho(x). \tag{1}$$

Definition 12 A representation ρ of a multiplicative color Hom-Lie algebra $(\mathfrak{g}, \alpha, \varepsilon)$ on a Γ-graded vector space (M, β) is a representation of color Hom-Lie algebra (i.e. a linear map satisfying (1)) such that in addition the following condition holds:

$$\beta\big(\rho(x)(m)\big) = \rho(\alpha(x))(\beta(m)), \quad \forall\, m \in M \text{ and } \forall x \in \mathfrak{g}. \tag{2}$$

Now, let $(\mathfrak{g}, [\cdot,\cdot], \alpha, \varepsilon)$ be a color Hom-Lie algebra and (M, β) be a pair of a Γ-graded vector space M and an even homomorphism β. If ρ is a representation of $\mathfrak{g}$ on (M, β), then we can see easily that M is a $\mathfrak{g}$-module via $[x, m] = \rho(x)(m)$, $\forall x \in \mathfrak{g}$ and $\forall m \in M$. Conversely, if M is a $\mathfrak{g}$-Hom-module, then the linear map $\rho : \mathfrak{g} \longrightarrow End(M)$ defined by $\rho(x)(m) = [x, m]$, $\forall x \in \mathfrak{g}$ and $\forall m \in M$, is a representation of $\mathfrak{g}$ on (M, β).

Two representations ρ and ρ' of $\mathfrak{g}$ on (M, β) and (M', β') respectively are *equivalent* if there exists an even isomorphism of vector spaces $f : M \to M'$ such that $f \circ \beta = \beta' \circ f$ and $f \circ \rho(x) = \rho'(x) \circ f$, $\forall x \in \mathfrak{g}$.

Example 8 Let $(\mathfrak{g}, [\cdot,\cdot], \alpha, \varepsilon)$ be a color Hom-Lie algebra. The even linear map

$$ad : \mathfrak{g} \longrightarrow End(\mathfrak{g}) \quad \text{defined by} \quad ad(x)(y) := [x, y]\,, \quad \forall x, y \in \mathfrak{g}$$

is a representation of $\mathfrak{g}$ on $(\mathfrak{g}, \alpha)$. This representation is called adjoint representation.

Proposition 5 *Let* $(\mathfrak{g}, [\cdot,\cdot], \alpha, \varepsilon)$ *be a color Hom-Lie algebra and* ρ *be a representation of* $\mathfrak{g}$ *on* (M, β). *Let us consider* M^* *the dual space of* M *and* $\tilde{\beta} : M^* \longrightarrow M^*$ *an even homomorphism defined by* $\widetilde{\beta}(f) = f \circ \beta$, $\forall f \in M^*$. *Then, the even linear map* $\widetilde{\rho} : \mathfrak{g} \to End(M^*)$ *defined by* $\widetilde{\rho} := -^t\rho$, *that is* $\widetilde{\rho}(x)(f) = -\varepsilon(x, f) f \circ \rho(x)$, $\forall f \in M^*$ *and* $\forall x \in \mathfrak{g}$, *is a representation of* $\mathfrak{g}$ *on* $(M^*, \tilde{\beta})$ *if and only if*

$$\rho(x) \circ \rho(\alpha(y)) - \varepsilon(x, y)\rho(y) \circ \rho(\alpha(x)) = \beta \circ \rho([x, y]), \ \forall\, x, y \in \mathfrak{g}. \tag{3}$$

Proof Let $f \in M^*$, $x, y \in \mathfrak{g}$ and $m \in M$. We compute the right hand side of the identity (1), we have

$$\begin{aligned}
&\big(\widetilde{\rho}(\alpha(x)) \circ \widetilde{\rho}(y) - \varepsilon(x, y)\widetilde{\rho}(\alpha(y)) \circ \widetilde{\rho}(x)\big)(f)(m)\\
&= -\varepsilon(x, y + f)\widetilde{\rho}(y)(f) \circ \rho(\alpha(x))(m) + \varepsilon(y, f)\widetilde{\rho}(x)(f) \circ \rho(\alpha(y))(m)\\
&= \varepsilon(y, f)\varepsilon(x, y + f) f(\rho(y) \circ \rho(\alpha(x))(m)) - \varepsilon(x + y, f) f(\rho(x) \circ \rho(\alpha(y))(m))\\
&= -\varepsilon(x + y, f) f\big(-\varepsilon(x, y)\rho(y) \circ \rho(\alpha(x))(m) + \rho(x) \circ \rho(\alpha(y))(m)\big).
\end{aligned}$$

On the other hand, we set that the twisted map $\widetilde{\beta}$ is $\widetilde{\beta} ={}^t \beta$, then the left hand side (1) writes

$$\begin{aligned}\widetilde{\rho}([x,y]) \circ \widetilde{\beta}(f)(m) &= -\varepsilon(x+y,f)\widetilde{\beta}(f) \circ \rho([x,y])(m)\\ &= -\varepsilon(x+y,f)f\big(\beta \circ \rho([x,y])\big)(m).\end{aligned}$$

Thus (3) is satisfied. □

Corollary 2 *Let ad be the adjoint representation of a color Hom-Lie algebra* $(\mathfrak{g}, [\cdot,\cdot], \alpha, \varepsilon)$ *and let us consider the even linear map* $\pi : \mathfrak{g} \to End(\mathfrak{g}^*)$ *defined by* $\pi(x)(f)(y) = -\varepsilon(x,f)(f \circ ad(x))(y)$, $\forall x, y \in \mathfrak{g}$. *Then* π *is a representation of* $\mathfrak{g}$ *on* $(\mathfrak{g}^*, \widetilde{\alpha})$ *if and only if*

$$ad(x) \circ ad(\alpha(y)) - \varepsilon(x,y)ad(y) \circ ad(\alpha(x)) = \alpha \circ ad([x,y]),\ \forall\, x, y \in \mathfrak{g}. \quad (4)$$

We call the representation π *the* coadjoint representation *of* $\mathfrak{g}$.

Proposition 6 *Let* (M, ρ, γ) *be a representation of color Hom-Lie algebra* $(\mathfrak{g}, [\cdot,\cdot]_{\mathfrak{g}}, \alpha)$. *Then the space* $\mathfrak{g} \oplus M$ *is a* Γ*-graded color Hom-Lie algebra where* $(\mathfrak{g} \oplus M)_a = \mathfrak{g}_a \oplus M_a$, *with structure* $([\cdot,\cdot], \eta)$ *given, for* $x, y \in \mathfrak{g}$ *and* $u, v \in M$ *by*

$$[x+u, y+v] = [x,y]_{\mathfrak{g}} + \rho(x)(v) - \varepsilon(x,y)\rho(y)(u), \quad (5)$$

$$\eta(x+u) = \alpha(x) + \gamma(u), \quad (6)$$

we call the direct sum $\mathfrak{g} \oplus M$ *semidirect product of* $\mathfrak{g}$ *and* M. *It is denoted by* $\mathfrak{g} \ltimes M$.

Proof Observe that, for any three homogeneous elements $x+u$, $y+v$ and $z+w$ of $\mathfrak{g} \oplus M$

$$\begin{aligned}&\varepsilon(z,x)[\eta(x+u), [y+v, z+w]]\\ &= \varepsilon(z,x)[\alpha(x)+\gamma(u), [y,z]_{\mathfrak{g}} + \rho(y)(w) - \varepsilon(y,z)\rho(z)(v)]\\ &= \varepsilon(z,x)[\alpha(x), [y,z]_{\mathfrak{g}}]_{\mathfrak{g}} + \rho(\alpha(x)) \circ \rho(y)(w)\\ &-\varepsilon(z,x)\varepsilon(y,z)\rho(\alpha(x)) \circ \rho(z)(v) - \varepsilon(x,y)\rho([y,z]_{\mathfrak{g}})(\gamma(u)).\end{aligned}$$

And similarly

$$\begin{aligned}\varepsilon(x,y)[\eta(y+v), [z+w, x+u]] &= \varepsilon(x,y)[\alpha(y), [z,x]_{\mathfrak{g}}]_{\mathfrak{g}} + \rho(\alpha(y)) \circ \rho(z)(u)\\ &-\varepsilon(x,y)\varepsilon(z,x)\rho(\alpha(y)) \circ \rho(x)(w) - \varepsilon(y,z)\rho([z,x]_{\mathfrak{g}})(\gamma(v)).\end{aligned}$$

$$\begin{aligned}\varepsilon(y,z)[\eta(z+w), [x+u, y+v]] &= \varepsilon(y,z)[\alpha(z), [x,y]_{\mathfrak{g}}]_{\mathfrak{g}} + \rho(\alpha(z)) \circ \rho(x)(v)\\ &-\varepsilon(y,z)\varepsilon(x,y)\rho(\alpha(z)) \circ \rho(y)(u) - \varepsilon(z,x)\rho([x,y]_{\mathfrak{g}})(\gamma(w)).\end{aligned}$$

Therefore the identities (5), (1) induce that $(\mathfrak{g} \oplus M, [\cdot,\cdot], \eta)$ is a color Hom-Lie algebra. □

5 Extensions

In this section, we generalize to the case of color Hom-Lie algebras the notion of central extension and T^*-extension introduced in [34] for color Lie algebras. These two types of extensions require some elements of cohomology of color Hom-Lie algebras.

5.1 Elements of Cohomology

The cohomology of color Lie algebras was introduced in [30]. In the following, we define the first and the second cohomology groups of color Hom-Lie algebras, with values in a $\mathfrak{g}$-Hom-module (M, β).

Let $(\mathfrak{g}, [\cdot, \cdot], \alpha, \varepsilon)$ be a color Hom-Lie algebra and (M, β) be a Γ-graded $\mathfrak{g}$-Hom-module. we set

$$\begin{aligned}
&\mathscr{C}^n(\mathfrak{g}, M) = \left\{\varphi : \wedge^n \mathfrak{g} \longrightarrow M, \quad \text{such that} \quad \varphi \circ \alpha^{\otimes n} = \beta \circ \varphi\right\}, \quad \text{for } n \geq 1.\\
&\mathscr{C}^n(\mathfrak{g}, M) = \{0\}, \quad \text{for } n \leq -1,\\
&\mathscr{C}^0(\mathfrak{g}, M) = M.
\end{aligned}$$

A homogeneous element φ in $\mathscr{C}^n(g, M)$ is called an *n-cochain*. Next, we define the coboundary operators $\delta^n : \mathscr{C}^n(\mathfrak{g}, M) \longrightarrow \mathscr{C}^{n+1}(\mathfrak{g}, M)$ for $n = 1, 2$ by

$$\begin{aligned}
\delta^1 : \mathscr{C}^1(\mathfrak{g}, M) &\longrightarrow \mathscr{C}^2(\mathfrak{g}, M)\\
\varphi &\longmapsto \delta^1(\varphi)
\end{aligned}$$

such that, for homogeneous elements $x_0, x_1 \in \mathfrak{g}$,

$$\delta^1\varphi(x_0, x_1) = \varepsilon(\varphi, x_0)\rho(x_0)(\varphi(x_1))) - \varepsilon(\varphi + x_0, x_1)\rho(x_1)(\varphi(x_0)) - \varphi([x_0, x_1]).$$

and

$$\begin{aligned}
\delta^2 : \mathscr{C}^2(\mathfrak{g}, M) &\longrightarrow \mathscr{C}^3(\mathfrak{g}, M)\\
\varphi &\longmapsto \delta^2(\varphi)
\end{aligned}$$

such that, for homogeneous elements $x_0, x_1, x_2 \in \mathfrak{g}$,

$$\begin{aligned}
\delta^2\varphi(x_0, x_1, x_2) = {}& \varepsilon(\varphi, x_0)\rho(\alpha(x_0))(\varphi(x_1, x_2))) - \varepsilon(\varphi + x_0, x_1)\rho(\alpha(x_1))(\varphi(x_0, x_2))\\
& +\varepsilon(\varphi + x_0 + x_1, x_2)\rho(\alpha(x_2))(\varphi(x_0, x_1)) - \varphi([x_0, x_1], \alpha(x_2))\\
& +\varepsilon(x_1, x_2)\varphi([x_0, x_2], \alpha(x_1)) + \varphi(\alpha(x_0), [x_1, x_2]).
\end{aligned}$$

Proposition 7 *Let* $(\mathfrak{g}, [\cdot,\cdot], \alpha, \varepsilon)$ *be a color Hom-Lie algebra and* $\delta^n : \mathscr{C}^n(\mathfrak{g}, M) \longrightarrow \mathscr{C}^{n+1}(\mathfrak{g}, M)$ *be the coboundary operator defined above. Then, the composite* $\delta^2 \circ \delta^1 = 0$.

Remark 3 In order to define a cohomology complex, we may set $(\delta^0 m)(x) = [m, x]$, $\forall x \in \mathfrak{g}$ and $\forall m \in M$, and $\delta^n = 0$, $\forall n \geq 3$. Deeply constructions will be given in a forthcoming paper.

We denote the kernel of δ^n by $\mathscr{Z}^n(\mathfrak{g}, M)$, whose elements are called *n-cocycles*, and the image of δ^{n-1} by $\mathscr{B}^n(\mathfrak{g}, M)$, whose elements are called *n-coboundaries*. Moreover $\mathscr{Z}^n(\mathfrak{g}, M)$ and $\mathscr{B}^n(\mathfrak{g}, M)$ are two graded submodules of $\mathscr{C}^n(\mathfrak{g}, M)$. Following Proposition 7 we have

$$\mathscr{B}^n(\mathfrak{g}, M) \subset \mathscr{Z}^n(\mathfrak{g}, M).$$

Consequently, we construct the so-called *cohomology groups*

$$\mathscr{H}^n(\mathfrak{g}, M) = \mathscr{Z}^n(\mathfrak{g}, M)/\mathscr{B}^n(\mathfrak{g}, M).$$

Two elements of $\mathscr{Z}^n(\mathfrak{g}, M)$ are said to be *cohomologous* if their residue classes modulo $\mathscr{B}^n(\mathfrak{g}, M)$ coincide, that is if their difference lies in $\mathscr{B}^n(\mathfrak{g}, M)$.

5.2 Central Extensions

Let $(\mathfrak{g}_i, [\cdot,\cdot]_i, \alpha_i, \varepsilon)$, $i = 1, 2$, be two color Hom-Lie algebras. A color Hom-Lie algebra $\mathfrak{g}$ is an extension of $\mathfrak{g}_2$ by $\mathfrak{g}_1$ if there exists an exact sequence

$$0 \longrightarrow \mathfrak{g}_1 \stackrel{\mu_1}{\longrightarrow} \mathfrak{g} \stackrel{\mu_2}{\longrightarrow} \mathfrak{g}_2 \longrightarrow 0$$

such that μ_1 and μ_2 are two morphisms of color Hom-Lie algebras. The kernel of μ_2 is said to be the kernel of the extension. Two extensions $\mathfrak{g}$ and $\mathfrak{g}'$ of $\mathfrak{g}_2$ by $\mathfrak{g}_1$ are said to be equivalent if there exists a morphism of color Hom-Lie algebras $f : \mathfrak{g} \longrightarrow \mathfrak{g}'$ such that the following diagram commutes.

$$\begin{array}{ccccccccc} 0 & \longrightarrow & \mathfrak{g}_1 & \stackrel{\mu_1}{\longrightarrow} & \mathfrak{g} & \stackrel{\mu_2}{\longrightarrow} & \mathfrak{g}_2 & \longrightarrow & 0 \\ & & \downarrow & & \downarrow & & \downarrow & & \\ 0 & \longrightarrow & \mathfrak{g}_1 & \stackrel{\mu'_1}{\longrightarrow} & \mathfrak{g}' & \stackrel{\mu'_2}{\longrightarrow} & \mathfrak{g}_2 & \longrightarrow & 0 \end{array}$$

Definition 13 An extension $\mathfrak{g}$ of $\mathfrak{g}_2$ by $\mathfrak{g}_1$ is said to be central if the kernel of the extension lies in the center of $\mathfrak{g}$.

The following theorem gives a characterization of the bracket of the central extension of a color Hom-Lie algebra $\mathfrak{g}$ and Γ-graded vector space M.

Theorem 3 *Let* $(\mathfrak{g}, [\cdot,\cdot]_{\mathfrak{g}}, \alpha_{\mathfrak{g}}, \varepsilon)$ *be a color Hom-Lie algebra,* M *be a* Γ*-graded vector space and* $\Psi : \mathfrak{g} \times \mathfrak{g} \longrightarrow M$ *be an even bilinear map. Then, the* Γ*-graded vector space* $\mathfrak{g} \oplus M$*, where* $(\mathfrak{g} \oplus M)_\gamma = \mathfrak{g}_\gamma \oplus M_\gamma$ *for* $\gamma \in \Gamma$*, provided with the following bracket and even linear map defined respectively by*

$$[x+m, y+n] = [x,y]_{\mathfrak{g}} + \Psi(x,y), \tag{1}$$

$$\alpha(x+m) = \alpha_{\mathfrak{g}}(x) + m, \quad \forall\, x, y \in \mathfrak{g}, \ m, n \in M, \tag{2}$$

is a color Hom-Lie algebra central extension of $\mathfrak{g}$ *by* M *if and only if*

$$\circlearrowleft_{x,y,z} \varepsilon(z,x)\Psi(\alpha_{\mathfrak{g}}(x), [y,z]_{\mathfrak{g}}) = 0.$$

In particular, if $\mathfrak{g}$ *is multiplicative and* Ψ *satisfies* $\Psi(\alpha(x), \alpha(y)) = \Psi(x,y)$*,* $\forall x, y \in \mathfrak{g}$*, then the color Hom-Lie algebra* $\mathfrak{g} \oplus M$ *is also multiplicative.*

Proof We have for homogeneous elements $x, y, z \in \mathfrak{g}$ and $m, n, l \in M$,

$$\varepsilon(z,x)\big[\alpha_{\mathfrak{g}}(x) + l, [y+m, z+n]\big] = \varepsilon(z,x)\big[\alpha_{\mathfrak{g}}(x) + l, [y,z]_{\mathfrak{g}} + \Psi(y,z)\big]$$
$$= \varepsilon(z,x)\big[\alpha_{\mathfrak{g}}(x), [y,z]_{\mathfrak{g}}\big]_{\mathfrak{g}} + \varepsilon(z,x)\Psi(\alpha_{\mathfrak{g}}(x), [y,z]_{\mathfrak{g}}).$$

Consequently,

$$\circlearrowleft_{x,y,z} \varepsilon(z,x)\big[\alpha_{\mathfrak{g}}(x) + l, [y+m, z+n]\big] = 0$$

if and only if

$$\circlearrowleft_{x,y,z} \varepsilon(z,x)\Psi(\alpha_{\mathfrak{g}}(x), [y,z]_{\mathfrak{g}}) = 0.$$

Corollary 3 *Let* $(\mathfrak{g}, [\cdot,\cdot]_{\mathfrak{g}}, \alpha_{\mathfrak{g}}, \varepsilon)$ *be a multiplicative color Hom-Lie algebra. Then, the* Γ*-graded vector space* $\mathfrak{g} \oplus \mathbb{K}$ *provided with the product (1) and the map (2) is a multiplicative color Hom-Lie algebra if and only if* $\Psi \in \mathscr{Z}^2(\mathfrak{g}, \mathbb{K})$*.*

We denote by $\mathcal{E}(\mathfrak{g}, \mathbb{K})$ the set of all equivalent classes of central extensions of a multiplicative color Hom-Lie algebra $\mathfrak{g}$ by $\mathbb{K}$.

Corollary 4 *There exists a one-to-one correspondence between* $\mathcal{E}(\mathfrak{g}, \mathbb{K})$ *and the second cohomology group* $\mathscr{H}^2(\mathfrak{g}, \mathbb{K})$ *of a color Hom-Lie* $\mathfrak{g}$*.*

Following [33], we define α^k-derivations.

Definition 14 Let $(\mathfrak{g}, [\cdot,\cdot], \alpha, \varepsilon)$ be a multiplicative color Hom-Lie algebra. A homogeneous bilinear map $D : \mathfrak{g} \longrightarrow \mathfrak{g}$ of degree d is said to be an α^k-derivation, where $k \in \mathbb{N}$, if it satisfies

$$D \circ \alpha = \alpha \circ D,$$

$$D[x,y] = [D(x), \alpha^k(y)] + \varepsilon(d,x)[\alpha^k(x), D(y)],$$

for any homogeneous element x and for any $y \in \mathfrak{g}$.

The set of all α^k-derivations is denoted by $Der_\alpha^k(\mathfrak{g})$. The space $Der(\mathfrak{g}) = \bigoplus_{k\geq 0} Der_{\alpha^k}(\mathfrak{g})$ provided with the color-commutator is a color Lie algebra. The fact that $Der_{\alpha^k}(\mathfrak{g})$ is Γ-graded implies that $Der(\mathfrak{g})$ is Γ-graded

$$(Der(\mathfrak{g}))_\gamma = \bigoplus_{k\geq 0} (Der_{\alpha^k}(\mathfrak{g}))_\gamma, \quad \forall\, \gamma \in \Gamma.$$

Moreover, similarly to the non-graded case [33], equipped with the color commutator and the following even map

$$\widetilde{\alpha} : Der(\mathfrak{g}) \longrightarrow Der(\mathfrak{g})\ ;\ D \longmapsto \widetilde{\alpha}(D) = D \circ \alpha,$$

$Der(\mathfrak{g})$ is a color Hom-Lie algebra.

Definition 15 Let $(\mathfrak{g}, [\cdot,\cdot], \alpha, \varepsilon, B)$ be a quadratic color Hom-Lie algebra and $D : \mathfrak{g} \longrightarrow \mathfrak{g}$ be a homogeneous derivation of degree d. The derivation D is said to be ε-skew-symmetric with respect to B (or B-skew-symmetric) if it satisfies

$$B(D(x), y) = -\varepsilon(d, x)B(x, D(y)), \quad \forall x \in \mathfrak{g}_x, \forall y \in \mathfrak{g}.$$

In the following we characterize scalar cocycles of quadratic multiplicative color Hom-Lie algebras using ε-skew-symmetric derivations.

Lemma 1 *Let $(\mathfrak{g}, [\cdot,\cdot], \alpha, B, \varepsilon)$ be a quadratic multiplicative color Hom-Lie algebra.*

(i) If ω is a homogeneous scalar 2-cocycle, then there exists a homogeneous ε-skew-symmetric α-derivation D of $\mathfrak{g}$ such that

$$\omega(x, y) := B(D(x), y), \ \forall x, y \in \mathfrak{g}. \tag{3}$$

(ii) Conversely, if D a homogeneous ε-skew-symmetric α-derivation of $\mathfrak{g}$, then ω defined in (3) is a homogeneous scalar 2-cocycle on $\mathfrak{g}$, that is

$$\circlearrowleft_{x,y,z} \omega\big(\alpha(x), [y, z]\big) = 0, \quad \forall x, y, z \in \mathfrak{g}.$$

Proof (i) The fact that B is nondegenerate implies that $\phi : \mathfrak{g} \longrightarrow \mathfrak{g}^*$ defined by $\phi(x) = B(x, .)$ is an isomorphism of Γ-graded vector spaces. Let $y \in \mathfrak{g}$, since $\omega(y, .) \in \mathfrak{g}^*$ then there exists a unique $Y_y \in \mathfrak{g}$ such that $\phi(Y_y) = \omega(y, .)$, namely $B(Y_y, z) = \omega(y, z), \forall z \in \mathfrak{g}$. Set $D(y) = Y_y$, the map D is well-defined by the uniqueness of Y_y. By using the fact that B is even and ω is homogeneous, we deduce that D is homogeneous with the same degree as ω. The B-skew-symmetry of D comes from the ε-symmetry of B and the ε-skew-symmetry of ω. Finally, for homogeneous elements $x, y, z \in \mathfrak{g}$, we have

$$\omega(\alpha(x), [y, z]) = \omega([x, y], \alpha(z)) + \varepsilon(x, y)\omega(\alpha(y), [x, z]),$$

which is equivalent to

$$B(D(\alpha(x)), [y, z]) = B(D([x, y]), \alpha(z)) + \varepsilon(x, y)B(D(\alpha(y)), [x, z]).$$

Thus

$$B([D(x), \alpha(y)], \alpha(z)) = B(D([x, y]), \alpha(z)) + \varepsilon(x, y)B([D(y), \alpha(x)], \alpha(z)).$$

Nondegeneracy of B leads to

$$D([x, y]) = [D(x), \alpha(y)] - \varepsilon(x, y)[D(y), \alpha(x)].$$

Condition (ii) is obtained by straightforward computations. □

The following corollary, which is an immediate consequence of Theorem 4 and Lemma 1, defines the first extension. It is used to define the double extension of quadratic color Hom-Lie algebras.

Corollary 5 *Let* $(\mathfrak{g}, [\cdot, \cdot]_{\mathfrak{g}}, \alpha_{\mathfrak{g}}, B)$ *be a quadratic color Hom-Lie algebra and* $D : \mathfrak{g} \longrightarrow \mathfrak{g}$ *be an even skew-symmetric* α*-derivation, then the* Γ*-graded vector space* $\mathfrak{g} \oplus \mathbb{K}$ *provided with the even linear map* $\alpha_{\mathfrak{g}} \oplus id$ *and the following product*

$$[x + \lambda, y + \eta] = [x, y]_{\mathfrak{g}} + B(D(x), y)$$

is a color Hom-Lie algebra.

5.3 T^*-Extensions of Color Hom-Lie Algebras

We have the following main Theorem.

Theorem 4 *Let* $(\mathfrak{g}, [\cdot, \cdot]_{\mathfrak{g}}, \alpha, \varepsilon)$ *be a color Hom-Lie algebra and* $\omega : \mathfrak{g} \times \mathfrak{g} \longrightarrow \mathfrak{g}^*$ *be an even bilinear map. Assume that the coadjoint representation exists. The* Γ*-graded vector space* $\mathfrak{g} \oplus \mathfrak{g}^*$, *where* $(\mathfrak{g} \oplus \mathfrak{g}^*)_\gamma = \mathfrak{g}_\gamma \oplus \mathfrak{g}^*{}_\gamma$ *for* $\gamma \in \Gamma$, *provided with the following bracket and linear map defined respectively by*

$$[x + f, y + g] = [x, y]_{\mathfrak{g}} + \omega(x, y) + \pi(x)g - \varepsilon(x, y)\pi(y)f, \tag{4}$$

$$\Omega(x + f) = \alpha(x) + f \circ \alpha,$$

where π *is the coadjoint representation of* $\mathfrak{g}$, *is a color Hom-Lie algebra if and only if* $\omega \in \mathscr{Z}^2(\mathfrak{g}, \mathfrak{g}^*)$. *In this case, we call* $\mathfrak{g} \oplus \mathfrak{g}^*$ *the* T^**-extension of* $\mathfrak{g}$ *by means of* ω.

If $B_{\mathfrak{g}}$ *is an invariant scalar product on* $\mathfrak{g}$, *then*

1. *the coadjoint representation exists,*
2. *the T^*-extension of $\mathfrak{g}$ admits also an invariant scalar product B defined by*

$$\begin{aligned} B : (\mathfrak{g} \oplus \mathfrak{g}^*)^{\otimes 2} &\to \mathbb{K} \\ (x + f, y + g) &\mapsto B_{\mathfrak{g}}(x, y) + f(y) + \varepsilon(x, y)g(x). \end{aligned} \tag{5}$$

Proof For any homogeneous elements $(x, f),\ (y, g),\ (z, h)$ in $\mathfrak{g} \oplus \mathfrak{g}^*$ we have

$$\begin{aligned} &\varepsilon(z, x)[\Omega(x + f), [y + g, z + h]] \\ &= \varepsilon(z, x)[\alpha(x) + f \circ \alpha, [y, z]_{\mathfrak{g}} + \omega(y, z) + \pi(y)h - \varepsilon(y, z)\pi(z)g] \\ &= \varepsilon(z, x)[\alpha(x), [y, z]_{\mathfrak{g}}]_{\mathfrak{g}} + \varepsilon(z, x)\omega(\alpha(x), [y, z]_{\mathfrak{g}}) \\ &+\varepsilon(z, x)\pi(\alpha(x))\omega(y, z) + \varepsilon(z, x)\pi(\alpha(x))(\pi(y)h) \\ &-\varepsilon(z, x)\varepsilon(y, z)\pi(\alpha(x))(\pi(z)g) - \varepsilon(x, y)\pi([y, z]_{\mathfrak{g}})f \circ \alpha. \end{aligned}$$

Consequently,

$$\circlearrowleft_{x,y,z} \varepsilon(z, x)\big[\Omega(x + f), [y + g, z + h]\big] = 0$$

if and only if

$$\begin{aligned} 0 &= \pi(\alpha(x_0))\big(\omega(x_1, x_2)\big) - \varepsilon(x_0, x_1)\pi(\alpha(x_1))\big(\omega(x_0, x_2)\big) \\ &+ \varepsilon(x_0 + x_1, x_2)\pi(\alpha(x_2))\big(\omega(x_0, x_1)\big) \\ &+ \omega(\alpha(x_0), [x_1, x_2]) + \varepsilon(x_1, x_2)\omega([x_0, x_2], \alpha(x_1)) - \omega([x_0, x_1], \alpha(x_2)). \end{aligned}$$

That is $\omega \in \mathscr{Z}^2(\mathfrak{g}, \mathfrak{g}^*)$. The proof for the second item is similar to the proof for non-graded case in [11]. □

6 Color Hom-Leibniz Algebras Arising from Faulkner Construction

Faulkner construction was introduced in [17]. Recently in [18], it was shown how Faulkner construction gives rise to a quadratic Leibniz algebras. In the following, we generalize this result to the case of color Hom-Leibniz algebras. Color Hom-Leibniz algebras are a non ε-skew-symmetric generalization of color Hom-Lie algebras.

Definition 16 A color Hom-Leibniz algebra is a tuple $(\mathfrak{g}, [\cdot, \cdot], \alpha, \varepsilon)$ consisting of a Γ-graded vector space $\mathfrak{g}$, an even bilinear mapping $[\cdot, \cdot]$, an even linear map $\alpha : \mathfrak{g} \longrightarrow \mathfrak{g}$ and a bicharacter ε such that

$$\varepsilon(z, x)\big[\alpha(x), [y, z]\big] + \varepsilon(x, y)\big[\alpha(y), [z, x]\big] + \varepsilon(y, z)\big[\alpha(z), [x, y]\big] = 0, \ \ \forall x, y, z \in \mathfrak{g}.$$

Proposition 8 *Let $(\mathfrak{g}, [\cdot, \cdot], \alpha, \varepsilon)$ be a multiplicative color Hom-Lie algebra. If ρ_1 and ρ_2 are two representations of $\mathfrak{g}$ on (M_1, β_1) and (M_2, β_2) respectively, then*

$\rho_1 \otimes \rho_2$ is a representation of $\mathfrak{g}$ on $(M_1 \otimes M_2, \beta_1 \otimes \beta_2)$ defined for homogeneous elements $m_1 \otimes m_2$ by

$$(\rho_1 \otimes \rho_2)(x)(m_1 \otimes m_2) = \rho_1(x)(m_1) \otimes \beta_2(m_2) + \varepsilon(x, m)\, \beta_1(m_1) \otimes \rho_2(x)(m_2).$$

Proof We check for all x, y in $\mathfrak{g}$ the following two identities:

$$\begin{aligned}&(\beta_1 \otimes \beta_2)\big((\rho_1 \otimes \rho_2)(x)(m_1 \otimes m_2)\big) = (\rho_1 \otimes \rho_2)\big(\alpha(x)\big)\big((\beta_1 \otimes \beta_2)(m_1 \otimes m_2)\big)\\ &(\rho_1 \otimes \rho_2)(\alpha(x)) \circ (\rho_1 \otimes \rho_2)(y) - \varepsilon(x, y)\, (\rho_1 \otimes \rho_2)(\alpha(y)) \circ (\rho_1 \otimes \rho_2)(x)\\ &= (\rho_1 \otimes \rho_2)([x, y]) \circ (\beta_1 \otimes \beta_2).\end{aligned}$$

Let $m_1 \in M_1$ and $m_2 \in M_2$ be two homogeneous elements.

$$\begin{aligned}&(\beta_1 \otimes \beta_2)\big((\rho_1 \otimes \rho_2)(x)(m_1 \otimes m_2)\big)\\ &= (\beta_1 \otimes \beta_2)\big(\rho_1(x)(m_1) \otimes \beta_2(m_2) + \varepsilon(x, m)\, \beta_1(m_1) \otimes \rho_2(x)(m_2)\big)\\ &= \rho_1(\alpha(x))(\beta_1(m_1)) \otimes {\beta_2}^2(m_2) + \varepsilon(x, m)\, {\beta_1}^2(m_1) \otimes \rho_2(\alpha(x))(\beta_2(m_2))\\ &= (\rho_1 \otimes \rho_2)(\alpha(x))((\beta_1 \otimes \beta_2)(m_1 \otimes m_2)).\end{aligned}$$

And

$$\begin{aligned}&\big[(\rho_1 \otimes \rho_2)(\alpha(x)) \circ (\rho_1 \otimes \rho_2)(y) - \varepsilon(x, y)\, (\rho_1 \otimes \rho_2)(\alpha(y)) \circ (\rho_1 \otimes \rho_2)(x)\big](m_1 \otimes m_2)\\ &= \rho_1(\alpha(x))\big(\rho_1(y)(m_1)\big) \otimes \beta_2^2(m_2) + \varepsilon(x, y + m_1)\, \beta_1\big(\rho_1(y)(m_1)\big) \otimes \rho_2(\alpha(x))(\beta_2(m_2))\\ &+\varepsilon(y, m_1)\, \rho_1(\alpha(x))(\beta_1(m_1)) \otimes \beta_2(\rho_2(y)(m_2))\\ &+\varepsilon(x + y, m_1)\, {\beta_1}^2(m_1) \otimes \rho_2(\alpha(x))\big(\rho_2(y)(m_2)\big)\\ &-\varepsilon(x, y)\, \rho_1(\alpha(y))\big(\rho_2(x)(m_1)\big) \otimes {\beta_2}^2(m_2) - \varepsilon(y, m)\, \beta_1\big(\rho_1(x)(m_1)\big) \otimes \rho_2(\alpha(y))(\beta_2(m_2))\\ &-\varepsilon(x, y + m)\, \rho_1(\alpha(y))(\beta_1(m_1)) \otimes \beta_2\big(\rho_2(x)(m_2)\big)\\ &-\varepsilon(x, y + m_1)\varepsilon(y, m_1)\, {\beta_1}^2(m) \otimes \rho_2(\alpha(y))\big(\rho_2(x)(m_2)\big).\end{aligned}$$

By virtue of (1) and (2), we obtain

$$\begin{aligned}&\big[(\rho_1 \otimes \rho_2)(\alpha(x)) \circ (\rho_1 \otimes \rho_2)(y) - \varepsilon(x, y)\, (\rho_1 \otimes \rho_2)(\alpha(y)) \circ (\rho_1 \otimes \rho_2)(x)\big](m_1 \otimes m_2)\\ &= \rho_1([x, y])(\beta_1(m_1)) \otimes {\beta_2}^2(m_2) + \varepsilon(x + y, m_1){\beta_1}^2(m_1) \otimes \rho_2([x, y])(\beta_2(m_2))\\ &= \big[(\rho_1 \otimes \rho_2)([x, y]) \circ (\beta_1 \otimes \beta_2)\big](m_1 \otimes m_2).\end{aligned}$$

□

Now, let $(\mathfrak{g}, [\cdot, \cdot], \alpha, \varepsilon, B)$ be a quadratic multiplicative color Hom-Lie algebra and ρ be a representation of $\mathfrak{g}$ on (M, β), such that both α and β are two involutions. Assume that $\tilde{\rho}$ is a representation of $\mathfrak{g}$ on $(M^*, \tilde{\beta})$, where M^* is the dual space of M. We denote the dual pairing between M and M^* by $\langle -, - \rangle$. Recall that $\tilde{\rho}(x)(f) := -\varepsilon(x, f) f \circ \rho(x)$, $\forall x \in \mathfrak{g}$, $f \in M^*$ and which means that $\langle m, \tilde{\rho}(x)(f)\rangle = -\varepsilon(m, x)\langle \rho(x)(m), f\rangle$, $\forall m \in M$. For all $m \in M$ and $f \in M^*$, we define the even linear map $\mathscr{D} : M \otimes M^* \longrightarrow \mathfrak{g}$ as follows

$$B(x, \mathcal{D}(m \otimes f)) = \langle \rho(\alpha(x))(m), f\rangle = \varepsilon(x+m, f)\, f(\rho(\alpha(x))(m)), \quad \text{for all } x \in \mathfrak{g}. \quad (1)$$

Lemma 2 *The even linear map $\mathcal{D} : M \otimes M^* \longrightarrow \mathfrak{g}$ is a morphism of $\mathfrak{g}$-Hom-modules, that is*

$$[x, \mathcal{D}(m \otimes f)] = \mathcal{D}(\rho(x)(m) \otimes \tilde{\beta}(f) + \varepsilon(x, m)\, \beta(m) \otimes \tilde{\rho}(x)(f)), \ \forall x \in \mathfrak{g}, m \in M, f \in M^*.$$

Proof Let m and f be homogeneous elements in respectively M and M^* and y be a homogeneous element in $\mathfrak{g}$.

$$\begin{aligned}
&B(y, \mathcal{D}\big(\rho(x)(m) \otimes \tilde{\beta}(f) \ + \ \varepsilon(x, m)\, \beta(m) \otimes \tilde{\rho}(x)(f)\big))\\
&= B(y, \mathcal{D}\big(\rho(x)(m) \otimes \tilde{\beta}(f)\big)) \ + \ \varepsilon(x, m)\ B(y, \mathcal{D}\big(\beta(m) \otimes \tilde{\rho}(x)(f)\big))\\
&= \langle \rho\big(\alpha(y)\big)\big(\rho(x)(m)\big), \tilde{\beta}(f)\rangle \ + \ \varepsilon(x, m)\ \langle \rho\big(\alpha(y)\big)\big(\beta(m)\big), \tilde{\rho}(x)(f)\rangle\\
&= \langle \beta\big[\rho\big(\alpha(y)\big)\big(\rho(x)(m)\big)\big], f\rangle \ - \ \varepsilon(x, m)\varepsilon(y+m, x)\ \langle \rho(x)\big[\rho\big(\alpha(y)\big)\big(\beta(m)\big)\big], f\rangle\\
&= \langle \rho(y)\big(\rho(\alpha(x))(\beta(m))\big) \ - \ \varepsilon(y, x)\ \rho(x)\big(\rho(\alpha(y))(\beta(m))\big), f\rangle\\
&= \langle \big(\rho(\alpha[y, x]) \circ \beta\big)(\beta(m)), f\rangle\\
&= B([y, x], \mathcal{D}(m \otimes f))\\
&= B(y, [x, \mathcal{D}(m \otimes f)]).
\end{aligned}$$

Consequently, $\mathcal{D}$ is a morphism of $\mathfrak{g}$-Hom-modules. □

In [15], it was proved that if (ρ, M) is a faithful representation, then $\mathcal{D}$ is surjective. In the sequel, we assume that ρ on (M, β) is a faithful representation. Since α is invertible, then $\mathcal{D}$ is still surjective. Using this argument, it follows that the fact that $\mathcal{D}$ is a morphism of $\mathfrak{g}$-Hom-modules is equivalent to

$$\begin{aligned}
&[\mathcal{D}(m \otimes f), \mathcal{D}(m' \otimes f')] = \qquad (2)\\
&\mathcal{D}(\rho\big(\mathcal{D}(m \otimes f)\big)(m') \otimes \tilde{\beta}(f')) \ + \ \varepsilon(m+f, m')\ \mathcal{D}(\beta(m') \otimes \tilde{\rho}\big(\mathcal{D}(m \otimes f)\big)(f')).
\end{aligned}$$

Proposition 9 *The Γ-graded vector space $M \otimes M^*$ provided with the following bracket*

$$[m \otimes f, m' \otimes f'] = \rho\big(\mathcal{D}(m \otimes f)\big)(m') \otimes \tilde{\beta}(f') \ + \ \varepsilon(m+f, m')\, \beta(m') \otimes \tilde{\rho}\big(\mathcal{D}(m \otimes f)\big)(f') \quad (3)$$

and the even linear map $\beta \otimes \tilde{\beta}$ is a color Hom-Leibniz algebra.

Proof For homogeneous elements $m \otimes f$, $m' \otimes f'$ and $m'' \otimes f''$ in $M \otimes M^*$, we have:

$$\begin{aligned}
&[\beta(m) \otimes \tilde{\beta}(f), [m' \otimes f', m'' \otimes f'']]\\
&= \rho\big(\mathcal{D}(\beta(m) \otimes \tilde{\beta}(f))\big)\big[\rho\big(\mathcal{D}(m' \otimes f')\big)(m'')\big] \otimes \tilde{\beta}^2(f'')\\
&\quad + \varepsilon(m+f, m'+m''+f')\beta\big[\rho\big(\mathcal{D}(m' \otimes f')\big)(m'')\big] \otimes \tilde{\rho}\big(\mathcal{D}(\beta(m) \otimes \tilde{\beta}(f))\big)(\tilde{\beta}(f''))\\
&\quad + \varepsilon(m'+f', m'')\ \rho\big(\mathcal{D}(\beta(m) \otimes \tilde{\beta}(f))\big)(\beta(m'')) \otimes \tilde{\beta}\big[\tilde{\rho}\big(\mathcal{D}(m' \otimes f')\big)(f'')\big]\\
&\quad + \varepsilon(m+f+m'+f', m'')\beta^2(m'') \otimes \tilde{\rho}\big(\mathcal{D}(\beta(m) \otimes \tilde{\beta}(f))\big)\big[\tilde{\rho}\big((\mathcal{D}(m' \otimes f')\big)(f')\big].
\end{aligned}$$

$$\begin{aligned}
&[\beta(m')\otimes\tilde\beta(f'),[m\otimes f,m''\otimes f'']]=-\rho\Big(\mathscr{D}(\beta(m')\otimes\tilde\beta(f'))\Big)\Big[\rho\Big(\mathscr{D}(m\otimes f)\Big)(m'')\Big]\otimes\tilde\beta^2(f'')\\
&\quad-\varepsilon(m'+f',m+f+m'')\,\beta\Big[\rho\Big(\mathscr{D}(m\otimes f)\Big)(m'')\Big]\otimes\tilde\rho\Big(\mathscr{D}(\beta(m')\otimes\tilde\beta(f'))\Big)(\tilde\beta(f''))\\
&\quad-\varepsilon(m+f,m'')\,\rho\Big(\mathscr{D}(\beta(m')\otimes\tilde\beta(f'))\Big)(\beta(m''))\otimes\tilde\beta\Big[\tilde\rho\Big(\mathscr{D}(m\otimes f)\Big)(f'')\Big]\\
&\quad-\varepsilon(m+f+m'+f',m'')\,\beta^2(m'')\otimes\tilde\rho\Big(\mathscr{D}(\beta(m')\otimes\tilde\beta(f'))\Big)\Big[\tilde\rho\Big(\mathscr{D}(m\otimes f)\Big)(f'')\Big].
\end{aligned}$$

$$\begin{aligned}
&[[m\otimes f,m'\otimes f'],\beta(m'')\otimes\tilde\beta(f'')]=\rho\Big[\mathscr{D}\Big(\rho(\mathscr{D}(m\otimes f))(m')\otimes\tilde\beta(f')\Big)\Big](\beta(m''))\otimes\tilde\beta^2(f'')\\
&\quad+\varepsilon(m+f+m'+f',m'')\beta^2(m'')\otimes\tilde\rho\Big[\mathscr{D}\Big(\rho(\mathscr{D}(m\otimes f))(m')\otimes\tilde\beta(f'))\Big)\Big]\tilde\beta(f'')\\
&\quad+\varepsilon(m+f,m')\,\rho\Big[\mathscr{D}\Big(\beta(m')\otimes\tilde\rho(\mathscr{D}(m\otimes f))(f')\Big)\Big]\beta(m'')\otimes\tilde\beta^2(f'')\\
&\quad+\varepsilon(m+f,m')\varepsilon(m+f+m'+f',m'')\beta^2(m'')\otimes\tilde\rho\Big[\mathscr{D}\Big(\beta(m')\otimes\tilde\rho(\mathscr{D}(m\otimes f))(f')\Big)\Big]\tilde\beta(f'')
\end{aligned}$$

Using condition (2), we deduce that

$$\begin{aligned}
&\big[\beta(m)\otimes\tilde\beta(f),[m'\otimes f',m''\otimes f'']\big]\\
&-\varepsilon(m+f,m'+f')\big[\beta(m')\otimes\tilde\beta(f'),[m\otimes f,m''\otimes f'']\big]\\
&-\big[[m\otimes f,m'\otimes f'],\beta(m'')\otimes\tilde\beta(f'')\big]=0.
\end{aligned}$$

This finishes the proof. □

Corollary 6 *Consider the Hom-Leibniz algebra* $(M\otimes M^*,[,],\beta\otimes\tilde\beta)$ *defined above. If the morphism of* $\mathfrak{g}$*-Hom-modules* $\mathscr{D}$ *is bijective, then* $(M\otimes M^*,[,],\beta\otimes\tilde\beta)$ *is a quadratic Hom-Leibniz algebra such that the quadratic structure is given by* $B: M\otimes M^*\times M\otimes M^*\longrightarrow\mathbb{K}$ *such that*

$$B(m\otimes f,m'\otimes f')=B_{\mathfrak{g}}(\mathscr{D}(m\otimes f),\mathscr{D}(m'\otimes f')).$$

Proof By straightforward computations, we prove that B is ε-symmetric, $\beta\otimes\tilde\beta$ is B-symmetric and B is invariant. □

References

1. Aizawa, N., Sato, H.: q-deformation of the Virasoro algebra with central extension. Phys. Lett. B **256**, 185–190 (1991)
2. Albuquerque, H., Barreiro, E., Benayadi, S.: Odd-quadratic Lie superalgebras. J. Geom. Phys. **60**, 230–250 (2010)
3. Ammar, F., Ejbehi, Z., Makhlouf, A.: Cohomology and deformations of Hom-algebras. J. Lie Theory **21**(4), 813–836 (2011)
4. Ammar, F., Mabrouk, S., Makhlouf, A.: Constructions of quadratic n-ary Hom-Nambu algebras. In: Algebra, Geometry and Mathematical Physics, Springer Proceedings in Mathematics & Statistics, vol. 85, pp. 193–224 (2014)

5. Ammar, F., Makhlouf, A.: Hom-Lie superalgebras and Hom-Lie admissible superalgebras. J. Algebr. **324**, 1513–1528 (2010)
6. Ayadi, I., Benamor, H., Benayadi, S.: Lie superalgebras with some homogeneous structures. J. Algebr. Appl. **11**(05), 1250095 (2012)
7. Bajo, I., Benayadi, S., Bordemann, M.: Generalized double extension and descriptions of quadratic Lie superalgebras (2007). arXiv:0712.0228
8. Benamor, H., Benayadi, S.: Double extension of quadratic Lie superalgebras. Commun. Algebr. **27**, 67–88 (1999)
9. Benayadi, S.: Quadratic Lie superalgebras with the completely reducible action of the even part on the odd part. J. Algebr. **223**, 344–366 (2000)
10. Benayadi, S.: Socle and some invariants of quadratic Lie superalgebras. J. Algebr. **261**(2), 245–291 (2003)
11. Benayadi, S., Makhlouf, A.: Hom-Lie algebras with symmetric invariant nondegenerate bilinear forms. J. Geom. Phys. **76**, 38–60 (2014)
12. Bordemann, M.: NonDegenerate invariant bilinear forms in nonassociative algebras. Acta Math. Univ. Comenian **LXVI**(2), 151–201 (1997)
13. Chaichian, M., Ellinas, D., Popowicz, Z.: Quantum conformal algebra with central extension. Phys. Lett. B **248**, 95–99 (1990)
14. Chaichian, M., Kulish, P., Lukierski, J.: q-Deformed Jacobi identity, q-oscillators and q-deformed infinite-dimensional algebras. Phys. Lett. B **237**, 401–406 (1990)
15. De Medeiros, P., Figueroa-O'Farrill, J., Méndez-Escobar, E., Ritter, P.: On the Lie-algebraic origin of metric 3-algebras. Commun. Math. Phys. **290**(3), 871–902 (2009)
16. Elhamdadi, M., Makhlouf, A.: Deformations of Hom-Alternative and Hom-Malcev algebras. Algebr. Groups Geom. **28**(2), 117–145 (2011)
17. Faulkner, J.R.: On the geometry of inner ideals. J. Algebr. **26**(1), 1–9 (1973)
18. Figueroa-O'Farrill, J.: Deformations of 3-algebras. J. Math. Phys. **50**, 113514 (2009)
19. Hartwig, J., Larsson, D., Silvestrov, S.: Deformations of Lie algebras using σ -derivations. J. Algebr. **295**, 314–361 (2006)
20. Hegazi, A.: Classification of nilpotent Lie superalgebras of dimension five. I. Int. J. Theor. Phys. **38**(6), 1735–1739 (1999)
21. Larsson, D., Silvestrov, S.: Quasi-Hom-Lie algebras, central extensions and 2-cocycle-like identities. J. Algebr. **288**, 321–344 (2005)
22. Makhlouf, A., Silvestrov, S.: Notes on formal deformations of Hom-associative and Hom-Lie algebras. Forum Math. **22**(4), 715–759 (2010)
23. Makhlouf, A.: Hom-alternative algebras and Hom-Jordan algebras. Int. Electron. J. Algebr. **8**, 177–190 (2010)
24. Makhlouf, A., Silvestrov, S.: On Hom-algebra structures. J. Gen. Lie Theory Appl. **2**(2), 51–64 (2008)
25. Medina, A., Revoy, P.: Algèbres de Lie et produit scalaire invariant. Ann. Sci. Ecole Norm. Sup. (4) **18**, 553–561 (1985)
26. Qingcheng, Z., Yongzheng, Z.: Derivations and extensions of Lie color algebra. Acta Math. Sci. **28**, 933–948 (2008)
27. Rittenberg, V., Wyler, D.: Generalized superalgebras. Nucl. Phys. B **139**, 189–202 (1978)
28. Rittenberg, V., Wyler, D.: Sequences of graded $\mathbb{Z} \otimes \mathbb{Z}$ Lie algebras and superalgebras. J. Math. Phys. **19**, 2193 (1978)
29. Scheunert, M.: The theory of Lie Superalgebras: An Introduction. Springer (1979)
30. Scheunert, M., Zhang, R.B.: Cohomology of Lie superalgebras and their generalizations. J. Math. Phys. **39**, 5024 (1998)
31. Scheunert, M.: Generalized Lie algebras. In: Group Theoretical Methods in Physics, vol. 450–450 (1979)
32. Scheunert, M.: Graded tensor calculus. J. Math. Phys. **24**, 2658 (1983)
33. Sheng, Y.: Representations of Hom-Lie algebras. Algebr. Represent. Theory **15**(6), 1081–1098 (2012)

34. Wang, S., Zhu, L., Su, Y.: Non-degenerate invariant bilinear forms on Lie color algebras. Algebr. Colloq. **17**, 365–374 (2010)
35. Yau, D.: Enveloping algebra of Hom-Lie algebras. J. Gen. Lie Theory Appl. **2**, 95–108 (2008)
36. Yau, D.: Hom-algebras and homology. J. Lie Theory **19**, 409–421 (2009)
37. Yau, D.: Hom-Maltsev, hom-alternative and hom-Jordan algebras. Int. Electron. J. Algebr. **11**, 177–217 (2012)
38. Yuan, L.: Hom-Lie color algebra structures. Commun. Algebr. **40**(2), 575–592 (2012)
39. Zhang, R., Hou, D., Bai, C.: A Hom-version of the affinizations of Balinskii-Novikov and Novikov superalgebras. J. Math. Phys. **52**, 023505 (2011)

The Extension Property in the Category of Direct Sum of Cyclic Torsion-Free Modules over a BFD

Seddik Abdelalim, Abdelhak Chaichaa and Mostafa El garn

Abstract It is known that, in the category of vector space, all automorphisms satisfy the extension property. However, Schupp (Proc. A.M.S. **101**(2):226–228, 1987 [11]) proved that the automorphisms having the property of extension in the category of groups, characterize the inner automorphisms. Ben Yakoub (Port. Math. **51**(2):231–233, 1994, [5]) proved that this result is not true in the category of algebras. But there are some very important results in the category of groups. Then, in order to generalize the result of Schupp, Abdelalim and Essannouni (Port. Math. (Nova) **59**(3):325–333, 2002, [1]) characterized the automorphisms having the property of extension, in the category of abelian groups. Let A be an integral bounded factorization domain and M a direct sum of cyclic torsion-free modules over A. This work aims to prove that the automorphisms of M that satisfy the property of the extension are none other than the homotheties of invertible ratio.

Keywords Abelian · Category · Enough injective · Epic · Monic · Integral domain · Factorization (BFD) · Module · Automorphism · Torsion and torsion-free

1 Introduction

Let us assume that we are working in an abelian ambience, that is, our objects 'belong' to an abelian category. Consider an automorphism $\alpha : M \hookrightarrow M$ in C, that is said to satisfy the extension property, provided that, for any monomorphism $\lambda : M \hookrightarrow N$ in C, there exists an automorphism $\widetilde{\alpha} : N \hookrightarrow N$ in C such that $\widetilde{\alpha} \circ \lambda = \lambda \circ \alpha$.

S. Abdelalim (✉) · A. Chaichaa · M. El garn
Department of Mathematical and Computer Sciences, Faculty of Sciences Ain Chock, Laboratory of Topology, Algebra, Geometry and Discrete Mathematics, Hassan II University of Casablanca, BP 5366 Maarif, Casablanca, Morocco
e-mail: seddikabd@hotmail.com

A. Chaichaa
e-mail: abdelchaichaa@gmail.com

M. El garn
e-mail: elgarnmostafa@gmail.com

M. Siles Molina et al. (eds.), *Associative and Non-Associative Algebras and Applications*, Springer Proceedings in Mathematics & Statistics 311,
https://doi.org/10.1007/978-3-030-35256-1_17

A first reasonable question, which perhaps should appears here, is the subsequent one. In which category every automorphism satisfies the extension property? The answer is given by any semisimple abelian category. In particular the category of (left or right) vector space over a division ring (or a field), or the category of (left) module over a (left) semisimple algebra. In fact this holds true in any C_{Sp} abelian category where every monomorphism splits. The extension property can be relaxed to a certain class of monomorphism like the ones with codomain injective objects, or even to one specific monomorphism, as the one acquired by the injective envelope of a given object or when C has enough injective objects, which is the case of Grothendieck categories. Let C_0 denote the category of all left (or right) modules over a ring A which is a Grothendieck categories. This paper extends the result in [1] to a special category of modules, the category of direct sums of cyclic torsion-free modules over BFD rings [3]. In other words, the full subcategory additively generated by these cyclic modules. In the first time, we will brievelly prove that all automorphism, satisfy the extension property, in the category C_{Sp}. In the second part, let M_n be a finite direct sum of cyclic modules over A. We will prove that an automorphism $\alpha : M \longrightarrow M$, satisfy the extension property, in the category C_0, if and only if there exists a unit k in A such that $\alpha = k.1_M$. And in the last part, let M be an infinite direct sum of cyclic torsion-free modules over A. We will prove that an automorphism $\alpha : M \longrightarrow M$, satisfy the extension property, in the category C_0, if and only if there exists a unit k in A such that $\alpha = k.1_M$.

2 The Extension Property in the Category C_{Sp}

Definition 1 ([10]) A category C consists of:

- a collection $ob(C)$ of objects
- for each M, N in $ob(C)$, a collection $Hom(M, N)$ of maps or arrows or morphisms from M to N;
- for each M, N, P in $ob(C)$, a function:

$$Hom(M, N) \times Hom(N, P) \longrightarrow Hom(M, P)$$

$$(f, g) \longmapsto g \circ f$$

  called composition
- for each M in $ob(C)$, an element 1_M of $Hom(M, M)$, called the identity on M, satisfying the following axioms:
  - associativity: for each M, N, P and Q in $ob(C)$: $\forall (f, g, h) \in Hom(M, N) \times Hom(N, P) \times Hom(P, Q) \quad (h \circ g) \circ f = h \circ (g \circ f)$
  - identity laws: for each f in $Hom(M, N)$, we have $f \circ 1_M = f = 1_N \circ f$

Definition 2 ([10]) Let C be a category.

A map $f : M \longrightarrow N$ in C is monic (or a monomorphism) if for all objects P, maps $g : P \longrightarrow M$ and $g' : P \longrightarrow M$: $f \circ g = f \circ g' \Longrightarrow g = g'$.

A map $f : M \longrightarrow N$ in C is epic (or a epimorphism) if for all objects P, maps $g : N \longrightarrow P$ and $g' : N \longrightarrow P$: $g \circ f = g' \circ f \Longrightarrow g = g'$.

Definition 3 ([10]) Let C be a category.

A map $f : M \longrightarrow N$ in C is split monic if there exists a maps $g : N \longrightarrow M$ such that $g \circ f = 1_M$ and g is then called split epic.

Definition 4 ([8]) Let C be a category.

A map $f : M \longrightarrow N$ in C is an isomorphism if there exists a maps $g : N \longrightarrow M$ such that $g \circ f = 1_M$ and $f \circ g = 1_N$.

We write: $g = f^{-1}$ and $M \cong N$.

Definition 5 ([8]) A category C is abelian if:

- C has a zero objects.
- for every pair of objects there is a product and a sum
- Every map has a kernel and a cokernel.
- Every monomorphism is a kernel of a map.
- Every epimorphism is a cokernel of a map.

Definition 6 ([7]) Let C be a category.

If M is an object in the category C, then a subobject of M is a monomorphism $M' \longrightarrow M$ in C.

Definition 7 ([9]) Let C be an abelian category.

- A nonzero object M in C is simple if its only subobjects are 0 and M.
- C is semisimple if every object in it is a direct sum of finite simple objects.

Remark 1 ([2]) In an abelian category, if $f : N \longrightarrow M$ is a split epimorphism with split monomorphism $g : M \longrightarrow N$, then N is isomorphic to the direct sum of M and the kernel of f.

Definition 8 ([12]) Let C be a category.

- An object Q in C is said to be injective if for every monomorphism $f : M \longrightarrow P$ and every morphism $g : M \longrightarrow Q$ there exists a morphism $h : P \longrightarrow Q$ extending g to P, i.e. such that $h \circ f = g$.
- Moreover, if C is abelian, we say that it has enough injective objects (or enough injectives), if for every object M in C there is an injective object Q and there exists a monomorphism $f : M \longrightarrow Q$.

Proposition 1 *In the category C_{Sp}, where every monomorphisme splits, all automorphisms satisfy the extension property.*

Proof We have C_{Sp} an abelian category where every monomorphism splits. Assume we have an automorphism $\alpha : M \longrightarrow M$ and take any monomorphism $\lambda : M \longrightarrow N$ in C. Therefore, there is an isomorphism $N \cong M \oplus M'$, for some object M' in C, as λ splits. Up to this isomorphism, one can choose $\widetilde{\alpha}$ to be the following 2×2- matrix:

$$\widetilde{\alpha} = \begin{pmatrix} \alpha & 0 \\ 0 & id_{M'} \end{pmatrix} : N \longrightarrow N$$

Then $\widetilde{\alpha} \circ \lambda = \lambda \circ \alpha$, so α satisfy the extension property. □

Remark 2 Let C_{Ss} denote a semisimple abelian Grothendieck category (see [7]). We know, that in C_{Ss} every monomorphism splits (see [6]), therefore according the proposition 1, all automorphisms in C_{Ss}, satisfy the extension property.

3 The Extension Property in the Category C_0

3.1 Preliminary and Notations

Definition 9 ([3]) Let A be a integral domain. A is a bounded factorization domain (BFD), if for every nonzero nonunits a in A, the number of irreducible factors in a is finite. The Nœtherian domain or a Krull domain is a BFD.

Note 1 Let A be a bounded factorization domain (BFD) and M a module over A.

- $A^{\times}$ denote the units group of A.
- Let I an infinite part of $\mathbb{N}$. Let $(x_i)_{i \in I}$ a family of torsion-free elements. Without restriction, we can suppose that $I = \mathbb{N}^*$.
- We consider $M = \sum_{i \in \mathbb{N}^*} Ax_i$ a direct infinite A module. Let $M_n = \sum_{i=1}^{i=n} Ax_i$, where $n \in \mathbb{N}^*$.
- $E_p(M)$ denote the set of all A-automorphisms of M satisfying the extension property.
- $E_{n,p}(M_n)$ denote the set of all A-automorphisms of M_n satisfying the extension property.
- Let $n \in \mathbb{N}^*$, for $i \in \{1, 2, ..., n\}$:
  - E_i will denote an injective envelope of Ax_i. (exists because C_0 has enough injective objects).
  - $M_{i,n} = E_i + \sum_{k=1, k \neq i}^{k=n} A.x_k$.
  - $\overline{M_i} = E_i + \sum_{k=1, k \neq i}^{k=+\infty} A.x_k$.

Definition 10 Consider an automorphism $\alpha : M \hookrightarrow M$ in C_0, that is said to satisfy the extension property, provided that, for any monomorphism $\lambda : M \hookrightarrow N$ in C, there exists an automorphism $\widetilde{\alpha} : N \hookrightarrow N$ in C such that $\widetilde{\alpha} \circ \lambda = \lambda \circ \alpha$.

Definition 11 Let A be an integral domain, M an A-module and x an element of M. The annihilator of x is the ideal $Ann_A(x) = \{a \in A/a.x = 0\}$. The annihilator of M, is the ideal $Ann_A(M) = \{a \in A/a.y = 0\, for\ all\ y \in M\}$.

Definition 12 Let A be a ring and M a left A-module

- Let x be an element of M.
  x is a torsion-free element if $Ann(x) = \{0_A\}$ and x is a torsion element if $Ann(x) \neq \{0_A\}$.
- We say that M is a torsion-free module if all its elements are torsion-free.
- We say that M is a torsion module if all its elements are torsion.

Lemma 1 *For all $n \in \mathbb{N}^*$, the set $\{x_1, x_2, ..., x_n\}$ is A-free.*

Proof Let $n \in \mathbb{N}^*$ and $(a_1, a_2, ..., a_n) \in A^n$. Suppose that $a_1x_1 + a_2x_2 + ... + a_nx_n = 0_M$, as M is a direct sum of the family $(Ax_k)_{k \in \mathbb{N}^*}$ then $a_1x_1 = a_2x_2 = ... = a_nx_n = 0_M$. And since all the x_k are torsion-free then $a_1 = a_2 = ... = a_n = 0_A$. Which proves that $\{x_1, x_2, ..., x_n\}$ is A-free. As a consequence: If $(i, j) \in (\mathbb{N}^*)^2$ such that $i \neq j$ then $Ax_i \cap Ax_j = 0_M$. □

Lemma 2 *Let A be a ring and M a left A-module. For p an irreducible element in A. The set $I_p = \{p^n/n \in N\}$ is infinite.*

Proof Suppose that I_p is finite, necessarily $\exists (n, m) \in N^2$ such that $n \prec m$ and $p^n = p^m$. Then $p^n(1 - p^{m-n}) = 1 \Longrightarrow p^{m-n} = 1$ (integral domain). So p is a unit in A, which is not true. □

Proposition 2 *Let $\alpha : M \longrightarrow M$ be an automorphism in C_0. If α satisfies the extension property then α^{-1} also satisfies the extension property.*

Proof In any category, if we have a commutative diagram satisfied by the equality: $\lambda \circ \alpha = \widetilde{\alpha} \circ \lambda$, where both $\widetilde{\alpha}$ and α are isomorphism, then automatically, we have in the other way around a commutative diagram given by the equality: $\lambda \circ \alpha^{-1} = \widetilde{\alpha^{-1}} \circ \lambda$. □

Remark 3 Every isomorphism of modules has a bijective underlying maps between the underlying sets. Thus, in set theoretical setting any injective map commutes with intersections. Thus if N is an A module, $F = \bigcap_{a \in A^\times} a.N$ a submodule of N and $\widetilde{\alpha} \in Aut_A(N)$ then $\widetilde{\alpha}\,(F) = F$.

3.2 Characterization of the Set $E_{n,p}(M_n)$

In this section, let $n \in \mathbb{N}^*$. For all i in $\{1, 2, ..., n\}$, let $\mu_i : Ax_i \longrightarrow E_i$ be a monomorphism of A-modules.

Lemma 3 *For all $i \in \{1, 2, ..., n\}$, we have*

$$M_{i,n} = E_i \bigoplus \bigoplus_{k=1,k\neq i}^{k=n} A.x_k$$

and

$$E_i = \bigcap_{a\in A^*} a.M_{i,n}.$$

Proof We know that $E_1 \oplus E_2 \oplus ... \oplus E_n$ is an injective envelope of $Ax_1 \bigoplus Ax_2 \bigoplus ... \bigoplus Ax_n$ (see [4]). Then

$$M_{i,n} = E_i + \sum_{k=1,k\neq i}^{k=n} A.x_k = E_i \bigoplus \bigoplus_{k=1,k\neq i}^{k=n} Ax_k$$

Let $b \in A^\times$. As E_i is an injective envelope of Ax_i then it is a divisible module. So

$$E_i = b.E_i \subset b.M_i \subset \bigcap_{a\in A^\times} a.M_{i,n}.$$

Reciprocally: Let

$$t = e + \sum_{j=1,j\neq i}^{j=n} m_j.x_j \in \bigcap_{a\in A^\times} a.M_i \subset \bigcap_{k\in N} p^k.M_{i,n},$$

where $e \in E_i$. Then, there exist $e_k \in E_i$ and $m_{i,k} \in A$, such that

$$t = p^k \left(e_k + \sum_{j=1,j\neq i}^{j=n} m_{j,k}.x_i \right).$$

Moreover, as $M_{i,n}$ is a direct sum of A-modules, by identification, we have

$$e = p^k(e_k) \;\; and \;\; \sum_{j=1,j\neq i}^{j=n} m_j.x_j = p^k.(\sum_{j=1,j\neq i}^{j=n} m_{j,k}.x_j).$$

Now we have that

$$\sum_{j=1, j\neq i}^{j=n} m_j.x_j \neq 0 \Longrightarrow \exists j \in \{1, 2, ..., n\}/\{i\} : m_j \neq 0.$$

Therefore, $m_j = p^k.m_{j,k}$, so that $p^k | \, m_j$ for all integer k. And while according to Lemma 2, m_j will have an infinity of divisors, which is false in a BFD. Hence $t = p^k(e_k) \in E_i$, that concludes the proof. □

Lemma 4 *For all i and j in* $\{1, 2, ..., n\}$. *If* $i \neq j$ *then* $M_n = A\left(x_i + x_j\right) \bigoplus \bigoplus_{k=1, k\neq i}^{k=n} A.x_k$

Proof Firstly, we will proved that $A\left(x_i + x_j\right) \cap Ax_j = \{0_M\}$. Let $(a, b) \in A^2$:

- $a.\left(x_i + x_j\right) = b.x_j \Longrightarrow (b - a).x_j - a.x_i = 0_M$. And as $\left\{x_i, x_j\right\}$ is A-free then $a = b = 0_A$.
- $a.\left(x_i + x_j\right) = b.x_k$ *for* $1 \leq k \leq n$ *such that* $k \neq i$ *and* $k \neq j \Longrightarrow a.x_i + ax_j - b.x_k = 0_M$. And as $\left\{x_i, x_j, x_k\right\}$ is A-free then $a = b = 0_A$

It is clear that $A\left(x_i + x_j\right) \oplus Ax_i \subset Ax_i \oplus Ax_j$. Let $x \in Ax_i \oplus Ax_j$ then $x = a.x_i + b.x_j$, where a and b are in A, so $x = b.\left(x_i + x_j\right) + (a - b).x_i \in A\left(x_i + x_j\right) \oplus Ax_i$. □

Lemma 5 *Let* $\alpha \in E_{n,p}(M_n)$ *then* $(\forall i \in \{1, 2, ..., n\})(\exists k_i \in A^{\times})(\alpha(x_i) = k_i.x_i)$.

Proof Let us consider the following map:

$$\begin{array}{ccc} M_n & \xrightarrow{\quad\lambda\quad} & M_{i,n} \\ x = \sum_{j=1}^{j=n} m_j.x_j & \longmapsto & \lambda(x) = m_i.\mu_i(x_i) + \sum_{j=1, j\neq i}^{j=n} m_j.x_j, \end{array}$$

where $m_j \in A$ for all $j \in \{1, 2, ..., n\}$. As defined λ is a morphism of A-modules and we have: If

$$\lambda(x) = m_i.\mu_i(x_i) + \sum_{j=1, j\neq i}^{j=n} m_j.x_j = 0_M$$

then

$$m_i.\mu_i(x_i) = 0_M \; and \; m_j.x_j = 0_M$$

for all $j \in \{1, 2, ..., n\}/\{i\}$. However μ_i is injective then $m_i = 0_A$. As $\{x_1, x_2, ..., x_n\}/\{x_i\}$ is A-free then $m_j = 0$ for all $j \in \{1, 2, ..., n\}/\{i\}$. Then, we obtain $x = 0_M$. Therefore, λ is a monomorphism of A-modules.

We know that $\alpha(x_i) = k_i.x_i + \sum_{j=1, j\neq i}^{j=n} k_j.x_j$, where k_1,k_2,...,k_{n-1} and k_n are in A. As $\lambda o \alpha = \widetilde{\alpha} o \lambda$ and by lemma 4 we have

$$\lambda[\alpha(x_i)] = k_i.\mu(x_i) + \sum_{j=1, j\neq i}^{j=n} k_j\lambda\left(x_j\right) = k_i.\mu_i(x_i) + \sum_{j=1, j\neq i}^{j=n} k_j.x_j = \widetilde{\alpha}[\lambda(x_i)] \in E_i.$$

Since, by Lemma 3 E_i is a direct factor in $M_{i,n}$, so we have that $\alpha\,(x_i) = k_i.x_i$. □

The main result in this section is the following theorem:

Theorem 1 *We have* $E_{n,p}(M_n) = \left\{k\, id_{M_n}/k \in A^{\times}\right\}$.

Proof Applying Lemma 5, $\forall i \in \{1, 2, ..., n\}$ $\exists k_i \in A^{\times} : \alpha\,(x_i) = k_i.x_i$. We must find $k_i = k_j \in A^{\times}$, for all i and j in $\{1, 2, ..., n\}$ such that $i \neq j$. Applying now the theorem 1, for $A(x_i + x_j)$ there exists $k_{i,j} \in A^{\times}$ such that $\alpha\left(x_i + x_j\right) = k_{i,j}.\left(x_i + x_j\right) = k_{i,j}.x_i + k_{i,j}.x_j = k_i.x_i + k_j.x_j \Longrightarrow k_{i,j} = k_i = k_j$, since $\left\{x_i, x_j\right\}$ is A-free. Then, $\exists k \in A^*$ such that $\forall i \in \{1, 2, ..., n\}$ $\alpha\,(x_i) = k.x_i$. Therefore $\alpha = k.id_M$. We must prove that k is a unit in A. And as α satisfies the extension property then α^{-1} also satisfies the extension property by Proposition 1. Then we have $\exists r \in A^*\ : \alpha^{-1} = r.id_{M_n} \Longrightarrow \alpha \circ \alpha^{-1} = \alpha^{-1} \circ \alpha = k.r.id_{M_n}$. Then $r.k = 1_A$ which proves that k is a unit in A. □

3.3 Characterization of the Set $E_p(M)$

For all i in $\mathbb{N}^*$, let $\mu_i : Ax_i \longrightarrow E_i$ a monomorphism of A modules.

Lemma 6 *For* $i \geq 1$ *and* $j \geq 1$ *such that* $i \neq j$, *we have*

$$\bigoplus_{k=1}^{k=\infty} Ax_k = A(x_i + x_j) \bigoplus \bigoplus_{k=1,k\neq i}^{k=\infty} Ax_k$$

Proof Let $n \geq \sup(i, j)$, we must prove that $\{x_1, x_2, ..., x_n, x_i + x_j\}/\{x_i\}$ is A free. We suppose that $a_1x_1 + a_2x_2 + ... + a_{i-1}x_{i-1} + a_{i+1}x_{i+1} + ... + a_nx_n + a_{i,j}(x_i + x_j) = 0_M$. Then

$$\sum_{k=1,k\neq i,k\neq j}^{k=n} a_kx_k + a_{i,j}x_i + (a_j + a_{i,j})x_j \;=\; 0$$

implies that $a_k = 0_A$, for all $k \in \{1, 2, ..., n\}/\{i, j\}$, $a_{i,j} = 0_A$ and $a_j + a_{i,j} = 0_A$. This means that $a_k = 0_A$, for $k \in \{1, 2, ..., n\}/\{i\}$ and $a_{i,j} = 0_A$. □

Lemma 7 *For every* $i \in \mathbb{N}^*$, *we have that* $E_i = \bigcap_{a\in A^*} a\overline{M_i}$.

Proof Let $i \in \mathbb{N}^*$. As E_i is an injective enveloppe then a divisible module. So for all $a \in A^{\times}$ we have, $E_i = aE_i \subset \bigcap_{a\in A^*} a\overline{M_i}$. Consider $x \in \bigcap_{a\in A^*} a\overline{M_i}$. Then there exists $e_x \in E_i$, I_x a finite subset in $\mathbb{N}^*$ and a family element $(a_i)_{i\in I_x}$ in A such that $x = e_x + \sum_{i\in I_x} a_ix_i$ (*).

We know that $x \in \bigcap_{n\in N^*} p^n\overline{M_i}$. So $\forall n \geq 1$ $\exists e_x^n \in E_i$, I_x^n a finite subset in $\mathbb{N}^*$, and a family element $(a_i^n)_{i\in I_x^n}$ in A such that $x = p^n e_x^n + \sum_{i\in I_x^n} p^n a_i^n x_i$. From (*) and since M_i is a direct sum, we assume that $x = p^n e_x^n + \sum_{i\in I_x^n} p^n a_i^n x_i$, $p^n e_x^n = e_x$ and $\sum_{i\in I_x} a_i x_i = \sum_{i\in I_x^n} p^n a_i^n x_i$. If $\sum_{i\in I_x} a_i x_i \neq 0_M \Longrightarrow \exists i \in I_x \ : \ a_i$ is non-zero. So, $a_i = p^n a_i^n \Longrightarrow p^n | a_i$. Therefore, by Lemma 2 there exists an infinite nonzero divisors of a_i in A^*, which is not true in a BFD. Consequently, $x \in E_i$. Finally, we have that $E_i = \bigcap_{a\in A^*} a\overline{M_i}$.

□

Lemma 8 *Let α an automorphism, if α satisfies the extension property then*

$$\widetilde{\alpha}(E_i) = E_i.$$

Proof We have

$$E_i = \bigcap_{a\in A^\times} a.\overline{M_i} \Longrightarrow \widetilde{\alpha}(E_i) \subset \bigcap_{a\in A^\times} a.\widetilde{\alpha}(\overline{M_i}) = E_i$$

Let $y \in E_i$ then

$$\exists x = e + \sum_{k=1,k\neq i}^{k=n} a_k.x_k \in \overline{M_i} \ : \ y = \widetilde{\alpha}(x)$$

Let $a \in A^*$ $\exists y_a \in E_i \ : \ y = ay_a \Longrightarrow \exists x_a \in \overline{M_i} \ : \ y = a\widetilde{\alpha}(x_a) = \widetilde{\alpha}(ax_a)$
$\Longrightarrow x = ax_a \in a\overline{M_i} \Longrightarrow x \in E_i$. Therefore, $E_i = \widetilde{\alpha}(E_i)$. □

Lemma 9 *Let α an automorphism. If α satisfies the extension property then there exists $k_i \in A^\times$ such that $\alpha\,(x_i) = k_i.x_i$.*

Proof We consider $\lambda : M \longrightarrow \overline{M_i}$ the homomorphism of A modules, defined by $\lambda\,(x_i) = \mu_i\,(x_i)$ and $\lambda\,(x_j) = x_j$ for all $j \geq 1$ such that $j \neq i$. Let $t = \sum_{j=1}^{j=n} m_j.x_j \in M$, we can suppose that $n \geq i$. So, If $\lambda\,(t) = 0_M$ then $m_i.\mu_i\,(x_i) = 0_M$ and $\sum_{j=1,j\neq i}^{j=n} m_j.x_j = 0_{M_i} \Longrightarrow m_i = 0_A$, because μ_i is a monomorphism of A modules. As $\{x_1, x_2, ..., x_n\}$ is A-free, then $m_j = 0_A$ for j in $\{1, 2, ..., n\}$ such that $j \neq i \Longrightarrow t = 0_M$. So λ is a monomorphism of A modules. We know that there exist $n \in N$, $\alpha(x_i) = \sum_{j=1}^{j=n} a_j.x_j$, where $a_j \in A$, for all $j \in \{1, 2, ..., n\}$. As $\lambda\alpha = \widetilde{\alpha}\lambda$, then we have that

$$\lambda\left[\alpha\left(x_i\right)\right] = a_i.\mu_i\left(x_i\right) + \sum_{j=1,j\neq i}^{j=n} a_j.x_j = \widetilde{\alpha}(\mu_i)(x_i) \in E_i$$
$$\Longrightarrow \sum_{j=1,j\neq i}^{j=n} a_j.x_j = 0_M$$
$$\Longrightarrow \alpha(x_i) = a_i x_i,$$

and this finishes the proof. □

The main result in this section is the folowing theorem:

Theorem 2 *Keep the above notations. Then, we have that*

$$E_p(M, F) = \left\{k\, id_M / k \in A^{\times}\right\}.$$

Proof By Lemma 9, we know that $\alpha(x_k) = a_k.x_k$ (1) for all $k \geq 1$. Let $i \geq 1$ and $j \geq 1$. Now, we must prove that $a_i = a_j$. However, using Lemma 1, we have that

$$\bigoplus_{k=1}^{k=\infty} Ax_k = A(x_i + x_j) \bigoplus \bigoplus_{k=1,k\neq i}^{k=\infty} Ax_k$$

Applying the previous argument to $x_i + x_j$, there exists $a_{i,j} \in A^* : \alpha(x_i + x_j) = a_{i,j}.(x_i + x_j) = a_i.x_i + a_j.x_j$. As $\{x_i, x_j\}$ is A free, then $a_{i,j} = a_i = a_j \Longrightarrow \exists k \in A^* : \alpha = kid_M$. Since α^{-1} satisfies to the extension property, there exists $r \in A^*$ such that $\alpha^{-1} = r\, id_M \Longrightarrow kr = 1_A \Longrightarrow k$ is a unit in A. □

Acknowledgements We thank the referee for his suggestions and comments in the revision of this paper.

References

1. Abdelalim, S., Essannouni, H.: Characterization of the automorphisms of an Abelian group having the extension property. Port. Math. (Nova) **59**(3), 325–333 (2002)
2. Aluffi, P.: Algebra: chapter 0. Am. Math. Soc. **104**, 177–178 (2009)
3. Anderson, D.D., Anderson, D.F., Zafrullah, M.: Factorization in integral domains. J. Pure Appl. Algebr. **69**, 1–19 (1990)
4. Barry, M.: Caractérisation des anneaux commutatifs pour lesquels les modules vérifant (*I*) sont de types finis. Dissertation Doctorale, Université Cheikh anta diop de Dakar, faculté des sciences et techniques. Département de Mathématiques et Informatique (1998)
5. Ben Yakoub, L.: Sur un Théorème de Schupp. Port. Math. **51**(2), 231–233 (1994)
6. Dăscălescu, S., Năstăsescu, C., Raianu, Ş.: Hopf Algebras. An Introduction. Monographs Textbooks in Pure Applied Mathematics, vol. 235. Marcel Dekker, New York (2001)
7. Faith, C.: Algebra: Rings, Modules and Categories I. Die Grundlehren der mathematischen Wissenschaften in Einzeldarstellungen Band, vol. 190, xxiii+551p. Springer, Berlin (1973)

8. Freyd, P.: Abelian Categories. An Introduction to the Theory of Functors. Harper's Series in Modern Mathematics. Harper & Row Publishers, New York (1964): Reprints in Theory and Applications of Categories. No. **3**, 23–164 (2003)
9. Kuperberg, G.: Finite, connected, semisimple, rigid tensor categories are linear. arXiv preprint math/0209256 (2002)
10. Leinster, T.: Basic Category Theory, vol. 143. Cambridge University Press, Cambridge (2014)
11. Schupp, P.E.: A characterizing of inner automorphisms. Proc. A.M.S. **101**(2), 226–228 (1987)
12. Vialar, T.: Handbook of Mathematics. BoD-Books on Demand (2015)

History of Mathematics

This part contains only one contribution, which offers a brief survey on the Arabic scientific heritage in Morocco, specially focussing on the filed of mathematics.

Arabic Scientific and Technical Heritage in Morocco

Abdelmalek Azizi

Abstract During the first centuries of the birth of the Muslim empire, many scientific disciplines were well developed. Subsequently, several new areas of knowledge had emerged. Thus, there was the birth of algebra and combinatorial analysis in mathematics and the birth of cryptanalysis in cryptology. There were some very important discoveries in several scientific fields such as in physics, in chemistry and in medicine. In this paper, we will overfly the development of certain scientific disciplines in Morocco by highlighting the application of the results of these disciplines in the socio-economic world.

Keywords Moroccan scientific heritage · Algebra and combinatorial analysis · Cryptanalysis · Arab technical heritage

1 Introduction

The birth of the Muslim Empire had been accompanied by an intellectual development in all ancient scientific fields as well as in the new knowledge which had begun to be formed at that time.

For socio-economic needs encountered, among others, in certain problems of heritage, commerce, linguistics and astronomy, the Arabs had approached several scientific questions in all fields and left us a golden Heritage in all the scientific and technical fields as in Mathematics, in Cryptography and communication systems, in medicine, in physic and chemistry and in theirs technical applications (Clocks; astrolabes; water pumps, ...).

A. Azizi (✉)
ACSA Laboratory, Faculty of Sciences, Universite Mohammed Premier, Oujda, Morocco
e-mail: abdelmalekazizi@yahoo.fr

M. Siles Molina et al. (eds.), *Associative and Non-Associative Algebras and Applications*, Springer Proceedings in Mathematics & Statistics 311,
https://doi.org/10.1007/978-3-030-35256-1_18

2 Mathematics

After the work of Al Khawarizmi (783–850) in Algebra and Arithmetic's, and especially his contributions in the study of 2nd degree equations, in the development of calculus techniques using the decimal system and in his manner of resolving problems steps by steps in a clear order, what makes his name used famously in computer science as algorithm; other works have been recorded in the Middle East, in Egypt, in the Maghreb and in Andalusia. In particular, we find the remarkable work of Thabit Ibno Qurra (836–901) on the amicable numbers, the works of Abu Kamil (m.930) and those of al-Karaji (1029) in Algebra and the works of Al-Khayam (1048–1131) on the equations of the 3rd degree.

In the 10th century and after the birth of the first dynasty in Morocco, the Idrissid dynasty, Moroccans (traders, military, and politicians), had been interested in developing Algebra, Arithmetic, the writing of the integers using new symbols clear or not, ... Then, many results were established by Ibn Al Yassamin (...–1204), Ibn Muneim (...–1228), Al-Hassan al-Marrakchi (...–1262), Ibn Al Banna (1256–1321), Ibn Ghazi al-Meknassi (1437–1513) and others. In particular, Ibn al-Banna's book "Tanbih Al Albab" include the following results:

Combination of n letters p to p	$C_n^p = \sum_{p-1}{}^{n-1} C_k^{p-1}$
Switching n letters	$P_n = n!$
Permutation of n letters with repetition of p letters $k_1, k_2, ..., k_p$ times	$P_n^r = \frac{n!}{k_1!k_2!\cdots k_p!}$
Different readings of a word of n letters by swapping vowels and sukun	$S_n = 4S_{n-1} - 3S_{n-3}$
Arrangement of n letters k to k with vowels and sukun and their arrangement in tables	$A_n^k = S_k.P_k.C_n^k$
The formula long attributed to Pascal (1623–1662)	$C_n^p = \frac{n-p+1}{p} C_n^{p-1}$

Ibn Albanna applied the mathematical results to the socio-economic problems of his time, as the following applications:

- Calculations regarding the drop in irrigation canal levels,
- Arithmetical explanation of the [[Muslim]] laws of inheritance,
- Determination of the hour of the [[Asr]] prayer,
- Explanation of frauds linked to instruments of measurement,
- Enumeration of delayed prayers which have to be said in a precise order, and calculation of legal tax in the case of a delayed payment.

We find also several other notions and questions that are studied such:

- approximation of cubic roots, Pascal's triangle and other results on combinatory, the first proof using the principle of mathematical induction that are in Ibno Munim Book 'Fikh Al Hissab',

Table 1 Examples of Arabic contributions in mathematics

 Al Khawarizmi (783–850)	 Ibn Al-Banna (1256–1321) Kitab Raf'e al-hijab an wujuh a'mal al-hisab	 Ibn Al-Banna (1256–1321) Calculus Tables	 Ibn Al Yassamin Poetry in Algebra

- the Algebraic Symbolism used by Ibn al-Yassamin (–1204), Ibn Qunfudh (1339–1407),
- And the classification of mathematics established by Ibn Rachiq (1275) from Ceuta.

For more information on this contributions, see [10–12] (Table 1).

3 Cryptography and Communication Systems

The use of difficult or unconventional notions to establish Cryptographical algorithms was a tradition among Arab Cryptographers. They used, among other things, poetry as transmission means and used, for example, the difficulty of writing verses of poetry (or pieces of verses) according to a given model or verses that can be read from right to left and at the same time from left to right as the basis of Cryptographic Algorithms. Thus, Arabic poetry was a means of transmission, information, advertising and cryptography.

The Arabs used cryptography even before Islam; but the pillars of Arab Cryptography were built by Al Khalil (718–786) and Al Kindi (801–873). Other Arab scholars had written important documents on Cryptography, including Ibn Dunainir (1187–1229), Ibn Adlan (1187–1268), Ibn Ad-Duraihim (1312–1361) and Al-Qalqashandi (1355–1418), see [9, 14–16]).

The Moroccans, following the study of certain linguistic, mathematical and astronomic questions, had developed several cryptographical methods for sending secret messages (military, diplomatic, scientific, distractions). In particular, they had used the following methods:

(i) Methods of substitution and transposition: among these, we find the method which consists of coding the letters by names of birds and after coding the letters, we put the coded text in a poetry.

(ii) The use of the function h, "hissab Al Jommal", to encrypt short messages: the message to be encrypted is transformed by the function h; we obtain a number which is decomposed into a product of two numbers n and m. Then we look for sentences P_1 and P_2 such that $h(P_1) = n$ and $h(P_2) = m$ and the multiplication symbol is replaced by its equivalent in Arabic. We thus obtain a text that can constitute the encrypted text (Cryptography of gold invented by the Sultan Ahmed El Mansour at the end of the 16th century, see [6–8, 16]).

(iii) The third method consists in using the numerical coding by position: it is based on the use of a text that is inserted in a grid, and then a letter will be replaced by three digits which represent a position of the letter in the grid (on Thinks that this method was used to encrypt messages in the 18th century, we have no proof but we find examples of coding of clear texts by this method in [1]).

(iv) The fourth method is Telegram secret writing which was used towards the end of the 19th century: it consisted in giving numerical values to the different letters of the alphabet and then transforming the clear text by coding numerically the letters and separating them by a point and the digits are coded by the same numbers with a bar at the top. In addition, some important names or words or phrases have been coded by numbers (see [16]). This method had been used by the Moroccans, at the end of the 19th century, to write messages of telegram.

(v) Use of the function "Hissab Al Jommal" to sign, to leave a digital imprint or to code the name of the author; it was used by some poets especially in the poems of al-Malhoun.

(vi) Use of signatures by steganographical methods: hide the letters of the author's name in a poetry as the first letters of verses of poetry or as the second letters of the words of a verse...

(vii) Use of special coding of numbers (Al Kalam al-Fassi) by judges and by notaries for safeguarding financial or inheritance acts against the possible forgeries.

For more information on his methods, see [5–8] (Table 2).

Table 2 Examples of Moroccan manuscripts on coding and cryptography

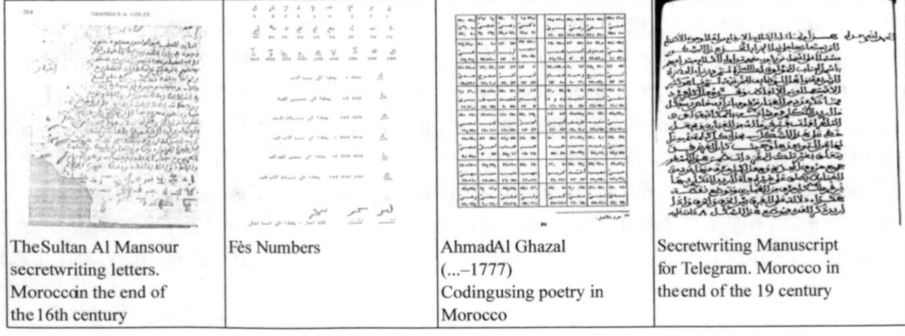

The Sultan Al Mansour secretwriting letters. Moroccan the end of the 16th century	Fès Numbers	AhmadAl Ghazal (...–1777) Codingusing poetry in Morocco	Secretwriting Manuscript for Telegram. Morocco in theend of the 19 century

Table 3 Examples of Arabic contributions in medicine and pharmacology

Ibn Al Nafis (1213–1288)	Kitâb al-Tasrîf li-man 'ajiza 'an al-ta'alîf Abû al-Qâsim Khalaf ibn al-'Abbâs al-Ansari al-Zahrâwî(936–1013)	Kitab al-Gami' li-mufradatal-adwiya wa-l-agáÿŔiya Ibn al-Baytar (1197–1248)	Kitab al-Kafi Ibn al-Baytar (1197–1248)

4 Medicine

Many theoretical and practical achievements by many scholars, such by al-Farabi (950) and Ibn Sina (1037) in the Mashreq, or by Ibno-Anafis (1211–1288) in Egypt and by Zahraoui (10th–11th centuries) In Andalusia. For more information on these achievements, see [3–5, 13].

In particular, the discovery of the pulmonary circulation by Ibno al-Nafis is a remarkable contribution, see Table 3.

5 Physics, Alchemy and Their Applications

The Arab world had made a remarkable development in physics (Optics, mechanics, ...) and Alchemy. This development had yielded its fruits mainly in the technological applications throughout the Arab world from the Mashreq to the Maghreb. For more information on these works, see [3–5, 13].
This was reflected in the following findings:

- The still and the hydrodistillation in Alchemy,
- The light ray by Ibn Al Haytham (965–1041),
- (hydro) water pump of al-Jazari (m 1206), (see Table 4),
- The astrolabe from al-Khawirzmi to al-Zarqali (11th c.), Al-Hassan al-Marrakchi (...–1262) and the Moroccan al-Roudani (17th c.), (see Table 5),
- The marvelous clocks: from the Elephant clock of al-Jazari from the 13th century to the water clock of Madrassa Albouanania from the 14th century through the Al-Qarawiyyin clock from the 13th century, (see Table 6).

Table 4 Examples of Arabic contributions in physics, Alchemy and their applications

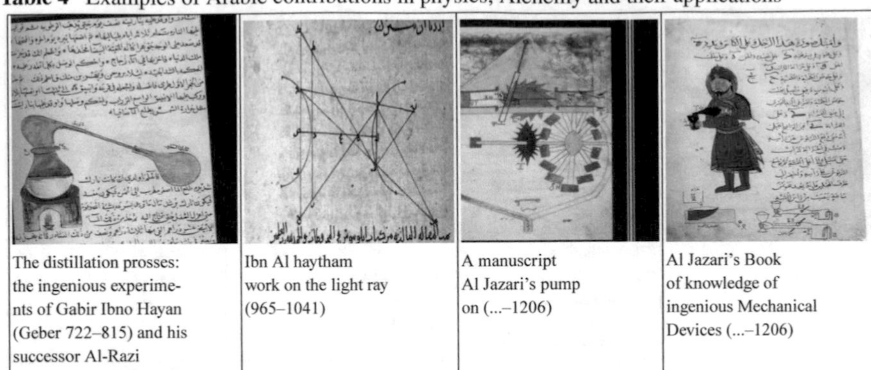

The distillation prosses: the ingenious experiments of Gabir Ibno Hayan (Geber 722–815) and his successor Al-Razi (Rhazes 864–925)	Ibn Al haytham work on the light ray (965–1041)	A manuscript Al Jazari's pump on (...–1206)	Al Jazari's Book of knowledge of ingenious Mechanical Devices (...–1206)

Table 5 Examples of Moroccan contributions in astronomy

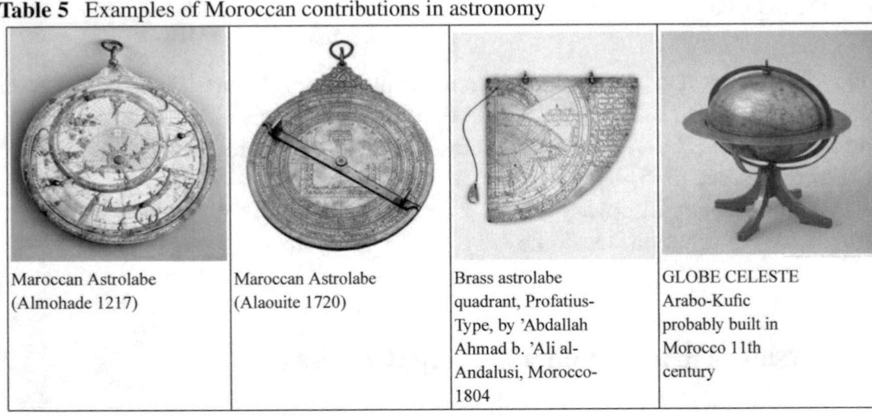

Maroccan Astrolabe (Almohade 1217)	Maroccan Astrolabe (Alaouite 1720)	Brass astrolabe quadrant, Profatius-Type, by 'Abdallah Ahmad b. 'Ali al-Andalusi, Morocco-1804	GLOBE CELESTE Arabo-Kufic probably built in Morocco 11th century

Table 6 Examples of Arabic ingenious clocks

Al-Jazari's Elephant Clock 13th century	The Al-Qarawiyyin Clock Fez Morocco 1286/87	Dar al-Magana (water clock) - Fez Morocco 1310–1331

Table 7 Examples of other distinguish Moroccan scientist

 Al Idrissi (1100–1165 or 1175)	 Ibno Rushd-Averroes (1126–1198)	 Ibn Battouta (1304–1377)	 Ibn Khaldoun (1332–1406)

6 Other Disciplines

Several other achievements or discoveries had been made in the Muslim world; we mention for examples:

- Realization of the first geographical map by the Moroccan Al Idrissi (m 1165),
- Volumes of philosophy and religion of Ibno Rushd (Averroes (1198)),
- Volumes of sociology (introduction) of Ibn Khaldoun (1406),
- Etc.

For more information on these achievements, see [3–5, 13] (Table 7).

7 Conclusions

Several other scientific and technical realizations were registered in the Muslim world and in particular in Morocco, as the notion of a symmetry in geometry which had been used in pavements or Zeliges in several places as in Alhambra of Grenada (pavement of the space plan by in most 17 different manners and which were all found in Alhambra), the sundial that Al-Hassan Al Marakchi had studied and he had written, in his book "Jamae al-Mabade' wal Ghayate fi Ilm al-Miqate", a chapter on his practical construction.

Since Saadiens, the Moroccans had acquired some technical Material and the knowledge of the use of that material, in several industrial domains and from several Arab or European countries.

A delay was noticed concerning the acquisition of the printing technology in the Arab world; in spite of its appearance in Lebanon since 1583. For Morocco, which had expressed a wish to be equipped with this technology well before on 1800 from French and Muslim world, he had been able to have this technology only by 1864 through a Judges after his return from the pilgrimage.

Unfortunately, we find few scientific or technical museums, in the Muslim world, which take care with getting back our scientific and technical inheritance, with repairing it or with restoring it. It is true that there are attempts of recovery or restoration as Al Karaouiyine's clock and the clock Al Bouenania in Fes, but this remains insufficient; because our manuscripts are still in European libraries, many of our astrolabes are in other centers.

Our scientific and technical Inheritance is a very important part of the history of our intellectual development which we have to integrate into the programs of our fundamental and technological education and so allow the young people to touch the theoretical ideas and the practices of their ancestors to understand them and be inspired there and do better.

Our scientific and technical inheritance contains many ideas, methods and brilliant practices. With the discoveries of the twentieth century, the old ideas, the methods and the practices can be developed to give extraordinary results. For example, we took certain ideas of cryptography of our old manuscripts and by means of the computing one found brilliant methods of cryptography.

It is the same for the technical instructions which were used to the manufacturing of certain instruments and devices; we can reconstruct them to put them in museums or in scientific and technical centers or improve them and put them in the market.

References

1. Al Iraki, A.: Ahmed El-Ghazal literary papers: Imprimerie Info-Print, Fès Maroc (1999)
2. Al-Hassani Salim, T.S. (ed.): 1001 Inventions: The Enduring Legacy of Muslim Civil, 3rd edn, p. 351. National Geographic Society, Washington (2012)
3. Firas, A.: Lost Islamic History. HURST and Company, London (2014)
4. Al-Manouni, M.: The Aspects of the Beginning of the revival of the new Morocco. Imrimerie Omnia Rabat (1973)
5. Azizi, A., Azizi et M., Ismaili, M.C.: Livre de recherche "Cryptographie : de l'histoire aux applications". In: Proceeding de cours, École CIMPA de Cryptographie Oujda 2009. Travaux en cours 80, Hermann Editions (2012)
6. Azizi, A.: Histoire de la Cryptographie Arabe au Maroc. Chapitre du livre "Cryptographie : de l'histoire aux applications". Travaux en cours 80, Hermann Editions (2012)
7. Azizi, A., Azizi, M.: Instance of arabic cryptography in Morocco. Cryptologia **35**(1), 47–57 (2011)
8. Azizi, A., Azizi, M.: Instance of arabic cryptography in Morocco II. Cryptologia **37**(4), 328–337 (2013)
9. Al-Kadi, I.A.: Origins of cryptology: The Arab contributions. Cryptologia **16**, 97–126 (1992)
10. Djebbar, A.: Les Mathématiques au Maghreb à l'époque d'Ibn Al-Banna. Actes du Congrès International Mathématiques et philosophie, Rabat 1982. Editeurs l'ARMATTAN Paris et OKAD Rabat (1987)
11. Djebbar, A.: Une histoire de la science arabe. Editions du Seuil (2001)
12. Djebbar, A.: L'Algèbre arabe, genèse d'un art. VUIBERT-ADAPT (2005)
13. Djebbar, A.: Les découvertes en pays d'Islam. Editions Le Pommier Paris (2009)
14. Mrayati, M., Meer Alam, Y., At-Tayyan, M.H.: Arabic Origins of Cryptology, Vols. 1, 2, 3 and 4. Published by KFCRIS and KACST, Riyadh (2003)
15. Kahn, D.: The Codebreakers: The Story of Secret Writind. Macmillan, New York (1967)
16. Tazi, A.: Les Codes Secrets des correspondances Marocaines á travers l'histoire. Librairie Almaarif Aljadida, Rabat (1983)

Printed in the United States
By Bookmasters